Multiple Access Techniques for 5G Wireless Networks and Beyond

Mojtaba Vaezi · Zhiguo Ding
H. Vincent Poor
Editors

Multiple Access Techniques for 5G Wireless Networks and Beyond

Springer

Editors
Mojtaba Vaezi
Villanova University
Villanova, PA
USA

H. Vincent Poor
Princeton University
Princeton, NJ
USA

Zhiguo Ding
The University of Manchester
Manchester
UK

ISBN 978-3-030-06360-3 ISBN 978-3-319-92090-0 (eBook)
https://doi.org/10.1007/978-3-319-92090-0

Printed on acid-free paper

This Springer imprint is published by the registered company Springer Nature Switzerland AG
The registered company address is: Gewerbestrasse 11, 6330 Cham, Switzerland

Preface

Waveform design, multiple access, and random access techniques for fifth-generation (5G) wireless networks and beyond are cutting-edge research topics that motivate a very wide range of research problems. Despite being different, these three areas are intertwined and lie at the heart of wireless communication systems. They allow multiple users to effectively share a communication medium. The previous generations of cellular networks have adopted radically different multiple access techniques with one common theme in mind: to have orthogonal signals for different users at the receiver side. As an example, the fourth-generation (4G) cellular networks have adopted orthogonal frequency division multiplexing (OFDM). In view of emerging applications such as the Internet of Things (IoT), and in order to fulfill the need for massive numbers of connections with diverse requirements in terms of latency and throughput, 5G and beyond cellular networks are experiencing a paradigm shift in design philosophy: shifting from *orthogonal* to *non-orthogonal* design in waveform, multiple access, and random access techniques.

This book provides a comprehensive and intensive examination of multiple access, random access, and waveform design techniques for 5G and beyond systems. It contains numerous state-of-the-art techniques and experimental results to address the challenges in building 5G and beyond wireless networks. The book will be of interest to readers from the *communications*, *signal processing*, and *information theory* communities. It will serve as a reference for graduate students, researchers, and engineers involved in the design and standardization of wireless communication systems. It can also serve as a reference book for graduate-level courses for students in electrical engineering.

The book is organized into four parts and twenty-one chapters, each meant to be self-contained. The contents of different chapters in each part are chosen so that they reinforce and complement each other. Part I is focused on waveform design for 5G and beyond and includes four chapters outlining several advanced multicarrier waveforms designs. Parts II through IV cover several related topics in multiple access and random access. These parts include various non-orthogonal multiple access (NOMA) techniques in the power domain, code domain, and other domains as well as topics on random access. Power domain NOMA is mainly covered in

Part II. Several code domain NOMA and other NOMA techniques as well as random access schemes are discussed in Part III. Part IV includes experimental trials and applications of NOMA in certain fields other than cellular communications.

Part I (Chaps. 1–4) addresses waveform design. Chapter 1 introduces the reader to 1G to 4G cellular systems and motivates the need for new multiple access and waveforms for 5G and beyond systems. Chapter 2 presents various waveform designs envisioned for the 5G new radio (NR). It first outlines the shortcomings of conventional OFDM in serving the diverse use cases envisioned for 5G systems. It then discusses design principles for new waveforms and introduces three new waveforms considered in the standardization process of 5G NR, all developed from OFDM. The new variants of OFDM are focused on reducing the out-of-band (OOB) emission of cyclic-prefix OFDM by signal processing techniques such as time-domain windowing and subband-based filtering. Chapter 3 introduces another advanced multicarrier waveform, namely filter bank multicarrier modulation (FBMC). FBMC can perform much better than windowed and filtered OFDM in reducing the OOB emissions of conventional OFDM and can be considered as a potential waveform for next generation wireless networks. Chapter 4 studies yet another advanced multicarrier waveform, generalized frequency-division multiplexing (GFDM). GFDM is a multicarrier waveform technique that encapsulates windowed and filtered OFDM techniques of 5G while providing an additional design space reserved for forward comparability beyond 5G.

Part II (Chaps. 5–11) is dedicated to NOMA relying on the power domain. Chapter 5 studies NOMA from an information-theoretic perspective. This chapter reviews the basic premise behind NOMA in single- and multi-cell networks both in the downlink and uplink. It also introduces various information-theoretic channels that can be used to model physical layer security in NOMA. Chapter 6 investigates power allocation for downlink NOMA under different performance metrics, such as fairness, sum rate, and energy efficiency. The design principles of multiple-antenna NOMA systems, including user clustering, channel state information acquisition, and transmit beamforming are studied in Chap. 7. Chapter 8 is focused on applying NOMA to millimeter wave networks with three transmissions schemes, namely unicast, multicast and cooperative multicast. Chapter 9 is dedicated to full-duplex NOMA, a technology that has the potential to double the spectral efficiency via simultaneous transmissions in the uplink and downlink. Resource allocation in heterogeneous NOMA with energy cooperation is discussed in Chap. 10. Chapter 11 evaluates the performance of NOMA in vehicle-to-vehicle massive MIMO channels.

Part III (Chaps. 12–17) introduces several code domain NOMA schemes as well as non-orthogonal *random access*. All multiple access techniques presented in this part have a common philosophy: to exploit efficient and low-complexity multiuser detection. In particular, Chap. 12 studies sparse code multiple access (SCMA), a code domain NOMA scheme that exploits the sparsity of the multi-dimensional codewords to apply the low-complexity message passing algorithm for multiuser detection. Chapter 13 discusses interleave division multiple access (IDMA).

IDMA applies a low-complexity iterative technique for multiuser detection and can achieve near-capacity sum rate with proper power allocation. Chapter 14 introduces pattern division multiple access (PDMA) which is a NOMA technique in which a pattern defines the mapping of transmitted data to a group of time, frequency, and spatial resources. In Chap. 15, low-density spreading (LDS), a variant of code division multiple access (CDMA) in which spreading sequences have low density, is studied. Owing to this, a near optimal message passing algorithm receiver with practically feasible complexity can be exploited. Chapter 16 discusses a grant-free multiple access scheme that enables lower transmission latency and savings in device energy. Chapter 17 provides a comprehensive survey of random access schemes that are suited to support IoT. These schemes are commonly based on time, frequency, and code division multiple access techniques and different variants of NOMA described in the previous chapters.

Part IV (Chaps. 18–21) outlines experimental trials, challenges, and future trends of NOMA. To evaluate the performance of NOMA using real-world hardware and in realistic radio environments, a test-bed is described in Chap. 18. Indoor and outdoor experimental trials are then conducted which confirm that NOMA improves user throughput as compared to orthogonal multiple access. Applications and extension of NOMA to visible light communication networks and terrestrial-satellite networks are studied in Chaps. 19 and 20, respectively. Finally, Chap. 21 provides future research directions for NOMA in 5G wireless networks and beyond as well as other fields.

We would like to extend our thanks to the people and organizations who made this book possible. Our sincere thanks go to the chapter authors; it has been an honor and a privilege to work with such a dedicated and talented group of authors and researchers. Princeton University and the universities, research laboratories, and corporations with which the authors are affiliated deserve credit for providing facilities and intellectual environments for this project. It is also a pleasure to acknowledge Springer and its team: Mary James, Brian Halm, and Zoe Kennedy. Finally, we offer our deepest appreciation and gratitude to our families for their patience and support during the months we were immersed in this project.

Villanova, PA, USA Mojtaba Vaezi
Manchester, UK Zhiguo Ding
Princeton, NJ, USA H. Vincent Poor

Contents

About the Editors

Mojtaba Vaezi received the Ph.D. degree in Electrical Engineering from McGill University in 2014. From 2015 to 2018, he was with Princeton University as a Postdoctoral Research Fellow and Associate Research Scholar. He is currently an Assistant Professor of ECE at Villanova University and a Visiting Research Collaborator at Princeton University. Before joining Princeton, he was a researcher at Ericsson Research in Montreal, Canada. His research interests include the broad areas of information theory, wireless communications, and signal processing, with an emphasis on physical layer security and radio access technologies. Among his publications in these areas is the book Cloud Mobile Networks: From RAN to EPC (Springer, 2017). He has served as the president of McGill IEEE Student Branch during 2012–2013. He is an Associate Editor of *IEEE Communications Magazine* and *IEEE Communications Letters*. He has co-organized four international NOMA workshops at VTC-Spring'17, Globecom'17, ICC'18, and Globecom'18. He is a recipient of a number of academic, leadership, and research awards, including the McGill Engineering Doctoral Award, IEEE Larry K. Wilson Regional Student Activities Award in 2013, the Natural Sciences and Engineering Research Council of Canada (NSERC) Postdoctoral Fellowship in 2014, and Ministry of Science and ICT of Korea's best paper award in 2017.

Zhiguo Ding received his B.Eng. in Electrical Engineering from the Beijing University of Posts and Telecommunications in 2000, and the Ph.D. degree in Electrical Engineering from Imperial College London in 2005. From July 2005 to April 2018, he was working in Queen's University Belfast, Imperial College, Newcastle University and Lancaster University. Since April 2018, he has been with the University of Manchester as a Professor in Communications. From September 2012 to September 2018, he has also been an academic visitor in Princeton University. His research interests are 5G networks, game theory, cooperative and energy harvesting networks and statistical signal processing. He is serving as an Editor for *IEEE Transactions on Communications and IEEE Transactions on Vehicular Technology*. He served as an Editor for *IEEE Wireless Communication Letters*, *IEEE Communication Letters*, and *Journal of Wireless Communications* *and Mobile Computing*. He received the best paper award in IET International Communication Conference on Wireless Mobile and Computing, 2009, and the IEEE WCSP 2015, IEEE Transactions on Vehicular Technologies Top Editor 2017, and the EU Marie Curie Fellowship 2012–2014.

H. Vincent Poor received the Ph.D. degree in EECS from Princeton University in 1977. From 1977 until 1990, he was on the faculty of the University of Illinois at Urbana-Champaign. Since 1990, he has been on the faculty at Princeton, where he is currently the Michael Henry Strater University Professor of Electrical Engineering. During 2006 to 2016, he served as Dean of Princeton's School of Engineering and Applied Science. His research interests are in the areas of information theory and signal processing, and their applications in wireless networks, energy systems, and related fields. He is a member of the National Academy of Engineering and the National Academy of Sciences and is a foreign member of the Chinese Academy of Sciences, the Royal Society, and other national and international academies. Other recognition of his work includes the 2017 IEEE Alexander Graham Bell Medal, and honorary doctorates and professorships from a number of universities.

Acronyms

1G	First generation
2G	Second generation
3G	Third generation
3GPP	3rd Generation Partnership Project
4G	Fourth generation
5G	Fifth generation
5GC	5G core
ACEP	Average codeword error probability
ACK	Acknowledgement
ACLR	Adjacent channel leakage rejection
ACO-OFDM	Asymmetrically clipped optical OFDM
ACRDA	Asynchronous contention resolution diversity ALOHA
ADC	Analog-to-digital converter
AF	Application function
AGC	Automatic gain control
AMC	Adaptive modulation and coding
AMF	Access and mobility management function
AMI	Averagemutual information
AMPS	Advanced mobile phone services
AN	Artificial noise
AoA	Angle of arrival
AoD	Angle of departure
APD	Avalanche photodiode
APM	Amplitude-phase modulation
APP	A posteriori probability
APSK	Amplitude phase-shift keying
AR	Augmented reality
AWGN	Additive white Gaussian noise
BC	Broadcast channel
BEP	Bit error probability

BER	Bit error rate
BF	Beamforming
BICM	Bit-interleaved coded modulation
BICM-ID	BIMC with iterative decoding
BLER	Block error rate
BP	Belief propagation
BPSK	Binary phase-shift keying
BRAM	Block RAM
BRU	Basic resource unit
BS	Base station
BSA	Binary switching algorithm
BSC	Base station controller
BSS	Base station subsystem
BTS	Base transceiver stations
C2X	Car to anything
CA	Carrier aggregation
CB	Coordinated beamforming
CBRS	Citizens broadband radio service
CCDF	Complementary cumulative distribution function
CCI	Co-channel interference
CDD	Cyclic delay diversity
CDF	Cumulative distribution function
CDMA	Code division multiple access
CEP	Channel estimation preamble
CFO	Carrier frequency offset
CIM	Color intensity modulation
CIR	Channel impulse response
CLT	Central limit theorem
CM	Cubic metric
CMT	Cosine-modulated multitone
CN	Core network
CN	Channel observation node
CNR	Channel-to-noise ratio
CoF	Cycle-of-four
CoMP	Coordinated multipoint
CP	Cyclic prefix
CPICH	Common pilot channel
CP-OFDM	Cyclic prefix OFDM
CPRI	Common public radio interface
CQI	Channel quality indicator
CR	Cognitive radio
CRA	Contention resolution ALOHA
C-RAN	Cloud radio access network
CRC	Cyclic redundancy check
CRDSA	Contention resolution diversity slotted ALOHA

CR-NOMA	Cognitive radio inspired NOMA
CRS	Cell-specific reference signal
CS	Circuit switched
CS	Cyclic suffix
CS	Coordinated scheduling
CSA	Coded Slotted ALOHA
CsDMA	Code-shift division multiple access
CSI	Channel state information
CSIR	Channel state information at the receiver
CSIT	Channel state information at the Transmitter
CSK	Color shift keying
CSMA	Carrier sense multiple access
CSMA/CA	Carrier sense multiple access/collision avoidance
CSMA/CD	Carrier sense multiple access/collision detection
CTS	Clear to send
CU	Centralized unit
CW-SIC	Codeword level SIC
D/A	Digital-to-analog
D2D	Device-to-device
DA	Deferred acceptance
DAC	Digital-to-analog converter
DACE	Data-aided channel estimation
DC	Direct current
DC	Difference of convex functions
DCCC	Discrete codebook-constrained capacity
DCI	Downlink control information
DCO-OFDM	Direct current-biased optical OFDM
DCS	Dynamic cell selection
DD	Direct detection
DEC	Decoder
DFE	Decision feedback equalizer
DL	Downlink
DMRS	Demodulation reference signal
DoF	Degree of freedom
DPC	Dirty-paper coding
DPD	Digital predistortion
DRAM	Dynamic random access memory
DS-ALOHA	Diversity slotted ALOHA
DS-CDMA	Direct-sequence CDMA
DSP	Digital signal processor
DS-SS	Direct-sequence spread spectrum
DTFT	Discrete-time Fourier transform
DU	Distributed unit
DVB	Digital video broadcasting
DVB-T	Digital video broadcasting terrestrial

DZT	Discrete Zak transform
EDGE	Enhanced data rates for GSM evolution
EE	Energy efficiency
eMBB	Enhanced mobile broadband
eMBMS	Evolved multimedia broadcast multicast service
eMTC	Enhanced machine type communications
eNB	Enhanced NodeB
EP	Expectation propagation
EPC	Equal power control
ES	Exhaustive search
ESE	Elementary signal estimation
E-SSA	Enhanced spread spectrum ALOHA
ETU	Extended typical urban
EV-DO	Evolution-data optimized
EVM	Error vector magnitude
EXIT	Extrinsic information transfer
FBMC	Filter bank multicarrier modulation
FD	Full-duplex
FDD	Frequency division duplex
FDE	Frequency domain equalization
FDM	Frequency division multiplex
FDMA	Frequency division multiple access
FEC	Forward error control
FER	Frame error rate
FET	Field-effect transistor
FFT	Fast Fourier transform
FIFO	First in first out
FIR	Finite impulse repose
FMT	Filtered multitone
FN	Function Node
FO	Frequency offset
f-OFDM	Filtered OFDM
FOV	Field of view
FPA	Fixed power allocation
FPGA	Field programmable gate array
FSPA	Full search power allocation
FTPA	Fractional transmission power allocation
GA	Genetic algorithm
GA	Gaussian approximation
GB	Grant-based
GEO	Geosynchronous earth orbit
GEVD	Generalized eigenvalue decomposition
GF	Grant-free
GFDM	Generalized frequency division multiplexing
GGSN	Gateway GPRS support node

GMSC	Gateway mobile switching center
GMSK	Gaussian minimum shift keying
gNB	Next generation NodeB
GOCA	Group-orthogonal coded access
GP	Guard period
GPRS	General packet radio services
GSM	Global system for mobile communications
GSVD	Generalized singular value decomposition
GT	Guard tone
HARQ	Hybrid automatic repeat request
HD	Half-duplex
HetNet	Heterogeneous network
HK	Han-Kobayashi
HLR	Home location register
HOM	Higher-order modulation
HOS	Hierarchy of orthogonal sequences
HPPP	Homogeneous Poisson point process
HSDPA	High speed downlink packet access
HSPA	High speed packet access
HSTRN	Hybrid satellite terrestrial relay network
HSUPA	High speed uplink packet access
i.i.d.	identically independently distributed
IA	Interference alignment
IAI	Inter-antena interference
IBFD	In-band full-duplex
IBI	Inter-block interference
IC	Interference cancellation
IC	Interference channel
ICI	Inter-carrier interference
ICI	Inter-cell interference
ICT	Information and communication technology
IDD	Iterative detection and decoding
IDFT	Inverse discrete Fourier transform
IDMA	Interleave division multiple access
IFFT	Inverse fast Fourier transform
IFPI	Interference-free pilot insertion
IFS	Inter-frame spacing
IGCH	Information-guided channel hopping
IGMA	Interleave grid multiple access
IM	Intensity modulation
I-MRC	Iterative maximum ratio combining
INI	Inter-numerology interference
IoT	Internet of Things
IR	Infrared
IRSA	Irregular repetition slotted ALOHA

ISI	Inter-symbol interference
ISM	Industrial scientific and medical
ITS	Intelligent transportation systems
ITU	International telecommunications union
IUI	Inter-user interference
JP	Joint processing
JRA	Joint resource allocation
JT	Joint transmission
J-TACS	Japan TACS
KKT	Karush-Kuhn-Tucker
KPI	Key performance indicator
LDM	Layer division multiplex
LDPC	Low-density parity-check
LDS	Low-density spreading
LED	Light-emitting diode
LEO	Low Earth orbit
LiFi	Light fidelity
LLR	Log-likelihood ratio
LMMSE	Linear minimum mean squared error
LOS	Line-of-sight
LP	Linear program
LPWAN	Low power wide area networks
LS	Least squares
LTE	Long term evolution
LTE-A	LTE advanced
M2M	Machine to machine
MA	Multiple access
MAC	Multiple access channel
MAC	Medium access control
MAC	Medium access layer
MACA	Multiple access collision avoidance
MAI	Multiple access interference
MAP	Maximum a posterior
MARSALA	Multi-replica decoding using correlation based Iocalization
MBS	Macro base station
MC-CDMA	Multicarrier CDMA
MC-LDSMA	Multicarrier low density spreading multiple access
MC-NOMA	Multi-channel NOMA
MCS	Modulation coding scheme
MEC	Multi-access edge computing
MEO	Medium Earth orbit
ME-SSA	MMSE enhanced spread spectrum ALOHA
MF	Matched filter
MF-CRDSA	Multi-frequency CRDSA
MI	Mutual information

MIIT	Ministry of industrial and information technology
MIMO	Multiple-input and multiple-output
MINLP	Mixed integer nonlinear programming
MISO	Multiple-input and single-output
ML	Maximum likelihood
MM	Metameric modulation
MMF	Maximin fairness
MMSE	Minimum mean squared error
mMTC	Massive machine type communication
mmWave	Millimeter wave
MPA	Message passing algorithm
MPF	Marginal product of functions
MPR	Multi-packet reception
MRC	Maximum ratio combining
MRT	Maximum ratio transmission
MSC	Mobile switching center
MSE	Mean squared error
MTC	Machine type communication
MTSO	Mobile telephone switching office
MU	Mobile user
MUD	Multi-user detection
MUI	Multi-user interference
MU-MIMO	Multi-user MIMO
MUSA	Multi-user shared access
MuSCA	Multi-slots coded ALOHA
MUST	Multi-user superposition transmission
NAICS	Network-assisted interference cancellation and suppression
N-AMPS	Narrowband AMPS
NB	NodeB
NB-IoT	Narrow band IoT
NCMA	Non-orthogonal coded multiple access
NEF	Noise enhancement factor
NFV	Network function virtualization
NLOS	Non-line-of-sight
NMSE	Normalized mean-squared error
NMT	Nordic mobile telephone
NOCA	Non-orthogonal coded access
NOMA	Non-orthogonal multiple access
NP	Non-deterministic polynomial-time
NR	New radio
O/E	Optical to electrical
OFDM	Orthogonal frequency division multiplexing
OFDMA	Orthogonal frequency division multiple access
OMA	Orthogonal multiple access
OOB	Out-of-band

OOK	On-off keying
OQAM	Offset quadrature amplitude modulation
O-QPSK	Offset quadrature phase shift keying
OSS	Operation and support subsystem
OSTBC	Orthogonal space-time block coding
OVSF	Orthogonal variable spreading factor
P/S	Parallel-to-serial
PA	Power amplifier
PAM	Pulse amplitude modulation
PAPR	Peak-to-average power ratio
PBS	Pico BS
PC	Power control
PCCC	Parallel concatenated convolutional code
PCM	Policy control function
PD	Photodiode
PDCCH	Physical downlink control channel
pdf	Probability density function
PDMA	Pattern division multiple access
PDR	Packet drop rate
PEP	Pairwise error probability
PER	Packet error rate
PHY	Physical layer
PIA-ASP	Prior-information aided adaptive subspace pursuit
PIC	Parallel interference cancellation
p-i-n	Positive-intrinsic-negative
PL	Primary layer
PLR	Packet loss rate
PN	Pseudo noise
PPM	Pulse position modulation
PPP	Poisson point process
PRB	Physical resource block
PS	Packet switched
PS	Phase shifter
PSD	Power spectral density
PSK	Phase-shift keying
PSTN	Public switched telephone network
PWM	Pulse width modulation
QAM	Quadrature amplitude modulation
QoS	Quality of service
QPSK	Quadrature phase shift keying
RA	Random access
RACH	Random access channel
RAN	Radio access network
RAR	Random access response
RAT	Radio access technology

RB	Resource block
RC	Raised-cosine
RDMA	Repetition division multiple access
RE	Resource element
RF	Radio frequency
RMS	Root mean square
RNC	Radio network controller
RNTI	Radio network temporary identifier
RPC	Randomized power control
RPMA	Random phase multiple access
RR	Round robin
RRC	Radio resource control
RRC	Root-raised cosine
RRH	Remote radio head
RS	Rate-splitting
RSMA	Resource spread multiple access
RSRP	Reference signal received power
RTS	Request to send
RV	Redundancy version
RV	Random variable
RX	Receiver
S-ALOHA	Slotted ALOHA
SAMA	Successive interference cancellation amenable multiple access
SA-SCMA	Spread asynchronous scrambled coded multiple access
SC	Sub-carrier
SC	Superposition coding
SCA	Successive convex approximation
SC-FDMA	Single carrier frequency division multiple access
SCM	Superposition coded modulation
SCMA	Sparse code multiple access
SCR	Signal-to-clipping-noise ratio
SCS	Sub-carrier spacing
SDMA	Space division multiple access
SDN	Software defined networking
SDR	Software defined radio
SE	Spectral efficiency
SER	Symbol error rate
SFBC	Space frequency block coding
SG	Scheduling grant
SGSN	Serving GPRS support node
SIC	Successive interference cancellation
SINR	Signal-to-interference-plus-noise ratio
SIR	Signal-to-interference ratio
SISO	Single-input and single-output
SL	Secondary layer

SLA	Side lobe attenuation
SLL	Side lobe level
SLS	System level simulation
SL-SIC	Symbol level SIC
SM	Spatial modulation
SMF	Session management function
SMS	Short message service
SMT	Staggered multitone
SNR	Signal-to-noise ratio
SPS	Semi-persistent scheduling
SR	Scheduling request
SR	Sum rate
SrCMA	Scrambled coded multiple access
SS-ALOHA	Spread-spectrum ALOHA
SSD	Signal-space diversity
SSK	Space shift keying
STC	Space-time coding
STO	Symbol time offset
SUD	Single-user detection
SVD	Singular value decomposition
Tx	Transmit antenna
TACS	Total access communication system
TB	Transport block
TCP	Transmission control protocol
TDD	Time division duplex
TDL	Tapped delay line
TDM	Time division multiplexing
TDMA	Time division multiple access
TO	Transmission occasion
TO	Time offset
TR	Technical report
TR-STC	Time-reversal space-time coding
TS	Time-sharing
TTI	Transmission time interval
TX	Transmitter
UA	User association
UDM	Unified data management
UDN	Ultra dense network
UE	User equipment
UFMC	Universal filtered multicarrier
UF-OFDM	Universal filtered OFDM
UL	Uplink
UMTS	Universal mobile telephone system
UN	User node
UNB	Ultra narrow band

U-OFDM	Unipolar OFDM
UPC	Unequal power control
UPF	User plane function
URLLC	Ultra-reliable low latency communications
UTRAN	UMTS terrestrial radio access network
V2I	Vehicle-to-infrastructure
V2N	Vehicle-to-network
V2P	Vehicle-to-pedestrian
V2V	Vehicle-to-vehicle
V2X	Vehicle-to-everything
VANET	Vehicular ad hoc network
VBLAST	Vertical Bell laboratories layered space-time
VLC	Visible light communication
VLR	Visitor location register
VN	Variable node
VR	Virtual reality
WAVE	Wireless access for vehicular environments
WBE	Welch-bound equality
WCDMA	Wideband CDMA
WOLA	Weighted overlap and add
WSR	Weighted sum rate
ZF	Zero-forcing
ZFBF	Zero-forcing beamforming
ZP	Zero-padding
ZP-OFDM	Zero prefix OFDM
ZT-DFT	Zero tail DFT

Part I
Orthogonal Multiple Access Techniques and Waveform Design

Chapter 1
Introduction to Cellular Mobile Communications

Joseph Boccuzzi

1.1 Introduction

This chapter provides an overview of the evolution of the cellular mobile communication systems. We begin with a quote from a conversation held over a mobile cellular network from Martin Cooper on 3 April 1973 [1].

I'm calling you from a cell phone, a real handheld portable cell phone.

The mobile device used during this conversation was a Motorola DynaTAC weighing approximately 2.5 lbs with a cost of approximately $9,000 USD. This historic event ignited a movement which would change the lives of so many people. This life change is much, much more than supporting mobile users, it *snowballed* into creating highly complex devices (presently called smart phones) that help us remain connected to the world. These devices not only perform our much needed voice and data communication needs, but they also take on a very wide array of supporting applications such as keeping our friends informed on social media, competing with online gaming, consuming and producing video content, performing medical measurements, utilizing location-based services, etc.

As these wireless devices benefited from Moore's law, the cellular mobile technologies were able to remain a focal point to introduce such new and exciting features, and benefits, to the end user.

This chapter is intended to address important driving technologies behind the 5G *new radio* (NR) system designs, which focuses on solutions to supporting 5G new services in uplink (UL) transmissions with requirements such as low-latency and high-reliability, energy-saving, and small packet applications. Grant-free (GF) resources in NR UL is termed as "a configured grant," which means that the

J. Boccuzzi (✉)
Intel Corporation, San Diego, CA, USA
e-mail: Joseph.Boccuzzi@intel.com

M. Vaezi et al. (eds.), *Multiple Access Techniques for 5G Wireless Networks and Beyond*, https://doi.org/10.1007/978-3-319-92090-0_1

pre-configured UE-specific resources will be used for UE UL transmission without dynamic scheduling/grant. Also, the base station (BS) in 5G NR network is referred to as "next generation NodeB" or "gNB."

1.2 Cellular Mobile Communication: A Primer

The cellular standards use a variety of multiple access (MA) techniques, which we highlight in Table 1.1. These techniques include frequency division multiple access (FDMA), time division multiple access (TDMA), code division multiple access (CDMA), and orthogonal frequency division multiple access (OFDMA). We also describe the relevant duplex method used for two-way communication and the actual physical resources available to be assigned to each user. The duplex methods are time division duplex (TDD) and frequency division duplex (FDD).

All of the above multiple access techniques can be viewed as a form of "orthogonal" multiple access (OMA), where the access of users, theoretically, do not interfere with one another as they share the wireless medium. They are, however, limited to the number of resources available that make them orthogonal to each other. An exception to this would be CDMA, where the transmission from the wireless device to the base station is inherently non-orthogonal.

In FDMA, the frequency is divided into channels to be utilized by various users. In TDMA, time is divided into time slots as a means to allow various users to access the cellular system. In CDMA, users are separated by PN codes and transmit over the entire frequency channel, all at the same time. In OFDMA, users are allocated to various frequency channels (groups of sub-carriers) at different time slots. For the next generation digital cellular system called 5G, OFDMA is still used where the sub-carrier spacing and time slot durations are flexible and scalable to support wide-varying requirements and use cases. It is also expected to utilize NOMA in 5G. In Fig. 1.1, we provide an overview to showcase various multiple access techniques that will be discussed in this section. They are compared in three dimensions or domains: power, time, and frequency.

Table 1.1 Multiple access in different generations of cellular networks

Cellular generation	MA technique	Duplex method	Physical resources	Notable examples
1G	FDMA	FDD	Frequency	AMPS, NMT
2G	TDMA	FDD	Time slots	GSM, IS-54
3G	CDMA	FDD/TDD	Time slots/PN Codes	WCDMA
4G	OFDMA	FDD/TDD	Time/Frequency	LTE, LTE-A
5G	OFDMA	FDD/TDD	Time/Frequency	5G-NR

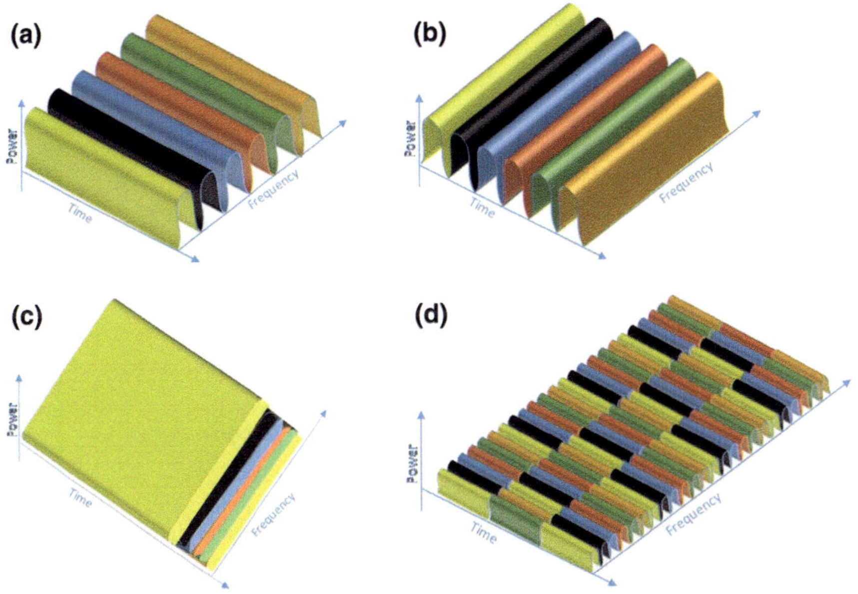

Fig. 1.1 Overview of various multiple access techniques: **a** FDMA, **b** TDMA, **c** CDMA, **d** OFDMA-4G

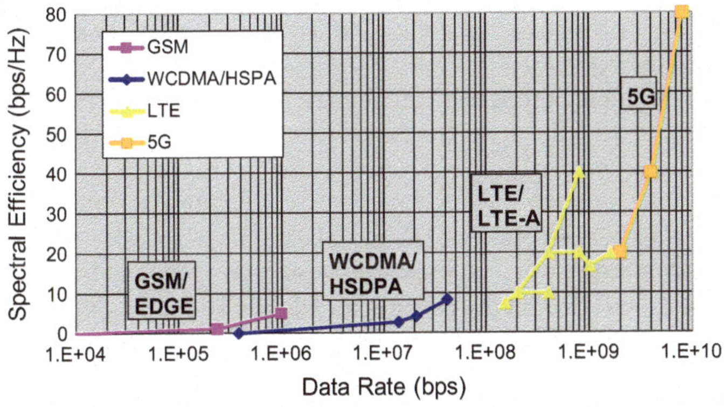

Fig. 1.2 Spectral efficiencies (bps/Hz) of the digital cellular evolution

A system performance metric that continues to be improved in every generation is spectral efficiency, bps/Hz. Figure 1.2 shows the DL spectral efficiencies of the 2G, 3G, 4G, and 5G digital cellular standards versus the peak theoretical data rates. Notice that with each new standard the demand for higher and higher data rates along with an increased demand for spectral efficiency becomes more pronounced.

With every cellular generation, there is not only an expectation of increased performance, but also the addition of new features. Figure 1.3 shows how the cellular

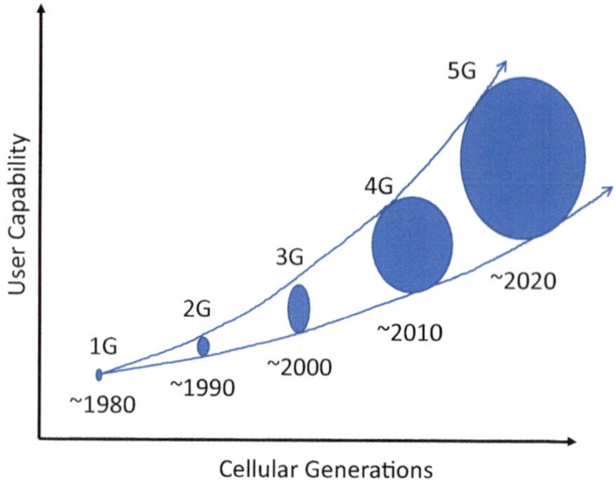

Fig. 1.3 User capabilities as a function of cellular generations

user capabilities (and expected features) have increased exponentially over the evolution of the cellular generations. We started with voice only and then moved on to voice and short message service (SMS) capabilities in 2G. The data capabilities were improved in 3G to include packet-switched services. 4G provided mobile Internet with expanded use cases for the Internet of things (IoT), vehicle-to-everything (V2X), device-to-device (D2D), etc. The next generation cellular system, 5G, is expected to only increase the use case possibilities, thus opening many doors for innovative products to be delivered.

The DL is the communication direction from the BS to the handset or user equipment (UE). The UL is the communication direction from the UEs to the BS. The UL also consists of random access where UEs attempt to access the communication systems resources, either from power on state or initiating a new transaction.

The method used to separate the DL and UL communication is called duplex. For example, this operation can be performed in the time (TDD) and/or the frequency (FDD) domains. In TDD, certain time slots are allocated to the DL and other time slots to the UL. In FDD, the UL and DL transmission occur simultaneously in different frequency bands. The benefits of TDD are a single spectrum is needed and shared (no paired spectrum is necessary), and there are symmetrical channel views (UL measurements can be used for DL communications and vice versa). The benefit of FDD is the need for less timing synchronization requirements; however, due to the frequency separation between the DL and UL, the UL measurements may not be useful for DL communications as reciprocity cannot be guaranteed. Whichever method is used, latency (time duration to access the networks resources) is becoming more and more critical as a system performance indicator.

1.2.1 *The Evolution of Mobile Technologies*

In this section, we will introduce the mobile radio access technologies (RATs) and comprehend their evolutionary benefits and advantages. Figure 1.1 shows the cellular standard evolution from 1G to 4G. We notice as 2G and 3G evolved, there was an increase in system complexity across multiple standards. This changed as the industry converged to a single 4G standard, where now there is an increase in complexity within a single standard.

Orthogonal Multiple Access Techniques

- **FDMA** (frequency division multiple access)
  - ✓ Difficult to assign multiple carriers in the same channel
  - ✓ Narrowband channels (less than the coherence bandwidth of the wireless channel) are desirable
  - ✓ Guard bands in frequency domain are needed to reduce spectral emissions into adjacent frequency bands
  - ✓ Finite number of orthogonal resources.
- **TDMA** (time division multiple access)
  - ✓ Inter-symbol interference compensation (equalization) is needed
  - ✓ Uses guard bands in time domain to allow for time delay variations of UL transmissions
  - ✓ Synchronization of time slots across all uses is critical to not destroy the OMA principle
  - ✓ Finite number of orthogonal resources.
- **CDMA** (code division multiple access)
  - ✓ Uses the entire bandwidth at the same time utilizing spreading codes
  - ✓ Finite number of orthogonal resources.
- **OFDMA** (orthogonal frequency division multiple access)
  - ✓ Assigns different sub-carriers to different users (at different time slots)
  - ✓ Finite number of orthogonal resources.

Spectrum is very precious to the operators and remains necessary to deliver increased system and user throughput. There is an industry-wide movement to not only use the traditional licensed spectrum, but also embrace the unlicensed (traditionally used by WiFi devices) and the shared spectrum whenever and wherever possible.

1.2.2 First-Generation Cellular Systems

The first-generation (1G) mobile cellular system was created to enable voice communications and support mobile users when a voice call would "hand off" to another base station (or cell) as the mobile user physically traversed the cellular environment. The technology used was analog frequency modulation (FM), and the spectrum was divided into 30 kHz segments, called channels. A single user utilized the entire channel for the duration of its call. This system is called advanced mobile phone services (AMPS) and is referred to as 1G [2].

In order to support a wide coverage area, a frequency reuse technique was introduced. Here the same frequency channels were allowed to be reused by other users, at the same time, as long as the distance was large enough to cause minimal interference. This interference is called co-channel interference or inter-cell interference.

In an effort to increase overall system capacity, a new technology was introduced called narrowband AMPS (N-AMPS). Here channel spacing was reduced to 10 kHz. Similarly, in an effort to introduce data services (which were not supported in AMPS), cellular digital packet data was proposed which utilized frequency channels when voice users were not present. However, it was quickly determined that an integrated voice and data wireless network is needed to effectively and efficiently deliver such services. Simple and robust discriminator detectors were used which were implementable, yet susceptible to random FM and deep fades from multipath observed in the radio environment. Forcing the mobile cellular system community to move to a different modulation technique [3, 4].

A typical cellular network architecture for 1G is provided in Fig. 1.4. Where a cell is denoted as a hexagonal shape. To be able to increase capacity, the cells can be divided into smaller cells, also called sectors. The mobile telephone switching office (MTSO) connects to base transceiver stations (BTS) and the public switched telephone network (PSTN). It also controls handovers, call routing, registration, authentication, etc. This was a circuit switched (CS)-based network. The network

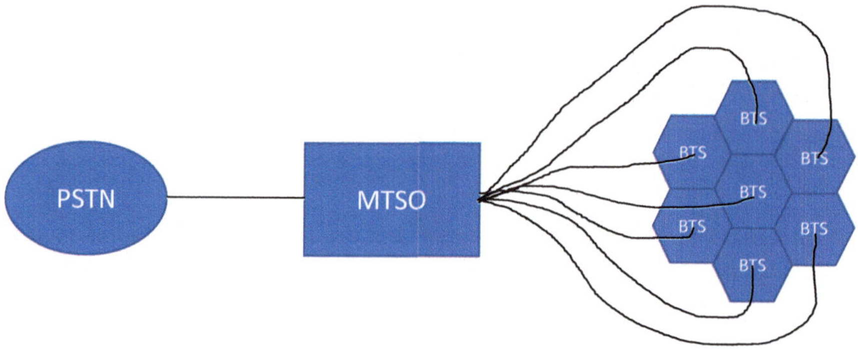

Fig. 1.4 1G network architecture block diagram

used licensed spectrum to deliver the voice services, spectrum that operators purchased from the relevant governing bodies.

The 1G analog cellular standards globally deployed are listed below. Note a single global standard did not exist.

- Advanced mobile phone services (AMPS)—*US based*
  - ✓ Analog FM modulation, FDD duplex, FDMA-based multiple access
  - ✓ Supports N-AMPS for narrowband, the channel bandwidth was decreased from 30 to 10 kHz.
- Nordic mobile telephone (NMT)—*The Nordic countries*
  - ✓ Analog FM modulation, FDD duplex, FDMA-based multiple access
  - ✓ Channel bandwidth was dependent on the frequency band deployed: either 25 kHz or 12.5 kHz
  - ✓ Supported roaming in European countries.
- Total access communication system (TACS)—*UK based*
  - ✓ Variant for Japan available (J-TACS)
  - ✓ Analog FM modulation, FDD duplex, FDMA-based multiple access
  - ✓ Channel bandwidth 30 kHz.

1.2.3 Second-Generation Cellular Systems

The second-generation (2G) mobile cellular systems were created to expand the voice user capacity as well as to offer an integrated data services capability. The technology moved away from analog and toward digital modulation. This shift to digital enabled better quality voice communications via usage of voice coders (vocoders), support of data services, initially through short messaging services (SMS), enabled encryption to support security, and increased system capacity.

This generation created a shift from FDMA to TDMA and CDMA. These were very interesting times the cellular users were facing; by this we mean being exposed to incompatible 2G cellular systems. The European community was backing global system for mobile communications (GSM), while the USA was struggling with two competing standards: IS-54 (later renamed IS-135) based on TDMA and IS-95 (later renamed CDMA-One) based on CDMA. All three of these cellular standards had technical merit.

In order to increase system capacity, not only was the frequency band divided into channels, but also time was divided into time slots for TDMA. In the CDMA case, each user's information was scrambled and frequency spread by a pseudo-noise (PN) sequence; all users transmitted at the same time over the entire channel.

These standards used licensed spectrum purchased by network operators from the local spectrum governing body. Receiver complexity was growing exponentially especially when considering data rates, modulation scheme, and number of antennas involved have increased.

The 2G digital cellular standards globally deployed are listed below. Note a single global cellular standard did not exist.

- GSM—*single standard in Europe*
  - ✓ TDMA based
  - ✓ Digital modulation (GMSK), FDD duplex
  - ✓ Channel bandwidth = 200 kHz
  - ✓ Frame duration = 4.615 ms
  - ✓ Time slot duration = 0.557 ms (8 slots/frame)
  - ✓ Data Rate = 270.833 Kbps
  - ✓ Evolved to general packet radio services (GPRS), also considered 2.5G
  - ✓ Evolved to enhanced data rates for GSM evolution (EDGE), also considered 2.75G.
- IS-54 (also called IS-136)—*standard in US*
  - ✓ TDMA based
  - ✓ Digital modulation ($\pi/4$-DQPSK), FDD duplex
  - ✓ Channel bandwidth = 30 kHz
  - ✓ Frame duration = 40 ms
  - ✓ Time slot duration = 6.67 ms (6 slots/frame)
  - ✓ Data rate = 48.6 Kbps.
- IS-95 (also called CDMA-One)—*standard in US and Korea*
  - ✓ CDMA based, developed by Qualcomm
  - ✓ Digital modulation (QPSK, O-QPSK), FDD duplex
  - ✓ Frame duration = 20 ms
  - ✓ Data rate = 115 Kbps.

These standards were all circuit switched (CS)-based networks, which over time, had extensions (e.g., evolving from 2G → 2.5G → 2.75G) which allowed interfacing to packet switched (PS)-based networks. Due to economies of scale, deployment costs, patent policies, and global backing, GSM held the largest piece of the cellular market share. The users' appetite increased thus forcing 2G to take incremental evolutionary steps such as 2.5G (GPRS) and 2.75 (EDGE). Both of which were created to increase the user data rate beyond the baseline GSM capability as well as add packet services capability. These systems are very much in use today [5].

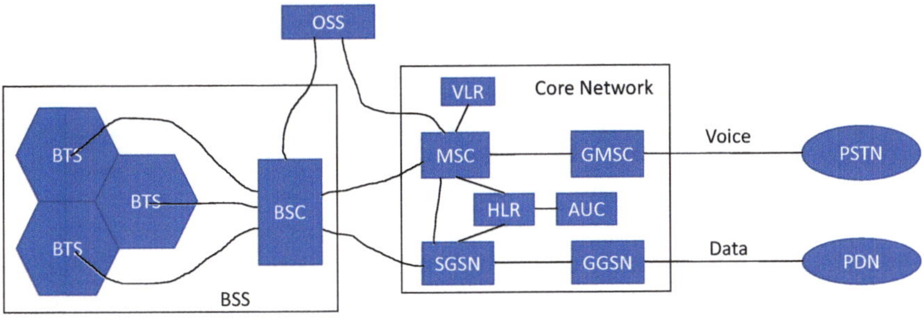

Fig. 1.5 2G GSM network architecture block diagram

The GSM network architecture block diagram is shown in Fig. 1.5 and is made up of the following network elements:

- Base station subsystem (BSS) which is composed of two parts: BTSs and base station controller (BSC)
- Operation and support subsystem (OSS) which controls and monitors the overall GSM network
- Mobile switching center (MSC) which provides registration, authentication, call location, call routing, etc.
- Home/visitor location register (HLR/VLR), a database of subscriber information
- Gateway mobile switching center (GMSC) obtains subscriber information from HLR to route calls to correct MSC
- Serving GPRS support node (SGSN) for packet routing and mobility management
- Gateway GPRS support node (GGSN) organizes the GPRS network and external packet-switched internetworking.

1.2.4 Third-Generation Cellular Systems

This third-generation (3G) digital cellular system was created to increase system user capacity and satisfy the increasing data rate appetite. This generation provided users the ability to surf the Internet and have simultaneous voice and data services. It also was the ecosystem catalyst to introduce video applications to the cellular user's devices. Both CS and PS services were supported from its initial definition. At this point, in the cellular evolution, mobile access to the Internet was becoming more and more important. The MA technique shifted from using both TDMA and CDMA to standardizing on CDMA. CDMA-One evolved into CDMA2000, and GSM/IS-136 evolved into Wideband CDMA (WCDMA).

CDMA is a multiple access technique where multiple users are separated by PN codes and transmit at the same time over the whole bandwidth allocated. It is well known as more users transmit, intra-cell interference grows called multiple access interference. A *power control* mechanism was used in the system to not only improve performance in a multipath fading environment, but also control the interference introduced by each additional user in the system. Power control was the solution to the *near-far* problem, with its goal of having the UE transmission flexible so that all users received by the NodeB would have comparable energy. This created a solution where all users were able to equally interfere with each other.

The international telecommunications union (ITU) provided 3G goals in the form of IMT-2000 requirements. The 3GPP standards body was formed and created specifications to support implementations which satisfied these ITU requirements. The 3G cellular system continued to use the licensed spectrum. The *small cells* concept was introduced in the standard and was called HomeNB. *Carrier aggregation* (CA) was a seed planted into the 3G system as a method to evolve and support higher user data rates. This seed grew and is presently benefiting the modern 4G systems. Multiple-input and multiple-output (MIMO) *spatial multiplexing* was also a seed planted into 3G where multiple streams or layers were transmitted to the user (provided the channel matrix rank requirement was satisfied). *Higher-order modulation* (HOM) was also standardized; a movement from 16-QAM to 256-QAM in a land mobile cellular system was very new during these times.

An example of a *rake receiver*, designed to counter the effects of multipath fading, used for the reception of the WCDMA downlink signal is provided in Fig. 1.6. A key WCDMA system design parameter is to have the transmission bandwidth be larger than the *coherence bandwidth* of the wireless channel so that multipath (or echos) can be used to exploit time diversity of the channel. The rake receiver consists of N fingers which individually track multipath and demodulate the respective waveforms. Each finger is assumed to demodulate the common pilot channel (CPICH) to support channel estimation [6, 7].

Receiver complexity grows linearly with data rate, modulation scheme used and number of antennas supported. In the WCDMA standard, both FDD and TDD duplex options were provided for paired and un-paired spectrum, respectively. To aid receiver digital signal processing, both common and dedicated pilot symbols were inserted into the waveforms. A complete shift from non-coherent detection to coherent detection was recognized by the cellular industry.

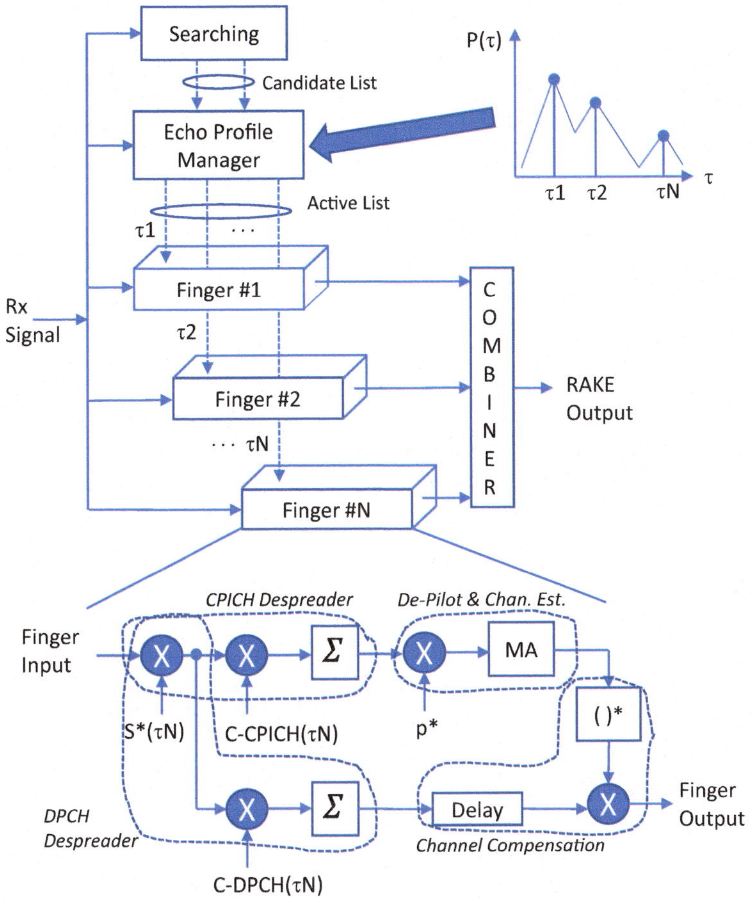

Fig. 1.6 3G WCDMA rake receiver block diagram

The 3G WCDMA network architecture is shown in Fig. 1.7. The NodeB replaced the BTS functions, and the radio network controller (RNC) replaced the BSC functions. WCDMA is also called universal mobile telephone system (UMTS). The UMTS terrestrial radio access network (UTRAN) consists of NodeB and RNC groupings [8].

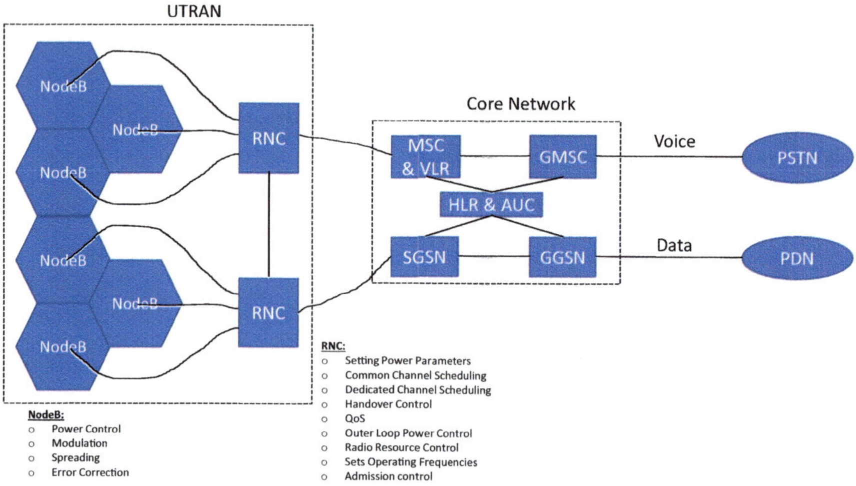

Fig. 1.7 3G WCDMA network architecture block diagram

The 3G cellular standards globally deployed are listed below. Note that a single global cellular standard did not exist.

- WCDMA (also called UMTS)
  - ✓ Digital modulation (QPSK, 16-QAM, 64-QAM, etc.), FDD/TDD duplex
  - ✓ Channel bandwidth = 5 MHz (with a chip rate = 3.84 Mcps)
  - ✓ Frame duration = 10 ms
  - ✓ Time slot duration = 0.667 ms (15 time slots/frame)
  - ✓ Data rates up to 1 Mbps
  - ✓ Defined by the 3GPP standards body.
- CDMA2000
  - ✓ Digital modulation (QPSK, 16-QAM, 64-QAM), FDD duplex
  - ✓ Channel bandwidth = 1.25 MHz × 3
  - ✓ Frame duration = 10 ms
  - ✓ Time slot duration = 0.667 ms (15 time slots/frame)
  - ✓ Data rates up to 1 Mbps
  - ✓ Defined by 3GPP2 standards.

A WCDMA high-level functional block diagram of the downlink transmitter is shown in Fig. 1.8. Each cell has a unique scrambling code, whereas the same spreading codes (orthogonal variable spreading factor (OVSF)) are reused in every cell. The spreading codes were also called channelization codes. The block diagram of the uplink transmitter is also shown in Fig. 1.9. Each cell has a unique scrambling

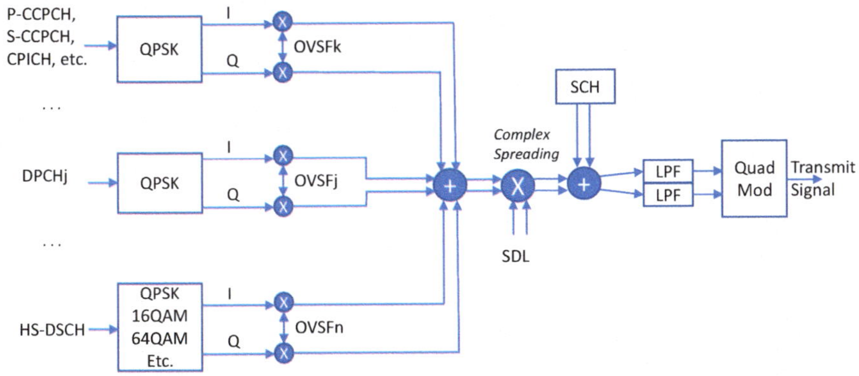

Fig. 1.8 WCDMA downlink transmitter block diagram

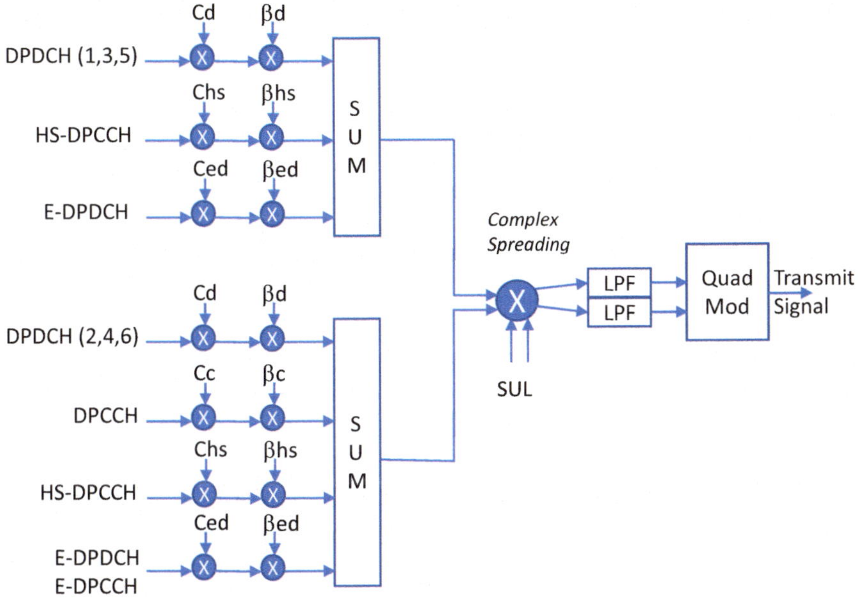

Fig. 1.9 WCDMA uplink transmitter block diagram

code, spreading codes are also reused in every cell. The difference is the uplink also uses quadrature multiplexing between the I and Q channels [9, 10].

The WCDMA cellular system evolved to what is called high speed packet access (HSPA)[1] which consisted of both the downlink (HSDPA) and uplink (HSUPA) components. HSPA was created because an efficient way to deliver packet services was

[1]The CDMA2000 cellular system evolved to what is called evolution-data optimized (EV-DO) to support data only extension.

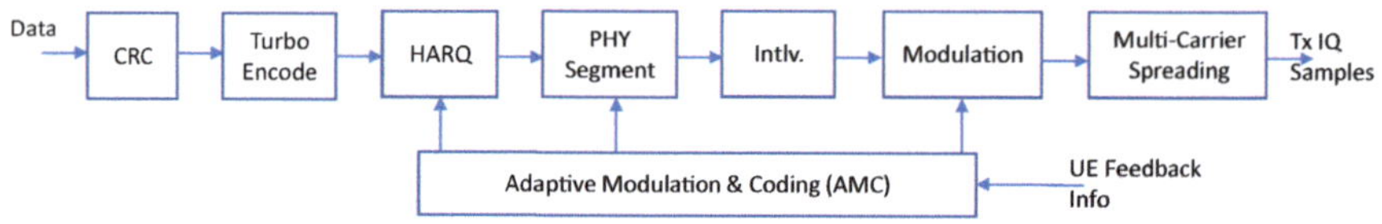

Fig. 1.10 HSDPA transmitter block diagram

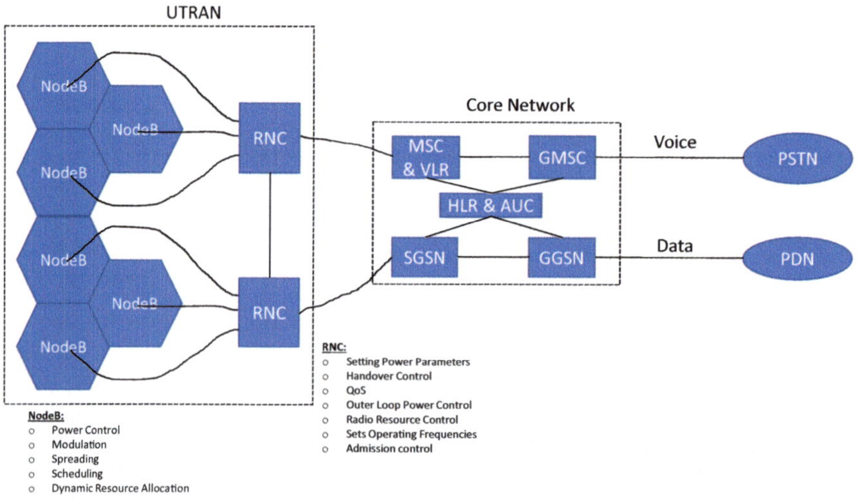

Fig. 1.11 HSDPA network architecture block diagram

needed. HSPA introduced the shared channel concept and adaptive modulation and coding (AMC) supporting hybrid automatic repeat request (HARQ). Also, in HSPA

- Whole frequency band was used (no frequency reuse greater than 1).
- Users scheduled on time slot (referred to a transmission time interval (TTI) with a duration of 2 ms) basis and used PN codes as physical resources.
- Network architecture flattening concept was introduced to support low-latency communications.

An HSDPA block diagram is shown in Fig. 1.10. Each user packet is protected and transmitted to the UE; an acknowledgement (ACK) is expected to ensure error-free communication. In the event of negative acknowledgements (NACK), the scheduler will decide which combination of coding, modulation, and physical resources should be used to increase the likelihood of error-free communication [11]. The 3G HSDPA network architecture block diagram is shown in Fig. 1.11. Certain functionality (highlighted in the figure) previously performed in the RNC are now performed closer to the edge of the access network within the NodeB—supporting the network flattening initiative.

Evolving WCDMA further became a great concern to cellular system designers. Every known tool was being used to increase the user data rate. HOM was used to increase the data rate within an allowed spectral bandwidth. MIMO, in the form of spatial multiplexing, was used to increase the data rate within an allowed spectral bandwidth. The spectral bandwidth was also increased in the form of aggregating carriers, to increase the data rate; however, the spectral efficiency remained unchanged. Increasing the single-carrier bandwidth brought along increased concerns. The baseline WCDMA system used a rake receiver which performs better when the processing gain is larger, rather than smaller. This bandwidth expansion factor (signal spreading operation), coupled with the desired high user data rates, prohibited the conventional WCDMA system from evolving further.

1.2.5 Fourth-Generation Cellular Systems

This fourth-generation (4G) digital cellular system was created to support the exponential system capacity and data rate appetite. Much higher data rates were required to enable mobile Internet access and video applications. Long-term evolution (LTE) is also known as 4G and only supports PS-based networking. The standard is also evolving to use licensed, unlicensed, and shared spectrum options—all with a common goal of increasing the user data rate, increasing system capacity, lowering latency, and improving the user experience. The ITU provided 4G goals in the form of IMT-2010 requirements.

At this point in the cellular evolution, the industry converged to a single standard, LTE. The LTE cellular system is based on OFDMA where the TTI has been reduced from 2 ms (used in the 3G cellular system) to 1 ms. This TTI reduction improved performance by being able to more quickly react to changing channel conditions, so more efficient scheduling algorithms can be used. The reduced TTI also provided a reduction in the end-to-end latency. The frequency bandwidth options have also increased: 1.4, 3, 5, 10, 15, and 20 MHz to provide flexible bandwidth deployments. To efficiently support FDMA multiple access, OFDMA (via inverse fast Fourier transform (iFFT) and FFT operations) was chosen which divided the frequency band into sub-channels (or sub-carriers) of 15 kHz spacing. To keep receiver signal processing complexity to a minimum, it was desirable to have the sub-carrier spacing less than the coherence bandwidth of the wireless channel. To deliver higher data rate services, MIMO support is mandatory to accommodate multiple layers through spatial multiplexing [12].

Recall that with a TDMA system, the increased data rate (or decreased symbol time duration) caused the receiver to use an equalizer to combat inter-symbol interference (ISI). The higher data rates, higher-order modulation, and longer delay spreads caused a significant increase in equalizer complexity. With a WCDMA system, the increased data rate (or decreased chip time duration) forced the receiver to make use of the time diversity of the wireless channel but required a large processing gain to adequately combat ISI. With WCDMA, the motivation was to have the transmission bandwidth larger than the coherence bandwidth of the wireless channel; however, for OFDM, the opposite holds true. OFDM addresses the higher data rate demand

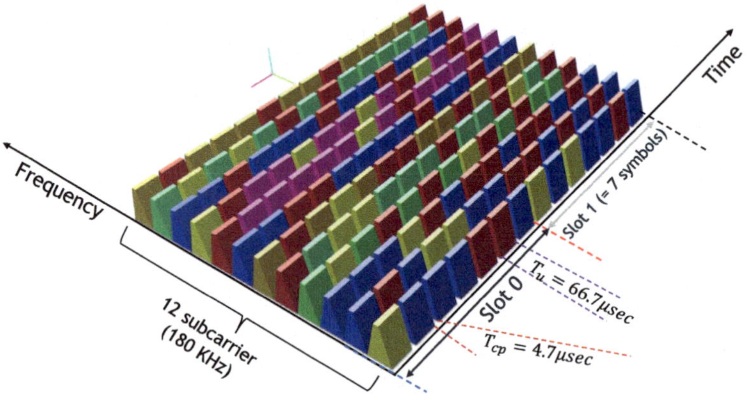

Fig. 1.12 Time/frequency representation of the OFDM signal for LTE standard. There are four different symbols (QPSK) each represented by one color

by generating many narrowband channels, where each narrowband channel can be seen to experience frequency flat disturbance. This observation coupled with the fact the frequency domain signal processing is possible, made OFDM a very attractive multi-carrier technique to mitigate a frequency selective fading environment.

In OFDMA, users are multiplexed in both the frequency and time domains, as depicted in Fig. 1.12 for LTE system. On the LTE air interface, the unit of allocation is a physical resource block (PRB). A PRB is 12 sub-carriers by 7 OFDM symbols which is equal to 84 modulation symbols. The minimum allocation to a single UE during a subframe (1 ms) is 2 PRBs with one PRB in each slot of the subframe. Thus, a UE will get a total of 2 PRBs/subframe which equals to 168 modulation symbols/subframe. Note that not all these 168 modulation symbols can be used to transmit user information, but some of these modulation symbols are used for synchronization or as pilot for channel estimation. Each PRB contains 12 sub-carriers, and thus have a bandwidth of 12×15 kHz $= 180$ kHz. Figure 1.12 represents 2 PRB (2×7 symbols $\times$ 12 sub-carrier). Assuming QPSK modulation, there are four different symbols represented by four different colors. Each color represents one resource element (RE) and carries two bits with QPSK modulation.

A block diagram providing an example of the OFDMA waveform generation is provided in Fig. 1.13. We also highlight the various points on the processing chain that can significantly impact system performance. The number of sub-carriers (SCs) has a direct impact on the data rate and user capacity of the system. From a system perspective, this value should be as large as possible; however, the occupied bandwidth needs to be controlled via the spectral shaping function. The OFDM symbol's peak to average power also impacts the occupied bandwidth and imposes linearity requirements to be met to minimize any increased spectral growth. Lastly, note the addition of *cyclic prefix* (CP) removes ISI from the wireless channel. The CP time duration should be large enough to exceed the length of the

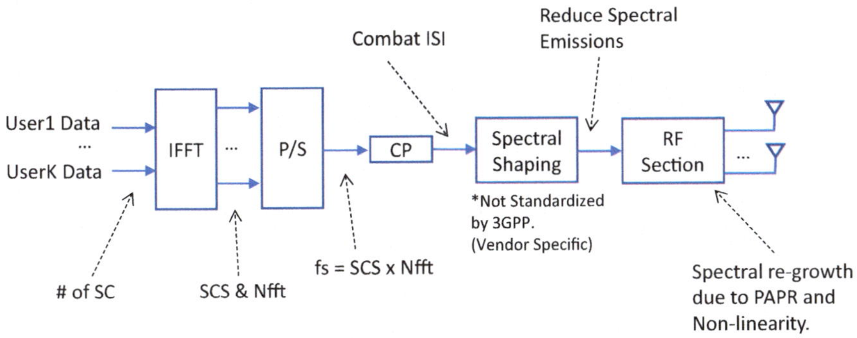

Fig. 1.13 OFDMA waveform generation with K sub-carrier (SC)

wireless channel time dispersion, but also as small as possible to maximize the user data information during the subframe duration.

There are certain disadvantages with OFDM that should be addressed in future systems, such as:

- **CP overhead:** The need for adding the CP introduces redundancy to the transmitted and thus results in a loss in spectral efficiency. This loss is larger when long CP is used or when the sub-carrier spacing (SCS) is small.
- **Sensitivity to frequency and timing offsets:** In order to keep the orthogonality in OFDM, the transmitter and receiver must have exactly the same reference frequency. Any frequency offsets will ruin the orthogonality, causing sub-carrier leakage known as inter-carrier interference (ICI).
- **High out-of-band (OOB) emission:** OFDM assumes rectangular pulse in time domain which is equivalent to *sinc* in the frequency domain which has infinite bandwidth theoretically and cause relatively high (OOB) emissions. The lack of spectral shaping (either filtering or windowing) is creating large spectral side lobes in the transmit spectrum.
- **High peak-to-average power ratio (PAPR):** The envelope of the OFDM waveform has a large variation which causes problems when encountering a nonlinear device such as a transmit power amplifier. The high PAPR in OFDM compared to the single-carrier transmission technique is due to the summation of the many individual sub-carriers with different phases which can results in a high PAPR when added together.

In terms of occupied bandwidth, 3GPP did not specify any spectral shaping technique in LTE and such each equipment and device vendor implements their own solution. OFDM sub-carriers are treated as $\sin(x)/x$, so applying spectral shaping will help produce a more spectrally efficient waveform with minimal or no impact to orthogonality performance. These spectral side lobes are relatively high in power due

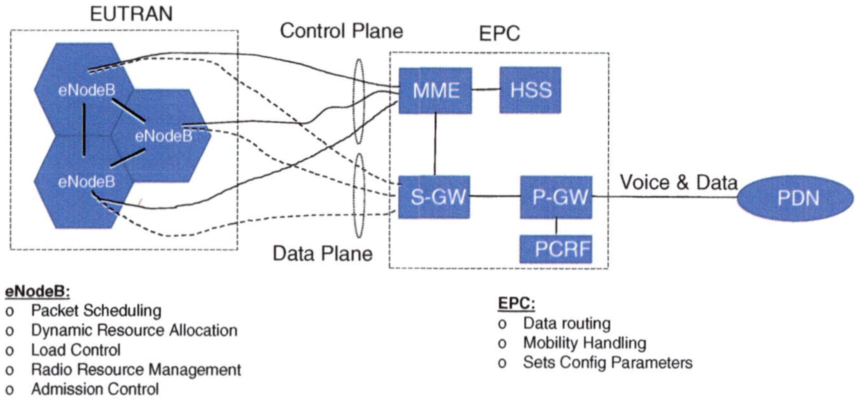

Fig. 1.14 4G LTE network architecture block diagram

to the assumed rectangular shaping. These high side lobes require a large guard band to reduce the out of band interference. Applying spectral shaping techniques, such as filter bank multi-carrier, universal filtered multi-carrier, etc., which are discussed in Chap. 2 through Chap. 4 of this book, will help reduce the side lobes. The other component can be found when viewing adjacent transmitted OFDM symbols in the time domain; there will be phase discontinuities that also cause spectral emissions.

The peak to average power concern of OFDM can be viewed as a weighted sum of sinusoids, which helps explain the large PAPR of the generated OFDM symbol (as high as 12 dB). A high PAPR can be problematic if the waveform encounters nonlinarites. Crest factor reduction is a technique used to reduce PAPR and a technique used to compensate for nonlinear distortion is digital pre-distortion. The LTE uplink waveform uses the single-carrier FDMA (SC-FDMA) method to reduce the PAPR impact on portable devices.

Lastly, to minimize ISI and provide the property of cyclic convolution, a small part of the end of each symbol is added to the beginning of each transmitted OFDM symbol. CP size depends on delay spread and LTE uses a short and a long CP. For LTE the short CP has a value of 4.7 μs, which is approximately 8% of the symbol time. Generally speaking, if a large delay spread is not expected to be encountered in a particular deployment, then a lower or shortened duration CP should be used.

The 4G LTE network architecture block diagram is shown in Fig. 1.14. Note that we now have a single global cellular standard. The evolved packet core (EPC) replaced the core network (CN) functions, and the eNodeB replaced the NodeB functions. The EUTRAN consists of eNodeB and EPC groupings. The EUTRAN to EPC connection consists of both *control plane* and *user plane* signaling. This was the beginning of an effort to separate user and control planes to allow for different evolution rates and network deployment scenarios/options [13].

In LTE, the SCS is set to 15 kHz which translates to an OFDM symbol duration of 66.67 μs. There are 14 data symbols per time slot (1 ms), and every OFDM symbol

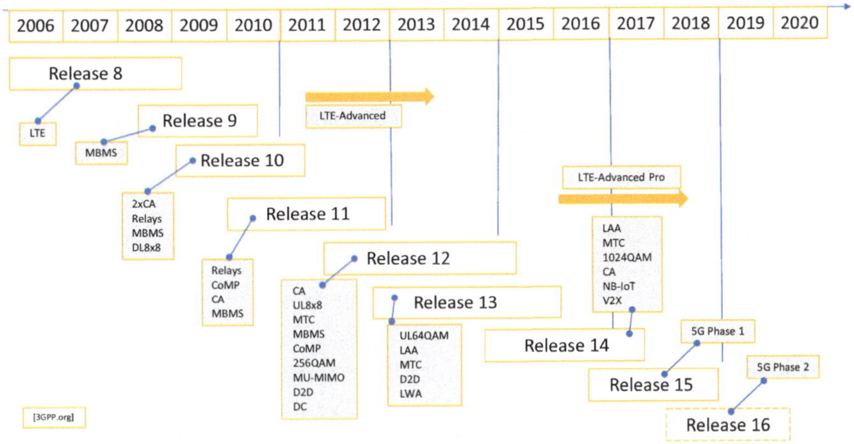

Fig. 1.15 3GPP release feature timeline

requires a CP. Including the time durations of all the CPs, results in an equivalence of 15 symbols (data + CP) in a time slot. The largest FFT size is 2048 which creates a sampling frequency of 30.72M samples/s. Below we list some LTE features in the 3GPP standards body release schedule.

- Data rates increased by employing HOM, MIMO, and CA
- New features are added: DC, V2X, IoT, D2D, etc.
- Advanced technology to support performance improvement: coordinated multi-point (CoMP), latency reduction, beamforming
- Spectral and RAT flexibility: licensed, shared, unlicensed and LTE-WiFi aggregation.

As discussed earlier, certain seeds were planted in the 3G cellular system to observe how beneficial they would become to the later generations. For example, CA continues to be useful, MIMO has become more and more essential, and HOM is effective. In fact, all three techniques have been successfully tested and are commercially deployed; they are required to achieve the greater than 1 Gbps data rate in LTE [14].

Figure 1.15 also reveals a departure from the typical cellular system evolution which has been given to increasing data rates, increasing user capacity, and lowering latency. This new trend clearly shows the additions of new services (or features, use cases) that the industry is recognizing are required as society demands. These new services were not really intended to be addressed when 4G was created in 2006. The width of new expected services is growing very rapidly (as depicted in Fig. 1.3).

The network is also experiencing its own evolutionary growth. The software-defined networking (SDN) and network function virtualization (NFV)-based "wave front" is departing the data center [15], making its way through the core network and on to the wireless access network. These CN and RAN workloads have started to be implemented on a homogeneous, general-purpose CPU-based platform (instead of the traditional dedicated logic + digital signal processor + microcontroller approaches). This has sparked the cloud-RAN movement utilizing the Information and communication technology (ICT) industry benefits. 4G will deploy these technologies, and when they are successful, the expectations are that 5G will be a network upgrade.

1.3 5G Drivers, Technologies, and Spectrum

The network architecture block diagram of fifth-generation (5G) of cellular is shown in Fig. 1.16. The 5G core (5GC) replaced EPC; next-generation radio access network (NG-RAN) consists of distributed unit (DU) and centralized unit (CU) grouping; the gNodeB replaced the eNodeB. As we will discuss, providing a flexible and scalable network architecture is essential for 5G. In this theme, the combination of DU and CU were introduced to support various RAN split options to extract the above-said benefits [16]. The 5GC elements consists of the following:

- Access and mobility management function (AMF): performs ciphering and integrity protection, mobility management, authentication, and authorization, etc.
- Session management function (SMF): performs UE IP address allocation and management, selection and control of UPF, roaming, etc.
- Unified data management (UDM): performs subscription management, user data, registration and mobility management, etc.
- Policy control function (PCF): performs policy rules for CP functions, etc.
- User plane function (UPF): performs the external interconnect point to data network, QoS handling of UP, etc.
- Application function (AF): interacts with policy framework for policy control, etc.

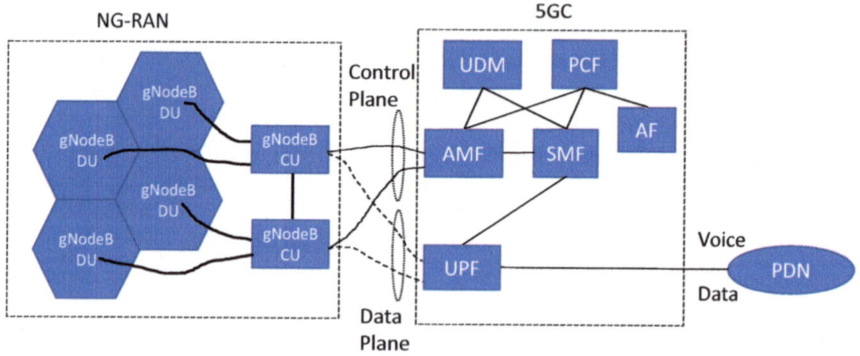

Fig. 1.16 5G network architecture block diagram

1.3.1 5G Drivers

5G cellular systems need to make a significant jump in features and performance over LTE as incremental improvements are not wanted and do not justify the significant capital investments operators need to commit to deploy 5G services. It is important to note that we have maintained a single global cellular standard. The 5G driving factors are [17]:

- Increased user data rate
- Increased system capacity
- Massive number of connections
- Reduction in end-to-end latency
- Heterogenous mix of services
- Flexible bandwidth deployments
- Network flexibility
- Move to more energy-efficient communications.

The ITU provided 5G goals in the form of IMT-2020 requirements, they are provided in Table 1.2. For comparison purposes, we have also included the IMT-advanced requirements.

5G NR will support both non-stand-alone and stand-alone modes of deployment. The NSA deployment will use LTE to provide wide area coverage, control and data planes and connection into an evolving EPC. 5G services will provide high-speed data via a dual connectivity scenario. The stand-alone deployment will provide control and data planes as well as a connection into a 5G CN.

The ITU published the diagram shown in Fig. 1.17 to identify 5G services. The three significant use cases (corners of the triangle) are meant to encapsulate the expected usages of 5G in the future:

- Enhanced mobile broadband (eMBB)
- Massive machine to machine communication (mMTC)
- Ultra-reliable low-latency communications (URLLC).

These 5G use cases range from smart home, connected drones, ehealth, connected energy, autonomous cars, real-time virtual reality/augmented reality gaming, etc. The introduction of low latency techniques has started in LTE to aid the transition to transforming the network in preparations for the widely varying 5G services [17]. The 5G cellular system is expected to support these usage scenarios by employing the following technologies:

- **Flexible spectrum deployments:** licensed, unlicensed and shared spectrum, larger and contiguous bandwidth, multi-RAT, etc.
- **Improved network architecture:** support of the ICT industry cloud trend, SDN/NFV, network slicing, multi-access edge computing, lower latency, etc.

Table 1.2 Comparison of IMT-2010 and IMT-2020 requirements

System metrics	IMT-2010	IMT-2020	Comments
Peak data rate	DL: 1 Gbps UL: 0.5 Gbps	DL: 20 Gbps UL: 10 Gbps	Maximum achievable data rate under ideal conditions
Area traffic capacity (Mbps/m^2)	0.1	10	Total traffic served per geographic area
Network energy efficiency (bit/Joule)	1x	100x (less)	Quantity of info bits per unit of energy consumption
Connection density (devices/km^2)	10^4	10^6	Total number of connected devise per unit area
Latency (ms)	10	1	Time from when source sends a packet to when the destination receives it (end-to-end one way)
Mobility (kmph)	350	500	Maximum speed a defined QoS can be achieved
Spectral efficiency (bps/Hz)	1x	3x (more)	Average data throughput per unit of spectrum and per cell
User expected data rate (Mbps)	10	100	Achievable data rate ubiquitously available across the coverage area

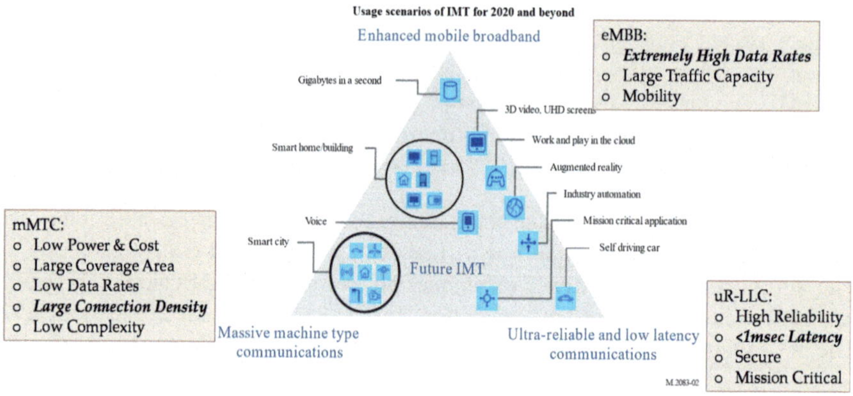

Fig. 1.17 IMT-2020 usage scenarios

- **Flexible numerology:** Support a wide variety of use cases and spectral deployments (below and above 6 GHz bands), flexible time slots and transmission bandwidths, etc.
- **Modulation and coding:** QAM modulation continues to provide a reasonable spectral and power efficiency trade-off, polar and other forward error correcting codes.
- **Advanced techniques:** NOMA, full-duplex, spectral shaping, etc.

1.3.2 5G Technologies

Expected to be commercialized around 2019/2020 time frame, 5G mobile networks are under intense reach and development activities. Compared to the current 4G mobile networks, 5G networks are expected to support enormous system capacity, much less latency, and about 1000 times more devices per squared kilometer, among other requirements. To satisfy these requirements, several new technologies have been suggested and are being developed for 5G networks. These technologies include but are not limited to: massive MIMO, software-defined networking, mm-Wave, cloud radio access network (cloud-RAN), non-orthogonal multiple access, M2M communications, mobile edge computing, wireless caching, ultra-dense networks, and full-duplex communication. In the following, we briefly describe some of these technologies.

1.3.2.1 Massive MIMO

In discussing massive MIMO, let us first address the term "massive." It is used to denote the large number of antenna elements that are used in the antenna signal processing. The number of antennas to be considered massive should be greater than 64 elements. Massive MIMO relies on the *law of large numbers* to make sure that the channel and hardware imperfections (e.g., noise, fading, and hardware) average out when signals from a large number of antennas are combined in the air together [18]. Multiple antennas afford two options in which the antennas can be used: First is to provide an array gain by focusing energy in desired directions and nulling in unwanted signal directions (forming a beam). Second, is to provide spatial multiplexing gain by sending independent data streams on each antenna. Either technique can be used to increase the overall user or system data rate. Both options are shown in Fig. 1.18 [18].

First, consider using massive MIMO for *beamforming*; here the antenna arrays can be arranged in either linear, rectangular, or circular arrays that can also be stacked. Massive MIMO will be deployed for 4G and 5G; in fact, high-frequency bands lead to more compact, large-scale antenna arrays due to the smaller wavelength. Massive MIMO can be deployed in either FDD or TDD duplex methods, TDD systems allow

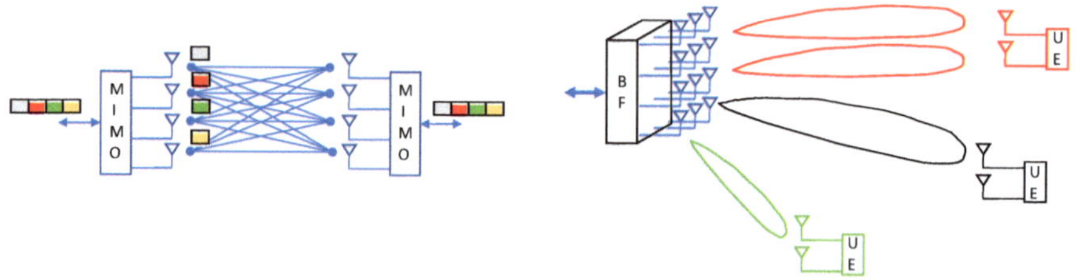

Fig. 1.18 Massive MIMO examples: spatial multiplexing (left) and single-/multi-user beamforming (right)

the users to invoke the theorem of reciprocity to apply what is observed on the UL to the DL.

Next, consider using massive MIMO for *spatial multiplexing*, which has been widely used for 4G and will continue in 5G deployments. Spatial multiplexing can be achieved provided the rank of the channel matrix between the transmit and receive antennas is greater than 1. In fact, for a 4×4 MIMO system, full capacity is only observed if the channel matrix rank is full (in this case, a value of 4).

Due to the success of spatial multiplexing in LTE, it would be logical to assume this continues for 5G and on a larger scale. This is true; however, should come with a warning. The larger the MIMO dimension, the less likely one would experience full rank. This means designing a 256×256 MIMO array and expecting to send 256 layers to a single user, all the time is a difficult assumption to make. This is one of the reasons 5G has limited the number of DL layers per user to 8. The implementation complexity involved in implementing massive MIMO in the digital domain is significant. Hybrid beamforming has been proposed to provide a compromise in performance/capability with complexity.

This brings forth an interesting question: Assuming a maximum number of layers of 8, what can one do with the remaining degrees of freedom? Some can be used to create (or form) beams and some can be used to multiplex other users over the antenna array. This last comment is known as multi-user MIMO (MU-MIMO). Here, multiple users transmit and their collective transmissions are treated as though they came from a single source of multiplexing. The beamforming weights can create a beam in the azimuth and elevation directions.

When considering beamforming, array gain can be used in a variety of ways. It can be used to the extend coverage area, reduce the transmit power of devices on the UL, improve signal-to-interference-plus-noise ratio (SINR) resulting in high user throughput and to reduce the transmit power on the DL thus improving overall power efficiency.

The number of antenna elements needed depends on a few items:

- Array gain (coverage area, power relief, etc.)
- Multiplexing layers needed
- Multi-users expected to be serviced
- Frequency band used (form factor, etc.)

- Signal processing complexity (CSI estimation, analog vs. digital domain, etc.)
- System performance gains (SINR, capacity, data rate, etc.).

One of the benefits of using multiple antenna techniques, for either transmitting or receiving, is the significant reduction in channel variation. This behavior is essential in combating multipath fading, and having at least 64 antennas in the antenna array significantly reduces the channel variations. Multiple 5G deployment scenarios proposed by 3GPP have varying use cases for eMBB, URLLC, and mMTC services. In these deployment scenarios, the maximum number of DL antennas discussed was 256 and maximum number of UL antennas discussed was 32.

1.3.2.2 Software-Defined Networking

Network functions virtualization (NFV) and software-defined networking (SDN) are supporting the movement to a software-centric network. These capabilities offer great technical (in the form of system performance) and financial (in the form of CAPEX and OPEX) improvements to the network operators. This movement provides the network operators with tremendous benefits such as: a more manageable means to monitor the network, better support of new feature roll-outs, network relocation, etc. However, it also opens the doors for new market players (such as Internet service giants, cable service providers, etc.) who wish to establish wireless network presence. The adoption has been to initially virtualize the less timing critical functions, such as in the EPC (also called vEPC) and then transition down the protocol software stack toward the physical layer [15].

Moving to a SDN allows network operators to become nimble in deployments of various use cases. One benefit is called network slicing. Here the network will be able to dynamically pull together the access and core network functions necessary to satisfy the requirements of a specific use case (latency, bandwidth, etc.). We have seen a trend that started in 4G where a diverse set of services have emerged, and 3GPP is addressing this demand as part of LTE's evolution. We expect this demand to increase and continue to create diverse requirements. The LTE network architecture (at its conception) has been called monolithic and needs to be more flexible and scalable as we introduce 5G services. Network slicing is a technique proposed to support these wide variety of use cases.

Network slicing creates virtual network architectures based on SDN and NFV principles. These virtual networks (or slices) are created on top of a common shared physical infrastructure and can be "optimized" to meet requirements of applications, services, or operators. The virtual networks consist of a set of network functions instantiated to provide a complete end-to-end logical (or virtual) network to meet the targeted performance requirements. For example, mMTC communications rely on user capacity and not necessarily low latency, whereas autonomous cars rely on low latency and not necessarily the highest throughput eMBB services would require. Figure 1.19 provides a block diagram example of how the network may be sliced to support the various 5G services discussed above.

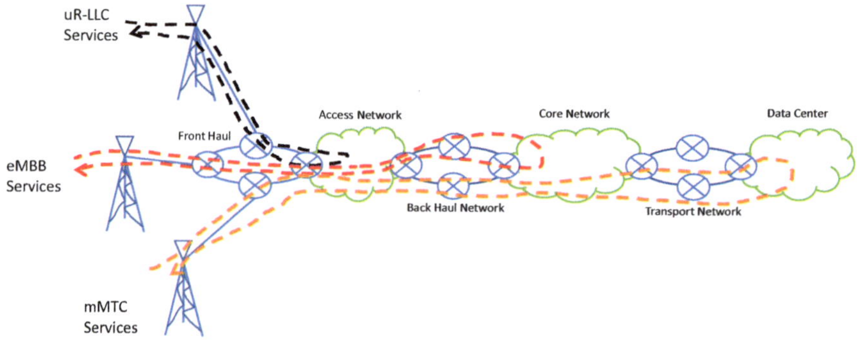

Fig. 1.19 Network slicing example supporting UR-LLC, eMBB and mMTC services

1.3.2.3 Multi-Access Edge Computing

To support demands for lower latency, optimizations in the 5G air interface alone are not sufficient, we must also optimize the network. Multi-access edge computing (MEC) is a method of moving core network or data center centric functions closer to the edge of the network (toward the antenna) where the data will be operated upon. It has been shown using this Principle of Relocation; the user end-to-end latency can be significantly reduced. Additionally, the backhaul traffic can also be reduced since the "back-and-forth" traffic has been significantly minimized by this move [19].

MEC enables cloud computing capability to be within the access network, which is closer to the user devices. This will also be supported by *fog computing*. The edge of the network is considered to be the antenna within the remote radio heads (RRHs) which are connected to the radio access network (RAN). There are a number of reasons to place the computing capability at the edge of the network. The most significant reason is to reduce latency (or delay) a mobile application encounters when trying to connect to a server. This eliminates the time a packet needs to enter the wireless network before being acted upon. The closer the MEC server is to the edge, the smaller the delay the applications would encounter. Examples of the expected delays are: latencies <1 ms are needed to support industrial robots and autonomous driving applications, latencies <10 ms are needed to support augmented reality applications, and latencies <100 ms are needed to support-assisted driving applications.

Figure 1.20 shows the concept of distributing the functionality which is typically located in the CN and data center (cloud computing) to the edge (fog computing). Besides lower application latency, we can observe lower backhaul traffic by not sending large packets all the way into the network to be processed and then sent all the way back to the edge [20].

MEC will perform compute and storage functionality with some market drivers for MEC deployments being:

- Reduce total cost of ownership (OPEX and CAPEX)

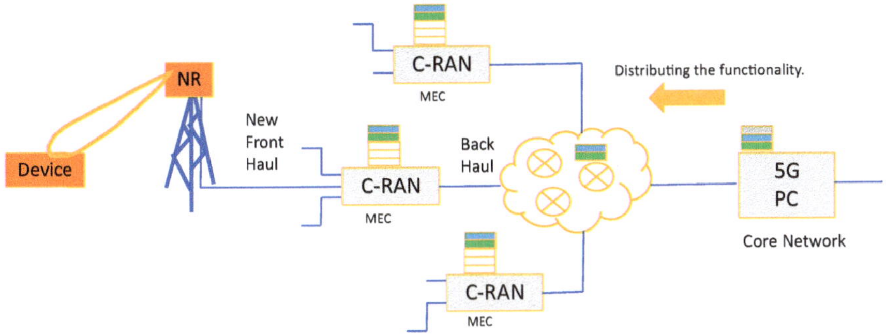

Fig. 1.20 Network diagram displaying distributing computing functionality to the edge

- Increase revenue by providing ability to create new services utilizing new technology such as artificial intelligence, content distribution network, etc.
- Natural migration as virtualization proliferates out to the access network (edge and fog)
- Improve performance (lower latency, reduce backhaul traffic).

A point also worth discussing. Why does the network edge need to be at the antennas? We should move away from the black and white viewpoint of network/device (also known as a cell-centric view) toward a more colorful viewpoint (also known as a user-centric view) where the edge is more blurred. Many reports reveal the total wireless devices are expected to be greater than 20B devices around the 2025 time frame. We should be cognizant that the number of devices is exceeding the number of people in the world. Also, since the computing performance of devices (handheld, laptop, etc.) is becoming more and more complex and capable, devices should be considered as an extension of the network—in other words the network edge.

1.3.2.4 RAN Split

The traditional and most commonly deployed fronthaul technology is based on fiber using the common public radio interface (CPRI) protocol. CPRI carries the IQ samples between the RAN and RRHs [21]. The CPRI capabilities are being stressed to support the evolution of LTE, especially when CA and massive MIMO deployments are required. This stress is due to the larger bandwidth required to transport the IQ waveform samples to the RRH, and only becomes more problematic when 5G enters the picture. Hence, next generation front haul technology is needed to support the expected 5G services [16].

A few front haul options exist: One solution is to standardize on another protocol that can use higher bandwidth technologies such as ethernet-based protocols (e.g., 25, 100 GB) while another component of the solution is to use a different RAN split options (with lower bandwidth requirements). A few RAN split options exist (proposed by the 3GPP) that can reduce the front haul bandwidth requirements as well as latency, and potentially trade-off performance [16].

One RAN split option transports modulated symbols, which is a point in the processing chain that is prior to being converted to the time domain by the iFFT operation on the transmit side. The frequency domain sampling rate is much lower, thus allowing more carrier-antenna combinations to be supported. This technique still maintains a centralized processing capability to allow for more complicated scheduling across cells. Another RAN split option transports user data packets, for example PDCP packets. These packets have had their headers compressed and properly ciphered and protected to address any security concerns. This results in a much lower data rate, but loses the centralized processing ability.

In addition to splitting the RAN functions, control and user planes are migrating to become separable to allow for separate evolution rates, lower latency, and support new deployment scenarios. This will, for example, provide the ability to have a control plane supplied by a wide area LTE macrocell while the user plane supplied by a small cell 5G. The 5G cellular system will also be based on OFDMA where the time slots have been defined to be variable to handle the widely varying requirements across all the expected services. As noticed in 4G, spectrum is extremely important to provide higher data rates. The OFDMA parameters (sub-carrier spacing, time slot duration, iFFT/FFT size, etc.) have been made flexible to support various spectral deployments.

1.3.3 5G Spectrum and mm-Wave Band

LTE has a maximum bandwidth of 20 MHz, as previously discussed user data rates have been increasing due to the use of HOM, MIMO spatial layers, and CA techniques. While present solutions support up to 5 CA, it is worth mentioning the 3GPP LTE specifications can support up to 32 carriers. This means if we sacrifice the complexity in supporting many carriers, there is plenty of room to further increase the data rates. In many cases, operators need to aggregate licensed and unlicensed spectrum (via license-assisted access) to reach the Gbps data rates. In fact, band number 46 (B46), whose spectral range is 5.15–5.925 GHz. is defined for that intention [22].

5G is defined to have a maximum bandwidth of 100 MHz for frequency bands below 6 GHz. Note that large bandwidth delivers high data rates, but lower bandwidth can also provide 5G services. This coupled with the fragmented spectral band allocations is a reason to support the need for flexibility in the OFDMA parameters discussed above. Another option besides traditional licensed and unlicensed (5–5.9, 64–71 GHz) spectrum usage is to use the citizens broadband radio service (CBRS) spectrum. The CBRS spectrum range is 3.55–3.7 GHz (totaling 150 MHz of bandwidth) and is governed by a three-tiered spectrum authorization framework to accommodate users on a shared basis with incumbent federal and non-federal users of this band. A summary of the items that need to be considered in using 5G frequency bands is provided in Fig. 1.21. A point worth mentioning, in these new

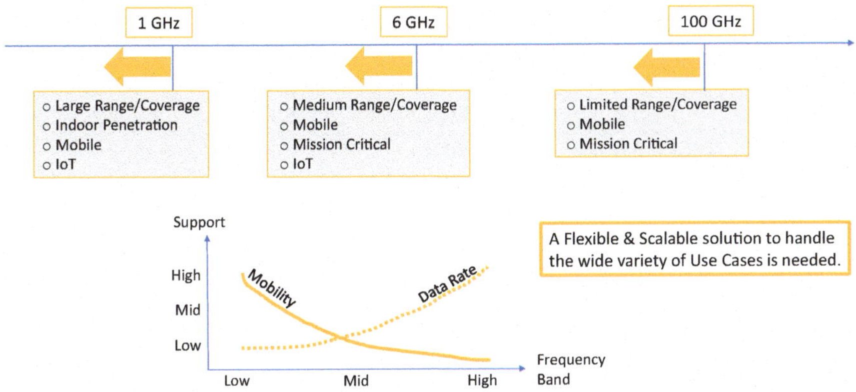

Fig. 1.21 5G frequency band considerations

frequency bands the availability of paired spectrum to support FDD is minimal forcing the industry to focus more on TDD deployments. Hence, not only do we expect bandwidth availability to vary across the low (<1 GHz), medium (<6 GHz), and high (>6 GHz) frequency bands, but we should also expect the duplex method to also vary.

Some operators are focusing on fixed wireless access to deliver high-speed 5G services (approximately 1Gbps) in place of cable/fiber deployments as initial 5G deployments in mm-Wave bands instead of, and in addition to, supporting mobile broadband applications. This approach will help develop a mm-wave-based ecosystem that will enable 5G technologies which need to be used for battery operated devices.

The heterogeneous spectrum usages discussed so far assumed the licensed spectrum is always used; there is an initiative to support services which only use the unlicensed spectrum (like WiFi today). The MulteFire alliance allows LTE technology (and 5G) to be exclusively used (in a stand-alone fashion) in shared and unlicensed spectrum to enable private services, neutral host network architecture, industrial networks, etc.

Spectrum for 5G service will be challenging. Some of the new frequency bands being considered in 5G NR by region is provided in Table 1.3. Operators and equipment manufacturers are faced with various options to identify spectrum (re-farm, acquire new, partner, etc.). We see reasonable convergence (toward Global Harmonization) around the 3–4 GHz frequency bands around the world and, at the moment, less so in the USA.

Table 1.3 New 5G frequency bands

Region	Freq. band (<6 GHz)	<6 GHz bandwidth	Freq. band (>6 GHz)	>6 GHz bandwidth
Europe	3.4–3.8	400 MHz	24.25–27.35	3.1 GHz
China	3.3–3.6	300 MHz		
Japan	3.6–4.2	800 MHz	27.5–29.5	2 GHz
Korea	3.4–3.7	300 MHz	26.5–29.5	3 GHz
United States	3.55–3.7	150 MHz	27.5–28.35	0.85 GHz

1.4 Waveform Design for 5G

As discussed in Sect. 1.2.5, CP-OFDM has certain limitations that makes it not the most suitable waveform for all 5G applications. However, due to its advantages and for backward compatibility reasons, OFDM will still be the main waveform for 5G systems. On the other hand, due to its limitations, certain modifications have been proposed in the literature to make it suitable for 5G application. Among these limitations, fixed SCS (in 4G LTE), CP overhead, and high OOB emission are the most important. Before listing these new waveforms, in the following, we discuss these limitations one by one.

Internet of Things (IoT) is a main contributor to the exponential growth of users in 5G. IoT devices, e.g., sensors, usually send sporadic short data packets and have a limited power. On the other hand, for eMBB a large volume of data should be transmitted in a short amount of time. Such varying characteristics of the bursts to be transported makes CP-OFDM with a fixed SCS an inefficient waveform. For IoT applications, 5G waveform is required to support a transmission mode with very low air interface latency enabled by very short frames [23]. To enable low-latency transmissions, very short TTIs are required, for energy-efficient communications by minimizing on times of low-cost devices. OOB emission can be reduced by applying time domain windowing that smooths the transition from one symbol to another.

As discussed earlier, the OFDM parameters have been made flexible to support various spectral deployments. Specifically, the SCS numerology is now 15, 30, 60, 120, 240, and 480 kHz. The maximum FFT size is now set to 4096, and the maximum number of resource blocks (RBs) that can be transmitted was also increased to 275 (or 3300 sub-carriers). Besides spectral deployment advantages, these options allow more spectrally efficient transmissions to occur. For example, in LTE, we utilize 18 MHz of the available 20 MHz of spectrum, with the adoption of the new numerology we are capable to utilize up to 99 MHz of the available 100 MHz of spectrum. In considering an example 100 MHz deployment, a set of parameters can include SCS = 30 kHz and FFT size = 4096 thus resulting in a sampling frequency of 122.88 MHz (which is 4 times greater than LTE while utilizing 5 times greater spectrum).

Having a flexible OFDMA system is critical to efficiently deploying the wide variety of 5G services [24]. Based on propagation characteristics, it is expected the lower frequency bands will be used for large-area deployments with smaller SCS and the associated larger subframe time durations, while higher frequency bands are expected to be used for the dense deployments with larger SCS and their associated smaller subframe time durations. These are examples and other others can surely exist. As can be seen, this deployment capability can be easily derived from a flexible numerology system.

To reduce OOB emission, various filtering and windowing-based solutions are applied to OFDM [23]. Filtered OFDM (F-OFDM), windowed OFDM (also known as weighted overlap and add or WOLA-OFDM), universal filtered OFDM (UF-OFDM), filter bank multi-carrier (FBMC), and other candidates have been suggested for new waveform in 5G and beyond. These candidates will be studied Chap. 2 through Chap. 4.

1.5 Multiple Access Techniques in 1G to 5G

Let us recall the multiple access techniques deployed in the cellular systems so far. In the first-generation, cellular systems employed FDMA where the frequency band was divided up into frequency channels and users were assigned channels. In the second-generation, TDMA and CDMA were used and in both cases the frequency band was divided into smaller frequency channels. In TDMA, the new dimension of time was used as a resource (time slot), and in CDMA, the new dimension in the code domain (PN sequence) was used. TDMA receiver complexity grew exponentially as the data rate was increased, modulation order increased and number of antennas increased. In the third-generation, CDMA was deployed which utilized larger bandwidth and more importantly introduced the concept of a shared channel. Here the physical resources allocated to users are: time slots and PN codes. CDMA technology complexity increased as the data rate increased. The resulting WCDMA spread bandwidth required a larger processing gain to have reasonable inter-path interference suppression capabilities.

The fourth-generation of cellular systems deployed OFDMA and kept the shared channel concept. Here the physical resources were time slots and frequency subcarriers. OFDMA technology maintained the flexibility of resources and kept the available information bandwidth at the desired value. Due to the use of the cyclic prefix and frequency domain signal processing, the receiver complexity is manageable. It is also a reason why the fifth-generation has decided to continue with OFDMA.

We would like to briefly discuss the differences between the DL and UL communication links; this is shown in Fig. 1.22. The DL starts with a common signal transmitted which consists of the aggregate sum of all UEs in that cell. Each UE is physically located in a different cell position and thus experiences different multipath fading, denoted by h_i. Each UE has its own additive noise, denoted by n_i. The UL starts with individual signal transmissions that encounter different fading, due to the physical locations within a cell. These individual signals are summed at the base

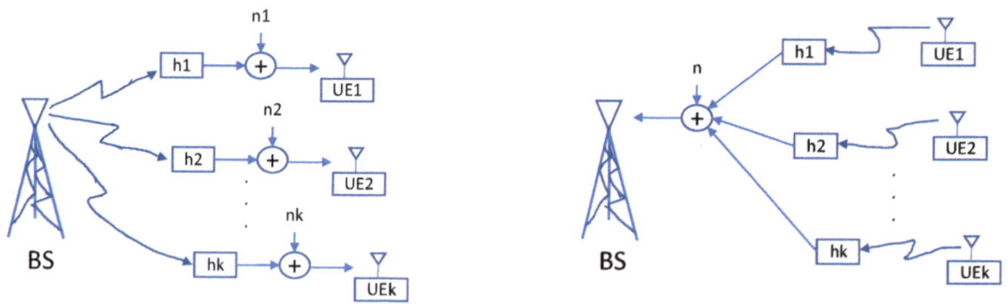

Fig. 1.22 Downlink and uplink communications

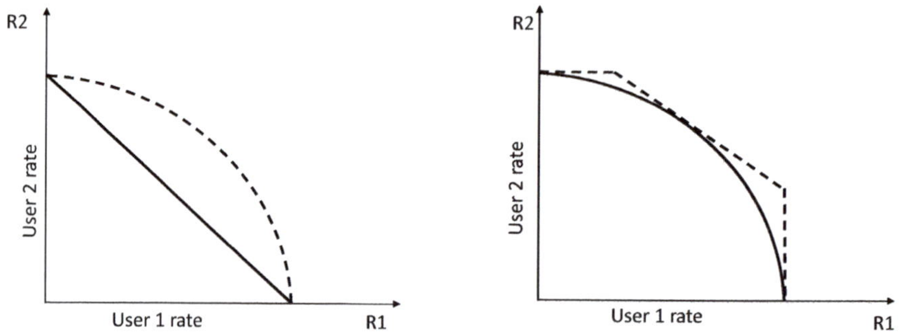

Fig. 1.23 The rate region for two-user DL and UL

station receive antenna, where the base station adds its additive noise. The rate regions of the DL and UL multiple access are shown in Fig. 1.23, for the two-user cases. The figures compare OMA, denoted by the solid line, against superposition coding, denoted by the dashed line. The left curve is used to showcase the DL capacity while the right curve is used to showcase the UL capacity [25–27].

1.6 What is Non-Orthogonal Multiple Access?

In an orthogonal multiple access (OMA) system, such as TDMA and FDMA, orthogonal resource allocation is used among users to avoid intra-cell (inter-user) interference. The number of users that can be supported is then limited by the number of orthogonal resources available. Non-orthogonal multiple access (NOMA) allows and utilizes intra-cell interference in the resource allocation of users. Interference cancelation techniques, such as success interference cancelation (SIC) or multi-user detection (MUD) are used to mitigate this interference. NOMA is a technique being considered by 3GPP in Release 16.

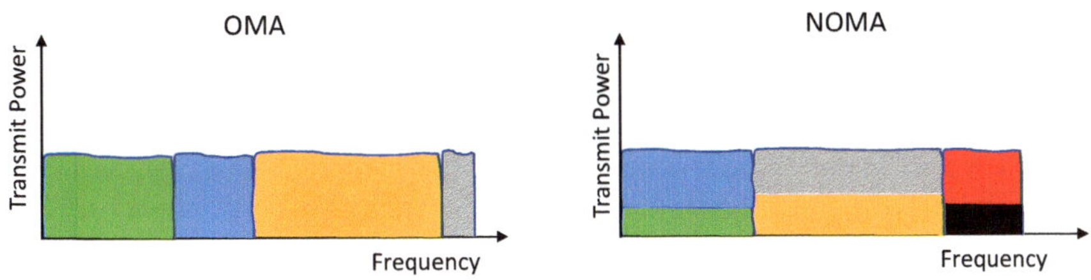

Fig. 1.24 OMA and NOMA power and spectral allocations

NOMA refers to *non-orthogonal* MA that can support multiple users within a single resource and thus can improve user and overall system throughput. It can be realized within the power domain, code domain or other domains.

Power domain NOMA (see Part II of this book) exploits the channel strength differences between users and is the optimal capacity-achieving multiple access technique in a single-cell network, as shown in Figs. 1.22 and 1.23. More details and the multi-cell cases can be found in Chap. 5 and [25–27]. The spectrum and power allocation for power domain NOMA is graphically compared with that of OMA in Fig. 1.24. In the NOMA-based systems, two users can share the same spectral band, where each user has a different power allocated to it.

Code domain NOMA schemes (see Part III of this book) usually exploit low-complexity multi-user detection schemes. Sparse code multiple access (SCMA), interleave division multiple access (IDMA), and low-density spreading (LDS)-CDMA are notable examples of code domain NOMA.

Some of the possible benefits when using NOMA are [27]:

- **Massive connectivity:** While OMA is limited by the number of orthogonal resources, NOMA is not. Theoretically, NOMA can support an unlimited number of users.
- **Lower latency:** OMA waits for available resource blocks to transmit which is accomplished by waiting for an access grant whereas NOMA can support a flexible scheduling and grant-free transmission.
- **Improved spectral efficiency (bps/Hz):** Every NOMA user can utilize the entire bandwidth, whereas an OMA user can utilize a limited amount. The data rates of properly grouped users can be increased when compared to OMA.

The NOMA cellular system components are

- Multi-user grouping, i.e., deciding which users should be grouped together to deploy NOMA.
- Resource allocation (power, code, etc.), e.g., for the power domain NOMA case, users with large power differences are favorable.
- SIC or MUD interference cancelation techniques to remove the controlled NOMA additions.

With SIC or MUD, NOMA can support this multiple access concept. As we have seen, we expect to support an exponential growth in system capacity and user throughput in future systems. This amount of growth presents challenges that forces us to investigate new solutions. The choice of radio access technology plays an important role. NOMA is a proposed scheme to address the future system demands.

1.7 Conclusion

In this chapter, we have reviewed the evolution of 1G to 5G cellular networks. A special emphasis has been placed on orthogonal and non-orthogonal multiple access techniques and network architectures in different generations of cellular technologies. The IMT-2020 requirements for 5G including enhanced mobile broadband, massive machine to machine communications and ultra-reliable and low-latency communications has been discussed and possible modifications such as flexible OFDM, required to address these requirements have been briefly reviewed. A few key technical components for 5G wireless network, including massive MIMO, cloud-RAN, and SDN, have been addressed. Advantages and issues of CP-OFDM have been listed and possible direction for new waveform design has been outlined.

References

1. The story behind the first cell phone call ever made (Online), https://www.bloomberg.com/news/articles/2015-04-24/the-story-behind-the-first-cell-phone-call-ever-made
2. V.H. McDonald, The cellular concept. Bell Syst. Tech. J. (1979)
3. W. Lee, *Mobile Communications Engineering* (McGraw-Hill, 1982)
4. J.G. Proakis, *Digital Communications* (McGraw-Hill, 2001)
5. J. Romero, T. Halonen, J. Melero, *GSM, BPRS and EDGE Performance: Evolution Towards 3G/UMTS* (Wiley, 2002)
6. F. Adachi, M. Sawahashi, H. Suda, Wideband ds-cdma for next generation mobile communications systems. IEEE Commun. Mag. **36**(9), 56–69 (1998)
7. H. Andoh, M. Sawahashi, F. Adachi, Channel estimation using time multiplexed pilot symbols for coherent Rake combining for DS-CDMA mobile radio, in *Proceedings of the 8th IEEE International Symposium on Personal, Indoor and Mobile Radio Communications (PIMRC)*, vol. 3 (1997) pp. 954–958
8. H. Holma, A. Toskala, *WCDMA for UMTS Radio Access for Third Generation Mobile Communications* (Wiley, 2004)

9. The 25 series of technical specification (TS) for 3G (Online), www.3gpp.org
10. J. Boccuzzi, *Signal Processing for Wireless Communications* (McGraw-Hill, 2008)
11. H. Holma, A. Toskala, *HSDPA/HSUPA for UMTS* (Wiley, 2006)
12. A. Ghosh, J. Zhang, J.G. Andrews, R. Muhamed, *Fundamentals of LTE* (Pearson Education, 2010)
13. S.P.E. Dahlman, J. Skold, *4G: LTE-Advanced Pro and The Road to 5G* (Elsevier, 2016)
14. The 36 series of technical specification (TS) for LTE (Online), www.3gpp.org
15. M. Vaezi, Y. Zhang, *Cloud Mobile Networks: From RAN to EPC* (Springer, 2017)
16. The 38 series of technical specification (TS) for 5G-NR (Online), www.3gpp.org
17. ITU-R, IMT vision–framework and overall objectives of the future development of IMT for 2020 and beyond (2015)
18. T.L. Marzetta, E.G. Larsson, H. Yang, H.Q. Ngo, Fundamentals of Massive MIMO (Cambridge University Press, 2016)
19. H. Zhang, N. Liu, X. Chu, K. Long, A.-H. Aghvami, V.C. Leung, Network slicing based 5G and future mobile networks: mobility, resource management, and challenges. IEEE Commun. Mag. **55**(8), 138–145 (2017)
20. D. Bubley, Mobile/Multi-access edge computing: how can telcos monetise this cloud? (2017) (Online), https://stlpartners.com/research/mobilemulti-access-edge-computing-how-can-telcos-monetise-this-cloud/
21. M. Vaezi, Y. Zhang, Radio access network evolution, in *Cloud Mobile Networks* (Springer, 2017), pp. 77–86
22. B. Ayvazian, H. Sarkissian, Spectrum Strategies for 5G. Wireless 20/20 Report (2017) (Online), http://www.wireless2020.com/media/articles.html
23. F. Schaich, T. Wild, Waveform contenders for 5GOFDM vs. FBMC vs. UFMC, in *Proceedings of 6th International Symposium on Communications, Control and Signal Processing (ISCCSP)*
24. Y. Liu, X. Chen, Z. Zhong, B. Ai, D. Miao, Z. Zhao, J. Sun, Y. Teng, H. Guan, Waveform design for 5g networks: analysis and comparison. IEEE Access **5**, 19282–19292 (2017)
25. D. Tse, P. Viswanath, *Fundamentals of Wireless Communication* (Cambridge University Press, 2005)
26. A. Goldsmith, *Wireless Communications* (Cambridge University Press, 2005)
27. W. Shin, M. Vaezi, B. Lee, D.J. Love, J. Lee, H.V. Poor, Non-orthogonal multiple access in multi-cell networks: theory, performance, and practical challenges. IEEE Commun. Mag. **55**(10), 176–183 (2017)

Chapter 2
OFDM Enhancements for 5G Based on Filtering and Windowing

Rana Ahmed, Frank Schaich and Thorsten Wild

2.1 Motivation

One of the main drivers of new radio (NR) is the huge market opportunity in the Internet of Things (IoT) applications [1]. It is predicted that such verticals will evolve as well as their needs. New services will emerge which cannot be efficiently served with any of the available dedicated solutions. The killer application (service) in 2030 cannot be easily predicted 10 years in advance; hence, forward compatibility is a design principle in NR. Such a killer application will not be most efficiently served by a fixed design, a configurable multiservice air interface is therefore the solution.

In contrast to previous long-term evolution (LTE) releases, which mainly target serving broadband users and to which serving verticals came only as an afterthought, e.g., narrow band IoT (NB-IoT) in Releases 13 and 14 [2, 3], 5G NR aims at serving verticals as a basic system capability in addition to broadband users. Consequently, the use cases considered by 5G NR are more diverse. Beyond enhanced mobile broadband (eMBB), massive machine communication (mMTC) and ultra-reliable low latency communication (URLLC) have to be supported. For example, NR targets applications with limited battery capability, which demand less stringent time synchronization requirements, and at the same time NR targets applications which are very sensitive to time delay and thus require shorter symbol transmission time. To make it possible, various considerations for the radio access in general, and for the design of the waveform in particular, have to be accounted for, as will be discussed in the upcoming sections.

R. Ahmed (✉) · F. Schaich · T. Wild
Nokia Bell Labs Stuttgart, Lorenzstrasse 10, Stuttgart, Germany
e-mail: rana.ahmed_salem@nokia-bell-labs.com

F. Schaich
e-mail: frank.schaich@nokia-bell-labs.com

T. Wild
e-mail: thorsten.wild@nokia-bell-labs.com

M. Vaezi et al. (eds.), *Multiple Access Techniques for 5G Wireless Networks and Beyond*, https://doi.org/10.1007/978-3-319-92090-0_2

2.1.1 Multi-carrier Transmission

The selection of the applied waveform is one of the most fundamental design decisions to be taken. It defines the temporal and spectral characteristics of the transmit signal. The resulting time domain peak-to-average power ratio will impact the power amplifier design and hence is responsible for the energy efficiency of communication. The resulting transmit signal spectrum impacts spectral efficiency and the coexistence with other communication systems. Waveforms will carry modulated symbols, so their design will impact the multiple access possibilities and frame structure design for multiplexing data symbols, pilot symbols, and blocks of control symbols.

With conventional single carrier transmission, each transmitted symbol occupies the whole transmission bandwidth. As the transmission bandwidth increases, and thus the symbol length T_s gets shorter, the channel delay spread τ_{max} becomes more significant. This leads to a distortion caused by inter-symbol interference (ISI), where the sum of several delayed replicas of the transmitted symbol is received. To compensate for this effect, multi-tap equalizers have to be employed at the receiver side, e.g., nonlinear decision feedback equalizer (DFE) [4].

In general, if the symbol duration T_s is much larger than the maximum delay spread $T_s \gg \tau_{max}$ or alternatively the signal bandwidth is much smaller than the coherence bandwidth $B_s \ll B_c$, the channel is considered as a "flat" channel; i.e., the received signal can be considered to be a version of the transmitted signal weighted by a complex scalar factor. Thus very low equalization effort is required. Indeed, this is exactly the basic idea of a multi-carrier system. The idea is to divide the total bandwidth of the signal B_s into smaller subchannels, referred to as subcarriers. Such that each subchannel bandwidth is equal to $B_{sc} = \frac{B_s}{N}$, referred to as subcarrier spacing Δf. The information is transmitted in parallel over these subcarriers. If the number of subcarriers, N, is large enough for a given overall bandwidth such that $\Delta f \ll B_c$, the channel experienced over every subcarrier is "flat," and hence, a single-tap equalizer is sufficient to compensate the channel distortion in the frequency domain.

The idea of multi-carrier transmission has emerged a long time ago as early as the 1966 paper of Chang [5]. However, it was only considered for practical implementation when an efficient implementation using the fast Fourier transform (FFT) [6] was proposed.

Multi-carrier modulation is therefore a favorable choice in channels with long delay spread, since it avoids the high computational complexity needed with the single carrier equalization. The inherent serial-to-parallel conversion of multi-carrier modulation naturally offers a basic delay spread protection, which can be extended by using a cyclic prefix or zero postfix.

Furthermore, multi-carrier modulation allows frequency-selective channel access, which exploits high gain links while avoiding fading dips. The decoupling into narrowband subchannels is very appealing for multiple input multiple output (MIMO) antenna processing techniques. The multi-carrier flexibility in time–frequency multiplexing allows good design properties for frame structures, including the multiplexing of pilot symbols and control information. Pilot symbols design can be flexibly tailored to the coherence bandwidth and coherence time of the radio propagation channel.

All those benefits made multi-carrier modulation the technique of choice in 4G LTE, which uses orthogonal frequency division multiplexing (OFDM) in the downlink (DL). The merits of OFDM, however, come at the price of an increased peak-to-average power ratio (PAPR), compared to single carrier transmission. In fact, as the number of subcarriers increases, the PAPR increases as well. That is why single carrier frequency division multiple access (SC-FDMA) is used in the uplink (UL) transmission of LTE, instead of OFDM. Note, however, that via a discrete Fourier transform (DFT) precoding, OFDM can be transformed into SC-FDMA, so the established multi-carrier processing techniques can be reapplied to single carrier modulation.

To provide the required orthogonality[1] between the different subcarriers:

- Every OFDM symbol is appended at the beginning by a guard interval of N_{GI} samples, where N_{GI} is designed to be larger than the channel delay spread. The guard interval contains either:

  1. A duplicate of the last N_{GI} samples of the OFDM symbol and hence is referred to as cyclic prefix (CP).
  2. Or N_{GI} zero samples and hence is referred to as zero prefix (ZP).
     The guard interval is important to avoid inter-block interference (IBI) between successive OFDM symbols. Therefore, the symbol time (in samples) is equal to $N_s = N + N_c$ in case of CP-OFDM or $N_s = N + N_z$ in case of ZP-OFDM, where N_c and N_z are the number of samples in the cyclic prefix and the zero prefix, respectively.

- The subcarriers are arranged in the frequency domain such that the frequency spacing between the subsequent subcarriers is

$$\Delta f = \frac{1}{T_u} = \frac{1}{NT} \tag{2.1}$$

  where T is the sampling period between two successive samples in the time domain and $T_u = NT$ is the useful symbol period in time domain. Such an arrangement guarantees that, at the frequency sampling point of any subcarrier q, no other contribution from any other subcarrier $q' \neq q$ exists after removal of the CP, i.e., orthogonality between the different subcarriers.

Assuming an UL transmitter to which Q subcarriers are allocated, transmitting at a symbol rate $\frac{1}{T_s}$, where $T_s = N_s T$, the output of a multi-carrier CP-OFDM transmitter

[1]Orthogonality here means that no crosstalk occurs in the detection process between the different subcarriers.

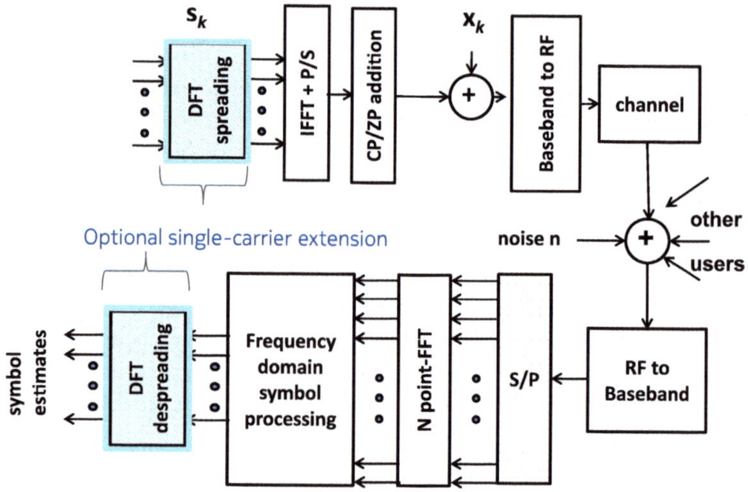

Fig. 2.1 Transmit–receive chain for one subband in an OFDM system

at time instant n can be written as [7] (after dropping the DFT spreading for ease of representation)

$$x(n) = \sum_{i} \sum_{q=0}^{Q-1} s_{q,i} w(n - iN_s) e^{j2\pi f_q (n - N_c - iN_s)}, \tag{2.2}$$

where $w(n)$ is a rectangular window function, holds the value of 1 over the interval $[0, N + N_c]$. $s_{q,i}$ are the i.i.d. complex-valued symbols transmitted at subcarrier frequency f_q and symbol i. Figure 2.1 shows the block diagram of the OFDM transceiver for one subband.

In frequency domain, the window function for one subcarrier in Eq. 2.2, $w(n)$, is written as

$$W(f) = \sqrt{(N + N_c)} e^{-j2\pi f (N+N_c)} \frac{sin(\pi f (N + N_c))}{\pi f (N + N_c)}, \tag{2.3}$$

As shown in Fig. 2.2, the nulls of $W(f)$ for subcarrier q occur at $f = f_q + \frac{m}{(N+N_c)}$, where m is an integer not equal to zero.

It is worth noting that in ZP-OFDM, the nulls of $W(f)$ for subcarrier q occur at $f = f_q + \frac{m}{N}$. Therefore, one can find that the power spectral density of ZP-OFDM-based waveforms have "true" nulls in the frequency domain, since the subcarrier spacing of the transmitted subcarriers, according to Eq. 2.1, coincides exactly with the nulls of $W(f)$.

The first notable multi-carrier modulation technique came up even before OFDM; filter-bank multi-carrier (FBMC) [8]. FBMC is a consequent application of Gabor signaling [9] generated by an orthogonal train of time–frequency shifted pulses. In

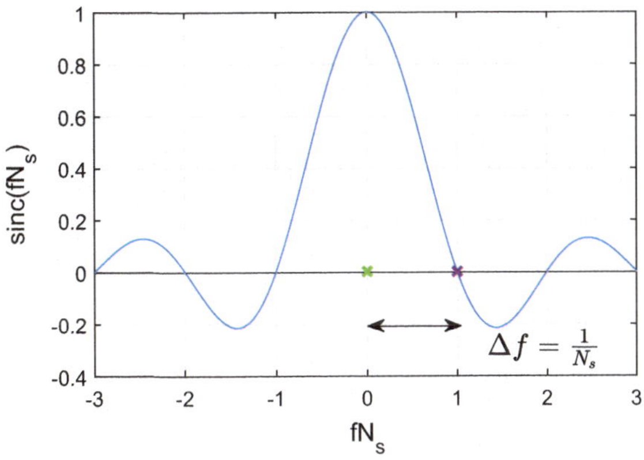

Fig. 2.2 Windowing function in frequency domain $W(f)$ with CP-OFDM

order to make it spectrally efficient, offset-QAM is used [10]. The overlapping symbols generated by long filters create an impressive spectral containment of signals. However, this comes at the price of suitability for short bursts and loss in multiplexing flexibility [11].

2.2 5G Waveform Requirements and Scenarios

NR is targetting a diverse set of use cases [12, 13]. It is foreseen that NR will have to support a wide range of user velocities, data rates, reliability, and power efficiency requirements. In order to be able to configure the waveform parameters to match the requirement of every use case, NR is supposed to support subcarrier spacing scaling principle of $\Delta f = 15 \times 2^S$ kHz, where S is an integer, unlike LTE which supports only $\Delta f = 15$ KHz.

The coexistence of all these services (waveform configurations) together is essential for an efficient use of resources and to be able to adapt to traffic load changes. Since the waveform is a fundamental component in the design of the air interface, NR waveform should be designed to facilitate that coexistence. In other words, NR waveform should be robust enough against possible inter-carrier interference (ICI) distortions caused by the support of different services, which will be detailed in Sects. 2.2.1 and 2.2.2

As mentioned earlier, in CP-OFDM with perfect synchronization, only a frequency domain one tap zero-forcing (ZF) equalizer is sufficient at the receiver side to equalize the effect of the channel. However, in reality, ICI can occur at the receiver side due to Doppler distortions caused by temporal channel variations or synchronization errors, etc. In such case, the relatively slow decay of the sinc waveform in Fig. 2.2 is especially problematic and the overall performance is not adequate. Therefore, in NR, it is desirable to design waveforms with faster decay rates in frequency, i.e., better frequency localization. In addition, waveforms with higher frequency localization than baseline OFDM can achieve higher spectrum utilization than 90%, which was the maximum achieved in LTE [14]. A further driver for higher spectral confinement is the created forward compatibility; i.e., any kind of signals/waveforms which are favorable for a future use case can be inserted in the evolution of the 5G standard, as well-defined in-band requirements allow for "cleaning-up" the spectrum from sidelobes much better than regular OFDM could do. As will be shown in Sects. 2.3.1–2.3.3, the design is a trade-off between time and frequency localization.

2.2.1 Mixed Numerology

In LTE, OFDM parameters, namely cyclic prefix length and subcarrier spacing, are selected as a reasonable compromise for different transmission scenarios (e.g., Doppler spread vs channel delay spread). In NR, because of the extreme use cases, more configuration options are available to serve each use case most efficiently [15]. For example, on one hand with use cases requiring URLLC, in order to save on the latency part, one option is to reduce the symbol lengths. This corresponds to a wider subcarrier spacing. Similarly, users who travel at a very high speed (e.g., high-speed trains), can benefit from having a large subcarrier spacing, reducing the interference arising from Doppler spread. On the other hand, for low-end devices, to enhance the coverage, or for users in a high channel spread environments, longer symbol durations (consequently longer CP) and smaller subcarrier spacings are more favored.

For carrier frequencies below 6 GHz, Release 15 supports subcarrier spacings of 15, 30 and 60 kHz [14]. NR supports the multiplexing of these numerologies in UL and DL [16]. On one carrier, NR supports mixed numerologies in frequency division multiplex (FDM) or time domain multiplex manner. As discussed in [15], FDM of different numerologies in neighboring subbands generates ICI at the edge between the two different numerologies, which can be explained as follows:

Assuming two neighboring allocations in FDM with two numerologies, namely $(\Delta f_1, T_{s1})$ and $(\Delta f_2, T_{s2})$, such that $\Delta f_2 = 2\Delta f_1$ and $T_{s2} = \frac{T_{s1}}{2}$.

1. As shown in Fig. 2.3, allocation 1 suffers from ICI because the neighboring allocation (with larger subcarrier spacing) have now nonzeros contributions at its own subcarrier positions.
2. The ICI on the larger numerology can be better understood in time domain, where the symbol of allocation 2 is half the duration of the symbol of allocation 1, $T_{s2} = \frac{T_{s1}}{2}$. Therefore, the demodulator of allocation 2 collects only half the samples of allocation 1. The effect is equivalent to multiplying with a time domain window, which is equivalent to a convolution with the frequency response of the window [15].

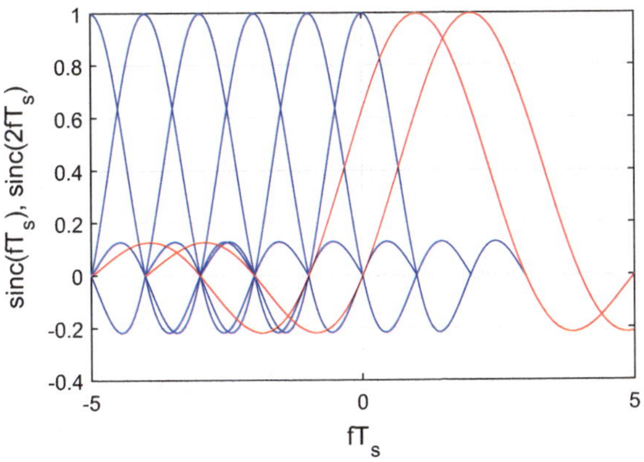

Fig. 2.3 ICI on smaller numerology

2.2.2 *Asynchronous Uplink Transmission*

The OFDM orthogonality in LTE UL requires that uplink transmissions from various users are synchronized at the BS; i.e., the different symbols from the UEs arrive at the BS within a certain time window which does not exceed the CP length. To compensate for different propagation delays, a user equipment (UE) would apply a timing advance [2] depending on how far it is from the BS, which means that UE devices which are far from the BS send their UL signals earlier than those close to the BS. Therefore, any device wanting to transmit a few bits of data has to enter the network via the random access procedure, wherein Msg 2 of the random access procedure, the BS informs the UE about its timing advance value (Fig. 2.4).

The whole random access procedure includes a closed loop timing advance control, which in some cases may not be suitable for low-end machine-type communication (MTC) devices, dealing with sporadic traffic and stringent requirements on energy efficiency. The lack of UL synchronization, as shown in Fig. 2.5, creates ICI between neighboring UL users as explained in [17, 18]. The effect is actually similar to the windowing effect mentioned in Sect. 2.2.1.

2.3 Candidate 5G Waveforms

As discussed in Sects. 2.1–2.2.2, there is a strong need for a waveform design in NR, which is robust against time–frequency misalignments. NR has selected CP-OFDM/DFT-s-OFDM as the baseline waveform including the optional addition of a windowing or a filtering functionality [16, 19]. However, such improvements should

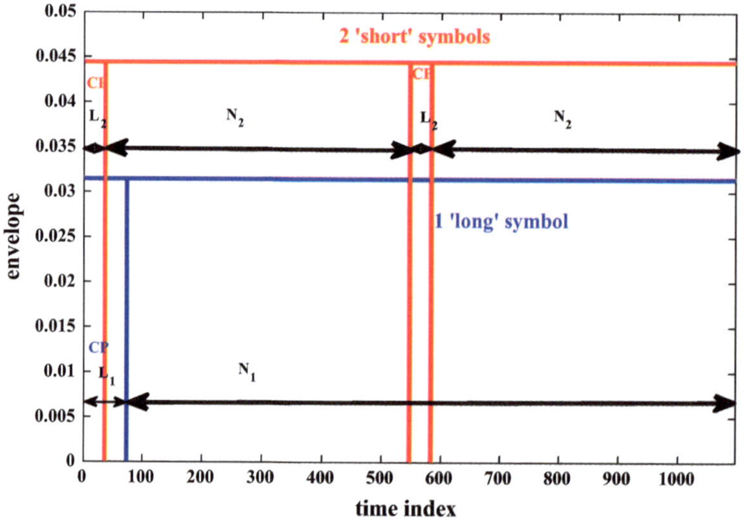

Fig. 2.4 ICI on larger numerology

Fig. 2.5 Multi-user scenario (FDMA)

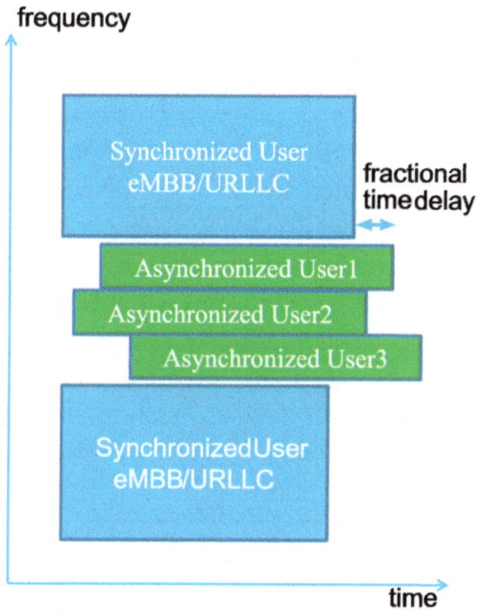

be agnostic to the UE/BS in Release 15 [20], meaning that a baseline CP-OFDM/DFT-s-OFDM receiver should work seamlessly without prior knowledge of the method used at the transmitter to reduce the out-of-band leakage (OOB). To enable a seamless multiplexing of different services, the used waveform should abide by in-band requirements, which are currently discussed in RAN4 [21]. These requirements can be met either by using gaps between the different subband allocations (guard

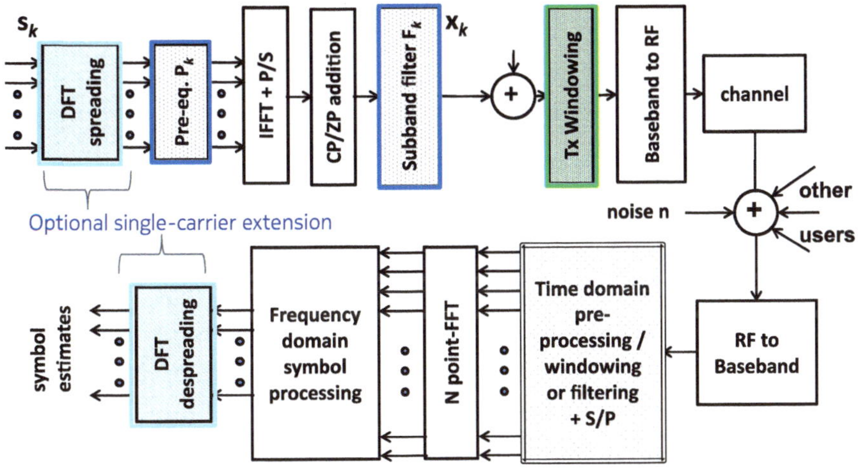

Fig. 2.6 Options for modifying the OFDM transceiver

subcarriers) or by using a frequency localized waveform. Figure 2.6 shows the options for modifying the baseline CP-OFDM transceiver in Fig. 2.1 to achieve this goal.

In this section, we discuss waveform examples which can be used in NR. Two main classes appear in this context; subband filtering and windowing, where the latter can be mapped to subcarrier filtering [7]. Subband filtering is motivated by the fact that ICI typically occur at the edge between neighboring subband allocations (blocks of subcarriers). For example, different uplink users having different waveform configurations/requirements. Therefore, the idea is to apply a well frequency localized filter, the bandwidth of which is close to the subbands bandwidth. As a result, only a few subcarriers close to the edges of the subband in frequency are affected by the filter, as the filter suppresses their out-of-subband sidelobes. By adapting filter parameter, the distortion on subband edges can be alleviated with little negative impact, depending on the use case. We discuss two waveforms which belong to this category, namely UF-OFDM (a.k.a UFMC) and f-OFDM.

Windowing, on the other hand, is applied in the time domain, by modifying the rectangular pulse shape of CP-OFDM waveform to have smoother transitions in time at both ends. As an example of this category, we discuss weighted overlap and add (WOLA) waveform. Other candidate windowing techniques exist [22], but will not be discussed here.

Other waveforms which gained a lot of attention, but were not considered for Release 15 due to incompatibility with CP-OFDM, are FBMC, mentioned at the beginning of this chapter, and zero tail discrete Fourier transform (ZT-DFT-s) [23].

All the methods mentioned in this chapter are compatible with DFT spreading as shown in Fig. 2.6.

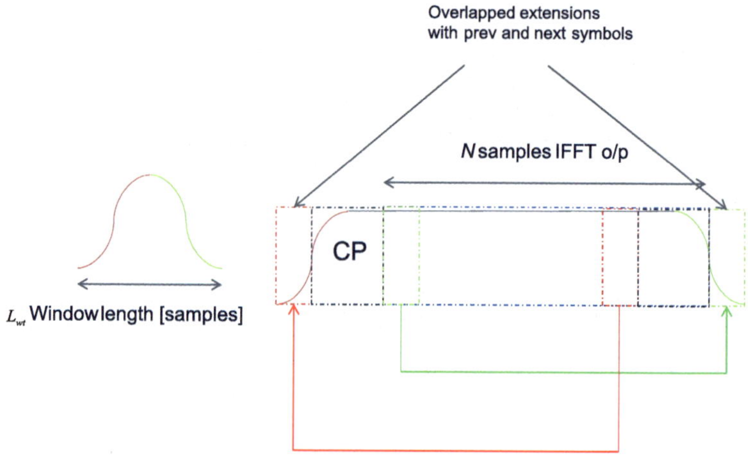

Fig. 2.7 WOLA waveform

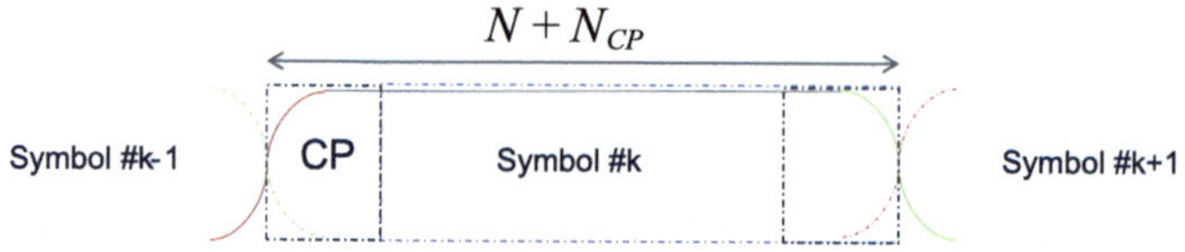

Fig. 2.8 WOLA transmitter operation

2.3.1 *Weighted Overlap and Add (WOLA)*

In CP-OFDM with WOLA, the windowing function $w(n)$ in (2.2) is replaced by a pulse function with soft edges at both sides, the length of the window is extended to $\left[-\frac{L_{wt}}{2}, N + \frac{L_{wt}}{2}\right]$ [24, 25], where L_{wt} is the length of the extension beyond the CP-OFDM length. The soft edges at the beginning and end of the window function result in better localization of the WOLA waveform in the frequency domain.

In [24, 25], the CP-OFDM symbol is first extended by a cyclic extension in the time domain, and both edges are shaped by a weighting function, as shown in Fig. 2.7. As shown in Fig. 2.8, the resulting symbol is overlapped and added to the next symbol, and hence the overhead remains the same as in CP-OFDM.

It is worth noting that, although the WOLA symbols overlap within one burst, when considering TDD, the tails from the end of the last WOLA symbol of one burst and from the first WOLA symbol of the following burst would already extend into the guard period (GP). In TDD, the GP is placed between two successive bursts when DL/UL switching is made. Therefore, the window length should be chosen carefully so as not to hinder TDD transmission.

At the receiver side, a WOLA receiver can optionally be applied to suppress ICI, leaked from a neighboring allocation [24, 25]. The WOLA receiver processing is shown in Fig. 2.9.

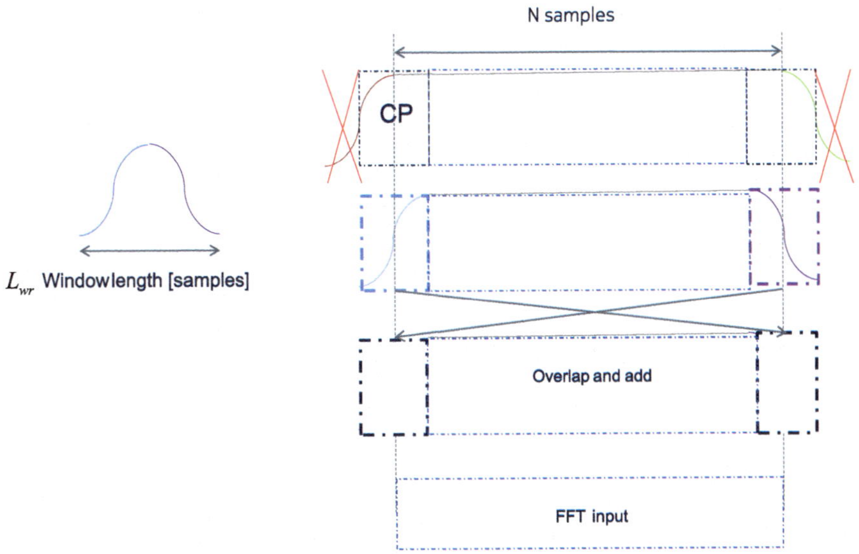

Fig. 2.9 WOLA receiver operation

The design of the soft edges of $w(n)$, including the overlap length L_{wt}, determines the frequency domain behavior of the WOLA symbol. In [24, 25], a raised cosine window design is used, but this does not exclude other possible window shapes. In general, the design is a trade-off between time and frequency localization, i.e., between ISI and ICI. The longer the window length L_{wt}, the better the ICI localization of the WOLA waveform is, but the longer the overlap between the successive WOLA symbols, and the less robust the WOLA symbol is to channels with long delay spread. Figure 2.10 shows the power spectral density (PSD) of WOLA for a subband of 2 PRBs with two different window lengths $L_{wt} = 144, 72$ samples. Compared to CP-OFDM ($L_{wt} = 0$), we can see that windowing reduced the OOB leakage of the waveform.

2.3.2 *Universal Filtered OFDM (UF-OFDM)*

UF-OFDM is a 5G candidate waveform, also known as universal filtered multi-carrier (UFMC), where blocks of subcarriers (subbands) are filtered. As shown in Fig. 2.6, this is done by passing the OFDM signal output for user k through a subband filter $f_k(n)$ with filter order L_f. $f_k(n)$ is built by shifting a prototype filter $f(n)$ to the center of subband of user k. The modified subband filtered OFDM signal for user k can be written as

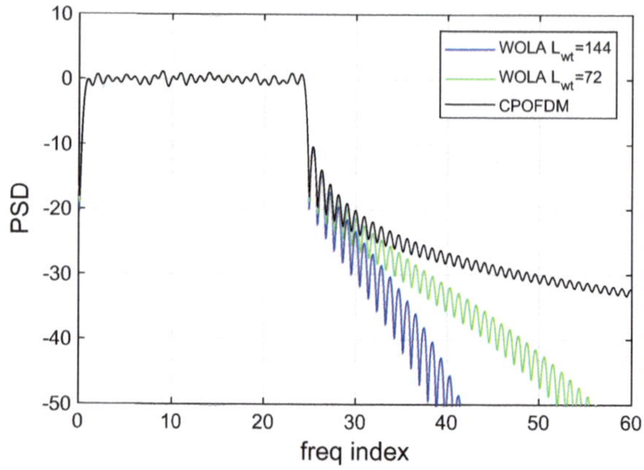

Fig. 2.10 PSD of WOLA waveform for different window lengths at a subband allocation size of 2 PRBs

$$x_k(n) = \sum_i f_k(n) * \left[\sum_{q=0}^{Q-1} s_{k,q,i} w(n - iN_s) e^{j2\pi f_q (n - N_c - iN_s)} \right]. \tag{2.4}$$

The input to the UF-OFDM subband filter can be either a ZP-OFDM signal or a CP-OFDM signal (depending on the value of N_c in Eq. 2.4). The advantage of applying the subband filter on a ZP-OFDM signal is that the resulting overall symbol can be limited in time domain, such that no overlapping between the successive symbols occur in an ISI-free environment (if the subband filter order is smaller than the guard interval $L_f < N_{GI}$). The prototype filter of choice in UF-OFDM is the Dolph–Chebychev filter, but it is not restricted to this selection.

If the subband filter is applied at the transmitter on a ZP-OFDM signal, at the receiver side, the L_f samples from the tail of the received signal are simply added to the beginning of the symbol before applying the FFT. Hence, even if a ZP-OFDM signal is used, no extra complexity is needed at the receiver side to demodulate the signal as explained in [26]. In [7], it was shown that subband filtering applied on CP-OFDM signal has near identical rate-versus-SNR performance to subband filtering applied on a ZP-OFDM signal.

In [27], it was proposed to use Dolph–Chebyshev-based subband filtering in combination with variable CP/ZP. The summation of the CP plus ZP parts was assumed to be constant, and the filter order matches the CP length. Each part (CP/ZP) can be tuned depending on the channel environment, i.e., to optimize the trade-off between ISI and ICI rejection. However, in Release 15, the NR waveform is based on CP-OFDM [16], therefore only CP-based UF-OFDM implementation can be supported in Release 15.

Dolph–Chebychev filters are optimal in the sense that for a given side lobe level (SLL) the main lobe width is minimized. They are adjustable by the tuning parameter for the side lobe attenuation (SLA) as well as by the filter length L. The optimum filter choice of L and SLA depends on the use case. For example, on the one hand, in high ICI use cases with asynchronous transmission, it makes sense to use filters

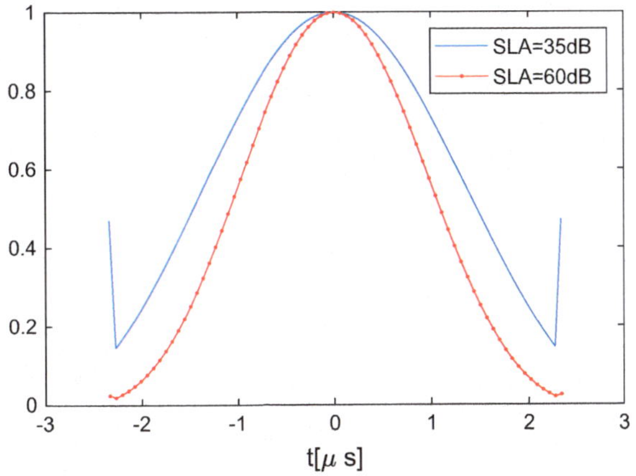

Fig. 2.11 Time domain impulse response for Dolph–Chebyshev filter with $L = 72$ and $SLA = 35, 60$ dB

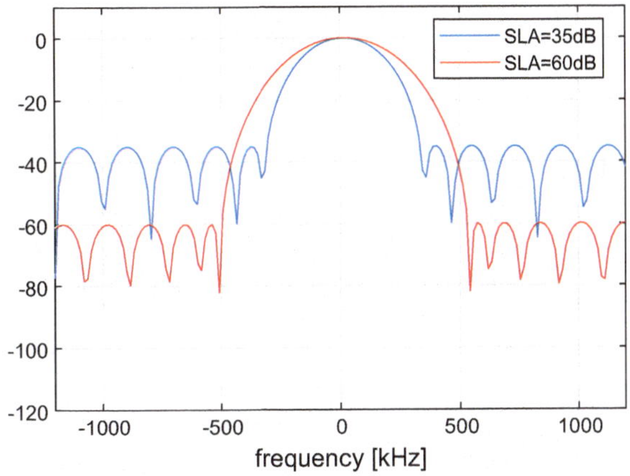

Fig. 2.12 Frequency domain response for Dolph–Chebyshev filter with $L = 72$ and $SLA = 35, 60$ dB

which are longer than the guard interval $L > N_{GI}$, at the price of higher vulnerability to delay spreads [7]. On the other hand, in environments with high delay spread, a shorter filter length is used to protect against ISI. The SLA controls the trade-off between the main lobe width and the SLL. As shown in Figs. 2.11 and 2.12, as the SLA increases, the main lobe width increases and the SLL decreases.

In general, for any subband filtering technique, when the allocation size is larger than a single PRB, a subband filter which has a broader passband offers a steeper side lobe level decay at the pass-band edge.

For equalization at the receiver side, the receiver has to be aware of the filter coefficients. In order to avoid that, pre-equalization of a UF-OFDM signal is proposed in [28] for reference signals, but can also be applied to data symbols [26]. In addition, equal transmit power per subcarrier prevents an increased number of error events at the pass-band edge, as those errors from weaker edge subcarriers, would be detrimental for bit error rate (BER) performance.

To further reduce the effect of ICI leaked from a neighboring allocation due to mixed numerology or asynchronous transmission, etc., windowing or matched filtering can be used at the receiver side of the UF-OFDM [7, 18, 29].

2.3.3 Filtered OFDM (f-OFDM)

Filtered OFDM (f-OFDM) is a 5G candidate waveform based on subband filtering of a CP-OFDM signal [30, 31]. Therefore, the f-OFDM signal can also be built according to (2.4). Compared to UF-OFDM, the key property of f-OFDM is that the filter length, L_f, can well exceed the guard interval length, which enables it to provide very good frequency localization. As shown in [30], soft truncation of a prototype filter is used, which in this case is a sinc impulse response $p_B(n)$. The sinc impulse response should have a bandwidth B in the frequency domain equal to the subband allocation size. The subband filter is obtained by applying a time windowing mask to $p_B(n)$

$$f_k(n) = w(n)p_B(n), \tag{2.5}$$

where $w(n)$ is the windowing mask with duration T_w. The windowing mask has to have smooth transitions to zero on its both ends so that it avoids abrupt jumps at the beginning and end of the truncated filter. An example of such a windowing function is the Hanning window as proposed in [30] or the raised cosine window as proposed in [32]. T_w is usually chosen as $T_w = \frac{T_u}{2}$.

In Figs. 2.13 and 2.14, the frequency domain response and the time domain impulse response of the designed filter for f-OFDM with bandwidth equal to 2 PRBs (360 kHz) and 3 PRBs (540 ‘kHz) are depicted, where the Hanning window function is used [30]. The filter length should be long in order to achieve desirable frequency localization. With half OFDM symbol filter length for example, one f-OFDM symbol extends into 25% of each of the previous and following f-OFDM symbols. However, most of the energy of the time domain impulse response, for allocation sizes greater than 3 PRBS, is limited to the CP part which is eventually dropped at the receiver side.

Similar to WOLA, the long tails of the f-OFDM is also a problem in TDD transmission. Especially with long filter lengths, $T_w = \frac{T_u}{2}$. That is why in [32], signal burst tail reduction is proposed, where a hard truncation is suggested to reduce the tail overhead at both ends.

In [32], one design criteria is to choose the sinc filters bandwidth B to be larger than the subband bandwidth W by a small excess bandwidth ∂W, called tone offset (TO), on each side, i.e., $B = W + 2 \times \partial W$. The motivation behind using the TO is to guarantee a flat passband across all used subcarriers. The TO in f-OFDM is then analogous in use to the pre-equalization used on UF-OFDM, as mentioned in Sect. 2.3.2. The benefits of the TO in f-OFDM and the pre-equalization in UF-OFDM come at the expense of a larger mainlobe (or higher SLL), and thus, a weaker frequency localization.

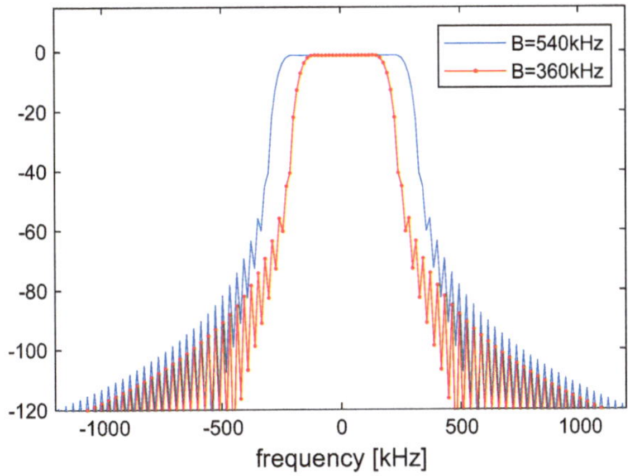

Fig. 2.13 Frequency domain response of designed filter for f-OFDM with 2 PRB and 3 PRBs

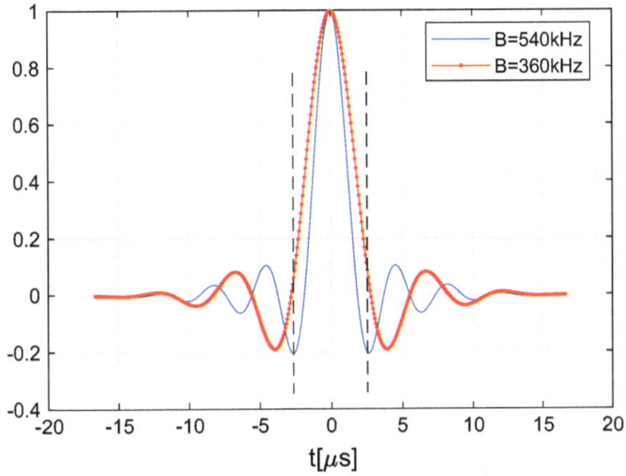

Fig. 2.14 Impulse response of designed filter for f-OFDM with 2 PRB and 3 PRBs versus sampling time

f-OFDM is often described in conjunction with a matched filter, which suppresses the ICI leaked into the UE subband from neighboring UEs. In addition, it maximizes the received signal-to-noise ratio (SNR) of the UE.

2.3.4 Comparison Between the Different Waveforms

In this section, we compare the performance of the four different waveforms discussed in this chapter, namely CP-OFDM, WOLA, UF-OFDM (CP based), and f-OFDM. Table 2.1 depicts the waveform parameters which were assumed for the different preprocessing approaches applied on top of the baseline CP-OFDM signal.

Table 2.1 Waveform parameters

Waveform	Parameters
Baseline CP-OFDM	$N_{GI} = 72$ samples, FFT size $N = 1024$, one subband = 1 and 4 PRBs
WOLA	$L_{wt} = 144$
CP-UFOFDM	SLA=25dB and 75dB for subband sizes 1 PRB and 4 PRBs, respectively, $L_f = 73$ samples, pre-equalization applied
f-OFDM	$L_f = 513$, TO=0 and 2 and 4, Hanning window used for soft truncation of sinc impulse response

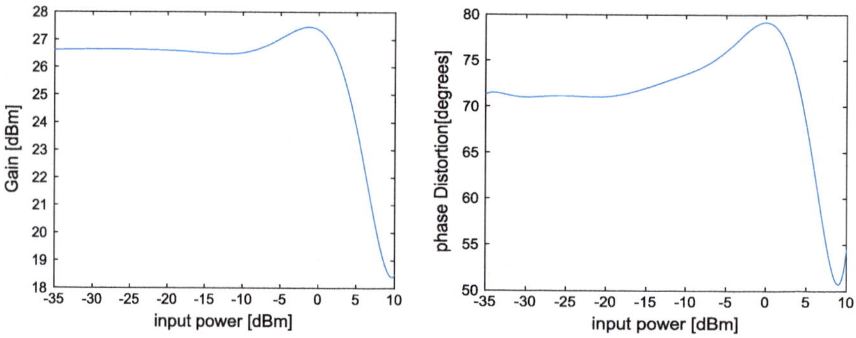

Fig. 2.15 UL polynomial power amplifier model: gain and phase distortion

2.3.4.1 Power Amplifier Model

The power amplifier, at the RF frontend, introduces signal nonlinearities in the final transmitted signal. Hence, the effectiveness of the OOB emission reduction by subband processing is eventually limited by the spectral regrowth due to these nonlinearities. As the PAPR increases, the spectral regrowth increases for the same power amplifier efficiency.

Third-generation partnership project (3GPP) uses modified Rapp model and polynomial model to model the effect of the power amplifier in NR DL and UL, respectively [33]. Figure 2.15 shows the UL polynomial power amplifier model gain and phase distortion curves. In order to operate at an output power of 22 dBm, an input power operating point of −5.12 dBm is assumed, as shown in Fig. 2.16.

As shown in Fig. 2.17, the error vector magnitude (EVM), used to describe the signal quality [34], reaches a minimum point at a phase compensation value of −77.1°. As mentioned earlier, one of the drawbacks of multi-carrier OFDM is the increase in the PAPR. The subband processing, based on windowing or filtering, leads to a further increase in the PAPR of the input signal to the power amplifier, which can also be observed in Fig. 2.17, where CP-OFDM has the lowest EVM value.

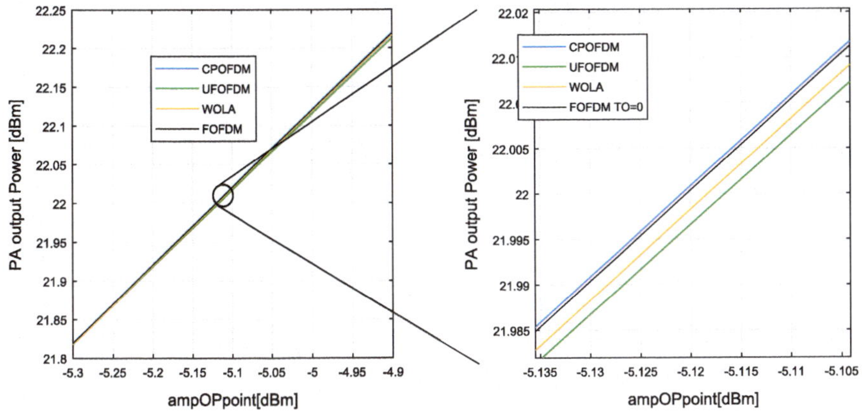

Fig. 2.16 Output power in dBm of amplifier versus input power in dBm for UL power amplifier

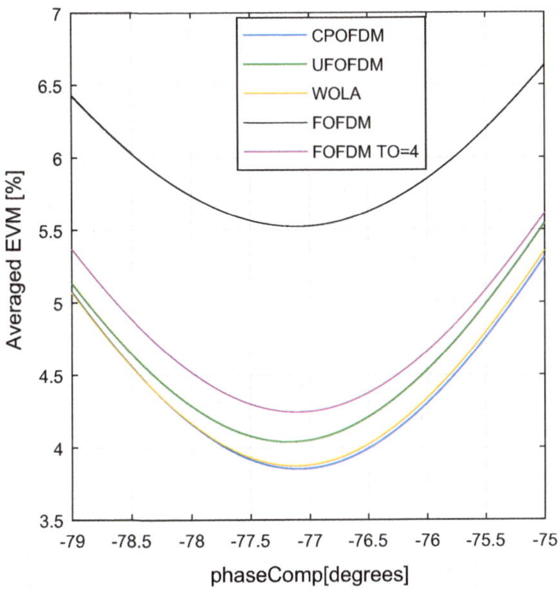

Fig. 2.17 EVM versus phase compensation value for all waveforms for UL power amplifier

2.3.4.2 Performance Comparison

In Figs. 2.18 and 2.19, we can see the PSD of all waveforms before and after the UL power amplifier. As we can see, with the given waveform configurations, sub-band filtering (f-OFDM followed by CP-UFOFDM) have better spectral localization compared to WOLA and is therefore more effective in dealing with frequency misalignments. After the power amplifier, the relative performance of the different waveforms remain the same, but the gap between them is significantly reduced. A waveform with high spectral localization translates into higher spectral efficiency,

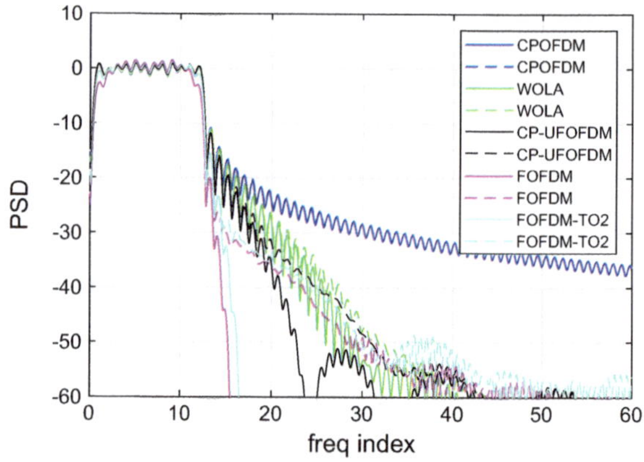

Fig. 2.18 PSD with 1 PRB allocation: solid lines are PSD without PA, dashed lines are PSD with PA

since with high spectral localization, relatively smaller number of in-band guards are needed to satisfy the in-band requirements of NR.

Figures 2.18 and 2.19, however, do not show the performance in a synchronous environment, especially with high delay spread. Results for this case in [35, 36] show an opposite ranking for the three waveforms in a synchronous environment with high delay spread, where WOLA shows the best performance followed by CP-UFOFDM and then f-OFDM. In [37], it is shown that by taking into consideration, the reduced guard band overhead required in case of well spectrally localized f-OFDM, f-OFDM without using a matched filter can have a higher spectral efficiency than than of WOLA. No comparison was made in [37] against UF-OFDM. In [7], it is shown that both windowing and filtering techniques are comparably effective in combating channel time–frequency misalignments depending on the selected waveform configuration, with filtering techniques showing a slightly superior performance.

2.3.4.3 Implementation Aspects

WOLA requires an additional complexity over baseline CP-OFDM depending on the window length, which is L_{2wt} multiplications, as shown in Fig. 2.7. Similarly, if a WOLA receiver L_{2wr} is applied to suppress ICI from a neighboring allocation, extra L_{2wr} multiplications are needed on top [36]. Therefore, the total added complexity is that of the overlap operation and $L_{2wt}+L_{2wr}$.

For subband filtering, higher computational complexity is required, and therefore, a range of low-complex solutions exists in the literature. Two main directions that will be discussed here are:

- Multi-window approximation discussed in [38, 39]
- Fast convolution discussed in [40, 41]

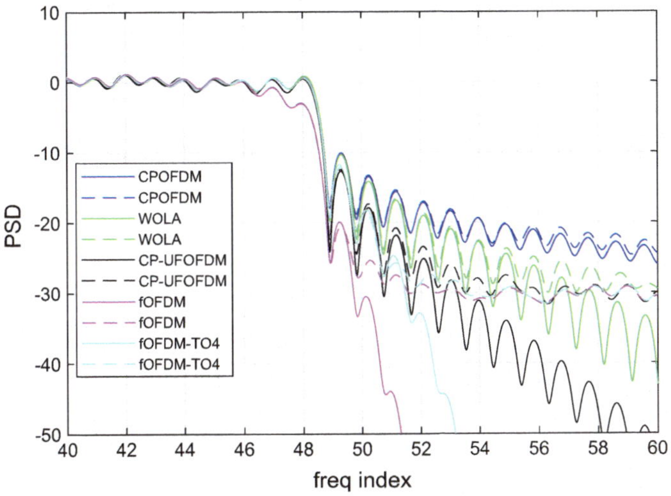

Fig. 2.19 PSD with 4 PRB allocation: solid lines are PSD without PA, dashed lines are PSD with PA

Recalling Eq. 2.4, the overall signal $x(n)$ composed of all subbands can be written as

$$x(n) = \sum_{k=0}^{K-1} x_k(n) e^{\frac{j2\pi k Q n}{N}}. \quad (2.6)$$

The complexity of generating the overall signal (e.g., in the DL case) for one symbol scales linearly with the number of subcarriers per subband Q (to generate one subband k) and with the number of allocated subbands K.

The multi-window approximation is based on the observation that subband filtering is equivalent to subcarrier windowing, with the result that Eq. 2.4 can be rewritten, for one symbol duration, as [38]

$$x_k(n) = \sum_{q=0}^{Q-1} f_q(n) s_{k,q} w(n) \quad (2.7)$$

where $f_q(n)$ is the effective filter that is used to modulate the qth subcarrier in each subband. One can observe that for close subcarriers, the value of $f_q(n)$ is not so different, depending mainly on the bandwidth for which the prototype filter $f(n)$ was designed. Therefore, the idea is to divide the subband allocation into multiple subcarrier groups, every subcarrier group g consists of Q_g subcarriers and is windowed using one effective filter $f_g(n)$, with the result that, the complexity scales only with the number of subcarrier groups G [39]. The overall signal for one symbol duration can be approximated as

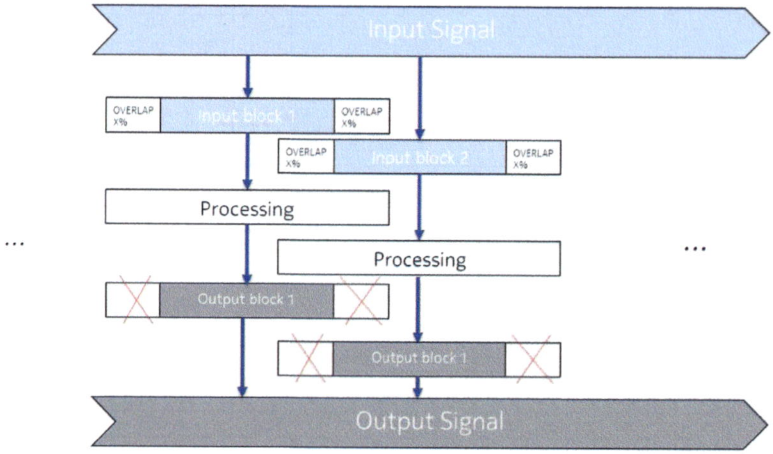

Fig. 2.20 Overlap-save processing flow used for fast convolution

$$x(n) \approx \sum_{g=0}^{G-1} f_g(n) \sum_{k=0}^{K-1} \sum_{q=0}^{Q-1} s_{k,q} w(n) e^{\frac{j2\pi k Qn+qn}{N}}, \tag{2.8}$$

where the strength of the approximation thus depends on the number of subcarrier groups G. In the extreme case of using $G = 1$, this approximation has the same complexity as a windowing operation. With $G = 3$, the spectral localization is improved, coming at the price of roughly 3 times baseline OFDM complexity [38].

The main idea behind fast convolution algorithms is that filtering through a higher-order impulse response can be implemented effectively through multiplication in FFT-domain. This is done by taking the DFT of the input sequence as well as the DFT of the filter impulse response. The time domain output signal is finally obtained by IDFT [40]. As shown in Fig. 2.20, overlap-save processing is applied for long sequences to combine all processed blocks into the output signal.

2.4 Summary

In this chapter, we have discussed the requirements on the waveform design in 5G NR is driven by the new and challenging use cases in NR. To enable a seamless multiplexing of different services and to fulfill the forward comparability vision for NR, it is highly desirable to have a frequency localized waveform design in NR without sacrificing too much time localization. CP-OFDM has been proven as a powerful and flexible waveform already for 4G, and hence in 5G, it is the dominant candidate solution, while dealing with its weak points such as frequency localization by some modifications To this end, three candidate waveform preprocessing techniques for the 5G NR are discussed, namely WOLA, UF-OFDM, and f-OFDM, where each is based on either time domain windowing or subband filtering. The design principle and implementation aspects of each waveform are outlined.

The design parameters of each waveform can be tuned to fit the target use case, e.g., the window length or the window shape in WOLA and the subband filter length in subband filtering techniques, mainly optimizing a trade-off between time domain localization and frequency domain localization. Depending on the choice of these parameters, the performance of windowing and subband filtering techniques is found to be comparable, with subband filtering offering a slightly better performance, at the price of higher complexity.

Several techniques can be used to reduce the implementation complexity of the subband filtering technique. One interesting technique, besides frequency domain-based fast convolution, is to approximate the subband filtering by a multi-windowing operation, which in essence yields a hybrid between a subband filtering technique and a windowing technique, with a trade-off between frequency localization (and hence performance) versus implementation complexity cost.

References

1. Machina research, Technical report, Aug 2016
2. 3rd Generation Partnership Project; TS 36.211, E-UTRA, Physical Channels and Modulation (Release 13) (2016)
3. 3rd Generation Partnership Project; TS 36.211, E-UTRA, Physical Channels and Modulation (Release 14) (2017)
4. J. Proakis, *Digital Communications*. 4th edn. (Mc Graw-Hill Book Company, 2001)
5. R.W. Chang, High-speed multichannel data transmission with bandlimited orthogonal signals. Bell Sys. Tech. **45**, 1775–96 (1966)
6. S.B. Weinstein, The history of orthogonal frequency-division multiplexing [History of Communications]. Commun. Mag. IEEE **47**(11), 26–35 (2009)
7. S. Venkatesan, R.A, Valenzuela, OFDM for 5G: cyclic prefix versus zero postfix, and filtering versus windowing, in *2016 IEEE International Conference on Communications (ICC)*, pp. 1–5, May 2016
8. B. Farhang-Boroujeny, OFDM versus filter bank multicarrier. IEEE Signal Process. Mag. **28**(3), 92–112 (2011)
9. G. Wunder, P. Jung, M. Kasparick, T. Wild, F. Schaich, Yejian Chen, S. Brink, I. Gaspar, N. Michailow, A. Festag, L. Mendes, N. Cassiau, D. Ktenas, M. Dryjanski, S. Pietrzyk, B. Eged, P. Vago, F, Wiedmann, 5GNOW: non-orthogonal, asynchronous waveforms for future mobile applications. IEEE Commun. Mag. **52**(2), 97–105 (2014)
10. M. Bellanger, *FBMC Physical Layer: A Primer* (2010)
11. F. Schaich, T. Wild, Y. Chen, Waveform Contenders for 5G—suitability for short packet and low latency transmissions, in *2014 IEEE 79th Vehicular Technology Conference (VTC Spring)*, pp. 1–5, May 2014
12. NGMN Alliance, *NGMN 5G White Paper* (2015), http://www.ngmn.org/5g-white-paper.html

13. A. Osseiran, V. Braun, T. Hidekazu, P. Marsch, H. Schotten, H. Tullberg, M.A. Uusitalo, M. Schellman, The foundation of the mobile and wireless communications system for 2020 and beyond: challenges, enablers and technology solutions, in *2013 IEEE 77th Vehicular Technology Conference (VTC Spring)*, pp. 1–5, June 2013
14. 3GPP TR 38.912 Study on New Radio (NR) access technology (Release 14) (2017)
15. F. Schaich, T. Wild, Subcarrier spacing–a neglected degree of freedom? in *2015 IEEE 16th International Workshop on Signal Processing Advances in Wireless Communications (SPAWC)*, pp. 56–60, June 2015
16. R1-167963 Way forward on waveform RAN1#86 (2016)
17. Y. Chen, F. Schaich, T. Wild, Multiple access and waveforms for 5G: IDMA and universal filtered multi-carrier, in *2014 IEEE 79th Vehicular Technology Conference (VTC Spring)*, pp. 1–5, May 2014
18. F. Schaich, T. Wild, Relaxed synchronization support of universal filtered multi-carrier including autonomous timing advance, in *2014 11th International Symposium on Wireless Communications Systems (ISWCS)*, pp. 203–208, Aug 2014
19. R1-163615 WF on overview of NR RAN1#84bis (2016)
20. RAN1#86bis chairmans notes
21. R4-1610921 Way forward on in-band requirements for NR RAN4#81, Nov 2016
22. Ericsson. R1-163224 Waveform candidates RAN1#84bis (2016)
23. G. Berardinelli, F.M.L. Tavares, T.B. Strensen, P. Mogensen, K. Pajukoski, Zero-tail DFT-spread-OFDM signals, in *2013 IEEE Globecom Workshops (GC Wkshps)*, pp. 229–234, Dec 2013
24. Qualcomm, *5G Waveform & Multiple Access Techniques*, https://www.qualcomm.com/documents/5g-research-waveform-and-multiple-access-techniques (2015)
25. Qualcomm, R1-162199 Feasibility of Mixing Numerology in an OFDM System RAN1#84bis (2016)
26. Nokia, R1-165014 Subband-wise filtered OFDM for New Radio below 6 GHz RAN1#85 (2016)
27. LG. R1-162516 Flexible CP-OFDM with variable ZP RAN1#84bis, Apr 2016
28. X. Wang, T. Wild, F. Schaich, S. ten Brink, Pilot-aided channel estimation for universal filtered multi-carrier, in *2015 IEEE 82nd Vehicular Technology Conference (VTC2015-Fall)*, pp. 1–5, Sept 2015
29. R. Ahmed, T. Wild, F. Schaich, Coexistence of UF-OFDM and CP-OFDM, in *2016 IEEE 83rd Vehicular Technology Conference (VTC Spring)*, pp. 1–5, May 2016
30. J. Abdoli, M. Jia, J. Ma, Filtered OFDM: a new waveform for future wireless systems, in *2015 IEEE 16th International Workshop on Signal Processing Advances in Wireless Communications (SPAWC)*, pp. 66–70, June 2015
31. X. Zhang, M. Jia, L. Chen, J. Ma, J. Qiu, Filtered-OFDM—enabler for flexible waveform in the 5th generation cellular networks, in *2015 IEEE Global Communications Conference (GLOBECOM)*, pp. 1–6, Dec 2015
32. Huawei, R1-164033 f-OFDM scheme and filter design RAN1#85 (2016)
33. R1-166004, R4-164542, Response LS on realistic power amplifier model for NR waveform evaluation, May 2016
34. E. Dahlman, S. Parkvall, J. Skold, *4G: LTE/LTE-Advanced for Mobile Broadband*. Academic Press, 1st ed. (2011)
35. R1-1609564 Implementation-specific UF-OFDM for New Radio (2016)
36. R1-164685, OFDM based waveform single user evaluation RAN1#85 (2016)
37. R1-166093, Waveform evaluation updates for case 1a and case 1b, Aug 2016
38. M. Matthe, D. Zhang, F. Schaich, T. Wild, R. Ahmed, G. Fettweis, A reduced complexity time-domain transmitter for UF-OFDM, in *2016 IEEE 83rd Vehicular Technology Conference (VTC Spring)*, pp. 1–5, May 2016

39. Samsung, R1-166746 Discussion on multi-window OFDM for NR waveform RAN1#86 (2016)
40. M. Renfors, J. Yli-Kaakinen, T. Levanen, M. Valkama, T. Ihalainen, J. Vihriala, Efficient fast-convolution implementation of filtered CP-OFDM waveform processing for 5G, in *2015 IEEE Globecom Workshops (GC Wkshps)*, pp. 1–7, Dec 2015
41. J. Yli-Kaakinen, T. Levanen, S. Valkonen, K. Pajukoski, J. Pirskanen, M. Renfors, M. Valkama, Efficient fast-convolution-based waveform processing for 5G Physical Layer. *CoRR*, abs/1706.02853 (2017)

Chapter 3
Filter Bank Multicarrier Modulation

Ronald Nissel and Markus Rupp

3.1 Why FBMC?

Future mobile systems will be characterized by a large range of different use cases, ranging from enhanced mobile broadband (eMBB) over enhanced machine type communications (eMTC) to ultra-reliable low-latency communications (URLLC) [2, 41, 53, 63]. To efficiently support such diverse use cases, a flexible time–frequency allocation becomes necessary. In particular, the out-of-band (OOB) emissions must be sufficiently low in order to efficiently support different use cases within the same band. Furthermore, low OOB emissions reduce the synchronization requirements. Conventional orthogonal frequency-division multiplexing (OFDM) with cyclic prefix (CP) performs poorly in this context because of the underlying rectangular prototype filter, which causes large OOB emissions. To improve spectral properties in OFDM, the 3rd Generation Partnership Project (3GPP) is therefore considering windowing and filtering [45, 52, 63]. The windowed OFDM scheme is called OFDM with weighted overlap and add (WOLA) and the filter-based methods are called universal filtered OFDM (UF-OFDM) and filtered OFDM (f-OFDM). While windowing and filtering can indeed reduce the OOB emissions of conventional OFDM, filter bank multicarrier modulation (FBMC) with offset quadrature amplitude modulation (OQAM) [41] still performs much better, as shown in Fig. 3.1. Additionally, FBMC-OQAM has a maximum symbol density, that is, a time–frequency spacing of $TF = 1$ for complex-valued symbols. In OFDM-based schemes, on the other hand, the symbol density is lower, as indicated by $TF > 1$, additionally worsening the spectral efficiency. Besides the better support of different use cases, FBMC also increases the throughput of legacy Long Term Evolution (LTE) transmissions because fewer

R. Nissel (✉) · M. Rupp
TU Wien, Gusshausstraße 25, 1040 Vienna, Austria
e-mail: rnissel@nt.tuwien.ac.at

M. Rupp
e-mail: mrupp@nt.tuwien.ac.at

M. Vaezi et al. (eds.), *Multiple Access Techniques for 5G Wireless Networks and Beyond*, https://doi.org/10.1007/978-3-319-92090-0_3

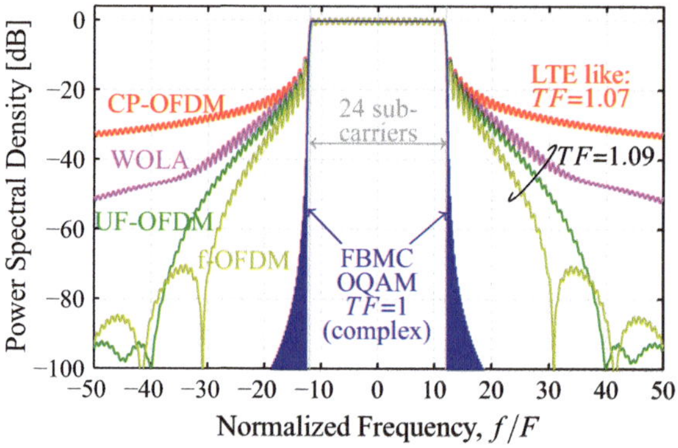

Fig. 3.1 FBMC has much better spectral properties compared with CP-OFDM. Windowing (WOLA) and filtering (UF-OFDM, f-OFDM) can improve the spectral properties of CP-OFDM. However, FBMC still performs much better and has the additional advantage of a maximum symbol density, $TF = 1$ (complex). ©2017 IEEE, [41]

Fig. 3.2 Real-world testbed measurements at 2.5 GHz show that FBMC has a higher throughput than OFDM (1.4 MHz LTE resembling SISO signal) because of a higher available bandwidth and no CP overhead [38, 41]. The channel estimation in FBMC is based on the data spreading approach [21, 35]

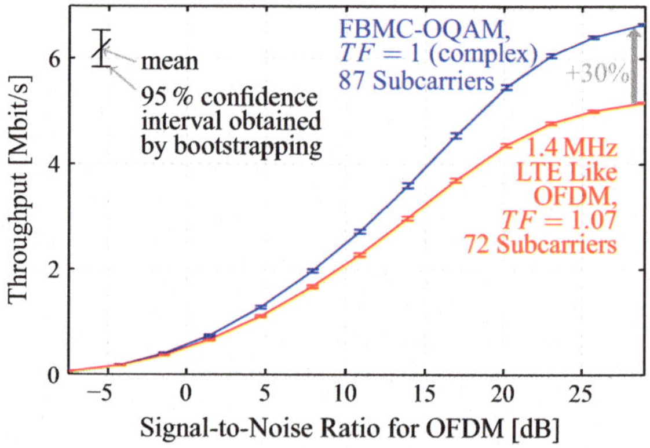

guard subcarriers are required and no CP is needed. Figure 3.2 shows real-world testbed measurements and compares FBMC with an 1.4 MHz LTE single-input and single-output (SISO) signal (including pilots but ignoring signaling overhead). For high signal-to-noise ratio (SNR) values, the throughput of FBMC is approximately 30% higher than for OFDM. Even compared with f-OFDM, FBMC would still be approximately 20% better, as indicated by the time–frequency efficiency calculations in [41]. However, one has to keep in mind that the potential improvement strongly depends on the number of subcarriers and the required guard band. In particular, once the number of subcarriers is very high, windowed OFDM and filtered OFDM will perform close to FBMC.

Unfortunately, all the nice features of FBMC-OQAM come at a price, namely, the complex orthogonality condition is replaced by the less strict real orthogonality condition. While this limitation has in many cases either no or only a minor influence on the performance, some important methods, such as channel estimation and some multiple-input and multiple-output (MIMO) techniques, become more challenging.

There exist different variants of FBMC, but we will mainly focus on OQAM-based schemes because they provide the highest spectral efficiency. Different names are used to describe OQAM, such as OFDM/OQAM [8], fbmc-pulse-amplitude modulation (PAM) [26], staggered multitone (SMT) or Cosine Modulated Multitone (CMT) [12], which, however, are essentially all the same. One can easily transform one of those schemes into another by appropriately tuning the underlying parameters. For example, FBMC-PAM is a conventional FBMC-OQAM scheme for which the subcarrier spacing is reduced by a factor of two, the number of subcarriers is increased by two, and the offset is applied in the frequency domain instead of the time domain. In general, all those "different" FBMC schemes are characterized by:

- A prototype filter which is localized in time and frequency.
- Only real-valued information symbols can be transmitted at a given time–frequency position.
- A time–frequency spacing of $TF = 0.5$ for real-valued symbols (equivalent to $TF = 1$ for complex-valued symbols).
- Intrinsic imaginary interference.

Although FBMC has been considered as a strong contender for replacing OFDM in the fifth-generation (5G) of wireless systems [4, 5, 59], 3GPP eventually decided that they will stick to OFDM [3]. While such decision makes sense in terms of backward compatibility to fourth-generation (4G) wireless systems, it is not the most efficient technique for all possible use cases, especially if the number of subcarriers is low. Thus, if the envisioned concept of different use cases within the same band turns out to be successful in 5G, we expect that FBMC will again gain momentum for beyond 5G communications.

3.2 Multicarrier Modulation

Multicarrier modulation has a long-standing history in wireless communications [9, 51, 58]; however, widespread practical applications have only been realized in the latest versions of wireless systems in the form of OFDM, enabled by advances in the field of integrated circuits. Current applications of OFDM include LTE, WiFi and digital video broadcasting-terrestrial (DVB-T).

In multicarrier systems [50], information is commonly transmitted over orthogonal pulses which overlap in time and frequency. The big advantage is that these pulses usually occupy only a small bandwidth, so that frequency-selective broadband channels transform into multiple, virtually frequency flat, sub-channels (subcarriers). Mathematically, the transmitted signal, $s(t)$, of a multicarrier system in the time domain can be expressed as

$$s(t) = \sum_{k=0}^{K-1} \sum_{l=0}^{L-1} g_{l,k}(t)\, x_{l,k}, \tag{3.1}$$

where $x_{l,k}$ denotes the transmitted symbol at subcarrier position l and time position k, and is chosen from a symbol alphabet, usually a QAM or a PAM signal constellation. The total number of subcarriers is denoted by L and the total number of symbols in time by K. The basis pulse $g_{l,k}(t)$ in (3.1) is defined by

$$g_{l,k}(t) = p(t - kT)\; \mathrm{e}^{\mathrm{j}2\pi\, lF\,(t-kT)}\; \mathrm{e}^{\mathrm{j}\theta_{l,k}}, \tag{3.2}$$

and is, essentially, a time and frequency shifted version of prototype filter $p(t)$, with T denoting the time spacing and F the frequency spacing (subcarrier spacing). The choice of phase shift $\theta_{l,k}$ becomes relevant later in the context of FBMC-OQAM. After transmission over a channel, the received symbols are decoded by projecting the received signal, $r(t)$, onto the basis pulses, $g_{l,k}(t)$, that is,

$$y_{l,k} = \langle r(t), g_{l,k}(t) \rangle = \int_{-\infty}^{\infty} r(t)\, g_{l,k}^{*}(t)\, \mathrm{d}t. \tag{3.3}$$

In (3.3), we implicitly apply a matched filter if the channel perturbation is additive white Gaussian noise (AWGN), maximizing the SNR. In a doubly selective channel, it might be better to choose the transmit and receive prototype filters slightly different, as, for example, suggested in pulse-shaped multicarrier transmissions [27] or in the practically more relevant case of CP-OFDM. However, employing an AWGN matched filter is usually close to the optimum because the channel induced interference is often dominated by noise, see Sect. 3.5. One of the biggest advantages of orthogonal multicarrier systems is that the transmission in (3.3) can be modeled by a one-tap channel, that is,

$$y_{l,k} = H(kT, lF)\, x_{l,k} + n_{l,k} + z_{l,k}, \tag{3.4}$$

where $H(t, f)$ denotes the time-variant transfer function and represents the one-tap channel. The noise in (3.4) is described by $n_{l,k}$ and the channel induced interference by $z_{l,k}$. Often, the delay spread and the Doppler spread are low enough so that the channel induced interference is dominated by noise, that is, $\mathbb{E}\{|n_{l,k}|^2\} \gg \mathbb{E}\{|z_{l,k}|^2\}$. Thus, the channel induced interference can be neglected and the employment of low-complexity one-tap equalizers correspond to the maximum likelihood symbol detection in case of Gaussian noise.

Multicarrier systems are mainly characterized by prototype filter $p(t)$ as well as time spacing T and frequency spacing F, so that the ambiguity function [12, 50],

$$A(\tau,\nu) = \int_{-\infty}^{\infty} p\left(t-\frac{\tau}{2}\right) p^*\left(t+\frac{\tau}{2}\right) e^{j2\pi\nu t}\, dt, \tag{3.5}$$

captures the main properties of a multicarrier system in a compact way. The projection of the transmitted basis pulses $g_{l_1,k_1}(t)$ onto the received basis pulses $g_{l_2,k_2}(t)$ can then be expressed by the ambiguity function according to

$$\langle g_{l_1,k_1}(t), g_{l_2,k_2}(t)\rangle = \underbrace{e^{-j\pi\, TF(l_1+l_2)(k_1-k_2)}\, e^{j(\theta_{l_1,k_1}-\theta_{l_2,k_2})}}_{\text{only a phase shift}}\ \underbrace{A(\,T(k_1-k_2),\, F(l_1-l_2)\,)}_{\text{ambiguity function}}. \tag{3.6}$$

There exist some fundamental limitations of multicarrier systems, as formulated by the Balian–Low theorem [13], which states that it is mathematically impossible that the following four desired properties are fulfilled at the same time:

1. Maximum symbol density,

$$TF = 1, \tag{3.7}$$

2. Time localization,

$$\sigma_t = \sqrt{\int_{-\infty}^{\infty} (t-\bar{t})^2\, |p(t)|^2\, dt} \; < \infty, \tag{3.8}$$

3. Frequency localization,

$$\sigma_f = \sqrt{\int_{-\infty}^{\infty} (f-\bar{f})^2\, |P(f)|^2\, df} \; < \infty, \tag{3.9}$$

4. Orthogonality,

$$\langle g_{l_1,k_1}(t), g_{l_2,k_2}(t)\rangle = \delta_{(l_1-l_2),(k_1-k_2)} \tag{3.10}$$

$$A(T(k_1-k_2), F(l_1-l_2)) = \delta_{(l_1-l_2),(k_1-k_2)}, \tag{3.11}$$

with δ denoting the Kronecker delta function. The pulse $P(f) = \int_{-\infty}^{\infty} p(t)\, e^{-j2\pi f t} dt$ in (3.9) represents the Fourier transform of $p(t)$ while $\bar{t} = \int_{-\infty}^{\infty} t\, |p(t)|^2\, dt$ corresponds to the mean time and $\bar{f} = \int_{-\infty}^{\infty} f\, |P(f)|^2\, df$ the mean frequency of the pulse. Furthermore, we assume that $p(t)$ is normalized to preserve unit energy. The localization measures in (3.8) and (3.9) can be interpreted as standard deviation, with $|p(t)|^2$ and $|P(f)|^2$ representing the probability density function (pdf). This relates the Balian-Low condition to the Heisenberg uncertainty relationship [57, Chap. 7].

The Balian-Low theorem implies that at least one of the four desired properties has to be sacrificed when designing multicarrier waveforms. For example, CP-OFDM sacrifices frequency localization while FBMC-OQAM the complex orthogonality condition.

3.2.1 CP-OFDM

CP-OFDM is the most prominent multicarrier technique and is applied, for example, in Wireless LAN and LTE [24, 29]. CP-OFDM employs a rectangular transmit and receive pulse, which greatly reduces the computational complexity. Furthermore, the CP guarantees orthogonality in frequency-selective channels. The transmitter (TX) and receiver (RX) prototype filter can be expressed by

$$p_{\mathrm{TX}}(t) = \begin{cases} \frac{1}{\sqrt{T_0}} & \text{if } -\left(\frac{T_0}{2} + T_{\mathrm{CP}}\right) \leq t < \frac{T_0}{2} \\ 0 & \text{otherwise} \end{cases} \tag{3.12}$$

$$p_{\mathrm{RX}}(t) = \begin{cases} \frac{1}{\sqrt{T_0}} & \text{if } \frac{T_0}{2} \leq t < \frac{T_0}{2} \\ 0 & \text{otherwise} \end{cases} \tag{3.13}$$

for which

$$\begin{array}{ll} \textbf{(Bi)-Orthogonal}: T = T_0 + T_{\mathrm{CP}}; & F = 1/T_0 \rightarrow TF = 1 + \frac{T_{\mathrm{CP}}}{T_0} \\ \textbf{Localization}: \quad \sigma_t = \frac{T_0 + T_{\mathrm{CP}}}{2\sqrt{3}}; & \sigma_f = \infty \end{array}, \tag{3.14}$$

with T_0 representing a time-scaling parameter which depends on the desired subcarrier spacing (or time spacing). Note that, in contrast to FBMC, the prototype filter is differently at the TX and RX side.

Figure 3.3 shows the ambiguity function, see (3.5), for CP-OFDM (without CP, that is, $T_{\mathrm{CP}} = 0$). Orthogonality is guaranteed for a time spacing of $T = T_0$ and a frequency spacing of $F = 1/T_0$, leading to $TF = 1$. This is also indicated by the rectangular grid (the small circles) inside of Fig. 3.3. The ambiguity function decays very slowly in frequency because of the underlying rectangular pulse. Additionally, the CP simplifies equalization in frequency-selective channels but also reduces the spectral efficiency. In order to reduce the OOB emissions in OFDM, 3GPP is currently considering windowing [45] and filtering [52, 63]. As already shown in Fig. 3.1, such

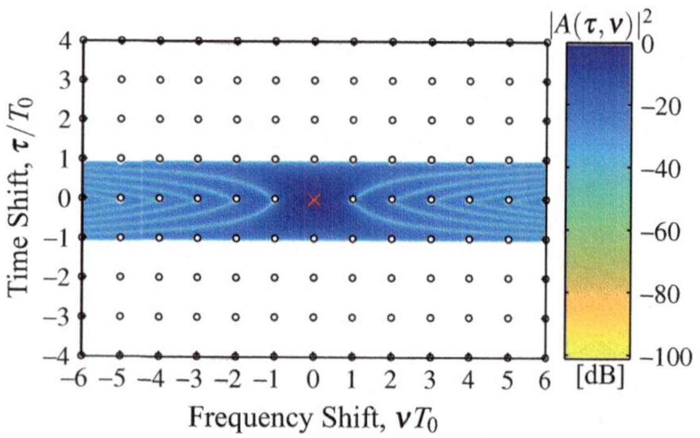

Fig. 3.3 The ambiguity function, $10\log_{10}|A(\tau,\nu)|^2$, for OFDM (without CP) shows a good localization in the time domain, but a poor localization in the frequency domain. The orthogonal time–frequency spacing is $TF = 1$, indicated by the small circles in the figure

methods can reduce the OOB emissions in OFDM, but still do not perform as good as FBMC and have the additional disadvantage of a reduced symbol density, that is, $TF > 1$, even in an AWGN channel. Note that for all the windowed- and filtered-based OFDM techniques, receive windowing and filtering is of utmost importance [41]. Very often, people forget this crucial aspect and only focus on reducing the OOB emissions at the transmitter.

3.3 FBMC-OQAM

FBMC-OQAM replaces the complex orthogonality condition with the less strict real orthogonality condition, $\Re\{\langle g_{l_1,k_1}(t), g_{l_2,k_2}(t)\rangle\} = \delta_{(l_2-l_1),(k_2-k_1)}$, and works, in principle, as follows:

1. Design a prototype filter with $p(t) = p(-t)$ which is orthogonal for a time spacing of $T = T_0$ and a frequency spacing of $F = 2/T_0$, leading to $TF = 2$.
2. Reduce the time–frequency spacing by a factor of two each, that is, $T = T_0/2$ and $F = 1/T_0$, leading to $TF = 0.5$.
3. The so induced interference is shifted to the purely imaginary domain by the phase shift $\theta_{l,k} = \frac{\pi}{2}(l+k)$ in (3.2).

Let us take a closer look at the intrinsic interference. With $\theta_{l,k} = \frac{\pi}{2}(l+k)$ and $TF = 0.5$, the inner product in (3.6) transforms to

$$\langle g_{l+\Delta l,k+\Delta k}(t), g_{l,k}(t)\rangle = \underbrace{\mathrm{e}^{\mathrm{j}\frac{\pi}{2}(\Delta l+\Delta k)}\mathrm{e}^{-\mathrm{j}\frac{\pi}{2}\Delta k(2l+\Delta l)}}_{\text{purely imaginary for odd } \Delta k,\, \Delta l} \quad \underbrace{A(\Delta k\, T, \Delta l\, F)}_{0 \text{ if both } \Delta k\neq 0,\, \Delta l\neq 0 \text{ are even}}. \tag{3.15}$$

The ambiguity function in (3.15) approaches zero if Δk and Δl are even because the prototype filter is designed to be orthogonal for those cases. If, on the other hand, either Δk or Δl is odd, $A(\cdot)$ no longer approaches zero, leading to interference. The main idea of FBMC is to shift this interference to the imaginary domain. To be specific, the exponential function in (3.15) becomes purely imaginary valued if either Δk or Δl is odd. Furthermore, the ambiguity function is always real-valued because of $p(t) = p(-t)$, so that it has no influence on the imaginary part. Note that we consider a phase shift of $\theta_{l,k} = \frac{\pi}{2}(l+k)$, but other phase shifts are also possible, for example, $\theta_{l,k} = \mathrm{j}\frac{\pi}{2}(l+k) - \mathrm{j}\pi lk$.

Similar as in orthogonal multicarrier systems, FBMC also allows the employment of low-complexity one-tap equalizers. To be specific, the transmission can be modeled by a one-tap channel $H(kT, lF)$, similar as in (3.4), according to

$$y_{l,k} = H(kT, lF)\,(x_{l,k} + \mathrm{j}\, v_{l,k}) + n_{l,k} + z_{l,k}. \tag{3.16}$$

Compared with orthogonal multicarrier systems, the data symbols $x_{l,k}$ are real-valued and there exists an imaginary interference term, described by $\mathrm{j}v_{l,k}$, which depends on the adjacent symbols. The big advantage of FBMC compared with other non-orthogonal schemes is that the imaginary interference can easily be canceled, simply by taking the real part after equalization.[1] Thus, computational demanding equalization and cancellation methods are not necessary. Note that the imaginary interference does not carry any useful information for $L \to \infty$ and $K \to \infty$, so that, by taking the real part, we do not lose any useful information; see Sect. 3.6 for more details.

Although the time–frequency spacing in FBMC-OQAM is equal to $TF = 0.5$, only real-valued information symbols can be transmitted in such as system, leading to an equivalent time–frequency spacing of $TF = 1$ for complex-valued symbols. Very often, the real part of a complex-valued symbol is mapped to the first time slot and the imaginary part to the second time slot, thus the name offset-QAM. However, such self-limitation is not necessary. We can equivalently perform this mapping over subcarriers or directly consider PAM symbols instead of "staggered" QAM symbols.

As already mentioned in the beginning of this section, the prototype filter has to be an even function and orthogonal for a time–frequency spacing of $TF = 2$. All prototype filters which satisfy this condition can be utilized in FBMC-OQAM. Let us discuss two prominent prototype filters, namely the Hermite filter and the PHYDYAS filter. The Hermite prototype filter is based on Hermite polynomials $H_n(\cdot)$, as proposed in [16], and can be expressed as

$$p(t) = \frac{1}{\sqrt{T_0}} \mathrm{e}^{-2\pi\left(\frac{t}{T_0}\right)^2} \sum_{\substack{i=\{0,4,8,\\12,16,20\}}} a_i H_i\left(2\sqrt{\pi}\frac{t}{T_0}\right), \tag{3.17}$$

for which the coefficients can be found to be [35]

$$\begin{aligned} a_0 &= 1.412692577 & a_{12} &= -2.2611 \times 10^{-9} \\ a_4 &= -3.0145 \times 10^{-3} & a_{16} &= -4.4570 \times 10^{-15} \\ a_8 &= -8.8041 \times 10^{-6} & a_{20} &= 1.8633 \times 10^{-16}, \end{aligned} \tag{3.18}$$

leading to the following properties of (3.17),

$$\begin{aligned} &\textbf{Orthogonal}: T = T_0; \qquad F = 2/T_0 \qquad \to TF = 2 \\ &\textbf{Localization}: \sigma_t = 0.2015\, T_0;\ \sigma_f = 0.403/T_0. \end{aligned} \tag{3.19}$$

Note that in practical systems, the pulse in (3.17) will be truncated so that it fits within the time interval $-\frac{OT_0}{2} \leq t < \frac{OT_0}{2}$, with O denoting the overlapping factor.

Figure 3.4 shows the ambiguity function for the Hermite prototype filter. The filter has the same shape in time and frequency, allowing us to exploit symmetries. Furthermore, the filter is based on a Gaussian function and therefore has a good joint time–frequency localization of $\sigma_t \sigma_f = 1.02 \times 1/4\pi$, almost as good as the bound of $\sigma_t \sigma_f \geq 1/4\pi \approx 0.08$ (attained by the Gaussian pulse). In Fig. 3.4, we also observe

[1]The channel must be known at the receiver. Channel estimation itself becomes more challenging in FBMC, as we will discuss later in this section.

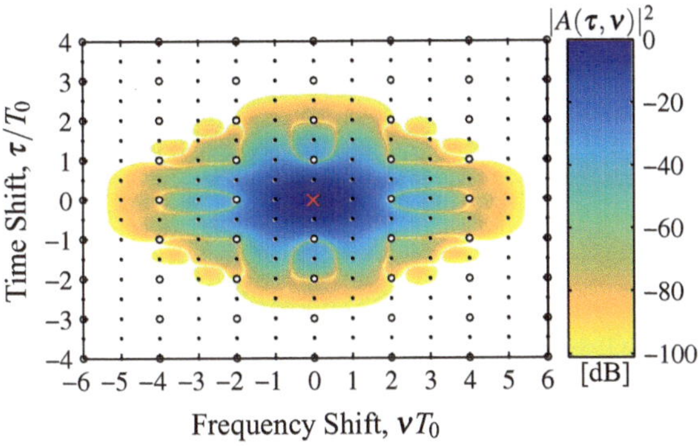

Fig. 3.4 The ambiguity function for the Hermite prototype filter shows good localization in both, time and frequency. The orthogonal time–frequency spacing is $TF = 2$. To improve the spectral efficiency, the time–frequency spacing is reduced to $TF = 0.5$, indicated by the black markers. The so induced interference is shifted to the purely imaginary domain

that the pulse is orthogonal for a time spacing of $T = T_0$ and a frequency spacing of $F = 2/T_0$, indicated by the small circles. In FBMC-OQAM, the time–frequency spacing is reduced to $T = T_0/2$ and $F = 1/T_0$. This causes interference, indicated by the black markers in Fig. 3.4, which, however, is purely imaginary valued, see (3.15).

Another prominent filter is the PHYDYAS prototype filter [6, 7], constructed by:

$$p(t) = \begin{cases} \frac{1+2\sum_{i=1}^{O-1} b_i \cos\left(\frac{2\pi t}{OT_0}\right)}{O\sqrt{T_0}} & \text{if } -\frac{OT_0}{2} \le t < \frac{OT_0}{2} \\ 0 & \text{otherwise} \end{cases}. \tag{3.20}$$

The coefficients b_i were calculated in [28] and depend on overlapping factor O. For example, for an overlapping factor of $O = 4$, the coefficients become,

$$b_1 = 0.97195983; \;\; b_2 = \sqrt{2}/2; \;\; b_3 = 0.23514695, \tag{3.21}$$

and lead to the following properties of (3.20)

$$\begin{aligned} &\textbf{Orthogonal}: \; T = T_0; \qquad\quad F = 2/T_0 \qquad \rightarrow TF = 2 \\ &\textbf{Localization}: \; \sigma_t = 0.2745\,T_0; \; \sigma_f = 0.328/T_0. \end{aligned} \tag{3.22}$$

Figure 3.5 shows the ambiguity function for the PHYDYAS prototype filter ($O = 4$). Compared to the Hermite prototype filter, the PHYDYAS filter has a better frequency localization but a worse time localization. The joint time–frequency localization of $\sigma_t\sigma_f = 1.13 \times 1/4\pi$ is also worse. Note that the PHYDYAS filter is not perfectly orthogonal, as shown in Fig. 3.5 for time position ± 3 and frequency position $\{-2, 0, 2\}$, However, this interference is very low and thus be neglected for typical working points of wireless communications.

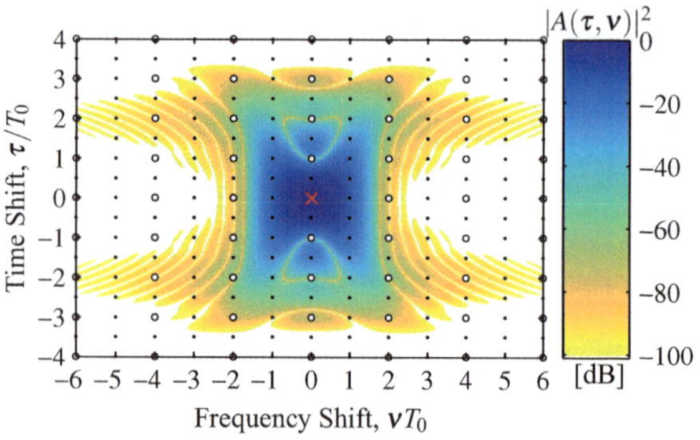

Fig. 3.5 The PHYDYAS prototype filter shows good localization in time and frequency. Compared with the Hermite prototype filter, the time localization is worse but the frequency localization is better. The small circles indicate the orthogonal time–frequency spacing, while the black markers correspond to the imaginary interference

3.3.1 Latency

For our latency considerations, we focus on the underlying pulse duration but ignore other sources of delays, such as channel delays and processing delays. The transmission time of FBMC-OQAM then depends on subcarrier spacing F, the number of symbols in time K and overlapping factor O, according to,

$$T_{\text{block}} = \frac{O}{F} + \frac{0.5}{F}(K-1). \tag{3.23}$$

In particular, the subcarrier spacing can be utilized to reduce the latency. This, however, comes at the expense of an increased sensitivity to frequency-selective channels (but improves the robustness in time-variant channels). When compared to OFDM, the overlapping factor plays an important role in FBMC. A common value is $O = 4$, leading to very low OOB emissions for the PHYDYAS prototype filter, see Fig. 3.1. However, FBMC allows more flexibility than that. For example, the Hermite prototype filter can be used with a lower overlapping factor. One can also design a prototype filter that is specifically tailored for low-latency scenarios, such as a time domain root-raised-cosine pulse with overlapping factor $O = 1$.

An LTE subframe requires a transmission time of 1 ms and consists of $K = 14$ OFDM symbols ($T_{\text{CP}} = \frac{1}{14F}$ and $F = 15$ kHz). Because each FBMC symbol only carries half the information of that of an OFDM symbol (same number of subcarriers), we need in total $K = 28$ FBMC symbols for a fair comparison to LTE. For an overlapping factor of $O = 1$, this implies that FBMC has a transmission time of $T_{\text{block}} \approx 0.97$ ms and is therefore faster than LTE. For an overlapping factor of $O = 1.5$, the transmission time is exactly 1 ms, same as in LTE. An overlapping factor of $O = 4$, on the other hand, performs relatively poor and requires 1.2 ms, 20% longer than LTE.

Note that the ramp-up and ramp-down period in FBMC increases the latency, but not necessarily the sum throughput of the whole system, because different blocks can overlap in time. This works as long as the required phase pattern, which shifts the intrinsic interference to the purely imaginary domain, is fulfilled, as typically the case in downlink transmissions. However, in multi-user uplink transmissions, different users experience different phase shifts. Thus, the necessary phase pattern is violated, leading to interference, even for a perfectly time and frequency synchronized system. One then has to include a guard time in order to avoid interference. In such cases, the ramp-up and ramp-down period not only increases the latency, but also reduces the sum throughput.

3.3.2 Channel Estimation

The main idea of FBMC-OQAM is to equalize the phase, followed by taking the real part in order to get rid of the imaginary interference. This, however, only works once the phase is known, thus only after channel estimation. The channel estimation itself has to be performed in the complex domain, affected by the imaginary interference, and one observes an SIR of 0 dB. Thus, additional processing becomes necessary. Preamble-based channel estimation was, for example, discussed in [20]. However, LTE employs pilot-aided channel estimation because it has a low overhead and allows a simple tracking of the channel in time. A straightforward method for pilot-aided channel estimation in FBMC was proposed in [19], where one data symbol per pilot, the so-called auxiliary symbol, is sacrificed to cancel the imaginary interference at the pilot position. The big disadvantage of such method is the high power of the auxiliary symbols, worsening the PAPR and wasting signal power. Subsequently, different methods have been proposed to mitigate these harmful effects [10, 21, 35, 60]. In particular, the data spreading approach of [21] is promising because no energy is wasted, there is no noise enhancement, and the performance is close to OFDM [33]. The idea of [21] is to spread data symbols over several time–frequency positions, close to the pilot symbol, in such a way, that the imaginary interference at the pilot position is canceled. The drawback is a slightly higher computational complexity [35]. To reduce the computational complexity and to improve the applicability in doubly selective channels, one can combine the data spreading approach with the auxiliary symbol method, as proposed in [11]. Note, however, that also the classical spreading approach performs well in doubly selective channels [43].

3.4 Discrete-Time System Model

The continuous-time representation, discussed so far, provides analytical insights and gives physical meaning to multicarrier systems. However, such representation becomes analytically hard to track in doubly selective channels because double integrals have to be solved. Furthermore, in practice, the signal is generated in the discrete-time domain. Thus, we will switch from the continuous-time domain to the discrete-time domain. In contrast to many other authors, we employ a matrix-based system model instead of a discrete-time filter representation because it simplifies analytical investigations and provides a more compact description. If one is interested in the conventional discrete-time filter representation, we refer to [47, 55].

In our matrix-based system model, the basis pulses in (3.2) are sampled at rate $f_s = 1/\Delta t = F N_{\text{FFT}}$ and stacked in a basis pulse vector $\mathbf{g}_{l,k} \in \mathbb{C}^{N\times 1}$ according to

$$[\mathbf{g}_{l,k}]_n = \sqrt{\Delta t}\; g_{l,k}(t)\Big|_{t=n\Delta t - OT}, \tag{3.24}$$

for $n = 0, 1, \ldots, N-1$, where the total number of samples is given by $N = ON_{\text{FFT}} + \frac{N_{\text{FFT}}}{2}(K-1)$. The interpretation of overlapping factor O and fast Fourier transform (FFT) size $N_{\text{FFT}} \geq L$ becomes more clear later in this section, when we discuss an efficient FFT implementation. Practical systems will never operate at a critically sampling rate ($N_{\text{FFT}} = L$) because this would lead to large OOB emissions, caused by the repetition of the spectrum in the frequency domain. We strongly advise to never use a critically sampled system, which is only useful for emulating the asymptotic case of infinitely many subcarriers, $L \to \infty$. Unfortunately, many authors ignore this important aspect.

By stacking all basis pulse vectors from (3.24) in a large transmit matrix $\mathbf{G} \in \mathbb{C}^{N\times LK}$,

$$\mathbf{G} = \left[\mathbf{g}_{0,0} \cdots \mathbf{g}_{L-1,0}\; \mathbf{g}_{0,1} \cdots \mathbf{g}_{L-1,K-1}\right], \tag{3.25}$$

and all data symbols in a large transmit symbol vector $\mathbf{x} \in \mathbb{C}^{LK\times 1}$,

$$\mathbf{x} = \text{vec}\left\{\begin{bmatrix} x_{0,0} & \cdots & x_{0,K-1} \\ \vdots & \ddots & \vdots \\ x_{L-1,0} & \cdots & x_{L-1,K-1} \end{bmatrix}\right\} \tag{3.26}$$

$$= \left[x_{0,0} \cdots x_{L-1,0}\; x_{0,1} \cdots x_{L-1,K-1}\right]^{\mathsf{T}}, \tag{3.27}$$

we can express the sampled transmit signal $\mathbf{s} \in \mathbb{C}^{N\times 1}$ in (3.1) by:

$$\mathbf{s} = \mathbf{G}\mathbf{x}. \tag{3.28}$$

Because of linearity, matrix $\mathbf{G}$ can easily be found even if the underlying modulation format is not known in detail. For that, all transmitted symbols have to be set to zero, except $x_{l,k} = 1$. Vector $\mathbf{s}$ then provides immediately the $l + Lk$-th column vector of $\mathbf{G}$. Repeating this step for each time–frequency position delivers transmit matrix $\mathbf{G}$.

At the receiver, we perform matched filtering by $\mathbf{G}^{\mathsf{H}}$, so that the whole transmission system simplifies to

$$\mathbf{y} = \mathbf{G}^{\mathsf{H}}\mathbf{H}\,\mathbf{G}\,\mathbf{x} + \mathbf{G}^{\mathsf{H}}\mathbf{n} \tag{3.29}$$

$$\approx \text{diag}\{\mathbf{h}\}\,\mathbf{G}^{\mathsf{H}}\mathbf{G}\,\mathbf{x} + \mathbf{G}^{\mathsf{H}}\mathbf{n}, \tag{3.30}$$

with $\mathbf{y} \in \mathbb{C}^{LK\times 1}$ denoting the received symbols, $\mathbf{H} \in \mathbb{C}^{N\times N}$ the banded time-variant convolution matrix ($[\mathbf{H}]_{i,j} = h_{\text{conv.}}[i, i-j]$ with time-variant impulse response $h_{\text{conv.}}[i, m_\tau]$) and $\mathbf{n} \sim \mathcal{CN}(\mathbf{0}, P_{\text{n}}\,\mathbf{I}_N)$ the additive white Gaussian noise in the time domain with zero mean and variance P_{n}. In most practical scenarios, the delay spread and the Doppler spread are low enough so that the channel induced interference can be neglected [36, 41]. This allows us to factor out the channel in (3.29) according to (3.30), for which $\mathbf{h} \in \mathbb{C}^{LK\times 1}$ describes the one-tap channels (frequency domain), that is, the diagonal elements of $\mathbf{G}^{\mathsf{H}}\mathbf{H}\,\mathbf{G}$. In particular, the $l + Lk$-th element of $\mathbf{h}$ is given by

$$h_{l,k} = \mathbf{g}_{l,k}^{\mathsf{H}}\mathbf{H}\,\mathbf{g}_{l,k} \approx H(kT, lF), \tag{3.31}$$

and represents the one-tap channel at subcarrier position l and time position k. FBMC experiences imaginary interference, described by the off-diagonal elements of $\mathbf{G}^{\mathsf{H}}\mathbf{G}$ and only the real orthogonality condition holds true, that is, $\Re\{\mathbf{G}^{\mathsf{H}}\mathbf{G}\} = \mathbf{I}_{LK}$.

3.4.1 IFFT Implementation

Practical systems must be much more efficient than the matrix multiplication in (3.28). It was, for example, shown in [55] that FBMC-OQAM can be efficiently implemented by an inverse FFT (IFFT) together with a polyphase network. Unfortunately, the authors of [55] rely on a filter bank representation which is very different to the conventional OFDM description. We therefore consider an alternative interpretation, more closely related to conventional OFDM systems. A similar representation was, for example, suggested in [27] for pulse-shaping multicarrier systems.

To simplify the exposition and without losing generality, we consider only time position $k = 0$. Any other time position can easily be obtained by time shifting this special case by $T = \frac{T_0}{2}$, respectively $\frac{N_{\text{FFT}}}{2}$. The main idea is to factor out the prototype filter $p(t)$ from (3.1), so that the sampled transmit signal can be expressed by

$$s_0(n\,\Delta t) = p(n\,\Delta t)\sum_{l=0}^{L-1} \mathrm{e}^{\mathrm{j}2\pi\, l\,\frac{n}{N_{\text{FFT}}}}\,\mathrm{e}^{\mathrm{j}\theta_{l,0}}\,x_{l,0}, \tag{3.32}$$

for $n = -\frac{O N_{\text{FFT}}}{2}, \ldots, \frac{O N_{\text{FFT}}}{2} - 1$. The summation in (3.32) corresponds to an N_{FFT} point IFFT with the input arguments $\{\mathrm{e}^{\mathrm{j}\theta_{0,0}}x_{0,0},\ \mathrm{e}^{\mathrm{j}\theta_{1,0}}x_{1,0},\ \ldots,\ \mathrm{e}^{\mathrm{j}\theta_{L-1,0}}x_{L-1,0},\ 0,\ 0,\ \ldots\}$. Furthermore, because l is an integer, the summation in (3.32) is N_{FFT} periodic with respect to n. Thus, the IFFT has to be calculated only for N_{FFT} samples. Those samples can then be copied O-times, followed by an element-wise multiplication with prototype filter $p(n\,\Delta t)$. By stacking the transmitted samples in a vector $\mathbf{s}_0 \in \mathbb{C}^{O N_{\text{FFT}}\times 1}$, we can therefore express (3.32) by

$$\mathbf{s}_0 = \mathbf{p} \circ \underbrace{\left(\underbrace{\mathbf{1}_{O\times 1} \otimes \underbrace{\mathbf{W}_{N_\mathrm{FFT}}^\mathrm{H} \begin{bmatrix} \mathrm{e}^{\mathrm{j}\theta_{0,0}}\, x_{0,0} \\ \vdots \\ \mathrm{e}^{\mathrm{j}\theta_{L-1,0}}\, x_{L-1,0} \\ 0 \\ \vdots \end{bmatrix}}_{\text{IFFT}}}_{\text{repeat } O\text{-times}} \right)}_{\text{element-wise multiplication}}, \tag{3.33}$$

where $\circ$ denotes the element-wise Hadamard product, $\otimes$ the Kronecker product and $\mathbf{W}_{N_\mathrm{FFT}} \in \mathbb{C}^{N_\mathrm{FFT}\times N_\mathrm{FFT}}$ a Discrete Fourier Transform (DFT) matrix. Note that, by circular shifting the IFFT input in (3.33), we can shift the signal in frequency by multiples of F. The sampled prototype filter $\mathbf{p} \in \mathbb{C}^{O N_\mathrm{FFT}\times 1}$ in (3.33) is given by,

$$[\mathbf{p}]_n = \sqrt{\Delta t}\; p(t)\Big|_{t=n\,\Delta t - OT} \quad \text{for} \quad n = 0, 1, \ldots, O\, N_\mathrm{FFT} - 1. \tag{3.34}$$

Figure 3.6 illustrates the efficient FBMC-OQAM implementation and compares it to windowed OFDM. Both modulation schemes apply the same basic operations, that is, IFFT, repetition and element-wise multiplications. However, windowed OFDM has overall a lower complexity because the element-wise multiplication is limited to a window of size $2\,T_\mathrm{W}$ and time symbols are further apart, that is, $T = T_\mathrm{W} + T_\mathrm{CP} + T_0$ in windowed OFDM versus $T = T_0/2$ in FBMC-OQAM. Thus, FBMC needs to apply the IFFT more than two times (exactly two times if $T_\mathrm{W} = T_\mathrm{CP} = 0$). Of course, the overhead $T_\mathrm{W} + T_\mathrm{CP}$ in windowed OFDM reduces the spectral efficiency. Because

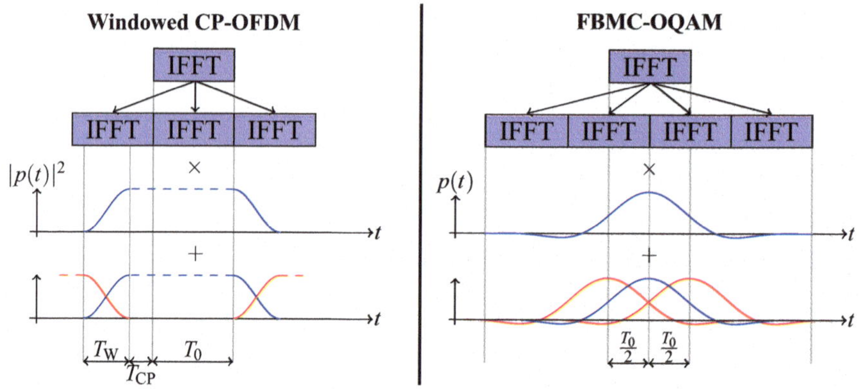

Fig. 3.6 From a conceptional point of view, the signal generation in windowed OFDM and FBMC-OQAM requires the same basic operations, namely, an IFFT, copying the IFFT output, element-wise multiplication with the prototype filter and, finally, overlapping. ©2017 IEEE, [41]

the signal generation for both modulation formats is very similar, FBMC-OQAM can utilize the same hardware components as windowed OFDM.

The receiver works in a similar way, but in reversed order, that is, element-wise multiplication, reshaping the received symbol vector to $N_{\mathrm{FFT}} \times O$ followed by a row-wise summation and, finally, an FFT. In matrix notation, this can be expressed by

$$\mathbf{y}_0 = \left[\mathrm{diag}\left\{\left[\mathrm{e}^{-\mathrm{j}\theta_{0,0}} \ldots \mathrm{e}^{-\mathrm{j}\theta_{L-1,0}}\right]\right\} \mathbf{0}_{L\times(N_{\mathrm{FFT}}-L)}\right] \mathbf{W}_{N_{\mathrm{FFT}}} \left(\mathbf{1}_{1\times O} \otimes \mathbf{I}_{N_{\mathrm{FFT}}}\right) \left(\mathbf{p} \circ \mathbf{r}_0\right), \tag{3.35}$$

where $\mathbf{r}_0 \in \mathbb{C}^{ON_{\mathrm{FFT}}\times 1}$ represents the received samples and $\mathbf{y}_0 \in \mathbb{C}^{L\times 1}$ the received symbols, both at time position $k = 0$. WOLA requires at the receiver the same basic operations as FBMC. However, in contrast to FBMC, WOLA employs a different transmit and receive prototype filter.

3.5 One-Tap Equalizers in Doubly Selective Channels

The biggest advantage of multicarrier systems is that the transmission over a doubly selective channel can be approximated by one-tap channels. In this section, we calculate the approximation error by considering the signal-to-interference ratio (SIR). In orthogonal multicarrier systems, the SIR can be calculated by

$$\mathrm{SIR}_{\mathrm{QAM}} = \frac{\mathbb{E}\{|h_{l,k}\, x_{l,k}|^2\}}{\mathbb{E}\{|z_{l,k}|^2\}} \tag{3.36}$$

with interference $z_{l,k}$ given by,

$$z_{l,k} = \mathbf{g}_{l,k}^{\mathsf{H}}\, \mathbf{H}\, \mathbf{G}\, \mathbf{x} - h_{l,k}\, x_{l,k}. \tag{3.37}$$

In FBMC-OQAM, on the other hand, the SIR cannot be calculated as easily as in (3.36) because directly applying (3.36) leads to an SIR of approximately 0 dB due to the inherent self-interference. We have to equalize the phase followed by taking the real part, before calculating the SIR. Thus, the SIR can be expressed by

$$\mathrm{SIR}_{\mathrm{OQAM}} = \frac{\mathbb{E}\{|\Re\{\mathrm{e}^{-\mathrm{j}\varphi_{l,k}}\, h_{l,k}\, x_{l,k}\}|^2\}}{\mathbb{E}\{|\Re\{\mathrm{e}^{-\mathrm{j}\varphi_{l,k}}\, z_{l,k}\}|^2\}}, \tag{3.38}$$

where $\mathrm{e}^{\mathrm{j}\varphi_{l,k}} = h_{l,k}/|h_{l,k}|$ represents the phase of the one-tap channel. The SIR is very helpful because it determines the point at which interference starts to dominate the noise, that is, SNR > SIR, leading to a saturation of the bit error rate (BER), see [36]. As long as the SNR is approximately 10 dB lower than the SIR, interference is completely dominated by noise and can thus be neglected. Once the SNR approaches

Fig. 3.7 The SIR depends on the subcarrier spacing. For a fair comparison of different modulation schemes, we consider an optimal subcarrier spacing

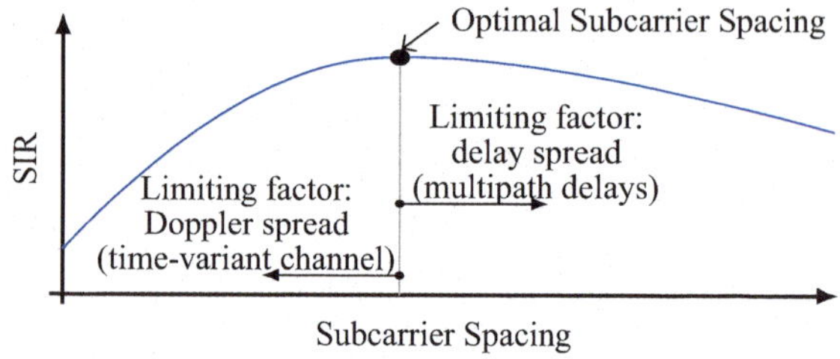

the SIR, that is, SNR = SIR, one observes a performance degeneration equivalent to an SNR shift of approximately 3 dB. The matrix representation in Sect. 3.4 can be utilized to calculate the SIR in (3.36). For example, the channel power can be calculated by $\mathbb{E}\{|h_{l,k}|^2\} = (\mathbf{g}_{l,k}^{\mathsf{T}} \otimes \mathbf{g}_{l,k}^{\mathsf{H}})\mathbf{R}_{\text{vec}\{\mathbf{H}\}}(\mathbf{g}_{l,k}^{\mathsf{T}} \otimes \mathbf{g}_{l,k}^{\mathsf{H}})^{\mathsf{H}}$, with channel correlation matrix $\mathbf{R}_{\text{vec}\{\mathbf{H}\}} = \mathbb{E}\{\text{vec}\{\mathbf{H}\}\text{vec}\{\mathbf{H}\}^{\mathsf{H}}\} \in \mathbb{C}^{N^2 \times N^2}$. For OQAM, on the other hand, additional processing becomes necessary, see [41] for more details. Note that the SIR can also be calculated with the ambiguity function, as, for example, demonstrated in [15, 50].

In [36], we showed that FBMC (Hermite prototype filter) outperforms CP-OFDM in high-velocity scenarios. This, however, was only true because interference from the Doppler spread dominated interference from the delay spread. By increasing the subcarrier spacing, the overall SIR could be improved, as illustrated in Fig. 3.7. The big question is then, does FBMC still outperform CP-OFDM if both modulation schemes apply an optimal subcarrier spacing? Because 5G will include a flexible subcarrier spacing [3], our considerations here are also relevant for future wireless systems. As a rule of thumb, the subcarrier spacing should be chosen so that [12]

$$\frac{\sigma_t}{\sigma_f} \approx \frac{\tau_{\text{rms}}}{\nu_{\text{rms}}}, \tag{3.39}$$

where time localization σ_t and frequency localization σ_f are given by (3.19) for the Hermite pulse and by (3.22) for the PHYDYAS pulse. For FBMC-OQAM, this leads to the following optimal subcarrier spacings:

$$F_{\text{opt,Hermite}} \approx 0.71 \times \sqrt{\frac{\nu_{\text{rms}}}{\tau_{\text{rms}}}}, \tag{3.40}$$

$$F_{\text{opt,PHYDYAS}} \approx 0.91 \times \sqrt{\frac{\nu_{\text{rms}}}{\tau_{\text{rms}}}}. \tag{3.41}$$

For a Jakes Doppler spectrum, the root mean square (RMS) Doppler spread is given by $\nu_{\text{rms}} = \frac{1}{\sqrt{2}} \frac{v}{c} f_{\text{c}}$, with v denoting the velocity, c the speed of light and f_{c} the carrier frequency. Note that (3.39) represents only an approximation. The exact relationship can be calculated, as, for example, done in [17] for the Gaussian pulse, and depends on the underlying channel model and the prototype filter. However, for our chosen numerical parameters, the differences between the optimal SIR (exhaustive search) and the SIR obtained by applying the rule in (3.39) is less than 0.1 dB for FBMC-OQAM and less than 1 dB for FBMC-QAM. For the rest of this section, we always find the optimal subcarrier spacing in FBMC through an exhaustive search.

As a reference, we also consider an optimal subcarrier spacing in CP-OFDM. The rule in (3.39), however, cannot be applied because the underlying rectangular pulse is not localized in frequency. Instead, we assume, for a fixed CP overhead of $\kappa = \frac{T_{\mathrm{CP}}}{T_0} = T_{\mathrm{CP}} F = TF - 1$, that the subcarrier spacing is chosen as high as possible, while still satisfying the condition of no Inter Symbol Interference (ISI), that is, $T_{\mathrm{CP}} = \tau_{\max}$. This leads to $F = \frac{\kappa}{\tau_{\max}}$. For a Jakes Doppler spectrum, the SIR can be expressed by a generalized hypergeometric function ${}_1F_2(\cdot)$ [48], which, together with an optimal subcarrier spacing, leads to [41],

$$\mathrm{SIR}^{\mathrm{CP-OFDM}}_{\mathrm{opt.,\,noISI}} = \frac{{}_1F_2\left(\frac{1}{2}; \frac{3}{2}, 2; -\left(\pi \frac{\nu_{\max}\tau_{\max}}{TF-1}\right)^2\right)}{1 - {}_1F_2\left(\frac{1}{2}; \frac{3}{2}, 2; -\left(\pi \frac{\nu_{\max}\tau_{\max}}{TF-1}\right)^2\right)}. \tag{3.42}$$

For our numerical example, we consider a TDL-B channel model, as proposed by 3GPP [1, Sect. 7.7.3], and a carrier frequency of 2.5 GHz. Furthermore, we assume a long delay spread of 300 ns. We expect that in future wireless systems, the "typical" delay spread will be much lower than 300 ns [41]. Nonetheless, such a long delay spread allows robustness considerations. The optimal subcarrier spacing for a velocity of zero approaches $F \rightarrow 0$ Hz, not feasible in practice. We therefore assume that the subcarrier spacing is lower bounded by $F \geq 15$ kHz. For our channel parameters, the SIR is illustrated in Fig. 3.8 and allows the following conclusions:

1. The SIR in FBMC is high enough, so that the channel induced self-interference is usually dominated by noise. Thus, self-interference can be neglected.[2]
2. For an optimal subcarrier spacing, the Hermite prototype filter outperforms the PHYDYAS prototype filter, but only by approximately 0.6 dB.
3. For low velocities, on the other hand, the PHYDYAS prototype filter becomes better than the Hermite filter because of a better frequency localization in combination with a fixed subcarrier spacing (lower bound).
4. For a maximum symbol density of $TF = 1$ (complex), FBMC performs much better than OFDM without CP, especially for low velocities.
5. CP-OFDM, see (3.42), performs best, but also has a lower data rate than FBMC because of the CP overhead.
6. WOLA ($T_{\mathrm{W,TX}} = T_{\mathrm{W,RX}} = \frac{\kappa}{2F}$ and $T_{\mathrm{CP}} = 0$) has a lower SIR than CP-OFDM and performs close to FBMC in high-velocity scenarios.
7. In general, OFDM-based schemes perform good in low-velocity scenarios, while in high-velocity scenarios, they lose most of their advantages when compared with FBMC.

[2]The SNR is often below 20 dB. Wireless systems are interference limited and what we call "noise" is in practice often interference from other users and real-world hardware effects.

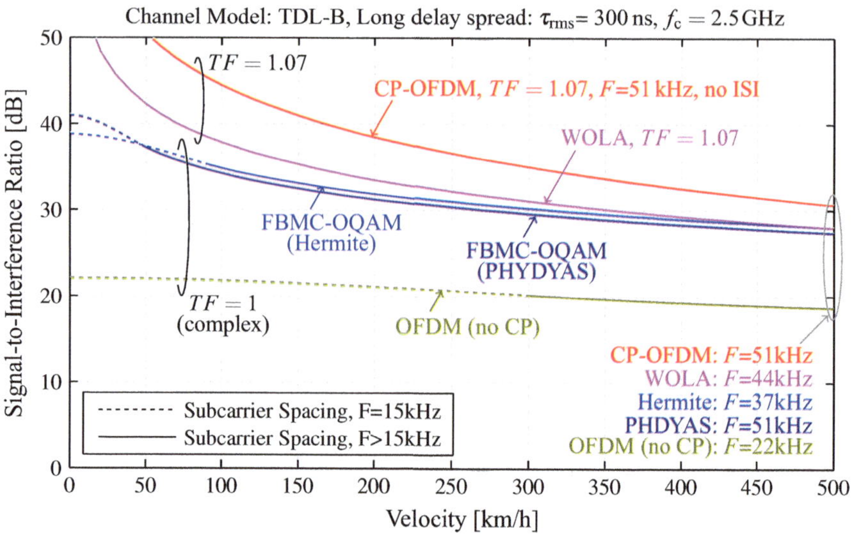

Fig. 3.8 The SIR is high enough so that the channel induced interference can usually be neglected in FBMC because it is dominated by noise

Note that Fig. 3.8 represents a typical behavior. Different channel models and RMS delay spread values only shift the curves to some other point [41]. The main conclusion here is that one-tap equalizers are in most practical cases sufficient. For all of our testbed measurements in [34, 38, 42], low-complexity one-tap equalizers were sufficient. Feedback delays and repeated handovers are usually more problematic than a small, channel induced, interference.

For the sake of completeness, we also consider a highly double-selective channel, where we assume an RMS delay spread of 720 ns and a carrier frequency of 60 GHz, as suggested in [1]. However, we want to emphasize that such extreme channel condition will rarely happen in practice and a system should therefore not be optimized for it, but it should be able to cope with those scenarios, at least to some extent. There exist several ways of dealing with such harsh channel environments: Firstly, the employment of computationally demanding equalizers, as, for example, proposed in [40, 49]. Secondly, we can treat interference as noise and accept a (small) throughput loss, see Fig. 3.2. Thirdly, spectral efficiency can be sacrificed in order to gain robustness. Let us discuss the last method in more detail, where we utilize the underlying orthogonality of the prototype filter in FBMC to transmit complex-valued symbols. This is enabled by setting some symbols, corresponding to the black markers in Figs. 3.4 and 3.5, to zero, so that a time–frequency spacing of $TF = 2$ is achieved. We call this transmission scheme FBMC-QAM but there does not exist a unique definition and different authors use this term differently. In FBMC-QAM, complex-valued symbols are transmitted and no intrinsic imaginary interference appears, allowing us to straightforwardly apply all known methods of

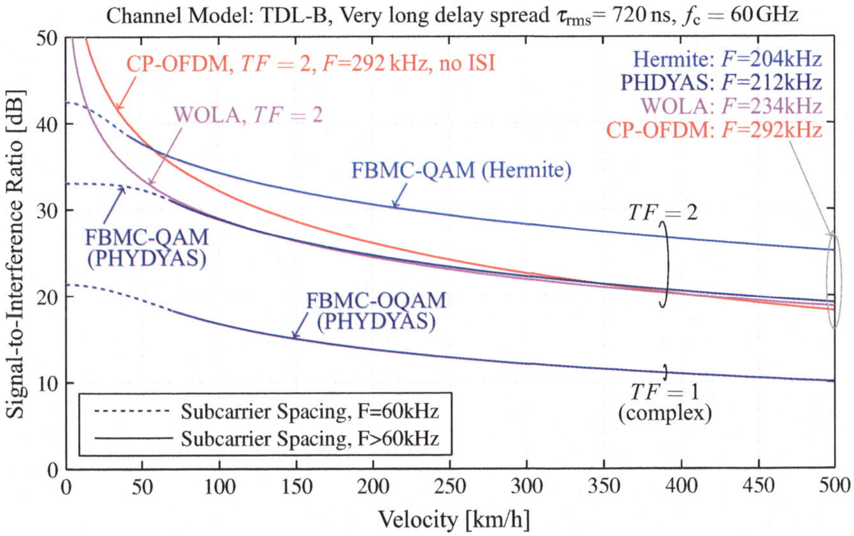

Fig. 3.9 In the rare case of a highly double-selective channel, it is possible to sacrifice spectral efficiency ($TF = 2$) in order to improve robustness. FBMC-QAM then even outperforms CP-OFDM

OFDM. Figure 3.9 shows the SIR for a highly double-selective channel. The following important observations can be identified.

1. The Hermite prototype filter performs much better than the PHYDYAS prototype filter thanks to a better joint time–frequency localization. In FBMC-OQAM, this effect is largely lost because of the time–frequency squeezing.
2. FBMC with a Hermite prototype filter outperforms even CP-OFDM for velocities larger than 60 km/h.

3.6 Block Spread FBMC: Enabling All MIMO Methods

The loss of complex orthogonality is the main obstacle in FBMC-OQAM and seriously hampers some important transmission techniques, such as channel estimation [35], Alamouti's space-time block code [23] or maximum likelihood MIMO detection [62]. In particular, the limited MIMO compatibility[3] is a major issue, preventing a widespread application of FBMC.

In this section, we investigate a method which restores complex orthogonality in FBMC-OQAM, so that all known techniques for OFDM can be straightforwardly

[3]Only some specific MIMO techniques become more challenging in FBMC. Many other MIMO methods, such as receive diversity or spatial multiplexing based on Zero-Forcing (ZF) or MMSE equalization, can be straightforwardly employed in FBMC.

employed in FBMC. This is enabled by adding an additional code dimension (besides time and frequency). In contrast to conventional FBMC, the data symbols no longer belong to a certain time–frequency position, but are rather spread over several time or frequency positions. Such spreading typically increases the sensitivity to doubly selective channels. However, if the delay spread and the Doppler spread are sufficiently low, the channel induced interference can still be neglected.

Spreading is very beneficial in FBMC because it can solve the underlying problem of Alamouti's space-time block code and maximum likelihood (ML) MIMO detection in FBMC. For example, authors in [46] proposed a block-Alamouti scheme (over time) which can be seen as a special kind of spreading (distributing symbols in time). The same method was recently applied by [30] in the frequency domain. However, Walsh–Hadamard spreading [23, 34] offers more flexibility because it restores complex orthogonality, so that it not only works for Alamouti transmissions (as in [30, 46]), but additionally allows to straightforwardly employ all other methods known in OFDM, such as channel estimation, other space-time block codes or low-complexity maximum likelihood symbol detection. Similar to Walsh–Hadamard spreading, authors in [62] propose FFT spreading in time to restore (quasi)-orthogonality.

Let us describe the spreading approach in more detail. In a first step, we assume an AWGN channel, that is, $\mathbf{H} = \mathbf{I}_N$, so that (3.30) transforms to

$$\mathbf{y} = \mathbf{G}^{\mathsf{H}}\mathbf{G}\,\mathbf{x} + \mathbf{G}^{\mathsf{H}}\mathbf{n}. \tag{3.43}$$

Note that (3.43) describes a block transmission of L subcarriers and K symbols in time. Several of those blocks must be concatenated in time and frequency to achieve a desired bandwidth and transmission time.

In spread FBMC, complex-valued data symbols $\tilde{\mathbf{x}} \in \mathbb{C}^{\frac{LK}{2}\times 1}$ are precoded by a coding/spreading matrix $\mathbf{C} \in \mathbb{C}^{LK\times\frac{LK}{2}}$, so that the transmitted symbols $\mathbf{x} \in \mathbb{C}^{LK\times 1}$ can be expressed by

$$\mathbf{x} = \mathbf{C}\,\tilde{\mathbf{x}}. \tag{3.44}$$

A priori, the size of $\mathbf{C}$ and $\tilde{\mathbf{x}}$ is unknown. We will explain later in this section why the size was chosen that way. The received data symbols $\tilde{\mathbf{y}} \in \mathbb{C}^{\frac{LK}{2}\times 1}$ are obtained by decoding of the received symbols according to

$$\tilde{\mathbf{y}} = \mathbf{C}^{\mathsf{H}}\,\mathbf{y}. \tag{3.45}$$

To restore complex orthogonality, the coding matrix must be chosen so that the following condition is fulfilled,

$$\mathbf{C}^{\mathsf{H}}\,\mathbf{G}^{\mathsf{H}}\mathbf{G}\,\mathbf{C} = \mathbf{I}_{LK/2}. \tag{3.46}$$

A straightforward way to find coding matrix $\mathbf{C}$ is based on an eigenvalue decomposition of $\mathbf{G}^{\mathsf{H}}\mathbf{G} = \mathbf{U}\boldsymbol{\Lambda}\mathbf{U}^{\mathsf{H}}$, so that coding matrix $\mathbf{C} \in \mathbb{C}^{LK\times\frac{LK}{2}}$ becomes,

$$\mathbf{C} = \mathbf{U}\begin{bmatrix} \Lambda_1^{-1/2} & 0 & 0 \\ 0 & \ddots & 0 \\ 0 & 0 & \Lambda_{LK/2}^{-1/2} \\ 0 & \dots & 0 \\ \vdots & \vdots & \vdots \\ 0 & \dots & 0 \end{bmatrix}, \tag{3.47}$$

where Λ_i represents the i-th eigenvalue (sorted) of $\mathbf{G}^{\mathsf{H}}\mathbf{G}$ and $\mathbf{U}$ the unitary eigenvector matrix. Figure 3.10 shows the eigenvalues of a typical FBMC-OQAM transmission matrix $\mathbf{G}^{\mathsf{H}}\mathbf{G}$. For the limit case of $K \rightarrow \infty$ and $L \rightarrow \infty$, the eigenvalues are $\Lambda_1 = \Lambda_2 = \cdots = \Lambda_{LK/2} = 2$ and $\Lambda_i = 0$ for $i > \frac{LK}{2}$. Thus, (3.47) implicitly applies water filling [56], so that $\mathbf{C}$ becomes the optimal spreading matrix in terms of maximizing the information rate. In particular, it shows that the optimal size of the spreading matrix is $LK \times \frac{LK}{2}$ and that any matrix, $\mathbf{C} \in \mathbb{C}^{LK \times \frac{LK}{2}}$, which satisfies (3.46), is optimal for $K \rightarrow \infty$ and $L \rightarrow \infty$ (the SNR is always the same). Moreover, it also shows that the intrinsic imaginary interference does not consist of any useful information and can thus be canceled by taking the real part. For a limited number of subcarriers and time symbols, the spreading matrix in (3.47) no longer corresponds to the optimal solution. Instead, water filling could improve the performance, where the column size of matrix $\mathbf{C}$ will usually be larger than $\frac{LK}{2}$. However, a spreading matrix of size $LK \times \frac{LK}{2}$, which satisfies (3.46), still performs close to the optimum, as indicated by the eigenvalues in Fig. 3.10. For example, for $L = 36$ and $K = 30$, the suboptimal spreading matrix performs only 3.6% worse in terms of achievable rate than the optimal spreading matrix (water filling) for SNR values smaller than 20 dB. Furthermore, the optimal spreading matrix requires different code rates for layers close to the eigenvalue of $\Lambda_{LK/2}$. This increases the overall complexity, while the possible improvement is rather low, so that, employing a slightly suboptimal spreading matrix makes sense in practical systems. Note that precoding reduces the average transmit power by a factor of two, that is, $\operatorname{tr}\left\{\mathbf{G}\mathbf{C}\mathbf{C}^{\mathsf{H}}\mathbf{G}^{\mathsf{H}}\right\} = \operatorname{tr}\left\{\mathbf{C}^{\mathsf{H}}\mathbf{G}^{\mathsf{H}}\mathbf{G}\mathbf{C}\right\} = \frac{LK}{2} = \frac{1}{2}\operatorname{tr}\left\{\mathbf{G}^{\mathsf{H}}\mathbf{G}\right\}$. Thus, for the same SNR as without precoding, the data symbol power has to be increased by a factor of two. It is also interesting that the noise after despreading is white, $\tilde{\mathbf{n}} \sim \mathcal{CN}(\mathbf{0}, P_{\mathrm{n}}\, \mathbf{I}_{LK/2})$,

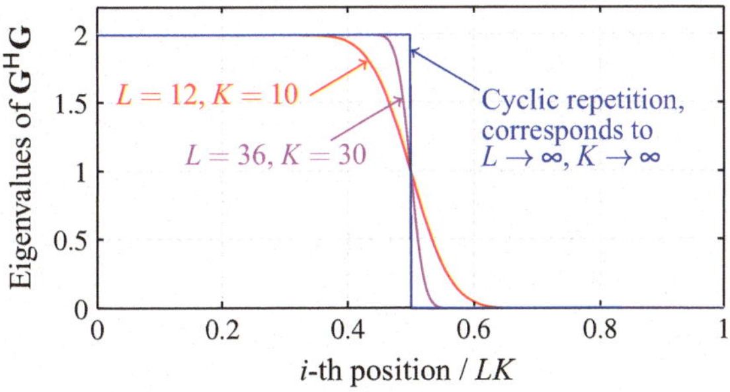

Fig. 3.10 The eigenvalues of $\mathbf{G}^{\mathsf{H}}\mathbf{G}$ for an FBMC-OQAM system. Similar as for the derivation of the MIMO channel capacity, the eigenvalues in combination with eigenvector precoding can be utilized to determine the optimal precoding matrix

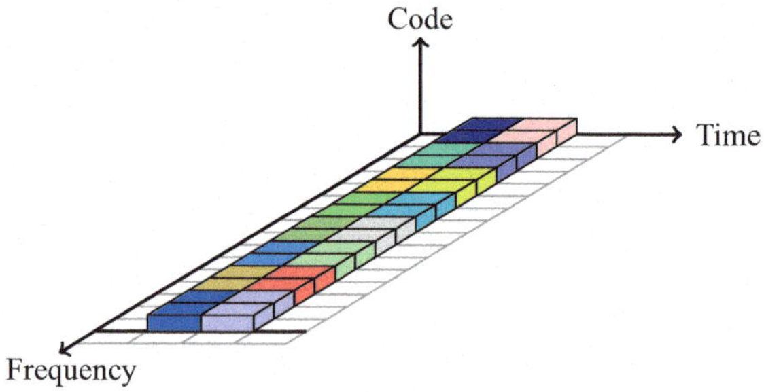

Fig. 3.11 In conventional FBMC-OQAM, real-valued symbols are transmitted over a rectangular time–frequency grid ($TF = 0.5$). Two real-valued symbols are required to transmit one complex-valued symbol. Thus, the name "offset"-QAM, where we apply the offset not in time (as often in literature) but in frequency. ©2017 IEEE, [39]

even though the spreading matrix itself is not necessarily semi-unitary. The reason behind this is again the orthogonalization condition in (3.46), implying that $\mathbf{R}_{\tilde{\mathbf{n}}} = \mathbf{C}^{\mathsf{H}}\mathbf{G}^{\mathsf{H}}\mathbf{R}_{\mathbf{n}}\mathbf{G}\mathbf{C} = P_{\mathrm{n}}\,\mathbf{I}_{LK/2}$.

While the spreading matrix in (3.47) provides analytical insight, it is not very practical because of a high computational complexity and the fact that the spreading is performed in both, time and frequency, which only works for a doubly flat channel. Walsh–Hadamard spreading [22, 23, 34, 39], on the other hand, is a much more practical solution because it requires almost no additional complexity and the spreading is performed in only one dimension, either in time or in frequency. In conventional FBMC-OQAM, see Fig. 3.11, each time–frequency position can only carry real-valued symbols, so that two time–frequency positions are required to transmit one complex-valued data symbol, indicated by the color in Fig. 3.11. In block spread FBMC-OQAM, on the other hand, data symbols no longer belong to a specific time–frequency position, but are spread over several subcarriers, see Fig. 3.12. To keep the spectral efficiency the same as in FBMC-OQAM (ignoring possible guard symbols), several data symbols are transmitted over the same time–frequency resources, but differentiated by their spreading/coding sequence. To be specific, $L/2$ complex-valued data symbols are spread over L subcarriers. This leads to the same information rate as in conventional FBMC-OQAM (again, ignoring possible guard symbols). Although complex orthogonality can be perfectly restored within one block, there still exists interference between different blocks. Thus, if we spread in frequency, a guard subcarrier might be necessary. If we spread in time, on the other hand, a guard symbol in time might be required. Such guard symbols typically reduce the spectral efficiency by a few percent, see [34, 39].

Spreading in frequency can be described by frequency spreading matrix $\mathbf{C}_f \in \mathbb{R}^{L\times\frac{L}{2}}$ for which we take every second column out of a sequency-ordered [25] Walsh–Hadamard matrix $\boldsymbol{\mathcal{H}} \in \mathbb{R}^{L\times L}$, that is,

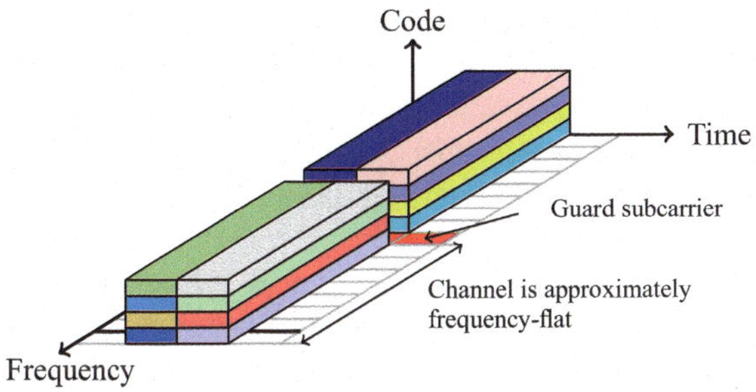

Fig. 3.12 In block spread FBMC-OQAM, complex-valued symbols are spread over several subcarriers (or time positions), allowing to restore complex orthogonality within one block. To improve the SIR between different blocks, a guard symbol might be necessary. ©2017 IEEE, [39]

$$[\mathbf{C}_f]_{l,m} = [\mathcal{H}]_{l,2m-1} \quad ; \text{for} \quad l = 1, 2, \ldots, L; \; m = 1, 2, \ldots, \frac{L}{2}. \tag{3.48}$$

Note that matrix $\mathbf{C}_f$ in (3.48) could equivalently be defined by $[\mathcal{H}]_{l,2m}$. Utilizing the underlying structure of our matrix notation (vectorization) and the fact that we spread in frequency only, allows us to express overall spreading matrix $\mathbf{C} \in \mathbb{R}^{LK \times \frac{LK}{2}}$ by,

$$\mathbf{C} = \mathbf{I}_K \otimes \mathbf{C}_f, \tag{3.49}$$

where Kronecker product $\otimes$ together with identity matrix $\mathbf{I}_K$ map coding matrix $\mathbf{C}_f$ to the correct time slots.

Spreading in time can be described in a similar way except that we have to alternate between spreading with $\mathbf{C}_{t'} \in \mathbb{R}^{K \times \frac{K}{2}}$ and spreading with $\mathbf{C}_{t''} \in \mathbb{R}^{K \times \frac{K}{2}}$ for adjacent subcarriers. The spreading matrices itself are again found by taking every second column out of a sequency-ordered [25] Walsh–Hadamard matrix $\mathcal{H} \in \mathbb{R}^{K \times K}$, that is,

$$\begin{aligned} [\mathbf{C}_{t'}]_{k,m} &= [\mathcal{H}]_{k,2m-1} \\ [\mathbf{C}_{t''}]_{k,m} &= [\mathcal{H}]_{k,2m} \end{aligned} \quad ; \text{for} \quad k = 1, 2, \ldots, K; \; m = 1, 2, \ldots, \frac{K}{2}. \tag{3.50}$$

To find the overall spreading matrix $\mathbf{C} \in \mathbb{R}^{LK \times \frac{LK}{2}}$, we have to map the individual spreading matrices $\mathbf{C}_{t'}$ and $\mathbf{C}_{t''}$ to the correct subcarrier positions. For the vectorized system model in (3.26), this implies that,

$$\mathbf{C} = \mathbf{C}_{t'} \otimes \mathbf{I}_{L/2} \otimes \begin{bmatrix} 1 & 0 \\ 0 & 0 \end{bmatrix} + \mathbf{C}_{t''} \otimes \mathbf{I}_{L/2} \otimes \begin{bmatrix} 0 & 0 \\ 0 & 1 \end{bmatrix}, \tag{3.51}$$

where the matrices $\left[\begin{smallmatrix} 1 & 0 \\ 0 & 0 \end{smallmatrix}\right]$ and $\left[\begin{smallmatrix} 0 & 0 \\ 0 & 1 \end{smallmatrix}\right]$ are necessary to alternate between spreading with $\mathbf{C}_{t'}$ and $\mathbf{C}_{t''}$ for adjacent subcarriers.

It can easily be checked by numerical evaluations that (3.49) and (3.51) satisfy the complex orthogonalization condition in (3.46). For a formal proof that Walsh–Hadamard spreading restores complex orthogonality in FBMC, we refer to [22]. Authors of [22] left the question open whether it is possible to find a spreading matrix that has more than $\frac{LK}{2}$ columns while still satisfying (3.46). Our investigations in (3.47) show that this is not possible (ignoring any edge effects which become negligible for a large number of K and L). A small disadvantage of Walsh–Hadamard spreading is the fact that the spreading length has to be a power of two. This makes the integration into existing systems problematic, but has almost no impact if a system is designed from scratch. The big advantage of Walsh–Hadamard spreading, on the other hand, is that only additions, but no multiplications are needed. Thus, the additional computational complexity is very low. Moreover, a fast Walsh–Hadamard transformation can be used, further reducing the computational complexity. For example, spreading in time only requires $\log_2(K) - 1$ extra additions/subtractions per data symbol at the transmitter and $\log_2(K)$ extra additions/subtractions per data symbol at the receiver. For spreading in frequency, it is $\log_2(L) - 1$, respectively $\log_2(L)$.

Similar as in (3.29), it is possible to include a doubly selective channel into our transmission model. The input output relationship between the transmitted data symbols $\tilde{\mathbf{x}}$ and the received data symbols $\tilde{\mathbf{y}}$ can then be modeled by

$$\tilde{\mathbf{y}} = \mathbf{C}^{\mathsf{H}}\mathbf{G}^{\mathsf{H}}\,\mathbf{H}\,\mathbf{G}\mathbf{C}\,\tilde{\mathbf{x}} + \mathbf{C}^{\mathsf{H}}\mathbf{G}^{\mathsf{H}}\mathbf{n} \tag{3.52}$$

$$\approx \operatorname{diag}\{\tilde{\mathbf{h}}\}\,\tilde{\mathbf{x}} + \mathbf{C}^{\mathsf{H}}\mathbf{G}^{\mathsf{H}}\mathbf{n}. \tag{3.53}$$

Similar as in (3.29), if the delay spread and the Doppler spread are sufficiently low, the transmission can be approximated by a one-tap channel, see (3.53). If we spread in time, block spread FBMC becomes more sensitive to time-variant channels. At the same time, it becomes slightly more robust to multipath delays. For spreading in frequency, the opposite holds true. In [34, 39], we discuss the effect of doubly selective channels on block spread FBMC.

Precoding by $\mathbf{C}$ can also be interpreted as transforming the underlying basis pulses according to $\tilde{\mathbf{G}} = \mathbf{G}\mathbf{C} = \left[\tilde{\mathbf{g}}_1 \cdots \tilde{\mathbf{g}}_{LK/2}\right]$. Thus, instead of modulating data symbols with $g_{l,k}(t)$, as in (3.1), we modulate them with $\tilde{g}_i(t)$. In contrast to conventional multicarrier systems, however, the transformed basis pulses $\tilde{g}_i(t)$ no longer all employ the same underlying prototype filter $p(t)$. Instead, many basis pulses have their own, unique, prototype filter $p_i(t)$. Thus, we cannot directly implement $\tilde{\mathbf{G}}$ in an efficient way. However, by interpreting $\tilde{\mathbf{G}}$ as a precoded FBMC system, the advantage of an efficient signal generation are preserved. Moreover, such interpretation offers overall a high flexibility.

We have validated the block spread FBMC approach by real-world testbed measurements in [34] for outdoor-to-indoor scenarios (150 m link distance, 2.5 GHz carrier frequency) and in [42] for indoor-to-indoor scenarios (5 m link distance, 60 GHz carrier frequency). For both measurements, the assumption of a low delay spread and a low Doppler spread was fulfilled, so that (3.53) accurately described

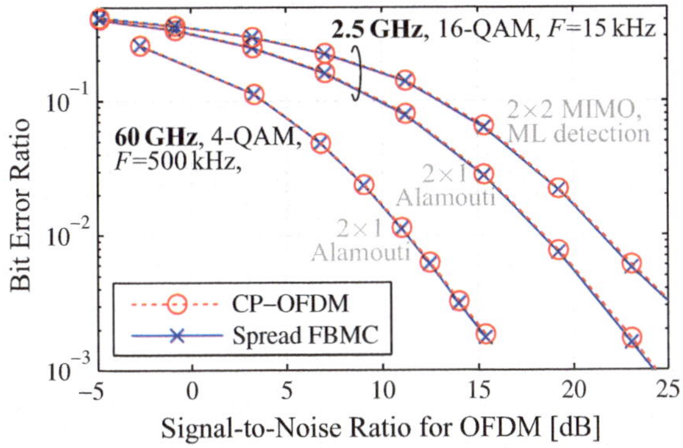

Fig. 3.13 Real-world testbed measurements [34, 42] show that MIMO works in FBMC once symbols are spread in time. The spreading process itself has a low computational complexity because of a fast Walsh–Hadamard transformation. FBMC and OFDM experience both the same BER, but FBMC has lower OOB emissions. ©2017 IEEE, [41]

the true physical behavior. Because of a time-invariant channel, we spread in time instead of frequency, see (3.50) and (3.51). For the 60 GHz measurement setup, we employed a high subcarrier spacing of $F = 500\,\text{kHz}$, as often considered in millimeter wave transmissions [44]. This implicitly reduces the latency, so that even though we spread symbols in time, the overall transmission time was less than $40\,\mu\text{s}$, satisfying the low-latency condition of $100\,\mu\text{s}$ [14].

Figure 3.13 shows the measured BER over SNR. Alamouti's space-time block code and low-complexity ML MIMO detection performs in FBMC as good as in OFDM, but FBMC has the advantage of much lower OOB emissions.

3.7 Pruned DFT-Spread FBMC-OQAM: Reducing the PAPR

Besides the intrinsic imaginary interference, nonlinearities, such as a limited Digital-to-Analog Converter (DAC) resolution or a nonlinear power amplifier, are even more problematic in practical systems because they destroy the superior spectral properties of FBMC [41, 50]. Thus, the concept of sharp digital filters to enable a flexible time–frequency allocation, as discussed in [41], only works as long as FBMC operates in the linear regime. In multicarrier systems, this is, in general, hard to achieve because of the poor peak-to-average power ratio (PAPR). To reduce the PAPR in practical systems, LTE employs single carrier frequency division multiple access (SC-FDMA) in the uplink [54], a DFT precoded OFDM system. The same technique will also be used in 5G [3]. Unfortunately, simply combining FBMC and a DFT, as done in SC-FDMA for OFDM, performs poorly in FBMC [18, 31, 61]. This motivated us to develop pruned DFT-spread FBMC [32, 37], a novel transmission technique

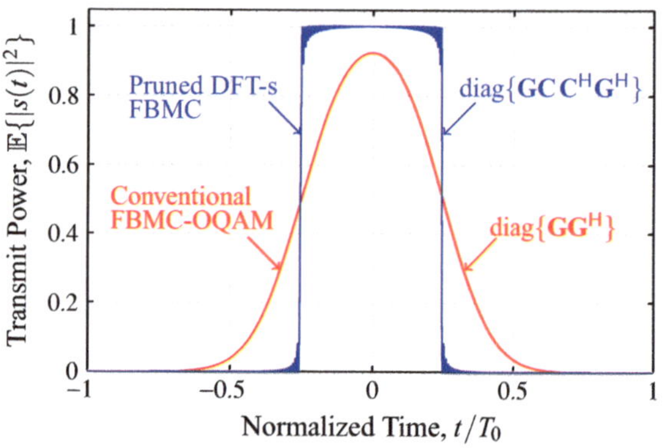

Fig. 3.14 Precoding matrix $\mathbf{C}$ shapes the transmitted signal in such a way, that the average transmit power (diagonal elements of $\mathbb{E}\{\mathbf{s}\,\mathbf{s}^{\mathsf{H}}\} = \mathbf{G}\mathbf{C}\,\mathbf{C}^{\mathsf{H}}\mathbf{G}^{\mathsf{H}}$) shows an almost perfect rectangular shape, with many beneficial properties. For the illustration we consider only one FBMC symbol, that is, $K = 1$

with the remarkable properties of a low PAPR, low-latency transmissions and a high spectral efficiency. The idea of pruned DFT-spread FBMC is closely related to Walsh–Hadamard spreading, see Sect. 3.6. In particular, we again spread $\frac{L}{2}$ data symbols over L subcarriers, described by frequency spreading matrix $\mathbf{C}_f \in \mathbb{C}^{L\times\frac{L}{2}}$,

$$\mathbf{C}_f = \mathbf{W}_{L\times L/2}\,\mathrm{diag}\{\mathbf{b}\}, \tag{3.54}$$

where $\mathbf{W}_{L\times L/2} \in \mathbb{C}^{L\times\frac{L}{2}}$ describes a pruned DFT matrix, that is, a conventional DFT matrix for which $\frac{L}{2}$ column vectors are canceled. Scaling vector $\mathbf{b} \in \mathbb{R}^{\frac{L}{2}\times 1}$, on the other hand, guarantees that the diagonal elements of $\mathbf{C}^{\mathsf{H}}\,\mathbf{G}^{\mathsf{H}}\mathbf{G}\,\mathbf{C}$ are exactly one, with $\mathbf{C} = \mathbf{I}_K \otimes \mathbf{C}_f$.

Figure 3.14 shows the expected transmit power over time for one FBMC symbol. In conventional FBMC, there exists a large overlapping of symbols in time and the transmission requires a long ramp-up and ramp-down period. In pruned DFT-spread FBMC, on the other hand, precoding by $\mathbf{C}_f$ shapes the transmitted signal in such a way, that the overlapping in time is very low and the ramp-up and ramp-down period dramatically reduced. This reduces the overall latency. The pruned DFT matrix in (3.54) is found by canceling those column vectors of a conventional DFT matrix, so that the main energy is concentrated within the time interval $-\frac{T_0}{4} \le t \le \frac{T_0}{4}$, see Fig. 3.14. The OOB emissions of pruned DFT-spread FBMC are comparable to conventional FBMC transmissions, leading to a high spectral efficiency.

From a conceptional point of view, the key difference between pruned DFT-spread FBMC and block spread FBMC, discussed in Sect. 3.6, is that pruned DFT-spread FBMC spreads the data symbols over the whole bandwidth, while for block spread FBMC the bandwidth is split into smaller chunks. Those small chunks can then be equalized by a simple one-tap equalizer, so that Alamouti's space-time block code and ML MIMO detection become feasible. In pruned DFT-spread FBMC, on the other hand, low-complexity ML detection is often not possible and one has to rely on minimum mean squared error (MMSE) equalization before despreading by $\mathbf{C}_f^{\mathsf{H}}$, same as in SC-FDMA. Another small drawback of pruned DFT-spread FBMC is

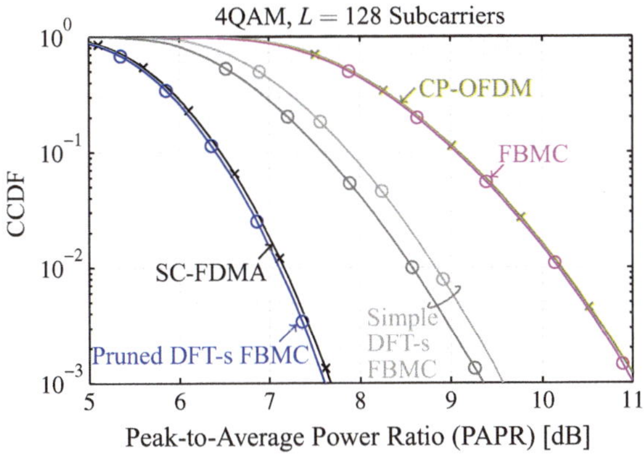

Fig. 3.15 Pruned DFT-spread FBMC [32, 37] has the same PAPR as SC-FDMA, but the additional advantages of a higher spectral efficiency. Note that a simple DFT-spread FBMC transmission scheme performs relatively poor [18, 31]

that orthogonality is only approximately restored, that is, $\mathbf{C}^{\mathsf{H}}\,\mathbf{G}^{\mathsf{H}}\mathbf{G}\,\mathbf{C} \approx \mathbf{I}_{LK/2}$, leading to some (small) interference. By restricting the time domain of $p(t)$ to approximately $-\frac{3}{4}T_0 \leq t \leq \frac{3}{4}T_0$, the interference can be reduced, so that it becomes neglectable in most cases. Moreover, a frequency CP can further reduce the interference [32, 37], if necessary.

Figure 3.15 shows the complementary cumulative distribution function (CCDF) of the PAPR for a 4-QAM signal constellation and $L = 128$ subcarriers. Conventional FBMC has the same poor PAPR as OFDM. A simple DFT-spread FBMC transmission scheme, as proposed in [18], only slightly improves the PAPR. Even an optimal phase condition [31], that is, $\mathrm{e}^{\mathrm{j}\frac{\pi}{2}(l+k)} \rightarrow \mathrm{e}^{\mathrm{j}\frac{\pi}{2}(l+k)}\mathrm{e}^{-\mathrm{j}\pi lk}$, hardly reduces the PAPR. In pruned DFT-spread FBMC, on the other hand, the PAPR is as good as in SC-FDMA and approximately 3 dB better than in OFDM and FBMC.

3.8 Summary

FBMC has the best spectral properties among all 5G waveform candidates. This is especially useful if the number of subcarriers is low, for example, in eMTC, and to support different use cases within the same band. To efficiently generate an FBMC signal, many hardware components from windowed OFDM can be reused. The main drawback of FBMC is that orthogonality only holds in the real domain, which makes some techniques, such as channel estimation or some MIMO transmission methods, more challenging. However, there exist several solutions to overcome those limitations.

Future wireless systems will be characterized by a relatively low delay spread, so that the channel induced self-interference in FBMC is very low and can often be neglected. This is even more true if an optimal subcarrier spacing is employed. To restore complex orthogonality in FBMC, one can spread data symbols over several time or frequency positions. In this context, we presented block spread FBMC which allows to straightforwardly employ all known methods from OFDM in FBMC (if the delay spread and the Doppler spread are sufficiently low). If, on the other hand, the focus lies on reducing the PAPR, pruned DFT-spread FBMC is a better option.

References

1. 3GPP, TR 38.900: study on channel model for frequency spectrum above 6GHz (release 14) (2016a), http://www.3gpp.org/DynaReport/38900.htm
2. 3GPP, TR 38.913: study on scenarios and requirements for next generation access technologies (release 14) (2016b), http://www.3gpp.org/DynaReport/38913.htm
3. 3GPP, TR 38.802: study on new radio access technology; physical layer aspects (release 14) (2017), http://www.3gpp.org/DynaReport/38802.htm
4. J.G. Andrews, S. Buzzi, W. Choi, S.V. Hanly, A. Lozano, A.C. Soong, J.C. Zhang, What will 5G be? IEEE J. Sel. Areas Commun. **32**(6), 1065–1082 (2014)
5. P. Banelli, S. Buzzi, G. Colavolpe, A. Modenini, F. Rusek, A. Ugolini, Modulation formats and waveforms for 5G networks: who will be the heir of OFDM? an overview of alternative modulation schemes for improved spectral efficiency. IEEE Signal Process. Mag. **31**(6), 80–93 (2014)
6. M. Bellanger, D. Le Ruyet, D. Roviras, M. Terré, J. Nossek, L. Baltar, Q. Bai, D. Waldhauser, M. Renfors, T. Ihalainen et al., *FBMC Physical Layer: A Primer*. PHYDYAS (2010)
7. M.G. Bellanger, Specification and design of a prototype filter for filter bank based multicarrier transmission, in *IEEE International Conference on Acoustics, Speech, and Signal Processing*, vol. 4, (2001), pp. 2417–2420
8. H. Bölcskei, Orthogonal frequency division multiplexing based on offset QAM, in *Advances in Gabor analysis* (Springer, 2003), pp. 321–352
9. R.W. Chang, Synthesis of band-limited orthogonal signals for multichannel data transmission. Bell Syst. Tech. J. **45**(10), 1775–1796 (1966)
10. J.M. Choi, Y. Oh, H. Lee, J.S. Seo, Pilot-aided channel estimation utilizing intrinsic interference for FBMC/OQAM systems. IEEE Trans. Broadcast. **63**(4), 644–655 (2017)
11. W. Cui, D. Qu, T. Jiang, B. Farhang-Boroujeny, Coded auxiliary pilots for channel estimation in FBMC-OQAM systems. IEEE Trans. Veh. Technol. **65**(5), 2936–2946 (2016)
12. B. Farhang-Boroujeny, OFDM versus filter bank multicarrier. IEEE Signal Process. Mag. **28**(3), 92–112 (2011)
13. H.G. Feichtinger, T. Strohmer, *Gabor Analysis and Algorithms: Theory and Applications* (Springer Science & Business Media, 2012)
14. G.P. Fettweis, The tactile internet: applications and challenges. IEEE Veh. Technol. Mag. **9**(1), 64–70 (2014)
15. M. Fuhrwerk, J. Peissig, M. Schellmann, Channel adaptive pulse shaping for OQAM-OFDM systems, in *IEEE European Signal Processing Conference (EUSIPCO)* (2014), pp. 181–185
16. R. Haas, J.C. Belfiore, A time-frequency well-localized pulse for multiple carrier transmission. Wirel. Pers. Commun. **5**(1), 1–18 (1997)
17. F.M. Han, X.D. Zhang, Wireless multicarrier digital transmission via Weyl-Heisenberg frames over time-frequency dispersive channels. IEEE Trans. Commun. **57**(6) (2009)
18. T. Ihalainen, A. Viholainen, T.H. Stitz, M. Renfors, M. Bellanger, Filter bank based multi-mode multiple access scheme for wireless uplink, in *IEEE European Signal Processing Conference (EUSIPCO)* (2009), pp. 1354–1358
19. J.P. Javaudin, D. Lacroix, A. Rouxel, Pilot-aided channel estimation for OFDM/OQAM, in *IEEE Vehicular Technology Conference (VTC)*, vol. 3 (2003), pp. 1581–1585
20. E. Kofidis, D. Katselis, A. Rontogiannis, S. Theodoridis, Preamble-based channel estimation in OFDM/OQAM systems: a review. Signal Proces. **93**(7), 2038–2054 (2013)

21. C. Lélé, R. Legouable, P. Siohan, Channel estimation with scattered pilots in OFDM/OQAM, in *IEEE Workshop on Signal Processing Advances in Wireless Communications (SPAWC)* (2008), pp. 286–290
22. C. Lélé, P. Siohan, R. Legouable, M. Bellanger, CDMA transmission with complex OFDM/OQAM. EURASIP J. Wirel. Commun. Netw. Article ID **748063**, 1–12 (2008)
23. C. Lélé, P. Siohan, R. Legouable, The Alamouti scheme with CDMA-OFDM/OQAM. EURASIP J. Adv. Signal Process. Article ID **703513**, 1–13 (2010)
24. Y.G. Li, G.L. Stuber, *Orthogonal Frequency Division Multiplexing for Wireless Communications* (Springer Science & Business Media, 2006)
25. J. Manz, A sequency-ordered fast Walsh transform. IEEE Trans. Audio Electroacoustic **20**(3), 204–205 (1972)
26. D. Mattera, M. Tanda, M. Bellanger, Filter bank multicarrier with PAM modulation for future wireless systems. Signal Process. **120**, 594–606 (2016)
27. G. Matz, D. Schafhuber, K. Grochenig, M. Hartmann, F. Hlawatsch, Analysis, optimization, and implementation of low-interference wireless multicarrier systems. IEEE Trans. Wirel. Commun. **6**(5), 1921–1931 (2007)
28. S. Mirabbasi, K. Martin, Design of prototype filter for near-perfect-reconstruction overlapped complex-modulated transmultiplexers, in *IEEE International Symposium on Circuits and Systems* (2002)
29. A.F. Molisch, *Wireless Communications*, vol. 34 (Wiley, 2012)
30. D. Na, K. Choi, Intrinsic ICI-free Alamouti coded FBMC. IEEE Commun. Lett. **20**(10), 1971–1974 (2016)
31. D. Na, K. Choi, Low PAPR FBMC. IEEE Trans. Wirel. Commun. **17**(1), 182–193 (2018)
32. R. Nissel, Filter bank multicarrier modulation for future wireless systems. Dissertation, TU Wien, 2017
33. R. Nissel, M. Rupp, Bit error probability for pilot-symbol aided channel estimation in FBMC-OQAM, in *IEEE International Conference on Communications (ICC)* (2016a)
34. R. Nissel, M. Rupp, Enabling low-complexity MIMO in FBMC-OQAM, in *IEEE Globecom Workshops (GC Wkshps)* (2016b)
35. R. Nissel, M. Rupp, On pilot-symbol aided channel estimation in FBMC-OQAM, in *IEEE International Conference on Acoustics, Speech and Signal Processing (ICASSP)* (2016c), pp. 3681–3685
36. R. Nissel, M. Rupp, OFDM and FBMC-OQAM in doubly-selective channels: Calculating the bit error probability. IEEE Commun. Lett. **21**(6), 1297–1300 (2017)
37. R. Nissel, M. Rupp, Pruned DFT spread FBMC-OQAM: low-PAPR, low latency, high spectral efficiency. IEEE Trans. Commun. (2018)
38. R. Nissel, S. Caban, M. Rupp, Experimental evaluation of FBMC-OQAM channel estimation based on multiple auxiliary symbols, in *IEEE Sensor Array and Multichannel Signal Processing Workshop (SAM)* (2016)
39. R. Nissel, J. Blumenstein, M. Rupp, Block frequency spreading: a method for low-complexity MIMO in FBMC-OQAM, in *IEEE Workshop on Signal Processing Advances in Wireless Communications (SPAWC)* (2017a)
40. R. Nissel, M. Rupp, R. Marsalek, FBMC-OQAM in doubly-selective channels: a new perspective on MMSE equalization, in *IEEE Workshop on Signal Processing Advances in Wireless Communications (SPAWC)* (2017b)
41. R. Nissel, S. Schwarz, M. Rupp, Filter bank multicarrier modulation schemes for future mobile communications. IEEE J. Sel. Areas Commun. **35**(8), 1768–1782 (2017c)
42. R. Nissel, E. Zöchmann, M. Lerch, S. Caban, M. Rupp, Low-latency MISO FBMC-OQAM: it works for millimeter waves! in *IEEE International Microwave Symposium* (2017d)
43. R. Nissel, E. Zöchmann, M. Rupp, On the influence of doubly-selectivity in pilot-aided channel estimation for FBMC-OQAM, in *IEEE Vehicular Technology Conference (VTC Spring)* (2017e)
44. Z. Pi, F. Khan, An introduction to millimeter-wave mobile broadband systems. IEEE Commun. Mag. **49**(6), 101–107 (2011)

45. Qualcomm Incorporated, Waveform candidates, in *3GPP TSG-RAN WG1 84b* (Busan, Korea, 2016)
46. M. Renfors, T. Ihalainen, T. H. Stitz, A block-Alamouti scheme for filter bank based multicarrier transmission, in *European Wireless Conference (EW)* (2010)
47. M. Renfors, X. Mestre, E. Kofidis, F. Bader, *Orthogonal Waveforms and Filter Banks for Future Communication Systems* (Academic Press, 2017)
48. P. Robertson, S. Kaiser, The effects of Doppler spreads in OFDM(A) mobile radio systems, in *IEEE Vehicular Technology Conference (VTC Fall)* (1999), pp. 329–333
49. F. Rottenberg, X. Mestre, D. Petrov, F. Horlin, J. Louveaux, Parallel equalization structure for MIMO FBMC-OQAM systems under strong time and frequency selectivity. IEEE Trans. Signal Process. **65**(17), 4454–4467 (2017)
50. A. Sahin, I. Guvenc, H. Arslan, A survey on multicarrier communications: prototype filters, lattice structures, and implementation aspects. IEEE Commun. Surv. Tutor. **16**(3), 1312–1338 (2012)
51. B. Saltzberg, Performance of an efficient parallel data transmission system. IEEE Trans. Commun. Technol. **15**(6), 805–811 (1967)
52. F. Schaich, T. Wild, Y. Chen, Waveform contenders for 5G-suitability for short packet and low latency transmissions, in *IEEE Vehicular Technology Conference (VTC Spring)* (2014), pp. 1–5
53. F. Schaich, T. Wild, R. Ahmed, Subcarrier spacing-how to make use of this degree of freedom, in *IEEE Vehicular Technology Conference (VTC Spring)* (2016), pp. 1–6
54. S. Sesia, M. Baker, I. Toufik, *LTE-the UMTS Long Term Evolution: from Theory to Practice* (Wiley, 2011)
55. P. Siohan, C. Siclet, N. Lacaille, Analysis and design of OFDM/OQAM systems based on filterbank theory. IEEE Trans. Signal Process. **50**(5), 1170–1183 (2002)
56. E. Telatar, Capacity of multi-antenna gaussian channels. Trans. Emerg. Telecommun. Technol. **10**(6), 585–595 (1999)
57. M. Vetterli, J. Kovačević, V.K Goyal, *Foundations of Signal Processing* (Cambridge University Press, 2014)
58. S. Weinstein, P. Ebert, Data transmission by frequency-division multiplexing using the discrete fourier transform. IEEE Trans. Commun. Technol. **19**(5), 628–634 (1971)
59. G. Wunder, P. Jung, M. Kasparick, T. Wild, F. Schaich, Y. Chen, S. Brink, I. Gaspar, N. Michailow, A. Festag et al., 5GNOW: non-orthogonal, asynchronous waveforms for future mobile applications. IEEE Commun. Mag. **52**(2), 97–105 (2014)
60. B. Yu, S. Hu, P. Sun, S. Chai, C. Qian, C. Sun, Channel estimation using dual-dependent pilots in FBMC/OQAM systems. IEEE Commun. Lett. **20**(11), 2157–2160 (2016)
61. C.H. Yuen, P. Amini, B. Farhang-Boroujeny, Single carrier frequency division multiple access (SC-FDMA) for filter bank multicarrier communication systems, in *IEEE International Conference on Cognitive Radio Oriented Wireless Networks (CROWNCOM)* (2010), pp. 1–5
62. R. Zakaria, D. Le Ruyet, A novel filter-bank multicarrier scheme to mitigate the intrinsic interference: application to MIMO systems. IEEE Trans. Wirel. Commun. **11**(3), 1112–1123 (2012)
63. X. Zhang, M. Jia, L. Chen, J. Ma, J. Qiu, Filtered-OFDM-enabler for flexible waveform in the 5th generation cellular networks, in *IEEE Global Communications Conference (GLOBECOM)* (2015), pp. 1–6

Chapter 4
Generalized Frequency Division Multiplexing: A Flexible Multicarrier Waveform

Ahmad Nimr, Shahab Ehsanfar, Nicola Michailow, Martin Danneberg, Dan Zhang, Henry Douglas Rodrigues, Luciano Leonel Mendes and Gerhard Fettweis

4.1 Introduction to GFDM Modulator

In this section, we give an overview of generalized frequency division multiplexing (GFDM) waveform. We start from the continuous model representation, from which an analytical expression of power spectral density (PSD) is derived. Based on that, a discrete signal model representation is derived and expressed in terms of matrix model. We show the structure of the modulation matrix by mean of decomposition.

A. Nimr (✉) · S. Ehsanfar · M. Danneberg · D. Zhang · G. Fettweis
Vodafone Chair Mobile Communication Systems, Technische Universität Dresden, Dresden, Germany
e-mail: ahmad.nimr@ifn.et.tu-dresden.de

S. Ehsanfar
e-mail: shahab.ehsanfar@ifn.et.tu-dresden.de

M. Danneberg
e-mail: martin.danneberg@ifn.et.tu-dresden.de

D. Zhang
e-mail: dan.zhang@ifn.et.tu-dresden.de

G. Fettweis
e-mail: gerhard.fettweis@ifn.et.tu-dresden.de

N. Michailow
National Instruments Corp., 11500, Mopac Expwy, Austin, TX 78759, USA
e-mail: nicola.michailow@ni.com

H. D. Rodrigues · L. L. Mendes
Intituto Nacional de Telecomunicações (Inatel), Santa Rita do Sapucai, MG, Brazil
e-mail: henry@inatel.br

L. L. Mendes
e-mail: lucianol@inatel.br

M. Vaezi et al. (eds.), *Multiple Access Techniques for 5G Wireless Networks and Beyond*, https://doi.org/10.1007/978-3-319-92090-0_4

From this model, the design requirements and performance indicators are derived. Finally, we show the ability of GFDM representation to generate other state-of-the-art waveforms.

4.1.1 Continuous Signal Model

GFDM is a block-based multicarrier modulation technique that employs circular filtering [1]. For a better understanding of the GFDM structure, we represent the modulation technique using the continuous time model. Consider a time-frequency resource block defined by time duration T and frequency bandwidth B. The target is to use this resource to convey a data message of maximum length of N data symbols. For this purpose, the available bandwidth is divided into K equally spaced subcarriers with subcarrier spacing $\Delta f = \frac{B}{K}$, and the available time is divided into M subsymbols with subsymbol spacing $T_{\text{sub}} = \frac{T}{M}$. The subcarrier spacing is related to the subsymbol spacing with the relation $\Delta f T_{\text{sub}} = 1$. Hence, $T = \frac{N}{B} = \frac{\Delta f}{M}$, where $N = KM$. Each pair (k, m)-(subcarrier, subsymbol) can be used to transmit one data symbol $d_{k,m}$ modulated by a pulse shape $g_{k,m}(t)$ given by

$$g_{k,m}(t) = w_T(t) g_T(t - mT_{\text{sub}}) e^{j2\pi k \Delta f t}, \tag{4.1}$$

where $w_T(t)$ is a rectangular window of duration T, namely $w_T(t) = 1, t \in [0, T]$ and 0 elsewhere. $g_T(t)$ is a prototype periodic pulse shape of period T, which can be expressed using Fourier series. This means that the pulse shapes are generated by time and frequency shifts of a periodic prototype pulse shape in addition to multiplication with a finite time window to form the GFDM block. In conventional GFDM, the number of the frequency components of g_T is limited to $2M$ such that,

$$g_T(t) = \sum_{q=-M}^{M-1} \tilde{g}_T[q] e^{j2\pi \frac{q}{M} \Delta f t}, \tag{4.2}$$

where $\tilde{g}_T[q]$ are the nonzero coefficients of the Fourier series. This allows each subcarrier to span at maximum two subcarrier spacing. The frequency response of the prototype pules shape is given by

$$G_T(f) = \sum_{q=-M}^{M-1} \tilde{g}_T[q] \delta(f - \tfrac{q}{M} \Delta f). \tag{4.3}$$

Here, $\delta(\cdot)$ is the Dirac pulse. GFDM commonly adopts a cyclic prefix (CP) of duration $T_{\text{cp}} \geq \tau_{\max}$ to tackle the impact of fading channel with maximum excess delay spread $\tau_{\max}$. In addition, a cyclic suffix (CS) with duration T_{cs} may be added to the end of the block. The CP and CS can be simply introduced by extending the window to

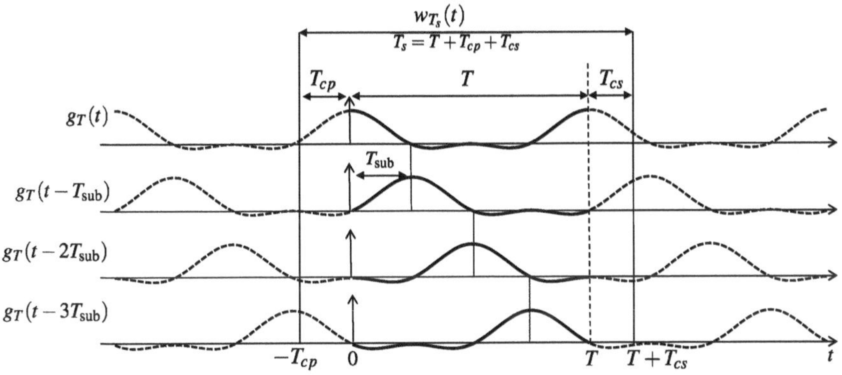

Fig. 4.1 Pulse shape generation using periodic Raised-Cosine prototype pulse shape, with roll-off factor $\alpha = 0, M = 4$ for $k = 0, m = 0, \ldots, 3$

$w_{T_s}(t)$ where $T_s = T + T_{cp} + T_{cs}$, as depicted in Fig. 4.1. Mathematically, it can be written as

$$g_{k,m}^{(t)}(t) = w_{T_s}(t - T_{\text{cp}})g_T(t - mT_{\text{sub}})e^{j2\pi k\Delta ft}. \tag{4.4}$$

Therefore, the frequency domain representation is given by

$$G_{k,m}^{(t)}(f) = \left(W_{T_s}(f - k\Delta f)e^{-j2\pi T_{\text{cp}}(f-k\Delta f)}\right) * \left(G_T(f)e^{-j2\pi mT_{\text{sub}}f}\right), \tag{4.5}$$

where $*$ denotes the convolution operator. Replacing $G_T(f)$ from (4.3), we get

$$G_{k,m}^{(t)}(f) = e^{-j2\pi T_{\text{cp}}(f-k\Delta f)} \sum_{q=-M}^{M-1} \tilde{g}_T[q]W_{T_s}(f - (kM + q)\tfrac{\Delta f}{M})e^{-j2\pi m\frac{q}{M}}e^{+j2\pi T_{\text{cp}}\Delta f\frac{q}{M}}. \tag{4.6}$$

In practice, not all subcarriers and subsymbols are used, thus, we define $\mathcal{K}_{\text{on}}$ and $\mathcal{M}_{\text{on}}$ as the sets of active subcarriers and subsymbols, respectively. Therefore, the signal corresponding to the i-th GFDM block that modulates the data symbols $\{d_{k,m,i}\}$ is generated as

$$x_i(t) = \sum_{m\in\mathcal{M}_{\text{on}}} \sum_{k\in\mathcal{K}_{\text{on}}} d_{k,m,i}g_{k,m}^{(t)}(t). \tag{4.7}$$

Moreover, the signal of a frame that contains N_s blocks can be expressed as

$$x(t) = \sum_{i=0}^{N_s-1} x_i(t - iT_s) = \sum_{i=0}^{N_s-1} \sum_{m\in\mathcal{M}_{\text{on}}} \sum_{k\in\mathcal{K}_{\text{on}}} d_{k,m,i}g_{k,m}^{(t)}(t - iT_s). \tag{4.8}$$

From (4.8) and (4.6), the PSD of the GFDM signal assuming uncorrelated data symbols with unit power, i.e., $\mathbb{E}\left[d_{k_1,m_1,i_1}^* d_{k_2,m_2,i_2}\right] = \delta(k_1 - k_2)\delta(m_1 - m_2)\delta(i_1 - i_2)$, can be computed as,

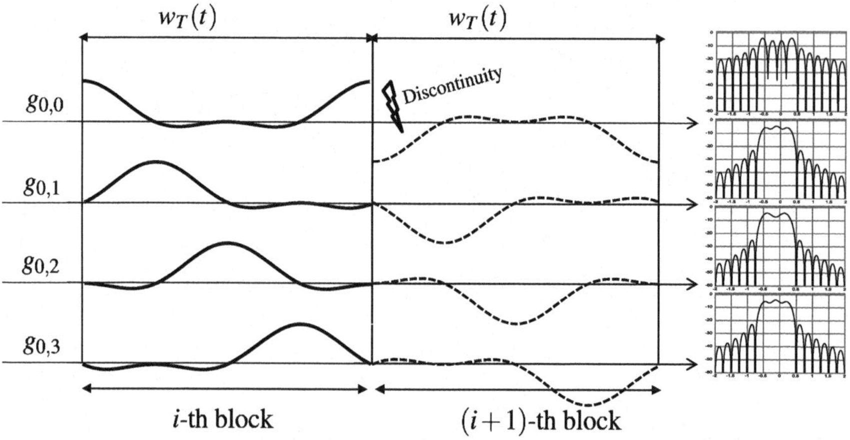

Fig. 4.2 PSD of subsymbols using a periodic raised-cosine (RC) prototype pulse with roll-off factor $\alpha = 0$, $M = 4$ for $k = 0$, $m = 0, \ldots, 3$

$$\begin{aligned} S_x(f) &= \frac{1}{T_s} \sum_{m \in \mathcal{M}_{\text{on}}} \sum_{k \in \mathcal{K}_{\text{on}}} \left| G^{(t)}_{k,m}(f) \right|^2 \\ &= \frac{1}{T_s} \sum_{m \in \mathcal{M}_{\text{on}}} \sum_{k \in \mathcal{K}_{\text{on}}} \left| \sum_{q=-M}^{M-1} \tilde{g}_T[q] W_{T_s}(f - (kM + q)\tfrac{\Delta f}{M}) e^{-j2\pi m \frac{q}{M}} e^{+j2\pi T_{\text{cp}} \Delta f \frac{q}{M}} \right|^2 . \end{aligned} \tag{4.9}$$

According to (4.9), the PSD is obviously influenced by the prototype pulse shape coefficients $\{\tilde{g}_T[q]\}$, the window $W_{T_s}(t)$ and the number of subsymbols M. In addition, the active subsymbol set $\mathcal{M}_{\text{on}}$ plays an important role throughout the phase term $e^{-j2\pi m \frac{q}{M}}$, which depends on the subsymbol index m. To clarify that, Fig. 4.2 shows the individual PSD of each subsymbol using a rectangular window and no CP nor CS. As can be seen, the first subsymbol, namely $m = 0$, is the source of high out-of-band (OOB) emission for the selected prototype pulse shape. This can be intuitively understood via the discontinuity between successive blocks, which happens when $d_{k,0,i}$ and $d_{k,0,i+1}$ are not identical.

4.1.2 Discrete Signal Model

The discrete time signal representation can be derived from the sampling of the analog signal with frequency $F_s = B$. With that, we get K, N, L_{cp}, and L_{cs} samples per subsymbol, symbol, CP and CS, respectively. The discrete prototype pulse shape results from (4.2) with

$$g[n] = \sum_{q=-M}^{M-1} \tilde{g}_T[q] e^{j2\pi \frac{qn}{N}}, n = 0, \ldots, N-1. \quad (4.10)$$

Let $\tilde{g}$ = N-DFT $\{g\}$ be the N-point finite discrete Fourier transform (DFT) of g such that

$$g[n] = \frac{1}{N} \sum_{q=0}^{N-1} \tilde{g}[q] e^{j2\pi \frac{qn}{N}}, \quad (4.11)$$

then, the relation between the Fourier series coefficient $\tilde{g}_T[q]$ of the continuous model and the frequency bins $\{\tilde{g}[q]\}$ of the discrete model can be expressed as,

$$\tilde{g}[q] = \frac{1}{N} \tilde{g}_T[< q >_N], \quad (4.12)$$

where $< \cdot >_N$ is the modulo-N operator. Assuming a rectangular window as in (4.1), we get

$$g_{k,m}[n] = g[< n - mK >_N] e^{j2\pi \frac{k}{K} n}. \quad (4.13)$$

Thus, the subcarrier–subsymbol pulse shapes are generated from the circular shift of the prototype pulse shape in the time and frequency domains. In fact, the circularity in time results from the design with periodic pulse shape and in frequency from the sampling. Consequently, the design of the prototype pulse shape $g[n]$ can be carried out in the frequency domain such that only the first and last M samples of $\tilde{g}$ have nonzero values. This ensures limited inter-carrier interference (ICI) to only adjacent subcarriers under the assumption of perfect synchronization. However, in the case of asynchronous subcarriers, we resort to the continuous discrete-time Fourier transform (DTFT) $G_{k,m}(\nu) = \text{DFT}(g_{k,m}[n])$, which takes into account the frequency response of the window as well. Actually, $\tilde{g}_{k,m}[q] = G_{k,m}(\nu = \frac{q}{N})$, as illustrated in Fig. 4.3.

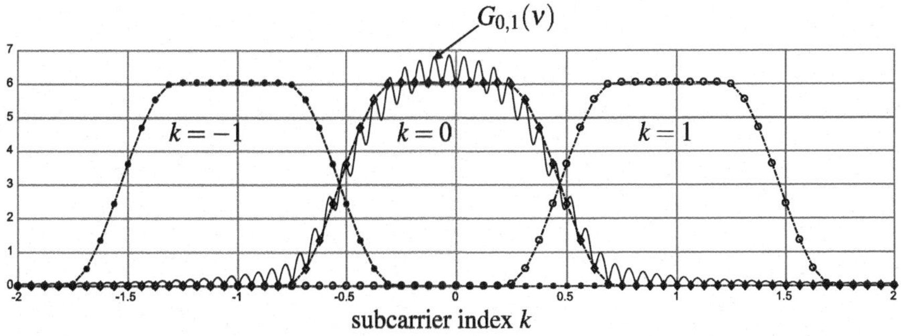

Fig. 4.3 ICI of adjacent subcarriers using periodic RC prototype pulse with roll-off factor $\alpha = 0.5$, $M = 16$ for $k = -1, 0, 1, m = 0$. The sampling points stand for $\tilde{g}_{k,m}[q]$, and the solid line represents $G_{k,m}(\nu)$

Next, we focus on the matrix representation of the GFDM block showing that both modulation and demodulation matrices follow the same structure. This structure is determined with the decomposition of the GFDM matrix.

4.1.2.1 Modulation Matrix Model

One GFDM block can be represented in a vector $\mathbf{x} \in \mathbb{C}^{N\times 1}$ such that,

$$[\mathbf{x}]_{(n)} = \sum_{m\in\mathcal{M}_{\text{on}}} \sum_{k\in\mathcal{K}_{\text{on}}} d_{k,m} g[< n - mK >_N] e^{j2\pi \frac{k}{K} n}. \tag{4.14}$$

In the frequency domain, $\tilde{\mathbf{x}} = \text{N-DFT}\{x\}$ can be written as

$$\left[\tilde{\mathbf{x}}\right]_{(n)} = \sum_{m\in\mathcal{M}_{\text{on}}} \sum_{k\in\mathcal{K}_{\text{on}}} d_{k,m} \tilde{g}[< n - kM >_N] e^{-j2\pi \frac{m}{M} n}. \tag{4.15}$$

Adding CP and CS is done by copying the last L_{cp} to the beginning and the first L_{cs} samples to the end of $\mathbf{x}$. Let $\mathbf{D} \in \mathbb{C}^{K\times M}$ be the matrix representing the data symbols, with $[\mathbf{D}]_{(k,m)} = d_{k,m}$ and for $(k, m) \notin \mathcal{K}_{\text{on}} \times \mathcal{M}_{\text{on}}$, $d_{k,m} = 0$. We define the data vector $\mathbf{d} = \text{vec}\{\mathbf{D}\}$, namely, $[\mathbf{d}]_{(k+mK)} = [\mathbf{D}]_{(k,m)}$. In addition, the modulation matrix $\mathbf{A} \in \mathbb{C}^{N\times N}$ is defined by

$$[\mathbf{A}]_{(n,k+mK)} = g[< n - mK >_N] e^{j2\pi \frac{k}{K} n}. \tag{4.16}$$

Thereby,

$$[\mathbf{x}]_{(n)} = \sum_{m=0}^{M-1} \sum_{k=0}^{K-1} [\mathbf{A}]_{(n,k+mK)} [\mathbf{d}]_{(k+mK)}, \tag{4.17}$$

and thus,

$$\mathbf{x} = \mathbf{A}\mathbf{d}. \tag{4.18}$$

Taking into account the active sets of subcarriers and subsymbols, we can define a compact representation for the active resource set

$$\mathcal{N}_{\text{on}} = \{n = k + mK, (k, m) \in \mathcal{K}_{\text{on}} \times \mathcal{M}_{\text{on}}\} \tag{4.19}$$

as

$$\mathbf{x} = \mathbf{A}^{(\text{on})}\mathbf{d}^{(\text{on})}, \tag{4.20}$$

where $\mathbf{A}^{(\text{on})} = [\mathbf{A}]_{(:,\mathcal{N}_{\text{on}})}$ stands for the active modulation matrix and $\mathbf{d}^{(\text{on})} = [\mathbf{d}]_{(\mathcal{N}_{\text{on}})}$ is the vector of the active data symbols.

4.1.2.2 Demodulation Matrix Model

An equalized received signal $y[n]$, represented by the vector $\mathbf{y} \in \mathbb{C}^{N\times 1}$, is to be demodulated using a receive prototype filter $\gamma[n]$. The estimated data symbols can be expressed as

$$\begin{aligned}\hat{d}_{k,m} &= \gamma^*[-n] \circledast \left(y[p]e^{-j2\pi\frac{k}{K}n}\right)|_{n=mK}\\ \left[\hat{\mathbf{d}}\right]_{(k+mK)} &= \sum_{n=0}^{N-1} y[n]\gamma^*[<n-mK>_N]e^{-j2\pi\frac{k}{K}n}.\end{aligned} \tag{4.21}$$

Then,

$$\hat{\mathbf{d}} = \mathbf{B}^H\mathbf{y}, \tag{4.22}$$

where

$$\begin{aligned}\left[\mathbf{B}^H\right]_{(k+mK,n)} &= \gamma^*[<n-mK>_N]e^{-j2\pi\frac{k}{K}n}, \text{ so that}\\ [\mathbf{B}]_{(n,k+mK)} &= \gamma[<n-mK>_N]e^{j2\pi\frac{k}{K}n}.\end{aligned} \tag{4.23}$$

Comparing with (4.16), we conclude that the demodulation matrix has the same structure as the modulation matrix.

4.1.3 GFDM Matrix Decomposition

As the GFDM matrix is generated by circular shift of a prototype pulse shape in the time and frequency domains, it has a well-defined structure and its properties can be derived from the prototype pulse shape. To investigate the structure, first, we define several auxiliary matrices to facilitate the derivation. For a vector $\mathbf{a} \in \mathbb{C}^{PQ\times 1}$, we define the following matrices

$$\mathbf{V}_{P,Q}^{(\mathbf{a})} = \left(\text{unvec}_{Q\times P}\{\mathbf{a}\}\right)^T, \tag{4.24}$$

$$\mathbf{Z}_{P,Q}^{(\mathbf{a})} = \mathbf{F}_P\mathbf{V}_{P,Q}^{(\mathbf{a})} = \tilde{\mathbf{V}}_{P,Q}^{(\mathbf{a})}, \tag{4.25}$$

where $\text{unvec}_{Q\times P}\{\mathbf{a}\}$ denotes the inverse of vectorization operation. Thus,

$$\left[\mathbf{V}_{P,Q}^{(\mathbf{a})}\right]_{(p,q)} = [\mathbf{a}]_{(q+pQ)}\,. \tag{4.26}$$

$\mathbf{F}_P$ is the P-point DFT matrix, where $[\mathbf{F}_P]_{(i,j)} = e^{-j2\pi\frac{ij}{P}}$. The matrix $\mathbf{V}_{P,Q}^{(\mathbf{a})}$ represents the polyphase components generated by the sampling of $\mathbf{a}$ with factor Q, while $\mathbf{Z}_{P,Q}^{(\mathbf{a})}$ is known as discrete Zak transform (DZT) transform [2]. Figure 4.4 visualizes these matrices by mean of an example.

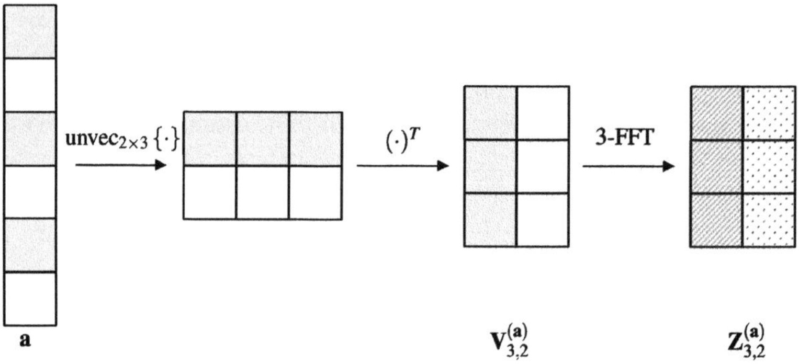

Fig. 4.4 DZT example. **a** of size 6×1, $P = 3$, $Q = 2$

Recall the GFDM block Eq. (4.14) with the consideration of full allocation and defining $n = q + pK$, $q = 0, \ldots, K - 1$ and $p = 0, \ldots, M - 1$, then

$$[\mathbf{x}]_{(q+pK)} = \sum_{m=0}^{M-1}\sum_{k=0}^{K-1} d_{k,m} g[< q + pK - mK >_N] e^{j2\pi \frac{k}{K} q}. \tag{4.27}$$

Based on (4.26), $[\mathbf{x}]_{(q+pK)} = \left[\mathbf{V}_{M,K}^{(\mathbf{x})}\right]_{(p,q)}$ and $\left[\mathbf{V}_{M,K}^{(\mathbf{g})}\right]_{(p,q)} = g[< q + pK >_N]$. Following this, it is easy to show that $g[< q + pK - mK >_N] = \left[\mathbf{V}_{M,K}^{(\mathbf{g})}\right]_{(<p-m>_M, q)}$. Putting all notations together we get

$$\begin{aligned} \left[\mathbf{V}_{M,K}^{(\mathbf{x})}\right]_{(p,q)} &= \sum_{m=0}^{M-1} \left[\mathbf{V}_{M,K}^{(\mathbf{g})}\right]_{(<p-m>_M, q)} \sum_{k=0}^{K-1} d_{k,m} e^{j2\pi \frac{k}{K} q} \\ &= \sum_{m=0}^{M-1} \left[\mathbf{V}_{M,K}^{(\mathbf{g})}\right]_{(<p-m>_M, q)} \left[\mathbf{D}^T \mathbf{F}_K^H\right]_{(m,q)}. \end{aligned} \tag{4.28}$$

The second line represents circular convolution between the q-th column of $\mathbf{V}_{M,K}^{(\mathbf{g})}$ and the q-th column of $\mathbf{D}^T \mathbf{F}_K^H$. By representing the circular decomposition in the frequency domain we get,

$$\left[\mathbf{V}_{M,K}^{(\mathbf{x})}\right]_{(:,q)} = \frac{1}{M} \mathbf{F}_M^H \text{diag}\left\{\left[\tilde{\mathbf{V}}_{M,K}^{(\mathbf{g})}\right]_{(:,q)}\right\} \mathbf{F}_M \left[\mathbf{D}^T \mathbf{F}_K^H\right]_{(:,q)}. \tag{4.29}$$

Finally, by stacking the columns according to the q index and using (4.25), we get

$$\mathbf{V}_{M,K}^{(\mathbf{x})} = \frac{1}{M} \mathbf{F}_M^H \left(\mathbf{Z}_{M,K}^{(\mathbf{g})} \odot \left[\mathbf{F}_M \mathbf{D}^T \mathbf{F}_K^H\right]\right). \tag{4.30}$$

Here, $\odot$ denotes the element-wise multiplication operator. Thus, $\mathbf{x} = \text{vec}\left\{\left(\mathbf{V}^{(\mathbf{x})}_{M,K}\right)^T\right\}$. Following the same approach on $\tilde{\mathbf{x}}$ defined in (4.15), we get

$$\left[\mathbf{V}^{(\tilde{\mathbf{x}})}_{K,M}\right]_{(q,p)} = \sum_{k=0}^{K-1}\left[\mathbf{V}^{(\tilde{\mathbf{g}})}_{K,M}\right]_{(<q-k>_K,p)}\sum_{m=0}^{M-1} d_{k,m}e^{-j2\pi\frac{m}{M}p}. \tag{4.31}$$

Thereby,

$$\mathbf{V}^{(\tilde{\mathbf{x}})}_{K,M} = \frac{1}{K}\mathbf{F}^H_K\left(\mathbf{Z}^{(\tilde{\mathbf{g}})}_{K,M}\odot[\mathbf{F}_K\mathbf{D}\mathbf{F}_M]\right). \tag{4.32}$$

Similarly, $\tilde{\mathbf{x}} = \text{vec}\left\{\left(\mathbf{V}^{(\tilde{\mathbf{x}})}_{K,M}\right)^T\right\}$. Finally, using the vectorized output of (4.30) and (4.32) with respect to (4.14), we can express the modulation matrix $\mathbf{A}$ as

$$\mathbf{A} = \frac{1}{M}\mathbf{P}^T_{M,K}\mathbf{U}^H_{K,M}\mathbf{L}^{(g)}\mathbf{U}_{K,M}\mathbf{P}_{M,K}\mathbf{U}^H_{M,K}, \tag{4.33}$$

$$= \frac{1}{N}\frac{1}{K}\mathbf{F}^H_N\mathbf{P}_{M,K}\mathbf{U}^H_{M,K}\mathbf{L}^{(\tilde{g})}\mathbf{U}_{M,K}\mathbf{P}^T_{M,K}\mathbf{U}_{K,M}\mathbf{P}_{M,K}. \tag{4.34}$$

Here, $\mathbf{P}_{P,Q} \in \Re^{PQ\times PQ}$ is the permutation matrix that fulfills for any $Q \times P$ matrix $\mathbf{X}$

$$\text{vec}\left\{\mathbf{X}^T\right\} = \mathbf{P}_{P,Q}\text{vec}\left\{\mathbf{X}\right\}, \tag{4.35}$$

$$\mathbf{U}_{P,Q} = \mathbf{I}_P \otimes \mathbf{F}_Q, \tag{4.36}$$

where $\otimes$ is the Kronecker product, and

$$\mathbf{L}^{(g)} = \text{diag}\left\{\text{vec}\left\{\mathbf{Z}^{(\mathbf{g})}_{M,K}\right\}\right\}, \tag{4.37}$$

$$\mathbf{L}^{(\tilde{g})} = \text{diag}\left\{\text{vec}\left\{\mathbf{Z}^{(\tilde{\mathbf{g}})}_{M,K}\right\}\right\}. \tag{4.38}$$

Either of these diagonal matrices can be used to analyze the GFDM performance indicators and derive the demodulator matrix. In other words, $\mathbf{Z}^{(\mathbf{g})}_{M,K}$ or $\mathbf{Z}^{(\tilde{\mathbf{g}})}_{M,K}$ is the key of GFDM design and analysis.

4.1.3.1 Demodulation Prototype Filter

While the GFDM-based demodulator matrix has the same structure as the modulation matrix, by using the decomposition (4.33) or (4.34), the prototype filter of the demodulator can be easily computed from the inverse of DZT as

$$\gamma = \frac{1}{M}\text{vec}\left\{\left(F^H_M\mathbf{Z}^{(\gamma)}_{M,K}\right)^T\right\}. \tag{4.39}$$

First, we need to find $\mathbf{B}$. Using the representation (4.33), the product $\mathbf{B}^H\mathbf{A}$ can be calculated as

$$\begin{aligned}\mathbf{B}^H\mathbf{A} &= \frac{1}{M}\mathbf{U}_{K,M}\mathbf{P}_{M,K}^T\mathbf{U}_{K,M}^H\mathbf{L}^{(\gamma)H}\mathbf{L}^{(g)}\mathbf{U}_{K,M}\mathbf{P}_{M,K}\mathbf{U}_{M,K}^H, \\ &= K\mathbf{U}^H\mathbf{L}^{(\gamma)H}\mathbf{L}^{(g)}\mathbf{U},\end{aligned} \tag{4.40}$$

where

$$\mathbf{U} = \frac{1}{\sqrt{N}}\mathbf{U}_{K,M}\mathbf{P}_{M,K}\mathbf{U}_{M,K}^H, \tag{4.41}$$

is a unitary matrix. Thus, the diagonal elements of $K\mathbf{L}^{(\gamma)H}\mathbf{L}^{(g)}$ represent the eigenvalues of $\mathbf{B}^H\mathbf{A}$. Obviously, the matched filter demodulator uses $\gamma[n] = g[n]$. On the other hand, the zero forcing (ZF) is determined when $K\mathbf{L}^{(\gamma)H}\mathbf{L}^{(g)} = \mathbf{I}_N$. Consequently, γ_{zf} can be computed form the DZT inverse of $\mathbf{Z}_{M,K}^{(\gamma_{\text{zf}})}$, defined by

$$\left[\mathbf{Z}_{M,K}^{(\gamma_{\text{zf}})}\right]_{(m,k)} = \frac{1}{K\left[\mathbf{Z}_{M,K}^{(g)}{}^*\right]_{(k,m)}}. \tag{4.42}$$

The minimum mean square error (MMSE) demodulator, assuming additive white Gaussian noise (AWGN) of noise power P_N and uncorrelated data symbols with power P_D, can be computed via the matrix

$$\begin{aligned}\mathbf{B}_{\text{MMSE}} &= \left(\mathbf{A}\mathbf{A}^H + \tfrac{P_N}{P_D}\mathbf{I}_N\right)^{-1}\mathbf{A} \\ &= \frac{1}{M}\mathbf{P}_{M,K}^T\mathbf{U}_{K,M}^H\left(K\mathbf{L}^{(g)}\mathbf{L}^{(g)H} + \tfrac{P_N}{P_D}\right)^{-1}\mathbf{L}^{(g)}\mathbf{U}_{K,M}\mathbf{P}_{M,K}\mathbf{U}_{M,K}^H.\end{aligned} \tag{4.43}$$

Comparing with (4.33), we find that

$$\left[\mathbf{Z}_{M,K}^{(\gamma_{\text{MMSE}})}\right]_{(m,k)} = \left[\mathbf{Z}_{M,K}^{(g)}\right]_{(k,m)}\left(K\left|\left[\mathbf{Z}_{M,K}^{(g)}\right]_{(k,m)}\right|^2 + \tfrac{P_N}{P_D}\right)^{-1}. \tag{4.44}$$

Then, γ_{MMSE} can be found form the DZT inverse using (4.39). It is worth noting that in the case of non-full allocation, an MMSE or least squares (LS) receiver can be derived based on the compact model $\mathbf{x} = \mathbf{A}^{(\text{on})}\mathbf{d}^{(\text{on})}$. However, the obtained matrix does not necessary have a GFDM structure and may complicate the implementation.

4.1.4 Performance Indicators

In order to evaluate the design with a given prototype pulse shape, we study three performance indicators of the modulation matrix $\mathbf{A}$ under the full allocation assumption.

All these indicators can be computed from $\mathbf{Z}_{K,M}^{(\check{g})}$ using the decomposition represented in (4.34), which can be reformulated as

$$\mathbf{A} = \frac{1}{\sqrt{K}}\mathbf{W}^H\mathbf{L}^{(\check{g})}\mathbf{V}, \tag{4.45}$$

with

$$\mathbf{W}^H = \frac{1}{\sqrt{NK}}\mathbf{F}_N^H\mathbf{P}_{M,K}\mathbf{U}_{M,K}^H, \tag{4.46}$$

$$\mathbf{V} = \frac{1}{\sqrt{N}}\mathbf{U}_{M,K}\mathbf{P}_{M,K}^T\mathbf{U}_{K,M}\mathbf{P}_{M,K}, \tag{4.47}$$

are unitary matrices.

Conditional Number

Defining the short-hand notation $z_{k,m} = \left[\mathbf{Z}_{K,M}^{(\check{g})}\right]_{(k,m)}$, then $\{\sigma_{k,m}^2 = \frac{|z_{k,m}|^2}{K}\}$ correspond to the singular values of $\mathbf{A}$. The conditional number of $\mathbf{A}$ is given by

$$cond\ (\mathbf{A}) = \frac{\max_{k,m}\{\sigma_{k,m}\}}{\min_{k,m}\{\sigma_{k,m}\}} = \frac{\max_{k,m}\{|z_{k,m}|\}}{\min_{k,m}\{|z_{k,m}|\}}. \tag{4.48}$$

When $cond\ (\mathbf{A}) = 1$, i.e., $|z_{k,m}| = 1, \forall(k, m)$, then $\mathbf{A}$ is orthogonal, and when there is at least (k_0, m_0) such that $z_{m_0,k_0} = 0$, $\mathbf{A}$ becomes singular. The rank of $\mathbf{A}$ is reduced by the number of zero elements in $\mathbf{Z}_{K,M}^{(\check{g})}$. The conditional number is important in all receiver processing steps that require the computation of the inverse of $\mathbf{A}$. Thus, a well-conditioned modulation matrix with smaller conditional number is preferred. Although we can always design GFDM with an orthogonal matrix, some other requirements cannot be achieved. More details on that are introduced in Sect. 4.1.5.

Noise Enhancement Factor

Considering the received signal in AWGN channel,

$$\mathbf{y} = \mathbf{A}\mathbf{d} + \mathbf{w}, \tag{4.49}$$

with $\mathbb{E}\left[\mathbf{d}\mathbf{d}^H\right] = P_D\mathbf{I}_N$ and $\mathbb{E}\left[\mathbf{w}\mathbf{w}^H\right] = P_N\mathbf{I}_N$, the noise enhancement factor (NEF) is defined by the ratio of the average signal-to-noise ratio (SNR) before and after applying the ZF demodulator;

$$\begin{aligned}\xi &= \frac{\text{trace}\left\{\mathbb{E}\left[\mathbf{A}\mathbf{d}\mathbf{d}^H\mathbf{A}^H\right]\right\}}{\text{trace}\left\{\mathbb{E}\left[\mathbf{v}\mathbf{v}^H\right]\right\}}\left(\frac{\text{trace}\left\{\mathbb{E}\left[\mathbf{d}\mathbf{d}^H\right]\right\}}{\text{trace}\left\{\mathbb{E}\left[\mathbf{A}^{-1}\mathbf{v}\mathbf{v}^H\mathbf{A}^{-1H}\right]\right\}}\right)^{-1}\\ &= \frac{1}{N^2}\text{trace}\left\{\mathbf{A}\mathbf{A}^H\right\}\text{trace}\left\{\mathbf{A}^{-1}\mathbf{A}^{-1H}\right\}\end{aligned}$$

$$
\begin{aligned}
&= \frac{1}{N^2} \left\| \mathbf{L}^{(\tilde{g})} \right\|_F^2 \left\| \mathbf{L}^{(\tilde{g})-1} \right\|_F^2 \\
&= \frac{1}{N^2} \left(\sum_{k,m} |z_{k,m}|^2 \right) \left(\sum_{k,m} \frac{1}{|z_{k,m}|^2} \right).
\end{aligned} \tag{4.50}
$$

Noting that after applying the ZF demodulator, we get $\mathbf{A}^{-1}\mathbf{y} = \mathbf{d} + \mathbf{A}^{-1}\mathbf{w}$ and because the rows of $\mathbf{A}^{-1}$ are generated from circular shift of the prototype filter $\boldsymbol{\gamma}_{zf}$, the noise enhancement is equal on each of the elements of $\mathbf{d}$ in the case of white noise. Contrariwise, it depends on the subcarrier–subsymbol index for colored noise.

Self-interference Ratio

Because the columns of $\mathbf{A}$ are generated from the prototype pulse shape $g[n]$ by circular shift in time and frequency, then

$$
\left[\mathbf{A}^H \mathbf{A}\right]_{(n,n)} = \|\mathbf{g}\|^2, \tag{4.51}
$$

and hence,

$$
\text{trace}\left\{\mathbf{A}^H \mathbf{A}\right\} = \frac{1}{K} \sum_{k,m} |z_{k,m}|^2 = N \|\mathbf{g}\|^2. \tag{4.52}
$$

After matched filter we get

$$
\begin{aligned}
\mathbf{A}^H \mathbf{y} &= \|\mathbf{g}\|^2 \mathbf{d} + \left(\mathbf{A}^H \mathbf{A} - \|\mathbf{g}\|^2 \mathbf{I}_N\right) \mathbf{d} + \mathbf{A}^H \mathbf{w} \\
&= \|\mathbf{g}\|^2 \mathbf{d} + \mathbf{V}^H \left(\tfrac{1}{K} \mathbf{L}^{(\tilde{g})H} \mathbf{L}^{(\tilde{g})} - \|\mathbf{g}\|^2 \mathbf{I}_N\right) \mathbf{V}\mathbf{d} + \mathbf{A}^H \mathbf{w}.
\end{aligned} \tag{4.53}
$$

The signal-to-interference ratio (SIR) is defined by the ratio of the signal power to the self-interface

$$
\begin{aligned}
\text{SIR} &= \frac{N \|\mathbf{g}\|^4}{\sum_{k,m} \left(\frac{1}{K} |z_{k,m}|^2 - \|\mathbf{g}\|^2\right)^2} \\
&= \frac{N}{\sum_{k,m} \left(\frac{|z_{k,m}|^2}{K\|\mathbf{g}\|^2} - 1\right)^2} \\
&= \left[\frac{1}{N} \sum_{k,m} \left(\frac{|z_{k,m}|^2}{\frac{1}{N} \sum_{k,m} |z_{k,m}|^2} - 1 \right)^2 \right]^{-1}.
\end{aligned} \tag{4.54}
$$

The computed self-interference is averaged over all data symbols. Nevertheless, when full allocation is considered and all data symbols have the same power, the SIR is identical for each data symbol. This is because $\left|[\mathbf{V}]_{(i,j)}\right| = \frac{1}{\sqrt{N}}, \forall i, k \in \{0, \ldots, N-1\}$.

The later discussion shows the dependency of the modulation on the prototype pulse shape, where all indicator can be expressed in terms of its DZT. Although we use DZT of $\tilde{\mathbf{g}}$, the same results hold with respect to the DZT of $\mathbf{g}$. In the next section, we introduce a method to design the prototype filter $\tilde{\mathbf{g}}$.

4.1.5 GFDM Pulse Shaping Filter Design

Starting from the DTFT of a preselected basis filter $h[n]$ of practical interests, e.g., RC or root-raised cosine (RRC), which is denoted as $H(\nu)$. Here, ν is the normalized frequency and thus the period of $H(\nu)$ is equal to 1. Then, we compute $\tilde{g}[n] = H(\frac{n}{N})$. With such design, it has been shown in [3] that $\mathbf{A}$ becomes singular for even M, K and a real symmetric filter $h[n]$. This is caused by $\left[\mathbf{Z}_{K,M}^{(\tilde{\mathbf{g}})}\right]_{(K/2,M/2)} = 0$. The requirement of odd M or K impedes an efficient implementation in terms of low-complexity radix-2 FFT operations. In [4], we present a design approach that overcomes this restriction for any basis filter $h[n]$ fulfilling the following conditions,

1. $h[n]$ is real-valued, i.e., $H(\nu) = H^*(1-\nu) = H^*(-\nu)$.
2. $H(\nu)$ spans two subcarriers within each period, namely $H(\nu) = 0, \forall \nu \in [\frac{1}{K}, \frac{1}{2}]$.
3. $|H(\nu)|$ is decreasing from 1 to 0 for $\nu \in [0, \frac{1}{K}]$.

The idea is to introduce a fractional shift $\lambda \in [0, 1]$ when sampling $H(\nu)$, as shown in Fig. 4.5. Accordingly, the samples of $\tilde{\mathbf{g}}$ are defined by

$$[\tilde{\mathbf{g}}]_n(\lambda) = \left\{ \begin{matrix} H\left(\frac{n+\lambda}{N}\right), & 0 \le n < M - \lambda \\ H^*\left(\frac{N-n-\lambda}{N}\right), & N - M - \lambda < n \le N-1 \\ 0, & \text{otherwise} \end{matrix} \right\}. \tag{4.55}$$

Moreover, $\tilde{\mathbf{g}}$ can be reshaped as in (4.24) to

$$\left[\mathbf{V}_{K,M}^{(\tilde{\mathbf{g}})}(\lambda)\right]_{(k,m)} = \left\{ \begin{matrix} H\left(\frac{m+\lambda}{N}\right), & k = 0 \\ H^*\left(\frac{M-m-\lambda}{N}\right), & k = K-1 \\ 0, & \text{elsewhere} \end{matrix} \right\}. \tag{4.56}$$

With this design restriction, the same frequency taps can be used with different values of $K \ge 2$. Then, we compute the DZT from (4.25), so that

$$z_{k,m}(\lambda) = H\left(\frac{m+\lambda}{N}\right) + H^*\left(\frac{M-m-\lambda}{N}\right) e^{j2\pi\frac{k}{K}}. \tag{4.57}$$

Due to the symmetry of $H(\nu)$, we have

$$\begin{aligned} z_{k,m}(1-\lambda) &= z_{k,M-1-m}^*(\lambda) e^{j2\pi\frac{k}{K}}, \\ |z_{k,m}(1-\lambda)|^2 &= |z_{k,M-1-m}(\lambda)|^2. \end{aligned} \tag{4.58}$$

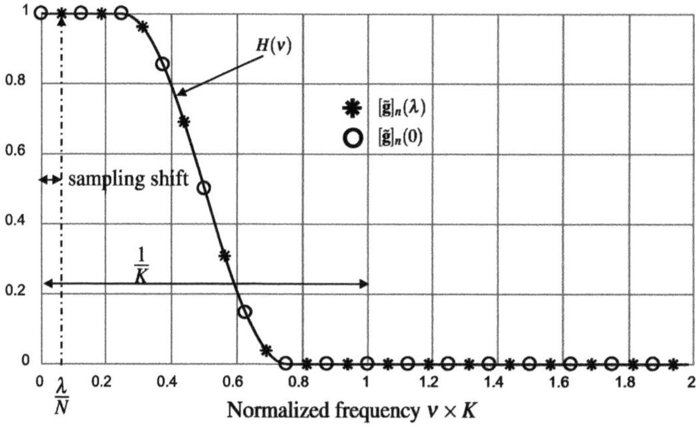

Fig. 4.5 Filter design with sampling. Basis filter is Raised-Cosine with roll-off factor $\alpha = 0.5$, $M = 8$. The samples are shown for $0 \leq n < M - \lambda$

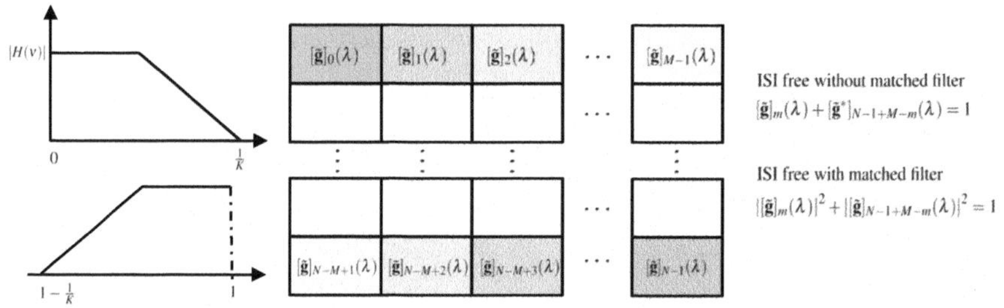

Fig. 4.6 Filter design with ISI free with and without matched filtering. Here, $\mathbf{V}^{(\tilde{\mathbf{g}})}_{K,M}(\lambda)$ is shown

Thereby, all results regarding conditional number, NEF and SIR are symmetric around $\lambda = 0.5$. Moreover, Eq. (4.57) shows that $z_{k,m}(1 + \lambda) = z_{k,m+1}(\lambda)$. Therefore, it suffices to study the range $0 \leq \lambda \leq 0.5$.

We subsequently focus on particular design cases for $K = 2^x, x > 1$ and different values of M. Namely, design with the families of $H(\nu)$ that fulfill the inter-symbol-interference (ISI)-free criterion without or with matched filtering, Fig. 4.6.

4.1.5.1 ISI-Free Without Matched Filter

In this case, $H(\nu)$ additionally satisfies the condition

$$\sum_{k=0}^{K-1} H\left(\nu - \frac{k}{K}\right) = 1. \tag{4.59}$$

From the symmetry and limited band of $H(\nu)$, it follows that

$$H(\nu) + H^*\left(\frac{1}{K} - \nu\right) = 1, \forall \nu \in \left[0, \frac{1}{K}\right]. \tag{4.60}$$

As a result, $H\left(\frac{m+\lambda}{N}\right) + H^*\left(\frac{M-m-\lambda}{N}\right) = 1$. From that, there exists a function $f(\nu) = r(\nu)e^{j\phi(\nu)}$ with $f(\nu) = -f^*\left(\frac{1}{K} - \nu\right)$ and

$$H(\nu) = \frac{1}{2}\left(1 + f(\nu)\right), \forall \nu \in \left[0, \frac{1}{K}\right]. \tag{4.61}$$

Let us assume a real-valued $f(\nu)$, i.e., $\phi(\nu) = 0$ and $f(\nu) = r(\nu)$. A complex-valued $f(\nu)$ as in Xia filters is treated in the following section. Due to the constraint of the decreasing amplitude of $H(\nu)$, $r(\nu)$ must be decreasing from 1 to -1 for $\nu \in [0, \frac{1}{K}]$. From (4.60) and (4.61) we get

$$|z_{A_{k,m}}(\lambda)|^2 = \frac{(1 + f_m^2(\lambda)}{2} + \frac{(1 - f_m^2(\lambda))}{2}\cos\left(2\pi\frac{k}{K}\right), \tag{4.62}$$

where $f_m(\lambda) = f\left(\frac{m+\lambda}{N}\right) = 2H\left(\frac{m+\lambda}{N}\right) - 1$. The singular values are symmetric with respect to k, and decreasing with $k = 0, \ldots, \frac{K}{2}$. Therefore, $|z_{A_{0,m}}(\lambda)|^2 = 1$ and $|z_{A_{\frac{K}{2},m}}(\lambda)|^2 = f_m^2(\lambda)$ are the maximum and minimum singular values with respect to k, respectively. Therefore, $|z_{A_{max}}(\lambda)|^2 = 1$, because $f_m^2(\lambda) \leq 1$, and $|z_{A_{min}}(\lambda)|^2$ is obtained from $\min_m\{f_m(\lambda)^2\}$. Since $f(\nu)$ is decreasing and antisymmetric around $\frac{1}{2K}$, $f(\nu)^2$ is decreasing $\forall \nu \in [0, \frac{1}{2K}]$ and increasing $\forall \nu \in [\frac{1}{2K}, \frac{1}{K}]$. As a result, when M is even and $0 \leq \lambda \leq 0.5$, $|z_{A_{min}}|^2$ is obtained at $m = M/2$, and when M is odd, it is obtained at $m = (M-1)/2$. Consequently,

$$|z_{A_{min}}|^2(\lambda) = f^2\left(\frac{1}{2K} + \frac{S(\lambda)}{2N}\right). \tag{4.63}$$

$$\text{where} \quad S(\lambda) = \begin{Bmatrix} 2\lambda, & M \text{ is even} \\ 1 - 2\lambda, & M \text{ is odd} \end{Bmatrix}. \tag{4.64}$$

From the increasing/decreasing intervals of $f^2(\nu)$, $|z_{A_{min}}|^2(\lambda)$ increases with $0 \leq \lambda \leq 0.5$ for even M and decreases when M is odd. Hence, the condition number can be expressed as

$$\text{cond}(\mathbf{A}_A)(\lambda) = \frac{1}{\left|f\left(\frac{1}{2K} + \frac{S(\lambda)}{2N}\right)\right|}. \tag{4.65}$$

Similarly, $\text{cond}(\mathbf{A}_A)(\lambda)$ is decreasing for even M and increasing for odd M. Hence, the best condition of $\mathbf{A}$ is attained at $\lambda = 0.5$ for even M and $\lambda = 0$ for odd M.

4.1.5.2 ISI-Free After Matched Filtering

A filter $H(\nu)$ is ISI-free after matched filtering if

$$\sum_{k=0}^{K-1}\left|H\left(\nu-\frac{k}{K}\right)\right|^2=1. \tag{4.66}$$

By exploiting the symmetry and band limit, we get

$$|H(\nu)|^2+\left|H^*\left(\frac{1}{K}-\nu\right)\right|^2=1, \forall\nu\in\left[0,\frac{1}{K}\right], \tag{4.67}$$

and hence, $\left|H\left(\frac{m+\lambda}{N}\right)\right|^2+\left|H^*\left(\frac{M-m-\lambda}{N}\right)\right|^2=1$. Furthermore, there exists a real-valued function $f(\nu)=-f\left(\frac{1}{K}-\nu\right)$, which is decreasing from 1 to -1 in the interval $\nu\in[0,\frac{1}{K}]$ with

$$|H(\nu)|^2=\frac{1}{2}(1+f(\nu)), \forall\nu\in\left[0,\frac{1}{K}\right]. \tag{4.68}$$

Adding an (arbitrary) phase $\phi(\nu)$ yields the original $H(\nu)$ by

$$H(\nu)=e^{j\phi(\nu)}\sqrt{\frac{1}{2}(1+f(\nu))}, \forall\nu\in\left[0,\frac{1}{K}\right]. \tag{4.69}$$

Using (4.67) and (4.69),

$$\begin{aligned}H\left(\frac{m+\lambda}{N}\right)&=e^{j\phi_{a,m}(\lambda)}\sqrt{\frac{1}{2}(1+f_m(\lambda))},\\ H^*\left(\frac{M-m-\lambda}{N}\right)&=e^{j\phi_{b,m}(\lambda)}\sqrt{\frac{1}{2}(1-f_m(\lambda))},\end{aligned} \tag{4.70}$$

where $\phi_m^a(\lambda)=\phi\left(\frac{m+\lambda}{N}\right)$, $\phi_m^b(\lambda)=-\phi\left(\frac{M-m-\lambda}{N}\right)$, and $f_m(\lambda)=f\left(\frac{m+\lambda}{N}\right)$. As special cases, we study the phase in the form $\phi(\nu)=-\phi\left(\frac{1}{K}-\nu\right)+\beta\frac{\pi}{2}$, $\beta=0,1,2,3$. Then, $e^{j\phi_{a,m}(\lambda)}=j^\beta e^{j\phi_{b,m}(\lambda)}$. The case of no ISI with and without (MF), as the Xia filters [5] provide, is obtained with $f(\nu)=\cos(2\phi(\nu))$ and $\beta=2$ or, equivalently, $\phi(\nu)=\frac{1}{2}\text{acos}(f(\nu)$. From (4.57), we get

$$\sigma^2_{B_{k,m}}(\lambda)=1+\sqrt{1-f_m^2(\lambda)}\cos\left(2\pi\frac{k-\beta\frac{K}{4}}{K}\right). \tag{4.71}$$

The maximum singular value with respect to k is located at $k_{max}=\beta\frac{K}{4}$ and the minimum one at $k_{min}=<\beta+2>_4\frac{K}{4}$. This requires that K is a multiple of 4 for $\beta=1,3$.

$$|z_{B_{k_{max},m}}(\lambda)|^2 = 1 + \sqrt{1 - f_m^2(\lambda)},$$
$$|z_{B_{k_{min},m}}(\lambda)|^2 = 1 - \sqrt{1 - f_m^2(\lambda)}. \tag{4.72}$$

Following the same arguments as in the previous subsection and based on the properties of $f(\nu)$, both $|z_{B_{min}}(\lambda)|^2$ and $|z_{B_{max}}(\lambda)|^2$ are obtained at $m = M/2$ for even M and $m = \frac{M-1}{2}$ for odd M. Accordingly,

$$|z^2_{B_{max}}(\lambda)|^2 = 1 + \sqrt{1 - f^2\left(\frac{1}{2K} + \frac{S(\lambda)}{2N}\right)},$$
$$|z^2_{B_{min}}(\lambda)|^2 = 1 - \sqrt{1 - f^2\left(\frac{1}{2K} + \frac{S(\lambda)}{2N}\right)}, \tag{4.73}$$

and the conditional number can then be written as

$$\text{cond}(\mathbf{A}_B)(\lambda) = \frac{\left|f\left(\frac{1}{2K} + \frac{S(\lambda)}{2N}\right)\right|}{1 - \sqrt{1 - f^2\left(\frac{1}{2K} + \frac{S(\lambda)}{2N}\right)}}. \tag{4.74}$$

It can be seen that $\text{cond}(\mathbf{A}_B)(\lambda)$ is decreasing for even M and increasing for odd M with $\lambda \in [0, 0.5]$. When using the same function $f(\nu)$ in cases A and B, it is notable that that $|z_{B_{max}}(\lambda)|^2 \geq 1 = |z_{A_{max}}|^2$ and $|z_{B_{min}}|^2 \leq f^2_{m_{max}}(\lambda) = |z_{A_{min}}|^2$, and hence,

$$\text{cond}(\mathbf{A}_A)(\lambda) \leq \text{cond}(\mathbf{A}_B)(\lambda), \tag{4.75}$$

proving that the condition number is smaller when using an ISI-free filter, compared to using its square root.

4.1.5.3 Numerical Example

In this section, we study the family of prototype filters with roll-off factor α, being obtained with the generator function Fig. 4.7,

$$f(\nu) = \begin{Bmatrix} 1, & 0 \leq \nu \leq \frac{1-\alpha}{2K} \\ f^a\left(\frac{2K}{\alpha}[\nu - \frac{1}{2K}]\right), & \frac{1-\alpha}{2K} < \nu \leq \frac{1+\alpha}{2K} \\ -1, & \frac{1+\alpha}{2K} < \nu \leq \frac{1}{K} \end{Bmatrix}. \tag{4.76}$$

Here, f^a is a real-valued antisymmetric ($f^a(x) = f^a(-x)$), and decreasing from 1 to -1 for $x \in [-1, 1]$, which produces $f(\nu) = -f(\frac{1}{K} - \nu)$. Hence, $f(\nu)$ can be used to construct pulse shapes that provide ISI free without matched filter used to generate $\mathbf{A}_A$

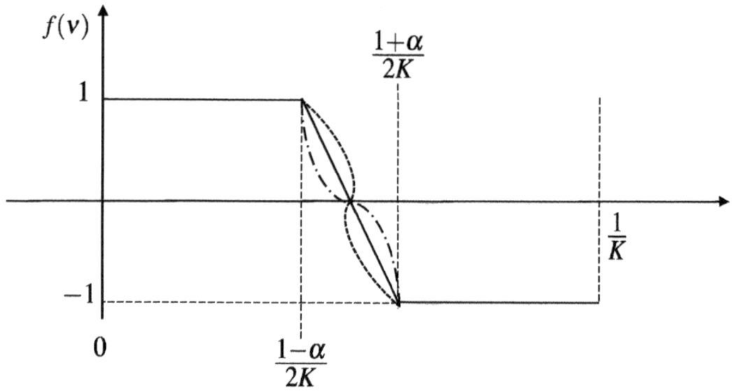

Fig. 4.7 Generator function of filters with roll-off factor α

(4.61) or ISI free with matched filter used to generate $\mathbf{A}_B$ (4.69). From (4.74), (4.65), and (4.76), we get for $M\alpha \leq S(\lambda)$, $\text{cond}(\mathbf{A}_A) = \text{cond}(\mathbf{A}_B) = 1$. For $S(\lambda) \leq M\alpha$,

$$\text{cond}(\mathbf{A}_A)(\lambda) = \frac{1}{\left|f^a\left(\frac{S(\lambda)}{\alpha M}\right)\right|},$$
$$\text{cond}(\mathbf{A}_B)(\lambda) = \frac{\left|f^a\left(\frac{S(\lambda)}{\alpha M}\right)\right|}{1-\sqrt{1-f^{a2}\left(\frac{S(\lambda)}{\alpha M}\right)}}. \tag{4.77}$$

The condition number is independent of K and, based on the properties of f^a, increases with αM. As a particular example, RC and RRC use the function $f^a(x) = -\sin(\frac{\pi}{2}x)$. Replacing in (4.77) we get,

$$\text{cond}(\mathbf{A}_{\text{RC}})(\lambda) = \left(\sin\left(\frac{\pi}{2}\frac{S(\lambda)}{\alpha M}\right)\right)^{-1},$$
$$\text{cond}(\mathbf{A}_{\text{RRC}})(\lambda) = \left(\tan\left(\frac{\pi}{4}\frac{S(\lambda)}{\alpha M}\right)\right)^{-1}. \tag{4.78}$$

Figure 4.8 illustrates the condition number of $\mathbf{A}$ for different sampling shift λ and validates the closed-form expressions (4.78) numerically. As shown, $\lambda = 0$ is optimal for odd M and $\lambda = \frac{1}{2}$ for even M when K is also even. In addition, as proven in (4.78), using RC yields a better conditioned $\mathbf{A}$ than RRC. Furthermore, numerically obtained values for the NEF as shown in Fig. 4.9 behave similarly as the condition number. This can be explained by the influence of the smaller singular value on the noise enhancement. In both cases, the condition number as well as the smallest singular value depends on $\left|f^a\left(\frac{S(\lambda)}{\alpha M}\right)\right| = \sin\left(\frac{\pi}{2}\frac{S(\lambda)}{\alpha M}\right)$. Considering the optimum λ, Fig. 4.10 shows the NEF and SIR with different M. The proper choice of λ with

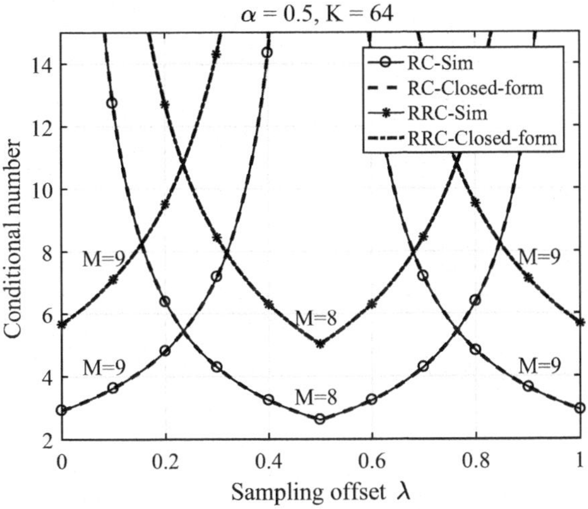

Fig. 4.8 Conditional number

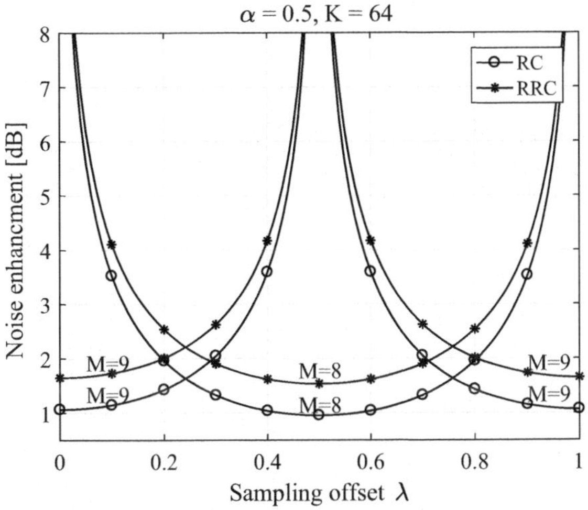

Fig. 4.9 NEF versus λ

respect to M preserves the trend of NEF which increases with M. On the other hand, the SIR is independent of M when M is big enough. In fact, the SIR approaches the interference value that can be directly obtained from $\text{SIR} = 2\int_{\frac{1}{2K}}^{\frac{1}{2}} |H(\nu)|^2 d\nu$, which is independent of λ and K but depends on α. Finally, Fig. 4.11 depicts the dependency of the NEF on K for the case of design with ISI free without matched filter. Although the condition number is independent of the even values of K, the

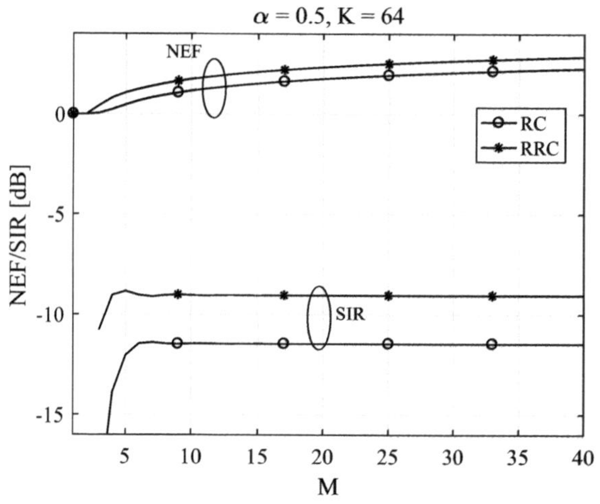

Fig. 4.10 NEF and SIR for optimal λ

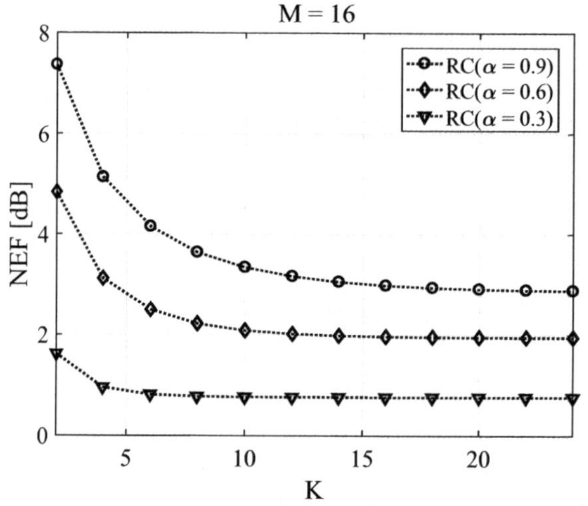

Fig. 4.11 NEF versus even values of K

NEF is higher for smaller values, especially when α is small, and converges to fixed value when K gets bigger. As a result, K should be bigger than the product of αM.

4.1.6 *Multicarrier Waveforms Generator*

In this section, we provide a general representation of linear multicarrier modulation techniques inspired from the GFDM model. Then, we show how different state-of-the-art waveforms can be generated from the GFDM block modulator, block multiplexing, windowing, and filtering. This enables the development of a universal waveform generator.

4.1.6.1 General Linear Modulation

In general multicarrier modulation, a stream of data symbols can be split into time-frequency substreams $\{d_{k,m,i}\}$, with k is the index in the frequency domain denoted as subcarrier, m the index in the time domain as subsymbol, and i stands for the block index. Each stream is modulated by a transmitter pulse shape $g^t_{k,m}[n]$ with finite length L_t, specifically, $\forall (k,m)$ and $\forall n \notin \{0,\ldots,L_t-1\}$, $g_{k,m}[n]=0$. The discrete transmitted signal can be written as

$$\begin{aligned} x^t[n] &= \sum_{i=0}^{\infty} \sum_{k\in\mathcal{K}_{\text{on}}} \sum_{m\in\mathcal{M}_{\text{on}}} d_{k,m,i} g^t_{k,m}[n-iL_s] \\ &= \sum_{i=0}^{\infty} x^t_i[n-iL_s], \end{aligned} \tag{4.79}$$

where L_s is the block spacing, $\mathcal{K}_{\text{on}}$ and $\mathcal{M}_{\text{on}}$ are the sets of active subcarriers and subsymbols, respectively, and $x^t_i[n]$ is the i-th multicarrier block, which is given by

$$x^t_i[n] = \sum_{k\in\mathcal{K}_{\text{on}}} \sum_{m\in\mathcal{M}_{\text{on}}} d_{k,m,i} g^t_{k,m}[n]. \tag{4.80}$$

Accordingly, $x^t_i[n]$ has a length of L_t samples. The number of available resources per block is denoted as $N = MK$. The difference between the block length and the block spacing $L_o = L_t - L_s$ determines the overlapping between successive blocks as illustrated in Fig. 4.12. When $L_o > 0$, the blocks overlap, which means that the last L_o samples of the previous block are added to the first L_o samples of the current block prior to transmission. On the other hand, for $L_o \leq 0$, there is a guard interval of L_o zero padding (ZP) samples between successive blocks. Therefore, the multicarrier waveform can be defined by knowing the set of pulse shapes $\{g_{k,m}\}$, the set of active resources $\mathcal{K}_{\text{on}} \times \mathcal{M}_{\text{on}}$, the resource dimensions K, M and the overlapping length L_o.

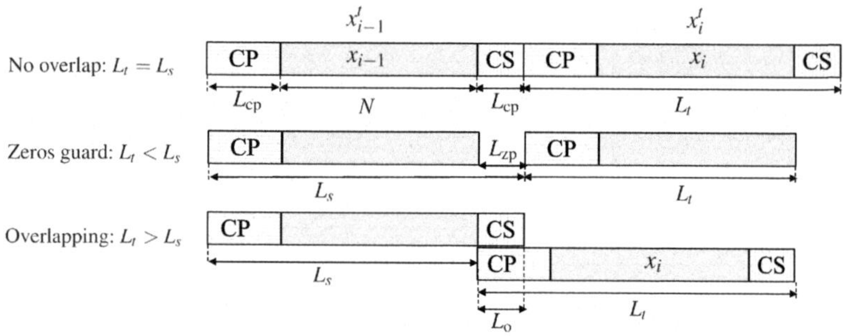

Fig. 4.12 Different cases of block multiplexing

4.1.6.2 Generic Waveforms Generator

In the common modulation techniques, $g^t_{k,m}[n]$ can be generated from a prototype pulse shape with shift in time and frequency. Moreover, CP and CS can be added afterward. In addition, windowing and subband filtering may be applied. Some waveforms involve more than one prototype pulse shape, which can be seen as superposition of different waveforms. As an example, we list the overall procedure for generating waveforms employing one prototype pulse shape $g[n]$ of length N samples.

- Shifting in time and frequency

$$g_{k,m}[n] = g[< n - mK >_N]e^{j2\pi \frac{k}{K}n}, n = 0, \ldots, N-1. \tag{4.81}$$

- CP and CS insertion

$$g^{cp}_{k,m}[n] = g_{m,k}[< n - mK - L_{\text{cp}} >_N], n = 0, \ldots, N + L_{\text{cp}} + L_{\text{cs}} - 1. \tag{4.82}$$

- Time domain windowing using a window function $w[n]$ of length $N + L_{\text{cp}} + L_{\text{cs}}$ samples

$$g^{cp,w}_{k,m}[n] = w[n] \cdot g^{cp}_{m,k}[n], n = 0, \ldots, N + L_{\text{cp}} + L_{\text{cs}} - 1. \tag{4.83}$$

- Filtering using a filter $f[n]$ with L_f samples

$$g^{cp,w,f}_{k,m}[n] = f[n] * g^{cp,w}_{k,m}[n], n = 0, \ldots, N + L_{\text{cp}} + L_{\text{cs}} + L_f - 2. \tag{4.84}$$

$g^t_{k,m}[n]$ results from one or more of that steps. In the most complicated case $g^t_{k,m}[n] = g^{cp,w,f}_{k,m}[n]$. Therefore, (4.80) becomes

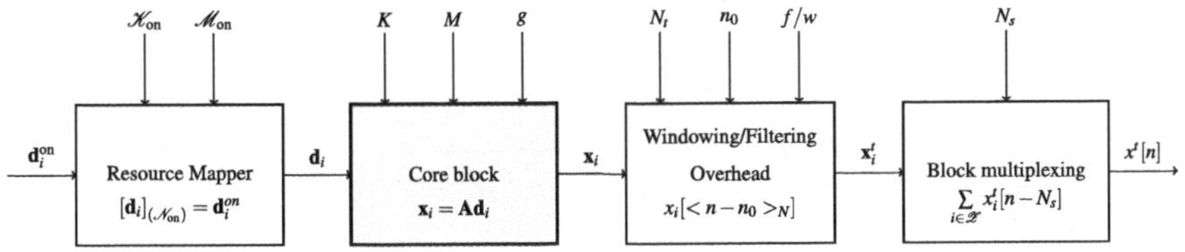

Fig. 4.13 Multicarrier waveform generator stages

$$
\begin{aligned}
x_i^t[n] &= f[n] * \left(w[n] \cdot \sum_{k \in \mathcal{K}_{\text{on}}} \sum_{m \in \mathcal{M}_{\text{on}}} d_{k,m,i} g_{m,k}[< n - mK - L_{\text{cp}} >_N] \right) \\
&= f[n] * \left(w[n] \cdot x_i[< n - L_{\text{cp}} >_N] \right),
\end{aligned} \tag{4.85}
$$

where

$$
\begin{aligned}
x_i[n] &= \sum_{k \in \mathcal{K}_{\text{on}}} \sum_{m \in \mathcal{M}_{\text{on}}} d_{k,m,i} g_{k,m}[n], \, n = 0, \ldots, N-1 \\
&= \sum_{k \in \mathcal{K}_{\text{on}}} \sum_{m \in \mathcal{M}_{\text{on}}} d_{k,m,i} g[< n - mK >_N] e^{j2\pi \frac{k}{K} n}.
\end{aligned} \tag{4.86}
$$

This means that the transmitted block can be obtained starting from a core block of length N samples, which can be generated using the GFDM modulator. Then, the CP and CS can be added to the core block followed by windowing and filtering. Finally, the blocks are multiplexed in the time domain to generate the waveform. This procedure is depicted in Fig. 4.13.

4.1.6.3 Example of the State-of-the-Art Waveforms

As discussed, the core block is the essential part of the waveform, and this can be generated with the GFDM modulator under proper setting of the related parameters.

OFDM Variants

Obviously, orthogonal frequency division multiplexing (OFDM) is a special case of GFDM when $M = 1$ and $g[n]$ is a rectangular pulse. In the simplest form of OFDM, only a CP is added. Thus $L_t = K + L_{\text{cp}}$ and no overlapping, i.e., $L_s = L_t$. In the windowed version, a window is applied after adding the CP and CS. Then, $L_t = K + L_{\text{cp}} + L_{\text{cs}}$, with $L_{\text{cp}} > L_{\text{cs}}$. In order to reduce the overhead, the blocks are overlapped with $L_o = L_{\text{cs}}$, so that $L_s = K + L_{\text{cp}}$, Fig. 4.12. In the filtered variants, either the ones based on subband filtering for multiple users or the others that apply filtering to the whole signal, a filter is applied on the generated CP symbols with the corresponding active subcarriers. Hence, $L_t = K + L_{\text{cp}} + L_f - 1$, $L_s = K + L_{\text{cp}}$, and due to the filtering $L_o = L_f - 1$.

DFT-spread OFDM

In this modulation, [6] a set of M data symbols are transferred into the frequency domain using $\mathbf{F}_M$ and then allocated to the subcarrier set of N subcarriers using $\frac{1}{N}\mathbf{F}_N^H$. Thus, the modulation matrix of this waveform is given by

$$\mathbf{A} = \frac{1}{N}\mathbf{F}_N^H \mathbf{U}_{K,M}. \tag{4.87}$$

Comparing with (4.34), this waveform can be generated with the GFDM modulator where the input vector is given by $\text{vec}\left\{\mathbf{D}^T\right\}$ and using a pulse shape $\tilde{g}$ with $\left[\mathbf{Z}_{K,M}^{(\tilde{g})}\right]_{(k,m)} = 1, \forall(k,m)$. This actually corresponds to the Dirichlet pulse given by

$$g[n] = e^{j\pi n\frac{M-1}{N}} \frac{\sin(\pi\frac{n}{K})}{\sin(\pi\frac{n}{N})}. \tag{4.88}$$

Filtered Multitone (FMT)

This waveform does not originally consider subsymbols representation [7]. However, it is easy to transform it to fit in this framework. The prototype pulse $g_{\text{FMT}}[n]$ is assumed to have a length of KM_o, where M_o is denoted as the overlapping factor.

$$\begin{aligned} x_{\text{FMT}}^t[n] &= \sum_{p=0}^{\infty}\sum_{k=0}^{K-1} d_{k,p}g_{\text{FMT}}[n-pK]e^{j2\pi\frac{k}{K}n} \\ &= \sum_{i=0}^{\infty}\sum_{k=0}^{K-1}\sum_{m=0}^{M_d-1} d_{k,m+iM_d}g_{\text{FMT}}[n-(m+iM_d)K]e^{j2\pi\frac{k}{K}n} \\ &= \sum_{i=0}^{\infty} x_i[n-iM_dK], \end{aligned} \tag{4.89}$$

where

$$x_i[n] = \sum_{k=0}^{K-1}\sum_{m=0}^{M_d-1} d_{k,m+iM_d}g_{\text{FMT}}[n-mK]e^{j2\pi\frac{k}{K}n}. \tag{4.90}$$

With respect to GFDM notations, we define the number of subsymbols $M = M_d + M_o - 1$, where $M_d \geq 1$ is the number of active subsymbols, the data symbol $d_{k,m,i} = d_{k,m+iM_d}$ and the pulse shape $g[n]$ as

$$g[n] = g_{\text{FMT}}[< n + \tfrac{KM_o}{2} >_N], n = 0, \ldots, N-1. \tag{4.91}$$

Therefore, we get the GFDM block

$$x_i[n] = \sum_{k=0}^{K-1}\sum_{m=M_o/2}^{M_d+M_o/2-1} d_{k,m,i}g[< n - mK >_N]e^{j2\pi\frac{k}{K}n}. \tag{4.92}$$

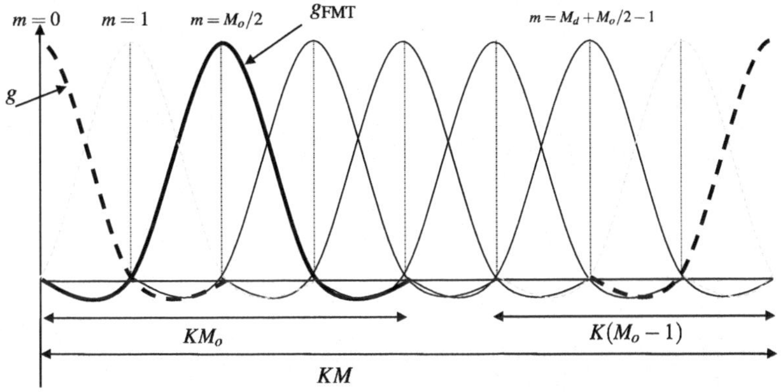

Fig. 4.14 FMT pulse shape to GFDM

Noting that $\mathcal{M}_{\text{on}} = \{M_o/2, \ldots, M_d + M_o/2 - 1\}$, in other words, there are $M_o - 1$ subsymbols turned off, Fig. 4.14. However, while $L_s = M_d K$ and $L_t = MK$, there is an overlapping of $L_o = (M_o - 1)K$ samples, which compensates the overhead results from the non-active subsymbols.

4.1.6.4 Further Degrees of Freedom

Recalling the compact equation

$$\mathbf{X} = \mathbf{V}_{M,K}^{(\mathbf{x})} = \frac{1}{M}\mathbf{F}_M^H \left(\mathbf{Z}_{M,K}^{(\mathbf{g})} \odot \left[\mathbf{F}_M \mathbf{D}^T \mathbf{F}_K^H\right]\right).$$

The GFDM core block is generated as $\mathbf{x} = \text{vec}\left\{\mathbf{X}^T\right\}$ and then further processing steps take place. Inspired from that, it is possible to generate three types of other signals;

1. By constructing the core block according to the columns, i.e., $\mathbf{x}_1 = \text{vec}\left\{\mathbf{X}\right\}$. This can be seen as a block of stacked K precoded OFDM symbols with M subcarriers.
2. Adding a CP to each column of $\mathbf{X}$ to get $\mathbf{X}^{(cp)}$, then $\mathbf{x}_2 = \text{vec}\left\{\mathbf{X}^{(cp)}\right\}$. This signal can be seen as transmitting precoded OFDM symbols successively.
3. Adding a CP to each column of $\mathbf{X}^T$ to get $\mathbf{X}_3^{(cp)}$, then $\mathbf{x}_3 = \text{vec}\left\{\mathbf{X}_3^{(cp)}\right\}$. This signal can be seen as inserting CPs within the GFDM symbol.

Moreover, the matrix $\mathbf{Z}_{M,K}^{(\mathbf{g})}$ can be populated with unit amplitude and variant phase elements, namely, $\left|\left[\mathbf{Z}_{K,M}^{(\tilde{g})}\right]_{(k,m)}\right| = 1$, so that the overall modulation matrix is orthogonal. For example, with $M = 1$ and $\left[\mathbf{Z}_{K,M}^{(\tilde{g})}\right]_{(k,1)} = e^{j2\pi c\frac{n^2}{N}} = \left[\mathbf{L}^{(\mathbf{c})}\right]_{(k,k)}$, we get

$$\mathbf{x} = \mathbf{L}^{(c)}\mathbf{F}_K^H\mathbf{d}, \tag{4.93}$$

which represents a chirp-based waveform [8].

In conclusion, the GFDM-inspired waveform generator is a very powerful tool for unified implementation of various standard waveforms, which makes it appropriate for mixed numerology approach.

4.1.7 Channel Estimation for GFDM Detection

Consider a multiple-input multiple-output (MIMO) transceiver with N_t transmit and N_r receive antennas where the complex-valued data symbols are spatially multiplexed. Combining the MIMO system with a GFDM-based modulation, we encounter a three-dimensional interference situation due to ISI, ICI as well as inter-antenna-interference (IAI). Such an interference-limited scenario challenges the MIMO-GFDM receiver design in different stages. A critical functional unit at the receiver side is the channel estimation. Due to broad subcarrier spacing in GFDM, the individual subcarriers become frequency selective and correct detection of the data symbols requires a reliable estimate of the wireless channel transfer function.

In the following, we focus on pilot-aided channel estimation techniques, where some reference signals (also referred as pilots), which are known to both transmitter and receiver, are multiplexed with the data symbols within the same time-frequency resource block. Given the knowledge of the pilots at the receiver, the frequency selective channel transfer function can be estimated and utilized for coherent detection of the data symbols. Here, we also assume that the interference autocorrelation matrix of the MIMO-GFDM transceiver is known at the receiver side, and based on such knowledge we derive the two well-known estimation techniques, namely least squares (LS) and linear minimum mean squared error (LMMSE). Later on, we also discuss an alternative approach for the pilot insertion of a MIMO-GFDM system in order to achieve interference-free channel estimation performance.

4.1.7.1 MIMO Wireless Channel

We assume an urban scenario with multipath Rayleigh fading MIMO channels. Further, we assume that the individual channels between each antenna pairs are independent and they are block fading, namely the channel impulse response (CIR) does not vary significantly within one GFDM symbol duration and therefore, it can be considered as constant. Hence, we model the CIR between the i_t-th and i_r-th antennas as a linear finite-impulse-response filter given by

$$h_{i_t i_r}[n] = \sum_{\ell=0}^{L-1} h_{\ell,i_t i_r}\delta[n - \tilde{n}_\ell], \tag{4.94}$$

where $\tilde{n}_\ell$ is the discrete-time-delay of the ℓ-th path, $h_{\ell,i_t i_r}$ is the complex-valued gain of the ℓ-th path, and it is independent and identically distributed (i.i.d.) zero mean Gaussian process parameterized by the power-delay profile (PDP) $\mathbb{E}\left[|h_{\ell,i_t i_r}|^2\right] = P_{i_t i_r}[\ell]$. Collecting all the L paths in a vector, we have

$$\mathbf{h}_{i_t i_r} = \sqrt{\text{diag}\left\{\mathbf{p}_{i_t i_r}\right\}}\mathbf{q}_{i_t i_r}, \tag{4.95}$$

where, $\mathbf{q}_{i_t i_r} \sim \mathbb{C}(\mathbf{0}, \mathbf{I}_L)$ and $\mathbf{p}_{i_t i_r} \in \mathbb{R}^L$ is the vector of normalized PDP.

Thanks to the utilization of CP in GFDM, the individual channels between each antenna pair are diagonal in the frequency domain. Thus, the receive (Rx) signal in DFT domain at the Rx antenna i_r is characterized by the linear expression

$$\tilde{\mathbf{y}}_{i_r} = \sum_{i_t=1}^{N_t}(\mathbf{D}^{(\tilde{x})}_{p,i_t} + \mathbf{D}^{(\tilde{x})}_{d,i_t})\mathbf{F}_{N,L}\mathbf{h}_{i_t i_r} + \tilde{\mathbf{w}}_{i_r}, \tag{4.96}$$

where $\tilde{\mathbf{w}}_{i_r}$ is the frequency counterpart of AWGN samples with variance σ_w^2, $\mathbf{F}_{N,L} \subseteq \mathbf{F}_N$ is the N-DFT matrix with the L columns associated to the discrete path delays $\tilde{n}_\ell$. Moreover, $\mathbf{D}^{(\tilde{x})}_{s,i_t} = \text{diag}(\tilde{\mathbf{x}}_{s,i_t})$, $s \in \{p, d\}$ and $\tilde{\mathbf{x}}_{s,i_t} = \mathbf{F}_N\mathbf{x}_{s,i_t}$, is the diagonal frequency domain transmitted signal associated with either pilots or data. For the time domain signal, we have $\mathbf{x}_{s,i_t} = \mathbf{A}\mathbf{d}_{s,i_t}$ where $\mathbf{d}_{p,i_t}$ and $\mathbf{d}_{d,i_t}$ are the N-dimensional vectors of the pilots and data symbols at data symbols at transmit (Tx) antenna i_t, respectively. In addition, the multiplexing of the pilots and data symbols satisfies $\mathbf{d}_{p,i_t} \circ \mathbf{d}_{d,i_t} = \mathbf{0}$.

If the number of pilot subcarriers is smaller than K, i.e., the spacing between the pilot subcarriers $\Delta k > 1$, only a subset of the observations in the frequency domain with $N_p = \lfloor N/\Delta k \rfloor$ samples, that contain the knowledge of pilots, are of interest. Therefore, the observed signal $\bar{\tilde{\mathbf{y}}}_{i_r}$ of size $N_p \times 1$ at the pilot-bearing frequency bins[1] follows

$$\bar{\tilde{\mathbf{y}}}_{i_r} = \sum_{i_t=1}^{N_t}(\mathbf{D}^{(\bar{\tilde{x}})}_{p,i_t} + \mathbf{D}^{(\bar{\tilde{x}})}_{d,i_t})\mathbf{F}'_{N,L}\mathbf{h}_{i_t i_r} + \bar{\tilde{\mathbf{w}}}_{i_r}, \tag{4.97}$$

where $\mathbf{D}^{(\bar{\tilde{x}})}_{s,i_t} = \text{diag}(\bar{\tilde{\mathbf{x}}}_{s,i_t})$, $s \in \{p, d\}$, $\bar{\tilde{\mathbf{x}}}_{s,i_t} = \mathbf{F}'_N\mathbf{x}_{s,i_t}$, $\mathbf{F}'_{N,L} \subseteq \mathbf{F}_{N,L}$ of size $N_p \times N$, and $\mathbf{F}'_N \subseteq \mathbf{F}_N$ of size $N_p \times N$ includes the rows of $\mathbf{F}_{N,L}$ and respectively $\mathbf{F}_N$ in which their inner product with the time domain signal $\mathbf{x}_{p,i_t}$ is nonzero.

We rearrange the expression (4.97) into a matrix form as

$$\mathbf{Y} = (\mathbf{X}_p + \mathbf{X}_d)\mathbf{F}'_{N_t}\mathbf{H} + \mathbf{W}, \tag{4.98}$$

herein, each of the above parameters is defined as $\mathbf{F}'_{N_t} = \mathbf{I}_{N_t} \otimes \mathbf{F}'_{N,L}$, $\mathbf{Y} = (\bar{\tilde{\mathbf{y}}}_1, \ldots, \bar{\tilde{\mathbf{y}}}_{N_r}) \in \mathbb{C}^{N_p \times N_r}$, $\mathbf{W} = (\bar{\tilde{\mathbf{w}}}_1, \ldots, \bar{\tilde{\mathbf{w}}}_{N_r}) \in \mathbb{C}^{N_p \times N_r}$, $\mathbf{X}_s = (\mathbf{D}^{(\bar{\tilde{x}})}_{s,1}, \ldots, \mathbf{D}^{(\bar{\tilde{x}})}_{s,N_t}) \in \mathbb{C}^{N_p \times N_p N_t}$

[1] We refer to each inner product of the rows of the DFT matrix with the time domain signal as a frequency bin.

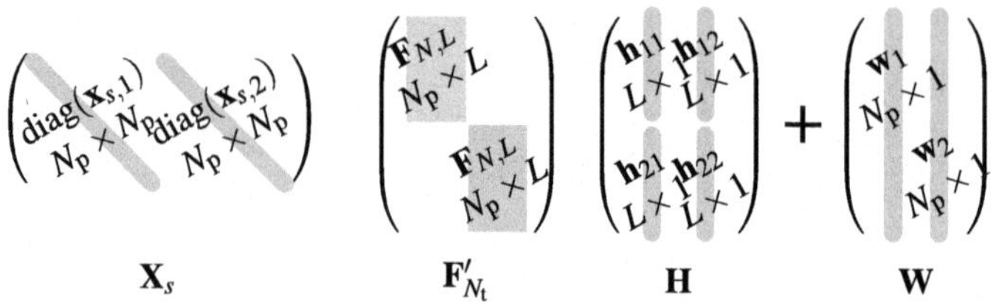

Fig. 4.15 Overview of the matrix structure for a 2 × 2 MIMO channel

with $s \in \{p, d\}$. The matrix $\mathbf{H} \in \mathbb{C}^{LN_t \times N_r}$ is the $N_t \times N_r$ block-matrix of channel impulse responses $\mathbf{h}_{i_t i_r}$. An example of the matrix structures for a 2×2 MIMO channel is depicted in Fig. 4.15. Note that due to the structure of $\mathbf{H}$, its vectorization $\mathbf{h} = \text{vec}\{\mathbf{H}\}$ consists of $N_t N_r$ independent column vectors of channel impulse responses. Thus, by assuming Rayleigh fading channels with no spatial correlation, the covariance matrix of all channel impulse responses $\mathbb{E}\left[\mathbf{h}\mathbf{h}^H\right]$ becomes diagonal.

Resorting to the matrix property $\text{vec}\{\mathbf{ABC}\} = (\mathbf{C}^T \otimes \mathbf{A})\text{vec}\{\mathbf{B}\}$, the associated vectorization of the observed matrix $\mathbf{Y}$ yields the expression

$$\mathbf{y} = \text{vec}\{\mathbf{Y}\} = \tilde{\mathbf{X}}_p\mathbf{h} + \tilde{\mathbf{X}}_d\mathbf{h} + \mathbf{w}, \tag{4.99}$$

where $\tilde{\mathbf{X}}_s = (\mathbf{I}_{N_r} \otimes \mathbf{X}_s\mathbf{F}'_{N_t}) \in \mathbb{C}^{N_p N_r \times LN_t N_r}$ and $\mathbf{w} = \text{vec}\{\mathbf{W}\}$.

4.1.7.2 Least Squares Estimation

The LS estimator of $\mathbf{H}$ minimizes $\|\mathbf{Y} - \mathbf{X}_p\mathbf{H}\|^2$ with respect to $\mathbf{H}$. This yields [9]

$$\hat{\mathbf{H}}_{\text{LS}} = \mathbf{Q}_{\text{LS}}\mathbf{Y} = \left((\mathbf{X}_p\mathbf{F}'_{N_t})^H(\mathbf{X}_p\mathbf{F}'_{N_t})\right)^{-1}(\mathbf{X}_p\mathbf{F}'_{N_t})^H\mathbf{Y}, \tag{4.100}$$

which subjects to estimation errors due to the enhancement of noise as well as interference. Subsequently, the mean squared error (MSE) of the LS estimation is given by

$$\begin{aligned}\text{MSE}_{\text{LS}} &= \mathbb{E}\left[\|\mathbf{H} - \hat{\mathbf{H}}_{\text{LS}}\|^2\right] \\ &= \frac{1}{\Delta k}\text{trace}\left\{\left(\mathbf{I}_{N_r} \otimes (\mathbf{Q}_{\text{LS}}^H\mathbf{Q}_{\text{LS}})\right)\mathbf{R}_{\Psi\Psi}\right\} + \frac{\sigma_w^2}{\Delta k}\text{trace}\left\{\left(\mathbf{I}_{N_r} \otimes (\mathbf{Q}_{\text{LS}}^H\mathbf{Q}_{\text{LS}})\right)\right\}\end{aligned} \tag{4.101}$$

where we calculate the interference covariance matrix as

$$\begin{aligned}\mathbf{R}_{\Psi\Psi} &= \mathbb{E}\left[\text{vec}\left\{\mathbf{X}_d\mathbf{F}'_{N_t}\mathbf{H}\right\}\text{vec}\left\{\mathbf{X}_d\mathbf{F}'_{N_t}\mathbf{H}\right\}^H\right] \\ &= \mathbb{E}_{\mathbf{X}_d}\left[(\mathbf{I}_{N_r} \otimes \mathbf{X}_d\mathbf{F}'_{N_t})\mathbb{E}\left[\mathbf{h}\mathbf{h}^H|\mathbf{X}_d\right](\mathbf{I}_{N_r} \otimes \mathbf{X}_d\mathbf{F}'_{N_t})^H\right] \\ &= \mathbb{E}_{\mathbf{X}_d}\left[(\mathbf{I}_{N_r} \otimes \mathbf{X}_d\mathbf{F}'_{N_t})\mathbf{R}_{\text{hh}}(\mathbf{I}_{N_r} \otimes \mathbf{X}_d\mathbf{F}'_{N_t})^H\right].\end{aligned} \tag{4.102}$$

In the above expression, $\mathbf{R}_{\mathrm{hh}}$ is diagonal because independent Rayleigh fading has been considered for the MIMO channels. Thus, the covariance matrix is given by

$$\mathbf{R}_{\mathrm{hh}} = \mathrm{diag}\left\{[\mathbf{p}_{11}^T, \ldots, \mathbf{p}_{N_\mathrm{t}1}^T, \ldots, \mathbf{p}_{(N_\mathrm{t}-1)N_\mathrm{r}}^T, \mathbf{p}_{N_\mathrm{t}N_\mathrm{r}}^T]^T\right\}. \tag{4.103}$$

Further, due to the block-diagonal structure of $(\mathbf{I}_{N_\mathrm{r}} \otimes \mathbf{X}_d \mathbf{F}'_{N_\mathrm{t}})$ in (4.102), the resulting interference covariance matrix $\mathbf{R}_{\Psi\Psi}$ follows the form

$$\mathbf{R}_{\Psi\Psi} = \mathbf{R}_{\Psi\Psi(1)} \oplus \cdots \oplus \mathbf{R}_{\Psi\Psi(i_\mathrm{r})} \oplus \cdots \oplus \mathbf{R}_{\Psi\Psi(N_\mathrm{r})}, \tag{4.104}$$

where $\oplus$ is the direct sum of matrices. Moreover, we calculate the interference covariance matrix $\mathbf{R}_{\Psi\Psi(i_\mathrm{r})}$ at Rx antenna i_r as

$$\mathbf{R}_{\Psi\Psi(i_\mathrm{r})} = \sum_{i_\mathrm{t}=1}^{N_\mathrm{t}} \mathbf{R}_{\Psi\Psi(i_\mathrm{t},i_\mathrm{r})}, \tag{4.105}$$

wherein for each antenna pair i_t-i_r, we have

$$\begin{aligned}\mathbf{R}_{\Psi\Psi(i_\mathrm{t},i_\mathrm{r})} &= \mathbb{E}_{\mathbf{X}_{d,i_\mathrm{t}}}\left[\mathbf{X}_{d,i_\mathrm{t}}\mathbf{F}'_{N,L}\mathbb{E}_{\mathrm{h}}\left[\mathbf{h}_{i_\mathrm{t}i_\mathrm{r}}\mathbf{h}_{i_\mathrm{t}i_\mathrm{r}}^H\right]\mathbf{F}'^H_{N,L}\mathbf{X}^H_{d,i_\mathrm{t}}\right] \\ &= \mathbf{R}_{\mathrm{hh_f}(\mathrm{i_ti_r})} \circ \mathbf{R}_{X_dX_d,i_\mathrm{t}},\end{aligned} \tag{4.106}$$

where $\mathbf{R}_{\mathrm{hh_f}(\mathrm{i_ti_r})} = \mathbf{F}'_{N,L}\mathrm{diag}\left\{\mathbf{p}_{i_\mathrm{t}i_\mathrm{r}}\right\}\mathbf{F}'^H_{N,L}$ is the channel covariance in the frequency domain. Furthermore, the covariance $\mathbf{R}_{X_dX_d,i_\mathrm{t}}$ is given by

$$\mathbf{R}_{X_dX_d,i_\mathrm{t}} = \mathbf{F}'_N\mathbf{A}\mathbf{R}_{dd,i_\mathrm{t}}\mathbf{A}^H\mathbf{F}'^H_N, \tag{4.107}$$

herein, $\mathbf{R}_{dd,i_\mathrm{t}} = \mathbb{E}\left[\mathbf{d}_{d,i_\mathrm{t}}\mathbf{d}^H_{d,i_\mathrm{t}}\right]$ is the covariance matrix of the data symbols transmitted on antenna i_t. Assuming the data symbols are i.i.d. with unit variance, $\mathbf{R}_{dd,i_\mathrm{t}}$ becomes diagonal with zero entries at the pilot positions.

4.1.7.3 Linear Minimum Mean Squared Error Estimation

The LMMSE estimation calculates the coefficients of a linear filter aiming at minimizing the MSE. In accordance with (4.98) and the corresponding vectorization in (4.99), we have

$$\hat{\mathbf{h}}_{\mathrm{LMMSE}} = \underbrace{\mathbf{R}_{\mathrm{hh}}\tilde{\mathbf{X}}_p^H}_{\mathbf{R}_{\mathrm{hy}}}\underbrace{(\tilde{\mathbf{X}}_p\mathbf{R}_{\mathrm{hh}}\tilde{\mathbf{X}}_p^H + \mathbf{R}_{\Psi\Psi} + \sigma_w^2\mathbf{I}_{N_\mathrm{p}N_\mathrm{r}})}_{\mathbf{R}_{\mathrm{yy}}}{}^{-1}\mathbf{y}. \tag{4.108}$$

Here, $\hat{\mathbf{h}}_{\mathrm{LMMSE}} \in \mathbb{C}^{LN_\mathrm{t}N_\mathrm{r}}$ is a column vector that contains $N_\mathrm{t}N_\mathrm{r}$ individual columns of size L associated to the LMMSE estimate of each channel impulse response.

The resulting MSE performance of the LMMSE estimation follows

$$\mathrm{MSE}_{\mathrm{LMMSE}} = \mathrm{trace}\left\{\mathbf{R}_{\mathrm{HH}} - \mathbf{R}_{\hat{\mathrm{H}}\hat{\mathrm{H}}}\right\}, \tag{4.109}$$

with

$$\mathbf{R}_{\mathrm{HH}} = (\mathbf{I}_{N_\mathrm{r}} \otimes \mathbf{F}'_{N_\mathrm{t}})\mathbf{R}_{\mathrm{hh}}(\mathbf{I}_{N_\mathrm{r}} \otimes \mathbf{F}'_{N_\mathrm{t}})^H, \tag{4.110}$$

$$\mathbf{R}_{\hat{\mathrm{H}}\hat{\mathrm{H}}} = (\mathbf{I}_{N_\mathrm{r}} \otimes \mathbf{F}'_{N_\mathrm{t}})\mathbf{R}_{\mathrm{hy}}\mathbf{R}_{\mathrm{yy}}^{-1}\mathbf{R}_{\mathrm{hy}}^H(\mathbf{I}_{N_\mathrm{r}} \otimes \mathbf{F}'_{N_\mathrm{t}})^H. \tag{4.111}$$

4.1.7.4 Interference-Free Pilot Insertion

In this section, we slightly modify the GFDM modulation at the pilot subcarriers[2] in order to insert orthogonal pilots. The low complexity frequency domain processing of the GFDM modulation can be written as in [10].

$$\mathbf{x} = \mathbf{F}_N^H \sum_{k=0}^{K-1} \mathbf{P}^{(k)}\mathbf{G}^{\delta}\mathbf{T}^{(\delta)}\mathbf{F}_M\mathbf{d}, \tag{4.112}$$

where $\mathbf{T}^{(\delta)}$ is δ-fold repetition matrix which concatenates δ identity matrices $\mathbf{I}_M$ of size M, i.e., $\mathbf{T}^{(\delta)} = (\mathbf{I}_M\mathbf{I}_M \ldots)^T$. The value of δ is based on the number of nonzero values in the filter frequency response, e.g., if a filter spans over two subcarriers δ is typically selected as $\delta = 2$. In (4.112), due to the circular filtering, the subcarrier filter $\mathbf{G}^{(\delta)} = \mathrm{diag}\left\{\mathbf{F}_{\delta M}\mathbf{g}^{(\delta)}\right\}$ is diagonal in frequency domain. The circulant filter $\mathbf{g}^{(\delta)}$ is the down-sampled version of $\mathbf{g} = \big(g[n]\big)_{n=0,\ldots,N-1}$ by factor K/δ. The permutation matrix $\mathbf{P}^{(k)}$ shifts the DC signals to their corresponding subcarriers (i.e., k) and is given by

$$\mathbf{P}^{(k)} = \mathbf{C}_{n'}\begin{bmatrix} \mathbf{0}_{M\delta/2} & \mathbf{I}_{M\delta/2} & \mathbf{0}_{M\delta/2\times(N-\delta M)} \\ \mathbf{I}_{M\delta/2} & \mathbf{0}_{M\delta/2} & \mathbf{0}_{M\delta/2\times(N-\delta M)} \end{bmatrix}^T, \tag{4.113}$$

where $n' = kM - M\delta/2$. The circulant matrix $\mathbf{C}_{n'}$ follows

$$\mathbf{C}_{n'} = \mathrm{circ}([\mathbf{0}^T_{n' \bmod N}, 1, \mathbf{0}^T_{(N-n'-1)\bmod N}]), \tag{4.114}$$

here, $\mathrm{circ}(\cdot)$ returns a circulant matrix associated to its input row vector.

Note that in (4.112) the M-point DFT matrix $\mathbf{F}_M$ can be considered as a special form of precoding. Hence, by slightly modifying such precoder we can reserve some frequency bins specifically for the pilots without any influence from the data symbols [11]. Thus, at the pilot subcarriers $k \in \mathcal{K}_p$, we modify the expression (4.112) by replacing $\mathbf{F}_M$ with $\mathbf{C}_\mathrm{P}$

$$\mathbf{x}_{\mathcal{K}_p} = \mathbf{F}_N^H \sum_{k\in\mathcal{K}_p} \mathbf{P}^{(k)}\mathbf{G}^{(\delta)}\mathbf{T}^{(\delta)}\mathbf{C}_\mathrm{P}\mathbf{d}_k, \tag{4.115}$$

[2]Pilot subcarrier is referred to a subcarrier in which M_p pilots are multiplexed with M_d data subsymbols while $M = M_p + M_d$.

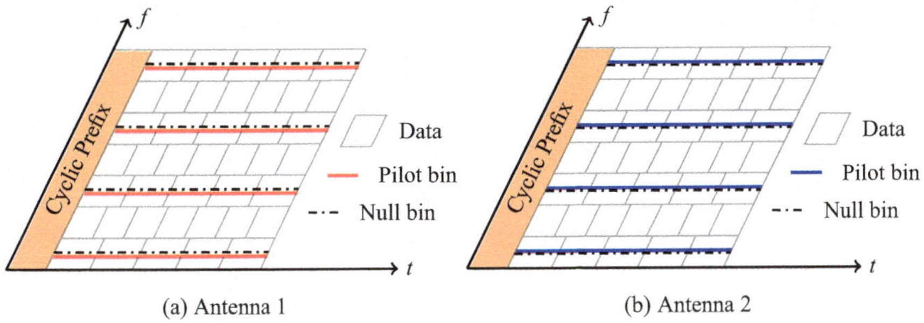

Fig. 4.16 Pilots and data subsymbols in time-frequency resources

Fig. 4.17 DFT domain of the signal for $M = 15$, $K = 4$, $\mathbf{P}' = \mathbf{I}_M$

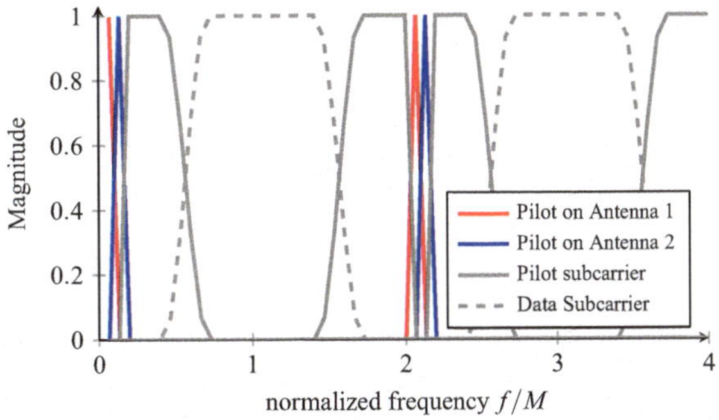

where $\mathbf{C}_\mathrm{P} = \mathbf{P}'(\lambda \mathbf{I}_{N_\mathrm{t}} \oplus \mathbf{F}_{(M-N_\mathrm{t})})$. Here, λ is a scaling factor that normalizes the pilots energy to one. $\mathbf{P}'$ can be any permutation matrix of compatible size which allocates the pilots to any frequency bin within the pilot subcarriers. For instance, the choice $\mathbf{P}' = \mathbf{I}_M$ allocates the pilots on the center frequency bins of the pilot subcarriers. Furthermore, $\mathbf{I}_{N_\mathrm{t}}$ in $\mathbf{C}_\mathrm{P}$ ensures that the first N_t subsymbols of the pilot subcarriers (i.e., $d_k[0], d_k[1], \ldots, d_k[N_\mathrm{t}-1]$ for $k \in \mathcal{K}_p$ which are filled with pilots) are processed directly in the frequency domain being orthogonal to the rest of subsymbols (i.e., $d_k[N_\mathrm{t}], \ldots, d_k[M-1]$ for $k \in \mathcal{K}_p$). Nevertheless, such orthogonality holds if and only if the pilots are located at the frequency bins where no inter-carrier interference is present. Moreover, reserving each orthogonal subsymbol for a specific Tx antenna, the $N_\mathrm{t} \times N_\mathrm{r}$ MIMO channel can be processed in terms of $N_\mathrm{t}N_\mathrm{r}$ single-input single-output (SISO) channels for channel estimation. The approach can be considered as a variation of cell-specific reference signal mapping in Long-Term Evolution (LTE) [12].

Figure 4.16 shows an example how the pilot subsymbols in the GFDM data block are mapped into the time-frequency grid of the resources for a 2×2 MIMO channel. Here, two frequency bins of the pilot subcarriers are reserved only for the pilots while at each Tx antenna only one pilot is being transmitted. Thus, the pilot is being transmitted during the whole GFDM symbol, while also the energies of the data subsymbols are no longer concentrated at equally spaced M peaks. Figure 4.17

shows an example of the signal filtering in frequency domain, where the pilots at different antennas are orthogonal to one another as well as to the data bins within the pilot subcarrier.

4.1.7.5 Simulation Results

In this section, we verify the validity of the closed-form expressions of the channel estimation MSE by simulation and numerical results while we also compare them with the channel estimation performance of OFDM. Later on, we evaluate the performance of MIMO-GFDM with an interference-free pilot design, where we adopt 2×2 MIMO block fading multipath channel with Rayleigh distribution. Since the interference-free pilot insertion (IFPI) in the GFDM block might modify the original signal characteristics, we analyze the Tx signal in terms of peak-to-average power ratio (PAPR) and OOB emission via Monte Carlo simulations.

Consider a sequence of 16-QAM symbols with energy per symbol E_s being transmitted through a multipath MIMO channel with noise energy N_0 and with $N_t = \{2, 3, 4\}$ and $N_r = \{2, 3, \ldots, 8\}$ antennas. A single block of GFDM contains $M = 7$ subsymbols, and it is filtered by an RC pulse with roll-off factor $\alpha = 0.3$. For comparison purpose, we configure OFDM to have $K' = MK$ subcarriers. Assuming both signals have an identical bandwidth, the subcarrier spacing of GFDM becomes M times broader with respect to the OFDM one and therefore, each GFDM subcarrier consists of M bins while OFDM has a single frequency bin per subcarrier.

Figure 4.18 illustrates the MSE evaluations for theoretical analysis as well as simulation results. Here, each channel is chosen to have $L = 9$ taps with exponential PDP. As expected from the theoretical expressions, the channel estimation for GFDM contains an error floor due to the interference from data symbols while for OFDM, the MSE decreases linearly with the increase of the SNR. Moreover, comparing the

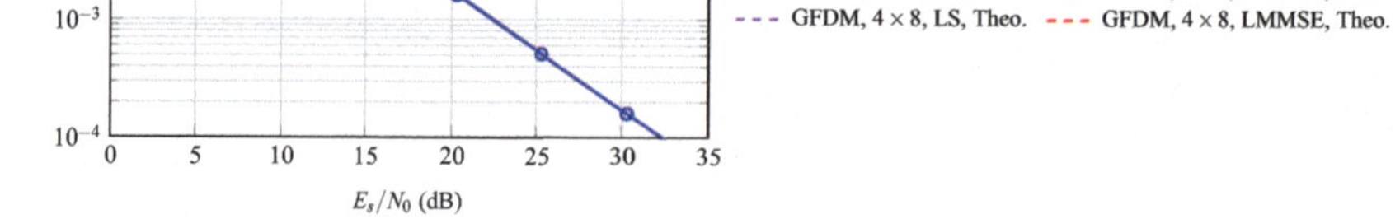

Fig. 4.18 MSE results of channel estimation versus E_s/N_0 for simulation and theoretical calculations in Rayleigh fading MIMO channel with a pilot spacing of $\Delta k = 2$ and $K = 128$ subcarriers

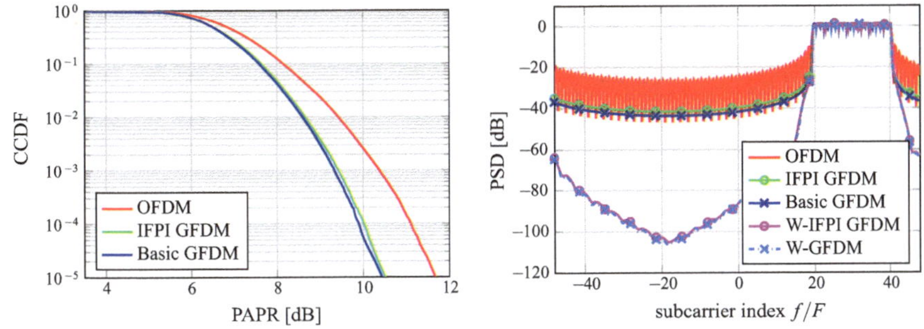

(a) Signal PAPR for $M = 21$, $K = 4$ vs. its CCDF

(b) OOB for $K = 96$, $M = 15$ and the total number of frequency samples F.

Fig. 4.19 Transmitted signal characteristics

GFDM channel estimation results for various number of Tx and Rx antennas, we notice that the error does not directly depend on the number of receive antennas, e.g., the MSE curves for 2×2 versus 2×8 antennas are overlapped (as well as 4×4 vs. 4×8). This is due to the fact that, by linearly increasing the number of Rx antennas we increase the number of observations while the number of estimation parameters (i.e., channel taps) also increases linearly, e.g., doubling the number of Rx antennas, we also double the number of channel taps while their ratio remains identical. As a consequence, no analytical difference should be expected in this case. On the other hand, as we increase the number of Tx-Rx antennas, the estimation performance for both LS and LMMSE estimators degrades, because by linearly increasing the number of Tx-Rx antennas, the number of channel taps increases quadratically and thus, the estimation performance degrades.

The PAPR of the IFPI GFDM is compared to the original GFDM beside OFDM in Fig. 4.19a. One can see that due to orthogonal pilot insertion, the PAPR of IFPI GFDM increases with respect to the basic GFDM. However, it still has more than one dB difference with the PAPR of an OFDM signal. On the other hand, comparing the power spectral densities of the signals in Fig. 4.19b, we observe that, despite IFPI GFDM has slightly larger OOB compared to the original GFDM signal, the windowed case achieves almost the same OOB radiation as in original windowed GFDM (W-GFDM). The window function is configured in form of an RC window with a ramp length of a quarter subsymbol. For further details regarding the windowing process, we refer the interested readers to [1].

The coded performances of the three receiver types are provided in Fig. 4.20. Here, a GFDM block has $M = 7$ subsymbols and $K = 96$ subcarriers. The received signal constellations are detected via GFDM zero forcing demodulation and MMSE frequency domain channel equalization. The channel codes are chosen as parallel concatenated convolutional codes (PCCCs) (1, 15/13), and they provide a gain in spectral efficiency leading to energy per bit $E_b/N_0 = E_s/N_0 - 10\log_{10}(\mu r)$ where μ and r denote the modulation order and the code rate, respectively. The detected data

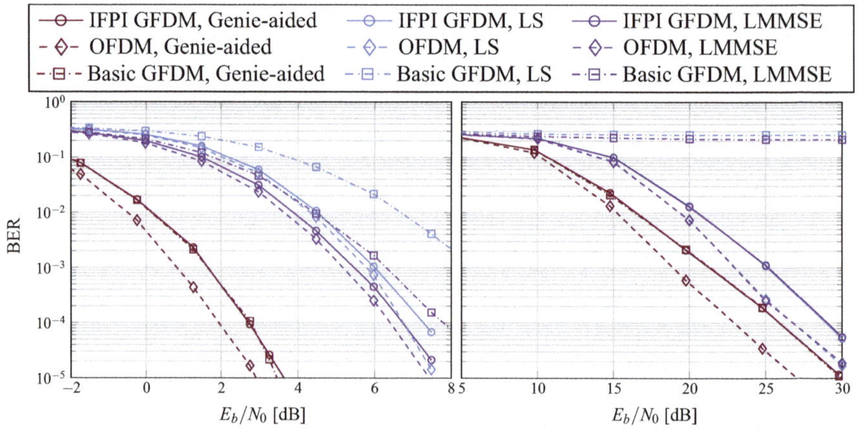

(a) 1/3 code-rate (b) 5/6 code-rate

Fig. 4.20 Bit error rate performance with 5% pilots overhead over 2×2 MIMO channel ($M = 7, K = 96$)

symbols are transferred into maximum likelihood (ML) symbol log-likelihoods and they are inserted into the soft demapper with 8 turbo decoder iterations. In Fig. 4.20, the channel estimation performance for IFPI GFDM and OFDM is identical, although the basic pilot insertion for GFDM has an error floor at high SNR regions as was shown before.

Note that in Fig. 4.20, employing a robust code rate of 1/3, OFDM, IFPI GFDM, and basic GFDM receivers obtain almost similar BER, though, basic GFDM with LS estimation has 2–3 dB worse BER performance than the rest of receivers with imperfect channel knowledge. Here, due to around 1 dB gap of the genie-aided receivers of OFDM and IFPI GFDM, the latter receiver stays around 0.5 dB behind OFDM when the channel is estimated through pilot transmission. On the other hand, Fig. 4.20b shows that the basic GFDM channel estimation with non-orthogonal pilots has appreciable performance loss for a high code rate of 5/6 which is due to its large error floor in the channel estimation. Comparing OFDM and IFPI GFDM, we observe that the performance loss in GFDM which is a non-orthogonal waveform is not significant compared to OFDM. Furthermore, in Fig. 4.20b the BER for LS and LMMSE estimations in OFDM as well as IFPI GFDM are identical due to identical channel estimation performances at high SNR regions.

In short, if the SNR region is low and a robust code rate is utilized, with normal pilot insertion in GFDM the performance has slight degradation in comparison with OFDM. Although for higher SNR regions when faster code rates are in favor, it is necessary to insert interference-free pilots in order to fully exploit the capacity of the GFDM system while the advantages of its signal characteristics are also preserved.

4.1.8 Transmission Diversity for GFDM

Robustness against time-variant and frequency-selective channel is an important feature for the fifth-generation (5G) networks. Transmit diversity can be exploited to improve coverage, reduce the need for retransmissions, and improve the reliability of the system. Although the space-time coding (STC) as proposed in [13] can be applied to GFDM with the help of wide linear equalization, the complexity of the receiver can hinder its application in some 5G scenarios, such as Internet of Things (IoT) and machine-type communication (MTC).

A simple and elegant solution for this issue is based on employing the time-reversal space-time coding (TR-STC) [14] applied to GFDM in order to achieve maximum diversity gain and low implementation complexity without any performance loss. In this case, two antennas are used to transmit two subsequent GFDM blocks $x_i[n]$ and $x_{i+1}[n]$, building the STC codeword as

$$\begin{array}{l|cc} & \text{Antenna 1} & \text{Antenna 2} \\ \hline \text{Block } i & x_i[n] & -x_{i+1}^*[- < n >_N] \\ \text{Block } i+1 & x_{i+1}[n] & x_i^*[- < n >_N] \end{array}. \tag{4.116}$$

After CP removal, the received signals at the j-th receiving antenna, in the frequency domain, for the i-th and $(i+1)$-th time instants are given by

$$\begin{aligned} \tilde{\mathbf{y}}_{i,i_r} &= \mathbf{D}^{(\tilde{h})}_{i_r,1}\tilde{\mathbf{x}}_i - \mathbf{D}^{(\tilde{h})}_{i_r,2}\tilde{\mathbf{x}}^*_{i+1} + \tilde{\mathbf{w}}_{i,i_r} \\ \tilde{\mathbf{y}}_{i+1,i_r} &= \mathbf{D}^{(\tilde{h})}_{i_r,1}\tilde{\mathbf{x}}_{i+1} + \mathbf{D}^{(\tilde{h})}_{i_r,2}\tilde{\mathbf{x}}^*_i + \tilde{\mathbf{w}}_{i+1,i_r}, \end{aligned} \tag{4.117}$$

where $\mathbf{D}^{(\tilde{h})}_{i_r,i_t} = \text{diag}(\tilde{\mathbf{h}}_{i_r,i_t})$ with $\tilde{\mathbf{h}}_{i_r,i_t} = \mathbf{F}_N\mathbf{h}_{i_r,i_t}$, $\tilde{\mathbf{x}}_i = \mathbf{F}_N\mathbf{x}_i$ and $\tilde{\mathbf{w}}_{i,i_r}$ is the noise vector on the i-th time instant and i_r-th receive antenna. Assuming that N_r receiving antennas are employed by the receiver, the received signals can be combined in the frequency domain as

$$\begin{aligned} \hat{\tilde{\mathbf{x}}}_i &= \mathbf{D}^{(\tilde{h})}_{eq}{}^{-1}\sum_{i_r=1}^{N_r}\mathbf{D}^{(\tilde{h})}_{i_r,1}{}^{*}\tilde{\mathbf{y}}_{i,i_r} + \mathbf{D}^{(\tilde{h})}_{i_r,2}\tilde{\mathbf{y}}^*_{i+1,i_r} \\ \hat{\tilde{\mathbf{x}}}_{i+1} &= \mathbf{D}^{(\tilde{h})}_{eq}{}^{-1}\sum_{i_r=1}^{N_r}\mathbf{D}^{(\tilde{h})}_{i_r,1}{}^{*}\tilde{\mathbf{y}}_{i+1,i_r} - \mathbf{D}^{(\tilde{h})}_{i_r,2}\tilde{\mathbf{y}}^*_{i,i_r}, \end{aligned} \tag{4.118}$$

where

$$\mathbf{D}^{(\tilde{h})}_{eq} = \sum_{i_t=1}^{2}\sum_{i_r=1}^{N_r}\mathbf{D}^{(\tilde{h})}_{i_r,i_t}{}^{*}\mathbf{D}^{(\tilde{h})}_{i_r,i_t}. \tag{4.119}$$

The combined GFDM blocks in the time domain are given by

$$\hat{\mathbf{x}}_i = \frac{1}{N}\mathbf{F}_N^{\text{H}}\hat{\tilde{\mathbf{x}}}_i, \tag{4.120}$$

which can be used to recover the data symbols, using the demodulation matrix $\mathbf{B}$, as presented in Sect. 4.1.

TR-STC can achieve full diversity gain of order 2J, which means that the approximated expression for the symbol error rate derived for the maximum-ratio combiner (MRC) can be adapted for the TR-STC-GFDM. Assuming that a V-QAM constellation is used to map the data bits into each subcarrier and a non-orthogonal prototype pulse that leads to an NEF of $\xi = \frac{1}{N}\text{tr}(\mathbf{B}^{\text{H}}\mathbf{B})$ is employed, the symbol error probability for the TR-STC-GFDM is approximately given by

$$p_e \approx 4\mu \sum_{i=0}^{2N_\text{r}-1} \binom{2N_\text{r} - 1 + i}{i} \left(\frac{1+\eta}{2}\right)^i, \tag{4.121}$$

where

$$\mu = \left(\frac{\sqrt{V} - 1}{\sqrt{V}}\right) \left(\frac{1-\eta}{2}\right)^{2J} \quad \text{and} \tag{4.122}$$

$$\eta = \sqrt{\frac{\frac{3\sigma_e^2}{V-1}\frac{E_s}{\xi N_0}}{2 + \frac{3\sigma_e^2}{V-1}\frac{E_s}{\xi N_0}}}, \tag{4.123}$$

with $\sigma_e^2 = \sum_n \text{E}[|h_n|^2]$, E_S and N_0 denote the average symbol energy and the noise power, respectively.

Figure 4.21 shows the TR-STC-GFDM symbol error rate (SER) performance assuming the parameters presented in Table 4.1 and the average channel impulse response based on the Extended Pedestrian A model from LTE, which is described

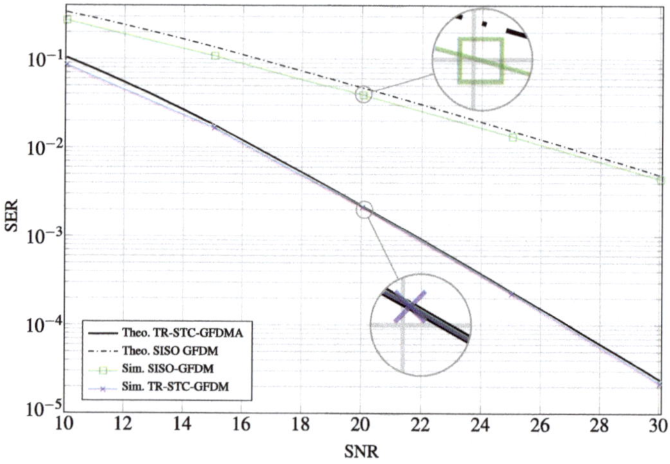

Fig. 4.21 TR-STC-GFDM SER performance under time-variant and frequency-selective channel

Table 4.1 Simulation parameters

Parameter	Symbol	GFDM	OFDM
Mapper	V-QAM	16-QAM	16-QAM
Transmit filter	$g[n]$	RC, $\alpha = 0.25$	Rect
# subcarriers	K	128	128
# subsymbols	M	7	1
CP length [samples]	L_{cp}	9	9
CS length [samples]	L_{cs}	3	3
Detector	–	ZF	ZF
Noise enhancement factor	ξ	1.02	1
# receiving antennas	J	1	1

Table 4.2 Channel power-delay profile used in the simulations

Tap (nth sample)	0	1	2	3	4	5	6
Tap gain h_n (dB)	0	−1	−2	−3	−8	−17.2	−20.8

by Table 4.2. The taps of the channel model are multiplied by i.i.d. complex Gaussian variable with zero mean and unitary variance, resulting in independent Rayleigh fading channels between each transmit and receive antennas. The results presented in Fig. 4.21 assumes two different situations. The first one considers SISO, where just one transmit antenna is employed to send data to the receiver. In the second scenario, both transmit antennas are active, providing full diversity gain. It is also assumed that the receiver knows the state information of all channels. We can see from Fig. 4.21 that the approximation presented in (4.121) can be used to predict the TR-STC-GFDM under time-variant frequency-selective channels. Also, it is possible to conclude that the TR-STC is able to provide full diversity gain to GFDM, introducing a considerable SER performance gain when compared to the SISO-GFDM.

Therefore, the high SER performance gain introduced by the TR-STC comes with a very small complexity increment on the receiver side, once the received signals can be easily combined in the frequency domain.

4.2 Link-Level Waveform Comparison

In this section, we compare the link-level performance of advanced multicarrier waveforms under MIMO wireless communication channels. The baseline waveform is CP-OFDM. In the last decade, it has evolved as a popular multicarrier scheme in different standards, including 3GPP LTE and Wi-Fi families. However, with new and even more stringent requirements in 5G and beyond, OFDM faces its limitations, such as sensitivity to time-frequency misalignments, high OOB emission, limited flexibility, and high PAPR [15, 16]. To overcome these limitations, advanced alternatives have been intensively investigated in recent years. Two groups will be examined in this section.

One group of waveforms attempts to improve OFDM while mostly keeping its orthogonality. Filtered OFDM (F-OFDM) [17] linearly filters a set of contiguous subcarriers that form a subband. It is evident that filtering is effective to limit the OOB emission. On the other hand, the presence of inter-block-interference (IBI) is a new issue as the linear filter tail will spread outside the duration of each OFDM block. One pragmatic solution is to insert one or several guard tones (GTs) between adjacent subbands. A widened subband helps reducing the filter length, thereby alleviating IBI. Another approach is to use ZP instead of CP as ZP has zero energy and minimizes IBI. This yields the second waveform in this group, namely universal-filtered OFDM (UF-OFDM) [18].

The other group of waveforms, consisting of filter bank multicarrier (FBMC) [19] and GFDM [1], completely discards the orthogonality requirement of OFDM to achieve better temporal and spectral characteristics. Comparing with OFDM and its variants in the first group, FBMC and GFDM have three core different features that are in common. They are: (1) filtering on a subcarrier basis, (2) permission of more than one data symbol per subcarrier, and (3) being subject to ISI and ICI arising from their non-orthogonality. Between FBMC and GFDM, they both also have several distinct features.[3] FBMC adopts linear filtering to achieve ultra-low OOB emission [19, 20]. On the other hand, the long filter length makes it more suitable for continuous rather than burst transmission, considering the usage efficiency of time resources. FBMC does not use a CP and relies on the soft transition of its filter tail to combat multipath fading. In GFDM, a unique feature initially adopted by it is circular filtering. This ensures a block-based waveform with no filter tails, but at the cost of an increased OOB emission. CP has been suggested as a default setting for GFDM, but a single one can protect multiple data symbols for the sake of temporal efficiency.

The self-introduced interference of non-orthogonal waveforms often results in much-increased receiver complexity, e.g., [21–23] and references therein. In MIMO communications, IAI will additionally takes place. To tackle such three-dimensional interference, we contributed an innovative way in [24] to perform MMSE equalization such that they can be jointly resolved with complexity in the same polynomial order as that of the (quasi-)orthogonal waveforms in the first group.

The focus of this section is to study the link-level performance of the above-mentioned waveforms, including OOB emission, PAPR and coded frame error rate (FER) achieved by the MMSE receiver mentioned before. Challenging channel conditions in terms of large delay spread and time-varying fading with imperfect synchronization and channel estimation are under the consideration.

[3] Here we consider the features that were suggested when FBMC and GFDM were invented. As the recent progress in waveform engineering, these features start becoming mutually usable. Therefore, they may no longer be regarded as distinct.

4.2.1 System Configurations

Taking OFDM as the baseline, we use K to denote the total number of subcarriers, where the indices of the active ones are recorded in the set $\mathscr{K}_{\text{on}}$. The symbol duration without CP and ZP is denoted as T. At the sampling rate K/T, the CP and ZP respectively contain L_{cp} and L_{zp} samples. Here we assume each frame carries a codeword, producing N_s OFDM blocks per frame. For the linear filter adopted by F-OFDM and UF-OFDM, we use L_f to denote its length that is normalized by the sampling period T/K. Since both GFDM and FBMC permit multiple data symbols per subcarrier, let M denote this number and T be the time spacing between two consecutive data symbols.

Table 4.3 summarizes the representative configurations of the waveform candidates, deriving from the baseline CP-OFDM. Their configurations attempt to fulfill the constraint that each frame uses the same bandwidth and carries the same number of data symbols.[4]

Exploiting the additional degree of freedom in the time domain, we particularly examine two configuration types of GFDM and FBMC. The first type sets the time spacing to be the same as the duration of the OFDM symbol, i.e., yielding identical subcarrier spacing $1/T$. Each frame consists of a single block to optimize the temporal efficiency. On contrary, the second type makes the subcarrier spacing of GFDM and FBMC M times wider than that of OFDM, i.e., M times shorter time spacing between two consecutive data symbols. In doing so GFDM-II has the same block length as OFDM, thereby yielding the same number of blocks per frame. For the sake of temporal efficiency, FBMC-II still has one block per framework, resulting M times more data symbols than FBMC-I per subcarrier.[5]

Now, let us set the parameters of the baseline CP-OFDM as: $K = 1536$, $|\mathscr{K}_{\text{on}}| = 36$, $T = 66.67\,\mu\text{s}$, and $N_s = 7$. With the subcarrier spacing $T^{-1} = 15\,\text{kHz}$, the occupied subband out of the total 23.04 MHz band has about 0.5 MHz bandwidth. In accordance with Table 4.3, the corresponding configurations for the other waveforms can be readily computed. For the type-II configuration of GFDM and FBMC in Table 4.3, we additionally set $M = 12$. Furthermore, following the suggestions in the literature, the filters adopted by the waveform candidates except CP-OFDM are set as follows. UF-OFDM adopts the Dolph–Chebyshev filter with length $L_f = 74$ and the side-lobe attenuation $-51\,\text{dB}$ [25]. The filter used by F-OFDM is a Hanning windowed sinc-function with length $L_f = K/2 + 1$ [26]. The PHYDYAS filter of FBMC has the longest filter length equal to 4 K [19]. GFDM adopts a periodic RC function with the roll-off factor $\alpha = 1$.

---

[4]Due to different choices of the filter, it is difficult to achieve the same frame duration without violating the bandwidth constraint. Considering the strict regulation on the spectrum, identical bandwidth is our primary constraint.

[5]For FBMC, the guard time interval to accommodate the filter tail between blocks can be too large. Therefore, in both configuration types, we only consider one block per framework to ensure a good temporal efficiency.

Table 4.3 Frame parameterization and modulation complexity

Waveform	Nr. dat. symbs per subcar.	Nr. blks per frame	Nr. subcar. per blk.	Nr. act. subcar.	Subcar. spacing.	Sampling rate	Frame length	Arithmetic complexity per data symb.
CP-OFDM	–	N_s	K	$\lvert\mathscr{K}_{on}\rvert$	$1/T$	K/T	$\frac{(K+L_{cp})N_s}{K}T$	$O\left(\frac{K}{\lvert\mathscr{K}_{on}\rvert}\log K\right)$
(F/UF)OFDM	–	N_s	K	$\lvert\mathscr{K}_{on}\rvert$	$1/T$	K/T	$\frac{(K+L_{cp})N_s+L_f-1}{K}T$	$O\left(\frac{K}{\lvert\mathscr{K}_{on}\rvert}\log K+\frac{KL_f}{\lvert\mathscr{K}_{on}\rvert}\right)$
GFDM-I	N_s	1	K	$\lvert\mathscr{K}_{on}\rvert$	$1/T$	K/T	$(N_s+\frac{L_{cp}}{K})T$	$O\left(\frac{K}{\lvert\mathscr{K}_{on}\rvert}\log K+\frac{KM}{\lvert\mathscr{K}_{on}\rvert}\right)$
FBMC-I	N_s	1	K	$\lvert\mathscr{K}_{on}\rvert$	$1/T$	K/T	$(N_s+\frac{7}{2})T$	$O\left(\frac{K}{\lvert\mathscr{K}_{on}\rvert}\log K+\frac{4K}{\lvert\mathscr{K}_{on}\rvert}\right)$
GFDM-II	M	N_s	K/M	$\lvert\mathscr{K}_{on}\rvert/M$	M/T	K/T	$\frac{(K+L_{cp})N_s}{K}T$	$O\left(\frac{K}{\lvert\mathscr{K}_{on}\rvert}\log(K/M)+\frac{KM}{\lvert\mathscr{K}_{on}\rvert}\right)$
FBMC-II	N_sM	1	K/M	$\lvert\mathscr{K}_{on}\rvert/M$	M/T	K/T	$(N_s+\frac{7}{2M})T$	$O\left(\frac{K}{\lvert\mathscr{K}_{on}\rvert}\log(K/M)+\frac{4K}{\lvert\mathscr{K}_{on}\rvert}\right)$

For UF-OFDM, its L_{zp} takes on the same value as L_{cp}
$O(1)$ represents the arithmetic complexity per multiplication

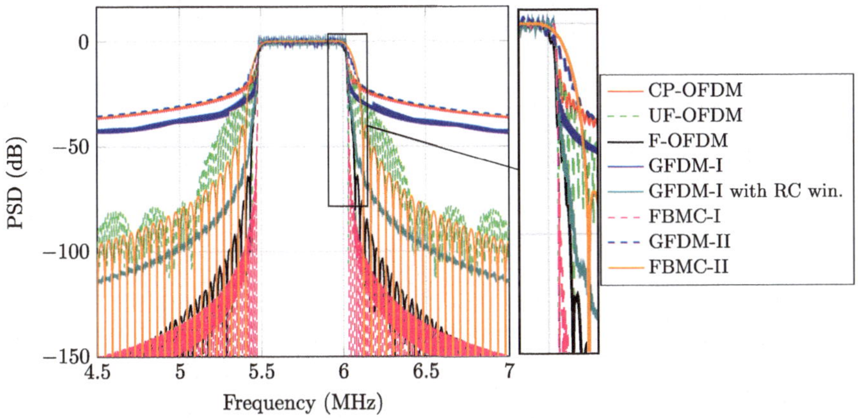

Fig. 4.22 Power spectral densities (PSDs) of the waveform candidates in their baseband signal form (per transmit antenna)

4.2.2 OOB Emission

Figure 4.22 depicts the PSD of the waveform candidates in their baseband signal form. Note that the impairment from the RF front-end is not considered here. As expected, the OOB emission of CP-OFDM, namely outside the allocated subband ranging from 5.5 to 6 MHz, is very high, because of the disruptive change from one OFDM block to another in the time domain. UF-OFDM, F-OFDM, and FBMC rely on linear filtering to smoothen the transition between blocks, thereby achieving lower OOB emission. The longer the filter is, the lower is the achieved OOB emission. For GFDM, circular filtering however keeps the disruptive change between blocks. GFDM-I achieves slightly improved OOB emission performance by having longer block duration and reducing the number of blocks per frame. On the other hand, GFDM-II with the same number of blocks per frame as OFDM achieves nearly identical OOB emission performance. Note that the soft transition of its PSD on the shoulder of the occupied spectrum is due to (1) its wider subcarrier spacing than CP-OFDM; and (2) the large roll-off factor. Such soft transition also appears with FBMC-II for the same reasons.

Compared to linear filtering, time domain windowing can be an attractive solution to reduce the OOB emission of block-based waveforms as well. GFDM-I uses the minimum time resource among all evaluated waveforms, even $6L_{\mathrm{cp}}$ shorter than the baseline CP-OFDM. Targeting the same temporal efficiency, we can extend the CP of GFDM-I by $3L_{\mathrm{cp}}$ samples and add a CS with identical length to window each GFDM block without impairing the data transmission plus one CP to combat the multipath fading channel. Here, the time domain window takes the frequency domain expression of an RC function, whose ramp up and down are contained within the extended part of the CP and CS. Figure 4.22 shows that such RC windowing is very efficient in reducing the OOB emission of GFDM-I, making it competent with

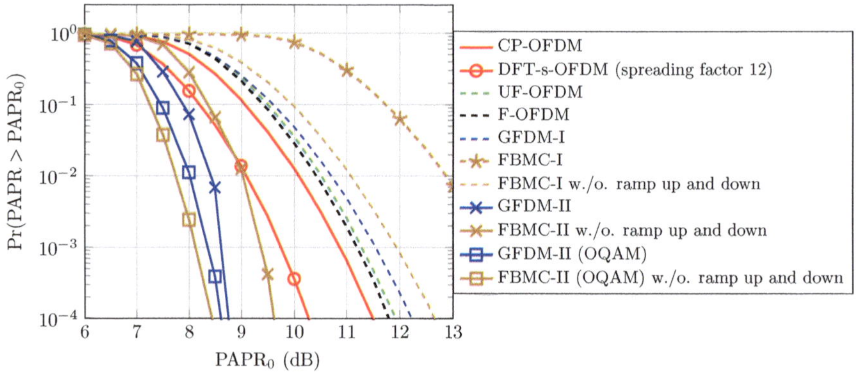

Fig. 4.23 PAPRs of the waveform candidates in their baseband signal form (per transmit antenna), where the CCDF is empirically constructed from 10^6 frames and the oversampling factor is 4. The default modulation scheme is 16-QAM (gray). The explicit label *OQAM* represents offset-16-QAM (gray)

the outer waveforms using linear filters. Besides filtering and windowing, it is also possible to reduce the OOB emission via preprocessing the transmitted data symbols, e.g., [27–29], but it is beyond the scope of our discussion here.

4.2.3 PAPR

Figure 4.23 depicts the PAPRs achieved by the waveforms. For those using linear filter, we note that the ramp up and down of the filtered signal reduce the average power without affecting the peak power. This causes the PAPR increment of UF-OFDM, F-OFDM, and FBMC-I compared to CP-OFDM. It is more noticeable when the filter length is longer. Note that such PAPR increment will not impose additional challenge on designing the power amplifier. So, it is not a concern. Since FBMC has the longest linear filter length among all waveforms, we compute its PAPR of FBMC excluding its ramp up and down phase on purpose.

For GFDM and FBMC, their additional degree of freedom in the time domain can be leveraged to improve the PAPR performance. By having fewer subcarriers and more data symbols per subcarrier, the configuration type-II achieves much lower PAPRs than the type-I. Figure 4.23 also shows that offset quadrature amplitude modulation (OQAM) is beneficial to further reduce the PAPR of GFDM-II and FBMC-II. If necessary, other PAPR reduction techniques, such as tone reservation and active constellation extension, are applicable on top of these waveforms.

4.2.4 FER Under a Doubly Dispersive Channel

In this following, the waveforms are evaluated in a 4×4 spatial multiplexing MIMO system with spatially uncorrelated multipath Rayleigh fading channels. On top of the waveforms, two modulation coding schemes (MCSs) are applied. Namely, the turbo code with the generator polynomial $\{1, 15/13\}_o$ can operate at rate 1/2 and 3/4,[6] which are respectively modulated with 16- and 64-QAM (gray). Unless otherwise stated, QAM is the default choice. Its comparison with OQAM is always under the same modulation order. The metric E_s/N_0 denotes the energy per data symbol to noise ratio.

As for the channel model, we choose the extended typical urban (ETU) model specified by 3GPP and with the total power of the path gains normalized to one. Given its large delay spread, we accordingly choose the long CP mode in LTE, specifically, $L_{cp} = 16.67\,\mu s$. Following the Jakes' model, the maximum Doppler frequency f_d reflects the channel varying rate. For coherently equalizing each block, the used CIR is obtained by averaging the continuously time-variant CIR over each block duration. Additionally, we assume perfect synchronization.

Generally speaking, the time-varying channel can affect the FER performance from two conflicting aspects. First, a continuous time-varying channel introduces ICI that increases along with the maximum Doppler frequency. Second, the time selectivity across the blocks is desirable for the decoder to exploit the code diversity in order to improve the decoding performance. When the additive noise dominates over ICI, the second aspect plays the determinant role. Therefore, lower FERs are attained at a higher maximum Doppler frequency, e.g., CP-OFDM, UF-OFDM, F-OFDM, and GFDM-II in Fig. 4.24a, c. However, when the ICI becomes the dominant factor, we observe higher FERs as the maximum Doppler frequency increases, e.g., in Fig. 4.24b, d.

For a similar reason, the IBI introduced by linear filtering of UF-OFDM and F-OFDM becomes particularly harmful at the higher MCS, which requires higher operating E_s/N_0 for a satisfactory FER performance. UF-OFDM benefits from the use of ZP and a shorter filter to suffer from less IBI than F-OFDM. As mentioned at the beginning of this section, we can alleviate the IBI issue encountered by F-OFDM through GT insertion. Figure 4.24b, d depict the performance achieved by having one GT on each side of the subband, which however costs spectral efficiency, i.e., $(2/38) \approx 5\%$ loss.

For FBMC and GFDM-I, they suffer from the need of a long block length. With long blocks, the time-variant CIR cannot be well approximated by its average value and the resulting mismatched channel knowledge can severely degrade the performance of equalization and subsequent decoding. Therefore, they perform poorly with $f_d = 300\,\text{Hz}$. Even at a lower maximum Doppler frequency $f_d = 70\,\text{Hz}$, such impairment is non-negligible at high SNRs, i.e., Fig. 4.24b.

---

[6]For every six information bits input to the turbo code, we keep all information bits plus two parity bits respectively generated by the two identical component convolutional codes.

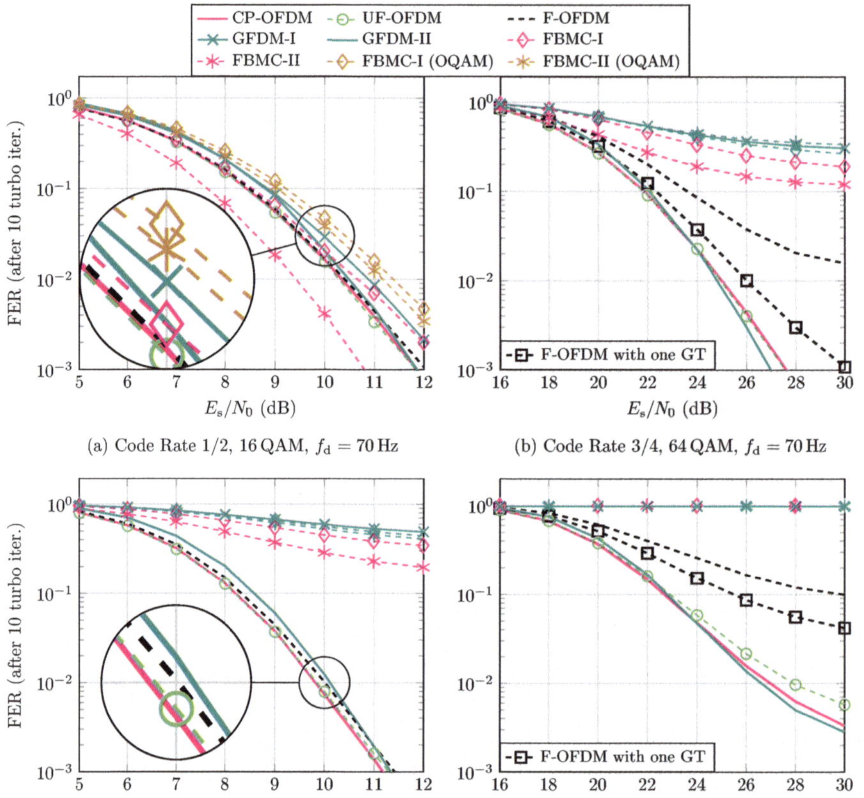

(a) Code Rate 1/2, 16 QAM, $f_d = 70$ Hz (b) Code Rate 3/4, 64 QAM, $f_d = 70$ Hz

(c) Code Rate 1/2, 16 QAM, $f_d = 300$ Hz (d) Code Rate 3/4, 64 QAM, $f_d = 300$ Hz

Fig. 4.24 FERs achieved by the waveforms under perfect synchronization and channel knowledge

Only in Fig. 4.24a, the benefit of FBMC becomes appreciable. Non-orthogonality not only introduces interference, but also spreads the information of each data symbol over more than one channel observations in the frequency domain. Particularly, the type-II has larger subcarrier spacing than the type-I to ensure the necessary frequency selectivity among the channel observations. Analogous to FBMC-II, GFDM-II is also equipped with such feature. Between them, FBMC-II in this case has more channel observations per data symbol to achieve a lower FER. Between GFDM-II and FBMC-II, the former permits short block lengths without considerably reducing the temporal efficiency. Therefore, it outperforms the latter in higher mobility case, e.g., Fig. 4.24c, d.

Last but not least, as shown in our work [24], QAM can more efficiently exploit the frequency selectivity of the channel than OQAM, therefore outperforming in Fig. 4.24.

4.2.5 FER with Imperfect Synchronization and Channel Estimation

This part investigates the impact of synchronization and channel estimation error on the FER performance of the waveforms. The channel is generated by following another 3GPP channel model termed extended vehicular A model (EVA) with the maximum Doppler frequency equal to 30 Hz and with the sum of the average path gains normalized to one. Due to the reduced maximum delay spread, the CP length accordingly decreases to 4.69 µs, namely the normal mode in LTE.

For data-aided channel estimation, we insert a preamble consisting of one baseline CP-OFDM block before each payload frame. It consists of N_t orthogonal pilot vectors that are periodically modulated onto the subcarriers belonging to the occupied subband plus N_t subcarriers on each side for achieving sufficient channel estimation quality also on the edge of the subband. Using such preamble, the receiver performs LMMSE channel estimation by assuming a uniform power-delay profile with maximum delay length equal to L_{cp} [30]. The obtained channel estimates will be used as the true one for coherently equalizing the whole frame. One important reason behind this setup is to assure that the MMSE equalizer of each waveform works with the same quality of channel knowledge.

Figure 4.25a shows that FBMC-II and GFDM-II both are robust against the channel estimation error, achieving up to 5 dB gain in comparison with CP-OFDM and its variants. This observation indicates that it is possible to harness the benefits of the waveform-induced interference instead of only suffering from it. To this end, we

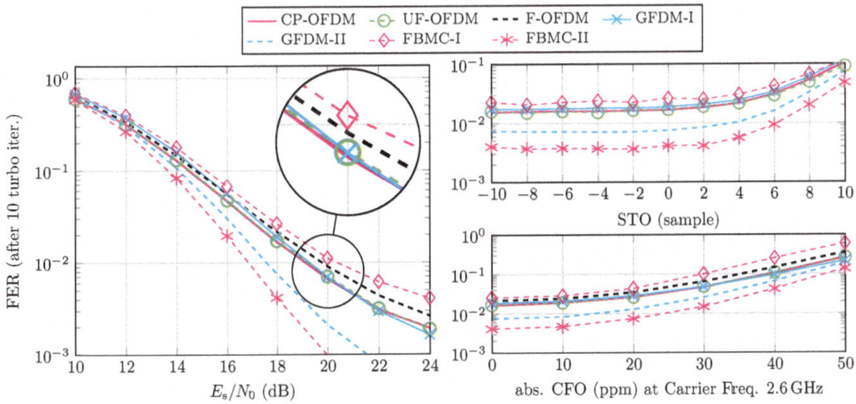

(a) Perf. Sync. and Imperf. Channel Knowledge

(b) Sensitivity to STO and CFO at $E_s/N_0 = 18$ dB and with Imperf. Channel Knowledge

Fig. 4.25 FERs achieved by the waveforms with the maximum Doppler frequency 30 Hz and relying on imperfect synchronization and channel knowledge, where the 16 QAM and code rate 1/2 are the default MCS and the contrastive OQAM has the same modulation order need to exploit the data information conveyed by the interference rather than treating it as a part of the background white noise.

We further investigate the sensitivity of the waveforms against symbol time offset (STO) and carrier frequency offset (CFO), respectively. Specifically, the use of guard intervals, no matter in the form of CP, ZP with overlap-and-add or soft termination of the filter, provides protection against negative STO estimates, i.e., the estimated frame arrival being earlier than the true one. The performance degradation appears once we have a positive STO estimate. How severe the degradation is depends on the power of the first non-negligible paths of the channel in its discrete-time model.[7] From Fig. 4.25b, we can infer that the initial 4 paths of the channel are insignificant. According to Fig. 4.25c, the waveforms can work under the CFO up to ± 20 ppm. For FBMC-II and GFDM-II, even with ± 30 ppm CFO, their performances are similar to that of OFDM working with ± 20 ppm CFO.

4.2.6 Section Summary

In this section, we have analyzed the link-level performance of advanced waveforms that are being intensively researched as alternatives to CP-OFDM for future systems. There is no single waveform that can outperform the others in all examined aspects, i.e., OOB emission, PAPR and FER under different channel conditions.

By observing the 3GPP RAN1 discussion through publicly available materials, the waveforms targeted by this section have all appeared in the proposals from different organizations. At this moment, the OFDM-based waveforms are more interested and supported by the main industrial players to ensure a good backward capability. Nevertheless, non-orthogonal waveforms, i.e., GFDM and FBMC, are definitely worth investigation. We believe their benefits can be exploited with a complexity that is affordable by today's hardware. Furthermore, non-orthogonality is not necessary to be a curse in the system design. Further research on non-orthogonal waveforms is no doubt valuable for a wide range of communications systems using multicarrier waveforms, including but not limited to mobile systems.

4.3 Multiple Access with GFDM

In multicarrier-based multiple access, the time and frequency resources are distributed among users. The basic resource element corresponds to transmitting one data symbol $d_{k,m}$ per block using the pulse shape $g^{t}_{k,m}[n]$. However, in practice, the

[7] Given the power-delay profile, the discrete-time channel model is obtained by sampling the low-pass filtered CIR, where the bandwidth equals the sampling rate. The discrete-time model very often have more resolvable paths than the power-delay profile. This is because the delays specified by the power-delay profile are not integer multiples of the sampling period.

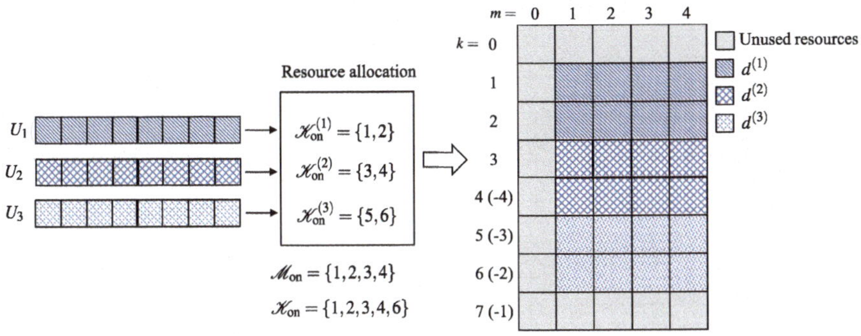

Fig. 4.26 GFDM resource allocation. In this example, $M = 5$, $K = 8$, $M_u = 4$ and $K_u = 2$

smallest physical resource block (PRB) consists of a number of basic resources and several blocks. For example, in LTE standard, which uses OFDM in the physical layer (PHY), the smallest PRB consists of 12 subcarriers and 7 or 6 OFDM symbols, for long and short CP, respectively. In addition, the scheduling, where the resources can be reassigned to different users, takes place after certain time, for example, in LTE after 12 or 14 symbols [31].

In GFDM, the resource allocation is done by setting the entries of the data matrix $\mathbf{D}$ corresponding to the allocated $\{(k, m)\}$ pairs and the other entries are set to zero. Let $\mathcal{K}_{\text{on}}^{(u)} \times \mathcal{M}_{\text{on}}^{(u)} \subset \mathcal{K}_{\text{on}} \times \mathcal{M}_{\text{on}}$ be the set of allocated $\{(k, m)\}$ to the u-th user. Practically, we fix $\mathcal{M}_{\text{on}}^{(u)} = \mathcal{M}_{\text{on}}$ and vary $\mathcal{K}_{\text{on}}^{(u)}$. Therefore, we define a PRB that have $M_u = |\mathcal{M}_{\text{on}}|$ subsymbols and K_u subcarriers. The scheduling can take place after certain number of GFDM blocks. Figure 4.26 shows an example of GFDM resource allocation.

4.3.1 Signal Model

We consider a network that consists of a base station (BS) and U users. In the downlink (DL), as seen in Fig. 4.27, the BS multiplexes the data symbols of all users in a data vector $\mathbf{d} = \text{vec}\{\mathbf{D}\}$ using the set of indexes $\mathcal{N}_{\text{on}}^{(u)} = \{n = k + mK,\ (k, m) \in \mathcal{K}_{\text{on}}^{(u)} \times \mathcal{M}_{\text{on}}^{(u)}\}$, with $\mathbf{d}^{(u)} = [\mathbf{d}]_{\left(\mathcal{N}_{\text{on}}^{(u)}\right)}$, and generates the modulated signal as discussed in Sect. 4.1.6.2. Each user receives the multiplexed signal given by

$$y^{(u)}[n] = e^{j2\pi f_u n} h^{(u)}[n] * x^{(u)}[n] + v[n], \tag{4.124}$$

where $h^{(u)}[n]$ is the fading channel and f_u represents the CFO between the u-th user and the BS. The STO is implicitly included in $h^{(u)}[n]$. Under perfect synchronization, i.e., f_u is perfectly estimated and the first channel tap is detected. By employing a sufficiently long CP, we get the following signal model after block demultiplexing,

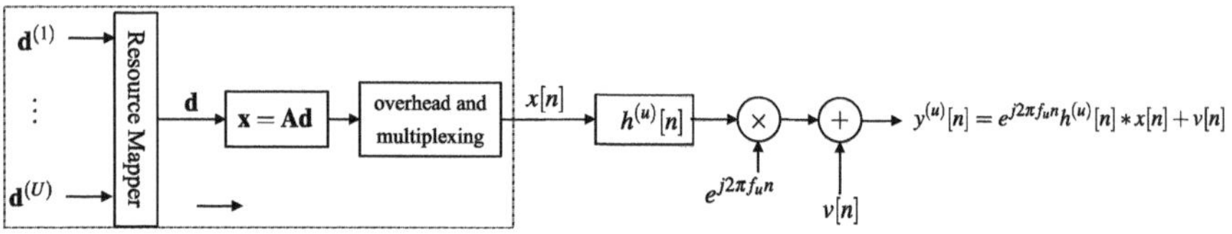

Fig. 4.27 DL signal model

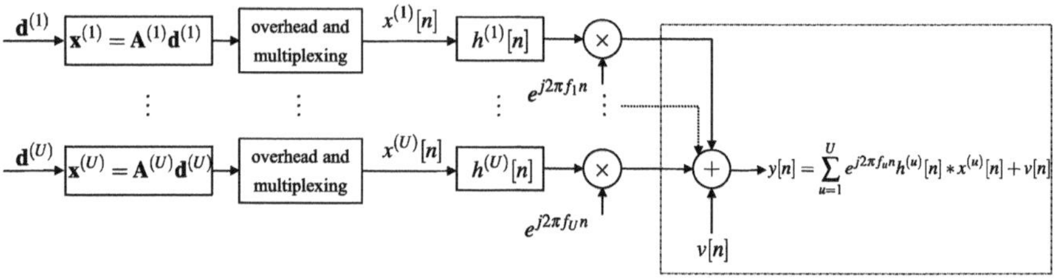

Fig. 4.28 UL signal model

$$\mathbf{y}^{(u)} = \mathbf{H}^{(u)}\mathbf{A}\mathbf{d} + \mathbf{v}, \tag{4.125}$$

where $\mathbf{H}^{(u)}$ is the circular channel matrix. This model can be used directly in a simple receiver, where the u-th user demodulates $\mathbf{y}^{(u)}$ to get $\hat{\mathbf{d}}$ and then applies resource demapping to extract its allocated data symbols. More advanced receiver can work on the model,

$$\mathbf{y}^{(u)} = \mathbf{H}^{(u)}\mathbf{A}^{(u)}\mathbf{d}^{(u)} + \mathbf{H}^{(u)} \sum_{v=1, v\neq u}^{U} \mathbf{A}^{(v)}\mathbf{d}^{(v)} + \mathbf{v}. \tag{4.126}$$

Here, the second term corresponds to the inter-user-interference (IUI). Actually, the interference in this case is inherited from the self-interference of the matrix $\mathbf{A}$. Thus, the study of the DL performance is similar to that of point-to-point link.

In the uplink (UL), Fig. 4.28, each user generates its modulated signal using the truncated modulation matrix $\mathbf{A}^{(u)} = [\mathbf{A}]_{\left(:,\mathcal{N}_{\text{on}}^{(u)}\right)}$, as discussed in Sect. 4.1.2.1. The BS receives a superposition of the signals from all users, which is expressed as

$$y[n] = \sum_{u=1}^{U} e^{j2\pi f_u n} h^{(u)}[n] * x^{(u)}[n] + v[n]. \tag{4.127}$$

In the UL scenario, we distinguish two multiple access (MA) schemes

- Synchronous MA: the users are strictly synchronized with the BS, i.e., $f_u = 0$ and no time offset, which can be achieved via closed-loop synchronization, then

$$\mathbf{y} = \mathbf{H}^{(u)}\mathbf{A}^{(u)}\mathbf{d}^{(u)} + \sum_{v=1,v\neq u}^{U} \mathbf{H}^{(v)}\mathbf{A}^{(v)}\mathbf{d}^{(v)} + \mathbf{v}. \tag{4.128}$$

In this case, the IUI is inherited from the self-interference properties of the GFDM waveform. For example, when the self-interference is limited to adjacent subcarriers, the IUI can be null if a sufficiently large guard subcarrier is used between the adjacent users.

- Asynchronous MA: the synchronization is coarse that there is a remaining CFO modeled by f_u and remaining STO which leads to the increase of the channel delay spread [32]. Nevertheless, the CP can be extended to take into account the maximum possible STO in additional to the channel excess delay. With that, and assuming the BS is able to perfectly estimate f_u and $\mathbf{H}^{(u)}$, we get the signal corresponding to each user after the CFO compensation as

$$\begin{aligned}\mathbf{y}^{(u)} &= \text{diag}\left\{\boldsymbol{\phi}^{(u)}\right\}^H \mathbf{y} \\ &= \mathbf{H}^{(u)}\mathbf{A}^{(u)}\mathbf{d}^{(u)} + \sum_{v=1,v\neq u}^{U} \text{diag}\left\{\boldsymbol{\phi}^{(v,u)}\right\} \mathbf{H}^{(v)}\mathbf{A}^{(v)}\mathbf{d}^{(v)} + \mathbf{v},\end{aligned} \tag{4.129}$$

where $\left[\boldsymbol{\phi}^{(u)}\right]_{(n)} = e^{j2\pi(f_u n + c_u)}$ and $\left[\boldsymbol{\phi}^{(v,u)}\right]_{(n)} = e^{j2\pi(\Delta f_{v,u} n + c_{v,u})}$ are constant phase and $\Delta f_{v,u} = f_v - f_u$ is the relative CFO. In this case, the IUI arises from the relative CFO among users.

In the next subsections, we study the IUI using the frequency domain processing.

4.3.2 *Frequency Domain Processing*

Consider GFDM with prototype pulse shape that has a maximum discrete frequency response within two subcarrier spacing. Without loss of generality, we let

$$\mathbf{V}_{K,M}^{(\tilde{\mathbf{g}})} = \begin{bmatrix} \tilde{\mathbf{g}}_1^T \\ \mathbf{0}_{K-2,M} \\ \tilde{\mathbf{g}}_2^T \end{bmatrix}, \tag{4.130}$$

where $\tilde{\mathbf{g}}_1 = \left[\tilde{\mathbf{g}}\right]_{(0:M-1)} \in \mathbb{C}^{M\times 1}$ and $\tilde{\mathbf{g}}_1 = \left[\tilde{\mathbf{g}}^*\right]_{(N-M:N-1)} \in \mathbb{C}^{M\times 1}$. Recalling (4.31),

$$\left[\mathbf{V}_{K,M}^{(\tilde{\mathbf{x}})}\right]_{(q,p)} = \sum_{m=0}^{K-1} \left[\mathbf{V}_{K,M}^{(\tilde{\mathbf{g}})}\right]_{(<q-k>_K,p)} \sum_{m=0}^{M-1} d_{k,m} e^{-j2\pi \frac{m}{M} p},$$

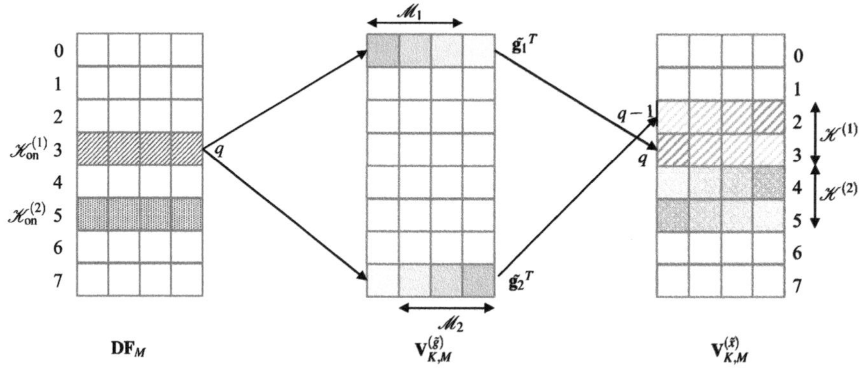

Fig. 4.29 Frequency domain signal processing model

we get

$$\left[\mathbf{V}_{K,M}^{(\tilde{\mathbf{x}})}\right]_{(q,:)} = \sum_{k=0}^{K-1}\left[\mathbf{V}_{K,M}^{(\tilde{\mathbf{g}})}\right]_{(<q-k>_K,:)} \odot \left[[\mathbf{D}]_{(k,:)}\mathbf{F}_M\right]$$
$$= \tilde{\mathbf{g}}_1 \odot \left[\tilde{\mathbf{D}}\right]_{(q,:)} + \tilde{\mathbf{g}}_2 \odot \left[\tilde{\mathbf{D}}\right]_{(<q+1>_K,:)}$$
$$\left[\mathbf{V}_{K,M}^{(\tilde{\mathbf{x}})}\right]_{(q+1,:)} = \tilde{\mathbf{g}}_1 \odot \left[\tilde{\mathbf{D}}\right]_{(q+1,:)} + \tilde{\mathbf{g}}_2 \odot \left[\tilde{\mathbf{D}}\right]_{(<q>_K,:)},$$

where

$$\tilde{\mathbf{D}} = \mathbf{D}\mathbf{F}_M \tag{4.131}$$

represents the M-DFT-spread data matrix. Thus, due to the M-DFT spreading, any allocated data symbol in the k-th subcarrier produces samples in the k-th and $(k+1)$-th subband, as demonstrated in Fig. 4.29. For the set of allocated subcarriers $\mathscr{K}_{\text{on}}^{(u)}$, the set of occupied frequency subbands can be defined by

$$\mathscr{K}^{(u)} = \mathscr{K}_1^{(u)} \cup \mathscr{K}_2^{(u)}, \tag{4.132}$$

with $\mathscr{K}_1^{(u)} = \mathscr{K}_{\text{on}}^{(u)}$ and $\mathscr{K}_2^{(u)} = <\mathscr{K}_{\text{on}}^{(u)} - 1>_K$. In addition, let $\mathscr{M}_1$ and $\mathscr{M}_2$ be the indexes of the nonzero elements of $\tilde{\mathbf{g}}_1$ and $\tilde{\mathbf{g}}_2$, respectively, we define the set of occupied frequency indexes as

$$\mathscr{S}^{(u)} = \{(k,m) \in \mathscr{K}_i \times \mathscr{M}_i, i = 1, 2\}. \tag{4.133}$$

Accordingly, we adapt a masking matrix $\mathbf{U}^{(u)}$ of size $|\mathscr{K}^{(u)}| \times M$ defined as

$$\left[\mathbf{U}^{(u)}\right]_{(k,m)} = 1, (k,m) \in \mathscr{S}^{(u)}, \text{ and 0 elsewhere.} \tag{4.134}$$

This matrix can be used as q frequency domain window at the receiver. Recalling that, $\tilde{\mathbf{x}} = \mathrm{vec}\left\{\mathbf{V}_{K,M}^{(\tilde{x})\,T}\right\}$, this window can be applied to the received vector using the form

$$\mathbf{w}_F^{(u)} = \mathrm{vec}\left\{\mathbf{U}^{(u)T}\right\}. \tag{4.135}$$

so that

$$\left[\mathbf{w}_F^{(u)}\right]_{(\mathscr{N}^{(u)})} = 1, \text{ and } 0, \text{ elsewhere}, \tag{4.136}$$

and

$$\mathscr{N}^{(u)} = \{n = m + kM \,:\, (k, m) \in \mathscr{S}^{(u)}\}. \tag{4.137}$$

Moreover, by taking the N-DFT of (4.129) and applying the frequency domain windowing we get

$$\begin{aligned}
\tilde{\mathbf{y}}^{(u)} &= \left[\mathbf{F}_N \mathrm{diag}\left\{\boldsymbol{\phi}^{(u)}\right\}^H \mathbf{y}\right] \odot \mathbf{w}_F^{(u)} \\
&= \mathbf{D}^{\left(\tilde{h}^{(l)}\right)} u]\tilde{\mathbf{A}}^{(u)} \mathbf{d}^{(u)} + \mathbf{w}_F^{(u)} \odot \left[\mathbf{F}_N \sum_{v=1, v\neq u}^{U} \mathrm{diag}\left\{\boldsymbol{\phi}^{(v,u)}\right\} \mathbf{H}^{(v)} \mathbf{A}^{(v)} \mathbf{d}^{(v)}\right] + \mathbf{w}_F^{(u)} \odot \tilde{\mathbf{v}} \\
&= \mathbf{D}^{\left(\tilde{h}^{(l)}\right)} u]\tilde{\mathbf{A}}^{(u)} \mathbf{d}^{(u)} + \sum_{v=1, v\neq u}^{U} \mathbf{Z}^{(v\to u)} \mathbf{d}^{(v)} + \mathrm{diag}\left\{\mathbf{w}_F^{(u)}\right\} \tilde{\mathbf{v}}.
\end{aligned} \tag{4.138}$$

where

$$\tilde{\mathbf{Z}}^{(v\to u)} = \mathrm{diag}\left\{\mathbf{w}_F^{(u)}\right\} \frac{1}{N} e^{j2\pi c_{v,u}} \mathbf{F}_N \mathrm{diag}\left\{\boldsymbol{\phi}^{(v,u)}\right\} \mathbf{F}_N^H \mathbf{D}^{\left(\tilde{h}^{(v)}\right)} \tilde{\mathbf{A}}^{(v)}. \tag{4.139}$$

and $\mathbf{D}^{\left(\tilde{h}^{(x)}\right)} = \mathrm{diag}\left\{\tilde{h}^{(x)}\right\}$. Equation (4.139) defines the signal model after synchronization, so we can compute the SIR assuming uncorrelated data as

$$\mathrm{SIR}^{(u)} = \frac{P_u \left\|\mathbf{D}^{\left(\tilde{h}^{(u)}\right)} \tilde{\mathbf{A}}^{(u)}\right\|_F^2}{\sum_{v=1, v\neq u}^{U} P_v \left\|\tilde{\mathbf{Z}}^{(v\to u)}\right\|_F^2}, \tag{4.140}$$

where $\|\cdot\|_F$ is the Frobenius norm, and $P_x = \mathbb{E}\left[|d^{(x)}|^2\right]$ is the x-th user power per symbol. The SIR depends on the modulation matrix, resource and power allocation, CFO and the CIR.

The SIR metric is useful to evaluate the IUI; however, it is also important to examine the overall performance after channel equalization and demodulation. Let $\tilde{\mathbf{B}}^{(u)}$ be the receiver matrix involving channel equalization and demodulation, then

$$\hat{\mathbf{d}}^{(u)} = \tilde{\mathbf{B}}^{(u)H}\tilde{\mathbf{y}}^{(u)}$$
$$= \tilde{\mathbf{B}}^{(u)H}\mathbf{D}^{\left(\tilde{h}^{(u)}\right)}\tilde{\mathbf{A}}^{(u)}\mathbf{d}^{(u)} + \tilde{\mathbf{B}}^{(u)H}\sum_{v=1,v\neq u}^{U}\tilde{\mathbf{Z}}^{(v\rightarrow u)}\mathbf{d}^{(v)} + \tilde{\mathbf{B}}^{(u)}\mathrm{diag}\left\{\mathbf{w}_F^{(u)}\right\}\tilde{\mathbf{v}}. \tag{4.141}$$

The average signal-to-interference-plus-noise ratio (SINR) is considered as a performance metric. It is given by

$$\mathrm{SINR}_B^{(u)} = \frac{P_u\left|\mathcal{N}_{\mathrm{on}}^{(u)}\right|}{P_u\left\|\tilde{\mathbf{B}}^{(u)H}\mathbf{D}^{\left(\tilde{h}^{(u)}\right)}\tilde{\mathbf{A}}^{(u)} - \mathbf{I}\right\|_F^2 + \sum_{v=1,v\neq u}^{U}P_v\left\|\tilde{\mathbf{B}}^{(u)}\tilde{\mathbf{Z}}^{(v\rightarrow u)}\right\|_F^2 + N_0\left\|\tilde{\mathbf{B}}^{(u)}\mathrm{diag}\left\{\mathbf{w}_F^{(u)}\right\}\right\|_F^2}. \tag{4.142}$$

Actually, $\left(\mathrm{SINR}_B^{(u)}\right)^{-1}$ is the normalized mean squared error (NMSE) of the data symbols.

4.3.3 Asynchronous MA Evaluation

For the purpose of comparing the asynchronous MA with different waveform parameters, we consider two users, i.e., $U = 2$, with identical power allocation $P_1 = P_2$ and evaluate the SIR (4.140) and the SINR (4.142) in different scenarios.

4.3.3.1 AWGN Channel

This evaluation is useful to comprehend the interference due to the CFO. Let

$$\mathbf{C}^{(1,2)} = \frac{1}{N}\mathbf{F}_N\mathrm{diag}\left\{\boldsymbol{\phi}^{(1,2)}\right\}\mathbf{F}_N^H, \tag{4.143}$$

then

$$\mathrm{SIR}^{(u)} = \frac{\sum_{k\in\mathcal{K}_{\mathrm{on}}^{(1)}}\sum_{m\in\mathcal{M}_{\mathrm{on}}}\left\|\tilde{\mathbf{g}}_{k,m}\right\|_2^2}{\sum_{k\in\mathcal{K}_{\mathrm{on}}^{(2)}}\sum_{m\in\mathcal{M}_{\mathrm{on}}}\left\|\mathrm{diag}\left\{\mathbf{w}_F^{(1)}\right\}\mathbf{C}^{(1,2)}\tilde{\mathbf{g}}_{k,m}\right\|_2^2}. \tag{4.144}$$

Noting that

$$\left[\mathbf{C}^{(1,2)}\tilde{\mathbf{g}}_{k,m}\right]_{(q)} = G_{k,m}(\nu = \tfrac{q}{N} - \Delta f_{1,2}), \tag{4.145}$$

where $G_{k,m}(\nu)$ is the DTFT of $g_{k,m}[n]$, then

$$\left\|\mathrm{diag}\left\{\mathbf{w}_F^{(1)}\right\}\mathbf{C}^{(1,2)}\tilde{\mathbf{g}}_{k,m}\right\|_2^2 = \sum_{q\in\mathcal{N}^{(1)}}\left|G_{k,m}(\tfrac{q}{N} - \Delta f_{1,2})\right|^2. \tag{4.146}$$

Therefore,

$$\left\| \tilde{\mathbf{Z}}^{(2\to 1)} \right\|_F^2 = N \sum_{q \in \mathcal{N}^{(1)}} S_2(\tfrac{q}{N} - \Delta f_{1,2}), \tag{4.147}$$

where $S_u(\nu)$ is the PSD of the signal of the u-th user without CP expressed as

$$S_x(\nu) = \frac{1}{N} \sum_{k \in \mathcal{K}_{\text{on}}^{(x)}} \sum_{m \in \mathcal{M}_{\text{on}}} \left| G_{k,m}(\nu) \right|^2 . \tag{4.148}$$

As a result, we can write the SIR in the form

$$\text{SIR}^{(1)} = \frac{\sum_{q \in \mathcal{N}^{(1)}} S_1(\tfrac{q}{N})}{\sum_{q \in \mathcal{N}^{(1)}} S_2(\tfrac{q}{N} - \Delta f_{1,2})}. \tag{4.149}$$

This form is intuitively comprehensive as $\mathcal{N}^{(1)}$ represents the frequency band of the first user. Additionally, the interference of the second user is the integral over the leakage within the first user's band, as illustrated in Fig. 4.30. From this, it can be shown how important to study the OOB of the waveform, which in the case of GFDM depends on the prototype filter. This equation can be simply extended to $U > 2$. Figure 4.31 justifies the closed-form solution with the numerical simulation. In addition, it can be seen that for the same allocated bandwidth and the same guard band, GFDM is significantly less sensitive to CFO compared to OFDM. It is important to highlight that the first subsymbol is turned off. Additionally, the worst situation for the used prototype filter happens when the CFO is half the frequency sample, which is exactly half the subcarrier spacing in OFDM. On the other hand, a shift by complete sample retains the orthogonality if a sufficient guard band is used.

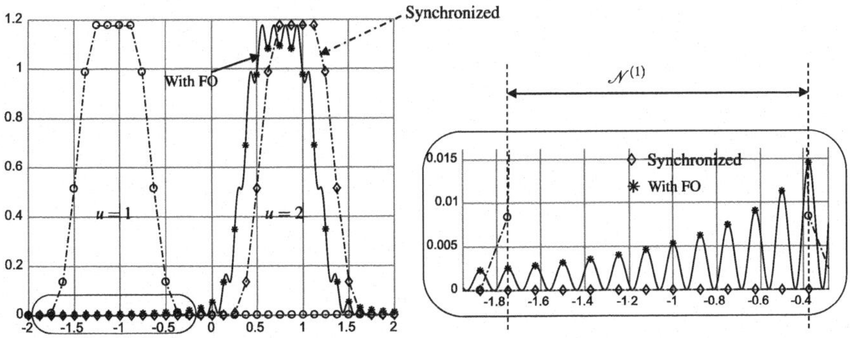

Fig. 4.30 IUI due to CFO using. In the synchronized case no interference as the sampling points are at zero crossing. With CFO, the samples are not equal to zeros

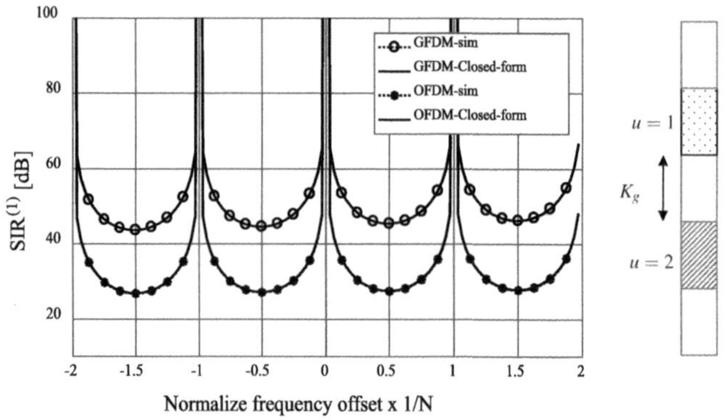

	GFDM	OFDM
K	64	1024
M	16	1
$\mathcal{M}_{\text{on}}$	$\{1,\cdots,15\}$	$\{1\}$
K_u	1	15
K_g	1	16
L_{cp}	64	64
g	RC (0.8)	rect

$N = 1024$

Fig. 4.31 $\text{SIR}^{(1)}$ simulation versus closed-form for GFDM and OFDM

4.3.3.2 Fading Channel

In the existence of fading channel, we define the modified pulse shape as

$$\mathbf{g}_{k,m|h^{(u)}} = \frac{1}{N}\mathbf{F}_N(\tilde{\mathbf{h}}^{(u)} \odot \tilde{\mathbf{g}}_{m,k}). \tag{4.150}$$

Following the same discussion, the PSD of the u-th user can be written as

$$S_{u|h^{(u)}}(\nu) = \frac{1}{N}\sum_{k\in\mathcal{K}_{\text{on}}^{(u)}}\sum_{m\in\mathcal{M}_{\text{on}}}\left|G_{k,m|h^{(u)}}(\nu)\right|^2, \tag{4.151}$$

and then

$$\text{SIR}^{(1)} = \frac{\sum_{q\in\mathcal{N}^{(1)}} S_{1|h^{(1)}}(\frac{q}{N})}{\sum_{q\in\mathcal{N}^{(1)}} S_{2|h^{(2)}}(\frac{q}{N} - \Delta f_{1,2})}. \tag{4.152}$$

By averaging over different channel realization, we achieve the SIR for certain PDP. Figure 4.32 shows the effect of the fading channel considering the TGn channel model in comparison with AWGN channel for different roll-off factors of RC filter. As expected, while larger roll-off factor produces well-localized filter and reduces the side lobes, the SIR is higher for larger roll-off. However, this gain may be reduced after the receive filter. For example, if ZF receiver is used, then the gain will be reduced by the NEF, which is higher for larger roll-off. On the other hand, with fading channel the average SIR increases, that is because the channel gain may attenuate the interfering signal. This can be beneficial if a proper equalization such as MMSE is used.

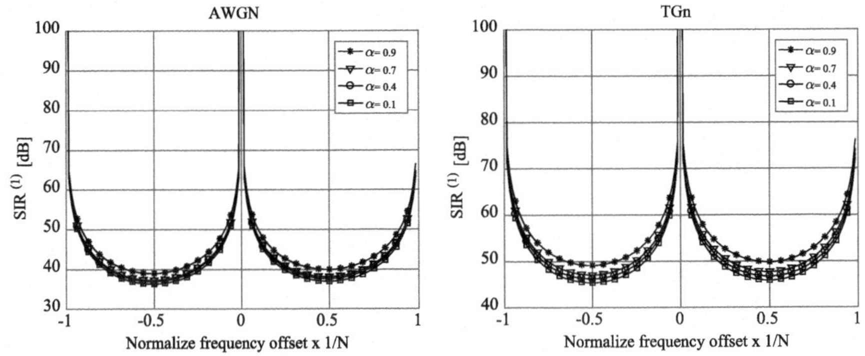

Fig. 4.32 SIR$^{(1)}$ for different roll-off in AWGN and TGn fading channel. $K = 64, M = 16, \mathcal{M}_{\text{on}} = \{1, \ldots, M-1\}$

4.3.4 *Mixed-Numerology with GFDM*

The supported services in the 5G are diverse, and each service has different requirements [33]. For example, to support the low latency of tactile Internet, shorter symbol duration needs to be used. To tackle the effect of Doppler shift in high speed vehicular communications, larger subcarrier spacing is required. The massive machine-type communications (mMTCs) employ fewer subcarriers to ensure lower PAPR. In order to multiplex all these services at one BS, the need of mixed-numerology arises. As discussed in Sect. 4.1.6, GFDM can be seen as a universal multicarrier waveform generator, with several reconfigurable parameters that can be altered on the fly to generate the corresponding signal. This feature promotes the GFDM framework to be a candidate for mixed-numerology.

The main issue arises here is the inter-numerology-interference (INI), which can be seen as IUI if we consider different users per service. In all cases, we need a model to evaluate the interference in order to optimize the design and the allocation of the services in the time and frequency domains.

4.3.4.1 General Inter-user-Interference Model

Without loss of generality, we consider two users $u = 1, 2$ transmitting their signals with GFDM modulation using K_u subcarriers, M_u subsymbols and a pulse shape $g^{(u)}[n]$ with subcarrier allocation indexes $\mathcal{K}_{\text{on},u}^{(u)}$, whose entries are subset of $\mathcal{K}_{\text{on},u}$ and $\mathcal{M}_{\text{on},u}^{(u)} = \mathcal{M}_{\text{on},u}$. Note that, the sub-index indicates the numerology index, while the sup-index represents the user index in that numerology. Let $y[n]$ be the discrete received signal

$$r[n] = r^{(1)}[n] + r^{(2)}[n] + v[n], \tag{4.153}$$

where $r^{(u)}[n]$ is the signal contribution from the u-th user and $v[n]$ is the additive noise. In order to decode the u-th signal, the receiver performs its operation based on $r^{(u)}[n]$ configurations, while the signal of the other user is the interference. After removing the CP, we get the block of the u-th user as

$$z_{u,i}[n] = r[n + L_{\text{cp},u} + iL_{\text{s},u}], n = 0, \ldots, N_u - 1, \tag{4.154}$$

where $L_{\text{cp},u}$ and $L_{\text{s},u}$ are the CP size and block spacing used by the u-th user, respectively. Afterward, N_u-DFT is computed, where $= K_u M_u$. Therefore, we get

$$\tilde{\mathbf{z}}_{1,i} = \tilde{\mathbf{z}}_i^{(1)} + \tilde{\mathbf{z}}_i^{(2\rightarrow 1)} + \tilde{\mathbf{v}}_{1,i} \in \mathbb{C}^{N_1 \times 1}, \tag{4.155}$$

$$\tilde{\mathbf{z}}_{2,i} = \tilde{\mathbf{z}}_i^{(2)} + \tilde{\mathbf{z}}_i^{(1\rightarrow 2)} + \tilde{\mathbf{v}}_{2,i} \in \mathbb{C}^{N_2 \times 1}. \tag{4.156}$$

Here, $\tilde{\mathbf{z}}_i^{(u_1)}$ is the i-th block of the u-th user containing all samples, $\tilde{\mathbf{z}}_i^{(u_1\rightarrow u_2)}$ is the interfering block from u_1 to u_2 and $\tilde{\mathbf{v}}_{u_1,i}$ the additive noise. First, we consider the case of AWGN. Using the multicarrier representation, we have

$$\tilde{\mathbf{z}}_i^{(u_1)} = \sum_{k\in\mathcal{K}_{\text{on},u_1}^{(u_1)}} \sum_{m\in\mathcal{M}_{\text{on},u_1}} \tilde{\mathbf{g}}_{k,m}^{(u_1)} d_{k,m,i}^{(u_1)}, \tag{4.157}$$

where $\tilde{\mathbf{g}}_{k,m}^{(u_1)}$ is the N_{u_1}-DFT of the pulse shapes of the u_1-th user as originally generated by the waveform. Moreover, the interfering signal can be expressed as

$$\tilde{\mathbf{z}}_i^{(u_2\rightarrow u_1)} = \sum_{j\in\mathcal{J}_i} \sum_{k\in\mathcal{K}_{\text{on},u_2}^{(u_2)}} \sum_{m\in\mathcal{M}_{\text{on},u_2}} \tilde{\mathbf{g}}_{k,m,j}^{(u_2\rightarrow u_1)} d_{k,m,\mathcal{I}_j(i)}^{(u_2)}. \tag{4.158}$$

In this notation, $\tilde{\mathbf{g}}_{k,m,j}^{(u_2\rightarrow u_1)}$ is the N_{u_1}-DFT of the pulse shapes seen by the receiver of u_1. This pulse shape can be computed from the original pulse shape $g^{(u_2)}$ and the multiplexing parameters, as seen in Fig. 4.33. The set $\mathcal{J}_i$ represents a set of indexes depending on the index i and $\mathcal{I}_j(i)$ is the block index of u_2 that contributes to the interference. Note that $\mathcal{I}_{j_1}(i_1) \neq \mathcal{I}_{j_2}(i_2)$, $i_1 \neq i_2$ and $j_1 \neq j_2$. This is the key difference in mixed-numerology. If the same numerology is used, then $\mathcal{J}_i$ disappears and $\mathcal{I}_j(i) = i$. But in mixed-numerologies, the interference may depend on the block index. However, by a proper design we can control the pattern, such that $\mathcal{J}_{Pi+p} = \mathcal{J}_p$, in order to have P interference patterns applied to different block indexes, such that

$$\tilde{\mathbf{z}}_{Pi+p}^{(u_2\rightarrow u_1)} = \sum_{j\in\mathcal{J}_p} \sum_{k\in\mathcal{K}_{\text{on},u_2}^{(u_2)}} \sum_{m\in\mathcal{M}_{\text{on},u_2}} \tilde{\mathbf{g}}_{k,m,j}^{(u_2\rightarrow u_1)} d_{k,m,\mathcal{I}_j(i)}^{(u_2)}. \tag{4.159}$$

Following the same approach in (4.149), we get the SIR for each pattern

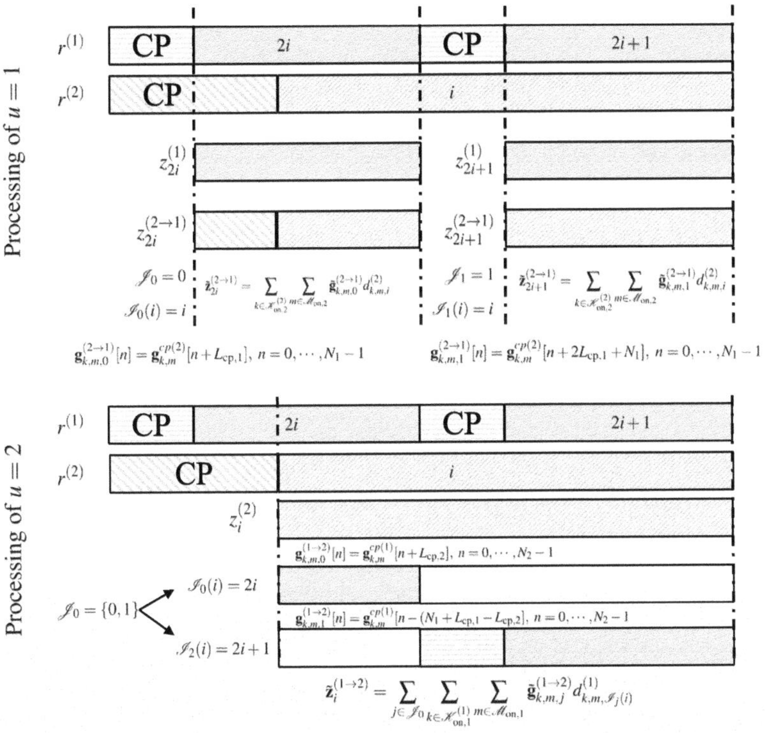

Fig. 4.33 INI example. $N_2 = 2N_1, L_{\text{cp},2} = 2L_{\text{cp},1}$. The i-th block of $u = 2$ interferes with the $2i$-th and $(2i+1)$-th blocks of $u = 1$ in two different patterns. Each pattern has one pulse shape. On the other side, two blocks of $u = 1$ interfere with one block of $u = 2$. In the latter case, there is only one pattern but with two pulse shapes

$$\text{SIR}_p^{(u_1)} = \frac{\sum_{q\in\mathcal{N}_{u_1}^{(u_1)}} S_1(\frac{q}{N})}{\sum_{q\in\mathcal{N}_{u_1}^{(u_1)}} S_{p,u_2\to u_1}(\frac{q}{N})}. \tag{4.160}$$

Here, $\mathcal{N}_{u_1}^{(u_1)}$ is the set of nonzero frequency samples of u_1 in its numerology, and

$$S_{p,u_2\to u_1}(\nu) = \frac{1}{N_{u_1}} \sum_{j\in\mathcal{J}_p} \sum_{k\in\mathcal{K}_{\text{on},u_2}^{(u_2)}} \sum_{m\in\mathcal{M}_{\text{on},u_2}} \left|G_{k,m,j}^{(u_2\to u_1)}(\nu)\right|^2. \tag{4.161}$$

The average SIR can be computed by averaging the SIR overall patterns. This can be extended in a similar way to the case of fading channel as in (4.152).

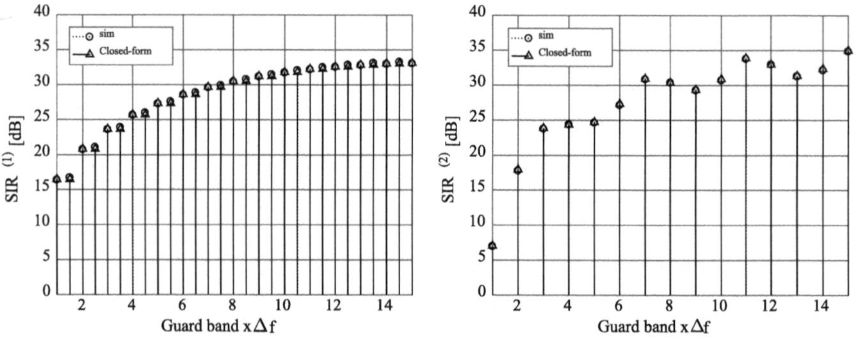

Fig. 4.34 INI for OFDM with $K_1 = 32$, $L_{cp,1} = 8$, $K_2 = 64$, $L_{cp,2} = 8$, $\Delta f = \Delta f_1 = 2\Delta f_2$. The users are allocated the same bandwidth, so that, $K_{u,1} = 1$, $K_{u,2} = 2$

4.3.4.2 Numerical Examples

To verify the closed-form solution in (4.160), consider OFDM systems with different subcarrier spacing. The first system uses K_1 subcarriers and the second system uses $K_2 = 2K_1$, in other word, the second system employs half the subcarrier spacing. The CP length in both cases is 1/4 of the symbol length, and both systems transmit data with unit power. Under perfect synchronization, we get the frame structure presented in Fig. 4.33. As illustrated in Fig. 4.34, we compare the SIR of one allocated subcarrier of the first system with two subcarriers from the other system, so we get the same bandwidth. The guard band is an integer number of subcarriers which is normalized to the largest subcarrier spacing Δf. As expected, the interference from user 2 to user 1 decreases with larger guard band. In addition, the guard band can be controlled by the smaller subcarrier spacing of system 2, where $\Delta f_2 = \Delta f/2$. Nevertheless, it can be observed that the behavior of the interference from user 1 to the user 2 fluctuates and is influenced by the guard interval.

Next, we compare GFDM with different settings. In this example, the target is to have short and long blocks as in Fig. 4.33. This can be attained by changing either the subcarrier spacing or the subsymbol spacing. The configurations and results are shown in Fig. 4.35, where the pair (U1, U2) corresponds to changing the subcarrier spacing, while the pair (U3, U2) realizes the different block lengths via different subsymbol spacing. It can be seen that with the used RC pulse shape, the interference in both directions follows the same behavior. Interestingly, it can be shown also that both settings achieve similar performance with slight gain when different subsymbol spacing is applied. Another advantage is that keeping smaller subcarrier spacing allows fine control of the guard band. Finally, GFDM outperforms OFDM by a gain up to 10 dB in the achieved SNR for the same subcarrier spacing.

	U1	U2	U3
K	32	64	64
M	16	16	8
$\mathcal{M}_{\text{on}}$	$\{0,\cdots,15\}$	$\{0,\cdots,15\}$	$\{0,\cdots,7\}$
K_u	1	2	2
L_{cp}	32	64	32
g	RC (0.8)	(0.8)	(0.8)

$\Delta f = \Delta f_1 = 2\Delta f_2$

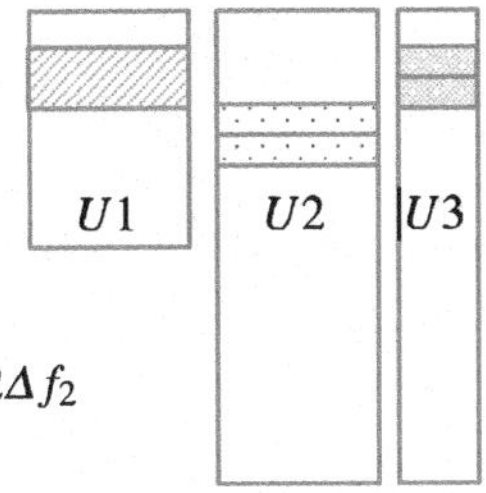

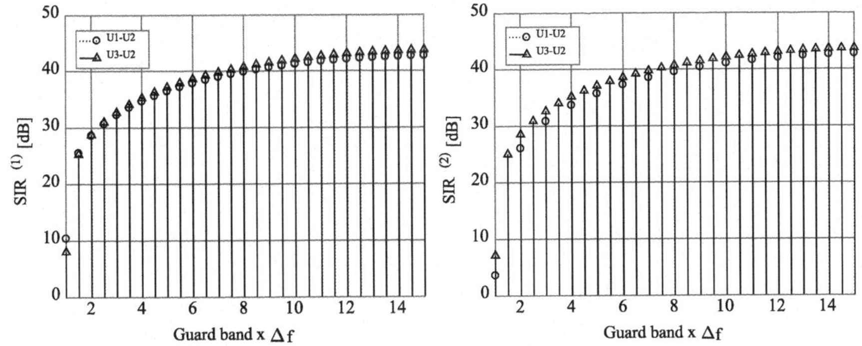

Fig. 4.35 INI for GFDM with different settings for user 1 and user 2

4.4 GFDM Implementation

The path to meeting all requirements of 5G is a challenging endeavor. The demand to achieve higher data rates for the enhanced media broadband (eMBB) scenario and novel use cases like ultra-reliable and low-latency communication (URLLC) and mMTC drive researchers and engineers to consider new concepts and technologies for future wireless communication systems. The goal is to identify promising candidate technologies among a vast number of new ideas and decide which are suitable to be implemented in future products. Figure 4.36 gives a rough overview of the development process.

New ideas and concepts typically first undergo extensive software simulations, which allow to make early predictions on the expected performance. After selection of the best candidates, individual aspects of the envisioned system can be implemented on a hardware-accelerated platform, e.g., software-defined radio (SDR), in order to learn about real-time behavior and over-the-air performance with real radio frequency (RF) components. Technologies that prove promising at this stage can be further evaluated in test beds, where the focus shifts toward the interaction of different technology building blocks and the realization of complete end-to-end applications. New concepts and technologies that have been proven to exhibit improved performance in practically relevant environments will ultimately find their way into new standards and lastly, industry will adopt them in future products.

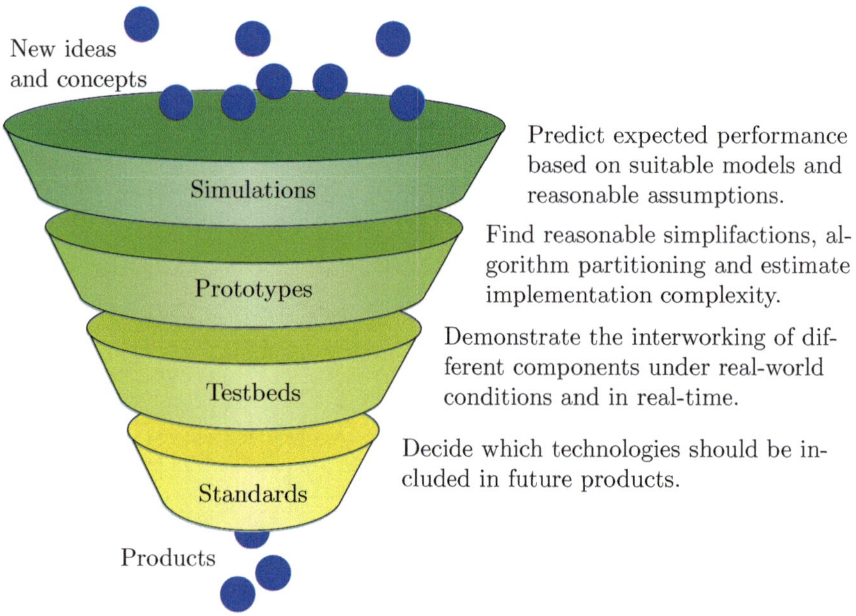

Fig. 4.36 From theory to practice

This section presents a field-programmable gate array (FPGA)-based, real-time implementation of a modulator/demodulator for multicarrier waveforms [34]. This proof-of-concept design provides a large number of degrees of freedom to the user and offers the flexibility for practical evaluation of new algorithms that aim to address various 5G aspects. Example applications include experiments with flexible numerology, which is a key differentiator of 5G new radio (NR) compared to fourth-generation (4G) LTE, as well as the design and implementation of a corresponding scheduler that utilizes the additional flexibility. The presented platform is a building block for test beds that will assist the design of 5G radio interface and network architecture [35, 36].

4.4.1 Modem Implementation

For the sake of simplicity, consider a communication system that consists of two nodes. Each node is realized by a control PC that is connected to a software-defined radio platform. This hardware setup is depicted in Fig. 4.37. Note that the design that is described in this section is tailored for the USRP-RIO hardware [37]. However, the basic principles of the implementation are valid for other platforms in general. The block diagram in Fig. 4.38 shows how the components of the overall system are mapped to the hardware platform.

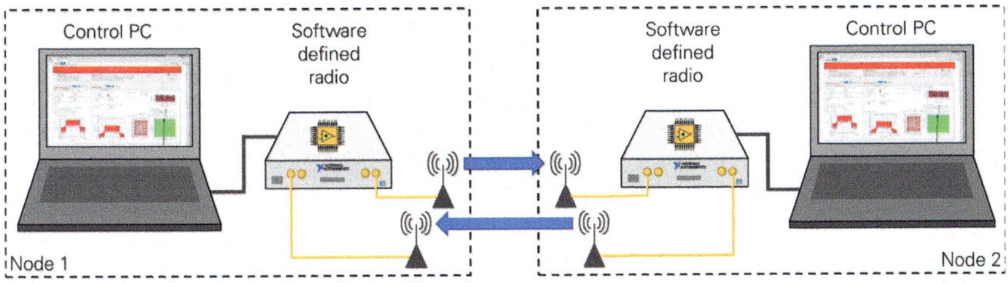

Fig. 4.37 Hardware setup

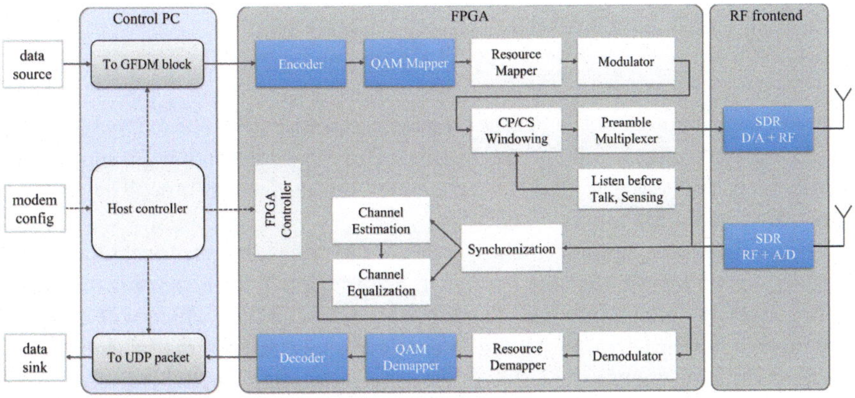

Fig. 4.38 Block diagram of the transceiver

In order to be able to support the various requirements of 5G wireless systems, the individual blocks need to be implemented with flexibility in mind. This section will focus on the PHY aspects of a 5G NR transceiver. The corresponding signal processing algorithms have to be implemented on FPGA, in order to be able to meet throughput and latency requirements. As medium access layer and higher layers have more relaxed requirements w.r.t. timing, when operating in sub-6 GHz bands, those components rarely require specialized hardware acceleration. Hence, the assumption is made that they are implemented using software running on standard PC hardware.

4.4.1.1 Baseband Modem

The baseband signal processing is performed by the resource mapper and modulator blocks on the transmit path, and demodulator and resource demapper on the receive path, respectively. Note that the block encoder, QAM mapper, decoder, and QAM demapper are standard implementations taken from a 4G LTE library [38], and hence will not be discussed here.

As seen in Fig. 4.39, the resource mapper takes complex symbols from various input sources, e.g., payload data, control channel data, and reference signals, and

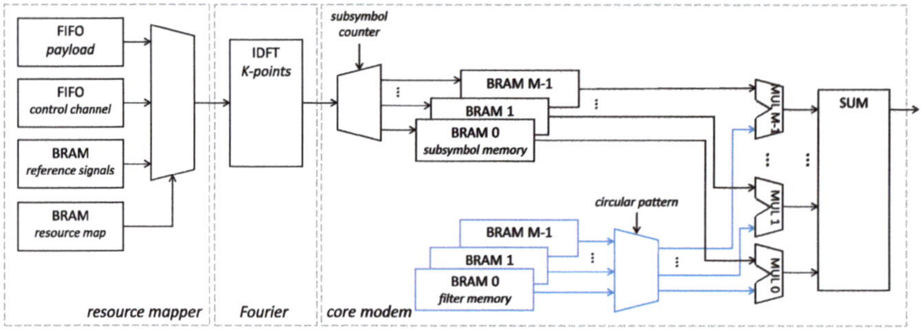

Fig. 4.39 Block diagram of the resource mapper and the modulator

maps them to a two-dimensional time-frequency resource grid. The mapping pattern, i.e., the resource map, is fully programmable. This mechanism allows to support any user-defined resource grid that can be adapted during the run-time. The resource demapper performs the inverse operation.

The modulation operation can be separated into two main functional blocks. The first block is an inverse discrete Fourier transform (IDFT) that transforms the data from frequency domain into time domain. The second block, which will be called core modem, applies a pulse shaping filter to each subcarrier of the transmit signal. The demodulation operation consists of the same processing blocks in reversed order, where the only differences are the direction of the Fourier transformation and the filter coefficients that are used in the core modem.

The first task of the core modem is to split the incoming IDFT output stream δ_m into M subsymbols with K samples in each, which are stored in block RAM (BRAM) $0, \ldots, M-1$. Each individual subsymbol has to be repeated M times, such that the filter g_m with N samples can be applied. Therefore, the K samples of each subsymbol are stored inside an independent subsymbol memory bank. All M memories are read sample-by-sample in parallel. The filter g_m is also stored in M parallel subsymbol filter banks. Each of the filter memory banks contains K coefficients that represent a different part of the pulse shaping filter in the time domain. The filter needs to be applied to the data in a circularly shifted way. This is implemented with a circular pattern that dynamically selects which filter BRAM is connected to which subsymbol BRAM. The last step in the core modem is to accumulate the contributions from all M parallel branches to get the transmit signal.

4.4.1.2 Post Modem Processing

After the signal is modulated, CP, CS, time-windowing, and preamble are added. Figure 4.40 depicts the general frame format. The CP and CS can be applied on both preamble and data block. $N_{\mathrm{P,CP}}$ defines the length of the CP for the preamble, N_{CP} for the data block, $N_{\mathrm{P,CS}}$ defines the length of the CS for the preamble, and N_{CS} for

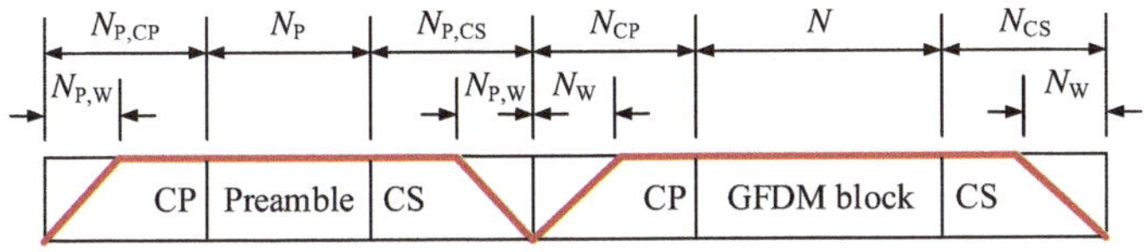

Fig. 4.40 Supported frame structure with one preamble and one data block

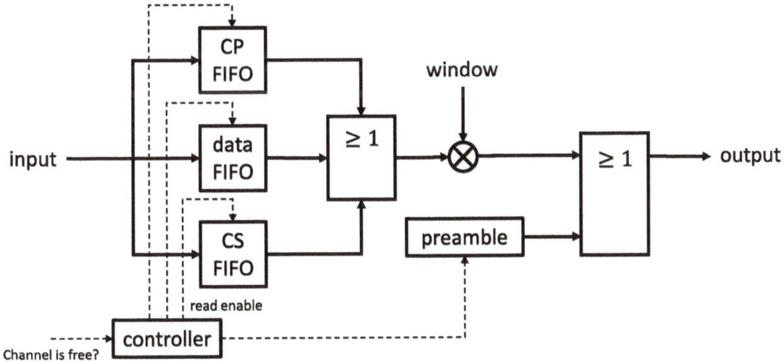

Fig. 4.41 CP, CS, windowing and preamble insertion

the data block. It is assumed that the window is symmetric, thus, the length of one half is given by $N_{P,W}$ and N_W. In addition, N_P denotes the length of the preamble.

The preamble is calculated in advanced on the host computer and written to the BRAM memory during the configuration phase of the transceiver. The CP and CS of the modulated GFDM data block are added via first in, first out (FIFO) memory as depicted in Fig. 4.41, which can be seen as a variable delay to shift the data samples into the correct output order.

The windowing unit follows, where only the rising half is stored inside a memory. An integrated counter in the control logic counts up until N_W is reached to trigger the memory for the appropriate samples. During the main data block, the unit is disabled. Finally, the same counter is decreased to create the falling flank.

Whenever the controller has finished reading in the first data block of a frame into the data-FIFO, the preamble insertion unit is triggered to push the preamble samples to the digital-to-analog (D/A) converter.

4.4.2 Complete Transceiver Chain and Extension for MIMO

A complete transceiver was implemented in FPGA as a proof of concept with its block diagram shown in Fig. 4.42. It includes all required processing functions for a real-world wireless communication system.

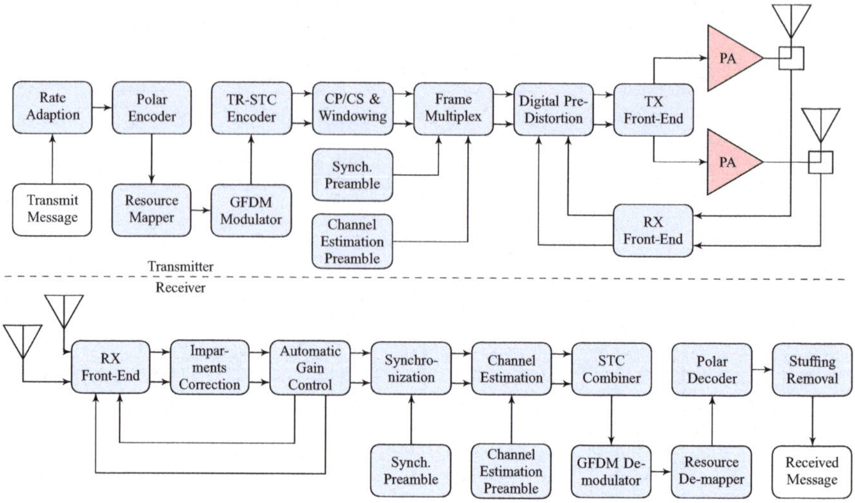

Fig. 4.42 Transceiver block diagram

A flexible frame structure was designed to cope with many design aspects such as synchronization, channel estimation, codeword length, resource allocation, MCS, channel multipath protection, OOB emission, MIMO operation, etc. Figure 4.43 depicts the adopted frame structure. The number of GFDM symbols within a frame, N_G, is chosen such that there is an integer number of codewords carried by an integer number of GFDM symbols. Therefore, the receiver can synchronously start decoding a codeword at the beginning of the frame. Synchronization and channel estimation were implemented using a preamble-based scheme, i.e., the transmitter inserts waveforms known by the receiver, multiplexed with the GFDM symbols. The periodicity of the synchronization preambles is proportional to frequency precision and stability difference between transmitter and receiver time base, likewise, the channel estimation preambles periodicity is inversely proportional to the mobile velocity.

The BSs operate in continuous transmission mode, where the carrier is always on the air independent if useful data is available for transmission. Considering the modulator is a wireless pipeline, it requires a constant input data rate. As the useful transmit message rate varies over time, a rate adaption scheme is required. The simplest solution is to fill dummy data in order to maintain a constant rate, and remove this stuffing data at the receiver.

Polar code was derived from the channel polarization theory and introduced in [39]. The code presents explicit construction, hardware efficient coding and decoding algorithms, and high flexibility, namely the code rate can continuously vary from $\frac{1}{N_C}$ to $\frac{N_C-1}{N_C}$, where N_C is the codeword length. It makes it a strong candidate for the future wireless networks, such as 5G. The implemented encoding algorithm is systematic [40] with semi-parallel architecture [41] for improved throughput. The implemented decoding algorithm is based on successive cancellation in the logarithm domain [42].

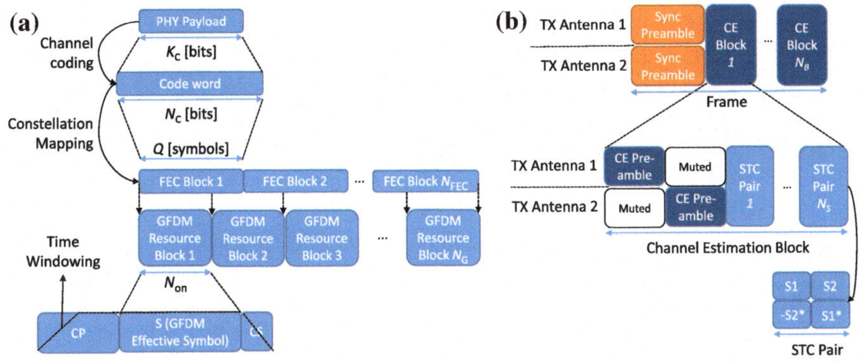

Fig. 4.43 Frame structure for **a** data encapsulation and; **b** waveforms multiplexing

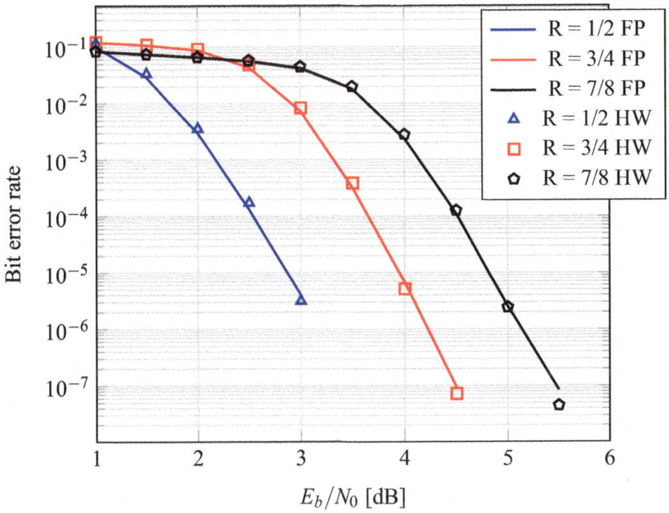

Fig. 4.44 Polar code bit error rate for different code rates in floating-point simulation (FP) and hardware fixed-point implementation (HW)

A comparison between a floating-point simulation and fixed-point implementation for the polar code is presented in Fig. 4.44.

The signal to be transmitted needs to be amplified by a power amplifier (PA). The majority of today's BSs employ the high electrical efficient Doherty topology [43], which is intrinsically nonlinear due to its class-C branch responsible for amplifying signal peaks. The nonlinearity distorts the amplified signal generating spectral regrowth, also known as intermodulation, presented at both in-band and OOB frequencies. The in-band intermodulation affects the signal quality, and, therefore, the receiving threshold. The OOB intermodulation results in OOB emissions, interfering in adjacent channels, and losing the benefit of using a waveform with low-OOB emissions such as GFDM. Linearization using digital predistortion (DPD) has been widely employed to mitigate the intermodulation [44]. The DPD system design procedure is subdivided into three distinct tasks: (1) to choose a behavioral model equation which is able to represent the PA characteristics, e.g., nonlinearity and memory effects, with the minimum number of coefficients; (2) to design a real-time DPD block which is able to generate the distortion according to the model equation and its coefficients; and (3) to select an algorithm which identifies the optimum values for the model coefficients in order to compensate for the PA distortion. Our chosen behavioral model equation is based on the memoryless orthogonal baseband polynomial for Gaussian distributed signals [45] and modified to include the memory as

$$y(n)=\sum_{m=0}^{M_{\mathrm{D}}-1}\sum_{k=1}^{K_{\mathrm{D}}}\sum_{l=1}^{k}\frac{h_{m,k}\sqrt{k}}{l!(-1)^{l-k}}\binom{k-1}{l-1}|x[n-m]|^{2(l-1)}x[n-m], \tag{4.162}$$

where $x[n]$ and $y[n]$ are the input and output model signals, respectively, K_{D} and M_{D} are the polynomial order and memory length, respectively, and $\{h_{m,k}\}$ are the DPD model coefficients at the m-th tap and k-th order.

The DPD performance was tested with a gallium nitride device operating at 3 W average, amplifying a 20 MHz wide GFDM signal centered at 723 MHz. The PA linearization results are shown in Fig. 4.45. Table 4.4 shows the DPD performance in terms of adjacent channel leakage rejection (ACLR).

There are impairments on the received signal caused by imperfections in the receiver front-end analog components. It causes interference and need to be digitally compensated in the real and imaginary parts of the received signal in three steps: (1) remove the average value, μ; (2) remove the average correlation between parts, β;

Fig. 4.45 DPD intermodulation reduction performance measured with a spectrum analyzer

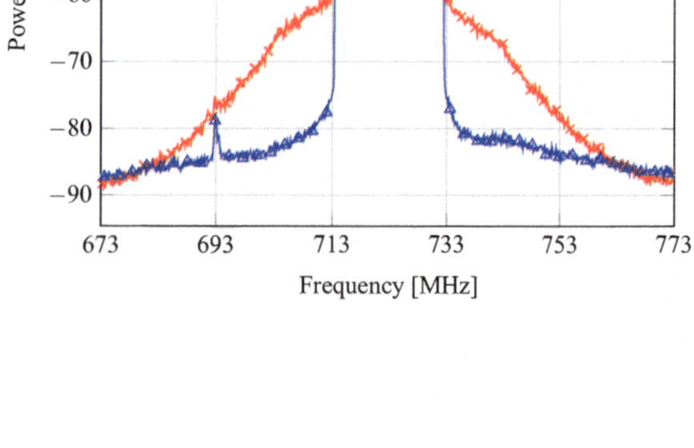

Table 4.4 DPD performance in terms of ACLR

	Lower ACLR (dB)	Upper ACLR (dB)
Without DPD	−36.6	−37.9
With DPD	−53.2	−53.1
Improvement	16.6	15.2

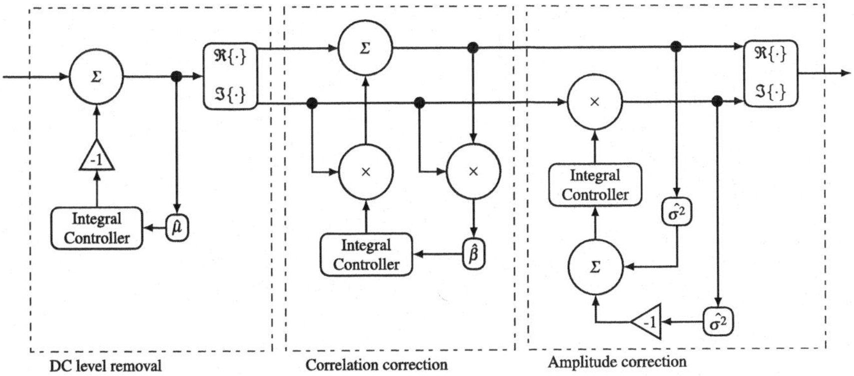

Fig. 4.46 Receiver impairments correction

and (3) equalize the power (or variance σ^2) difference between parts. This process is shown in Fig. 4.46.

At the receiver side, the signal peak-to-peak voltage needs to comply with the analog-to-digital converter (ADC) input range. If the voltage swing is too low, the SNR is compromised by the quantization error. If the voltage swing is too high, clipping effects will also affect the SNR. Assuming a Gaussian distribution and an ADC with a given number of bits, it is possible to find the optimum signal amplitude which maximizes the SNR [46]. The automatic gain control (AGC) goal is to keep the voltage level at this optimum point as shown in Fig. 4.47.

The existing preamble-based receiver synchronization techniques for OFDM are applicable for GFDM. For instance, a preamble with two repeated halves may be used, where the autocorrelation

$$\rho[n] = \sum_{k=n}^{n+N_\mathrm{P}-1} r[k]^* r[k + N_\mathrm{P}/2], \tag{4.163}$$

can be calculated between the halves, where N_P is the synchronization preamble length. Normalizing the autocorrelation by the signal energy leads to

$$\mu_S[n] = \frac{2\,|\rho[n]|^2}{\sum_{k=n}^{n+N_\mathrm{P}-1} |r[k]|^2}. \tag{4.164}$$

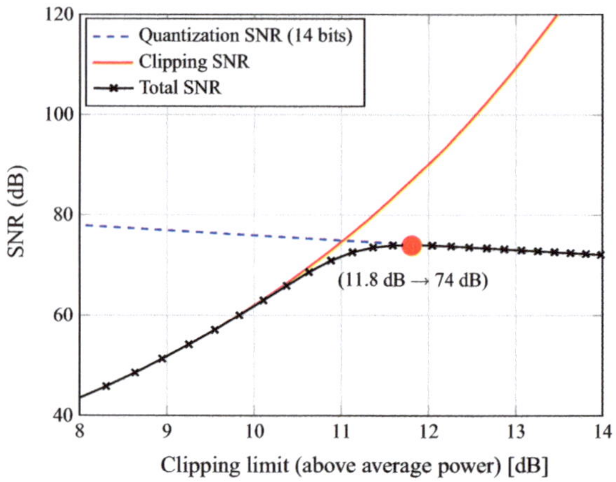

Fig. 4.47 SNR due to quantization and clipping, and optimal operating point

The presence of CP and CS produces the plateau effect which can be mitigated by integrating (4.164) and resulting in

$$\mu_M[n] = \frac{1}{L+1} \sum_{k=n-L}^{n} \mu_S[n], \tag{4.165}$$

where $L = N_{\text{CP}} + N_{\text{CS}}$.

The STO can also be estimated through the cross-correlation, which is given by

$$\rho_C[n] = \frac{1}{N_{\text{P}}} \sum_{k=0}^{N_{\text{P}}-1} r'[n+k] p_x^*[k], \tag{4.166}$$

where $p_x[n]$ is the known preamble waveform. Finally, (4.165) and (4.166) can combined for an enhanced performance given by

$$\mu_A[n] = |\rho_C[n]| \cdot \mu_M[n], \tag{4.167}$$

where the multiplication suppresses the cross-correlation side peaks, which appear due to the repeated halves. All discussed synchronization metrics are depicted in Fig. 4.48. The STO is estimated as the sample time index where the metric peak is.

The preamble-based channel estimation scheme is accomplished in the frequency domain. It is straightforward since the channel estimation preamble (CEP) in the frequency domain is known to the receiver. Considering the CEP length is usually shorter than the GFDM symbol, the estimated channel needs to be interpolated. However, when some subcarriers are muted, the IFFT/FFT interpolation method

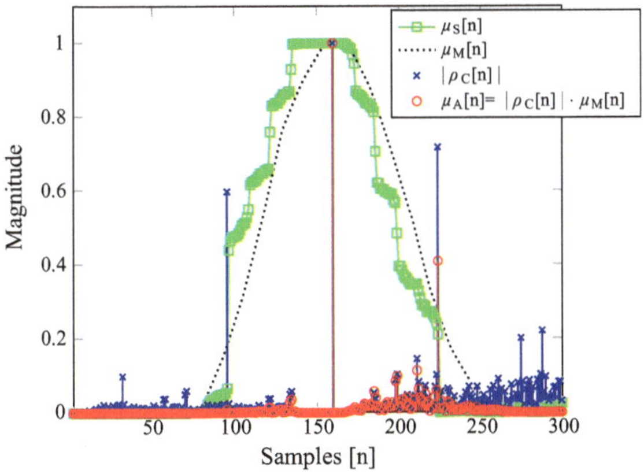

Fig. 4.48 Synchronization metrics

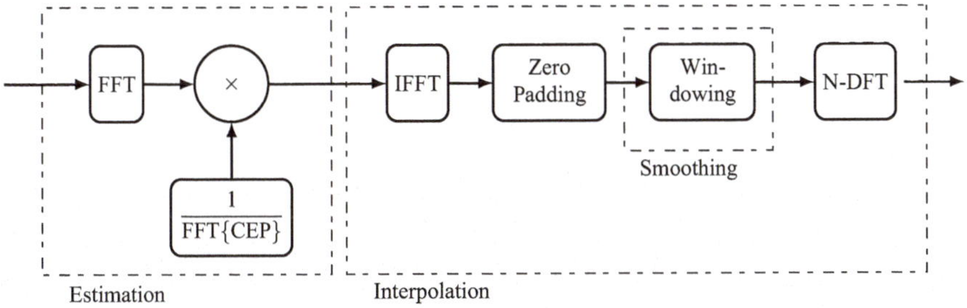

Fig. 4.49 Preamble-based channel estimation block diagram

fails because all the frequency components are required to calculate the interpolated response. In order to solve this problem, a time-windowing is done on the channel impulse response being interpolated. The detailed channel estimation block diagram is shown in Fig. 4.49.

References

1. N. Michailow, Gaspar I. Matthé et al., Generalized frequency division multiplexing for 5th generation cellular networks. IEEE Trans. Commun. **62**, 3045–3061 (2014)
2. H. Bölcskei, F. Hlawatsch, Discrete Zak transforms, polyphase transforms and applications. IEEE Trans. Signal Process. (1997), https://doi.org/10.1109/78.564174
3. M. Matthé et al., Generalized frequency division multiplexing in a Gabor transform setting. IEEE Commun. Lett. (2014), https://doi.org/10.1109/LCOMM.2014.2332155
4. A. Nimr et al., Optimal Radix-2 FFT compatible filters for GFDM. IEEE Commun. Lett. (2017), https://doi.org/10.1109/LCOMM.2017.2687926
5. X.-G. Xia, A family of pulse-shaping filters with ISI-free matched and unmatched filter properties. IEEE Trans. Commun. (1997), https://doi.org/10.1109/26.634674

6. G. Berardinelli, On the sensitivity of Zero-Tail DFT-spread-OFDM to small bandwidth allocations. IEEE Wirel. Commun. Lett. (2017), https://doi.org/10.1109/LWC.2017.2784825
7. G. Cherubini et al., Filtered multitone modulation for very high-speed digital subscriber lines. IEEE J. Sel. Areas Commun. (2002), https://doi.org/10.1109/JSAC.2002.1007382
8. K. Jung-Hyo et al., A novel OFDM chirp waveform scheme for use of multiple transmitters in SAR. IEEE Geosci. Remote Sens. Lett. (2013), https://doi.org/10.1109/LGRS.2012.2213577
9. S. Ehsanfar, M. Matthé, D. Zhang et al., A study of pilot-aided channel estimation in MIMO-GFDM systems, in *ITG/IEEE Workshop on Smart Antennas (WSA'16)* (2016)
10. I. Gaspar, N. Michailow, A. Navarro et al., Low complexity GFDM receiver based on sparse frequency domain processing, in *IEEE Vehicular Technology Conference (VTC)* (2013)
11. S. Ehsanfar, M. Matthé, D. Zhang et al., Interference-free pilot insertion for MIMO-GFDM channel estimation, in *IEEE Wireless Communication and Networking Conference (WCNC'17)* (2017)
12. E. U. T. R. Access, Physical channels and modulation, 3GPP TS, Vol. 36 (2009), p. V8
13. S. Alamouti, A simple transmit diversity technique for wireless communications. IEEE J. Sel. Areas Commun. **16**, 1451–1458 (1998)
14. M. Matthé, L. Mendes, I. Gaspar et al., Multi-user time-reversal STC-GFDMA for future wireless networks. EURASIP J. Wirel. Commun. Netw. **132**, 1–8 (2015)
15. L. Le, V. Lau, E. Jorswieck et al., Enabling 5G mobile wireless technologies. EURASIP J. Wirel. Commun. Netw. (2015), https://doi.org/10.1186/s13638-015-0452-9
16. P. Banelli, S. Buzzi, G. Colavolpe et al., Modulation formats and waveforms for 5G networks: who will be the heir of OFDM?: an overview of alternative modulation schemes for improved spectral efficiency. IEEE Signal Process. Mag. **31**, 80–93 (2014)
17. X. Zhang, M. Jia, L. Chen et al., Filtered-OFDM—enabler for flexible waveform in the 5th generation cellular networks, in *IEEE Global Communications Conference (GLOBECOM)* (2015), https://doi.org/10.1109/GLOCOM.2015.7417854
18. T. Wild, F. Schaich, Chen, 5G air interface design based on universal filtered (UF-)OFDM, in *International Conference on Digital Signal Processing* (2014), https://doi.org/10.1109/ICDSP.2014.6900754
19. M. Bellanger et al., FBMC physical layer: a primer, in *The PHYDYAS project* (2010), http://ict-phydyas.org
20. A. Aminjavaheri, A. Farhang, A. Rezazadeh Reyhani et al., Impact of timing and frequency offsets on multicarrier waveform candidates for 5G, in *IEEE Signal Processing and Signal Processing Education Workshop (SP/SPE)* (2015), https://doi.org/10.1109/DSP-SPE.2015.7369549
21. M. Matthé, I. Gaspar, D. Zhang et al., Reduced complexity calculation of LMMSE filter coefficients for GFDM, in *IEEE 82nd Vehicular Technology Conference (VTC2015-Fall)* (2015), https://doi.org/10.1109/VTCFall.2015.7391113
22. A. Farhang, N. Marchetti, L. Doyle, Low complexity GFDM receiver design: a new approach, in *IEEE International Conference on Communications (ICC)* (2015), https://doi.org/10.1109/ICC.2015.7249078
23. T. Ihalainen, A. Ikhlef, J. Louveaux et al., Channel equalization for multi-antenna FBMC/OQAM receivers. IEEE Trans. Veh. Technol. **60**, 2070–2085 (2011)
24. D. Zhang, M. Matthé, L. Mendes et al., A study on the link level performance of advanced multicarrier waveforms under MIMO wireless communication channels. IEEE Trans. Wirel. Commun. **16**, 2350–2365 (2017)
25. M. Matthé, D. Zhang, F. Schaich et al., A reduced complexity time-domain transmitter for UF-OFDM, in *IEEE 83rd Vehicular Technology Conference (VTC Spring)* (2016), https://doi.org/10.1109/VTCSpring.2016.7504101
26. J. Abdoli, M. Jia, J. Ma, Filtered OFDM: a new waveform for future wireless systems, in *IEEE 16th International Workshop on Signal Processing Advances in Wireless Communications (SPAWC)* (2015), https://doi.org/10.1109/SPAWC.2015.7227001
27. J. van de Beek, F. Berggren, N-continuous OFDM. IEEE Commun. Lett. **13**, 1–3 (2009)
28. Z. Yuan, A. Wyglinski, On sidelobe suppression for multicarrier-based transmission in dynamic spectrum access networks. IEEE Trans. Veh. Technol. **59**, 1998–2006 (2010)

29. P. Kryszkiewicz, H. Bogucka, Out-of-band power reduction in NCOFDM with optimized cancellation carriers selection. IEEE Commun. Lett. **17**, 1901–1904 (2013)
30. Y. Li, J. Cimini, N. Sollenberger, Robust channel estimation for OFDM systems with rapid dispersive fading channels. IEEE Trans. Commun. **46**, 902–915 (1998)
31. G. Ku, J.M. Walsh, Gross, Resource allocation and link adaptation in LTE and LTE advanced: a tutorial. IEEE Commun. Surv. Tutor. (2014), https://doi.org/10.1109/COMST.2014.2383691
32. M. Matthé et al., Asynchronous multi-user uplink transmission with generalized frequency division multiplexing, in *Communication Workshop (ICCW)* (2015), https://doi.org/10.1109/ICCW.2015.7247519
33. A.A. Zaidi, Gross et al., Waveform and numerology to support 5G services and requirements. IEEE Commun. Mag. (2016), https://doi.org/10.1109/MCOM.2016.1600336CM
34. GFDM on GitHub, Jan 2018, https://github.com/ewine-project/Flexible-GFDM-PHY
35. The ORCA Project, Jan 2018, https://www.orca-project.eu/
36. The eWINE Project, Jan 2018, https://ewine-project.eu/
37. National Instruments reconfigurable software defined radio (USRP-RIO), Jan 2018, http://sine.ni.com/nips/cds/view/p/lang/en/nid/212174
38. LabVIEW Communications LTE Application Framework, Jan 2018, http://sine.ni.com/nips/cds/view/p/lang/en/nid/213083
39. E. Arikan, Channel polarization: a method for constructing capacity-achieving codes for symmetric binary-input memoryless channels. IEEE Trans. Inf. Theory (2009), https://doi.org/10.1109/TIT.2009.2021379
40. E. Arikan, Systematic polar coding. IEEE Commun. Lett. (2011), https://doi.org/10.1109/LCOMM.2011.061611.110862
41. G. Sarkis, I. Tal, P. Giard, A. Vardy, C. Thibeault, W.J. Gross, Flexible and low-complexity encoding and decoding of systematic polar codes. IEEE Trans. Commun. (2016), https://doi.org/10.1109/TCOMM.2016.2574996
42. C. Leroux, I. Tal, A. Vardy, W.J. Gross, Hardware architectures for successive cancellation decoding of polar codes, in *IEEE International Conference on Acoustics, Speech and Signal Processing* (2011), https://doi.org/10.1109/ICASSP.2011.5946819
43. R. Darraji, P. Mousavi, F.M. Ghannouchi, Doherty goes digital: digitally enhanced Doherty power amplifiers. IEEE Microw. Mag. (2016), https://doi.org/10.1109/MMM.2016.2561478
44. J. Wood, System-level design considerations for digital pre-distortion of wireless base station transmitters. IEEE Trans. Microw. Theory Tech. (2017), https://doi.org/10.1109/TMTT.2017.2659738
45. R. Raich, G.T. Zhou, Orthogonal polynomials for complex Gaussian processes. IEEE Trans. Signal Process. (2004), https://doi.org/10.1109/TSP.2004.834400
46. C. Berger, Y. Benlachtar, R. Killey, Optimum clipping for optical OFDM with limited resolution DAC/ADC. Advanced Photonics, OSA Technical Digest, paper SPMB5 (2011)

Part II
Non-Orthogonal Multiple Access (NOMA) in the Power Domain

Chapter 5
NOMA: An Information-Theoretic Perspective

Mojtaba Vaezi and H. Vincent Poor

5.1 What Is Non-Orthogonal Multiple Access (NOMA)?

Multiple access lies at the heart of cellular communication systems. It refers to a technique that allows multiple users to share a communication channel. The first-generation (1G) to the fourth-generation (4G) of cellular networks have adopted radically different multiple access schemes with one common theme in mind—to have *orthogonal* signals for different users at the receiver side [1]. In particular, 1G to 4G cellular networks has adopted one or more of the following multiple access methods:

- Frequency division multiple access (FDMA)
- Time division multiple access (TDMA)
- Code division multiple access (CDMA)
- Orthogonal frequency division multiple access (OFDMA)
- Space division multiple access (SDMA)

For example, in OFDMA which has widely been used in 4G systems, different users' signals are orthogonal in the frequency and/or time domains. In other words, one orthogonal frequency division multiplexing (OFDM) *resource block* (180 kHz) cannot be allocated to more that one user.

Non-orthogonal multiple access (NOMA) [2], in contrast, allows multiple users to share the same resource elements, be it in the time, frequency, space, or code domain. NOMA is currently a hot research topic for 5G and beyond systems, both in academia and industry. While it is concerned with "non-orthogonality" of multiple access, it appears that the research community is perceiving this term in somewhat

M. Vaezi (✉)
Villanova University, Villanova, PA, USA
e-mail: mvaezi@villanova.edu

H. Vincent Poor
Princeton University, Princeton, NJ, USA
e-mail: poor@princeton.edu

M. Vaezi et al. (eds.), *Multiple Access Techniques for 5G Wireless Networks and Beyond*, https://doi.org/10.1007/978-3-319-92090-0_5

different ways. Due to the different interpretations, there is not a consensus about applying this term to some well-known existing techniques such a CDMA. While the majority of recent works see CDMA as an orthogonal multiple access (OMA) technique, there are a group of other papers that categorize it as a NOMA technique. In the following, we present and discuss different viewpoints used to define *non-orthogonality* in NOMA.[1]

NOMA Definitions and Viewpoints

- **Superposition Coding with Successive Interference Cancellation**: A large body of papers consider NOMA to be equivalent to superposition coding and successive interference cancellation (SC-SIC), respectively, at the transmitter and receivers. This is partly because the first paper using the term NOMA considered the problem of downlink transmission with SIC [2] and partly due to the fact that SC-SIC is the capacity-achieving technique for the downlink channel in single-cell single-input single-output (SISO) transmission, as we discuss later in this chapter. In fact, the key term is SIC as it also appears in the uplink transmission. With this in mind, SC-SIC is also applied to several different cases, such as multiple-input multiple-output (MIMO) networks and multi-cell networks (SISO or MIMO), in which SC-SIC is suboptimal. This method is also known as power domain NOMA.
- **Overloading**: A second important view is to distinguish NOMA and OMA based on the system loading. In this setting, NOMA refers to overloaded systems and "overloading" means to have more than one user per available resource element in the time, frequency, code, or space domain. This point of view is rooted in CDMA systems. With this definition, a CDMA system will be seen as a NOMA scheme if it is overloaded (i.e. when there are more users than the number of codes) and will be considered as an OMA scheme if there are more codes than the number of users in the systems. Examples of the NOMA schemes developed with this view are low-density spreading (LDS) CDMA, LDS-OFDM, sparse code multiple access (SCMA), and multi-user shared access (MUSA), which are collectively known as code domain NOMA. This definition (overloading) can be applied to other settings such as SDMA systems. That is, similar to the CDMA case, a multi-user MIMO (MU-MIMO) system can be seen as a NOMA or OMA scheme. The former happens when the number of transmit antennas n_t is less than the number of users K, whereas in the latter case $n_t \geq K$.
- **Linear Transform Decoding**: Some even define NOMA based on the complexity of multi-user detection. With this point of view, in an OMA scheme, the signals of different users can be separated in orthogonal subspaces using a linear transform. Then, any scheme that does meet this definition can be

[1] We should highlight that in this chapter we are discussing non-orthogonal *multiple access* methods. This should not be confused with non-orthogonal *random access* which is another related topic and will be discussed in Chap. 17 of this book.

categorized as NOMA. For example, CDMA systems *theoretically* can be constructed using independent random coding for different users. Such CDMA systems are naturally non-orthogonal. Since the 1990s, there has been a vast amount of research on the capacity of multiple access systems based on CDMA (involving power control, serial interference cancellation, dirty-paper coding, etc.)[2] However, historically, CDMA implies direct-sequence CDMA (DS-CDMA) operation as used in 3G wideband CDMA (WCDMA) systems. Since the 2000s, other types of CDMA schemes have been developed, such as interleave division multiple access (IDMA) [3]. In [3], NOMA is used to cover both DS-CDMA and IDMA. In this respect, NOMA has a broader meaning than DS-CDMA. Therefore, when comparing CDMA and NOMA, we need to distinguish between CDMA in the general sense, i.e. CDMA defined by information theory, and CDMA in a narrow sense such as DS-CDMA.

- **Information-Theoretic View**: Looking from an information-theoretic perspective, NOMA may refer to any technique in which concurrent transmission is allowed over the same resources in time/frequency/code/space. This achieves a better rate region when compared to orthogonalization of one or some of the resources and includes SC-SIC but is not limited to that. In this context, other techniques such as *rate-splitting* (RS) and *dirty-paper coding* (DPC) can be seen as NOMA. On the other hand, CDMA techniques that are based on orthogonality in the code domain will be seen as OMA techniques. This definition is very broad and includes many existing techniques as a subset, and in general, its concern is to promote optimal strategies in various uplink/downlink communication strategies.

In this chapter, NOMA refers to the last sense.[3] We should highlight that the theory of NOMA has been around for many years. In effect, the basic premise behind NOMA is to reap the benefits promised by information theory for the downlink and uplink transmission of wireless systems, modelled by the *broadcast channel* (BC) and *multiple access channel* (MAC), respectively. Capacity regions of the BC and MAC have been established several decades ago [1, 4, 5], and concurrent non-orthogonal transmission is the optimal transmission strategy in both cases. That is, in general, to achieve the capacity region, the users must transmit at the same time and frequency. In particular, the capacity region of the BC is achieved using *superposition coding* at the base station (BS). For decoding, the user with the stronger channel gain (usually the one closer to the BS) uses successive interference cancellation (SIC) to decode

---

[2]For example, CDMA with power control is capacity achieving for the single-antenna broadcast channel.

[3]However, in the SDMA case, we assume that the system is overloaded. This is relevant in the MIMO (and MU-MIMO) case where "space" comes in as a new resource. We note that when the number of transmit antennas n_t is greater than the number of users K, different users can be served in different (orthogonal) spaces, reducing the underlying system to an OMA one.

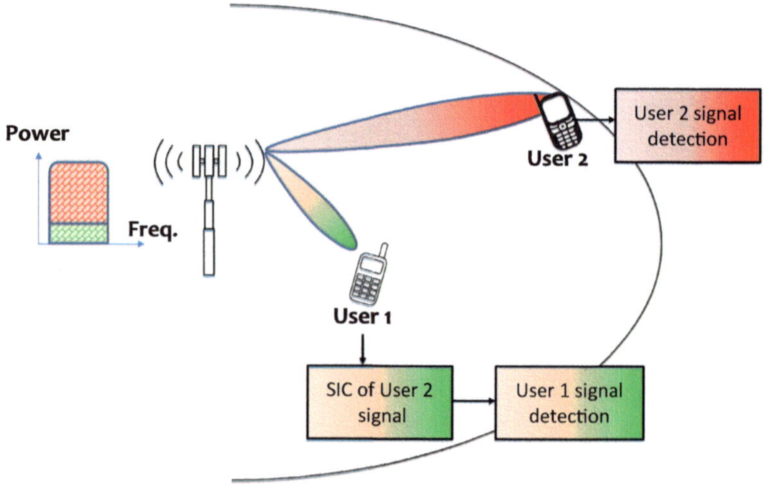

Fig. 5.1 Illustration of single-cell downlink NOMA using power domain multiplexing for two users

its signal free of interference, while the user with the weaker channel gain treats the signal of the stronger users as noise. This is illustrated in Fig. 5.1. Similarly, to get the highest achievable region in multi-cell systems, concurrent non-orthogonal transmission is still required [5–7], and orthogonal transmission is suboptimal.

5.2 What Drives NOMA?

The next generation of wireless networks must support very high throughput, low latency, and massive connectivity. According to the international telecommunication union (ITU) [8], 5G networks must fulfill several requirements including:

(1) a minimum peak data rate of 10 Gbps (100 times more than that in the 3rd Generation Partnership Project (3GPP) Long-Term Evolution (LTE))
(2) a latency of 1 ms (ten times lower than that in 4G networks), and
(3) a connection density of 1,000,000 devices per km^2 (100 times more than 4G networks).

Effective multiple access is a key enabler in achieving these requirements. As noted above, the first to the fourth-generations of cellular networks have adopted different multiple access schemes with one common theme in mind—to have orthogonal signals for different users at the receiver side. By allowing multiple users to share all domains (e.g. time, frequency, space), NOMA can address the above challenges of the next generation of wireless networks more efficiently than the conventional *orthogonal* multiple access schemes. NOMA can increase spectral efficiency and user-fairness by exploiting a capacity-achieving scheme in the downlink. It can support more connections in the uplink by letting multiple users simultaneously access the same wireless resources, which, in turn, can reduce latency [9–12].

Code domain NOMA[4] uses user-specific sequences for sharing the entire radio resource. On the other hand, power domain NOMA exploits the channel gain differences between the users for multiplexing via power allocation. In this chapter, we study NOMA in the power domain, a technique that can improve wireless communication through the following benefits:

- ***Massive Connectivity***: There appears to be a reasonable consensus that NOMA is essential for massive connectivity. This is because the number of served users in all OMA techniques is inherently limited by the number of resources (e.g. the number of codes in CDMA and the number of resource blocks in OFDMA). In contrast, by superimposing all users' signals, NOMA theoretically can serve an *arbitrary* number of users even within one resource block. In this sense, NOMA can be tailored to Internet of Things (IoT) applications where a large number of devices sporadically transmit a small number of packets. In fact, allocating an entire resource block (180 kHz in LTE) to one device, as is done in OMA, is extremely inefficient.
- ***Low Latency***: Latency requirements for 5G applications are rather diverse and very stringent in some cases. For example, ITU requires a user plan latency of 4ms and 1ms for enhanced mobile broadband (eMBB) and ultra-reliable and low-latency communications (URLLC), respectively [8]. With OMA, it is very difficult to guarantee such stringent delay requirements because no matter how many bits a device wants to transmit the device must wait until an unoccupied resource block becomes available. In contrast, NOMA supports flexible scheduling since it can accommodate a *variable* number of devices depending on the application that is being used and the perceived quality of service (QoS) of the device.
- ***High Spectral Efficiency***: According to ITU requirements for IMT-2020, downlink peak spectral efficiency should be 30 bits/s/Hz. NOMA offers a higher spectral efficiency and a better user-fairness compared to OMA. As will be seen in Sect. 5.3, NOMA is the theoretically *optimal* way of using the spectrum for both uplink and downlink communications in a single-cell network. Such better performance is achieved due to the fact that every NOMA user can enjoy the whole bandwidth, whereas OMA users are limited to a smaller fraction of the spectrum which is inversely proportional to the number of users. NOMA can also be combined with other emerging technologies, such as massive MIMO and millimeter wave (mmWave) technologies, to further improve spectral efficiency and support higher throughput.

[4] Sparse code multiple access (SCMA) is an important variant of NOMA in the code domain.

In consideration of the above benefits, NOMA has drawn significant attention from both academia and industry during past a few years. However, as we briefly mentioned earlier and will see in detail in this chapter, from an information-theoretic perspective the theory behind NOMA has been established for several decades. Nevertheless, despite this insight from information theory, orthogonal multiple access techniques have been used in the cellular networks from 1G to 4G. This was mainly to avoid inter-user *interference cancellation* which would have resulted in unacceptably complex receivers.

Today, with the advances in processing power, the implementation of interference cancellation at user equipment has been made practical. For example, a category of relatively complex user terminals, known as network-assisted interference cancellation and suppression (NAICS) terminals, has recently been adopted in 3GPP LTE-A. Such technological advances, in conjunction with the need to support exponentially increasing numbers of devices and better spectral efficiency, have motivated a new wave of research on NOMA. Saito et al. first showed that NOMA can improve system throughput and user-fairness over a SISO channel using OFDMA [2]. The spectral efficiency of NOMA-based systems is further boosted when combined with MIMO communication [13–15]. Successful operation of this technique, however, depends on knowledge of the channel state information (CSI) between the BS and the end-users. More practical solutions, e.g. those with limited and delayed CSI, are crucial in making NOMA workable. Today, a variation of NOMA, known as multi-user superposition transmission [12], is considered for the 3GPP LTE-A systems.

While recent advances in processor capabilities have made SIC, and consequently NOMA, feasible, significant research challenges remain to be addressed before NOMA can be deployed. In addition to the above practical issues, NOMA-based transmission introduces new security and privacy challenges. This is because in NOMA-based transmission a user with a better channel is able to decode the other user's signal. Even, a user with a weaker channel can also partly decode the stronger user's signal. On top of that, wireless transmission is naturally vulnerable to external eavesdroppers. Although upper-layer security approaches (e.g. cryptography) can be used to secure transmissions, there are numerous risks in cryptographic methods due to the rapid advancement of computing technologies. Also, cryptographic approaches require a key management infrastructure which should be secured, in turn. Moreover, traditional key agreement algorithms are not suitable for many existing and emerging wireless networks, such as ad hoc networks and IoT, since they consume scarce resources such as bandwidth and battery power.

5.3 Theory Behind NOMA

Analysis of cellular communication can generally be classified as either *downlink* or *uplink*. In the downlink channel, the BS simultaneously transmits signals to multiple users, whereas in the uplink channel multiple users transmit data to the same BS.

As noted above, from an information-theoretic perspective, the downlink and uplink are modelled by the broadcast channel and multiple access channel, respectively. The basic premise behind single-cell NOMA in the power domain is to reap the benefits promised by the theory of multi-user channels [1, 5]. As such, we review what information theory promises for these channels, both in the single-cell and multi-cell settings. In particular, we seek to answer the following two questions in this section: (1) What are the highest achievable throughputs for these multi-user channels? and (2) how can a system achieve such rates?

5.3.1 *Single-Cell NOMA*

For simplicity of illustration, we first consider a network consisting of a single cell. In addition, we assume that there are only two users in that cell. Later, we discuss the general case with multiple cells each consisting of multiple users. For OMA, we consider a TDMA technique[5] where a fraction α of the time ($0 \le \alpha \le 1$) is dedicated to user 1 and a fraction $\bar{\alpha} \triangleq 1 - \alpha$ of the time is dedicated to user 2.

The capacity regions of the two-user MAC and BC are achieved via NOMA, where both users' signals are transmitted at the same time and in the same frequency band [5]. To gain more insight, we describe how these regions are obtained.

Throughout the chapter, we use

$$\mathscr{C}(x) \triangleq \frac{1}{2}\log_2(1+x), \tag{5.1}$$

and $\gamma_i = |h_i|^2 P$ is the received signal-to-noise ratio (SNR) for user i, where h_i is the channel gain, P is the transmitter power, and the noise power is normalized to unity.

5.3.1.1 Two-User MAC (Uplink)

The capacity regions of the two-user MAC is achieved via non-orthogonal transmission in which both users' signals are transmitted at the same time and in the same frequency band [5]. The curves labelled NOMA in Fig. 5.2 represent the capacity regions of the MAC for different values of P_1 and P_2, which are the respective transmit powers of the two users. From these figures, it can be seen that except for a few points OMA is strictly suboptimal. One of these points is the sum capacity of

[5]FDMA has exactly the same performance [5].

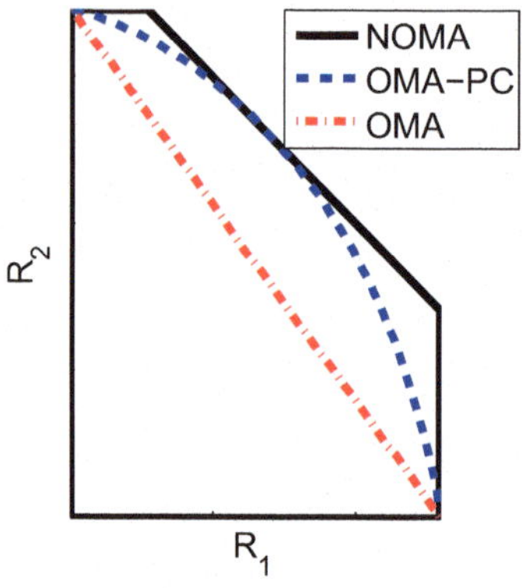

(a) $P_1 = 5$, $P_2 = 10$

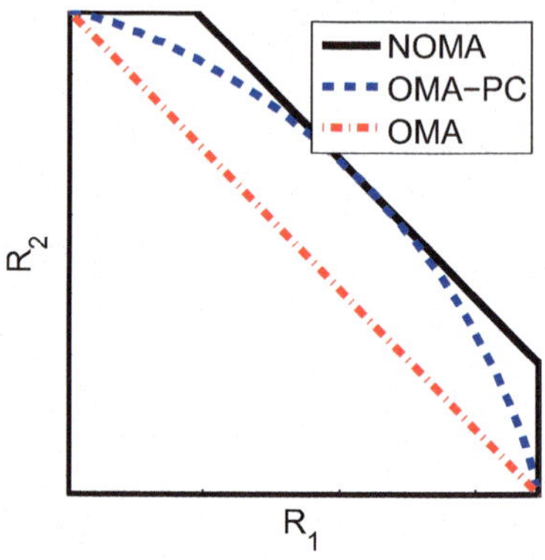

(b) $P_1 = 10$, $P_2 = 10$

Fig. 5.2 Best achievable regions by OMA and NOMA in the two-user MAC (uplink) for different values of the two users' transmit powers P_1 and P_2

the MAC. This means that both OMA and NOMA can achieve the sum capacity of MAC.

To gain more insight, we describe how the above regions are obtained. Using OMA, each user sees a single-user channel in its dedicated fraction of the time. As a result

$$R_1 = \alpha\mathscr{C}(\gamma_1), \tag{5.2a}$$
$$R_2 = \bar{\alpha}\mathscr{C}(\gamma_2), \tag{5.2b}$$

are achievable. If power control (PC) is applied, these rates can be increased to

$$R_1 = \alpha\mathscr{C}\left(\frac{\gamma_1}{\alpha}\right), \tag{5.3a}$$
$$R_2 = \bar{\alpha}\mathscr{C}\left(\frac{\gamma_2}{\bar{\alpha}}\right). \tag{5.3b}$$

This region is labelled NOMA-PC in Fig. 5.2a, b.

In the case of NOMA, both users concurrently transmit, and their signals interfere with each other at the BS. The BS can use SIC to achieve any point in the NOMA region, which is the capacity region of this channel [1].

Theorem 1 *The capacity region of the two-user MAC is the set of nonnegative* (R_1, R_2) *such that*

$$R_1 \leq \mathscr{C}(\gamma_1), \tag{5.4a}$$
$$R_2 \leq \mathscr{C}(\gamma_2), \tag{5.4b}$$
$$R_1 + R_2 \leq \mathscr{C}(\gamma_1 + \gamma_2). \tag{5.4c}$$

Particularly, to achieve the right corner point of the capacity region (in which $R_1 = \mathscr{C}(\gamma_1)$), the BS first decodes user 2's signal, treating the other signal as noise. This results in $R_2 = \mathscr{C}(\frac{\gamma_2}{\gamma_1+1})$. The BS then removes user 2's signal and decodes user 1's signal free of interference; i.e. $R_1 = \mathscr{C}(\gamma_1)$. As a result, the sum capacity $R_1 + R_2 = \mathscr{C}(\gamma_1 + \gamma_2)$ is achievable. To achieve the other corner point, the order of decoding needs to be changed. From Fig. 5.2, it is seen that the gap between the NOMA and OMA regions becomes larger if power control is not used in OMA.

5.3.1.2 Two-User BC (Downlink)

Similar to the two-user MAC, the capacity region of the two-user BC is known and is achieved via non-orthogonal transmission in which both users' signals are transmitted at the same time and in the same frequency band [5]. The curves in Fig. 5.3a–c represent the capacity regions of the BC as well as the best achievable regions obtained by OMA for different values of channel gains. From these figures, it can be seen that except for a few points, OMA is strictly suboptimal in the downlink. In fact, OMA can only achieve

$$R_1 = \alpha\mathscr{C}(\gamma_1), \tag{5.5a}$$

$$R_2 = \bar{\alpha}\mathscr{C}(\gamma_2). \tag{5.5b}$$

However, making use of a NOMA scheme can strictly increase this rate region as shown in Fig. 5.3. In particular, the capacity region of this channel is known and can be achieved using superposition coding at the BS and successive interference cancellation at the receiver. For decoding, the user with the stronger channel uses SIC to decode its signal free of interference at a rate of $R_1 = \mathscr{C}(\beta\gamma_1)$ while the other

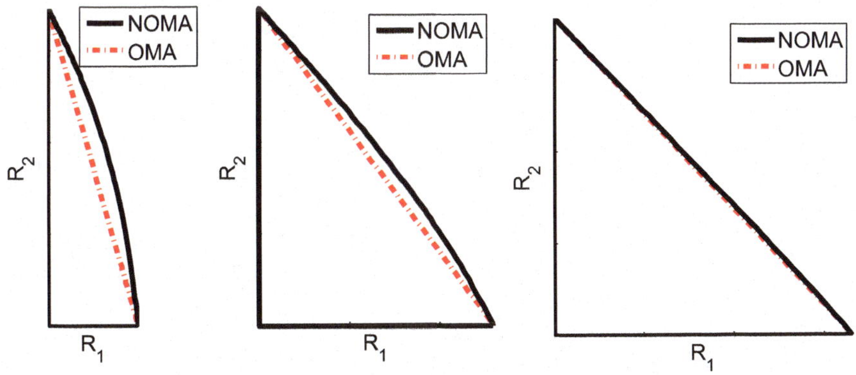

(a) $P_1 = 1, P_2 = 10$ (b) $P_1 = 5, P_2 = 10$ (c) $P_1 = 9, P_2 = 10$

Fig. 5.3 Best achievable regions by OMA and NOMA in the BC (download) for different values of P_1 and P_2

user is capable of decoding at a rate of $R_2 = \mathscr{C}(\frac{\bar{\beta}\gamma_2}{\beta\gamma_2+1})$ where β is the fraction of the BS power allocated to user 1's data and $\bar{\beta} = 1 - \beta$. By varying β from 0 to 1, any rate pair (R_1, R_2) on the boundary of the capacity region of the BC (NOMA region) can be achieved. That is,

Theorem 2 *The capacity region of the two-user BC is the set of nonnegative* (R_1, R_2) *such that*

$$R_1 \leq \mathscr{C}(\beta\gamma_1), \tag{5.6a}$$

$$R_2 \leq \mathscr{C}\left(\frac{\bar{\beta}\gamma_2}{\beta\gamma_2 + 1}\right), \tag{5.6b}$$

in which $\beta \in [0\ 1]$ *and* $\bar{\beta} = 1 - \beta$.

The fact that the capacity region of downlink NOMA is known enables us to find the optimum power allocation corresponding to any point (R_1, R_2) on the boundary of the capacity region. In fact, all we need to know to achieve such a rate pair is to find what fraction of the BS power should be allocated to each user. Corresponding to each (R_1, R_2), there is a $0 \leq \beta \leq 1$ such that βP and $\bar{\beta} P$ are the optimal powers for user 1 and user 2, respectively, where P is the BS power. Conversely, every β generates a point on the boundary of the capacity region.

The above argument implies that NOMA can improve *user-fairness* smoothly and in an optimal way by flexible power allocation. Suppose that a user has a poor channel or it has not been served for a long time (in OMA). To boost this user's rate and improve user-fairness, the BS can simply increase the fraction of power allocated to this user. We can look at this problem from yet another perspective. To increase the rate of such a user, we can maximize the weighted sum-rate $\mu R_1 + R_2$ where a high weight (μ) is given to such a user. This is because, to maximize $\mu R_1 + R_2$ for any $\mu \geq 0$, there exists an optimal power allocation strategy, determined by β. Seeing that $\mu > 1$ ($\mu < 1$) corresponds to the case where user 1 has higher (lower) weight than user 2, to improve user-fairness, we can assign an appropriate weight to the important user and find the corresponding β.

5.3.1.3 K-User Uplink/Downlink

In the above, we described coding strategies for the two-user uplink/downlink channels. Interestingly, very similar coding schemes are still capacity achieving for the K-user MAC and BC, as described in the following.

K-User MAC: To achieve the capacity region of the K-user MAC, the users transmit their signals concurrently and the BS applies SIC decoding. Specifically, we have [5].

Theorem 3 *The capacity region of the* K*-user MAC is the set of nonnegative* $(R_1, \ldots, R_K)$ *such that*

$$\sum_{j \in S} R_j \leq \mathscr{C}\Big(\sum_{j \in S} \gamma_j\Big) \qquad \textit{for every } S \subseteq [1:K]. \tag{5.7}$$

For example, for $K = 3$, the capacity region is the set of nonnegative (R_1, R_2, R_3) such that

$$R_1 \leq \mathscr{C}(\gamma_1), \tag{5.8a}$$
$$R_2 \leq \mathscr{C}(\gamma_2), \tag{5.8b}$$
$$R_3 \leq \mathscr{C}(\gamma_3), \tag{5.8c}$$
$$R_1 + R_2 \leq \mathscr{C}(\gamma_1 + \gamma_2), \tag{5.8d}$$
$$R_1 + R_3 \leq \mathscr{C}(\gamma_1 + \gamma_3), \tag{5.8e}$$
$$R_2 + R_3 \leq \mathscr{C}(\gamma_2 + \gamma_3), \tag{5.8f}$$
$$R_1 + R_2 + R_3 \leq \mathscr{C}(\gamma_1 + \gamma_2 + \gamma_3). \tag{5.8g}$$

The capacity-achieving scheme is based on non-orthogonal transmission that allows multiple users to transmit at the same time and frequency. To be specific, the capacity region is achieved by point-to-point codes, successive cancellation decoding, and time-sharing. Again, OMA is strictly suboptimal [1].

K-**User BC**: To achieve the capacity region of the K-user BC, the users' signals are superimposed at the BS and transmitted altogether. Without loss of generality assume that $\gamma_1 \geq \gamma_2 \geq \cdots \geq \gamma_K$. At the receiver sides, receiver $k \in [1, \ldots, K]$ first decodes the signal of users $k+1, \ldots, K$ and removes them from the received signals. It (receiver k) then decodes its own signal treating users' $1, \ldots, k-1$ signals as noise. As a result, we obtain

Theorem 4 *The capacity region of the K-user BC is the set of nonnegative $(R_1, \ldots, R_K)$ such that*

$$R_k \leq \mathscr{C}\left(\frac{\beta_k \gamma_k}{1 + \sum_{j=1}^{k-1} \beta_j \gamma_k}\right), \tag{5.9}$$

in which $k \in [1, \ldots, K]$, $\beta_j \geq 0\ \forall j$ and $\sum_{j=1}^{K} \beta_j = 1$.

Remark 1 To achieve the above rates, it is important to note that the single-antenna K-user Gaussian BC is a set of degraded channels. This implies that the users can be ordered based on their channel strengths (for example, in Theorem 4, we have assumed that $\gamma_1 \geq \gamma_2 \geq \cdots \geq \gamma_K$).

We note that due to the SIC decoding, the user with the strongest channel (user 1) is able to decode its own signal free of interference, whereas the user with the weakest channel (user K) has to treat all other users' signals as noise. For example, for $K = 3$ the capacity region is the set of nonnegative (R_1, R_2, R_3) such that

$$R_1 \leq \mathscr{C}(\beta_1\gamma_1), \tag{5.10a}$$

$$R_2 \leq \mathscr{C}\left(\frac{\beta_2\gamma_2}{1+\beta_1\gamma_2}\right), \tag{5.10b}$$

$$R_3 \leq \mathscr{C}\left(\frac{\beta_3\gamma_3}{1+(\beta_1+\beta_2)\gamma_3}\right). \tag{5.10c}$$

So far, we have assumed a network with a single cell. In fact, most of the work on NOMA is limited to single-cell analysis, where there is no co-channel interference caused by an adjacent BS. However, to verify the benefits of NOMA in a more realistic setting, it is necessary to consider a multi-cell network. Specifically, as wireless networks get denser and denser, inter-cell interference (ICI) becomes a major obstacle to achieving the benefits of NOMA. In the next section, we discuss the theory behind NOMA in multi-cell networks.

5.3.2 *Multi-Cell NOMA*

In a multi-cell setting, finding the best achievable region is much more involved than in the single-cell case, and simple channel models are insufficient. A few models, including the interference channel, interfering MAC, and interfering BC can be used to model multi-cell networks. Unfortunately, the capacities of these channels are unknown in general. However, the known achievable rate regions for these channels indicate the superiority of NOMA to OMA.

5.3.2.1 Interference Channel (IC)

The capacity region of the two-user IC is not known in general; however, it is known that OMA is strictly sub-optimal. Han and Kobayashi introduced an achievable region in 1981 [6], which is still the best known *inner bound* for the general interference channel. In the Han-Kobayashi (HK) scheme, each user can split its message to be sent into two submessages of smaller rates and power. These are known as *private* and *common* messages; the former is intended to be decoded only at the respective receiver, whereas the latter can be decoded at both receivers. The rationale behind this coding scheme is to decode part of the interference (the common message) and treat the rest as noise. The optimal input distributions are not known for the HK region. As such, commonly a subset of the HK region with Gaussian codebooks is used to represent the HK region for the Gaussian channel; see, e.g. [16–19].

The basic HK scheme employs *rate-splitting* (RS) and superposition coding (SC) at each transmitter and SIC at the receiver. Since both transmitters send their signals concurrently at the same frequency, the HK scheme is a NOMA scheme. Flexibility in splitting each user's transmission power into the common/private portions of information and *time-sharing* between them make the HK scheme very strong, but complicated. Not surprisingly, though, the optimal HK strategy is not well-understood, in general. Nevertheless, the HK scheme is known to be within $\frac{1}{2}$ bit of the capacity region of the two-user Gaussian interference channel [16]. Although this gap could be due to a suboptimal scheme, loose outer bounds, or both, the outer bounds seem to be the most crucial.

In general, the HK scheme applies time-sharing to improve the basic HK region and can be seen as a combination of NOMA and OMA [7]. Although the basic HK scheme is optimal for the strong and very strong interference regimes, when one interference link is strong and the other one is weak, time-sharing can enlarge the basic HK region in general. Specifically, we have [7, 20]

Lemma 1 *The HK achievable region for the one-sided IC is the set of rate pairs* (R_1, R_2) *satisfying*

$$R_1 \le \lambda_1 R_{11}, \tag{5.11a}$$

$$R_2 \le \lambda_1 R_{21} + \lambda_2 R_{22}, \tag{5.11b}$$

in which

$$R_{11} \le \gamma\left(\frac{\frac{P_1}{\lambda_1}}{1 + a\beta_1 P_{21}}\right), \tag{5.12a}$$

$$R_{21} \le \gamma\left(\frac{a\bar{\beta}_1 P_{21}}{1 + \frac{P_1}{\lambda_1} + a\beta_1 P_{21}}\right) + \gamma(\beta_1 P_{21}), \tag{5.12b}$$

$$R_{22} \le \gamma(P_{22}), \tag{5.12c}$$

where $\lambda_1 + \lambda_2 = 1$, $\lambda_1 P_{21} + \lambda_2 P_{22} = P_2$, $0 \le \beta_1 \le 1$, *and* $\bar{\beta}_1 = 1 - \beta_1$.

It is worth mentioning that Lemma 1 characterizes a strictly better region when compared with the HK region without time-sharing. In particular, for $\lambda_1 = 1$, this lemma reduces to the HK region without time-sharing (i.e., the basic HK region). These regions are compared in Fig. 5.4. Also, substituting $P_{21} = 0$, Lemma 1 reduces to

$$R_1 \le \lambda_1 \gamma\left(\frac{P_1}{\lambda_1}\right), \tag{5.13a}$$

$$R_2 \le (1 - \lambda_1)\gamma\left(\frac{P_2}{1 - \lambda_1}\right), \tag{5.13b}$$

for $0 \le \lambda_1 \le 1$. This region is known as the TDMA (or FDMA) region. The main difference between the TDMA and time-sharing regions is in the fact that in the TDMA approach only one user is allowed to transmit during each subband, while in time-sharing method, both users can transmit in the same subband (e.g. during λ_1 in the region defined by Lemma 1).

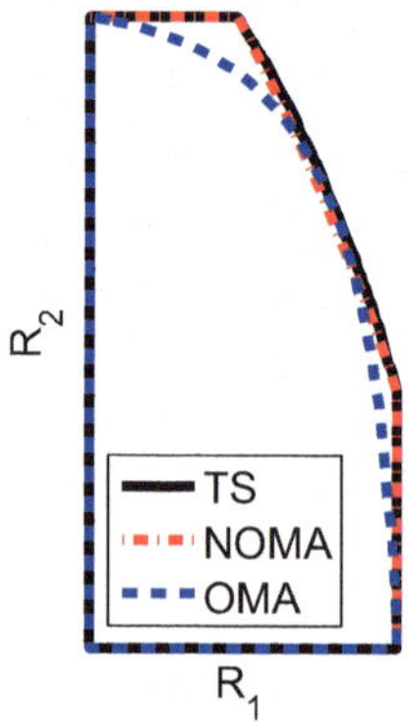

(a) $P_1 = 1$, $P_2 = 3$

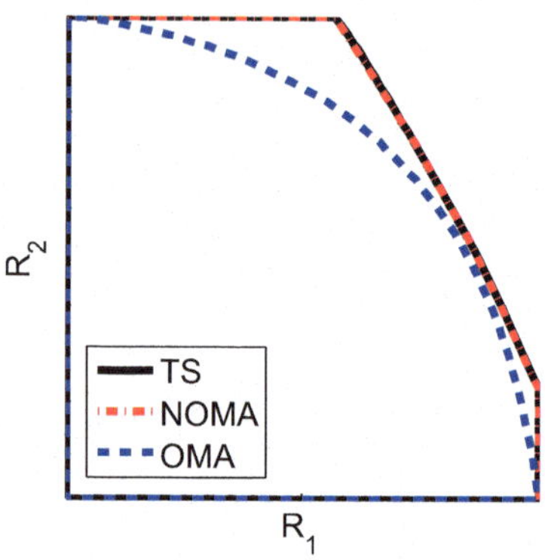

(b) $P_1 = 3$, $P_2 = 3$

Fig. 5.4 Best achievable regions by OMA, NOMA, TS (applying time-sharing to OMA and NOMA in different time slots) in the IC (multi-cell network) for different values of P_1 and P_2

In other words, the HK scheme that combines NOMA and OMA gives the largest rate region [7], as shown in Fig. 5.4 for different values of P_1 and P_2. In these plots, OMA refers to TDMA, whereas NOMA refers to the basic HK scheme in which time-sharing is not applied. In this case, both transmitters send their messages using the HK strategy but in one time-slot only. The third curve, labelled TS, is based on the HK scheme with time-sharing (TS) in which two time slots are used: in one time-slot, both users are active, while in the other time-slot, only one of them is transmitting. As can be seen from this figure, both NOMA and OMA are suboptimal when compared with the case where NOMA and OMA are combined. We should highlight that the rate region obtained via the HK scheme without time-sharing (NOMA) is however very close to that with time-sharing (OMA + NOMA). That is, NOMA yields a good approximation of the best achievable region for this channel, for the two-user IC.

5.3.2.2 Interfering MAC and BC

Consider a mutually interfering two-cell network in the uplink, where each cell includes one MAC. Assume that only one of the transmitters of each MAC (typically the closest one to the cell-edge) is interfering with the BS of the other MAC. In this network, the interfering transmitters can employ HK coding, similar to that used in the IC, while the non-interfering transmitters in each MAC employ single-user coding. This NOMA-based transmission results in an inner bound which is within a one-bit gap of the capacity region [21]. Likewise, one can use the interfering BC to model a mutually interfering two-cell downlink network.

Despite years of intensive research in information theory, finding optimal uplink and downlink transmit/receive strategies for multi-cell networks remains rather elusive. In fact, as discussed earlier, even for a much simpler case of the two-user IC, the optimal coding strategy is still unknown. Nonetheless, fundamental results from information theory as a whole suggest that NOMA-based techniques result in a superior rate region when compared with OMA.

It should be highlighted that, despite the above insight from information theory, OMA techniques have been used in the cellular networks from 1G to 4G, mainly to avoid interference and due to its simplicity. In addition, the lack of understanding of optimal strategies for multi-cell networks has motivated pragmatic approaches in which interference is simply treated as noise.

5.3.3 NOMA in MIMO Networks

With the rapid advancement of multi-antenna techniques, today wireless nodes (particularly BSs) are often equipped with more than one antenna. Multi-antenna systems create spatial dimension which, in turns, opens the door to SDMA. In SDMA, multiple users can communicate at the same time-frequency but are distinguished in space. That is, each user has a different beam. Then, in the downlink and uplink of single-cell networks, we will have the MIMO-BC and MIMO MAC. In addition, similar to the multi-cell SISO networks, we can use MIMO IC to model a multi-antenna multi-cell network. Although in some cases the capacity region of these channels is unknown in general, similar to the SISO case, it is known from information theory that TDMA (OMA) is suboptimal and concurrent transmission (NOMA) can result in a better rate region. We discuss this in the following sections.

5.3.3.1 MIMO BC

Unlike the Gaussian SISO BC, the Gaussian MIMO BC is *non-degraded* in general. That is, in general, the MIMO BC users cannot be ordered based on their channel strengths. This is because the users' channels are matrices (or vectors), and matrices cannot be ordered in general, unlike the scalars in the SISO BC. This has been the main difficulty in proving the capacity of the MIMO BC. However, the capacity region of the MIMO BC has recently been established in [22] in which it is proved that the capacity region can be achieved by DPC [23]. Because of the non-degradedness of the MIMO BC, SC-SIC which is the capacity-achieving technique for the SISO BC is not capacity achieving for the MIMO BC. While in SC-SIC interference cancellation is performed at the receiver side, DPC constitutes interference cancellation at the transmitter side and consequently the transmitter requires knowledge of CSI. The same strategy is also the only technique that achieves the capacity region of the multiple-input single-output (MISO) BC. DPC is a non-orthogonal transmission (NOMA) strategy, and it is well-understood that TDMA (OMA) is suboptimal for the MIMO BC and the MISO BC.

DPC is, however, prohibitively complex for practical systems. Practical MIMO systems usually use linear precoding to simplify the transmitter design. This technique creates different beams for different users (or sets of users) and allocates a fraction of the total transmit power to each beam. At the receiver side, the interference from other users is treated as noise. Due to its simplicity, SDMA is usually implemented using linear precoding. This access technique is the basic principle behind several well-established techniques in 4G and upcoming 5G networks. The examples include but are not limited to multi-user MIMO, network MIMO, coordinated multipoint (CoMP), millimeter-wave MIMO, and massive MIMO [24].

Another line of research seeks to extend the SISO NOMA principles to MIMO NOMA transmission. In fact, the performance of NOMA can be further boosted in multi-antenna networks. MIMO NOMA solutions exploit multiplexing and diversity gains to improve outage probability and throughput, by converting the MIMO channel into multiple parallel channels. One approach is to directly apply SISO NOMA methods by making the MIMO NOMA networks degraded. That is to order the users based on an effective scalar channel and decode their messages using SIC. A second approach combines SDMA with superposition coding at the receiver and SIC at the transmitter. Such solutions try to allocate different beams to each group of users and then to use SISO NOMA within each group. By allocating different beams to each cluster (group), the interference between the clusters can be managed and removed. Within each cluster, the SISO NOMA solutions are then applied.

SDMA and MIMO NOMA can both be viewed as superpositions in the power domain with different approaches at the receiver side. In the former, the users are separated spatial beamformers, whereas in the latter case SIC is used at the receivers to separate the users [14]. In other words, SDMA fully treats the interference as noise whereas NOMA with SIC fully decodes interference. Recalling the flexible rate-splitting in the HK scheme for the IC, in which a user's transmission power is split between the common/private portions of information and part of the interference (the common message) is decoded while treating the remainder as noise, makes the achievable region larger, researchers have applied the same concept to the MIMO BC channel. RS obviously can bridge between the SDMA and NOMA with SIC by enabling a receiver to decode part of the interference and treat the remaining part of that as noise. In this sense, SDMA and NOMA with SIC can be seen as two extreme cases of RS. All of these approaches suggested by information theory are based on the concurrent transmission of multiple users signal without making them orthogonal and thus can be seen as different variants of NOMA. To summarize, from information-theoretic results, it is clear that OMA is suboptimal in MIMO networks too. This is valid both for single-cell and multi-cell (network MIMO) cases. The best techniques are either based on NOMA or a combination of NOMA and OMA using time-sharing. Several other techniques such as superposition coding, SIC, RS, and time-sharing fall within these NOMA schemes.

5.3.3.2 MIMO IC

The capacity region of the MIMO IC, similar to many other multi-user networks, is unknown. Finding the exact capacity region has been quite challenging for those channels. Because of this, approximations are widely used to get insight into the behavior of these channels. One commonly used approximation metric is degrees of freedom (DoF), or multiplexing gain. The DoF gives the pre-log of the capacity of a given channel in the high SNR regime; i.e.

$$DoF = \lim_{SNR \to \infty} \frac{\mathscr{C}(SNR)}{\log(SNR)}. \tag{5.14}$$

For example, the DoF of the MIMO channel with M and N antennas at the transmitter and receiver is $\min(M, N)$ which is its multiplexing gain. The DoF of the MIMO broadcast channel with M antennas at the transmitter and N_1 and N_2 antennas at the receivers is $\min(M, N_1 + N_2)$. Interference alignment (IA) is the main technique used to achieve the degrees of freedom of interferences channels. IA was introduced by Maddah-Ali et al. in the context of the MIMO X channel [25, 26], where an iterative achievable scheme for this channel built upon dirty-paper coding and successive decoding was introduced. IA refers to a mechanism for aligning (overlapping) interference spaces. It was then applied to the K-user SISO IC in [27] leading to the surprising conclusion that wireless networks are not essentially interference limited.

The DoF of the K-user SISO IC is equal to $\frac{K}{2}$ [27], meaning that each user can enjoy half of the spectrum in the high SNR regime. The DoF of the K-user user MIMO Gaussian IC with M antennas at each transmitter and N antennas at each receiver is also known [28, 29] and is obtained using IA. All these results show the suboptimality of orthogonalization. For example, in the three-user MIMO IC with two antennas at each node, IA allows a total of three DoF, whereas the schemes based on orthogonalization can allow a maximum of two DoF per channel use. In the orthogonalized solutions, e.g. TDMA, only one user will be active at a time and can send two symbols on its 2×2 MIMO channel, whereas IA allows each user to send only one information symbol, but all the users are active at all times. These results show that the DoF of the K-user MIMO IC is achieved when all the users are active at all times and orthogonalization is suboptimal. In other words, NOMA performs better than OMA.

5.4 Moving from Theory to Practice

As explained in Sect. 5.3, the basic theory behind NOMA has been around for many years. Specifically, the capacity region of SISO NOMA in the single-cell setting has been known for several decades. Moreover, it is known that to get the highest achievable region in multi-cell systems concurrent non-orthogonal transmission is required [5–7], and orthogonal transmission is suboptimal. This conclusion is based on the exact and approximated (DoF) analysis, as discussed in the previous section. Furthermore, the theoretical results for MIMO networks (MIMO BC, MIMO MAC, and MIMO IC) on the whole indicate that OMA is suboptimal and NOMA, with sophisticated schemes such as DPC and IA, can achieve better rates.

Despite these insights from information theory, orthogonal multiple access techniques have been used in the cellular networks from 1G to 4G. This has mainly been to avoid inter-user interference cancellation which would have resulted in unacceptably complex receivers. In fact, schemes such as SIC and DPC used for interference cancellation are very complex to implement in user equipment. Today, with the advances in processing power driven by Moore's law,[6] the implementation of interference cancellation at user equipment has been made practical, as discussed in Sect. 5.4.1. Such technological advances, in conjunction with the need to support exponentially increasing numbers of devices and better spectral efficiency, have motivated a new wave of research on NOMA.

5.4.1 SIC in 4G Networks

Owing to advances in processing power, interference cancellation on user equipment has become more practical and is realized in LTE-A networks in a few different settings. Some of these are listed below.

- **Network-Assisted Interference Cancellation and Suppression (NAICS)**: NAICS refers to a category of relatively complex user terminals that has recently been adopted by 3GPP LTE-A. Network-assisted interference cancellation/suppression can enable more effective interference cancellation/suppression at the user-side with possible network coordination [30]. In developing NAICS, an extensive study was done on advanced receivers with various capabilities of interference cancellation/suppression. As an example, single-user MIMO with minimum mean square error successive interference cancellation (MMSE-SIC) has been designed in LTE Release 8 [31].
- **Multi-User Superposition Transmission (MUST)**: MUST is a recent proposal for 3GPP for downlink mobile broadband (MBB) services [12]. MUST has different categories corresponding to different transmitting schemes [32]. Although the interference scenarios are not the same in NAICS and MUST, many of the receivers proposed for NAICS can also be used for MUST.

[6]Moore's Law, hypothesized by Intel founder Gordon Moore in 1965, states that the number of transistors in a dense integrated circuit will double approximately every two years. This enables a larger number of transistors to be concentrated in a given area which, in turn, results in a faster processor that can operate at lower power.

Saito et al. first showed that NOMA can improve system throughput and user-fairness over a SISO channel using OFDMA [2]. The spectral efficiency of NOMA-based systems is further boosted when combined with MIMO communication [13–15]. Successful operation of this technique, however, depends on knowledge of the CSI between the BS and the end-users. More practical solutions, e.g. those with limited and delayed CSI, are crucial to making NOMA workable. Today, a variation of NOMA, known as MUST [12], is being considered for use in 3GPP LTE-A systems, as noted above.

5.4.2 Multi-Cell NOMA Solutions

Inter-cell interference reduces a cell-edge user's performance and is the main issue in multi-cell networks. ICI management approaches are used to improve cell-edge users' performance. Depending on the availability of the data messages desired at the users among multiple BSs, multi-cell techniques can be categorized into coordinated scheduling/beamforming (CS/CB) and joint processing (JP) [33]. Specifically, in CS/CB, data for a user is only available at and transmitted from a single BS, whereas in JP, the data is shared among multiple BSs. ICI management approaches can be combined with NOMA resulting in multi-cell NOMA solutions, e.g. NOMA-JP and NOMA-CS/CB.

In NOMA-JP, the users' data symbols are available at more than one BS. NOMA-JP has two main categories: NOMA-joint transmission (JT) and NOMA-dynamic cell selection (DCS). In the former, multiple BSs are active and simultaneously serve a cell-edge user using a shared wireless resource. NOMA-JP can significantly improve the quality of the signals received by cell-edge users as the two BSs cooperate instead of interfering with each other. In NOMA-DCS, although the user's data is shared among multiple BSs, this data is transmitted only from one BS at a time. The transmitting BS can be dynamically changed over time.

In NOMA CS/CB, the users' data is not shared among two or more cooperating BSs. Specifically, in NOMA CB, the beamforming decision is made with coordination of other BSs. As an example, IA-based CB is applied in [34] in which two BSs jointly optimize their beamforming vectors in order to improve the data rates of cell-edge users by removing ICI. On the other hand, in NOMA-CS multiple BSs coordinate scheduling to serve NOMA users with less ICI. The cooperating BSs in NOMA CS/CB need to exchange global CSI and cooperative scheduling information via a standardized interface named X2 which may result in considerable overhead especially when the users are highly mobile. A review of different multi-cell NOMA techniques can be found in [11].

While recent advances in processing capabilities have paved the way for SIC, and consequently NOMA, significant research challenges remain to be addressed before NOMA can be deployed. In addition to the above practical issues, NOMA-based transmission may introduce new security and privacy challenges. We introduce the channel modes relevant to the physical layer security of NOMA in the next section.

5.5 Physical Layer Security in NOMA

In a NOMA-based transmission, a user with a better channel is capable of decoding the other user's signal. Even, a user with a weaker channel can also partly decode the stronger user's signal. This may introduce new security and privacy challenges. Moreover, wireless transmission is naturally vulnerable to external eavesdroppers. Although upper-layer security approaches (e.g. cryptography) are still relevant since only the legitimate user has a key to decode its message, there are numerous risks in cryptographic methods due to the rapid advancement of computing technologies. Besides, cryptographic approaches require a key management infrastructure which should be secured, in turn. Moreover, traditional key agreement algorithms are not suitable for many existing and emerging wireless networks, such as ad hoc networks and IoT, since they consume scarce resources such as bandwidth and battery power. Considering these challenges, physical layer security schemes are of interest.

5.5.1 Description of the Channel Models

The goal of this section is to identify and leverage information-theoretic channel models for securing communication in the context of NOMA. Three basic channel models are considered as depicted in Figs. 5.5, 5.6 and 5.7. In these figures, n_t, n_1, n_2, and n_e are the number of antennas at the transmitter (Tx), legitimate receivers (Rx1 and Rx2), and eavesdropper (Eve), respectively. For the purpose of illustration, we only consider the case of two legitimate receivers. These models can be used for NOMA with two users. A more general setting is the case with K ($K \geq 2$) legitimate receivers. It should be highlighted that these channels also can be seen as multi-user MIMO BCs.

Strictly speaking, with these channel models, NOMA is relevant only in cases in which the number of transmit antennas (n_t) is less than the number of users (K), i.e. when the system is overloaded. This is because for $n_t < K$, there is more than one user per resource block (time/frequency/space), while for $n_t \geq K$, this figure is less than or equal to one, implying that each user's signal can be transmitted in a dedicated resource block orthogonal to the other users' resources. In the latter case, there is at least one spatial degree of freedom per user which is the meaningful operating regime for SDMA.

It should be highlighted that regardless of the relation between n_t and K, optimal beamforming provides interesting open problems. In the conventional multi-user MIMO BC, n_t users are selected out of K to satisfy the constraint $n_t \leq K$. However, in 5G networks, due to the explosion of connected devices, K is usually very large. In addition, many of these connections are from low-rate IoT devices, and a dedicated resource block may allocate more resources than needed.[7] In view of these factors,

[7] In LTE one resource block is 180 kHz.

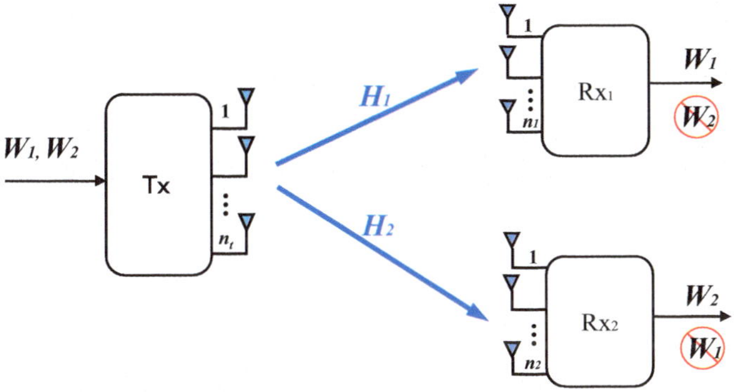

Fig. 5.5 MIMO BC with confidential messages

NOMA is a useful technology for accommodating a higher number of connections and improving spectral efficiency.

5.5.1.1 MIMO BC with Confidential Messages

A MIMO BC, similar to any other BC, has an inherent vulnerability in terms of security and privacy, even if there is no external eavesdropper. This is due to the fact that each legitimate user can, partly or wholly, decode the other legitimate user's message. To study this issue, the MIMO BC with confidential messages (see Fig. 5.5) has been introduced, in which independent messages W_1 and W_2 are intended for their respective receivers but need to be kept secret from the other receiver. This model is important in NOMA transmission as each legitimate user is a potential threat to security or privacy of its counterpart.

5.5.1.2 MIMO BC with an External Eavesdropper

In the above channel model, there is no external eavesdropper, but each legitimate user is seen as a potential adversary (eavesdropper) to the other legitimate user. In Fig. 5.6, another channel that models a class of BCs with an external eavesdropper is depicted. In this model, the messages are to be secured from the eavesdropper but not necessarily from other legitimate users. This channel is also known as the MIMO multi-receiver wiretap channel whose capacity was established in [35], under a matrix power constraint. This channel model is relevant for secure NOMA transmission when external eavesdropping is the only security issue but the legitimate users are not security threats to each other. It can be used, for example, for the case where the users are resource-limited single-antenna devices such as sensors and, thus, are not capable of or interested in decoding the other users' data. Therefore, the confidentiality of

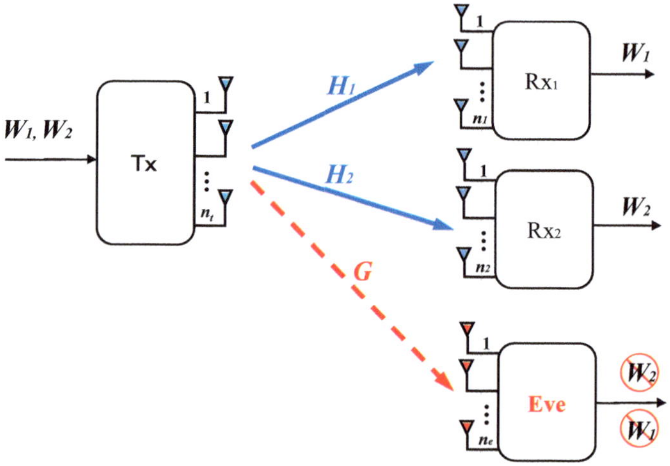

Fig. 5.6 MIMO BC with an external eavesdropper

a message from other legitimate users is not an issue. Obviously, in such cases, the channel model reduces to a MISO BC.

5.5.1.3 MIMO BC with Confidential Messages and an External Eavesdropper

A third and more general channel modeling secure communication in a NOMA-based transmission is the MIMO BC with confidential messages and an external eavesdropper, depicted in Fig. 5.7. This channel consists of a transmitter who wishes to communicate two messages to their respective receivers, each needing to be kept secret from the other legitimate receiver and a third unauthorized receiver (eavesdropper). This configuration models NOMA with two users and one external eavesdropper.

While the capacity region of the channel in Fig. 5.7 is unknown, the capacity regions of the channel models in Fig. 5.5 and Fig. 5.6 are established in [35] and [36], respectively. In [36], using a combination of DPC and stochastic encoding, known as secret DPC, it is shown that both messages can be simultaneously transmitted at their respective maximal secrecy rates under a matrix power constraint. Similar to the channel model in Fig. 5.5, secret DPC achieves the capacity region of the channel in Fig. 5.6, under a matrix power constraint. However, under the more practical total average power constraint, a computable secrecy capacity expression is not known for any of those channels, in general. Because of this and also due to the fact that the secret DPC is prohibitively complex for practical implementations, it is important to find simple solutions, e.g. based on linear precoding, that maximize the achievable secret rate and/or can achieve the secret capacity region in practice.

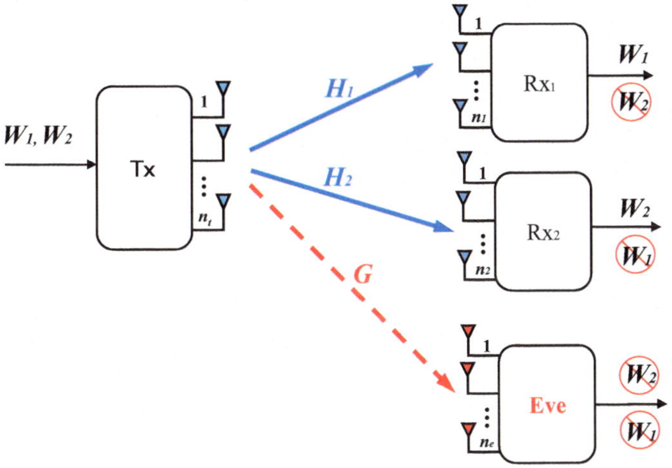

Fig. 5.7 MIMO BC with confidential messages and an external eavesdropper

5.5.2 *Physical Layer Security via Beamforming*

Developing practical physical layer security techniques that address the security issues of NOMA in the downlink is an important research topic. In particular, securing communication for the three basic channel models illustrated in Figs. 5.5, 5.6 and 5.7 and their extension to K-user cases leads to practically important open problems. Specifically, secure transmission strategies based on linear beamforming techniques are of interest. In the following, we summarize different types of beamforming that have been proposed for physical layer security.

Beamforming is one of the most widely studied approaches to physical layer security for multi-antenna systems. Beamforming approaches can be categorized as follows:

- *Zero-forcing beamforming*: One intuitive approach to secure transmission is to send the signal as orthogonal to the eavesdropper's channel as possible. When the transmitter has a larger number of antennas than the eavesdropper, it is possible to transmit the information in the *null-space* of the eavesdropper's channel. This approach, which is called zero-forcing beamforming, is a simple but suboptimal beamforming scheme for multi-user MIMO systems [37].
- *GSVD-based beamforming*: Generalized singular value decomposition (GSVD)-based precoding has been widely used for security and confidentiality purposes in the MIMO wiretap and BC channels [38, 39] as well as in multicasting [40]. Simplicity is the main advantage of GSVD-based transmission as it decouples the MIMO channel between the transmitter and the receivers into several parallel subchannels which can be selected independently and then be encoded separately. However, even in the case of the MIMO wiretap channel, GSVD-based precoding is neither capacity achieving nor very close to capacity in general [41]. Nevertheless,

this approach is useful in that it can usually result in reasonably high achievable rates with low complexity.

- *GEVD-based transmission*: Linear beamforming based on the generalized eigenvalue decomposition (GEVD) is known to be optimal for both the MISO wiretap channel [39] and MISO BC with confidential messages [42]. Despite being theoretically appealing for its optimality, in practice, there is no guarantee to have single-antenna eavesdroppers. It is worth noting that under a matrix power constraint, GEVD-based precoding also achieves the secrecy capacity of the MIMO wiretap channel [43].
- *Trigonometric approach*: This is another simple linear beamforming approach which was recently introduced in the context of the MIMO wiretap channel in [41, 44] and shown to be optimal for any numbers of antennas at the eavesdropper and the legitimate receiver when the transmitter has two antennas.
- *Convex-optimization-based precoding*: Numerical solutions based on convex optimization have also been proposed to compute a transmit covariance matrix for the secrecy capacity maximization in the MISO and MIMO channels [45, 46]. These methods solve the underlying non-convex optimization problem in an iterative way and thus are very complex because the objectives are matrices.
- *Artificial Noise (AN)-aided transmission*: When the eavesdropper's CSI is not available at the transmitter, an AN-aided transmission is useful for providing security at the physical layer. In this method, multiple antennas at the transmitter and legitimate receiver are used to inject *artificial noise* into directions orthogonal to those of the main channel [47]. Due to its simplicity and practicality against passive eavesdroppers, AN-based beamforming is widely used for secure transmissions [48, 49], particularly when the eavesdropper's CSI is not known. In some cases, e.g. for the fading MISO wiretap channel with no eavesdropper's CSI, it is shown that the optimal solution converges to transmitting AN in all null-space dimensions of the main channel [50].

5.5.3 Research Directions

It is of considerable interest to develop physical layer security for NOMA transmission with both in-network and external eavesdroppers. The former is specifically important in a NOMA-based transmission since a user with a stronger channel is able to decode a weaker user's signal, at the physical layer, and compromise their privacy. Therefore, security with respect to external eavesdroppers and confidentiality (privacy) with respect to legitimate users are both critical in NOMA. The case with multiple-antenna transmitters should be the main focus of such a study, as those are now ubiquitous in many wireless networks, including cellular networks. In this case, space-time signal processing such as linear beamforming with/without artificial noise can be used to put physical layer security into practice and to improve secure transmission rates. The key is to use signal processing techniques to increase the signal strength difference between the legitimate user and eavesdropper.

One set of problems in this setting is to identify linear beamforming techniques that maximize the secrecy rate region of the models shown in Figs. 5.5, 5.6 and 5.7 assuming the availability of the legitimate users' instantaneous perfect CSI at the transmitter (CSIT). Another set of problems may consider the above security problems with more realistic assumptions, i.e. developing linear beamforming techniques that maximize physical layer security with imperfect CSIT.

References

1. D. Tse, P. Viswanath, *Fundamentals of Wireless Communication* (Cambridge university press, 2005)
2. Y. Saito, Y. Kishiyama, A. Benjebbour, T. Nakamura, A. Li, K. Higuchi, Non-orthogonal multiple access (NOMA) for cellular future radio access, in *Proceedings of the 77th IEEE Vehicular Technology Conference (VTC Spring)* (2013), pp. 1–5
3. P. Wang, J. Xiao, P. Li, Comparison of orthogonal and non-orthogonal approaches to future wireless cellular systems. IEEE Veh. Technol. Mag. **1**(3), 4–11 (2006)
4. T.M. Cover, J.A. Thomas, *Elements of Information Theory* (Wiley, New York, 2006)
5. A. El Gamal, Y.H. Kim, *Network Information Theory* (Cambridge University Press, 2011)
6. T. Han, K. Kobayashi, A new achievable rate region for the interference channel. IEEE Trans. Inf. Theory **27**, 49–60 (1981)
7. M. Vaezi, H.V. Poor, Simplified Han-Kobayashi region for one-sided and mixed Gaussian interference channels, in *Proceedings of the IEEE International Conference on Communications (ICC)* (2016), pp. 1–6
8. ITU, Minimum requirements related to technical performance for IMT-2020 radio interface(s), February 2017, https://www.itu.int/md/R15-SG05-C-0040/en
9. S.R. Islam, N. Avazov, O.A. Dobre, K.-S. Kwak, Power-domain non-orthogonal multiple access (NOMA) in 5G systems: potentials and challenges. IEEE Commun. Surv. Tutor. (2017)
10. Z. Ding, Y. Liu, J. Choi, Q. Sun, M. Elkashlan, I. Chih-Lin, H.V. Poor, Application of non-orthogonal multiple access in LTE and 5G networks. IEEE Commun. Mag. **55**(2), 185–191 (2017)
11. W. Shin, M. Vaezi, B. Lee, D.J. Love, J. Lee, H.V. Poor, Non-orthogonal multiple access in multi-cell networks: theory, performance, and practical challenges. IEEE Commun. Mag. **55**(10), 176–183 (2017)
12. 3GPP TD RP-150496, Study on Downlink Multiuser Superposition Transmission, Mar 2015
13. Q. Sun, S. Han, C.-L. I, Z. Pan, On the ergodic capacity of MIMO NOMA systems. IEEE Wirel. Commun. Lett. **4**(4), 405–408 (2015)
14. Z. Ding, F. Adachi, H.V. Poor, The application of MIMO to non-orthogonal multiple access. IEEE Trans. Wirel. Commun. **15**(1), 537–552 (2016)
15. Z. Ding, R. Schober, H.V. Poor, A general MIMO framework for NOMA downlink and uplink transmission based on signal alignment. IEEE Trans. Wirel. Commun. **15**(6), 4438–4454 (2016)
16. R.H. Etkin, D.N. Tse, H. Wang, Gaussian interference channel capacity to within one bit. IEEE Trans. Inf. Theory **54**(12), 5534–5562 (2008)
17. A.S. Motahari, A.K. Khandani, Capacity bounds for the Gaussian interference channel. IEEE Trans. Inf. Theory **55**(2), 620–643 (2009)
18. X. Shang, G. Kramer, B. Chen, A new outer bound and the noisy-interference sum-rate capacity for Gaussian interference channels. IEEE Trans. Inf. Theory **55**(2), 689–699 (2009)
19. V.S. Annapureddy, V.V. Veeravalli, Gaussian interference networks: sum capacity in the low-interference regime and new outer bounds on the capacity region. IEEE Trans. Inf. Theory **55**(7), 3032–3050 (2009)

20. M. Vaezi, H.V. Poor, On limiting expressions for the capacity regions of Gaussian interference channels, in *Proceedings of the 49th Asilomar Conference on Signals, Systems and Computers*, Nov (2015), pp. 1292–1298
21. Y. Pang, M. Varanasi, Approximate capacity region of the MAC-IC-MAC (2016), arXiv:1604.02234
22. H. Weingarten, Y. Steinberg, S.S. Shamai, The capacity region of the Gaussian multiple-input multiple-output broadcast channel. IEEE Trans. Inf. Theory **52**(9), 3936–3964 (2006)
23. M. Costa, Writing on dirty paper. IEEE Trans. Inf. Theory **29**(3), 439–441 (1983)
24. V.O.L. Yijie Mao, B. Clerckx, Rate-splitting multiple access for downlink communication systems: bridging, generalizing and outperforming SDMA and NOMA, arXiv:1710.11018v3
25. M.A. Maddah-Ali, A.S. Motahari, A.K. Khandani, Signaling over MIMO multi-base systems: combination of multi-access and broadcast schemes, in *Proceedings of the IEEE International Symposium on Information Theory* (2006), pp. 2104–2108
26. M.A. Maddah-Ali, A.S. Motahari, A.K. Khandani, Communication over MIMO X channels: interference alignment, decomposition, and performance analysis. IEEE Trans. Inf. Theory **54**(8), 3457–3470 (2008)
27. V.R. Cadambe, S.A. Jafar, Interference alignment and degrees of freedom of the K-user interference channel. IEEE Trans. Inf. Theory **54**(8), 3425–3441 (2008)
28. C.M. Yetis, T. Gou, S.A. Jafar, A.H. Kayran, On feasibility of interference alignment in MIMO interference networks. IEEE Trans. Signal Process. **58**(9), 4771–4782 (2010)
29. T. Gou, S.A. Jafar, Degrees of freedom of the K user $M \times N$ MIMO interference channel. IEEE Trans. Inf. Theory **56**(12), 6040–6057 (2010)
30. 3GPP TR 36.866 V12.0.1, Study on Network-Assisted Interference Cancellation and Suppression (NAIC) for LTE (Release 12), Mar 2014
31. Q. Li, G. Li, W. Lee, M.-i. Lee, D. Mazzarese, B. Clerckx, Z. Li, MIMO techniques in WiMAX and LTE: a feature overview. IEEE Commun. Mag. **48**(5), (2010)
32. Y. Yuan, Z. Yuan, G. Yu, C.-H. Hwang, P.-K. Liao, A. Li, K. Takeda, Non-orthogonal transmission technology in LTE evolution. IEEE Commun. Mag. **54**(7), 68–74 (2016)
33. D. Lee, H. Seo, B. Clerckx, E. Hardouin, D. Mazzarese, S. Nagata, K. Sayana, Coordinated multipoint transmission and reception in LTE-advanced: deployment scenarios and operational challenges. IEEE Commun. Mag. **50**(2), 148–155 (2012)
34. W. Shin, M. Vaezi, B. Lee, D.J. Love, J. Lee, H.V. Poor, Coordinated beamforming for multi-cell MIMO-NOMA. IEEE Commun. Lett. **21**(1), 84–87 (2017)
35. E. Ekrem, S. Ulukus, The secrecy capacity region of the Gaussian MIMO multi-receiver wiretap channel. IEEE Trans. Inf. Theory **57**(4), 2083–2114 (2011)
36. R. Liu, T. Liu, H.V. Poor, S. Shamai, Multiple-input multiple-output Gaussian broadcast channels with confidential messages. IEEE Trans. Inf. Theory **56**(9), 4215–4227 (2010)
37. K. Wang, X. Wang, X. Zhang, SLNR-based transmit beamforming for MIMO wiretap channel. Wirel. Pers. Commun. **71**(1), 109–121 (2013)
38. S.A.A. Fakoorian, A.L. Swindlehurst, Optimal power allocation for GSVD-based beamforming in the MIMO Gaussian wiretap channel, in *Proceedings of the IEEE International Symposium on Information Theory* (2012), pp. 2321–2325
39. A. Khisti, G.W. Wornell, Secure transmission with multiple antennas I: the MISOME wiretap channel. IEEE Trans. Inf. Theory **56**(7), 3088–3104 (2010)
40. W. Mei, Z. Chen, J. Fang, GSVD-based precoding in MIMO systems with integrated services. IEEE Signal Process. Lett. **23**(11), 1528–1532 (2016)
41. M. Vaezi, W. Shin, V. Poor, Optimal beamforming for Gaussian MIMO wiretap channels with two transmit antennas. IEEE Trans. Wirel. Commun. **16**, 6726–6735 (2017)
42. R. Liu, H.V. Poor, Secrecy capacity region of a multiple-antenna Gaussian broadcast channel with confidential messages. IEEE Trans. Inf. Theory **55**(3), 1235–1249 (2009)
43. R. Bustin, R. Liu, H. V. Poor, S. Shamai, An MMSE approach to the secrecy capacity of the MIMO Gaussian wiretap channel. EURASIP J. Wirel. Commun. Netw. (1) (2009)
44. M. Vaezi, W. Shin, H. V. Poor, J. Lee, MIMO Gaussian wiretap channels with two transmit antennas: optimal precoding and power allocation, in *Proceedings of the IEEE International Symposium on Information Theory* (2017), pp. 1708–1712

45. Q. Li, W.-K. Ma, Optimal and robust transmit designs for MISO channel secrecy by semidefinite programming. IEEE Trans. Signal Process. **59**(8), 3799–3812 (2011)
46. Q. Li, M. Hong, H.-T. Wai, Y.-F. Liu, W.-K. Ma, Z.-Q. Luo, Transmit solutions for MIMO wiretap channels using alternating optimization. IEEE J. Sel. Areas Commun. **31**(9), 1714–1727 (2013)
47. S. Goel, R. Negi, Guaranteeing secrecy using artificial noise. IEEE Trans. Wirel. Commun. **7**(6), 2180–2189 (2008)
48. Z. Li, P. Mu, B. Wang, X. Hu, Optimal semiadaptive transmission with artificial-noise-aided beamforming in MISO wiretap channels. IEEE Trans. Veh. Technol. **65**(9), 7021–7035 (2016)
49. T.-X. Zheng, H.-M. Wang, J. Yuan, D. Towsley, M.H. Lee, Multi-antenna transmission with artificial noise against randomly distributed eavesdroppers. IEEE Trans. Commun. **63**(11), 4347–4362 (2015)
50. S. Gerbracht, C. Scheunert, E.A. Jorswieck, Secrecy outage in MISO systems with partial channel information. IEEE Trans. Inf. Forensics Secur. **7**(2), 704–716 (2012)

Chapter 6
Optimal Power Allocation for Downlink NOMA Systems

Yongming Huang, Jiaheng Wang and Jianyue Zhu

6.1 Introduction

With the popularity of smartphones and Internet of Things, there is an explosive demand for new services and data traffic for wireless communications. The capacity of the fourth-generation (4G) mobile communication system is insufficient to satisfy such a demand in the near future. The development of the fifth-generation (5G) mobile communication system has been placed on the agenda with higher requirements in data rates, latency, and connectivity [1]. In order to meet the new standards, some potential technologies, such as massive multiple-input–multiple-output (MIMO) [2], millimeter wave [3], and ultra densification [4, 5], will be introduced into 5G. Meanwhile, new multiple access technologies, which are flexible, reliable, and efficient in terms of energy and spectrum, are also considered for 5G communication.

Conventionally, cellular systems have adopted orthogonal multiple access (OMA) approaches, in which wireless resources are allocated orthogonally to multiple users. The common OMA techniques include frequency-division multiple access (FDMA), time division multiple access (TDMA), code-division multiple access (CDMA), and orthogonal frequency-division multiple access (OFDMA). Ideally, in OMA, the intra-cell interference does not exist as result of dedicated resource allocation. Also, for this reason, the information of multiple users can be retrieved at a low complexity. Nonetheless, the number of served users is limited by the number of orthogonal resources, which is generally small in practice. Consequently, it is difficult for OMA systems to support a massive connectivity.

Recently, non-orthogonal multiple access (NOMA) technologies are developed and proposed for 5G, which will contribute to disruptive design changes on radio access and alleviate the scarcity of suitable spectra. By using superposition coding

Y. Huang (✉) · J. Wang · J. Zhu
Southeast University, Nanjing 210096, China
e-mail: huangym@seu.edu.cn

J. Wang
e-mail: jhwang@seu.edu.cn

J. Zhu
e-mail: zhujy@seu.edu.cn

M. Vaezi et al. (eds.), *Multiple Access Techniques for 5G Wireless Networks and Beyond*, https://doi.org/10.1007/978-3-319-92090-0_6

at the transmitter with successive interference cancellation (SIC) at the receiver, downlink NOMA allows one (frequency, time, code, or spatial) channel to be shared by multiple users simultaneously [6, 7], thus leading to better performance in terms of spectral efficiency, fairness, or energy efficiency [8]. Therefore, NOMA has received much attention recently. Its combinations with MIMO and multi-cell technologies were studied in [9, 10] and [11], respectively. NOMA was also considered to be used in, e.g., visible light communication [12] and millimeter wave communication [13].

The principle of NOMA is to implement multiple access in the power domain [14]. Hence, allocation is critical for NOMA systems. In the literature, there are a number of works on power allocation for NOMA. In particular [15, 16] focused on power allocation in a two-user NOMA system and [17–21] investigated power allocation for multiple users (more than two) sharing one channel, which is referred as multi-user NOMA (MU-NOMA). There were also some works, e.g., [12, 22–28], studying the resource allocation problems in multi-channel NOMA (MC-NONA) systems, where multiple channels are available for NOMA. Different criteria, such as maximin fairness [18–20, 22], sum rate [15–17, 22, 28–30], and energy efficiency [21–23, 26], were considered.

This chapter focuses on power allocation for downlink NOMA. We first briefly review the basic concepts of downlink NOMA transmission and introduce the two-user NOMA, MU-NOMA, and MC-NOMA schemes. Then, we investigate the optimal power allocation strategies for these NOMA schemes under different performance criteria such as the maximin fairness, sum rate, and energy efficiency along with user weights and quality-of-service (QoS) constraints. We show that the optimal NOMA power allocation can be analytically characterized in most cases, otherwise it can be numerically computed via convex optimization methods.

This chapter is organized as follows. Section 6.2 introduces the fundamentals of downlink NOMA and the two-user NOMA, MU-NOMA, and MC-NOMA schemes. In Sects. 6.3–6.5, we investigate the optimal power allocation for two-user NOMA, MU-NOMA, and MC-NOMA schemes, respectively. The performance of the NOMA power allocation strategies is evaluated in Sect. 6.6 via simulations, and the conclusion is drawn in Sect. 6.7.

6.2 Fundamentals of Downlink NOMA

In this section, we review the basic concepts of downlink NOMA transmission in a single-cell network.[1] To begin with, we start from the simplest two-user case, where a base station (BS) serves two users, namely UE_1 and UE_2, on the same frequency band with bandwidth B. The BS transmits a signal s_n for user n ($UE_n, n = 1, 2$) with transmission power p_n. The total power budget of the BS is P, i.e., $p_1 + p_2 \leq P$. Such a simple downlink NOMA system is displayed in Fig. 6.1 [31].

[1] For multi-cell NOMA, the reader is referred to [11].

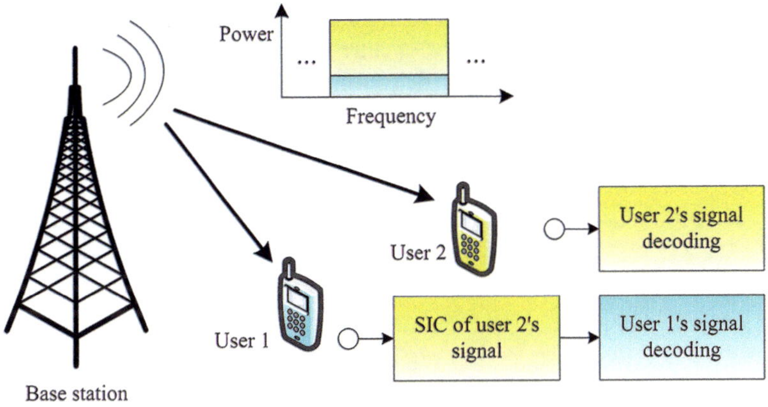

Fig. 6.1 A downlink NOMA system with one base station and two users

Superposition Coding: According to the NOMA principle, the BS exploits the superposition coding and broadcasts the signal

$$x = \sqrt{p_1}s_1 + \sqrt{p_2}s_2 \tag{6.1}$$

to both users. The received signal at UE_n is

$$y_n = h_n\left(\sqrt{p_1}s_1 + \sqrt{p_2}s_2\right) + z_n, \tag{6.2}$$

where $h_n = g_n d_n^{-\rho}$ is the channel coefficient from the BS to UE_n, g_n follows a Rayleigh distribution, d_n is the distance between the BS and UE_n, ρ is the path loss exponent, and z_n is the additive white Gaussian noise with zero mean and variance σ_n^2, i.e., $z_n \sim \mathcal{CN}\left(0, \sigma_n^2\right)$.

Successive Interference Cancellation (SIC): In NOMA systems, each user exploits SIC at its receiver. Let $\Gamma_n = |h_n|^2/\sigma_n^2$ be the channel-to-noise ratio (CNR) of UE_n. Assume without loss of generality (w.l.o.g.) that the users are ordered by their normalized channel gains as $\Gamma_1 \geq \Gamma_2$, i.e., UE_1 and UE_2 are regarded as the strong and weak users, respectively. It is expected that more power is allocated to the weak user UE_2 and less power is allocated to the strong user UE_1, i.e., $p_1 \leq p_2$ [14, 25]. Then, UE_1 first decodes the message of UE_2 and removes it from its received signal, while UE_2 treats the signal of UE_1 as interference and decodes its own message.

Achievable Rate: Suppose that the channel coding is ideal and UE_1 is able to decode the message of UE_2 successfully. Then, the achievable rates of UE_1 and UE_2 are given respectively by

$$R_1 = B\log\left(1 + p_1\Gamma_1\right), \; R_2 = B\log\left(1 + \frac{p_2\Gamma_2}{1 + p_1\Gamma_2}\right) \tag{6.3}$$

which are often used as the design objectives of NOMA systems.

Multi-User NOMA (MU-NOMA): Consider a more general case where a BS serves $N \geq 2$ users on the same spectrum, which are indexed by $n = 1, \ldots,$ N. The broadcast signal by the BS is then given by

$$x = \sum_{i=1}^{N} \sqrt{p_i} s_i \tag{6.4}$$

and then the received signal at each UE_n is given by

$$y_n = h_n \sum_{i=1}^{N} \sqrt{p_i} s_i + z_n. \tag{6.5}$$

Similarly, suppose that the users are ordered by their normalized channel gains as

$$\Gamma_1 \geq \Gamma_2 \geq \cdots \geq \Gamma_N \tag{6.6}$$

and the NOMA protocol allocates higher powers to the users with lower CNRs, leading to $p_1 \leq p_2 \leq \cdots \leq p_N$. Therefore, UE_n is able to decode the message of UE_l for $l > n$ and remove it from the received signal so that UE_n is only interfered by UE_j for $j < n$. Therefore, after SIC, the achievable rate of UE_n is

$$R_n = \log\left(1 + \frac{p_n \Gamma_n}{\sum_{j=1}^{n-1} p_j \Gamma_n + 1}\right) \tag{6.7}$$

for $n = 1, \ldots, N$.

Multi-Channel NOMA (MC-NOMA): The frequency band shared by the users could be viewed as a channel, which may also be a time slot, spread code, or resource block. In cellular systems, there are often multiple channels available, which leads to a more general NOMA scheme called multi-channel NOMA (MC-NOMA), where multiple users share multiple channels. Specifically, in a downlink MC-NOMA network, the BS serves N users through M channels and the total bandwidth B is equally divided to M channels so the bandwidth of each channel is $B_c = B/M$. Let $N_m \in \{N_1, N_2, ..., N_M\}$ be the number of users using channel m for $m = 1, 2, \ldots,$ M and $\mathrm{UE}_{n,m}$ denotes user n on channel m for $n = 1, 2, \ldots, N_m$. The signal transmitted by the BS on each channel m can be expressed as

$$x_m = \sum_{n=1}^{N_m} \sqrt{p_{n,m}} s_n, \tag{6.8}$$

where s_n is the symbol of $\mathrm{UE}_{n,m}$ and $p_{n,m}$ is the power allocated to $\mathrm{UE}_{n,m}$. The received signal at $\mathrm{UE}_{n,m}$ is

$$y_{n,m} = \sum_{i=1}^{N_m} \sqrt{p_{i,m}} h_{n,m} s_i + z_{n,m}. \tag{6.9}$$

It is easily seen that on each channel m is an MU-NOMA scheme. Similarly, assume w.l.o.g. that the CNRs of the users on channel m are ordered as

$$\Gamma_{1,m} \geq \cdots \geq \Gamma_{n,m} \geq \cdots \geq \Gamma_{N_m,m}, \tag{6.10}$$

which will lead to $p_{1,m} \leq \cdots \leq p_{n,m} \leq \cdots \leq p_{N_m,m}$. Then, the achievable rate of $\text{UE}_{n,m}$ using SIC is

$$R_{n,m} = B_c \log \left(1 + \frac{p_{n,m} \Gamma_{n,m}}{1 + \sum_{i=1}^{n-1} p_{i,m} \Gamma_{n,m}} \right). \tag{6.11}$$

The basic idea of NOMA is to implement multiple access in the power domain [14]. Hence, power allocation is the key to achieve the full benefit of NOMA transmission. In the following parts, we will investigate the optimal power allocation strategies for different NOMA schemes, including the simplest two-user case, the MU-NOMA scheme, and the MC-NOMA scheme, under different performance measures.

6.3 Two-User NOMA

In this section, we investigate the optimal power allocation for the two-user NOMA scheme. Although the two-user scheme is the simplest case of NOMA, the results and insights obtained in this case will serve the more complicated MU-NOMA and MC-NOMA schemes.

6.3.1 Optimal Power Allocation for MMF

The NOMA scheme enables a flexible management of users' achievable rates and provides an efficient way to enhance user fairness. A widely used fairness metric is the maximin fairness (MMF), which is achieved by maximizing the worst (i.e., minimum) user rate. According to (6.3), the power allocation to achieve the MMF is given by the solution to the following optimization problem:

$$TU^{\text{MMF}}: \begin{array}{ll} \max\limits_{p_1, p_2} & \min \{R_1(p_1, p_2),\ R_2(p_1, p_2)\} \\ \text{s.t.} & 0 \leq p_1 \leq p_2,\ p_1 + p_2 \leq P \end{array}$$

This problem admits a closed-form solution as follows.

Proposition 1 *Suppose that $\Gamma_1 \geq \Gamma_2$. Then, the optimal solution to TU^{MMF} is given by $p_1^\star = \Lambda$ and $p_2^\star = P - p_1^\star$, where $\Gamma_l \triangleq |h_l|^2/\sigma_l^2$ and*

$$\Lambda \triangleq \frac{-(\Gamma_1+\Gamma_2)+\sqrt{(\Gamma_1+\Gamma_2)^2+4\Gamma_1\Gamma_2^2 P}}{2\Gamma_1\Gamma_2}. \tag{6.12}$$

Proof Please refer to the proof of Proposition 1 in [22].

Remark 1 It can be verified that at the optimal point $R_1(p_1^\star, p_2^\star) = R_2(p_1^\star, p_2^\star)$, i.e., UE_1 and UE_2 achieve the same rate. This indicates that, under the MMF criterion, the NOMA system will provide absolute fairness for two users on one channel.

To elaborate another important insight, we introduce the following definition.

Definition 1 A NOMA system is called *SIC stable* if the optimal power allocation satisfies $p_1 < p_2$ on one channel.

Remark 2 In NOMA systems, SIC is performed according to the order of the CNRs of the users on one channel [14, 25], which is guaranteed by imposing an inverse order of the powers allocated to the users, i.e., $p_1 \leq p_2$. Specifically, UE_1 (the stronger user) first decodes the signal of UE_2 (the weaker user) and then subtracts it from the superposed signal. Therefore, from the SIC perspective, a large difference between the signal strengths of UE_2 and UE_1 is preferred [32]. However, even with the power order constraint $p_1 \leq p_2$, the power optimization may lead to $p_1 = p_2$; i.e., UE_1 and UE_2 have the same signal strength, which is the worst situation for SIC. In this case, SIC may fail or has a large error propagation and thus is unstable. Indeed, the authors in [33] pointed out that the power of the weak user must be strictly larger than that of the strong user, otherwise the users' outage probabilities will always be one. Definition 1 explicitly concretizes such a practical requirement in NOMA systems.

Lemma 1 *The NOMA system is SIC stable for TU^{MMF}.*

Proof Please refer to the proof of Lemma 1 in [22].

Indicated by Lemma 1, the two-user NOMA system is always SIC stable under the MMF criterion, as in this case the optimal power allocation always satisfies $p_1^\star < p_2^\star$. On the other hand, in the subsequent subsections, we will show that a NOMA system may not always be SIC stable under different criteria.

6.3.2 Optimal Power Allocation for SR Maximization

In this subsection, we seek the optimal power allocation for maximizing the sum rate (SR). In SR maximization, to take user priority or fairness into account, user weights or quality-of-service (QoS) constraints are often adopted.

6.3.2.1 Weighted SR Maximization (SR1)

According to (6.3), the problem of maximizing the weighted SR (WSR) is given by

$$TU^{\mathrm{SR1}}: \begin{array}{ll} \max\limits_{p_1, p_2} & W_1 R_1(p_1, p_2) + W_2 R_2(p_1, p_2) \\ \text{s.t.} & 0 \leq p_1 \leq p_2, \ p_1 + p_2 \leq P \end{array}$$

where W_i denotes the weight of UE_i for $i = 1, 2$. Note that TU^{SR1} is a nonconvex problem due to the interference between UE_1 and UE_2. Nevertheless, its optimal solution can be found as follows.

Proposition 2 *Suppose that* $\Gamma_1 \geq \Gamma_2$, $W_1 < W_2$ *and* $P > 2\Omega$, *with*

$$\Omega \triangleq \frac{W_2 \Gamma_2 - W_1 \Gamma_1}{\Gamma_1 \Gamma_2 (W_1 - W_2)}. \tag{6.13}$$

Then, the optimal solution to TU^{SR1} *is given by* $p_1^\star = \Omega$ *and* $p_2^\star = P - p_1^\star$.

Proof Please refer to the proof of Proposition 2 in [22].

Remark 3 In Proposition 2, the conditions $W_1 < W_2$ and $P > 2\Omega$ are both to avoid a failure of SIC. Indeed, if $W_1 \geq W_2$, the solution to TU^{SR1} is $p_1^\star = p_2^\star = P/2$; i.e., the NOMA system is unstable according to Definition 1. SIC may also fail if $P \leq 2\Omega$, which will lead to $p_1^\star = p_2^\star = P/2$ as well. Therefore, the two-user NOMA system is SIC stable for the WSR maximization if and only if $W_1 < W_2$ and $P > 2\Omega$.

6.3.2.2 SR Maximization with QoS (SR2)

Now, we consider maximizing the SR with QoS constraints. In this case, the power allocation problem is given by

$$TU^{\mathrm{SR2}}: \begin{array}{ll} \max\limits_{p_1, p_2} & R_1(p_1, p_2) + R_2(p_1, p_2) \\ \text{s.t.} & 0 \leq p_1 \leq p_2, \ p_1 + p_2 \leq P \\ & R_i \geq R_i^{\min}, i = 1, 2. \end{array}$$

where $R_i^{\min}$ is the QoS threshold of UE_i. The optimal solution to TU^{SR2} is provided in the following result.

Proposition 3 *Suppose that* $\Gamma_1 \geq \Gamma_2$, $A_2 \geq 2$, *and* $P \geq \Upsilon$, *with*

$$A_l = 2^{R_l^{\min}}, \quad \Upsilon \triangleq \frac{A_2(A_1 - 1)}{\Gamma_1} + \frac{A_2 - 1}{\Gamma_2}, \quad \Xi \triangleq \frac{\Gamma_2 P - A_2 + 1}{A_2 \Gamma_2}. \tag{6.14}$$

Then, the optimal solution to TU^{SR2} *is given by* $p_1^\star = \Xi$ *and* $p_2^\star = P - p_1^\star$.

Proof Please refer to the proof of Proposition 3 in [22].

Remark 4 Similarly, in Proposition 3, the conditions $A_2 \geq 2$ and $P \geq \Upsilon$ are to guarantee the SIC stability. Indeed, if $A_2 < 2$, then $\Xi > P/2$ and the optimal solution will be $p_1^\star = p_2^\star = P/2$, which may lead a failure of SIC. At the same time, SIC may also fail if $P < \Upsilon$, which will lead to $p_1^\star = p_2^\star = P/2$ as well. Therefore, the NOMA system is SIC stable in this case if and only if $A_2 \geq 2$ and $P \geq \Upsilon$.

According to Proposition 3, if the NOMA system is SIC stable, the optimal solution will be $p_1^\star = \Xi$ and $p_2^\star = P - p_1^\star$. Hence, we have $R_2(p_1^\star, p_2^\star) = R_2^{\min}$, implying that the user with a lower CNR (i.e., UE$_2$) receives the power to meet its QoS requirement exactly, while the remaining power is used to maximize the rate of the user with a higher CNR (i.e., UE$_1$).

6.3.3 Optimal Power Allocation for EE Maximization

In this subsection, we investigate the optimal power allocation for maximizing the energy efficiency (EE), which is defined as the ratio between the rate and the consumed power. Similarly, user weights and QoS constraints are considered.

6.3.3.1 Weighted EE Maximization (EE1)

The problem of maximizing the weighted EE is formulated as follows:

$$TU_a^{\mathrm{EE1}}: \begin{array}{l} \max\limits_{p_1, p_2} \ \eta = \frac{W_1 R_1(p_1, p_2) + W_2 R_2(p_1, p_2)}{P_T + p_1 + p_2} \\ \text{s.t.} 0 \leq p_1 \leq p_2, \ p_1 + p_2 \leq P \end{array}$$

where P_T is the power consumption of the circuits. Given the fraction form of the objective, TU_a^{EE1} is more complicated than TU^{SR1}. In the following, we show that this problem can also be optimally solved.

We introduce an auxiliary variable q with $p_1 + p_2 = q$. Then, TU_a^{EE1} can be equivalently written into

$$TU_b^{\mathrm{EE1}}: \begin{array}{l} \max\limits_{p_1, q} \ \eta = \frac{W_1 R_1(p_1) + W_2 R_2(p_1, q)}{P_T + q} \\ \text{s.t.} \ \ q \geq 2p_1, \ q \leq P \end{array}$$

where the rate of UE$_2$ can be expressed as

$$R_2(p_1, q) = B \log\left(\frac{1 + q\Gamma_2}{1 + p_1 \Gamma_2}\right). \tag{6.15}$$

To deal with the fractional form, let us introduce the following objective function:

$$H(p_1, q, \alpha) \triangleq W_1 R_1(p_1) + W_2 R_2(p_1, q) - \alpha(P_T + q), \tag{6.16}$$

where α is a positive parameter. Then, we consider the following problem for given α:

$$TU_c^{\text{EE1}}: \begin{array}{ll} \max\limits_{p_1, q} & H(p_1, q, \alpha) \\ \text{s.t.} & q \geq 2p_1, q \leq P. \end{array}$$

The relation between TU_c^{EE1} and TU_b^{EE1} is stated in the following result.

Lemma 2 *([34, pp. 493–494]) Let $H^\star(\alpha)$ be the optimal objective value of TU_c^{EE1} and $\mathbf{p}^\star(\alpha)$ be the optimal solution of TU_c^{EE1}. Then, $\mathbf{p}^\star(\alpha)$ is the optimal solution to TU_b^{EE1} if and only if $H^\star(\alpha) = 0$.*

According to Lemma 2, the optimal solution to TU_b^{EE1} can be found by solving TU_c^{EE1} parameterized by α such that $H^\star(\alpha) = 0$. Since $H^\star(\alpha)$ is monotonic in α, one can use any line search method, e.g., the bisection method, to find α such that $H^\star(\alpha) = 0$. Then, the left question is how to solve TU_c^{EE1} with given α.

Theorem 1 *Suppose that $\Gamma_1 \geq \Gamma_2$. Then, TU_c^{EE1} is a convex problem if one of the following conditions hold*

$$\begin{aligned} &C1: W_1 \geq W_2; \\ &C2: 1 < \tfrac{W_2}{W_1} \leq \tfrac{(\Gamma_1 + \Gamma_1\Gamma_2 P)^2}{(\Gamma_2 + \Gamma_1\Gamma_2 P)^2}. \end{aligned}$$

Proof See Appendix A.

Theorem 1 reveals that TU_c^{EE1} is in fact a convex problem if condition C1 or C2 holds. Consequently, TU_c^{EE1} can be efficiently solved via convex optimization methods. The optimal solution to TU_c^{EE1} can further be analytically characterized.

Proposition 4 *Suppose that $\Gamma_1 \geq \Gamma_2$, C2 holds, and $P \geq 2\Omega$ with*

$$\Omega = \frac{W_2\Gamma_2 - W_1\Gamma_1}{\Gamma_1\Gamma_2(W_1 - W_2)}. \tag{6.17}$$

Then, the optimal solution to TU_c^{EE1} is $p_1^\star = \Omega$ and

$$q^\star = \left[\frac{W_2 B}{\alpha \ln 2} - \frac{1}{\Gamma_2}\right]_{2\Omega}^{P} = \max\left\{2\Omega, \min\left\{\frac{W_2 B}{\alpha \ln 2} - \frac{1}{\Gamma_2}, P\right\}\right\}. \tag{6.18}$$

Proof See Appendix B.

Remark 5 Similarly, in Proposition 4, the conditions C2 and $P > 2\Omega$ are to avoid a failure of SIC. Although TU_c^{EE1} is convex under C1, condition C1 will lead to $p_1^\star = p_2^\star$, which, according to Definition 1, is SIC-unstable.

6.3.3.2 EE Maximization with QoS Constraints (EE2)

Then, we consider maximizing the EE with QoS constraints. In this case, the power allocation problem is given by

$$TU_a^{\mathrm{EE2}}: \begin{array}{ll} \max\limits_{p_1, p_2} & \frac{R_1(p_1,p_2)+R_2(p_1,p_2)}{P_T+p_1+p_2} \\ \text{s.t.} & 0 \le p_1 \le p_2,\ p_1 + p_2 \le P \\ & R_i \ge R_i^{\min},\ i = 1, 2. \end{array}$$

This problem can be optimally solved following the similar steps as in the previous subsection.

Specifically, using $p_1 + p_2 = q$, TU_a^{EE2} can be equivalently transformed into

$$TU_b^{\mathrm{EE2}}: \begin{array}{ll} \max\limits_{p_1, q} & \frac{R_1(p_1)+R_2(p_1,q)}{P_T+q} \\ \text{s.t.} & q \ge 2p_1,\ q \le P \\ & R_i \ge R_i^{\min},\ i = 1, 2. \end{array}$$

Then, we consider the following problem with given α:

$$TU_c^{\mathrm{EE2}}: \begin{array}{ll} \max\limits_{p_1, q} & Q\left(p_1, q, \alpha\right) \\ \text{s.t.} & q \ge 2p_1, q \le P \\ & R_i \ge R_i^{\min},\ i = 1, 2 \end{array}$$

where

$$Q\left(p_1, q, \alpha\right) \triangleq R_1(p_1) + R_2(p_1, q) - \alpha\left(P_T + q\right). \tag{6.19}$$

According to Lemma 2, the optimal solution to TU_b^{EE2} can be found by solving TU_c^{EE2} for a given α and updating α until the optimal objective value of TU_c^{EE2}, denoted by $Q^\star(\alpha)$, satisfies $Q^\star(\alpha) = 0$.

From Theorem 1, under condition C1, the objective of TU_c^{EE2} is concave and TU_c^{EE2} is a convex problem too. Therefore, TU_c^{EE2} can be efficiently solved. In fact, the optimal solution to TU_c^{EE2} can also be analytically characterized.

Proposition 5 *Suppose that $\Gamma_1 \ge \Gamma_2$, $A_2 \ge 2$, and $P \ge \Upsilon$, with*

$$A_l = 2^{\frac{R_l^{\min}}{B}}, \quad \Upsilon \triangleq \frac{A_2(A_1 - 1)}{\Gamma_1} + \frac{A_2 - 1}{\Gamma_2}. \tag{6.20}$$

Then, the optimal solution to TU_c^{EE2} *is*

$$p_1^\star = \frac{1 + q^\star \Gamma_2 - A_2}{A_2 \Gamma_2}, \tag{6.21}$$

$$q^\star = \left[\frac{1}{\alpha} - \frac{A_2}{\Gamma_1} + \frac{A_2 - 1}{\Gamma_2}\right]_{\Upsilon}^{P} = \max\left\{\Upsilon, \min\left\{\frac{1}{\alpha} - \frac{A_2}{\Gamma_1} + \frac{A_2 - 1}{\Gamma_2}, P\right\}\right\}. \tag{6.22}$$

Proof See Appendix C.

Similarly, it can be verified that the power allocation obtained from TU_c^{EE2} (or TU_a^{EE2}) is SIC stable if and only if $P \geq \Upsilon$ and $A_2 \geq 2$.

6.4 MU-NOMA

In this section, we consider the more general MU-NOMA scheme, where a BS serves $N \geq 2$ users on the same channel. Similarly, the optimal MU-NOMA power allocation is investigated under the MMF, SR, and EE criteria with user weights or QoS constraints.

6.4.1 Optimal Power Allocation for MMF

According to (6.7), the power allocation problem under the MMF criterion is formulated as

$$MU_a^{\mathrm{MMF}}: \begin{array}{ll} \max\limits_{\boldsymbol{p}} & \min\limits_{i=1,\ldots,N} \{R_i\} \\ \text{s.t.} & 0 < p_1 < \cdots < p_N, \ \sum\limits_{i=1}^{N} p_i \leq P \end{array}$$

where $\boldsymbol{p} = \{p_i\}_{i=1}^N$ denotes the powers allocated to the users. It has been shown in [20] that, though nonconvex, MU_a^{MMF} is a quasi-convex problem. Thus, the optimal solution to MU_a^{MMF} can be found by solving a sequence of convex problems.

Specifically, MU_a^{MMF} is equivalent to

$$MU_b^{\mathrm{MMF}}: \begin{array}{ll} \max\limits_{\boldsymbol{p},t} & t \\ \text{s.t.} & 0 < p_1 < \ldots < p_N, \ \sum\limits_{i=1}^{N} p_i \leq P \\ & R_i \geq t, i = 1, \ldots, N. \end{array}$$

The constraint $R_i \geq t$ can be rewritten into

$$p_i \geq \frac{2^t - 1}{\Gamma_i}\left(\sum_{j=1}^{i-1} p_j \Gamma_i + 1\right). \tag{6.23}$$

Hence, for fixed t, MU_b^{MMF} is a linear program (LP) and can be efficiently solved by a number of LP solvers. Then, one can exploit the bisection method to search the optimal t. Note that the optimal solution to MU_b^{MMF} for fixed t can be analytically characterized if there is no power order constraint.

Proposition 6 *In the absence of power order constraint, the solution to* MU_b^{MMF} *is given by*

$$p_i = \frac{2^t - 1}{\Gamma_i}\left(\sum_{j=1}^{i-1} p_j \Gamma_i + 1\right), i = 1, \cdots, N. \tag{6.24}$$

Proof Please refer to the proof of Theorem 1 in [20].

The solution in (6.24) implies that all users achieve the same data rate equal to t. Hence, in this case, the NOMA system will provide absolute fairness for all users. Note that, however, the solution in (6.24) is obtained without the power order constraint. One may wonder if this solution is still optimal if the power order constraint is not omitted. The following result provides a sufficient condition to characterize the optimality of (6.24).

Theorem 2 *The solution in* (6.24) *is optimal for* MU_b^{MMF} *if* $P \geq \chi$, *where* $\chi = \sum_{i=1}^{N} \frac{2^{N-i}}{\Gamma_i}$.

Proof See Appendix E.

Theorem 2 indicates that the power order constraints can be omitted under some conditions. In this case, the solution in (6.24) is indeed optimal for MU_b^{MMF}. On the other hand, it is unknown if the solution in (6.24) is optimal if the condition in Theorem 2 is not satisfied. Nevertheless, in this case, one can always numerically solve the linear problem MU_b^{MMF} for fixed t.

6.4.2 *Optimal Power Allocation for SR Maximization*

In this subsection, we investigate the SR maximization problems in MU-NOMA systems with user weights or QoS constraints.

6.4.2.1 Weighted SR Maximization (SR1)

The weighted SR maximization for MU-NOMA is formulated as

$$MU_a^{\mathrm{SR1}}: \quad \begin{aligned} \max_{p} \quad & R_{\mathrm{sum}} = \sum_{i=1}^{N} W_i R_i \\ \text{s.t.} \quad & \sum_{i=1}^{N} p_i \le P \\ & p_1 \le p_2 \le \cdots \le p_N. \end{aligned}$$

Unlike MU_a^{MMF} for MMF, MU_a^{SR1} in its original form is neither a convex nor quasi-convex problem, making it difficult to solve it. Nevertheless, we show that MU_a^{SR1} can be transformed into a convex problem via a linear transformation of the optimization variables.

Introduce the following variable transformation: $q_i = \sum_{j=1}^{i} p_j$ for $i = 1, 2, \ldots,$ N; and conversely $p_i = q_i - q_{i-1}$ for $i = 2, \ldots,$ N and $p_1 = q_1$. In this way, we have $R_1 = \log(q_1 \Gamma_1 + 1)$ and

$$R_i = \log\left(\frac{\sum_{j=1}^{i} p_j \Gamma_i + 1}{\sum_{j=1}^{i-1} p_j \Gamma_i + 1}\right) = \log\left(\frac{q_i \Gamma_i + 1}{q_{i-1}\Gamma_i + 1}\right) = \log(q_i \Gamma_i + 1) - \log\left(q_{i-1}\Gamma_i + 1\right) \tag{6.25}$$

for $i = 2, \ldots,$ N. Therefore, the weighted sum rate can be expressed as

$$\sum_{i=1}^{N} W_i R_i = W_1 \log(q_1 \Gamma_1 + 1) + \sum_{i=2}^{N} W_i \left(\log(q_i \Gamma_i + 1) - \log\left(q_{i-1}\Gamma_i + 1\right)\right) = \sum_{i=1}^{N} f_i(q_i), \tag{6.26}$$

where

$$f_i(q_i) = W_i \log(q_i \Gamma_i + 1) - W_{i+1} \log(q_i \Gamma_{i+1} + 1) \tag{6.27}$$

for $i = 1, \ldots, N-1$ and $f_N(q_N) = W_N \log(q_N \Gamma_N + 1)$. The power constraint $\sum_{i=1}^{N} p_i \le P$ is equal to $q_N \le P$. The power order constraint $p_1 \le p_2 \le \cdots \le p_N$ is equal to $q_1 \le q_2 - q_1 \le \cdots \le q_N - q_{N-1}$. Consequently, problem MU_a^{SR1} can be equivalently transformed into the following problem:

$$MU_b^{\mathrm{SR1}}: \quad \begin{aligned} \max_{q} \quad & \sum_{i=1}^{N} f_i(q_i) \\ \text{s.t.} \quad & q_N \le P \\ & 0 \le q_1 \le q_2 - q_1 \le \cdots \le q_N - q_{N-1} \end{aligned}$$

Then, the following result identifies the convexity of MU_a^{SR1} (or MU_b^{SR1}).

Theorem 3 *MU_a^{SR1} (or MU_b^{SR1}) is a convex problem if one of the following conditions hold for $i = 1, \ldots, N-1$:*

$$\begin{aligned} &T1: W_i \ge W_{i+1}; \\ &T2: 1 < \frac{W_{i+1}}{W_i} \le \frac{(\Gamma_i + \Gamma_i \Gamma_{i+1} P)^2}{(\Gamma_{i+1} + \Gamma_i \Gamma_{i+1} P)^2}. \end{aligned}$$

Proof Please refer to the proof of Theorem 1 in [17].

Remark 6 Theorem 3 indicates that MU_a^{SR1} (or MU_b^{SR1}) is a convex problem under some conditions of the user weights. From T1, if the user weights are in the same order as the channel gains, i.e., $W_1 \geq W_2 \geq \cdots \geq W_N$, then the objective function is concave and the problem is convex. Note that this situation includes the most common sum rate as a special case. On the other hand, the user weights can also be in the inverse order of the channel gains, i.e., $W_1 \leq W_2 \leq \cdots \leq W_N$, but in this case the ratio between W_{k+1} and W_k cannot be too large according to T2. Consequently, one can find the optimal power allocation via standard convex optimization methods, e.g., the interior point method.

6.4.2.2 SR Maximization with QoS (SR2)

Then, we consider the SR maximization problem with QoS constraints for MU-NOMA, which is given by

$$MU_a^{\text{SR2}}: \begin{array}{ll} \max\limits_{p} & \sum\limits_{i=1}^{N} R_i \\ \text{s.t.} & \sum\limits_{i=1}^{N} p_i \leq P,\ p_1 \leq p_2 \leq \cdots \leq p_N \\ & R_i \geq R_i^{\min}, i = 1, \ldots, N \end{array}$$

Similarly, although MU_a^{SR2} is nonconvex in its original formulation, it can be transformed into a convex problem.

In particular, we exploit the same variable transformation: $q_i = \sum_{j=1}^{i} p_j$ for $i = 1, 2, \ldots$, N. Then, MU_a^{SR2} is transformed into

$$MU_b^{\text{SR2}}: \begin{array}{ll} \max\limits_{q} & \sum_{k=1}^{N} g_i(q_i) \\ \text{s.t.} & q_N \leq P \\ & (a_1 - 1)/\Gamma_1 \leq q_1 \leq q_2 - q_1 \leq \cdots \leq q_N - q_{N-1} \\ & q_{i-1} \leq a_i q_i - \varepsilon_i, i = 2, \ldots, N \end{array}$$

where

$$g_i(q_i) = \log(q_i \Gamma_i + 1) - \log(q_i \Gamma_{i+1} + 1) \tag{6.28}$$

for $i = 1, \ldots, N - 1$ and $g_N(q_N) = \log(q_N \Gamma_N + 1)$, $a_i = 2^{-R_i}$ and $\varepsilon_i = (1 - a_i)/\Gamma_i$. According to condition T1 in Theorem 3, the objective in MU_b^{SR2} is concave and thus MU_b^{SR2} is a convex problem. Therefore, MU_b^{SR2} can also be efficiently solved via convex optimization methods. Moreover, we show that if the power order constraint $p_1 \leq p_2 \leq \cdots \leq p_N$ is absent, the optimal solution to MU_b^{SR2} can be analytically characterized.

Proposition 7 *Suppose that $P \geq \sum_{i=1}^{N} \varphi_i$, where $\varsigma_i = 2^{R_i} - 1/\Gamma_i$,*

$$\varphi_i = \begin{cases} \varsigma_1, & i = 1 \\ \max\left\{\varphi_{i-1}, \varsigma_i\left(1 + \Gamma_i \sum\limits_{j=1}^{i-1} \varphi_j\right)\right\}, & i = 2, \ldots, N \end{cases} \tag{6.29}$$

and the power order constraint is absent in MU_a^{SR2} *and* MU_b^{SR2}*. Then, the solution to* MU_b^{SR2} *is*

$$\tilde{q}_i = \begin{cases} a_{i+1}\tilde{q}_{i+1} - \varepsilon_{i+1}, & k = 1, \ldots, N-1 \\ P, & k = N \end{cases} \tag{6.30}$$

and the solution to problem MU_a^{SR2} *is*

$$\tilde{p}_i = \begin{cases} \tilde{q}_1, & k = 1 \\ (1 - a_i)\,\tilde{q}_i + \varepsilon_i, & k = 2, \ldots, N. \end{cases} \tag{6.31}$$

Proof Please refer to the proof of Proposition 3 and Lemma 2 in [17].

Then, a natural question is when the solution in Proposition 7 is indeed optimal with the power order constraint. The answer is given below.

Theorem 4 *The solution in* (6.31) *is optimal for problem* MU_a^{SR2} *with the power order constraint if T3:* $R_2^{\min} \geq 1$ *and*

$$R_i^{\min} \geq \log\left(2 - 2^{-R_{i+1}^{\min}}\right), \ i = 3, \ldots, N. \tag{6.32}$$

Proof Please refer to the proof of Theorem 2 in [17].

Corollary 1 *Condition T3 in Theorem 4 holds if* $R_i^{\min} \geq 1$ *for* $i = 2, \ldots, N$.

Theorem 4 indicates that the power order constraint can be omitted without loss of optimality if the QoS thresholds of the last $N - 1$ users are not small. Corollary 1 specifies that the QoS thresholds are only required to be no less than 1bps/Hz, which is usually satisfied in practice. Therefore, for the SR maximization with QoS constraints, the optimal power allocation is given by Proposition 7 in practical MU-NOMA systems.

6.4.3 Optimal Power Allocation for EE Maximization

In this subsection, we investigate the EE maximization for MU-NOMA systems.

6.4.3.1 Weighted EE Maximization (EE1)

The EE maximization with user weights in an MU-NOMA system is formulated as

$$MU_a^{\mathrm{EE1}}: \quad \begin{aligned} &\max_{p} \ \eta = \frac{\sum_{i=1}^{N} W_i R_i}{P_T + \sum_{i=1}^{N} p_i} \\ &\text{s.t.} \quad \sum_{i=1}^{N} p_i \leq P \\ &\qquad\ \ p_1 \leq p_2 \leq \cdots \leq p_N. \end{aligned}$$

To address this problem, we follow the similar steps for MU_a^{SR1} to simplify MU_a^{EE1}. Specifically, using the variable transformation: $q_i = \sum_{j=1}^{i} p_j$ for $i = 1, 2, \ldots$, N, MU_a^{EE1} can be reformulated as

$$MU_b^{\text{EE1}}: \begin{array}{ll} \max\limits_{\boldsymbol{q} } \eta = \frac{\sum_{k=1}^{N} f_i(q_i)}{P_T + q_N} & \\ \text{s.t.} \quad q_N \le P & \\ \quad\quad 0 \le q_1 \le q_2 - q_1 \le \cdots \le q_N - q_{N-1} & \end{array}$$

where $f_i(q_i)$ is defined in (6.27).

Then, we introduce the following objective function:

$$H(\boldsymbol{q}, \alpha) \triangleq \sum_{i=1}^{N} f_i(q_i) - \alpha \left(P_T + \sum_{i=1}^{N} p_i \right) \tag{6.33}$$

and consider the following problem parameterized by α:

$$MU_c^{\text{EE1}}: \begin{array}{ll} \max\limits_{\boldsymbol{q}} H(\boldsymbol{q}, \alpha) & \\ \text{s.t.} \quad q_N \le P & \\ \quad\quad 0 \le q_1 \le q_2 - q_1 \le \cdots \le q_N - q_{N-1}. & \end{array}$$

According to Lemma 2, the optimal solution to MU_b^{EE1} can be found by solving MU_c^{EE1} with α chosen such that $H^\star(\alpha) = 0$, where $H^\star(\alpha)$ is the optimal objective value of MU_c^{EE1}. The desirable α can be found via a line search method by exploring the monotonicity of $H^\star(\alpha)$. To solve MU_c^{EE1}, we provide the following result.

Theorem 5 *Given $\Gamma_1 \ge \Gamma_2 \ge \cdots \ge \Gamma_N$, MU_c^{EE1} is a convex problem if T1 or T2 in Theorem 3 holds for $i = 1, \ldots, N-1$.*

Proof Please refer to the proof of Theorem 1 in [17].

Theorem 5 indicates that, under the same condition in Theorem 3, MU_c^{EE1} is a convex problem. Therefore, one can efficiently compute its optimal solution via optimization tools, e.g., the interior method.

6.4.3.2 EE Maximization with QoS Constraints (EE2)

Then, we focus on maximizing EE with QoS constraints for MU-NOMA and the corresponding optimization problem is given by

$$MU_a^{\text{EE2}}: \begin{array}{ll} \max\limits_{\boldsymbol{p}} \eta = \frac{\sum_{i=1}^{N} R_i}{P_T + \sum_{i=1}^{N} p_i} & \\ \text{s.t.} \quad \sum\limits_{i=1}^{N} p_i \le P & \\ \quad\quad p_1 \le p_2 \le \cdots \le p_N & \\ \quad\quad R_i \ge R_i^{\min}, i = 1, \ldots, N. & \end{array}$$

By using the same variable transformation: $q_i = \sum_{j=1}^{i} p_j$ for $i = 1, 2, \ldots,$ N, MU_a^{EE2} can be transformed into

$$MU_b^{\text{EE2}}: \begin{array}{ll} \max\limits_{\boldsymbol{q}} & \eta = \frac{\sum_{k=1}^{N} g_i(q_i)}{P_T + q_N} \\ \text{s.t.} & q_N \leq P \\ & (a_1 - 1)/\Gamma_1 \leq q_1 \leq q_2 - q_1 \leq \cdots \leq q_N - q_{N-1} \\ & q_{i-1} \leq a_i q_i - \varepsilon_i, i = 2, \ldots, N \end{array}$$

where $g_i(q_i)$ is given in (6.28). Similarly, we introduce the following objective function

$$Q(\boldsymbol{q}, \alpha) \triangleq \sum_{k=1}^{N} g_i(q_i) - \alpha \left(P_T + \sum_{i=1}^{N} p_i \right), \tag{6.34}$$

and consider the problem parameterized by α:

$$MU_c^{\text{EE2}}: \begin{array}{ll} \max\limits_{\boldsymbol{q}} & Q(\boldsymbol{q}, \alpha) \\ \text{s.t.} & q_N \leq P \\ & (a_1 - 1)/\Gamma_1 \leq q_1 \leq q_2 - q_1 \leq \cdots \leq q_N - q_{N-1} \\ & q_{i-1} \leq a_i q_i - \varepsilon_i, i = 2, \ldots, N. \end{array}$$

Similarly, to obtain the optimal solution to MU_b^{EE2}, one can solve MU_c^{EE2} for given α and search α such that the optimal objective value of MU_c^{EE2} satisfies $Q^{\star}(\alpha) = 0$, for which we refer the reader to the previous subsection. To solve MU_c^{EE2}, we provide the following result.

Proposition 8 *Suppose that* $P \geq \sum_{i=1}^{N} \varphi_i$, *where* $\varsigma_i = 2^{R_i} - 1/\Gamma_i$ *and*

$$\varphi_k = \begin{cases} \varsigma_1, & i = 1 \\ \max\left\{ \varphi_{i-1}, \varsigma_i \left(1 + \Gamma_i \sum\limits_{j=1}^{i-1} \varphi_j \right) \right\}, & i = 2, \ldots, N \end{cases}, \tag{6.35}$$

then MU_c^{EE2} *is feasible and convex.*

Proof Please refer to the proof of Theorem 1 and Proposition 3 in [17].

Proposition 8 indicates that if the power budget of BS is not too small, MU_c^{EE2} can be solved by convex optimization methods, e.g., the interior point method.

6.5 MC-NOMA

In this section, we consider the MC-NOMA scheme, where multiple users share multiple channels. In this case, the resource optimization includes power allocation and channel assignment. However, the joint optimization results in a mixed integer problem and finding its solution requires exhaustive search [35], which causes prohibitive computational complexity. Therefore, in practice, power allocation and channel assignment are often separately and alternatively optimized [26, 28, 35]. In this section, we focus on seeking the optimal power allocation for given channel assignment.

Note that using SIC at each user's receiver causes additional complexity, which is proportional to the number of users on the same channel. Thus, in the multi-channel case, each channel is often restricted to be shared by two users [25, 26, 36], which is also beneficial to reduce the error propagation of SIC. In this section, we would also like to focus on this typical situation. In this case, suppose w.l.o.g. that the CNRs of $\mathrm{UE}_{1,m}$ and $\mathrm{UE}_{2,m}$ are ordered as $\Gamma_{1,m} \geq \Gamma_{2,m}$. Then, the rates of $\mathrm{UE}_{1,m}$ and $\mathrm{UE}_{2,m}$ on channel m are given, respectively, by

$$R_{1,m} = B_c \log\left(1 + p_{1,m}\Gamma_{1,m}\right), \; R_{2,m} = B_c \log\left(1 + \frac{p_{2,m}\Gamma_{2,m}}{p_{1,m}\Gamma_{2,m} + 1}\right). \tag{6.36}$$

6.5.1 Optimal Power Allocation for MMF

In MC-NOMA systems, the power allocation problem under the MMF criterion is given by

$$MC_a^{\mathrm{MMF}}: \begin{array}{ll} \max\limits_{\boldsymbol{p}_1, \boldsymbol{p}_2} & \min\limits_{m=1,\ldots,M} \left\{R_{1,m}(p_{1,m}, p_{2,m}),\; R_{2,m}(p_{1,m}, p_{2,m})\right\} \\ \text{s.t.} & 0 \leq p_{1,m} \leq p_{2,m}, m = 1, \ldots, M,\; \sum\limits_{m=1}^{M} p_{1,m} + p_{2,m} \leq P \end{array}$$

where $\boldsymbol{p}_1 = \{p_{1,m}\}_{m=1}^{M}$ and $\boldsymbol{p}_2 = \{p_{2,m}\}_{m=1}^{M}$. Note that MC_a^{MMF} is a nonconvex problem. To address it, we first introduce auxiliary variables $\boldsymbol{q} = \{q_m\}_{m=1}^{M}$, where q_m represents the power budget for channel m with $p_{1,m} + p_{2,m} = q_m$. Suppose that the channel power budgets $\{q_m\}_{m=1}^{M}$ are given. Then, MC_a^{MMF} is decomposed into a group of subproblems and each subproblem is same with TU^{MMF} in the two-user case with P replaced by q_m.

With given channel power budget q_m, the optimal power allocation for the two users on channel m has been provided in Proposition 1. We can use this result to further optimize the power budgets $\{q_m\}$. According to MC_a^{MMF} and TU^{MMF}, the corresponding power budget optimization problem is

$$MC_b^{\mathrm{MMF}}: \begin{array}{ll} \max\limits_{\boldsymbol{q}} & \min\limits_{m=1,\ldots,M} f_m^{\mathrm{MMF}\star}(q_m) \\ \text{s.t.} & \sum\limits_{m=1}^{M} q_m \leq P,\; \boldsymbol{q} \geq \boldsymbol{0} \end{array}$$

where $f_m^{\mathrm{MMF}\star}(q_m)$ is the optimal objective value of TU^{MMF} for each channel m. Using Proposition 1, we obtain

$$f_m^{\text{MMF}\star} \triangleq B_c \log \left(\frac{\Gamma_{2,m} - \Gamma_{1,m} + \sqrt{\left(\Gamma_{1,m} + \Gamma_{2,m}\right)^2 + 4\Gamma_{1,m}\Gamma_{2,m}^2 q_m}}{2\Gamma_{2,m}} \right). \tag{6.37}$$

Then, we show that MC_b^{MMF} has a closed-form solution.

Theorem 6 *The optimal solution to MC_b^{MMF} is given by*

$$q_m^\star = \frac{\left(Z(\lambda)\,\Gamma_{2,m} + \Gamma_{1,m}\right)\left(Z(\lambda) - 1\right)}{\Gamma_{1,m}\Gamma_{2,m}}, \ \forall m, \tag{6.38}$$

where

$$Z(\lambda) \triangleq X + \sqrt{X^2 + \frac{B_c}{2\lambda \sum_{m=1}^{M} 1/\Gamma_{1,m}}}, \quad X \triangleq \frac{\sum_{m=1}^{M} \left(\Gamma_{2,m} - \Gamma_{1,m}\right)/\left(\Gamma_{1,m}\Gamma_{2,m}\right)}{4\sum_{m=1}^{M} 1/\Gamma_{1,m}}, \tag{6.39}$$

and λ is chosen such that $\sum_{m=1}^{M} q_m^\star = P$.

Proof Please refer to the proof of Theorem 1 in [22].

Consequently, the optimal MC-NOMA power allocation under the MMF criterion is fully characterized by Theorem 6 and Proposition 1. It follows from (6.38) that $q_m^\star$ is monotonically decreasing in λ, so the optimal λ satisfying $\sum_{m=1}^{M} q_m^\star = P$ can be efficiently found via a simple bisection method.

6.5.2 *Optimal Power Allocation for SR Maximization*

In this subsection, we investigate the SR maximization problem with weights or QoS constraints in MC-NOMA systems.

6.5.2.1 Weighted SR Maximization (SR1)

With given channel assignment, the problem of maximizing the weighted sum rate is formulated as the following power allocation problem:

$$MC_a^{\text{SR1}}: \begin{array}{l} \max\limits_{p_1, p_2} \sum\limits_{m=1}^{M} \left(W_{1,m} R_{1,m}(p_{1,m}, p_{2,m}) + W_{2,m} R_{2,m}(p_{1,m}, p_{2,m})\right) \\ \text{s.t.}\ \ 0 \le p_{1,m} \le p_{2,m}, m = 1, \ldots, M, \ \sum\limits_{m=1}^{M} \left(p_{1,m} + p_{2,m}\right) \le P \end{array}$$

To solve it, similarly we introduce auxiliary variables $\boldsymbol{q} = \{q_m\}_{m=1}^{M}$ that represent the power budgets on each channel m with $p_{1,m} + p_{2,m} = q_m$. Then, MC_a^{SR1} is decomposed into a group of subproblems, where each subproblem is the same with TU^{SR1} except P replaced by q_m and its solution has been provided in Proposition 2.

Next, we further optimize the power budget q_m for each channel m. According to Remark 3, to guarantee that the NOMA system is SIC stable, it is reasonable to assume that $q_m \geq \Theta_m > 2\Omega_m$ and $P \geq \sum_{m=1}^{M} \Theta_m$ for some positive Θ_m. Then, from MC_a^{SR1} and TU^{SR1}, the corresponding power budget optimization problem is given by

$$MC_b^{\mathrm{SR1}}: \begin{array}{ll} \max\limits_{q} & \sum\limits_{m=1}^{M} f_m^{\mathrm{SR1}\star}(q_m) \\ \text{s.t.} & \sum\limits_{m=1}^{M} q_m \leq P, \; q_m \geq \Theta_m, \; \forall m \end{array}$$

where $f_m^{\mathrm{SR1}\star}(q_m)$ is the optimal objective value of each subproblem. Using Proposition 2, we obtain

$$f_m^{\mathrm{SR1}\star}(q_m) = W_{1,m} \log\left(1 + \Omega_m \Gamma_{1,m}\right) + W_{2,m} \log\left(\frac{q_m \Gamma_{2,m} + 1}{\Omega_m \Gamma_{2,m} + 1}\right). \tag{6.40}$$

It is easily seen that $f_m^{\mathrm{SR1}\star}(q_m)$ is a concave function, so MC_b^{SR1} is a convex problem, whose solution is provided in the following result.

Theorem 7 *The optimal solution to MC_b^{SR1} is given by*

$$q_m^\star = \left[\frac{W_{2,m} B_c}{\lambda} - \frac{1}{\Gamma_{2,m}}\right]_{\Theta_m}^{\infty}, \tag{6.41}$$

where λ is chosen such that $\sum_{m=1}^{M} q_m^\star = P$.

Proof The solution to MC_b^{SR1} is given by the well-known waterfilling form.

Consequently, the optimal power allocation for the weighted sum rate maximization in MC-NOMA systems is jointly characterized by Theorem 7 and Proposition 2 under the SIC stability.

6.5.2.2 SR Maximization with QoS (SR2)

Now, we consider maximizing the SR with QoS constraints. In this case, the power allocation problem is given by

$$MC_a^{\mathrm{SR2}}: \begin{array}{ll} \max\limits_{p_1, p_2} & \sum\limits_{m=1}^{M} \left(R_{1,m}(p_{1,m}, p_{2,m}) + R_{2,m}(p_{1,m}, p_{2,m})\right) \\ \text{s.t.} & 0 \leq p_{1,m} \leq p_{2,m}, m = 1, \ldots, M, \; \sum\limits_{m=1}^{M} (p_{1,m} + p_{2,m}) \leq P \\ & R_{n,m} \geq R_{n,m}^{\min}, \; n = 1, 2, \; m = 1, \ldots, M. \end{array}$$

We use the similar method to address MC_a^{SR2}. By introducing the power budget q_m on each channel m, MC_a^{SR2} decomposes into several subproblems and each of them has the same structure as TU^{SR2}. Thus, the optimal solution to each subproblem is given in Proposition 3 with P replaced by q_m.

Then, we focus on optimizing the power budget q_m for each channel. Similarly, according to Remark 4, to guarantee the NOMA system is SIC stable, we assume that $q_m \geq \Upsilon_m$ and $P \geq \sum_{m=1}^{M} \Upsilon_m$. According to MC_a^{SR2} and TU^{SR2}, the corresponding power budget optimization problem is as follows

$$MC_b^{\mathrm{SR2}}: \begin{array}{ll} \max\limits_{q} & \sum\limits_{m=1}^{M} f_m^{\mathrm{SR2}\star}(q_m) \\ \text{s.t.} & \sum\limits_{m=1}^{M} q_m \leq P, \ q_m \geq \Upsilon_m, \ \forall m \end{array}$$

where $f_m^{\mathrm{SR2}\star}(q_m)$ is the optimal objective value of each subproblem and given by

$$f_m^{\mathrm{SR2}\star}(q_m) = B_c \log\left(\frac{A_{2,m}\Gamma_{2,m} - A_{2,m}\Gamma_{1,m} + \Gamma_{1,m}\Gamma_{2,m}q_m + \Gamma_{1,m}}{A_{2,m}\Gamma_{2,m}}\right) + R_{2,m}^{\min}. \tag{6.42}$$

Since $f_m^{\mathrm{SR2}\star}(q_m)$ is a concave function, MC_b^{SR2} is a convex problem, whose solution is also given in a waterfilling form.

Theorem 8 *The optimal solution to MC_b^{SR2} is given by*

$$q_m^\star = \left[\frac{B_c}{\lambda} - \frac{A_{2,m}}{\Gamma_{1,m}} + \frac{A_{2,m}}{\Gamma_{2,m}} - \frac{1}{\Gamma_{2,m}}\right]_{\Upsilon_m}^{\infty}, \tag{6.43}$$

where λ is chosen such that $\sum_{m=1}^{M} q_m^\star = P$.

Proof The proof is simple and thus omitted.

Therefore, the optimal power allocation for the SR maximization with QoS constraints in MC-NOMA systems is jointly characterized by Proposition 3 and Theorem 8.

6.5.3 Optimal Power Allocation for EE Maximization

In this subsection, we investigate the optimal power allocation for maximizing the EE with weights or QoS constraints in MC-NOMA systems.

6.5.3.1 EE Maximization with Weights (EE1)

With given channel assignment, the problem of maximizing the weighted EE is formulated as the following power allocation problem:

$$MC_a^{\mathrm{EE1}}:\max_{\boldsymbol{p}_1,\boldsymbol{p}_2}\frac{\sum_{m=1}^{M}\left(W_{1,m}R_{1,m}(p_{1,m},p_{2,m})+W_{2,m}R_{2,m}(p_{1,m},p_{2,m})\right)}{P_T+\sum_{m=1}^{M}\left(p_{1,m}+p_{2,m}\right)}$$

$$\text{s.t. } 0\le p_{1,m}\le p_{2,m},\, m=1,\ldots,M,\ \sum_{m=1}^{M}\left(p_{1,m}+p_{2,m}\right)\le P.$$

The difficulties in solving MC_a^{EE1} lie in its nonconvex and fractional objective. In the following, we will show that this problem can also be optimally solved.

We use the similar trick to address this problem, i.e., introducing the auxiliary variables $\{q_m\}_{m=1}^{M}$ with $p_{1,m}+p_{2,m}=q_m$ for each channel m. Then, MC_a^{EE1} is decomposed into a group of subproblems. Each subproblem is the same with TU^{SR1} except P replaced by q_m, and thus, its solution is provided in Proposition 1.

Then, we concentrate on searching the optimal power budget q_m for each channel. Similarly, to guarantee the NOMA system is SIC stable, it is assumed that $q_m\ge\Theta_m>2\Omega_m$ and $P\ge\sum_{m=1}^{M}\Theta_m$ for some positive Θ_m. According to Proposition 1 and MC_a^{EE1}, the power budget optimization problem is formulated as

$$MC_b^{\mathrm{EE1}}:\max_{\boldsymbol{q}}\ \eta(\boldsymbol{q})\triangleq\frac{\sum_{m=1}^{M}f_m^{\mathrm{SR1}\star}(q_m)}{P_T+\sum_{m=1}^{M}q_m}\tag{6.44}$$

$$\text{s.t.}\ \sum_{m=1}^{M}q_m\le P,\ q_m\ge\Theta_m,\ \forall m$$

where $f_m^{\mathrm{SR1}\star}(q_m)$ is given in (6.40). Although $f_m^{\mathrm{SR1}\star}(q_m)$ is a concave function, MC_b^{EE1} is nonconvex due to the fraction form. To solve it, we introduce the following objective function:

$$\begin{aligned}H(\boldsymbol{q},\alpha)&\triangleq\sum_{m=1}^{M}f_m^{\mathrm{SR1}\star}(q_m)-\alpha\left(P_T+\sum_{m=1}^{M}q_m\right)\\&=\sum_{m=1}^{M}\left(\tilde{R}_{1,m}+W_{2,m}\log\left(\frac{q_m\Gamma_{2,m}+1}{\Omega_m\Gamma_{2,m}+1}\right)\right)-\alpha\left(P_T+\sum_{m=1}^{M}q_m\right),\end{aligned}\tag{6.45}$$

where $\tilde{R}_{1,m}\triangleq W_{1,m}\log\left(1+\Omega_m\Gamma_{1,m}\right)$ and α is a positive parameter. Then, we consider the following convex problem with given α:

$$MC_c^{\text{EE1}}: \begin{array}{l} \max\limits_{\boldsymbol{q}} H(\boldsymbol{q}, \alpha) \\ \text{s.t. } \sum\limits_{m=1}^{M} q_m \le P,\ q_m \ge \Theta_m,\ \forall m. \end{array}$$

According to Lemma 2, the optimal solution to MC_b^{EE1} can be found by solving MC_c^{EE1} with given α and then updating α until $H^\star(\alpha) = 0$. Hence, we first solve MC_c^{EE1} with given α, whose solution is provided in the following result.

Theorem 9 *The optimal solution to* MC_c^{EE1} *is*

$$q_m^\star = \left[\frac{W_{2,m} B_c}{\alpha + \lambda} - \frac{1}{\Gamma_{2,m}}\right]_{\Theta_m}^{\infty}, \tag{6.46}$$

where λ *is chosen such that* $\sum_{m=1}^{M} q_m^\star = P$.

Proof The solution is obtained by exploiting the KKT conditions of MC_c^{EE1}.

After the optimal solution to MC_c^{EE1} is obtained, we shall find an α such that $H^\star(\alpha) = 0$. Since $H^\star(\alpha)$ is monotonic in α, one can use the bisection method to find α. Thereby, the optimal power allocation for the weighted EE maximization in MC-NOMA systems is provided Proposition 2 and Theorem 9.

6.5.3.2 EE Maximization with QoS (EE2)

In this part, we consider maximizing the EE with QoS constraints. The corresponding power allocation problem is given by

$$MC_a^{\text{EE2}}: \begin{array}{l} \max\limits_{\boldsymbol{p}_1, \boldsymbol{p}_2} \dfrac{\sum_{m=1}^{M} \left(R_{1,m}\left(p_{1,m}, p_{2,m}\right) + R_{2,m}\left(p_{1,m}, p_{2,m}\right)\right)}{P_T + \sum_{m=1}^{M} \left(p_{1,m} + p_{2,m}\right)} \\ \text{s.t. } 0 \le p_{1,m} \le p_{2,m}, m = 1, \ldots, M, \sum\limits_{m=1}^{M} (p_{1,m} + p_{2,m}) \le P \\ \qquad R_{l,m} \ge R_{l,m}^{\min},\ l = 1, 2,\ m = 1, \ldots, M. \end{array}$$

We can use the similar method to solve MC_a^{EE2}. Briefly, we also adopt $\{q_m\}_{m=1}^{M}$ with $p_{1,m} + p_{2,m} = q_m$ and decompose MC_a^{EE2} into a group of subproblems, whose solution is coincided with TU^{SR2} and provided in Proposition 3.

Next, we optimize the channel power budget q_m for each channel. First, we assume that $q_m \ge \Upsilon_m$ and $P \ge \sum_{m=1}^{M} \Upsilon_m$ to guarantee the SIC stability. Then, according to Proposition 3 and MC_a^{EE2}, the power budget optimization problem is given by

$$MC_b^{\text{EE2}}: \max_{\boldsymbol{q}} \eta(\boldsymbol{q}) \triangleq \frac{\sum_{m=1}^{M} f_m^{\text{SR2}\star}(q_m)}{P_T + \sum_{m=1}^{M} q_m}$$

$$\text{s.t.} \sum_{m=1}^{M} q_m \le P,\ q_m \ge \Upsilon_m,\ \forall m$$

where $f_m^{\text{SR2}\star}(q_m)$ is given in (6.42). To solve MC_b^{EE2}, we introduce the objective function parameterized by α:

$$\begin{aligned} Q(\boldsymbol{q}, \alpha) &\triangleq \sum_{m=1}^{M} f_m^{\text{SR2}\star}(q_m) - \alpha \left(P_T + \sum_{m=1}^{M} q_m \right) \\ &= \sum_{m=1}^{M} \left(W_{1,m} \log \left(\frac{A_{2,m}\Gamma_{2,m} - A_{2,m}\Gamma_{1,m} + \Gamma_{1,m}\Gamma_{2,m}q_m + \Gamma_{1,m}}{A_{2,m}\Gamma_{2,m}} \right) + R_{2,m}^{\min} \right) \\ &-\alpha \left(P_T + \sum_{m=1}^{M} q_m \right), \end{aligned} \tag{6.47}$$

and formulate the following problem with given α:

$$MC_c^{\text{EE2}}: \begin{array}{l} \max\limits_{\boldsymbol{q}} \ Q(\boldsymbol{q}, \alpha) \\ \text{s.t.} \ \sum\limits_{m=1}^{M} q_m \le P,\ q_m \ge \Upsilon_m,\ \forall m. \end{array}$$

Then, from Lemma 2, we shall solve MC_c^{EE2}, which is a convex problem since $Q(\boldsymbol{q}, \alpha)$ is concave in $\boldsymbol{q}$. The optimal solution to MC_c^{EE2} is provided below.

Theorem 10 *The optimal solution to* MC_c^{EE2} *is*

$$q_m^\star = \left[\frac{W_{1,m}B_c}{\lambda + \alpha} - \frac{A_{2,m}}{\Gamma_{1,m}} + \frac{A_{2,m}}{\Gamma_{2,m}} - \frac{1}{\Gamma_{2,m}} \right]_{\Upsilon_m}^{\infty}, \tag{6.48}$$

where λ *is chosen such that* $\sum_{m=1}^{M} q_m^\star = P$ *.*

Proof The solution is obtained by exploiting the KKT conditions of MC_c^{EE2}.

Then, we can exploit the bisection method to find an α such that the optimal objective value of MC_c^{EE2} satisfies $Q^\star(\alpha) = 0$. Consequently, the optimal power allocation for the EE maximization with QoS constraints in MC-NOMA systems is obtained by using Theorem 10 and Proposition 3.

6.6 Numerical Results

This section evaluates the performance of the optimal power allocation investigated in this chapter. In simulations, the BS is located in the cell center and the users are randomly distributed in a circular range with a radius of 500 m. The minimum distance between users is set to be 40 m, and the minimum distance between the users and the BS is 50 m. Each channel coefficient follows an i.i.d. Gaussian distribution as $g \sim \mathcal{CN}(0, 1)$ and the path loss exponent is $\rho = 2$. The total power budget of the BS is P = 41 dBm and the circuit power consumption is P_T =30 dBm. The noise power is $\sigma^2 = BN_0/M$, where the bandwidth is B = 5 MHz and the noise power spectral density is $N_0 = -174$ dBm.

First, we evaluate the performance of the proposed optimal power solutions for two-user NOMA ($N = 2$) and MU-NOMA ($N = 6$) systems. The user weights satisfy $W_{i+1}/W_i = 0.5$ for $i = 1, \ldots, N - 1$ and the QoS thresholds to be $R_i^{\min} = 2$ bps/Hz for $i = 1, \ldots,$ N. In addition, we compare the NOMA schemes with OFDMA and the DC (difference of two convex functions) approach in [26], where the power allocation is optimized via waterfilling and via DC programming, respectively.

Figure 6.2 shows the minimum user rates of the two-user NOMA and MU-NOMA schemes using the optimal power allocation under the MMF criterion and the minimum user rate of the OFDMA scheme for different total power budgets and user numbers. The minimum user rate in the NOMA system is higher than that in the OFDMA system especially in the two-user case, implying that NOMA provides better fairness than OFDMA.

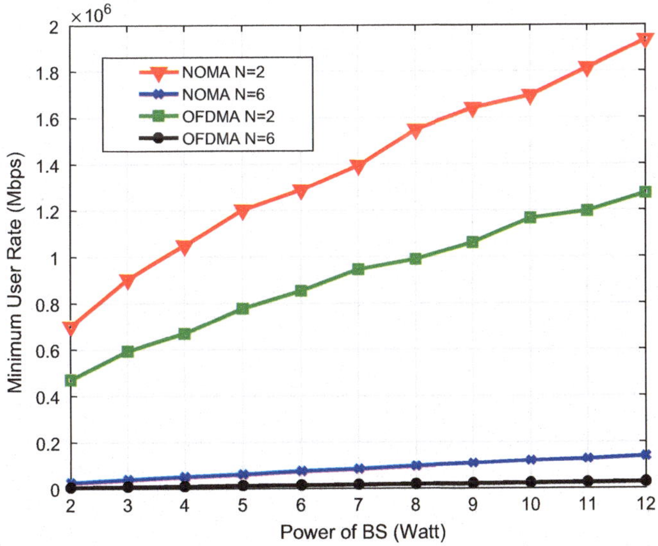

Fig. 6.2 Minimum user rate for different number of users versus BS power

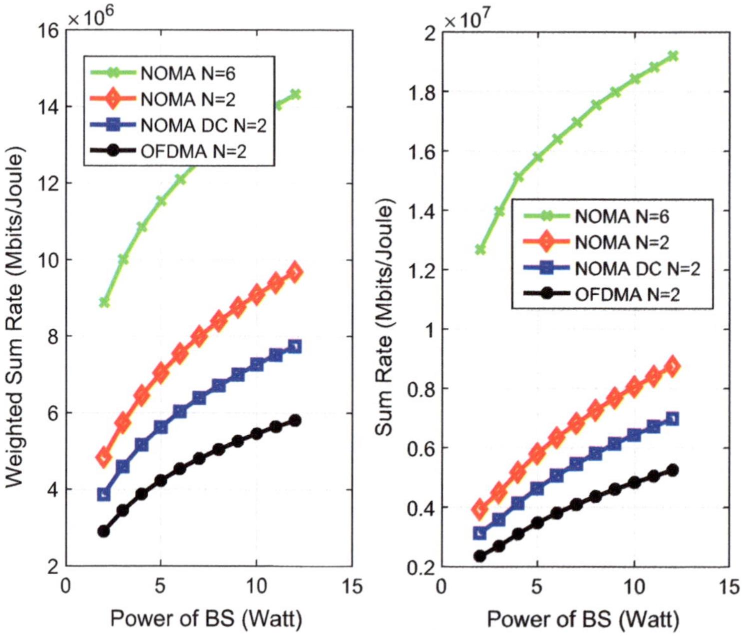

Fig. 6.3 Sum rate versus BS power

In Fig. 6.3 the left subfigure shows the weighted sum rate and the right subfigure shows the sum rate with QoS constraints. Here, in each subfigure, we compare the proposed methods with the OFDMA scheme and the NOMA scheme using DC programming in the two-user case. While NOMA outperforms OFDMA, NOMA with the optimal power allocation also achieves a higher sum rate than the DC approach, as the DC approach generally leads to a suboptimal power allocation. Meanwhile, as expected, the (weighted) sum rate increases with the user number, implying the potential of NOMA. In Fig. 6.4, the similar phenomenon can be observed, i.e., NOMA using the optimal power allocation outperforms OFDMA as well as the (suboptimal) DC approach in terms of energy efficiency.

Then, we show the performance of the optimal power allocation in MC-NOMA systems. The user weights are set to be $W_{1,m} = 0.9$ and $W_{2,m} = 1.1$ for $\forall m$ and the QoS thresholds are set to be $R_{l,m}^{\min} = 2$ bps/Hz for $l = 1, 2, \forall m$. In Fig. 6.5, we compare the joint resource allocation (JRA) method, which uses the optimal power allocation and the matching algorithm [22, 26] for channel assignment, with the exhaustive search (ES), which provides the jointly optimal solution but has high complexity. We set the number of users $N = 6$ and the power budget of the BS ranges from 2 to 12 W. From Fig. 6.5, the performance of JRA is very close to the

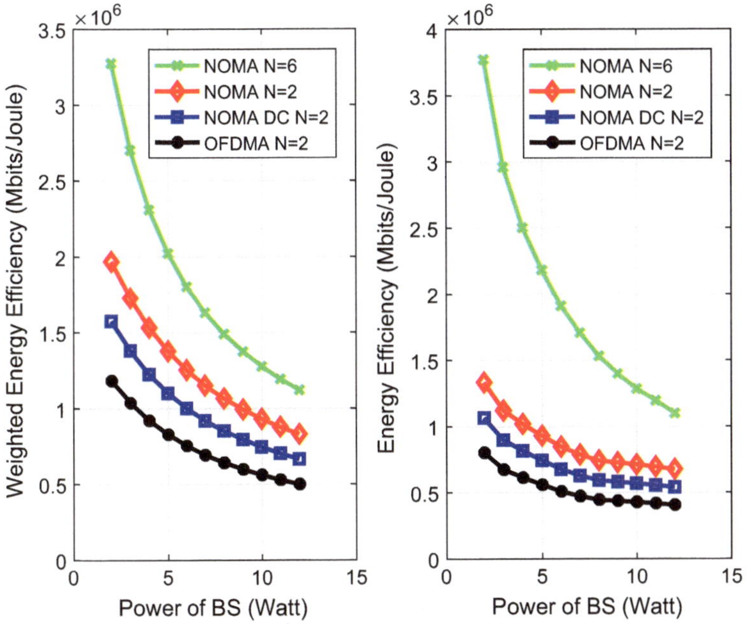

Fig. 6.4 Energy efficiency versus BS power

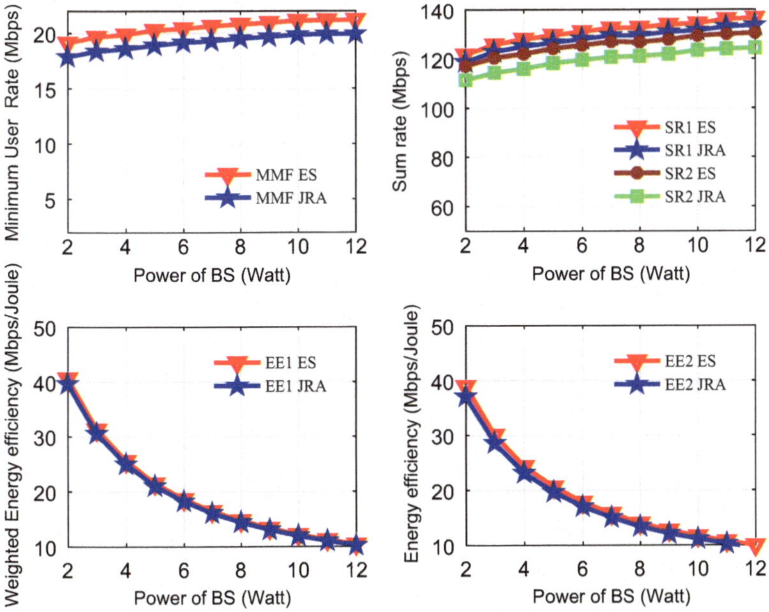

Fig. 6.5 Comparison with the exhaustive search (ES)

globally optimal value and the maximum gap is less than 5%. Therefore, the optimal power allocation method along with efficient (suboptimal) matching algorithm is able to achieve near-optimal performance with low complexity.

6.7 Conclusion

In this chapter, we discussed a promising multiple access technology, i.e., NOMA, for 5G networks and focused on the key problem of power allocation in NOMA systems. We have investigated the optimal power allocation for different NOMA schemes, including the two-user MU-NOMA, and MC-NOMA schemes. The optimal power allocation was derived under different performance measures, including the maximin fairness, weighted sum rate, and energy efficiency, wherein user weights or QoS constraints were also considered. We showed that in most cases the optimal NOMA power allocation admits an analytical solution, while in other cases it can be numerically computed via convex optimization methods.

Appendix

A. Proof of Theorem 1

Since the constraints in TU_c^{EE1} are all linear, it suffices to investigate the concavity of $H(p_1, q, \alpha)$. The second-order derivative of $H(p_1, q, \alpha)$ with respect to p_1 is

$$\frac{\partial^2 H}{\partial p_1^2} = \frac{1}{\ln 2}\left(\frac{\Upsilon \Theta}{(p_1\Gamma_1+1)^2 (p_1\Gamma_2+1)^2}\right), \tag{6.49}$$

where $\Upsilon = \sqrt{W_2 B}\Gamma_2 (p_1\Gamma_1 + 1) + \sqrt{W_1 B}\Gamma_1 (p_1\Gamma_2 + 1)$ and $\Theta = \sqrt{W_2 B}\Gamma_2 - \sqrt{W_1 B}\Gamma_1 + \sqrt{B}\Gamma_1\Gamma_2 p_1 \left(\sqrt{W_2} - \sqrt{W_1}\right)$. Given $\Gamma_1 \geq \Gamma_2$, if $W_1 \geq W_2$, then $\frac{\partial^2 H}{\partial p_1^2} \leq 0$. On the other hand, with $q \leq P$ and if C2 holds, we have

$$\sqrt{W_2 B}\Gamma_2 - \sqrt{W_1 B}\Gamma_1 + \sqrt{B}\Gamma_1\Gamma_2 p_1 \left(\sqrt{W_2} - \sqrt{W_1}\right) \sqrt{W_2 B}\Gamma_2 - \sqrt{W_1 B}\Gamma_1 + \left(\sqrt{W_2} - \sqrt{W_1}\right)\sqrt{B}\Gamma_1\Gamma_2 P \leq 0, \tag{6.50}$$

also implying $\frac{\partial^2 H}{\partial p_1^2} \leq 0$. Following the similar manner, it can be verified that $\frac{\partial^2 H}{\partial q^2} \leq 0$, $\frac{\partial^2 H}{\partial q \partial p_1} = 0$ and $\frac{\partial^2 H}{\partial p_1 \partial q} = 0$. Therefore, the Hessian matrix

$$\begin{pmatrix} \frac{\partial^2 H}{\partial p_1^2} & \frac{\partial^2 H}{\partial q \partial p_1} \\ \frac{\partial^2 H}{\partial p_1 \partial q} & \frac{\partial^2 H}{\partial q^2} \end{pmatrix}$$

is a negative semidefinite matrix, indicating that $H(p_1, q, \alpha)$ is a concave in (p_1, q).

B. Proof of Proposition 4

The Lagrange of TU_c^{EE1} is given by

$$L = W_1 R_1(p_1) + W_2 R_2(p_1, q) - \alpha\,(P_T + q) + \mu\,(q - 2p_1) - \lambda\,(q - P) \quad (6.51)$$

with Lagrange multipliers μ and $\lambda \geq 0$. According to Theorem 1, TU_c^{EE1} is a convex problem under condition C1 or C2. Therefore, its optimal solution is characterized by the following Karush–Kuhn–Tucker (KKT) conditions:

$$\frac{\partial L}{\partial p_1} = \frac{W_1 B \Gamma_1}{\ln 2\,(1 + p_1 \Gamma_1)} - \frac{W_2 B \Gamma_2}{\ln 2\,(1 + p_1 \Gamma_2)} - 2\mu = 0, \quad (6.52)$$

$$\frac{\partial L}{\partial q} = \frac{W_2 B \Gamma_2}{\ln 2\,(1 + q \Gamma_2)} - \alpha + \mu - \lambda = 0, \quad (6.53)$$

$$\mu\,(q - 2p_1) = 0, \quad (6.54)$$

$$\lambda\,(q - P) = 0. \quad (6.55)$$

According to Definition 1, if $p_1 = q/2$, then the NOMA system is SIC-unstable. Therefore, from (6.54), considering the SIC stability, we have $\mu = 0$. Hence, from (6.52) we obtain the optimal $p_1^\star = \Omega$. It follows from (6.55) that if $q < P$, then $\lambda = 0$. Then, from (6.53) we obtain

$$2\Omega \leq q = \frac{W_2 B}{\alpha \ln 2} - \frac{1}{\Gamma_2} < P. \quad (6.56)$$

On the other hand, if $q = P$, then from (6.53) we have

$$\lambda = \frac{W_2 B \Gamma_2}{\ln 2\,(1 + P \Gamma_2)} - \alpha \geq 0, \quad (6.57)$$

which leads to

$$\frac{W_2 B}{\alpha \ln 2} - \frac{1}{\Gamma_2} \geq P. \quad (6.58)$$

Therefore, the optimal q is given by $q^\star = \left[\frac{W_2 B}{\alpha \ln 2} - \frac{1}{\Gamma_2}\right]_{2\Omega}^{P}$.

C. Proof of Proposition 5

The Lagrange of TU_c^{EE2} is given by

$$L = R_1(p_1) + R_2(p_1, q) - \alpha\,(P_T + q) + \mu\,(q - 2p_1) - \lambda\,(q - P) \tag{6.59}$$
$$+ \sigma_1 \left(p_1 - \frac{A_1 - 1}{\Gamma_1} \right) + \sigma_2\,(1 + q\Gamma_2 - A_2 - A_2 p_1 \Gamma_2)\,,$$

where μ, λ, σ_1, and σ_2 are the Lagrange multipliers. The optimal solution is characterized by the following KKT conditions:

$$\frac{\partial L}{\partial p_1} = \frac{B\Gamma_1}{\ln 2\,(1 + p_1\Gamma_1)} - \frac{B\Gamma_2}{\ln 2\,(1 + p_1\Gamma_2)} - 2\mu + \sigma_1 - \sigma_2 A_2 \Gamma_2 = 0, \tag{6.60}$$

$$\frac{\partial L}{\partial q} = \frac{B\Gamma_2}{\ln 2\,(1 + q\Gamma_2)} - \alpha + \mu - \lambda + \sigma_2 \Gamma_2 = 0, \tag{6.61}$$

$$\mu\,(q - 2p_1) = 0, \tag{6.62}$$

$$\lambda\,(q - P) = 0, \tag{6.63}$$

$$\sigma_1 \left(p_1 - \frac{A_1 - 1}{\Gamma_1} \right) = 0, \tag{6.64}$$

$$\sigma_2\,(1 + q\Gamma_2 - A_2 - A_2 p_1 \Gamma_2) = 0. \tag{6.65}$$

In (6.62), considering the SIC stability, we have $q > 2p_1$ and hence $\mu = 0$. Note that $\sigma_2 \neq 0$. To see this, if $\sigma_2 = 0$, according to (6.60), we have

$$\frac{B\Gamma_1}{\ln 2\,(1 + p_1\Gamma_1)} - \frac{B\Gamma_2}{\ln 2\,(1 + p_1\Gamma_2)} + \sigma_1 = 0 \tag{6.66}$$

which, however, does not hold since $\frac{B\Gamma_1}{\ln 2(1+p_1\Gamma_1)} - \frac{B\Gamma_2}{\ln 2(1+p_1\Gamma_2)} + \sigma_1 > 0$ with $\Gamma_1 \geq \Gamma_2$.

We consider two cases: (1) $\sigma_1 \neq 0$, $\sigma_2 \neq 0$; and (2) $\sigma_1 = 0$, $\sigma_2 \neq 0$. First, if $\sigma_1 \neq 0$, $\sigma_2 \neq 0$, the optimal solution can be easily obtained as

$$p_1^\star = \frac{1 + q^\star \Gamma_2 - A_2}{A_2 \Gamma_2},\ q^\star = \Upsilon \tag{6.67}$$

from (6.64) and (6.65). Then, if $\sigma_1 = 0$, $\sigma_2 \neq 0$, according to (6.60) and (6.61), we have

$$\frac{A_2\Gamma_2}{(1 + q\Gamma_2)} + \left(\frac{1}{1/\Gamma_1 + p_1} - \frac{1}{1/\Gamma_2 + p_1} \right) = (\alpha + \lambda)\,A_2. \tag{6.68}$$

From (6.65), we obtain $p_1 = \frac{1+q\Gamma_2-A_2}{A_2\Gamma_2}$, which along with (6.68) leads to

$$q^\star = \frac{1}{\alpha+\lambda} - \frac{A_2}{\Gamma_1} + \frac{A_2-1}{\Gamma_2}. \tag{6.69}$$

It follows from (6.63) that if $q < P$, then $\lambda = 0$. From (6.69), we obtain

$$\Upsilon \le q = \frac{1}{\alpha} - \frac{A_2}{\Gamma_1} + \frac{A_2-1}{\Gamma_2} < P. \tag{6.70}$$

On the other hand, if $q = P$, then from (6.53) we have

$$\lambda = \frac{\Gamma_1\Gamma_2}{A_2\Gamma_2 - (A_2-1)\,\Gamma_1 + P\Gamma_2\Gamma_1} - \alpha \ge 0, \tag{6.71}$$

which leads to

$$\frac{1}{\alpha} - \frac{A_2}{\Gamma_1} + \frac{A_2-1}{\Gamma_2} \ge P. \tag{6.72}$$

Therefore, optimal q is given by $q^\star = \left[\frac{W_2 B}{\alpha \ln 2} - \frac{1}{\Gamma_2}\right]_\Upsilon^P$.

D. Proof of Theorem 2

Let $q_i = \sum_{j=1}^{i} p_j$, then $q_N = P$ and $p_i = \frac{2^t-1}{\Gamma_i}\left(\sum_{j=1}^{i-1} p_j\Gamma_i + 1\right)$ can be transformed into $q_i = q_{i-1}2^t + \frac{2^t-1}{\Gamma_i}$. Thus, we obtain $P = q_N = \sum_{i=1}^{N} \frac{(2^t-1)2^{(N-i)t}}{\Gamma_i} \ge \chi$, implying $t \ge 1$ and $p_i \ge p_{i-1}$ for $i = 2, \ldots, N$. Therefore, this solution satisfies the power order constraint.

References

1. J.G. Andrews, S. Buzzi, W. Choi, S.V. Hanly, A. Lozano, A.C. Soong, J.C. Zhang, What will 5G be? IEEE J. Sel. Areas Commun. **32**(6), 1065–1082 (2014)
2. V. Jungnickel, K. Manolakis, W. Zirwas, B. Panzner, V. Braun, M. Lossow, M. Sternad, R. Apelfrojd, T. Svensson, The role of small cells, coordinated multipoint, and massive MIMO in 5G. IEEE Commun. Mag. **52**(5), 44–51 (2014)
3. M. Xiao, S. Mumtaz, Y. Huang, L. Dai, Y. Li, M. Matthaiou, G.K. Karagiannidis, E. Bjrnson, K. Yang, I. Chih-Lin, A. Ghosh, Millimeter wave communications for future mobile networks. IEEE J. Sel. Areas Commun. **35**(9), 1909–1935 (2017)
4. J.G. Andrews, H. Claussen, M. Dohler, S. Rangan, M.C. Reed, Femtocells: past, present, and future. IEEE J. Sel. Areas Commun. **30**(3), 497–508 (2012)
5. J. Wang, W. Guan, Y. Huang, R. Schober, X. You, Distributed optimization of hierarchical small cell networks: a GNEP framework. IEEE J. Sel. Areas Commun. **35**(2), 249–264 (2017)

6. Y. Saito, Y. Kishiyama, A. Benjebbour, T. Nakamura, A. Li, K. Higuchi, Non-orthogonal multiple access (NOMA) for cellular future radio access, in *Proceeding of IEEE Vehicular Technology Conference (VTC Spring)* Dresden, Germany, June 2013, pp. 1–5
7. Z. Ding, Y. Liu, J. Choi, Q. Sun, M. Elkashlan, I. Chih-Lin, H.V. Poor, Application of non-orthogonal multiple access in LTE and 5G networks, IEEE Commun. Mag. **55**(2), 185–191 (2017)
8. L. Dai, B. Wang, Y. Yuan, S. Han, I. Chih-Lin, Z. Wang, Non-orthogonal multiple access for 5G: solutions, challenges, opportunities, and future research trends. IEEE Commun. Mag. **53**(9), 74–81 (2015)
9. Z. Ding, R. Schober, H.V. Poor, A general MIMO framework for NOMA downlink and uplink transmission based on signal alignment. IEEE Trans. Wirel. Commun. **15**(6), 4438–4454 (2016)
10. Z. Ding, F. Adachi, H.V. Poor, The application of MIMO to non-orthogonal multiple access. IEEE Trans. Wirel. Commun. **15**(1), 537–552 (2016)
11. W. Shin, M. Vaezi, B. Lee, D.J. Love, J. Lee, H.V. Poor, Non-orthogonal multiple access in multi-cell networks: theory, performance, and practical challenges. IEEE Commun. Mag. **55**(10), 176–183 (2017)
12. X. Zhang, Q. Gao, C. Gong, Z. Xu, User grouping and power allocation for NOMA visible light communication multi-cell networks. IEEE Commun. Lett. **21**(4), 777–780 (2017)
13. Y. Huang, C. Zhang, J. Wang, Y. Jing, L. Yang, X. You, Signal processing for MIMO-NOMA: present and future challenges. IEEE Wirel. Commun. **25**(2), 32–38 (2018)
14. L. Zhang, M. Xiao, G. Wu, M. Alam, Y.C. Liang, S. Li, A survey of advanced techniques for spectrum sharing in 5G networks. IEEE Wirel. Commun. **24**(5), 44–51 (2017)
15. C.-L. Wang, J.-Y. Chen, Y.-J. Chen, Power allocation for a downlink non-orthogonal multiple access system. IEEE Wirel. Commun. Lett. **5**(5), 532–535 (2016)
16. Z. Yang, Z. Ding, P. Fan, N. Al-Dhahir, A general power allocation scheme to guarantee quality of service in downlink and uplink NOMA systems. IEEE Trans. Wirel. Commun. **15**(11), 7244–7257 (2016)
17. J. Wang, Q. Peng, Y. Huang, H.M. Wang, X. You, Convexity of weighted sum rate maximization in NOMA systems. IEEE Signal Process. Lett. **24**(9), 1323–1327 (2017)
18. J. Choi, Power allocation for max-sum rate and max-min rate proportional fairness in NOMA. IEEE Commun. Lett. **20**(10), 2055–2058 (2016)
19. J. Cui, Z. Ding, P. Fan, A novel power allocation scheme under outage constraints in NOMA systems. IEEE Signal Process. Lett. **23**(9), 1226–1230 (2016)
20. S. Timotheou, I. Krikidis, Fairness for non-orthogonal multiple access in 5G systems. IEEE Signal Process. Lett. **22**(10), 1647–1651 (2015)
21. Y. Zhang, H.M. Wang, T.X. Zheng, Q. Yang, Energy-efficient transmission design in non-orthogonal multiple access. IEEE Trans. Veh. Technol. **66**(3), 2852–2857 (2017)
22. J. Zhu, J. Wang, Y. Huang, S. He, X. You, L. Yang, On optimal power allocation for downlink non-orthogonal multiple access systems. IEEE J. Sel. Areas Commun. **35**(12), 2744–2757 (2017)
23. F. Fang, H. Zhang, J. Cheng, S. Roy, V.C.M. Leung, Joint user scheduling and power allocation optimization for energy efficient NOMA systems with imperfect CSI. IEEE J. Sel. Areas Commun. **35**(12), 2874–2885 (2017)
24. B. Di, S. Bayat, L. Song, Y. Li, Radio resource allocation for downlink non-orthogonal multiple access (NOMA) networks using matching theory, in *Proc. of IEEE Global Communication Conference (GLOBECOM)*, pp. 1–6
25. Z. Ding, M. Peng, H.V. Poor, Cooperative non-orthogonal multiple access in 5G systems. IEEE Commun. Lett. **19**(8), 1462–1465 (2015)
26. F. Fang, H. Zhang, J. Cheng, V.C. Leung, Energy-efficient resource allocation for downlink non-orthogonal multiple access network. IEEE Trans. Commun. **64**(9), 3722–3732 (2016)
27. L. Lei, D. Yuan, C.K. Ho, S. Sun, Joint optimization of power and channel allocation with non-orthogonal multiple access for 5G cellular systems. in *Proceeding of IEEE Global Communication Conference (GLOBECOM)*, San Diego, CA, Dec 2015, pp. 1–6

28. P. Parida, S.S. Das, Power allocation in OFDM based NOMA systems: a DC programming approach, in *Proceeding of IEEE Globecom Workshops*, Dec 2014, pp. 1026–1031
29. Y. Sun, D.W.K. Ng, Z. Ding, R. Schober, Optimal joint power and subcarrier allocation for full-duplex multicarrier non-orthogonal multiple access systems. IEEE Trans. Commun. **65**(3), 1077–1091 (2017)
30. M.R. Hojeij, J.Farah, C.A. Nour, C. Douillard, Resource allocation in downlink non-orthogonal multiple access (NOMA) for future radio access, in *Proceeding of IEEE Vehicular Technology Conference (VTC Spring)*, Dresden, Germany, May 2015, pp. 1–6
31. Z. Wei, J. Yuan, D.W.K. Ng, M. Elkashlan, Z. Ding, A Survey of Downlink Non-orthogonal Multiple Access for 5G Wireless Communication Networks. CoRR. arXiv: 1609.01856, https://dblp.org/rec/bib/journals/corr/WeiYNED16 (2016)
32. M.S. Ali, H. Tabassum, E. Hossain, Dynamic user clustering and power allocation for uplink and downlink non-orthogonal multiple access (NOMA) systems. IEEE Access **4**, 6325–6343 (2016)
33. Z. Ding, Z. Yang, P. Fan, H.V. Poor, On the performance of non-orthogonal multiple access in 5G systems with randomly deployed users. IEEE Signal Process. Lett. **21**(12), 1501–1505 (2014)
34. W. Dinkelbach, On nonlinear fractional programming. Manag. Sci. **13**(7), 492–498 (1967)
35. S. Zhang, B. Di, L. Song, Y. Li, Radio resource allocation for non-orthogonal multiple access (NOMA) relay network using matching game," in *Proceeding of IEEE International Conference Communication (ICC)*, Kuala Lumpur, Malaysia, May 2016, pp. 1–6
36. Z. Ding, P. Fan, H.V. Poor, Impact of user pairing on 5G nonorthogonal multiple-access downlink transmissions. IEEE Trans. Veh. Technol. **65**(8), 6010–6023 (2016)

Chapter 7
On the Design of Multiple-Antenna Non-Orthogonal Multiple Access

Xiaoming Chen, Zhaoyang Zhang, Caijun Zhong and Derrick Wing Kwan Ng

7.1 Introduction

Current wireless communications in general adopt various types of orthogonal multiple access (OMA) technologies for serving multiple users, such as time division multiple access (TDMA), frequency division multiple access (FDMA), and code division multiple access (CDMA), where one resource block is exclusively allocated to one mobile user (MU) to avoid possible multiuser interference. In practice, the OMA technologies are relatively easy to implement, albeit at the cost of low spectral efficiency. Recently, with the rapid development of mobile Internet and proliferation of mobile devices, it is expected that future wireless communication systems should be able to support massive connectivity, which is an extremely challenging task for the OMA technologies with limited radio resources. Responding to this, non-orthogonal multiple access (NOMA) has been recently proposed as a promising access technology for the fifth-generation (5G) mobile communication systems, due to its potential in achieving high spectral efficiency and supporting massive access [1–4].

X. Chen (✉) · Z. Zhang · C. Zhong
College of Information Science and Electronic Engineering, Zhejiang University, Hangzhou, China
e-mail: chen_xiaoming@zju.edu.cn

Z. Zhang
e-mail: ning_ming@zju.edu.cn

C. Zhong
e-mail: caijunzhong@zju.edu.cn

D. W. K. Ng
School of Electrical Engineering and Telecommunications, The University of New South Wales, Sydney, NSW, Australia
e-mail: w.k.ng@unsw.edu.au

M. Vaezi et al. (eds.), *Multiple Access Techniques for 5G Wireless Networks and Beyond*, https://doi.org/10.1007/978-3-319-92090-0_7

The principle of NOMA is to exploit the power domain to simultaneously serve multiple MUs utilizing the same radio resources [5–7], with the aid of sophisticated successive interference cancellation (SIC) receivers [8, 9]. Despite the adoption of SIC, inter-user interference still exists except for the MU with the strongest channel gain, which limits the overall system performance [10]. To address this issue, power allocation has been considered as an effective method to harness multiuser interference [11, 12]. Since the overall performance is limited by the MUs with weak channel conditions, it is intuitive to allocate more power to the weak MUs and less power to the strong MU in order to enhance the effective channel gain and minimize the interference to the weak MUs [13]. For the specific two-user case, the optimal power allocation scheme was studied in [14], and [15] proposed two sub-optimal power allocation schemes exploiting the Karush–Kuhn–Tucker (KKT) conditions, while the issue of quality of service (QoS) requirements of NONA systems was investigated in [16]. For the case with arbitrary number of users, the computational complexity of performing SIC increases substantially and the design of the optimal power allocation becomes intractable. To facilitate an effective system design, clustering and user pairing have been proposed [17, 18]. Generally speaking, multiple MUs with distinctive channel gains are selected to form a cluster, in which SIC is conducted to mitigate the interference [19, 20]. In general, a small cluster consisting a small number of MUs implies low complexity of SIC, but leads to high inter-cluster interference. Thus, it makes sense to dynamically adjust the size of a cluster according to performance requirements and system parameters, so as to achieve a balance between implementation complexity and interference mitigation [21]. However, dynamic clustering is not able to reduce the inter-cluster interference, indicating the necessity of carrying out dynamic clustering in combination with efficient interference mitigation schemes.

It is well known that the multiple-antenna technology is a powerful interference mitigation scheme [22–25], hence, can be naturally applied to NOMA systems [26, 27]. In [28], the authors proposed a beamforming scheme for combating inter-cluster and intra-cluster interference in a NOMA downlink, where the base station (BS) was equipped with multiple antennas and the MUs have a single antenna each. A more general setup was considered in [29], where both the BS and the MUs are multiple-antenna devices. By exploiting multiple antennas at the BS and the MUs, a signal alignment scheme was proposed to mitigate both the intra-cluster and inter-cluster interference. It is worth pointing out that the implementation of the two above schemes requires full channel state information (CSI) at the BS, which is usually difficult and costly in practice. To circumvent the difficulty in CSI acquisition, random beamforming was adopted in [30], which inevitably leads to performance loss. Alternatively, the work in [31] suggested to employ zero-forcing (ZF) detection at the multiple-antenna MUs for inter-cluster interference cancelation. However, the ZF scheme requires that the number of antennas at each MU is greater than the number of antennas at the BS, which is in general impractical.

To effectively realize the potential benefits of multiple-antenna techniques, the amount and quality of CSI available at the BS play a key role [32, 33]. In practice, the CSI can be obtained in several different ways. For instance, in time duplex division (TDD) systems, the BS can obtain the downlink CSI through estimating the CSI of uplink by leveraging the channel reciprocity [34]. While in frequency duplex division (FDD) systems, the downlink CSI is usually first estimated and quantized at the MUs, and then is conveyed back to the BS via a feedback link [35]. For both practical TDD and FDD systems, the BS has access to only partial CSI. As a result, there will be residual inter-cluster and intra-cluster interference. To the best of the authors' knowledge, previous works only consider two extreme cases with full CSI or no CSI, the design, analysis and optimization of multiple-antenna NOMA systems with partial CSI remains an uncharted area. Motivated by this, we present a comprehensive study on the impact of partial CSI on the design, analysis, and optimization of multiple-antenna NOMA downlink communication systems.

The rest of this chapter is organized as follows: Sect. 7.2 gives a brief introduction of the considered NOMA downlink communication system and designs the corresponding multiple-antenna transmission framework. Section 7.3 first analyzes the average transmission rates in presence of imperfect CSI and then proposes three performance optimization schemes. Section 7.4 derives the average transmission rates in two extreme cases through asymptotic analysis and presents some system design guidelines. Section 7.5 provides simulation results to validate the effectiveness of the proposed schemes. Finally, Sect. 7.6 concludes this chapter.

7.2 System Model and Framework Design

Consider a downlink communication scenario in a single-cell system, where a base station (BS) broadcasts messages to multiple MUs, cf. Fig. 7.1. Note that the BS is equipped with M antennas, while the MUs have a single antenna each due to the size limitation.

7.2.1 User Clustering

To strike a balance between the system performance and computational complexity in NOMA systems, it is necessary to carry out user clustering. In particular, user clustering can be designed from different perspectives. For instance, a signal-to-interference-plus-noise ratio (SINR) maximization user clustering scheme was adopted in [36] and quasi-orthogonal MUs were selected to form a cluster in [37]. Intuitively, these schemes perform user clustering by the exhaustive search method, resulting in high implementation complexity. In this chapter, we design a simple user clustering scheme based on the information of spatial direction.[1] Specifically, the MUs in the same direction but with distinctive propagation distances are arranged

[1]The spatial direction of users can be found via various methods/technologies such as GPS or user location tracking algorithms.

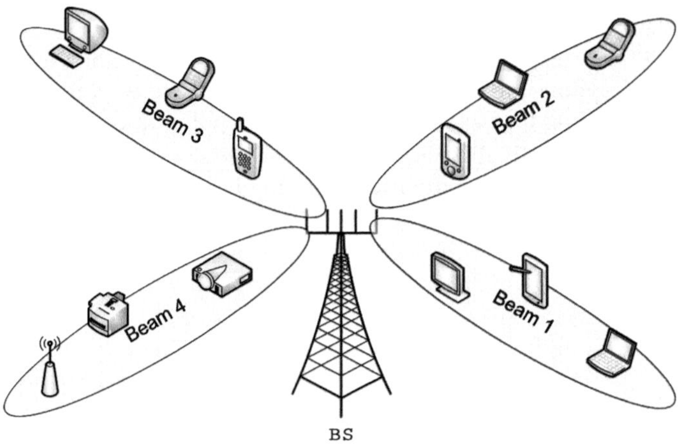

Fig. 7.1 A multiuser NOMA communication system with 4 clusters

into a cluster. On one hand, the same direction of the MUs in a cluster allows the use of a single beam to nearly align all MUs in such a cluster, thereby facilitating the mitigation of the inter-cluster interference and the enhancement of the effective channel gain. On the other hand, a large gap of propagation distances avoids severe inter-user interference and enables a more accurate SIC at the MUs [38–40]. If two MUs are close to each other with almost equal channel gains, it is possible to assign them in different clusters by improving the spatial resolution via increasing the number of spatial beams and the number of BS antennas. Without loss of generality, we assume that the MUs are grouped into N clusters with K MUs in each cluster. To facilitate the following presentation, we use $\alpha_{n,k}^{1/2}\mathbf{h}_{n,k}$ to denote the M-dimensional channel vector from the BS to the kth MU in the nth cluster, where $\alpha_{n,k}$ is the large-scale channel fading, and $\mathbf{h}_{n,k}$ is the small-scale channel fading following zero mean complex Gaussian distribution with unit variance. It is assumed that $\alpha_{n,k}$ remains constant for a relatively long period, while $\mathbf{h}_{n,k}$ keeps unchanged in a time slot but varies independently over time slots.

7.2.2 CSI Acquisition

For the TDD mode, the BS obtains the downlink CSI through uplink channel estimation. Specifically, at the beginning of each time slot, the MUs simultaneously send pilot sequences of τ symbols to the BS, and the received pilot at the BS can be expressed as

$$\mathbf{Y}_P = \sum_{n=1}^{N}\sum_{k=1}^{K}\sqrt{\tau P_{n,k}^{P}\alpha_{n,k}}\mathbf{h}_{n,k}\boldsymbol{\Phi}_{n,k} + \mathbf{N}_P, \tag{7.1}$$

where $P^P_{n,k}$ is the transmit power for the pilot sequence of the kth MU in the nth cluster, $\mathbf{N}_P$ is an additive white Gaussian noise (AWGN) matrix with i.i.d. zero mean and unit variance complex Gaussian distributed entries. $\boldsymbol{\Phi}_{n,k} \in C^{1\times\tau}$ is the pilot sequence sent from the kth MU in the nth cluster. It is required that $\tau > NK$, such that the pairwise orthogonality that $\boldsymbol{\Phi}_{n,k}\boldsymbol{\Phi}^H_{i,j} = 0$ and $\boldsymbol{\Phi}_{n,k}\boldsymbol{\Phi}^H_{n,k} = 1, \forall(n,k) \neq (i,j)$, can be guaranteed. By making use of the pairwise orthogonality, the received pilot can be transformed as

$$\mathbf{Y}_P\boldsymbol{\Phi}^H_{n,k} = \sqrt{\tau P^P_{n,k}\alpha_{n,k}}\mathbf{h}_{n,k} + \mathbf{N}_P\boldsymbol{\Phi}^H_{n,k}. \tag{7.2}$$

Then, by using minimum mean squared error (MMSE) estimation, the relation between the actual channel gain $\mathbf{h}_{n,k}$ and the estimated channel gain $\hat{\mathbf{h}}_{n,k}$ can be expressed as

$$\mathbf{h}_{n,k} = \sqrt{\rho_{n,k}}\hat{\mathbf{h}}_{n,k} + \sqrt{1-\rho_{n,k}}\mathbf{e}_{n,k}, \tag{7.3}$$

where $\mathbf{e}_{n,k}$ is the channel estimation error vector with i.i.d. zero mean and unit variance complex Gaussian distributed entries, and is independent of $\hat{\mathbf{h}}_{n,k}$. Variable $\rho_{n,k} = \frac{\tau P^P_{n,k}\alpha_{n,k}}{1+\tau P^P_{n,k}\alpha_{n,k}} = 1 - \frac{1}{1+\tau P^P_{n,k}\alpha_{n,k}}$ is the correlation coefficient between $\mathbf{h}_{n,k}$ and $\hat{\mathbf{h}}_{n,k}$. A large $\rho_{n,k}$ means a high accuracy for channel estimation. Thus, it is possible to improve the CSI accuracy by increasing the transmit power $P^P_{n,k}$ or the length τ of pilot sequence.

For the FDD mode, the CSI is usually conveyed from the MUs to the BS through a feedback link. Since the feedback link is rate-constrained, CSI at the MUs should first be quantized. Specifically, the kth MU in the nth cluster chooses an optimal codeword from a predetermined quantization codebook $\mathscr{B}_{n,k} = \{\tilde{\mathbf{h}}^{(1)}_{n,k}, \ldots, \tilde{\mathbf{h}}^{(2^{B_{n,k}})}_{n,k}\}$ of size $2^{B_{n,k}}$, where $\tilde{\mathbf{h}}^{(j)}_{n,k}$ is the jth codeword of a unit norm and $B_{n,k}$ is the number of feedback bits. Mathematically, the codeword selection criterion is given by

$$j^\star = \arg \max_{1\leq j\leq 2^{B_{n,k}}} \left|\mathbf{h}^H_{n,k}\tilde{\mathbf{h}}^{(j)}_{n,k}\right|^2. \tag{7.4}$$

Then, the MU conveys the index $j^\star$ to the BS with $B_{n,k}$ feedback bits, and the BS recoveries the quantized CSI $\tilde{\mathbf{h}}^{(j^\star)}_{n,k}$ from the same codebook. In other words, the BS only gets the phase information by using the feedback scheme based on a quantization codebook. However, as shown in below, the phase information is sufficient for the design of spatial beamforming. Similarly, the relation between the real CSI and the obtained CSI in FDD mode can be approximated as [41, 42]

$$\tilde{\mathbf{h}}_{n,k} = \sqrt{\rho_{n,k}}\tilde{\mathbf{h}}^\star_{n,k} + \sqrt{1-\rho_{n,k}}\tilde{\mathbf{e}}_{n,k}, \tag{7.5}$$

where $\tilde{\mathbf{h}}_{n,k} = \frac{\mathbf{h}_{n,k}}{\|\mathbf{h}_{n,k}\|}$ is the phase of the channel $\mathbf{h}_{n,k}$, $\tilde{\mathbf{h}}^\star_{n,k}$ is the quantized phase information, $\tilde{\mathbf{e}}_{n,k}$ is the quantization error vector with uniform distribution, and

$\rho_{n,k} = 1 - 2^{-\frac{B_{n,k}}{M-1}}$ is the associated correlation coefficient or CSI accuracy. Thus, it is possible to improve the CSI accuracy by increasing the size of quantization codebook for a given number of antennas M at the BS.

7.2.3 Superposition Coding and Transmit Beamforming

Based on the available CSI, the BS constructs one transmit beam for each cluster, so as to mitigate or even completely cancel the inter-cluster interference. To strike balance between system performance and implementation complexity, we adopt zero-force beamforming (ZFBF) at the BS. We take the design of beam $\mathbf{w}_i$ for the ith cluster as an example. First, we construct a complementary matrix $\bar{\mathbf{H}}_i$[2] as:

$$\bar{\mathbf{H}}_i = [\hat{\mathbf{h}}_{1,1}, \ldots, \hat{\mathbf{h}}_{1,K}, \ldots, \hat{\mathbf{h}}_{i-1,K}, \hat{\mathbf{h}}_{i+1,1}, \ldots, \hat{\mathbf{h}}_{N,K}]^H. \tag{7.6}$$

Then, we perform singular value decomposition (SVD) on $\bar{\mathbf{H}}_i$ and obtain its right singular vectors $\mathbf{u}_{i,j}$, $j = 1, \ldots, N_u$, with respect to the zero singular values, where N_u is the number of zero singular values. Finally, we can design the beam as $\mathbf{w}_i = \sum_{j=1}^{N_u} \theta_{i,j}\mathbf{u}_{i,j}$, where $\theta_{i,j} > 0$ is a weight such that $\sum_{j=1}^{N_u} \theta_{i,j} = 1$. Thus, the received signal at the kth MU in the nth cluster is given by

$$\begin{aligned} y_{n,k} &= \sqrt{\alpha_{n,k}}\mathbf{h}_{n,k}^H \sum_{i=1}^{N} \mathbf{w}_i s_i + n_{n,k} \\ &= \sqrt{\alpha_{n,k}}\mathbf{h}_{n,k}^H \mathbf{w}_n s_n + \sqrt{\alpha_{n,k}(1-\rho_{n,k})}\mathbf{e}_{n,k}^H \sum_{i=1, i \neq n}^{N} \mathbf{w}_i s_i + n_{n,k}, \end{aligned} \tag{7.7}$$

where $s_i = \sum_{j=1}^{K} \sqrt{P_{i,j}^S} s_{i,j}$ is the superposition coded signal with $P_{i,j}^S$ and $s_{i,j}$ being transmit power and transmit signal for the jth MU in the ith cluster, and $n_{n,k}$ is the AWGN with unit variance. In general, $P_{i,j}^S$ should be carefully allocated to distinguish the MUs in the power domain, which we will discuss in detail below. Note that Eq. (7.7) holds true due to the fact that $\mathbf{h}_{n,k}^H\mathbf{w}_i = \sqrt{\rho_{n,k}}\hat{\mathbf{h}}_{n,k}^H\mathbf{w}_i + \sqrt{1-\rho_{n,k}}\mathbf{e}_{n,k}^H\mathbf{w}_i = \sqrt{1-\rho_{n,k}}\mathbf{e}_{n,k}^H\mathbf{w}_i$ for ZFBF in TDD mode.[3] With perfect CSI at the BS, i.e., $\rho_{n,k} = 1$, the inter-cluster interference can be completely canceled.

---

[2]In FDD mode, the complementary matrix is given by $\bar{\mathbf{H}}_i = [\tilde{\mathbf{h}}_{1,1}^\star, \ldots, \tilde{\mathbf{h}}_{1,K}^\star, \ldots, \tilde{\mathbf{h}}_{i-1,1}^\star, \ldots, \tilde{\mathbf{h}}_{i-1,K}^\star, \tilde{\mathbf{h}}_{i+1,1}^\star, \ldots, \tilde{\mathbf{h}}_{N,K}^\star]^H$.

[3]In FDD mode, we have $\mathbf{h}_{n,k}^H\mathbf{w}_i = \sqrt{\rho_{n,k}}\|\mathbf{h}_{n,k}\|(\tilde{\mathbf{h}}_{n,k}^\star)^H\mathbf{w}_i + \sqrt{1-\rho_{n,k}}\|\mathbf{h}_{n,k}\|\tilde{\mathbf{e}}_{n,k}^H\mathbf{w}_i = \sqrt{1-\rho_{n,k}}\|\mathbf{h}_{n,k}\|\tilde{\mathbf{e}}_{n,k}^H\mathbf{w}_i \overset{d}{=} \sqrt{1-\rho_{n,k}}\mathbf{e}_{n,k}^H\mathbf{w}_i$, where $\overset{d}{=}$ denotes the equality in distribution. If $\rho_{n,k} = \rho_{n,k}$, Eq. (7.7) also holds true in FDD mode. In the sequel, without loss of generality, we no longer distinguish between TDD and FDD.

7.2.4 *Successive Interference Cancellation*

Although ZFBF at the BS can mitigate partial inter-cluster interference from the other clusters, there still exists intra-cluster interference from the same cluster. In order to improve the received signal quality, the MU conducts SIC according to the principle of NOMA. Without loss of generality, we assume that the effective channel gains in the ith cluster have the following order:

$$|\sqrt{\alpha_{i,1}}\mathbf{h}_{i,1}^H\mathbf{w}_i|^2 \geq \cdots \geq |\sqrt{\alpha_{i,K}}\mathbf{h}_{i,K}^H\mathbf{w}_i|^2. \tag{7.8}$$

It is reasonably assumed that the BS may know MUs' effective gains through the channel quality indicator (CQI) messages, and then determines the user order in (7.8). Thus, in the ith cluster, the jth MU can always successively decode the lth MU's signal, $\forall l > j$, if the lth MU can decode its own signal. As a result, the jth MU can subtract the interference from the lth MU in the received signal before decoding its own signal. After SIC, the signal-to-interference-plus-noise ratio (SINR) at the kth MU in the nth cluster can be expressed as

$$\gamma_{n,k} = \frac{\alpha_{n,k}|\mathbf{h}_{n,k}^H\mathbf{w}_n|^2 P_{n,k}^S}{\underbrace{\alpha_{n,k}|\mathbf{h}_{n,k}^H\mathbf{w}_n|^2\sum_{j=1}^{k-1}P_{n,j}^S}_{\text{Intra-cluster interference}} + \underbrace{\alpha_{n,k}(1-\rho_{n,k})\sum_{i=1,i\neq n}^{N}|\mathbf{e}_{n,k}^H\mathbf{w}_i|^2\sum_{l=1}^{K}P_{i,l}^S}_{\text{Inter-cluster interference}} + \underbrace{1}_{\text{AWGN}}}, \tag{7.9}$$

where the first term in the denominator of (7.9) is the residual intra-cluster interference after SIC at the MU, the second one is the residual inter-cluster interference after ZFBF at the BS, and the third one is the AWGN. For the 1st MU in each cluster, there is no intra-cluster interference, since it can completely eliminate the intra-cluster interference. Note that in this chapter, we assume that perfect SIC can be performed at the MUs. In practical NOMA systems, SIC might be imperfect due to a limited computational capability at the MUs. Thus, there exists residual intra-cluster interference from the weaker MUs even after SIC [43]. However, the study of the impact of imperfect SIC on the system performance is beyond the scope of this chapter and we would like to investigate it in the future work. Moreover, the transmit power has a significant impact on the SIC and the performance of NOMA [44]. Thus, we will quantitatively analyze the impact of transmit power and then aim to optimize the transmit power for improving the performance in the following sections.

7.3 Performance Analysis and Optimization

In this section, we concentrate on performance analysis and optimization of multi-antenna NOMA downlink with imperfect CSI. Specifically, we first derive closed-form expressions for the average transmission rates of the 1st MU and the other MUs, and then propose separate and joint optimization schemes of transmit power, feedback bits, and transmit mode, so as to maximize the average sum rate of the system.

7.3.1 Average Transmission Rate

We start by analyzing the average transmission rate of the kth MU in the nth cluster. First, we consider the case $k > 1$. According to the definition, the corresponding average transmission rate can be computed as

$$\begin{aligned}
R_{n,k} &= \mathrm{E}\left[\log_2\left(1+\gamma_{n,k}\right)\right] \\
&= \mathrm{E}\left[\log_2\left(\frac{\alpha_{n,k}|\mathbf{h}_{n,k}^H\mathbf{w}_n|^2\sum_{j=1}^{k}P_{n,j}^S+\alpha_{n,k}(1-\rho_{n,k})\sum_{i=1,i\neq n}^{N}|\mathbf{e}_{n,k}^H\mathbf{w}_i|^2\sum_{l=1}^{K}P_{i,l}^S+1}{\alpha_{n,k}|\mathbf{h}_{n,k}^H\mathbf{w}_n|^2\sum_{j=1}^{k-1}P_{n,j}^S+\alpha_{n,k}(1-\rho_{n,k})\sum_{i=1,i\neq n}^{N}|\mathbf{e}_{n,k}^H\mathbf{w}_i|^2\sum_{l=1}^{K}P_{i,l}^S+1}\right)\right] \\
&= \mathrm{E}\left[\log_2\left(\alpha_{n,k}|\mathbf{h}_{n,k}^H\mathbf{w}_n|^2\sum_{j=1}^{k}P_{n,j}^S+\alpha_{n,k}(1-\rho_{n,k})\sum_{i=1,i\neq n}^{N}|\mathbf{e}_{n,k}^H\mathbf{w}_i|^2\sum_{l=1}^{K}P_{i,l}^S+1\right)\right] \\
&\quad -\mathrm{E}\left[\log_2\left(\alpha_{n,k}|\mathbf{h}_{n,k}^H\mathbf{w}_n|^2\sum_{j=1}^{k-1}P_{n,j}^S+\alpha_{n,k}(1-\rho_{n,k})\sum_{i=1,i\neq n}^{N}|\mathbf{e}_{n,k}^H\mathbf{w}_i|^2\sum_{l=1}^{K}P_{i,l}^S+1\right)\right].
\end{aligned} \tag{7.10}$$

Note that the average transmission rate in (7.10) can be expressed as the difference of two terms, which have a similar form. Hence, we concentrate on the derivation of the first term. For notational convenience, we use W to denote the term $\alpha_{n,k}|\mathbf{h}_{n,k}^H\mathbf{w}_n|^2\sum_{j=1}^{k}P_{n,j}^S+\alpha_{n,k}(1-\rho_{n,k})\sum_{i=1,i\neq n}^{N}|\mathbf{e}_{n,k}^H\mathbf{w}_i|^2\sum_{l=1}^{K}P_{i,l}^S$. To compute the first expectation, the key is to obtain the probability density function (pdf) of W. Checking the first random variable $|\mathbf{h}_{n,k}^H\mathbf{w}_n|^2$ in W, since $\mathbf{w}_n$ of unit norm is designed independent of $\mathbf{h}_{n,k}$, $|\mathbf{h}_{n,k}^H\mathbf{w}_n|^2$ is χ^2 distributed with 2 degrees of freedom [45]. Similarly, $|\mathbf{e}_{n,k}^H\mathbf{w}_i|^2$ also has the distribution $\chi^2(2)$. Therefore, W can be considered as a weighted sum of N random variables with $\chi^2(2)$ distribution. According to [46], W is a nested finite weighted sum of N Erlang pdfs, whose pdf is given by

$$f_W(x)=\sum_{i=1}^{N}\Xi_N\left(i,\{\eta_{n,k}^q\}_{q=1}^N\right)g(x,\eta_{n,k}^i), \tag{7.11}$$

where

$$\eta_{n,k}^q=\begin{cases}\alpha_{n,k}\sum\limits_{j=1}^{k}P_{q,j}^S & \text{if}\, q=n\\ \alpha_{n,k}(1-\rho_{n,k})\sum\limits_{l=1}^{K}P_{q,l}^S & \text{if}\, q\neq n\end{cases},$$

$$g(x, \eta_{n,k}^{i}) = \frac{1}{\eta_{n,k}^{i}} \exp\left(-\frac{x}{\eta_{n,k}^{i}}\right),$$

$$\Xi_N\left(i, \{\eta_{n,k}^{q}\}_{q=1}^{N}\right) = \frac{(-1)^{N-1}\eta_{n,k}^{i}}{\prod_{l=1}^{N}\eta_{n,k}^{l}} \prod_{s=1}^{N-1}\left(\frac{1}{\eta_{n,k}^{i}} - \frac{1}{\eta_{n,k}^{s+\mathbf{U}(s-i)}}\right)^{-1},$$

and $\mathbf{U}(x)$ is the well-known unit step function defined as $\mathbf{U}(x \geq 0) = 1$ and zero otherwise. It is worth pointing out that the weights Ξ_N are constant for given $\{\eta_{n,k}^{q}\}_{q=1}^{N}$. Hence, the first expectation in (7.10) can be computed as

$$\begin{aligned}
\mathrm{E}[\log_2(1+W)] &= \int_0^{\infty} \log_2(1+x) f_W(x)dx \\
&= \sum_{i=1}^{N} \Xi_N\left(i, \{\eta_{n,k}^{q}\}_{q=1}^{N}\right) \int_0^{\infty} \log_2(1+x)\frac{1}{\eta_{n,k}^{i}} \exp\left(-\frac{x}{\eta_{n,k}^{i}}\right) dx \\
&= -\frac{1}{\ln(2)} \sum_{i=1}^{N} \Xi_N\left(i, \{\eta_{n,k}^{q}\}_{q=1}^{N}\right) \exp\left(\frac{1}{\eta_{n,k}^{i}}\right) \mathrm{E_i}\left(-\frac{1}{\eta_{n,k}^{i}}\right), \quad (7.12)
\end{aligned}$$

where $\mathrm{E_i}(x) = \int_{-\infty}^{x} \frac{\exp(t)}{t} dt$ is the exponential integral function. Equation (7.12) follows from [47, Eq. (4.3372)]. Similarly, we use V to denote $\alpha_{n,k}|\mathbf{h}_{n,k}^{H}\mathbf{w}_n|^2 \sum_{j=1}^{k-1} P_{n,j}^{S} + \alpha_{n,k}(1-\rho_{n,k}) \sum_{i=1, i\neq n}^{N} |\mathbf{e}_{n,k}^{H}\mathbf{w}_i|^2 \sum_{t=1}^{K} P_{i,t}^{S}$ in the second term of (7.10). Thus, the second expectation term can be computed as

$$\mathrm{E}[\log_2(1+V)] = -\frac{1}{\ln(2)} \sum_{i=1}^{N} \Xi_N\left(i, \{\beta_{n,k}^{v}\}_{v=1}^{N}\right) \exp\left(\frac{1}{\beta_{n,k}^{i}}\right) \mathrm{E_i}\left(-\frac{1}{\beta_{n,k}^{i}}\right), \quad (7.13)$$

where

$$\beta_{n,k}^{v} = \begin{cases} \alpha_{n,k} \sum\limits_{j=1}^{k-1} P_{v,j}^{S} & \text{if } v = n \\ \alpha_{n,k}(1-\rho_{n,k}) \sum\limits_{l=1}^{K} P_{v,l}^{S} & \text{if } v \neq n \end{cases}.$$

Hence, we can obtain the average transmission rate for the kth MU in the nth cluster as follows

$$\begin{aligned}
R_{n,k} = &\frac{1}{\ln(2)} \sum_{i=1}^{N} \Xi_N\left(i, \{\beta_{n,k}^{v}\}_{v=1}^{N}\right) \exp\left(\frac{1}{\beta_{n,k}^{i}}\right) \mathrm{E_i}\left(-\frac{1}{\beta_{n,k}^{i}}\right) \\
&-\frac{1}{\ln(2)} \sum_{i=1}^{N} \Xi_N\left(i, \{\eta_{n,k}^{q}\}_{q=1}^{N}\right) \exp\left(\frac{1}{\eta_{n,k}^{i}}\right) \mathrm{E_i}\left(-\frac{1}{\eta_{n,k}^{i}}\right). \quad (7.14)
\end{aligned}$$

Then, we consider the case $k = 1$. Since the first MU can decode all the other MUs' signals in the same cluster, there is no intra-cluster interference. In this case, the corresponding average transmission rate reduces to

$$R_{n,1} = \frac{1}{\ln(2)} \sum_{i=1}^{N-1} \Xi_{N-1}\left(i, \{\beta_{n,1}^{v}\}_{v=1}^{N-1}\right) \exp\left(\frac{1}{\beta_{n,1}^{i}}\right) \mathrm{E_i}\left(-\frac{1}{\beta_{n,1}^{i}}\right) \\ - \frac{1}{\ln(2)} \sum_{i=1}^{N} \Xi_{N}\left(i, \{\eta_{n,1}^{q}\}_{q=1}^{N}\right) \exp\left(\frac{1}{\eta_{n,1}^{i}}\right) \mathrm{E_i}\left(-\frac{1}{\eta_{n,1}^{i}}\right), \tag{7.15}$$

where

$$\eta_{n,1}^{q} = \begin{cases} \alpha_{n,1} P_{q,1}^{S} & \text{if } q = n \\ \alpha_{n,1}(1-\rho_{n,1}) \sum_{l=1}^{K} P_{q,l}^{S} & \text{if } q \neq n \end{cases},$$

and

$$\beta_{n,1}^{v} = \begin{cases} \alpha_{n,1}(1-\rho_{n,1}) \sum_{l=1}^{K} P_{v,l}^{S} & \text{if } v < n \\ \alpha_{n,1}(1-\rho_{n,1}) \sum_{l=1}^{K} P_{v+1,l}^{S} & \text{if } v \geq n \end{cases}.$$

Combing (7.14) and (7.15), it is easy to evaluate the performance of a multiple-antenna NOMA downlink with arbitrary system parameters and channel conditions. In particular, it is possible to reveal the impact of system parameters, i.e., transmit power, CSI accuracy, and transmission mode.

7.3.2 Power Allocation

From (7.14) and (7.15), it is easy to observe that with imperfect CSI, transmit power has a great impact on average transmission rates. On one hand, increasing the transmit power can enhance the desired signal strength. On the other hand, it also increases the interference. Thus, it is desired to distribute the transmit power according to channel conditions.

To maximize the sum rate of the considered multiple-antenna NOMA system subject to a total power constraint, we have the following optimization problem:

$$
\begin{aligned}
J_1 : \max_{P^S_{n,k}} & \sum_{n=1}^{N}\sum_{k=1}^{K} R_{n,k} \\
\text{s.t. C1} : & \sum_{n=1}^{N}\sum_{k=1}^{K} P^S_{n,k} \le P^S_{tot} \\
\text{C2} : & P^S_{n,k} > 0,
\end{aligned} \tag{7.16}
$$

where P^S_{tot} is the maximum total transmit power budget. It is worth pointing out that in certain scenarios, user fairness might be of particular importance. To guarantee user fairness, one can replace the objective function of J_1 with the maximization of a weighted sum rate, where the weights can directly affect the power allocation and thus the MUs' rates. Unfortunately, J_1 is not a convex problem due to the complicated expression for the objective function. Thus, it is difficult to directly provide a closed-form solution for the optimal transmit power. As a compromise solution, we propose an effective power allocation scheme based on the following important observation of the multiple-antenna NOMA downlink system:

Lemma 1 *The inter-cluster interference is dependent of power allocation between the clusters, while the intra-cluster interference is determined by power allocation among the MUs in the same cluster.*

Proof A close observation of the inter-cluster interference $\alpha_{n,k}(1-\rho_{n,k})\sum_{i=1,i\neq n}^{N} |\mathbf{e}^H_{n,k}\mathbf{w}_i|^2 \sum_{l=1}^{K} P^S_{i,l}$ in (7.9) indicates that $\sum_{l=1}^{K} P^S_{i,l}$ is the total transmit power for the ith cluster, which suggests that inter-cluster power allocation does not affect the inter-cluster interference. □

Inspired by Lemma 1, the power allocation scheme can be divided into two steps. In the first step, the BS distributes the total power among the N clusters. In the second step, each cluster individually carries out power allocation subject to the power constraint determined by the first step. In the following, we give the details of the two-step power allocation scheme. First, we design the power allocation between the clusters from the perspective of minimizing inter-cluster interference. For the ith cluster, the average aggregate interference to the other clusters is given by

$$
\begin{aligned}
I_i &= \mathrm{E}\left[\sum_{n=1,n\neq i}^{N}\sum_{k=1}^{K}\alpha_{n,k}(1-\rho_{n,k})|\mathbf{e}^H_{n,k}\mathbf{w}_i|^2\sum_{l=1}^{K}P^S_{i,l}\right] \\
&= \left(\sum_{n=1,n\neq i}^{N}\sum_{k=1}^{K}\alpha_{n,k}(1-\rho_{n,k})\right)P^S_i,
\end{aligned} \tag{7.17}
$$

where $P^S_i = \sum_{l=1}^{K} P^S_{i,l}$ is the total transmit power of the ith cluster. Equation (7.17) follows the fact that $\mathrm{E}[|\mathbf{e}^H_{n,k}\mathbf{w}_i|^2] = 1$. Intuitively, a large interference coefficient $\sum_{n=1,n\neq i}^{N}\sum_{k=1}^{K}\alpha_{n,k}(1-\rho_{n,k})$ means a more severe inter-cluster interference caused by the ith cluster. In order to mitigate the inter-cluster interference for improving the average sum rate, we propose to distribute the power proportionally to the reciprocal of interference coefficient. Specifically, the transmit power for the ith cluster can be computed as

$$P_i^S = \frac{\left(\sum_{n=1,n\neq i}^{N} \sum_{k=1}^{K} \alpha_{n,k}(1-\rho_{n,k})\right)^{-1}}{\sum_{l=1}^{N} \left(\sum_{n=1,n\neq l}^{N} \sum_{k=1}^{K} \alpha_{n,k}(1-\rho_{n,k})\right)^{-1}} P_{tol}^S. \tag{7.18}$$

Then, we allocate the power in the cluster for further increasing the average sum rate. According to the nature of NOMA techniques, the first MU not only has the strongest effective channel gain for the desired signal, but also generates a weak interference to the other MUs. On the contrary, the Kth MU has the weakest effective channel gain for the desired signal and also produces a strong interference to the other MUs. Thus, from the perspective of maximizing the sum of average rate, it is better to allocate the power based on the following criterion:

$$P_{n,1}^S \geq \cdots \geq P_{n,k}^S \geq \cdots \geq P_{n,K}^S. \tag{7.19}$$

On the other hand, in order to facilitate SIC, the NOMA in general requires the transmit powers in a cluster to follow a criterion below [31]:

$$P_{n,1}^S \leq \cdots \leq P_{n,k}^S \leq \cdots \leq P_{n,K}^S. \tag{7.20}$$

Under this condition, the MU performs SIC according to the descending order of the user index, namely the ascending order of the effective channel gain. Specifically, the kth MU cancels the interference from the Kth to the $(k+1)$th MU in sequence. Thus, the SINR for decoding each interference signal is the highest, which facilitates SIC at MUs [44].

To simultaneously fulfill the above two criterions, we propose to equally distribute the powers within a cluster, namely

$$P_{n,k}^S = P_n^S / K. \tag{7.21}$$

Substituting (7.18) into (7.21), the transmit power for the kth MU in the nth cluster can be computed as

$$P_{n,k}^S = \frac{\left(\sum_{i=1,i\neq n}^{N} \sum_{j=1}^{K} \alpha_{i,j}(1-\rho_{i,j})\right)^{-1}}{K\left(\sum_{l=1}^{N} \left(\sum_{i=1,i\neq l}^{N} \sum_{j=1}^{K} \alpha_{i,j}(1-\rho_{i,j})\right)^{-1}\right)} P_{tol}^S. \tag{7.22}$$

Thus, we can distribute the transmit power based on (7.22) for given channel statistical information and the CSI accuracy, which has a quite low computational complexity.

Remark 1 We note that path loss coefficient $\alpha_{n,k}, \forall n, k$, remain constant for a relatively long time, and it is easy to obtain at the BS via long-term measurement. Hence, the proposed power allocation scheme incurs a low system overhead and can be implemented with low complexity.

7.3.3 Feedback Distribution

For the FDD mode, the accuracy of quantized CSI relies on the size of codebook $2^{B_{n,k}}$, where $B_{n,k}$ is the number of feedback bits from the kth MU in the nth cluster. As observed in (7.14) and (7.15), it is possible to decrease the interference by increasing feedback bits. However, due to the rate constraint on the feedback link, the total number of feedback bits is limited. Therefore, it is of great importance to optimize the feedback bits among the MUs for performance enhancement.

According to the received signal-to-noise ratio (SNR) in (7.9), the CSI accuracy only affects the inter-cluster interference. Thus, it makes sense to optimize the feedback bits to minimizing the average sum of inter-cluster interference given by

$$\begin{aligned} I_{\text{inter}} &= \mathrm{E}\left[\sum_{n=1}^{N}\sum_{k=1}^{K}\alpha_{n,k}(1-\rho_{n,k})\sum_{i=1,i\neq n}^{N}|\mathbf{e}_{n,k}^{H}\mathbf{w}_i|^2\sum_{l=1}^{K}P_{i,l}^{S}\right] \\ &= \sum_{n=1}^{N}\sum_{k=1}^{K}\alpha_{n,k}\sum_{i=1,i\neq n}^{N}P_i^S 2^{-\frac{B_{n,k}}{M-1}}. \end{aligned} \tag{7.23}$$

Hence, the optimization problem for feedback bits distribution can be expressed as

$$\begin{aligned} J_2 : \min_{B_{n,k}} & \sum_{n=1}^{N}\sum_{k=1}^{K}\alpha_{n,k}\sum_{i=1,i\neq n}^{N}P_i^S 2^{-\frac{B_{n,k}}{M-1}} \\ \text{s.t. } C3 : & \sum_{n=1}^{N}\sum_{k=1}^{K}B_{n,k} \leq B_{\text{tot}}, \\ C4 : & B_{n,k} \geq 0, \end{aligned} \tag{7.24}$$

where B_{tot} is an upper bound on the total number of feedback bits. J_2 is an integer programming problem, hence is difficult to solve. To tackle this challenge, we relax the integer constraint on $B_{n,k}$. In this case, according to the fact that

$$\sum_{n=1}^{N}\sum_{k=1}^{K}\alpha_{n,k}\sum_{i=1,i\neq n}^{N}P_i^S 2^{-\frac{B_{n,k}}{M-1}} \geq NK\left(\prod_{n=1}^{N}\prod_{k=1}^{K}\alpha_{n,k}\sum_{i=1,i\neq n}^{N}P_i^S 2^{-\frac{B_{n,k}}{M-1}}\right)^{\frac{1}{NK}}$$
$$= NK\left(2^{-\frac{\sum_{n=1}^{N}\sum_{k=1}^{K}B_{n,k}}{M-1}}\right)^{\frac{1}{NK}}\left(\prod_{n=1}^{N}\prod_{k=1}^{K}\alpha_{n,k}\sum_{i=1,i\neq n}^{N}P_i^S\right)^{\frac{1}{NK}}$$
$$= NK\left(2^{-\frac{B_{tot}}{M-1}}\right)^{\frac{1}{NK}}\left(\prod_{n=1}^{N}\prod_{k=1}^{K}\alpha_{n,k}\sum_{i=1,i\neq n}^{N}P_i^S\right)^{\frac{1}{NK}}, \quad (7.25)$$

where the equality holds true only when $\alpha_{n,k}\sum_{i=1,i\neq n}^{N}P_i^S 2^{-\frac{B_{n,k}}{M-1}}$, $\forall n, k$ are equal. In other words, the objective function in (7.24) can be minimized while satisfying the following condition:

$$\alpha_{n,k}\sum_{i=1,i\neq n}^{N}P_i^S 2^{-\frac{B_{n,k}}{M-1}} = \left(2^{-\frac{B_{tot}}{M-1}}\right)^{\frac{1}{NK}}\left(\prod_{n=1}^{N}\prod_{k=1}^{K}\alpha_{n,k}\sum_{i=1,i\neq n}^{N}P_i^S\right)^{\frac{1}{NK}}. \quad (7.26)$$

Hence, based on the relaxed optimization problem, the optimal number of feedback bits for the kth MU in the nth cluster is given by

$$B_{n,k} = \frac{B_{\mathrm{tot}}}{NK} - \frac{1}{NK}\sum_{i=1}^{N}\sum_{j=1}^{K}\log_2\left(\alpha_{i,j}\sum_{l=1,l\neq i}^{N}P_l^S\right) + \log_2\left(\alpha_{n,k}\sum_{l=1,l\neq n}^{N}P_l^S\right). \quad (7.27)$$

Given channel statistical information and transmit power allocation, it is easy to determine the feedback distribution according to (7.27). Note that there exists an integer constraint on the number of feedback bits in practice, so we should utilize the maximum integer that is not larger than $B_{n,k}$ in (7.27), i.e., $\lfloor B_{n,k}\rfloor$, $\forall n, k$.

Remark 2 The number of feedback bits distributed to the kth MU in the nth cluster is determined by the average inter-cluster interference generated by the kth MU in the nth cluster with respect to the average inter-cluster interference of each MU. In other words, if one MU generates more inter-cluster interference, it would be allocated with more feedback bits, so as to facilitate a more accurate ZFBF to minimize the total interference.

7.3.4 Mode Selection

As discussed above, the performance of the multiple-antenna NOMA system is limited by both inter-cluster and intra-cluster interference. Although ZFBF at the BS and SIC at the MUs are jointly applied, there still exists residual interference. Intuitively, the strength of the residual interference mainly relies on the number of clusters N and the number of MUs in each cluster K. For instance, increasing the number of MUs in each cluster might reduce the inter-cluster interference, but also results in an increase in intra-cluster interference. Thus, it is desired to dynamically adjust the transmission mode, including the number of clusters and the number of MUs in each cluster, according to channel conditions and system parameters. For dynamic mode selection, we have the following lemma:

Lemma 2 *If the BS has no CSI about the downlink, it is optimal to set* $N = 1$. *On the other hand, if the BS has perfect CSI about the downlink,* $K = 1$ *is the best choice.*

Proof First, if there is no CSI, namely $\rho_{n,k} = 0, \forall n, k$, ZFBF cannot be utilized to mitigate the inter-cluster interference. If all the MUs belong to one cluster, interference can be mitigated as much as possible by SIC. In the case of perfect CSI at the BS, ZFBF can completely the interference. Thus, it is optimal to arrange one MU in one cluster. □

In above, we consider two extreme scenarios of no and perfect CSI at the BS, respectively. In practice, the BS has partial CSI through channel estimation or quantization feedback. Thus, we propose to dynamically choose the transmission mode for maximizing the sum of average transmission rate, which is equivalent to an optimization problem below:

$$\begin{aligned} J_3 : \max_{N,K} & \sum_{n=1}^{N} \sum_{k=1}^{K} R_{n,k} \\ \text{s.t. } & C5 : NK = N_u, \\ & C6 : N > 0, \\ & C7 : K > 0, \end{aligned} \tag{7.28}$$

where N_u is the number of MUs in the multiple-antenna NOMA system. J_3 is also an integer programming problem, so it is difficult to obtain the closed-form solution. Under this condition, it is feasible to get the optimal solution by numerical search and the search complexity is $O(N^K)$. In order to control the complexity of SIC, the number of MUs in one cluster is usually small, e.g., $K = 2$. Therefore, the complexity of numerical search is acceptable.

7.3.5 *Joint Optimization Scheme*

In fact, transmit power, feedback bits and transmission mode are coupled, and determine the performance together. Therefore, it is better to jointly optimize these variables, so as to further improve the performance of the multiple-antenna NOMA systems. For example, given a transmission mode, it is easy to first allocate transmit power according to (7.22), and then distribute feedback bits according to (7.27). Finally, we can select an optimal transmission mode with the largest sum rate. The complexity of the joint optimization is mainly determined by the mode selection. As mentioned above, if the number of MUs in one cluster is small, the complex of mode selection is acceptable.

7.4 Asymptotic Analysis

In order to provide insightful guidelines for system design, we now pursue an asymptotic analysis on the average sum rate of the system. In particular, two extreme cases are studied, namely interference limited and noise limited.

7.4.1 *Interference Limited Case*

With loss of generality, we let $P_{n,k}^S = \theta_{n,k} P_{tot}^S, \forall n, k$, where $0 < \theta_{n,k} < 1$ is a power allocation factor. For instance, $\theta_{n,k}$ is equal to $\frac{\left(\sum_{v=1,v\neq n}^N \sum_{j=1}^K \alpha_{v,j}(1-\rho_{v,j})\right)^{-1}}{K\left(\sum_{l=1}^N\left(\sum_{v=1,v\neq l}^N \sum_{j=1}^K \alpha_{v,j}(1-\rho_{v,j})\right)^{-1}\right)}$ in the proposed power allocation scheme in Sect. 7.3.2. If the total power P_{tot}^S is large enough, the noise term of SINR in (7.9) is negligible. In this case, with the help of [47, Eq. (4.3311)], the average transmission rate of the kth MU ($k > 1$) in the nth cluster reduces to

$$R_{n,k} = \frac{1}{\ln(2)} \sum_{i=1}^{N} \Xi_N\left(i, \{\eta_{n,k}^q\}_{q=1}^N\right) \ln(\eta_{n,k}^i) - \frac{1}{\ln(2)} \sum_{i=1}^{N} \Xi_N\left(i, \{\beta_{n,k}^v\}_{v=1}^N\right) \ln(\beta_{n,k}^i), \tag{7.29}$$

where we have also used the fact that

$$\sum_{i=1}^{N} \Xi_N\left(i, \{\eta_{n,k}^q\}_{q=1}^N\right) = \sum_{i=1}^{N} \Xi_N\left(i, \{\beta_{n,k}^v\}_{v=1}^N\right) = 1. \tag{7.30}$$

Similarly, the asymptotic average transmission rate of the 1st MU in the nth MU can be obtained as

$$R_{n,1} = \frac{1}{\ln(2)} \sum_{i=1}^{N} \Xi_N\left(i, \{\eta_{n,1}^q\}_{q=1}^N\right) \ln\left(\eta_{n,1}^i\right) - \frac{1}{\ln(2)} \sum_{i=1}^{N-1} \Xi_{N-1}\left(i, \{\beta_{n,1}^v\}_{v=1}^{N-1}\right) \ln\left(\beta_{n,1}^i\right). \tag{7.31}$$

Combining (7.29) and (7.31), we have the following important result:

Theorem 1 *In the region of high transmit power, the average transmission rate is independent of P_{tot}^S, and there exists a performance ceiling regardless of P_{tot}^S, i.e., once P_{tot}^S is larger than a saturation point, the average transmission rate will not increase further even the transmit power increases.*

Proof According to the definitions, $\eta_{n,k}^i$ and $\beta_{n,k}^i$ can be rewritten as $\eta_{n,k}^i = \omega_{n,k}^i P_{tot}^S$ and $\beta_{n,k}^i = \psi_{n,k}^i P_{tot}^S$, where

$$\omega_{n,k}^i = \begin{cases} \alpha_{n,k} \sum_{j=1}^{k} \theta_{i,j} & \text{if } i = n \\ \alpha_{n,k}(1-\rho_{n,k}) \sum_{l=1}^{K} \theta_{i,l} & \text{if } i \neq n \end{cases},$$

and

$$\psi_{n,k}^i = \begin{cases} \alpha_{n,k} \sum_{j=1}^{k-1} \theta_{i,j} & \text{if } i = n \\ \alpha_{n,k}(1-\rho_{n,k}) \sum_{l=1}^{K} \theta_{i,l} & \text{if } i \neq n \end{cases},$$

respectively. Thus, $\Xi_N\left(i, \{\eta_{n,k}^q\}_{q=1}^N\right)$ and $\Xi_N\left(i, \{\beta_{n,k}^v\}_{v=1}^N\right)$ are independent of P_{tot}^S. Hence, $R_{n,k}$ in (7.29) can be transformed as

$$\begin{aligned} R_{n,k} &= \frac{1}{\ln(2)} \sum_{i=1}^{N} \Xi_N\left(i, \{\eta_{n,k}^q\}_{q=1}^N\right) (\ln(P_{tot}^S) + \ln(\omega_{n,k}^i)) \\ &\quad - \frac{1}{\ln(2)} \sum_{i=1}^{N} \Xi_N\left(i, \{\beta_{n,k}^v\}_{v=1}^N\right) (\ln(P_{tot}^S) + \ln(\psi_{n,k}^i)) \\ &= \frac{1}{\ln(2)} \sum_{i=1}^{N} \Xi_N\left(i, \{\eta_{n,k}^q\}_{q=1}^N\right) \ln(\omega_{n,k}^i) - \frac{1}{\ln(2)} \sum_{i=1}^{N} \Xi_N\left(i, \{\beta_{n,k}^v\}_{v=1}^N\right) \ln(\psi_{n,k}^i), \end{aligned} \tag{7.32}$$

where Eq. (7.32) follows the fact that $\sum_{i=1}^N \Xi_N\left(i, \{\eta_{n,k}^q\}_{q=1}^N\right) = \sum_{i=1}^N \Xi_N\left(i, \{\beta_{n,k}^v\}_{v=1}^N\right) = 1$. Similarly, we can rewrite $R_{n,1}$ in (7.31) as

$$R_{n,1} = \frac{1}{\ln(2)} \sum_{i=1}^{N} \Xi_N \left(i, \{\eta_{n,1}^q\}_{q=1}^N\right) \ln\left(\omega_{n,1}^i\right)$$
$$- \frac{1}{\ln(2)} \sum_{i=1}^{N-1} \Xi_{N-1} \left(i, \{\beta_{n,1}^v\}_{v=1}^{N-1}\right) \ln\left(\psi_{n,1}^i\right), \quad (7.33)$$

where

$$\omega_{n,1}^i = \begin{cases} \alpha_{n,1}\theta_{i,1}^S & \text{if } i = n \\ \alpha_{n,1}(1-\rho_{n,1}) \sum_{l=1}^{K} \theta_{i,l}^S & \text{if } i \neq n \end{cases},$$

and

$$\psi_{n,1}^i = \begin{cases} \alpha_{n,1}(1-\rho_{n,1}) \sum_{l=1}^{K} \theta_{i,l}^S & \text{if } i < n \\ \alpha_{n,1}(1-\rho_{n,1}) \sum_{l=1}^{K} \theta_{i+1,l}^S & \text{if } i \geq n \end{cases}.$$

Note that both (7.32) and (7.33) are regardless of P_{tot}^S, which proves Theorem 1. □

Now, we investigate the relation between the performance ceiling in Theorem 1 and the CSI accuracy $\rho_{n,k}$. First, we consider $R_{n,k}$ with $k > 1$. As $\rho_{n,k}$ asymptotically approaches 1, the inter-cluster interference is negligible. Then, $R_{n,k}$ can be further reduced as

$$R_{n,k}^{\text{ideal}} = \mathrm{E}\left[\log_2\left(\alpha_{n,k}|\mathbf{h}_{n,k}^H \mathbf{w}_n|^2 \sum_{j=1}^{k} P_{n,j}^S\right)\right] - \mathrm{E}\left[\log_2\left(\alpha_{n,k}|\mathbf{h}_{n,k}^H \mathbf{w}_n|^2 \sum_{j=1}^{k-1} P_{n,j}^S\right)\right]$$
$$= \log_2\left(\frac{\sum_{j=1}^{k}\omega_{n,j}}{\sum_{j=1}^{k-1}\psi_{n,j}}\right). \quad (7.34)$$

It is found that even with perfect CSI, the average transmission rate for the $(k > 1)$th MU is still upper bounded. The bound $\log_2\left(\frac{\sum_{j=1}^{k}\omega_{n,j}}{\sum_{j=1}^{k-1}\psi_{n,j}}\right)$ is completely determined by channel conditions, and thus cannot be increased via power allocation. Differently, for the 1st MU, if the CSI at the BS is sufficiently accurate, the SINR $\gamma_{n,1}$ becomes high. As a result, the constant term 1 in the rate expression is negligible, and thus the average transmission rate can be approximated as

$$\begin{aligned} R_{n,1} &\approx \mathrm{E}\left[\log_2\left(\frac{\alpha_{n,1}|\mathbf{h}_{n,1}^H\mathbf{w}_n|^2 P_{n,1}^S}{\alpha_{n,1}(1-\rho_{n,1})\sum_{i=1,i\neq n}^N |\mathbf{e}_{n,1}^H\mathbf{w}_i|^2\sum_{l=1}^K P_{i,l}^S}\right)\right] \\ &= \underbrace{\mathrm{E}\left[\log_2\left(\alpha_{n,1}|\mathbf{h}_{n,1}^H\mathbf{w}_n|^2 P_{n,1}^S\right)\right]}_{\text{Ideal average rate}} \\ &\quad \underbrace{-\mathrm{E}\left[\log_2\left(\alpha_{n,1}(1-\rho_{n,1})\sum_{i=1,i\neq n}^N |\mathbf{e}_{n,1}^H\mathbf{w}_i|^2\sum_{l=1}^K P_{i,l}^S\right)\right]}_{\text{Rate loss due to imperfect CSI}}. \end{aligned} \tag{7.35}$$

In (7.35), the first term is the ideal average transmission rate with perfect CSI, and the second one is rate loss caused by imperfect CSI. We first check the term of the ideal average transmission rate, which is given by

$$\begin{aligned} R_{n,1}^{\text{ideal}} &= \mathrm{E}\left[\log_2\left(\alpha_{n,1}P_{tot}^S\theta_{n,1}|\mathbf{h}_{n,1}^H\mathbf{w}_n|^2\right)\right] \\ &= \log_2\left(\alpha_{n,1}P_{tot}^S\theta_{n,1}\right) - \frac{C}{\ln(2)}. \end{aligned} \tag{7.36}$$

Note that if there is perfect CSI at the BS, the average transmission rate of the 1st MU increases proportionally to $\log_2(P_{tot}^S)$ without a bound. However, as seen in (7.34), the $(k > 1)$th MU has an upper bounded rate under the same condition, which reconfirms the claim in Lemma 2 that it is optimal to arrange one MU in each cluster in presence of perfect CSI. Then, we investigate the rate loss due to imperfect CSI, which can be expressed as

$$\begin{aligned} R_{n,1}^{\text{loss}} &= \mathrm{E}\left[\log_2\left(\alpha_{n,1}(1-\rho_{n,1})P_{tot}^S\sum_{i=1,i\neq n}^N |\mathbf{e}_{n,1}^H\mathbf{w}_i|^2\sum_{t=1}^K\theta_{i,t}\right)\right] \\ &= \log_2\left(\alpha_{n,1}(1-\rho_{n,1})P_{tot}^S\right) - \frac{1}{\ln(2)}\sum_{i=1}^{N-1}\Xi_{N-1}\left(i,\{\mu_{n,1}^v\}_{v=1}^{N-1}\right)\left(C-\ln\left(\mu_{n,1}^i\right)\right), \end{aligned} \tag{7.37}$$

where

$$\mu_{n,1}^v = \begin{cases} \sum_{l=1}^K \theta_{v,l} & \text{if } v < n \\ \sum_{l=1}^K \theta_{v+1,l} & \text{if } v \geq n \end{cases}.$$

Given a $\rho_{n,1}$, the rate loss $R_{n,1}^{loss}$ enlarges as the total transmit power P_{tot}^S increases. In order to keep the same rate of increase to the ideal rate $R_{n,1}^{ideal}$, the CSI accuracy $\rho_{n,1}$ should satisfy the following theorem:

Theorem 2 *Only when $(1-\rho_{n,1})P_{tot}^S$ is equal to a constant ε, the average transmission rate of the 1st MU in the nth cluster with imperfect CSI remains a fixed gap with respect to the ideal rate. Specifically, the transmit power for training sequence* *should satisfy* $P^p_{n,1} = \frac{P^S_{tot}/\varepsilon - 1}{\alpha_{n,1}\tau}$ *in TDD systems, while the number of feedback bits should satisfy* $B_{n,1} = (M-1)\log_2(P^S_{tot}/\varepsilon)$ *in FDD systems.*

Proof The proof is intuitively. By substituting $\rho_{n,1} = 1 - \frac{1}{1+\tau P^P_{n,1}\alpha_{n,1}}$ into $(1-\rho_{n,1})P^S_{tot} = \varepsilon$ for TDD systems and $\rho_{n,1} = 1 - 2^{-\frac{B_{n,1}}{M-1}}$ into $(1-\rho_{n,1})P^S_{tot} = \varepsilon$ for FDD systems, we can get $P^P_{n,1} = \frac{P^S_{tot}/\varepsilon - 1}{\alpha_{n,1}\tau}$ and $B_{n,1} = (M-1)\log_2(P^S_{tot}/\varepsilon)$, which proves Theorem 2. □

Remark 3 For the CSI accuracy at the BS, $P^P_{n,1}\tau$ (namely transmit energy for training sequence) in TDD systems and $\frac{B_{n,1}}{M-1}$ (namely spatial resolution) in FDD systems are two crucial factors. Specifically, given a requirement on CSI accuracy, it is possible to shorten the length of training sequence by increasing the transmit power, so as to leave more time for data transmission in a time slot. However, in order to keep the pairwise orthogonality of training sequences, the length of training sequence τ must be larger than the number of MUs. In other words, the minimum value of τ is NK. Similarly, in FDD systems, it is possible to reduce the feedback bits by increasing the number of antennas M. Yet; in order to fulfill the spatial degrees of freedom for ZFBF at the BS, M must be not smaller than $(N-1)K+1$. This is because the beam $\mathbf{w}_i$ for the ith cluster should be in the null space of the channels for the $(N-1)K$ MUs in the other $N-1$ clusters.

Furthermore, substituting (7.36) and (7.37) into (7.35), we have

$$R_{n,1} \approx -\log_2(1-\rho_{n,1}) + \log_2(\theta_{n,1}) - \sum_{i=1}^{N-1} \Xi_{N-1}\left(i, \{\mu^v_{n,1}\}_{v=1}^{N-1}\right)\log_2\left(\mu^i_{n,1}\right). \quad (7.38)$$

Given a power allocation scheme, it is interesting that the bound of $R_{n,1}$ is independent of channel conditions. As analyzed above, it is possible to improve the average rate by improving the CSI accuracy. Especially, for FDD systems, we have the following lemma:

Lemma 3 *At the high power region with a large number of feedback bits, the average rate of the* 1*st MU increases linearly as the numbers of feedback bits increase.*

Proof Replacing $\rho_{n,1}$ in (7.38) with $\rho_{n,1} = 1 - 2^{-\frac{B_{n,1}}{M-1}}$, $R_{n,1}$ is transformed as

$$R_{n,1} \approx \frac{B_{n,1}}{M-1} + \log_2(\theta_{n,1}) - \sum_{i=1}^{N-1} \Xi_{N-1}\left(i, \{\mu^v_{n,1}\}_{v=1}^{N-1}\right)\log_2\left(\mu^i_{n,1}\right), \quad (7.39)$$

which yields Lemma 3. □

7.4.2 Noise-Limited Case

If the interference term is negligible with respect to the noise term due to a low transmit power, then the SINR $\gamma_{n,k}$, $\forall n, k$ is reduced as

$$\gamma_{n,k} = \alpha_{n,k} |\mathbf{h}_{n,k}^H \mathbf{w}_n|^2 P_{n,k}^S, \tag{7.40}$$

which is equivalent to the interference-free case. As discussed earlier, $|\mathbf{h}_{n,k}^H \mathbf{w}_n|^2$ is $\chi^2(2)$ distributed, then the average transmission rate can be computed as

$$\begin{aligned} R_{n,k} &= \int_0^\infty \log_2\left(1 + P_{n,k}^S \alpha_{n,k} x\right) \exp(-x) dx \\ &= -\exp\left(\frac{1}{P_{n,k}^S \alpha_{n,k}}\right) \mathrm{E_i}\left(-\frac{1}{P_{n,k}^S \alpha_{n,k}}\right). \end{aligned} \tag{7.41}$$

Note that Eq. (7.41) is independent of the CSI accuracy, thus it is unnecessary to carry out channel estimation or CSI feedback in this scenario. Since both intra-cluster interference and inter-cluster interference are negligible, ZFBF at the BS and SIC at the MUs are not required, and all optimization schemes asymptotically approach the same performance.

7.5 Simulation Results

To evaluate the performance of the proposed multiple-antenna NOMA technology, we present several simulation results under different scenarios. For convenience, we set $M = 6$, $N = 3$, $K = 2$, $B_{tot} = 12$, while $\alpha_{n,k}$ and $\rho_{n,k}$ are given in Table 7.1 for all simulation scenarios without extra specification. In addition, we use SNR (in dB) to represent $10 \log_{10} P_{tot}^S$.

First, we verify the accuracy of the derived theoretical expressions. As seen in Fig. 7.2, the theoretical expressions for both the 1st and the 2nd MUs in the 1st cluster well coincide with the simulation results in the whole SNR region, which confirms the high accuracy. As the principle of NOMA implies, the 1st MU performs better than

Table 7.1 Parameter Table for $(\alpha_{n,k}, \rho_{n,k})$, $\forall n \in [1, 3]$, and $k \in [1, 2]$

n	k	
	1	2
1	(1.00, 0.90)	(0.10, 0.70)
2	(0.95, 0.85)	(0.20, 0.75)
3	(0.90, 0.80)	(0.15, 0.80)

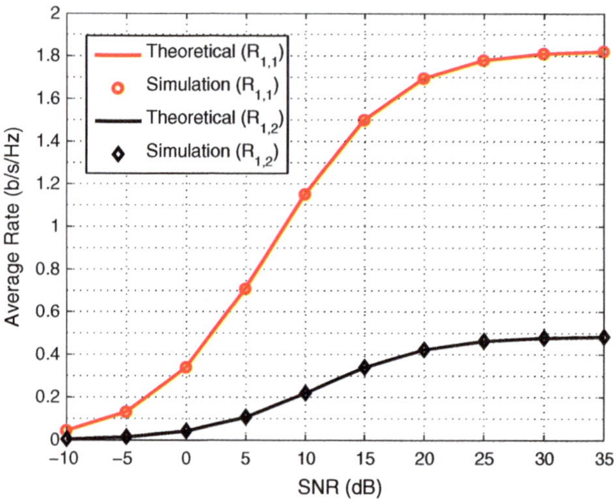

Fig. 7.2 Comparison of theoretical expressions and simulation results

the second MU. At high SNR, the average rates of the both MUs are asymptotically saturated, which proves Theorem 1 again.

Secondly, we compare the proposed power allocation scheme with the equal power allocation scheme and the fixed power allocation scheme proposed in [5]. Note that the fixed power allocation scheme distributes the power with a fixed ratio 1:4 between the two MUs in a cluster so as to facilitate the SIC. It is found in Fig. 7.3 that the proposed power allocation scheme offers an obvious performance gain over the two baseline schemes, especially in the medium SNR region. Note that practical communication systems, in general, operate at medium SNR, thus the proposed scheme is able to achieve a given performance requirement with a lower SNR. As the SNR increases, the proposed scheme and the equal allocation scheme achieve the same saturated sum rate, but the fixed allocation scheme has a clear performance loss.

Next, we examine the advantage of feedback allocation for the FDD-based NOMA system with equal power allocation, cf. Fig. 7.4. As analyzed in Sect. 7.4.2, at very low SNR, namely the noise-limited case, the average rate is independent of CSI accuracy, and thus the two schemes asymptotically approach the same sum rate. As SNR increases, the proposed feedback allocation scheme achieves a larger performance gain. Similarly, at high SNR, both the two schemes are saturated, and the proposed scheme obtains the largest performance gain. For instance, at SNR $= 30$ dB, there is a gain of more than 0.5 b/s/Hz. Furthermore, we investigate the impact of the total number of feedback bits on the average rates of different MUs at SNR $= 35$ dB. As shown in Fig. 7.5, the performance of the 1st MU is clearly better than that of the 2nd MU. Moreover, the average rate of the 1st MU is nearly a linear function of the number of feedback bits, which reconfirms the claims of Lemma 3.

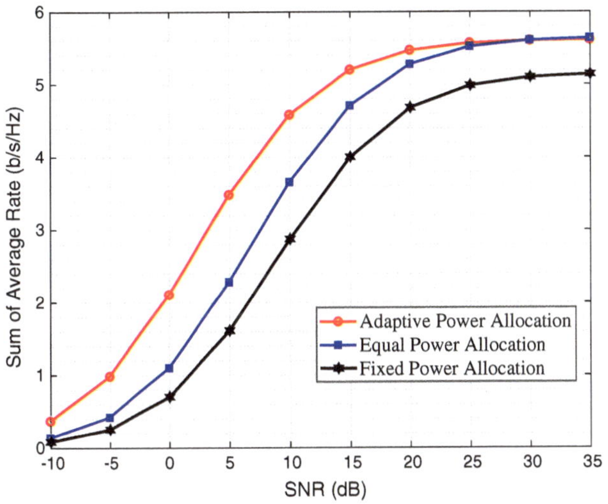

Fig. 7.3 Performance comparison of different power allocation schemes

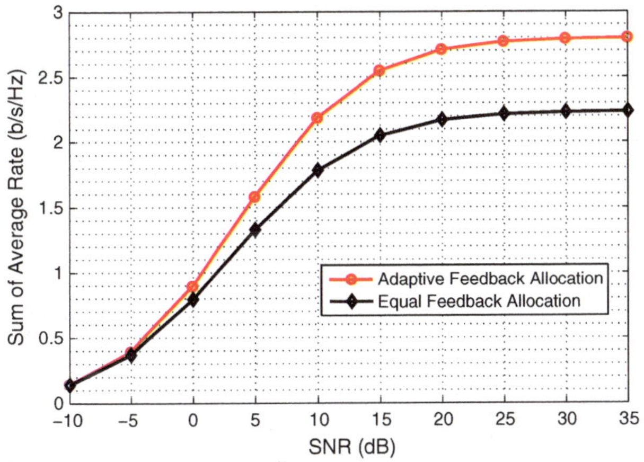

Fig. 7.4 Performance comparison of different feedback allocation schemes

Then, we investigate the impact of the transmission mode on the performance of the NOMA systems at SNR = 10 dB with equal power allocation in Fig. 7.6. To concentrate on the impact of transmission mode, we set the same CSI accuracy of all downlink channels as ρ. Note that we consider four fixed transmission modes under the same channel conditions in the case of six MUs in total. Consistent with the claims in Lemma 2, mode 4 with $N = 1$ and $K = 6$ achieves the largest sum rate at low CSI accuracy, while mode 1 with $N = 6$ and $K = 1$ performs best at high CSI accuracy. In addition, it is found that at medium CSI accuracy, mode 2 with $N = 3$ and $K = 2$ is optimal, since it is capable to achieve a best balance between intra-cluster

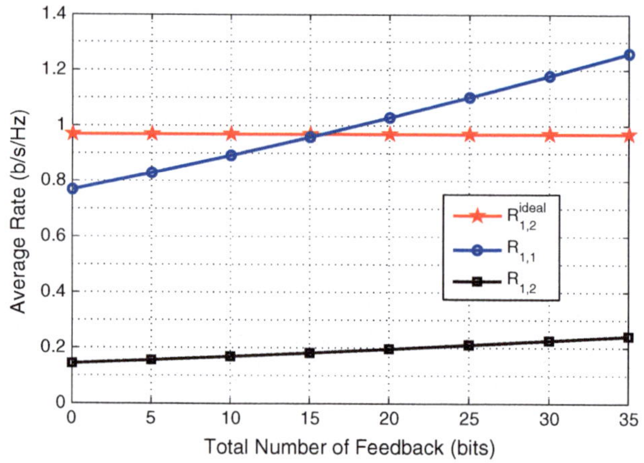

Fig. 7.5 Asymptotic performance with a large number of feedback bits

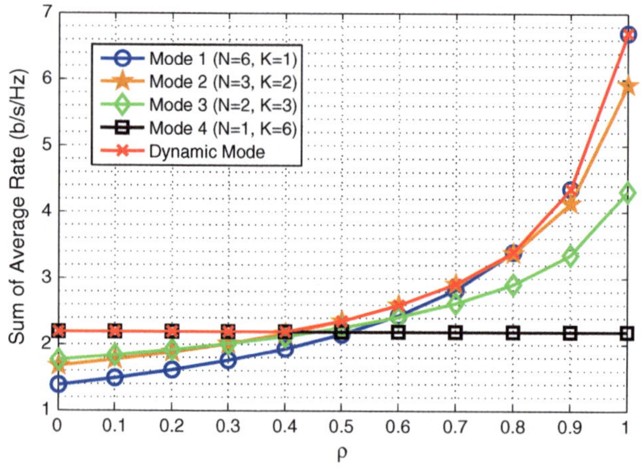

Fig. 7.6 Performance comparison of different transmission modes

interference and inter-cluster interference. Thus, we propose to dynamically select the transmission mode according to channel conditions and system parameters. As shown by the red line in Fig. 7.6, dynamic mode selection can always obtain the maximum sum rate.

Finally, we exhibit the superiority of the proposed joint optimization scheme for the NOMA systems at SNR = 10 dB. In addition, we take a fixed scheme based on NOMA and a time division multiple access (TDMA) based on OMA as baseline schemes. Specifically, the joint optimization scheme first distributes the transmit power with equal feedback allocation, then allocates the feedback bits based on the distributed power, finally selects the optimal transmission mode. The fixed scheme always adopts the mode 2 ($N = 3$, $K = 2$) with equal power and feedback allocation.

The TDMA equally allocates each time slot to the six MUs and utilizes maximum ratio transmission (MRT) based on the available CSI at the BS to maximize the rate. For clarity of notation, we use ρ to denote the CSI accuracy based on equal feedback allocation. In other words, the total number of feedback bits is equal to $B_{tot} = -K * N * (M - 1) * \log_2(1 - \rho)$. As seen in Fig. 7.7, the fixed scheme performs better than the TDMA scheme at low and high CSI accuracy, and slightly worse at the medium regime. However, the proposed joint optimization scheme performs much better than the two baseline schemes. Especially at high CSI accuracy, the performance gap becomes substantially large. For instance, there is a performance gain of about 3 b/s/Hz at $\rho = 0.8$, and up to more than 5 b/s/Hz at $\rho = 0.9$. As analyzed in Lemma 2 and confirmed by Fig. 7.6, when ρ is larger than 0.8, which is a common CSI accuracy in practical systems, mode 2 is optimal for maximizing the system performance. Thus, the joint optimization scheme is reduced to joint power and feedback allocation, which requires only a very low complexity. Thus, the proposed NOMA scheme with joint optimization can achieve a good performance with low complexity, and it is a promising technique for future wireless communication systems.

7.6 Conclusion

This chapter provided a comprehensive solution for designing, analyzing, and optimizing a NOMA technology over a general multiuser multiple-antenna downlink in both TDD and FDD modes. First, we proposed a new framework for multiple-antenna NOMA. Then, we analyzed the performance and derived exactly closed-form expressions for average transmission rates. Afterward, we optimized the three key

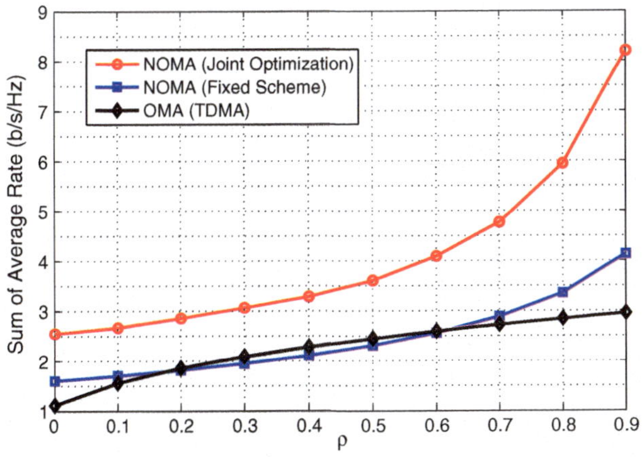

Fig. 7.7 Performance comparison of a joint optimization scheme and a fixed allocation scheme

parameters of multiple-antenna NOMA, i.e., transmit power, feedback bits, and transmission mode. Finally, we conducted asymptotic performance analysis and obtained insights on system performance and design guidelines.

References

1. V.W.S. Wong, R. Schober, D.W.K. Ng, L.-C. Wang, *Key Technologies for 5G Wireless Systems* (Cambridge University Press, Cambridge, U.K., 2017)
2. Y. Saito, Y. Kishiyama, A. Benjebour, T. Nakamura, A. Li, K. Higuchi, Non-orthogonal multiple access (NOMA) for cellular future radio access, in *Proceedings of the IEEE Vehicular Technology Conference (VTC-Spring)*, June 2013, pp. 1–5
3. Z. Ding, Y. Liu, J. Choi, Q. Sun, M. Elkashlan, C-L. I, H.V. Poor, Application of non-orthogonal multiple access in LTE and 5G networks. IEEE Commun. Mag. **55**(2), 185–191 (2017)
4. L. Dai, B. Wang, Y. Yuan, S. Han, C-L. I, Z. Wang, Non-orthogonal multiple access for 5G: solutions, challenges, opportunities, and future research trends. IEEE Commun. Mag. **53**(9), 74–81 (2015)
5. Z. Ding, Z. Yang, P. Fan, H.V. Poor, On the performance of non-orthogonal multiple access in 5G systems with randomly deployed users. IEEE Signal Process. Lett. **21**(12), 1501–1505 (2014)
6. Y. Yuan, Z. Yuan, G. Yu, C.-H. Hwang, P.-K. Liao, A. Li, K. Takeda, Non-orthogonal transmission technology in LTE evolution. IEEE Commun. Mag. **54**(7), 68–74 (2016)
7. W. Shin, M. Vaezi, B. Lee, D.J. Love, J. Lee, H.V. Poor, Non-orthogonal multiple access in multi-cell networks: theory, performance, and practical challenges. IEEE Commun. Mag. **55**(10), 176–183 (2017)
8. S. Timotheou, I. Krikidis, Fairness for non-orthogonal multiple access in 5G systems. IEEE Signal Process. Lett. **22**(10), 1647–1651 (2015)
9. S.-L. Shieh, Y.-C. Huang, A simple scheme for realizing the promised gains of downlink nonorthogonal multiple access. IEEE Trans. Commun. **64**(4), 1624–1635 (2016)
10. P. Xu, Y. Yuan, Z. Ding, X. Dai, R. Schober, On the outage performance of non-orthogonal multiple access with 1-bit feedback. IEEE Trans. Wirel. Commun. **15**(10), 6716–6730 (2016)
11. Z. Yang, Z. Ding, P. Fan, Z. Ma, Outage performance for dynamic power allocation in hybrid non-orthogonal multiple access systems. IEEE Commun. Lett. **20**(8), 1695–1698 (2016)
12. F. Fang, H. Zhang, J. Cheng, V.C.M. Leung, Energy-efficient resource allocation for downlink orthogonal multiple access (NOMA) network. IEEE Trans. Commun. **64**(9), 3722–3732 (2016)
13. H. Tabassum, M.S. Ali, E. Hossain, M.J. Hossain, D.I. Kim, Non-orthogonal multiple access (NOMA) in cellular uplink and downlink: challenges and enabling techniques, August 2016 [Online], http://128.84.21.199/abs/1608.05783
14. J. Choi, On the power allocation for a practical multiuser superposition scheme in NOMA systems. IEEE Commun. Lett. **20**(3), 438–441 (2016)
15. C.-L. Wang, J.-Y. Chen, Y.-J. Chen, Power allocation for downlink non-orthogonal multiple access system. IEEE Wirel. Commun. Lett. **5**(5), 532–535 (2016)
16. Z. Yang, Z. Ding, P. Fan, N. Al-Dhahir, A general power allocation scheme to guarantee quality of service in downlink and uplink NOMA systems. IEEE Trans. Wireless Commun. **15**(11), 7244–7257 (2016)
17. Y. Liu, M. Elkashlan, Z. Ding, G.K. Karagiannidis, Fairness of user clustering in MIMO non-orthogonal multiple access systems. IEEE Commun. Lett. **20**(7), 1465–1468 (2016)
18. H. Zhang, D.-K. Zhang, W.-X. Meng, C. Li, User pairing algorithm with SIC in non-orthogonal multiple access system, in *Proceedings of the IEEE International Conference on Communication (ICC)*, May 2016, pp. 1–6

19. J. Mei, L. Yao, H. Long, K. Zheng, Joint user pairing and power allocation for downlink non-orthogonal multiple access systems, in *Proceedings of the IEEE International Conference on Communication (ICC)*, May 2016, pp. 1–6
20. Z.Q. Al-Abbasi, D.K.C. So, User-pairing based non-orthogonal multiple access (NOMA) system, in *Proceedings of the IEEE Vehicular Technology Conference (VTC-Spring)*, April 2016, pp. 1–5
21. Md.S. Ali, H. Tabassum, E. Hossain, Dynamic user clustering and power allocation for uplink and downlink non-orthogonal multiple access (NOMA) systems. IEEE Access **4**, 6325–6343 (2016)
22. H. Weingarten, Y. Steinberg, S.S. Shamai, The capacity region of the Gaussian multiple-input multiple-output broadcast channel. IEEE Trans. Inf. Theory **52**(9), 3936–3964 (2006)
23. M.A. Maddah-Ali, M.A. Sadrabadi, A.K. Khandani, Broadcast in MIMO systems based on a generalized QR decomposition: signaling and performance analysis. IEEE Trans. Inf. Theory **54**(3), 1124–1138 (2008)
24. A.D. Dabbagh, D.J. Love, Precoding for multiple antenna Gaussian broadcast channels. IEEE Trans. Signal Process. **55**(7), 3837–3850 (2007)
25. X. Chen, C. Yuen, Performance analysis and optimization for interference alignment over MIMO interference channels with limited feedback. IEEE Trans. Signal Process. **62**(7), 1785–1795 (2014)
26. Q. Sun, S. Han, C-L. I, Z. Pan, On the ergodic capacity of MIMO NOMA systems. IEEE Wirel. Commun. Lett. **4**(4), 405–408 (2015)
27. J. Choi, On the power allocation for MIMO-NOMA systems with layered transmission. IEEE Trans. Wirel. Commun. **15**(5), 3226–3237 (2016)
28. Z. Chen, Z. Ding, X. Dai, Beamforming for combating inter-cluster and intra-cluster interference in hybrid NOMA systems. IEEE Access **4**, 4452–4463 (2016)
29. Z. Ding, R. Schober, H.V. Poor, A general MIMO framework for NOMA downlink and uplink transmission based on signal alignment. IEEE Trans. Wirel. Commun. **15**(6), 4438–4454 (2016)
30. K. Higuchi, Y. Kishyama, Non-orthogonal access with random beamforming and intra-beam SIC for cellular MIMO downlink, in *Proceedings of the IEEE Vehicular Technology Conference (VTC-Fall)*, September 2013, pp. 1–5
31. Z. Ding, F. Adachi, H.V. Poor, The application of MIMO to non-orthogonal multiple access. IEEE Trans. Wirel. Commun. **15**(1), 537–552 (2016)
32. X. Chen, Z. Zhang, H.-H. Chen, On distributed antenna system with limited feedback precoding-opportunities and challenges. IEEE Wirel. Commun. **17**(2), 80–88 (2010)
33. D.J. Love, R.W. Heath Jr., V.K.N. Lau, D. Gesbert, B.D. Rao, M. Andrews, An overview of limited feedback in wireless communication systems. IEEE J. Sel. Areas Commun. **26**(8), 1341–1365 (2008)
34. X. Chen, Z. Zhang, C. Zhong, R. Jia, D.W.K. Ng, Fully non-orthogonal communications for massive access. IEEE Trans. Commun. **66**(4), 1717–1731 (2018)
35. X. Chen, Z. Zhang, C. Zhong, D.W.K. Ng, Exploiting multiple-antenna techniques for non-orthogonal multiple access. IEEE J. Sel. Areas Commun. **35**(10), 2207–2220 (2017)
36. N. Nonaka, Y. Kishiyama, K. Higuchi, Non-orthogonal multiple access using intra-beam superposition coding and SIC in base station cooperative MIMO cellular downlink, in *Proceedings of the IEEE Vehicular Technology Conference (VTC-Spring)*, September 2014, pp. 1–5
37. S. Ali, E. Hossain, D.I. Kim, Non-orthogonal multiple access (NOMA) for downlink multiuser MIMO systems: user clustering, beamforming, and power allocation. IEEE Access **5**, 565–577 (2017)
38. M.B. Shahab, M. Irfan, M.F. Kader, S.Y. Shin, User pairing schemes for capacity maximization in nonorthogonal multiple access systems. Wirel. Commun. Mob. Comput. **16**(17), 2884–2894 (2016)
39. Z. Ding, P. Fan, V. Poor, Impact of user pairing on 5G non-orthogonal multiple access downlink transmissions. IEEE Trans. Veh. Technol. **65**(8), 6010–6023 (2016)
40. M.B. Shahab, M.F. Kader, S.Y. Shin, A virtual user pairing scheme to optimally utilize the spectrum of unpaired users in non-orthogonal multiple access. IEEE Signal Process. Lett. **23**(12), 1766–1770 (2016)

41. N. Jindal, MIMO broadcast channels with finite-rate feedback. IEEE Trans. Inf. Theory **52**(11), 5045–5060 (2006)
42. X. Chen, Z. Zhang, C. Yuen, Adaptive mode selection in multiuser MISO cognitive networks with limited cooperation and feedback. IEEE Trans. Veh. Technol. **63**(4), 1622–1632 (2014)
43. K. Saito, A. Benjebbour, Y. Kishiyama, Y. Okumura, T. Nakamura, Performance and design of SIC receiver for downlink NOMA with open-loop SU-MIMO, in *Proceedings of the IEEE International Conference on Communication Workshop (ICCW)*, June 2015, pp. 1161–1165
44. M.B. Shahab, M.F. Kader, S.Y. Shin, On the power allocation of non-orthogonal multiple access for 5G wireless networks, in *Proceedings of the International Conference on Open Source Systems Technology (ICOSST)*, December 2016, pp. 89–94
45. K.K. Mukkavilli, A. Sabharwal, E. Erkip, B. Aazhang, On beamforming with finite rate feedback in multiple-antenna systems. IEEE Trans. Inf. Theory **49**(10), 2562–2579 (2003)
46. G.K. Karagiannidis, N.C. Sagias, T.A. Tsiftsis, Closed-form statistics for the sum of squared Nakagami-m variates and its application. IEEE Trans. Commun. **54**(8), 1353–1359 (2006)
47. I.S. Gradshteyn, I.M. Ryzhik, *Tables of Integrals, Series, and Products* (Academic Press, USA, 2007)

Chapter 8
NOMA for Millimeter Wave Networks

Zhengquan Zhang and Zheng Ma

8.1 Introduction

Millimeter wave (mmWave) communications [1–3] are one of the most important technologies for 5G wireless networks and beyond, due to the rich spectrum resources in the mmWave band from 30 to 300 GHz. According to Shannon's capacity theorem, increasing the system bandwidth is an effective way to achieve high data rate transmissions. However, compared with conventional low-frequency cellular networks working in 900 MHz to 3.5 GHz, mmWave networks suffer from high path loss and directional transmissions, and are sensitive to blockage. To promote the application of mmWave communications in 5G and beyond, some efforts from both academic and industry have been devoted.

Conventionally, it is considered that mmWave communications will be mainly used to achieve high data rate transmissions for enhanced mobile broadband (eMBB) scenario. Recently, some works also discussed the potential that mmWave communications are used to other application scenarios, such as Internet of Things (IoT) cloud-enabled autonomous vehicles [4]. Therefore, to further improve spectrum efficiency and support massive connectivity in mmWave networks, the combination of mmWave communications with other key technologies, e.g., non-orthogonal multiple access (NOMA) [5–7] and multicast transmission [8], is still very important. The works [9–14] studied NOMA for mmWave networks. The multicast transmission for mmWave networks was also studied in [14, 15].

Z. Zhang (✉) · Z. Ma
Southwest Jiaotong University, West Section, High-tech Zone, Chengdu, Sichuan, China
e-mail: zhang.zhengquan@hotmail.com

Z. Ma
e-mail: zma@swjtu.edu.cn

M. Vaezi et al. (eds.), *Multiple Access Techniques for 5G Wireless Networks and Beyond*, https://doi.org/10.1007/978-3-319-92090-0_8

In this chapter, the application of NOMA to mmWave networks is studied, and unicast, multicast, and cooperative multicast transmissions for mmWave-NOMA networks are discussed. Their performance in terms of coverage probability, outage probability, and sum rate is also given. In Sect. 8.2, the fundamentals of mmWave communications including path loss model, directivity gain model, and user association are briefly discussed. Then, unicast transmissions for single-tier mmWave-NOMA networks are studied in Sect. 8.3, which enable a user pair by NOMA to be transmitted on the same radio resources. Next, to further improve the network performance, multicast transmissions for single-tier mmWave-NOMA networks are studied in Sect. 8.4, which distribute media data to all interested users in a non-orthogonal manner. Furthermore, considering two-tier mmWave heterogeneous networks (HetNets), cooperative multicast transmissions for mmWave-NOMA HetNets are further studied in Sect. 8.5, followed by summary in Sect. 8.6.

8.2 Fundamentals of mmWave Communications

8.2.1 Path Loss and Small-Scale Fading

mmWave communications exhibit obvious line-of-sight (LOS) and non-LOS (NLOS) propagations due to the very short wavelength, which is different from conventional low-frequency cellular networks. To simply characterize LOS and NLOS links, different path loss exponents can be used. Besides, it is assumed that the type of link observed by users is probabilistic. This means that for a link with length d, the probability that it is a LOS link is $p_L(d)$, while it is a NLOS link with probability (w.p) $p_N(d) = 1 - p_L(d)$. Therefore, the path loss model can be given by [16, 17]

$$L_S(d) = \begin{cases} C_{S,L} d^{-\alpha_{S,L}}, \text{ w.p } p_L(d), \\ C_{S,N} d^{-\alpha_{S,N}}, \text{ w.p } p_N(d), \end{cases} \tag{8.1}$$

where for $s \in \{\text{L (LOS), N (NLOS)}\}$, $\alpha_{S,s}$ is the path loss exponent for the s link. The intercept of path loss formula for the s link, $C_{S,s}$, is a function of reference distance and wavelength (or carrier frequency) and is $10^{-2\log_{10}(4\pi/\lambda_c)}$ for the close-in reference distance $d_{\text{ref}} = 1$ m [18].

According to [16, 17], the small-scale fading of mmWave link is assumed to be Nakagami-m fading. The Nakagami-m fading parameters, $N_{S,L}$ and $N_{S,N}$, are used to characterize the fading of the mmWave LOS and NLOS links, respectively. Let $h_{S,i}$ be the channel coefficient of the link between the i-th mmWave small cell and the user. Then, $H_{S,i} = |h_{S,i}|^2$ follows a normalized Gamma distribution. Similar to [16, 17], shadowing is ignored.

8.2.2 Directivity Gain

Beamforming [2] is a key technique to overcome high path loss in mmWave communications by using antenna arrays to form directional beams. Especially, analog beamforming is a simple way to form directional beams by using low-cost phase shifters (PSs). The deployment of antenna arrays at both BSs and user equipments (UEs) can be considered. According to 3GPP TR 38.913, up to 256 Tx and Rx BS antenna elements and 32 Tx and Rx UE antenna elements are assumed. To approximate the beamforming pattern for tractable analysis, sectored antenna model is available. In this case, the directivity gain can be given by [16, 17, 19]

$$G(\phi) = \begin{cases} G_M, & |\phi| \le \theta, \\ G_m, & |\phi| > \theta, \end{cases} \tag{8.2}$$

where G_M and G_m are the main and side lobe gains, respectively, ϕ is a certain angle, and θ is the beamwidth of the main lobe. Furthermore, according to [19], G_M and G_m can be equal to $\frac{2\pi-(2\pi-\theta)\varepsilon}{\theta}$ and ε, respectively. At the mmWave base station side, $G_{S,M}$, $G_{S,m}$, and θ_S denote the main lobe gain, side lobe gain, and beamwidth, respectively, which are denoted by $G_{U,M}$, $G_{U,m}$, and θ_U at the user side. Therefore, the directivity gain of the communication link between the user and the i-th mmWave base station is $G_i = G_S(\phi_S)G_U(\phi_U)$, where ϕ_S and ϕ_U are the angle of departure (AoD) and the angle of arrival (AoA) of the signal, respectively. Further, the user is always assumed to be aligned with its serving base station, B_0, such that the directivity gain is $G_0 = G_{U,M}G_{S,M}$. According to [16, 17], the directivity gain of the i-th interference mmWave link is assumed to be a discrete random variable (RV), whose probability distribution is $G_{S,i} = a_k$ with probability b_k, $k \in \{1, 2, 3, 4\}$, where a_k and b_k are constants defined in Table 8.1. $\bar{a}_k = \frac{a_k}{G_{U,M}G_{S,M}}$ is the normalized directional gain. Note that for macro BSs, the omnidirectional antennas are considered, i.e., $\theta = 2\pi$. As a result, there is no directivity gain.

8.2.3 User Association

Maximum average received power-based user association scheme enables users to be associated with a base station having the minimum average path loss, which averages the effect of fading and can provide robust performance. Note that this user

Table 8.1 Probability mass function of $G_{S,i} (i \ge 1)$

k	1	2	3	4
a_k	$G_{U,M}G_{S,M}$	$G_{U,M}G_{S,m}$	$G_{U,m}G_{S,M}$	$G_{U,m}G_{S,m}$
b_k	$\frac{\theta_U\theta_S}{(2\pi)^2}$	$\frac{\theta_U(2\pi-\theta_S)}{(2\pi)^2}$	$\frac{(2\pi-\theta_U)\theta_S}{(2\pi)^2}$	$\frac{(2\pi-\theta_U)(2\pi-\theta_S)}{(2\pi)^2}$

association strategy will not necessarily result in the maximum average user performance due to unbalanced load distribution. However, load balancing technique can be available to change connected users' distribution between adjacent base stations by optimizing their handover parameters.

In mmWave networks, each user can associate with one LOS or NLOS mmWave base station, which depends on the user location and the distance between user and mmWave base station. According to [16, 17], the probability that user associates with one LOS base station is

$$A_L = B_L \int_0^{\infty} e^{-2\pi\lambda_S \int_0^{\psi_L(x)} p_N(t)tdt} g_L(x)dx, \tag{8.3}$$

where the probability that the user has at least one LOS base station is

$$B_L = 1 - e^{-2\pi\lambda_S \int_0^{\infty} rp_L(r)dr}, \tag{8.4}$$

and given the user observes at least one LOS base station, the conditional probability density function (PDF) of the distance to its nearest LOS base station is

$$g_L(x) = 2\pi\lambda_S x p_L(x) e^{-2\pi\lambda_S \int_0^{x} rp_L(x)dr} / B_L. \tag{8.5}$$

The association probability to a NLOS base station is $A_N = 1 - A_L$. The probability that the user has at least one NLOS base station is

$$B_N = 1 - e^{-2\pi\lambda_S \int_0^{\infty} r(1-p_L(r))dr}, \tag{8.6}$$

and given the user observes one or more LOS base stations, the conditional PDF of the distance to its nearest LOS base station is

$$g_N(x) = 2\pi\lambda_S x(1 - p_L(x)) e^{-2\pi\lambda_S \int_0^{x} r(1-p_L(x))dr} / B_N. \tag{8.7}$$

8.3 Unicast Transmissions for mmWave-NOMA Networks

In this section, we will study unicast transmissions for mmWave-NOMA networks. Conventionally, unicasting employs point-to-point mechanism to achieve data transmissions, which has been widely used in cellular networks. However, with NOMA, the messages of multiple users can be multiplexed in the power domain by superposition coding at the transmitter side, and then each user decodes its desired message by successive interference cancelation (SIC). This non-orthogonality overcomes the loss of degree-of-freedom (DoF) caused by orthogonal transmission at the cost of increasing processing complexity.

8.3.1 System Model

Figure 8.1 illustrates the system model of single-tier mmWave-NOMA networks with downlink unicast transmissions. The mmWave base stations are located according to a homogeneous Poisson point process (HPPP) Φ_S with density λ_S and are assumed to have same transmit power P_S. We consider an M-user NOMA scenario, which means that a NOMA user pair with M users sorted by ascending order. The M users are randomly distributed in an mmWave service area. After channel ordering, we have $\frac{H_1 L(r_1)}{I_1+\sigma^2} \leq \frac{H_2 L(r_2)}{I_2+\sigma^2} \leq \cdots \leq \frac{H_M L(r_M)}{I_M+\sigma^2}$. The power allocated to the m-th user is $P_m > 0$ and satisfies $\sum_{m=1}^{M} P_m = P_S$. According to the principle of NOMA, the weak users are allocated to more power in order to ensure that they can decode successfully. Therefore, we have $P_1 \geq P_2 \geq \cdots \geq P_M$.

NOMA transmissions enable the messages of all users from a NOMA user pair to be multiplexed in the power domain to form a superposed signal, and then this superposed signal is transmitted on the same radio frequency. Therefore, the signal transmitted by the base station to users is

$$x = \sum_{m=1}^{M} \sqrt{P_m} x_m, \tag{8.8}$$

where x_m is the message of the m-th user and satisfies $\mathbb{E}[|x_m|] = 1$. The m-th user not only receives the signal from its serving base station, B_0, but also suffers from co-channel interference (CCI) from neighboring base stations and can be expressed as

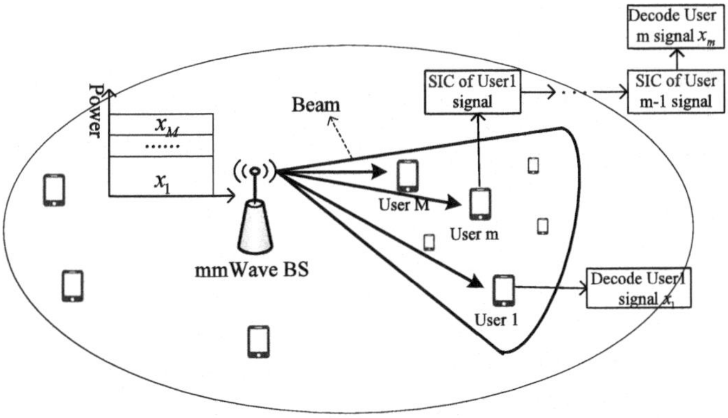

Fig. 8.1 System model of single-tier mmWave-NOMA networks with downlink unicast transmissions

$$y_m^1 = \frac{h_m \sqrt{G_0 L(r_m)} x}{\underbrace{\sum_{\mathrm{B}_i \in \Phi \backslash \mathrm{B}_0} h_{m,i} \sqrt{G_i L(r_{m,i})} x_i}_{I_{\mathrm{CCI},m}} + n_m}, \tag{8.9}$$

where n_m is the additive Gaussian noise with power σ^2 for the m-th user.

When the users receive signal from base stations, they employ successive SIC to decode their desired messages. According to the principle of SIC, the m-th user first decodes and cancels the messages of $\mathrm{UE}_i, i = 1, \ldots, m-1$ from the received sum signal orderly. The signal-to-interference-plus-noise ratio (SINR) that UE_m detects UE_i can be expressed as

$$\mathrm{SINR}_{m,i}^1 = \frac{H_m G_0 L(r_m) P_i}{\sum_{j=i+1}^{M} H_m G_0 L(r_m) P_j + I_{\mathrm{CCI},m} + \sigma^2}. \tag{8.10}$$

Then, UE_m decodes its own message after SIC with detecting SINR given by

$$\mathrm{SINR}_m^1 = \frac{H_m G_0 L(r_m) P_m}{\sum_{j=m+1}^{M} H_m G_0 L(r_m) P_j + I_{\mathrm{CCI},m} + \sigma^2}. \tag{8.11}$$

Note that for the M-th user, its SINR after SIC is given by

$$\mathrm{SINR}_M^1 = \frac{H_M G_0 L(r_M) P_M}{I_{\mathrm{CCI},M} + \sigma^2}. \tag{8.12}$$

8.3.2 Performance Analysis

In mmWave networks, for a random variable (RV), $X = \frac{H G_0 L(r)}{I_{\mathrm{CCI}} + \sigma^2}$, according to [16, 17], its complementary cumulative distribution function (CCDF) can be expressed as

$$\bar{F}(T) = A_L \bar{F}_L(T) + A_N \bar{F}_N(T), \tag{8.13}$$

while its CDF is $F(T) = 1 - \bar{F}(T)$. For $s \in \{L, N\}$, $\bar{F}_s(T)$ is the conditional CCDF given that the user is associated with a base station in Φ_s and can be written as

$$\bar{F}_L(T) \approx \sum_{n=1}^{N_L} (-1)^{n+1} \binom{N_L}{n} \int_0^\infty e^{-\left(\frac{n \eta_L x^{\alpha_{S,L}} T \sigma_S^2}{C_L G_{S,0}} + Q_n(T,x) + V_n(T,x)\right)} f_L(x) dx, \tag{8.14}$$

and

$$\bar{F}_N(T) \approx \sum_{n=1}^{N_N} (-1)^{n+1} \binom{N_N}{n} \int_0^\infty e^{-\left(\frac{n\eta_N x^{\alpha_{S,N}} T\sigma_S^2}{C_N G_{S,0}} + W_n(T,x) + Z_n(T,x)\right)} f_N(x)dx, \quad (8.15)$$

where

$$f_L(x) = \frac{B_L g_L(x)}{A_L} e^{-2\pi\lambda \int_0^{(C_N/C_L)^{1/\alpha_N} x^{\alpha_L/\alpha_N}} (1-p_L(t))tdt}, \quad (8.16)$$

$$f_N(x) = \frac{B_N g_N(x)}{A_N} e^{-2\pi\lambda \int_0^{(C_L/C_N)^{1/\alpha_L} x^{\alpha_N/\alpha_L}} p_L(t)tdt}, \quad (8.17)$$

$$Q_n(T,x) = 2\pi\lambda_S \sum_{k=1}^{4} b_k \int_x^\infty F(N_L, \frac{n\eta_L \bar{a}_k T x^{\alpha_{S,L}}}{N_L t^{\alpha_{S,L}}}) p_L(t)tdt, \quad (8.18)$$

$$V_n(T,x) = 2\pi\lambda_S \sum_{k=1}^{4} b_k \int_{\psi_L(x)}^\infty F(N_N, \frac{nC_N\eta_L \bar{a}_k T x^{\alpha_{S,L}}}{C_L N_N t^{\alpha_{S,N}}}) p_N(t)tdt, \quad (8.19)$$

$$W_n(T,x) = 2\pi\lambda_S \sum_{k=1}^{4} b_k \int_{\psi_N(x)}^\infty F(N_L, \frac{nC_L\eta_N \bar{a}_k T x^{\alpha_{S,N}}}{C_N N_L t^{\alpha_{S,L}}}) p_L(t)tdt, \quad (8.20)$$

and

$$Z_n(T,x) = 2\pi\lambda_S \sum_{k=1}^{4} b_k \int_x^\infty F(N_N, \frac{n\eta_N \bar{a}_k T x^{\alpha_{S,N}}}{N_N t^{\alpha_{S,N}}}) p_N(t)tdt. \quad (8.21)$$

8.3.2.1 Coverage Probability

The coverage probability can be used to characterize the quality of network coverage, which refers to the probability that the SINR received at user exceeds certain threshold T. According to SIC decoding, the coverage probability of the m-th ordered user in a NOMA user pair is defined as

$$P_{c,m}^1(T) = \mathbb{P}[\mathrm{SINR}_{m,1}^1 > T, \ldots, \mathrm{SINR}_{m,m}^1 > T]. \quad (8.22)$$

Since the maximum SINR that the m-th user detects the message of the i-th user from the superposed signal is $\lim_{H_m \to \infty} \mathrm{SINR}_{m,i}^1 = \frac{P_i}{\sum_{j=i+1}^M P_j}$, the coverage probability is equal to zero when the SINR threshold T is equal or greater than this maximum detecting SINR. When the SINR threshold T is below this maximum detecting SINR, substituting (8.10) and (8.11) into (8.22) and according to the law of total probability, the coverage probability can be rewritten as

$$P^1_{c,m}(T) = \mathbb{P}[X_m > b_1, \ldots, X_m > b_m]$$
$$\overset{(a)}{=} 1 - \sum_{i=m}^{n} \binom{n}{i} [1 - \bar{F}(\max(b_1, \ldots, b_m))]^i \bar{F}^{n-i}(\max(b_1, \ldots, b_m)), \tag{8.23}$$

where $X_m = \frac{H_m G_0 L(r_m)}{I_{\mathrm{CCI},m} + \sigma^2}$, $b_i = \frac{1}{\frac{P_i}{T} - \sum_{j=i+1}^{M} P_j}$, $i = 1, \ldots, m$, and (a) follows (8.13) and the property of order statistics [22].

Therefore, the coverage probability of the m-th ordered user in a NOMA user pair with M users in mmWave-NOMA networks can be finally expressed as

$$P^1_{c,m}(T) = \begin{cases} 0, \ T > \frac{p_1}{\sum_{j=2}^{M} P_j} \text{ or}, \ldots, \text{or } T > \frac{p_m}{\sum_{j=m+1}^{M} P_j}, \\ 1 - \sum\limits_{i=m}^{n} \binom{n}{i} [1 - \bar{F}(\max(b_1, \ldots, b_m))]^i \bar{F}^{n-i}(\max(b_1, \ldots, b_m)), \text{ otherwise.} \end{cases} \tag{8.24}$$

8.3.2.2 Outage Probability

The outage probability is used to characterize the probability that the user cannot achieve a target rate τ, and is defined as $P_o \triangleq \mathbb{P}[R < \tau]$. Further, due to $R = \log_2(1 + \mathrm{SINR}) < \tau$, its form related with SINR can be expressed as $P_o \triangleq \mathbb{P}[\mathrm{SINR} < 2^{\tau} - 1]$. We assume that the target rate for the m-th user is $\tau_m > 0$ and define the outage event $E_{m,i} = \{\mathrm{SINR}_{m,i} < \gamma_i\}$, where $\gamma_i = 2^{\tau_i} - 1$. This means that the m-th user failed to decode the message of the i-th user. Correspondingly, the complementary outage event is $\bar{E}_{m,i} = \{\mathrm{SINR}_{m,i} \geq \gamma_i\}$. According to the principle of NOMA, the outage probability of the m-th user can be expressed as

$$P^1_{o,m} = \mathbb{P}[E_{m,1} \cup E_{m,2} \cup \cdots \cup E_{m,m}] = 1 - \mathbb{P}[\bar{E}_{m,1} \cap \bar{E}_{m,2} \cap \cdots \cap \bar{E}_{m,m}]. \tag{8.25}$$

With SIC decoding, when the SINR that the m-th user detects the message of the i-th user is smaller than its maximum SINR, $\mathrm{SINR}^1_{m,i} = \frac{P_i}{\sum_{j=i+1}^{M} P_j}$, the SIC decoding failed. That is, when $\gamma_1 \geq \frac{P_1}{\sum_{j=2}^{M} P_j}$ or ... or $\gamma_m \geq \frac{P_m}{\sum_{j=m+1}^{M} P_j}$, the outage probability is equal to one. When $\gamma_1 < \frac{P_1}{\sum_{j=2}^{M} P_j}$ and ... and $\gamma_m < \frac{P_m}{\sum_{j=m+1}^{M} P_j}$, substituting (8.10) and (8.11) into (8.25), the outage probability can be written as

$$P^1_{o,m} = 1 - \mathbb{P}[X_m > c_1, \ldots, X_m > c_m]$$
$$\overset{(b)}{=} \sum_{i=m}^{n} \binom{n}{i} [1 - \bar{F}(\max(c_1, \ldots, c_m))]^i \bar{F}^{n-i}(\max(c_1, \ldots, c_m)), \tag{8.26}$$

where $c_i = \frac{1}{P_i \gamma_i^{-1} - \sum_{j=i+1}^{M} P_j}$, and (b) follows the property of order statistics and the complementary property of CCDF and CDF. Therefore, the outage probability of the m-th ordered user in a NOMA user pair with M users can be finally expressed as

$$P_{o,m}^{1} = \begin{cases} 1, \ \gamma_1 \geq \frac{P_1}{\sum_{j=2}^{M} P_j} \text{ or } \ldots \text{ or } \gamma_m \geq \frac{P_m}{\sum_{j=m+1}^{M} P_j}, \\ \sum\limits_{i=m}^{n} \binom{n}{i} [1 - \bar{F}(\max(c_1, \ldots, c_m))]^i \bar{F}^{n-i}(\max(c_1, \ldots, c_m)), \text{ otherwise.} \end{cases} \tag{8.27}$$

8.3.2.3 Sum Rate

To the m-th user's message, x_m, it should ensure that all users after it can also decode the message x_m in order to perform SIC successfully. Therefore, the data rate for the m-th user is equal to

$$R_m^1 = \min\left(R_{m,m}^1, R_{m+1,m}^1, \ldots, R_{M,m}^1\right), \tag{8.28}$$

where $R_{n,m}^1 = \log_2(1 + \mathrm{SINR}_{n,m}^1)$, $n = m, \ldots, M$ is the rate that the n-th user decodes the data x_m. Further, we have

$$\begin{aligned} R_{n,m}^1 &= \log_2\left(1 + \frac{H_n G_0 L(r_n) P_m}{\sum_{j=m+1}^{M} H_n G_0 L(r_n) P_j + I_{\mathrm{CCI},n} + \sigma^2}\right) \\ &= \log_2\left(1 + \frac{P_m}{\sum_{j=m+1}^{M} P_j + \frac{I_{\mathrm{CCI},n} + \sigma^2}{H_n G_0 L(r_n)}}\right). \end{aligned} \tag{8.29}$$

Due to $\frac{H_1 L(r_1)}{I_1} \leq \frac{H_2 L(r_2)}{I_2} \leq \cdots \leq \frac{H_M L(r_M)}{I_M}$, we have $R_{m,m}^1 \leq R_{m+1,m}^1 \leq \cdots \leq R_{M,m}^1$. Therefore, we have

$$R_m^1 = \min\left(R_{m,m}^1, R_{m+1,m}^1, \ldots, R_{M,m}^1\right) = R_{m,m}^1. \tag{8.30}$$

The sum rate for a typical NOMA user pair consisting of M users is defined as the sum of the average rate for each user and is

$$\tau_t^1 \triangleq \sum_{m=1}^{M} \mathbb{E}[R_m^1] = \frac{1}{\ln 2} \sum_{m=1}^{M} \mathbb{E}[\ln(1 + \mathrm{SINR}_m^1)]. \tag{8.31}$$

For the m-th ordered user, $m = 1, \ldots, M-1$, the average rate is taken over both the spatial PPP and the fading distribution. We have

$$\begin{aligned}
\bar{R}_m &= \frac{1}{\ln 2}\mathbb{E}[\ln(1+\mathrm{SINR}_m^1)] \\
&\stackrel{(c)}{=} \frac{1}{\ln 2}\Bigg\{A_L \int_{r>0}\int_0^{\ln\left(1+\frac{P_m}{\sum_{j=m+1}^M P_j}\right)} \mathbb{P}\left[X_m > \frac{1}{\frac{P_m}{e^t-1} - \sum_{j=m+1}^M P_j}\right] dt f_L(r) dr \\
&\quad + A_N \int_{r>0}\int_0^{\ln\left(1+\frac{P_m}{\sum_{j=m+1}^M P_j}\right)} \mathbb{P}\left[X_m > \frac{1}{\frac{P_m}{e^t-1} - \sum_{j=m+1}^M P_j}\right] dt f_N(r) dr\Bigg\} \\
&= \int_0^{\ln\left(1+\frac{P_m}{\sum_{j=i+1}^M P_j}\right)} \bar{F}_m\left(\frac{1}{\frac{P_m}{e^t-1} - \sum_{j=m+1}^M P_j}\right) dt,
\end{aligned} \tag{8.32}$$

where (c) follows $\mathbb{E}[W] = \int_{t>0}\mathbb{P}(W>t)dt$ for a positive RV, W [20], and

$$\lim_{I_{\mathrm{CCI},m}\to 0, H_m\to\infty} \ln\left(1+\frac{H_m G_0 L(r_m) P_m}{\sum_{l=m+1}^M H_M G_0 L(r_m) P_j + I_{\mathrm{CCI},m} + \sigma^2}\right) = \ln\left(1+\frac{P_m}{\sum_{j=m+1}^M P_j}\right). \tag{8.33}$$

For the M-ordered user, the integral domain of variable t in (8.32) is $(0, \infty)$. Therefore, the average rate can be written as

$$\bar{R}_M = \int_0^{\infty} \bar{F}_M\left(\frac{e^t-1}{P_M}\right) dt. \tag{8.34}$$

Combining (8.31), (8.32), and (8.34), the sum rate for a NOMA user pair in mmWave networks can be expressed as

$$\begin{aligned}
\tau_t^1 = \frac{1}{\ln 2}\Bigg(&\sum_{m=1}^{M-1}\int_0^{\ln\left(1+P_m(\sum_{j=i+1}^M P_j)^{-1}\right)} \bar{F}_m\left(\frac{1}{\frac{P_m}{e^t-1} - \sum_{j=m+1}^M P_j}\right) dt \\
&+ \int_0^{\infty}\bar{F}_M\left(\frac{e^t-1}{P_M}\right) dt\Bigg).
\end{aligned} \tag{8.35}$$

8.3.3 Numerical Results

To evaluate the performance of mmWave-NOMA networks, the parameters are used as follows: the carrier frequency is 28 GHz and the system bandwidth is 100 MHz; the base station transmit power is 30 dBm; the density of mmWave base station is $\lambda_S = \frac{1}{\pi 200^2}$; the path loss exponents for LOS and NLOS links are set as 2 and 4,

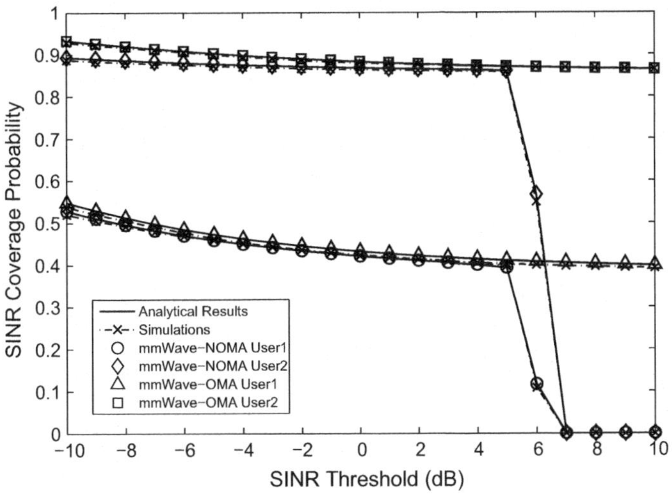

Fig. 8.2 SINR coverage probabilities of mmWave-NOMA networks with fixed power ratio (0.8, 0.2)

respectively, and a NOMA user pair consists of two random users with ascending order. To give a fair comparison to NOMA, OMA employs equal resource allocation for each user.

Figure 8.2 shows the coverage probabilities of mmWave-NOMA networks with fixed power ratio (0.8, 0.2). The results show that the coverage probabilities of both NOMA user1 and NOMA user2 are lower than that of OMA. This is because inter-user interference introduced by NOMA deteriorates the SINR received at users. The results also show that the coverage probabilities of NOMA users decline to zero when the SINR threshold exceeds a certain value (i.e., $\frac{P_1}{P_2}$), while OMA can still achieve some coverage. The reason is that the maximum SINR detecting NOMA user1 is limited by $\frac{P_1}{P_2}$, according to SIC decoding.

Figure 8.3 shows the outage probabilities of mmWave-NOMA networks with fixed power ratios (0.8, 0.2) and (0.9, 0.1). The results show that compared with OMA, the outage probability of the NOMA-weak user (i.e., User1) can be improved, while the NOMA-strong user (i.e., User2) suffers from some loss. The results also show that with the increase of power allocated to the NOMA-weak user, the outage probability of the NOMA-weak user can be further improved, while the NOMA-strong user's outage probability becomes worse.

Figure 8.4 shows the average rates for mmWave-NOMA networks with fixed power ratios (0.8, 0.2) and (0.9, 0.1). The results show that NOMA can achieve higher data rate than conventional OMA. This is because NOMA can enable all users in a NOMA user pair to occupy whole system bandwidth such that there is no loss in DoF, at the cost of the increase of processing complexity and the introduction

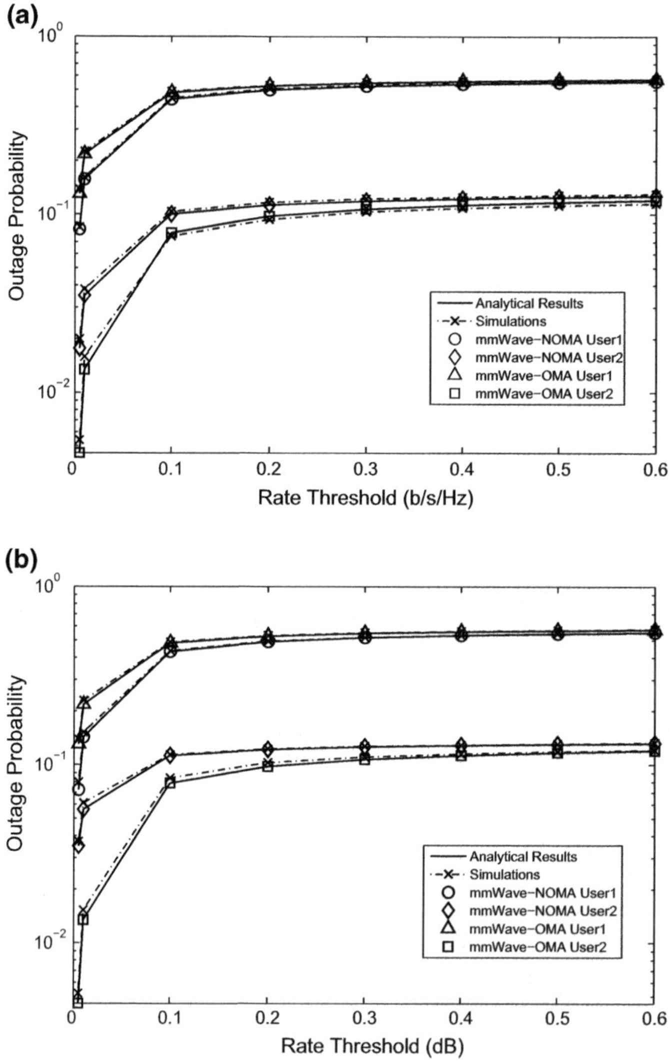

Fig. 8.3 Outage probabilities of mmWave-NOMA networks with fixed power ratios: **a** (0.8, 0.2); **b** (0.9, 0.1)

of inter-user interference. However, the results also show that for a certain transmit power range, e.g., [0, 5] dBm, NOMA with power ratio 0.9 can achieve higher data rate for the weak user (i.e, User1), while suffers from a lower data rate for the strong user (i.e., User2) and sum rate. This is because the fixed power ratio for NOMA is not optimal for all transmit powers. Comparing Fig. 8.4a, b, the weak user with power ratio 0.9 can achieve higher average rate than that of power ratio 0.8, while the opposite trend is for the strong user, which results in a lower sum rate.

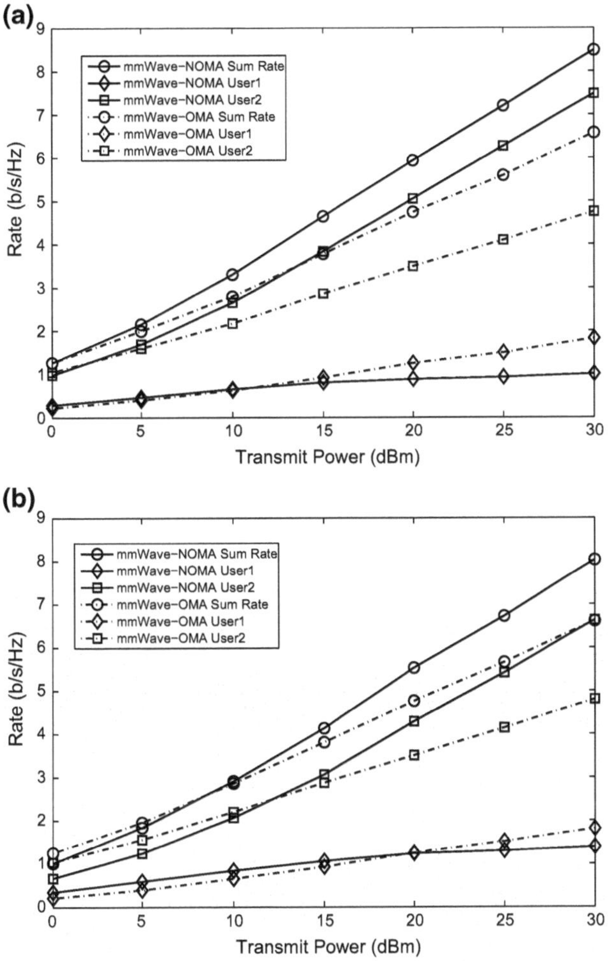

Fig. 8.4 Average rates for mmWave-NOMA networks with fixed power ratios: **a** (0.8, 0.2); **b** (0.9, 0.1)

8.4 Multicast Transmissions for mmWave-NOMA Networks

Compared with unicast transmissions discussed in Sect. 8.3, multicast transmissions employ the point-to-multipoint mechanism to distribute the same media data to multiple interested users on the same radio resources such that higher spectrum efficiency can be achieved. However, conventional multicast transmissions just deliver single data stream with a low data rate in order to ensure that most users can decode the media. With the help of NOMA, multicast transmissions can be enhanced by multiplexing multiple data streams with different data rates in the power domain. As a result, users can decode data streams according to their channel conditions. In this section, we will discuss multicast transmissions for mmWave-NOMA networks.

8.4.1 System Model

Figure 8.5 illustrates the system model of multicast transmissions in mmWave-NOMA networks. The mmWave base stations are located according to an HPPP Φ_S with density λ_S. The mmWave base stations are assumed to have same transmit power P_S. The users are located according to an HPPP Φ_U with density λ_U. With NOMA, N-layer superposition coded multicast transmission can be achieved by power allocation, which consists of one primary layer and $N-1$ secondary layers. The primary layer carries the basic data, while the secondary layers carry the corresponding enhanced data. The fixed data rate for each layer is also assumed.

Generally, the scalable media can be encoded into one basic data and $N-1$ enhanced data by source layered coding, where the basic data provides the basic service quality, while the enhanced data are used to improve the service quality. Note that the enhanced data cannot work without the basic data. With NOMA-enabled multicast transmissions, the basic data and enhanced data are multiplexed in the power domain to form a superposed signal. Then, this superposed signal is distributed on the same radio resources to all users who desire to receive the media. When users receive this superposed signal, they first decode the basic data to obtain the basic service quality directly, then try to decode the enhanced data to achieve better service quality. For the weak users, they can just decode the low-rate basic data, while the strong users can decode both the low-rate basic data and high-rate enhanced data. Therefore, NOMA-enabled multicast transmissions can fully utilize the channel difference between users to improve the performance of the strong users. This technique overcomes the shortage of conventional multicast transmissions that

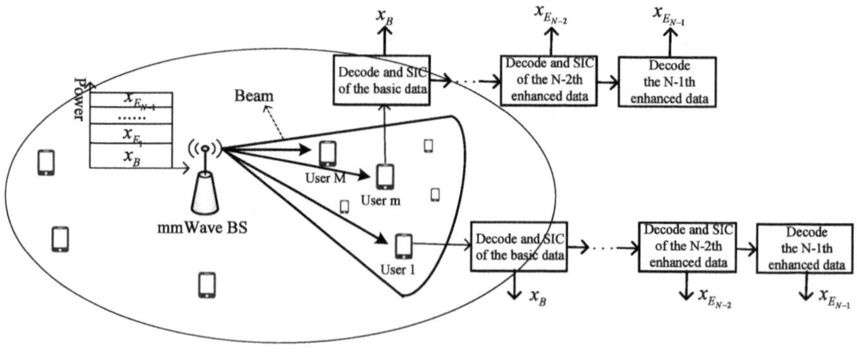

Fig. 8.5 System model of single-tier mmWave-NOMA networks with multicast transmissions

the strong users cannot fully utilize their good channel to obtain better service quality, as the conventional one just transmits single low-rate data to ensure that all users can decode the media successfully.

Without loss of generality, two-layer superposition coded multicast transmission is considered as an example. The mmWave base station transmits a superposed signal to all users within its coverage as

$$x = \sqrt{\alpha_p} x_B + \sqrt{1 - \alpha_p} x_E, \tag{8.36}$$

where $0 < \alpha_p < 1$ is the power allocation factor (PAF), x_B and x_E are the transmit messages of the primary and secondary layers, respectively. The signal received at the user with random distance, d_0, from its serving mmWave base station can be expressed as

$$y_S^2 = h_{S,0}\sqrt{G_{S,0} P_S L_S(d_0)} x + \underbrace{\sum_{X_i \in \Phi_S \setminus B_0} h_{S,i}\sqrt{G_{S,i} P_S L_S(d_i)} x_i}_{I_S} + n_S. \tag{8.37}$$

After receiving the superposed signal, the user first decodes the primary layer, and then cancel it from the received signal before decoding the secondary layer. Therefore, substituting (8.36) into (8.37), the SINRs of detecting the primary and secondary layers can be written, respectively, as

$$\mathrm{SINR}_{S,PL}^2 = \frac{\alpha_p H_S G_{S,0} L_S(d_0)}{(1 - \alpha_p) H_S G_{S,0} L_S(d_0) + \underbrace{\sum_{X_i \in \Phi_S \setminus B_0} H_{S,i} G_{S,i} L_S(d_i)}_{I_S} + \sigma_S^2}, \tag{8.38}$$

and

$$\mathrm{SINR}_{S,SL}^2 = \frac{(1 - \alpha_p) H_S G_{S,0} L_S(d_0)}{\underbrace{\sum_{X_i \in \Phi_S \setminus B_0} H_{S,i} G_{S,i} L_S(d_i)}_{I_S} + \sigma_S^2}. \tag{8.39}$$

Note that σ_S^2 is the thermal noise power, normalized by transmit power, P_S.

8.4.2 *Performance Analysis*

8.4.2.1 Coverage Probability

The coverage probability that users can decode the primary layer relative to the SINR threshold T_{PL} can be written as

$$P^2_{c,PL}(T_{PL}) = \mathbb{E}_R[\mathbb{P}[\text{SINR}^2_{PL} > T_{PL} \mid R = r]]. \tag{8.40}$$

Considering the maximum SINR for detecting the primary layer, $\lim_{H_s \to \infty} \text{SINR}^2_{PL} = \frac{\alpha_p}{1-\alpha_p}$, the integral domain is

$$\begin{aligned} D &= \left\{(H_S) \mid \text{SINR}^2_{PL} > T_{PL}\right\} \\ &= \left\{(H_S) \mid H_S > \frac{T_{PL}(I_S + \sigma_S^2)}{(\alpha_p - (1-\alpha_p)T_{PL})G_{S,0}L_S(r_0)} \middle| T_{PL} < \frac{\alpha_p}{1-\alpha_p}\right\}. \end{aligned} \tag{8.41}$$

Therefore, the coverage probability can be expressed as

$$P^2_{c,S,PL}(T_{PL}, \alpha_p) = \mathbb{E}_R\left[\mathbb{P}\left[H_S > \frac{T_{PL}(I_S + \sigma_S^2)}{(\alpha_p - (1-\alpha_p)T_{PL})G_{S,0}L_S(r_0)} \mid R = r\right]\right]. \tag{8.42}$$

According to [16, 17] and after some manipulations, the coverage probability of the primary layer can be obtained as

$$\begin{aligned} &P^2_{c,S,PL}(T_{PL}, \alpha_p) \\ &\approx \begin{cases} A_L \sum_{n=1}^{N_L} (-1)^{n+1} \binom{N_L}{n} \int_0^\infty e^{-\left(\frac{n\eta_L x^{\alpha_{S,L}} T_{PL}\sigma_S^2}{C_L(\alpha_p - (1-\alpha_p)T_{PL})G_S} + Q_n(T_{PL},x) + V_n(T_{PL},x)\right)} f_L(x)dx \\ +A_N \sum_{n=1}^{N_N} (-1)^{n+1} \binom{N_N}{n} \int_0^\infty e^{-\left(\frac{n\eta_N x^{\alpha_{S,N}} T_{PL}\sigma_S^2}{C_N(\alpha_p - (1-\alpha_p)T_{PL})G_S} + W_n(T_{PL},x) + Z_n(T_{PL},x)\right)} f_N(x)dx, & T_{PL} < \frac{\alpha_p}{1-\alpha_p}, \\ 0, & T_{PL} \geq \frac{\alpha_p}{1-\alpha_p}. \end{cases} \end{aligned} \tag{8.43}$$

The coverage probability of both the primary and secondary layers relative to the SINR thresholds T_{PL} and T_{SL} can be written as

$$P^2_{c,PSL}(T_{PL}, T_{SL}, \alpha_p) = \mathbb{E}_R[\mathbb{P}[\{\text{SINR}^2_{PL} > T_{PL} \cap \text{SINR}^2_{SL} > T_{SL}\} \mid R = r]]. \tag{8.44}$$

Its integral domain is

$$\begin{aligned} D &= \left\{(H_S) \mid \{\text{SINR}^2_{PL} > T_{PL} \cap \text{SINR}^2_{SL} > T_{SL}\}\right\} \\ &= \left\{(H_S) \mid \left\{H_S > \frac{T_{PL}(I_S + \sigma_S^2)}{(\alpha_p - (1-\alpha_p)T_{PL})G_{S,0}L_S(r_0)} \cap H_S > \frac{T_{SL}(I_S + \sigma_S^2)}{(1-\alpha_p)G_{S,0}L_S(r_0)}\right\}\right\}. \end{aligned} \tag{8.45}$$

Let $\frac{T_{PL}(I_S+\sigma_S^2)}{(\alpha_p-(1-\alpha_p)T_{PL})G_{S,0}L_S(r_0)} = \frac{T_{SL}(I_S+\sigma_S^2)}{(1-\alpha_p)G_{S,0}L_S(r_0)}$, we can obtain $\alpha_p = \frac{T_{PL}(1+T_{SL})}{T_{SL}+T_{PL}(1+T_{SL})}$.

Further, the coverage probability can be expressed as

$$P^2_{c,S,PSL}(T_{PL}, T_{SL}, \alpha_p) = \begin{cases} \mathbb{E}_R\left[\mathbb{P}\left[H_S > \frac{T_{PL}(I_S+\sigma_S^2)}{(\alpha_p-(1-\alpha_p)T_{PL})G_{S,0}L_S(r_0)} \mid R=r\right]\right], \ \alpha_p \le \frac{T_{PL}(1+T_{SL})}{T_{SL}+T_{PL}(1+T_{SL})}, \\ \mathbb{E}_R\left[\mathbb{P}\left[H_S > \frac{T_{SL}(I_S+\sigma_S^2)}{(1-\alpha_p)G_{S,0}L_S(r_0)} \mid R=r\right]\right], \alpha_p > \frac{T_{PL}(1+T_{SL})}{T_{SL}+T_{PL}(1+T_{SL})}. \end{cases} \tag{8.46}$$

Let

$$P^2_{c,S,SL}(T_{SL}, \alpha_p) = \mathbb{E}_R\left[\mathbb{P}\left[H_S > \frac{T_{SL}(I_S+\sigma_S^2)}{(1-\alpha_p)G_{S,0}L_S(r_0)} \mid R=r\right]\right]. \tag{8.47}$$

Therefore, the coverage probability, $P^2_{c,PSL}(T_{PL}, T_{SL}, \alpha_p)$, can finally be written as

$$P^2_{c,S,PSL}(T_{PL}, T_{SL}, \alpha_p) = \begin{cases} P^2_{c,S,PL}(T_{PL}, \alpha_p), \alpha_p \le \frac{T_{PL}(1+T_{SL})}{T_{SL}+T_{PL}(1+T_{SL})}, \\ P^2_{c,S,SL}(T_{SL}, \alpha_p), \alpha_p > \frac{T_{PL}(1+T_{SL})}{T_{SL}+T_{PL}(1+T_{SL})}, \end{cases} \tag{8.48}$$

where $P^2_{c,S,PL}(T_{PL}, \alpha_p)$ is expressed as in (8.43) and $P^2_{c,S,SL}(T_{SL}, \alpha_p)$ is approximated as

$$\begin{aligned} &P^2_{c,S,SL}(T_{SL}, \alpha_p) \\ &\approx A_L \sum_{n=1}^{N_L} (-1)^{n+1} \binom{N_L}{n} \int_0^\infty e^{-\left(\frac{n\eta_L x^{\alpha_{S,L}} T_{SL}\sigma_S^2}{C_L(1-\alpha_p)G_S} + Q_n(T_{SL},x) + V_n(T_{SL},x)\right)} f_L(x)dx \\ &+ A_N \sum_{n=1}^{N_N} (-1)^{n+1} \binom{N_N}{n} \int_0^\infty e^{-\left(\frac{n\eta_N x^{\alpha_{S,N}} T_{SL}\sigma_S^2}{C_N(1-\alpha_p)G_S} + W_n(T_{SL},x) + Z_n(T_{SL},x)\right)} f_N(x)dx. \end{aligned} \tag{8.49}$$

8.4.2.2 Average Number of Served Users

For the mmWave multicast cluster, B_o, the average number of served users by the primary layer can be expressed as

$$\mathbb{E}^o[N^2_{PL}] \triangleq \mathbb{E}^o\left[\sum_{y\in\Phi_{U,B_o}} \mathbb{I}(E^2_{PL}(y))\right], \tag{8.50}$$

where $E^2_{PL}(y) = \{\mathrm{SINR}^2_{S,PL} \ge 2^{R_{PL}} - 1\}$. Given the user density λ_U, the average number of users covered by a beam with width θ_S is $\lambda_U\theta_S(2\pi\lambda_S)^{-1}$. Further, considering the average coverage probability, $\mathbb{E}^o[N^2_{PL}]$ can be finally expressed as

$$\mathbb{E}^o[N^2_{PL}] = \lambda_U P^2_{c,S,PL}(T_{PL}, \alpha_p)\theta_S(2\pi\lambda_S)^{-1}, \tag{8.51}$$

The average number of served users, who can decode the data contained in both the primary and secondary layers, can be written as

$$\mathbb{E}^o[N_{PSL}^2] \triangleq \mathbb{E}^o\left[\sum_{y\in\Phi_{U,B_o}} \mathbb{I}(E_{PSL}^2(y))\right], \tag{8.52}$$

where $E_{PSL}^2(y) = \{\{\mathrm{SINR}_{S,PL}^2 \geq 2^{R_{PL}} - 1\} \cap \{\mathrm{SINR}_{S,SL}^2 \geq 2^{R_{SL}} - 1\}\}$. Similarly, $\mathbb{E}^o[N_{PSL}^2]$ can be finally expressed as

$$\mathbb{E}^o[N_{PSL}^2] = \lambda_U P_{c,S,PSL}^2(T_{PL}, T_{SL}, \alpha_p)\theta_S(2\pi\lambda_S)^{-1}. \tag{8.53}$$

8.4.2.3 Sum Rate

The sum rate for NOMA multicast is defined as the mean of the sum rate of all users in coverage of the multicast cluster, who successfully decode the primary layer with data rate, R_{PL}, or both the primary and secondary layers with data rate, $R_{PL} + R_{SL}$. This is given by

$$\begin{aligned}\bar{R}_{\text{sum}}^2 &= (\mathbb{E}^o[N_{PL}^2] - \mathbb{E}^o[N_{PSL}^2])R_{PL} + \mathbb{E}^o[N_{PSL}^2](R_{PL} + R_{SL})\\ &= \mathbb{E}^o[N_{PL}^2]R_{PL} + \mathbb{E}^o[N_{PSL}^2]R_{SL}.\end{aligned} \tag{8.54}$$

Note that $\mathbb{E}^o[N_{PL}^2] - \mathbb{E}^o[N_{PSL}^2]$ is the average number of served users by the mmWave multicast cluster, who can only decode the primary layer.

Combining (8.43), (8.49), (8.51), (8.53), and (8.54), the sum rate for the mmWave multicast cluster can be expressed as

$$\bar{R}_{\text{sum}}^2 = \begin{cases} \frac{(R_{PL}+R_{SL})\lambda_U P_{c,S,PL}^2(T_{PL},\alpha_p)\theta_S}{2\pi\lambda_S}, & \alpha_p \leq \frac{T_{PL}(1+T_{SL})}{T_{SL}+T_{PL}(1+T_{SL})} \text{ and } \mathrm{T_{PL}} < \frac{\alpha_p}{1-\alpha_p}, \\ \frac{R_{PL}P_{c,S,PL}^2(T_{PL},\alpha_p)\lambda_U\theta_S}{2\pi\lambda_S} + \frac{R_{SL}P_{c,S,SL}^2(T_{PL},T_{SL},\alpha_p)\lambda_U\theta_S}{2\pi\lambda_S}, & \\ \qquad \alpha_p > \frac{T_{PL}(1+T_{SL})}{T_{SL}+T_{PL}(1+T_{SL})} \text{ and } \mathrm{T_{PL}} < \frac{\alpha_p}{1-\alpha_p}, & \\ 0, \qquad T_{PL} \geq \frac{\alpha_p}{1-\alpha_p}, & \end{cases} \tag{8.55}$$

where, $T_{PL} = 2^{R_{PL}} - 1$ and $T_{SL} = 2^{R_{SL}} - 1$.

8.4.3 *Numerical Results*

Figure 8.6 depicts the coverage probabilities of multicast transmissions for mmWave-NOMA networks with fixed power ratio (0.8, 0.2). It can be observed that NOMA multicast can provide a similar primary coverage layer as the conventional one and achieve a secondary coverage layer as well. However, when the SINR threshold

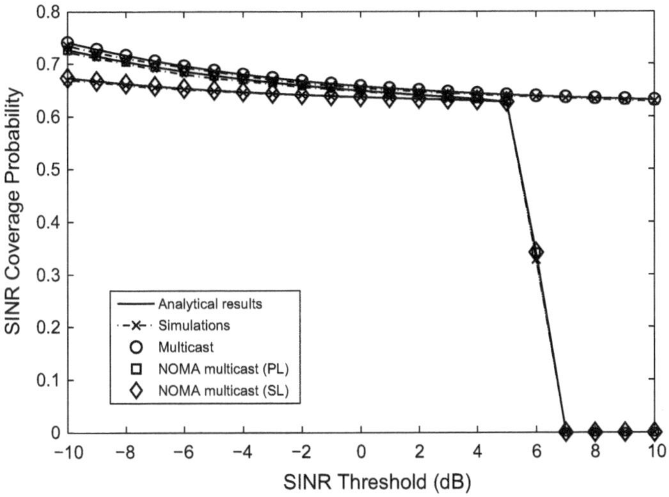

Fig. 8.6 SINR coverage probabilities of multicast transmission for mmWave-NOMA networks with fixed power ratio (0.8, 0.2)

T is larger than the maximum SINR, $\frac{\alpha_P}{1-\alpha_P}$, that users can detect the primary layer, the SINR coverage probabilities of the primary and secondary layers are equal to zero. This is because if users failed to decode the primary layer, they do not further decode the secondary layer through SIC.

Figure 8.7 depicts the sum rates for multicast transmission for mmWave-NOMA networks with fixed power ratios (0.8, 0.2) and (0.95, 0.05) and the secondary layer data rate, $R_{SL} = 4$ b/s/Hz. The results show that NOMA multicast can achieve a significant gain of sum rate, compared with the conventional one in the low multicast rate region. This is because NOMA multicast can fully utilize the channel conditions of strong users. However, the maximum SINR for detecting the primary layer is $\frac{\alpha_P}{1-\alpha_P}$ such that NOMA multicast cannot work when the multicast rate for the primary layer exceeds $\log_2(1 + \frac{\alpha_P}{1-\alpha_P})$. Comparing Fig. 8.7a, b, more power is allocated to the primary layer, a higher multicast rate for the primary layer can be provided, yet a lower sum multicast rate is achieved. This means that NOMA multicast gradually degrades to the conventional one, with the increase of power allocated to the primary layer.

Figure 8.8 depicts the sum rates for NOMA multicast with different power allocations, given 0.4, 1 b/s/Hz for the primary layer and 2, 4, 6 b/s/Hz for the secondary layer. It is shown that for each multicast rate pair, as the power ratio grows, the sum rate first experiences a sharp rise to the maximum, then it falls slowly in the medium power ratio region. Finally, it declines rapidly to the lowest point in the high power ratio region. Furthermore, with fixed multicast rate for the secondary layer, the sum rate of the primary layer with rate 1 b/s/Hz is higher than that of the primary layer with rate 0.4 b/s/Hz, which requires more power to be allocated to the primary layer. The sum rate of the secondary layer with rate 6 b/s/Hz is higher than that of the

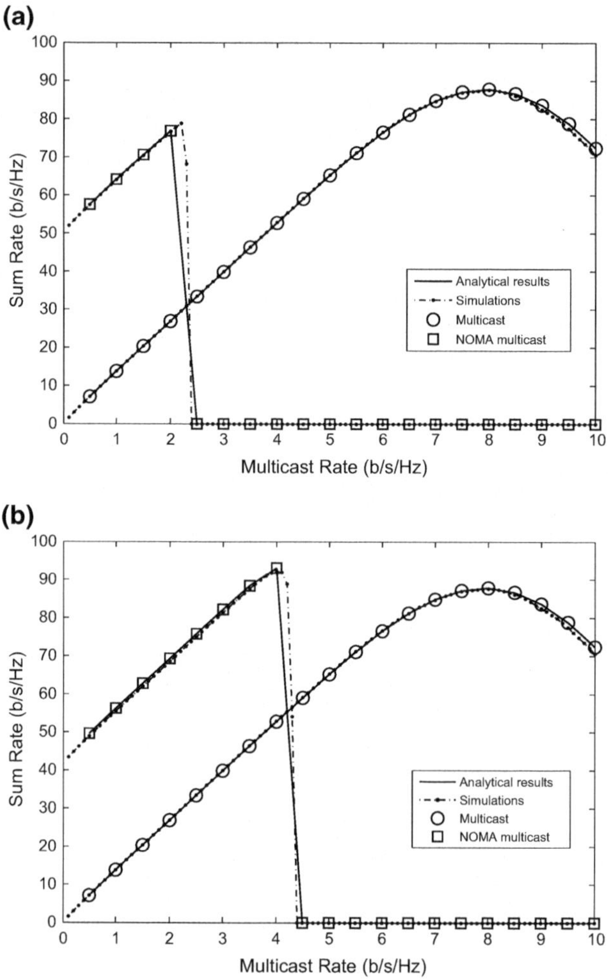

Fig. 8.7 Sum rates for multicast transmission for mmWave-NOMA networks with fixed power ratios. **a** (0.8, 0.2); **b** (0.95, 0.05)

secondary layer with rates 4 and 2 b/s/Hz, when the multicast rate for the primary layer is fixed.

8.5 Cooperative Multicast Transmissions for mmWave-NOMA HetNets

In this section, we will further discuss multicast transmissions in a two-tier mmWave-NOMA HetNet consisting of one low-frequency macro base station (MBS) tier and

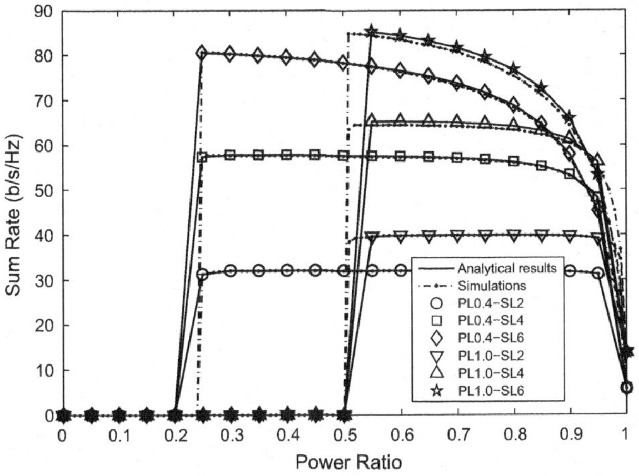

Fig. 8.8 Sum rate for multicast transmissions for mmWave-NOMA networks with different power ratios

one mmWave base station tier, which is the typical deployment for mmWave wireless networks, and introduce a cooperative multicast scheme for mmWave-NOMA networks to further improve the system performance. The cooperative multicast scheme can increase the success probability of decoding the primary layer with the help of cooperation from MBSs.

8.5.1 System Model

Figure 8.9 illustrates the system model of cooperative NOMA-enabled multicast transmissions for a two-tier mmWave HetNet, which consists of one low-frequency MBS tier with transmit power P_M and one mmWave base station tier with transmit power P_S. Cooperative NOMA-enabled multicast enables the MBS tier to cooperatively transmit the primary layer with a low data rate. With the cooperation of MBSs, the users who failed to decode the primary layer of the superposed signal from the mmWave base station tier will try to decode it from the macro BS tier. If the primary layer is decoded successfully, they cancel it from the superposed signal and further decode the secondary layer. Thus, this increases the success probability that decodes the primary and secondary layers, such that the NOMA multicast performance can be improved.

The signal received at the user with random distance, d_0, from its serving and interfering MBSs can be expressed as

$$y_M^3 = h_{M,0}\sqrt{P_M} d_0^{-\alpha_M/2} x_B + \underbrace{\sum_{X_i \in \Phi_M \setminus B_0} h_{M,i}\sqrt{P_M} d_i^{-\alpha_M/2} x_{i,B}}_{I_M} + n_M, \tag{8.56}$$

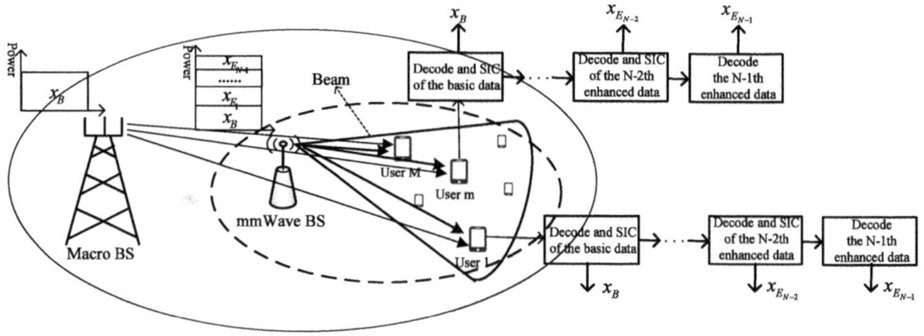

Fig. 8.9 System model of cooperative multicast transmissions for mmWave-NOMA HetNets

where n_M is the additive Gaussian noise with power σ_M^2. Therefore, the SINR of decoding the data can be written as

$$\mathrm{SINR}_{PL}^3 = \frac{H_{M,0} P_M d_0^{-\alpha_M}}{\underbrace{\sum_{\mathrm{X_i} \in \Phi_M \backslash \mathrm{B_0}} H_{M,i} P_M d_i^{-\alpha_M}}_{I_M} + \sigma_M^2}. \tag{8.57}$$

The SINR of decoding the primary layer, $\mathrm{SINR}_{S,PL}^3$, from the mmWave base station tier, can be written as in (8.38), while the SINR of the secondary layer, $\mathrm{SINR}_{S,SL}^3$, can be expressed as in (8.39).

8.5.2 *Performance Analysis*

8.5.2.1 Coverage Probability

With cooperative NOMA multicast in a two-tier mmWave HetNet, when users decode the primary layer from a macro or mmWave base station successfully, the basic data can be recovered. Therefore, the coverage probability of the primary layer can be expressed as

$$\begin{aligned} P_{c,PL}^3(T_{PL}, \alpha_p) &= \mathbb{P}\{\{\mathrm{SINR}_{M,PL}^3 > T_{PL}\} \cup \{\mathrm{SINR}_{S,PL}^3 > T_{PL}\}\} \\ &= \mathbb{P}\{\mathrm{SINR}_{M,PL}^3 > T_{PL}\} + \mathbb{P}\{\mathrm{SINR}_{S,PL}^3 > T_{PL}\} \\ &\quad - \mathbb{P}\{\mathrm{SINR}_{M,PL}^3 > T_{PL}\}\mathbb{P}\{\mathrm{SINR}_{S,PL}^3 > T_{PL}\}, \end{aligned} \tag{8.58}$$

where $\mathbb{P}\{\mathrm{SINR}_{S,PL}^3 > T_{PL}\}$ is obtained as in (8.43) due to $\mathrm{SINR}_{S,PL}^3 = \mathrm{SINR}_{S,PL}^2$. And according to [21], $P_{c,M,PL}^3(T_{PL})$ can be obtained as

$$\begin{aligned} P^{3}_{c,M,PL}(T_{PL}) &= \mathbb{P}\{\text{SINR}^{3}_{M,PL} > T_{PL}\} \\ &= \int_{r>0} \mathbb{P}\left[\frac{H_M P_M R^{-\alpha_M}}{I_M + \sigma_M^2} > T_{PL} \mid R = r\right] f_R(r)dr \\ &= \pi\lambda_M \int_0^\infty e^{-\pi\lambda_M(1+\rho(T_{PL},\alpha_M))x - T_{PL}(P_M/\sigma_M^2)^{-1}x^{\alpha_M/2}} dx, \end{aligned} \tag{8.59}$$

where

$$\rho(T_{PL}, \alpha_M) = T_{PL}^{2/\alpha_M} \int_{T_{PL}^{-2/\alpha_M}}^{\infty} (1 + t^{\alpha_M/2})^{-1} dt. \tag{8.60}$$

The coverage probability of the secondary layer can be expressed as

$$\begin{aligned} P^{3}_{c,PSL}(T_{PL}, T_{SL}, \alpha_p) = \mathbb{P}\{&\{\text{SINR}^{3}_{S,SL} > T_{SL} \cap \text{SINR}^{3}_{S,PL} > T_{PL}\} \\ &\cup \{\text{SINR}^{3}_{S,SL} > T_{SL} \cap \text{SINR}^{3}_{M,PL} > T_{PL}\}\}. \end{aligned} \tag{8.61}$$

After some manipulations,

$$\begin{aligned} P^{3}_{c,PSL}(T_{PL}, T_{SL}, \alpha_p) =& \mathbb{P}\{\text{SINR}^{3}_{S,SL} > T_{SL} \cap \text{SINR}^{3}_{S,PL} > T_{PL}\} \\ &+ \mathbb{P}\{\text{SINR}^{3}_{S,SL} > T_{SL}\}\mathbb{P}\{\text{SINR}^{3}_{M,PL} > T_{PL}\} \\ &- \mathbb{P}\{\text{SINR}^{3}_{S,SL} > T_{SL} \cap \text{SINR}^{3}_{S,PL} > T_{PL}\}\mathbb{P}\{\text{SINR}^{3}_{M,PL} > T_{PL}\}. \end{aligned} \tag{8.62}$$

Since $\text{SINR}^{3}_{S,SL} = \text{SINR}^{2}_{S,SL}$ and $\text{SINR}^{3}_{S,PL} = \text{SINR}^{2}_{S,PL}$, $\mathbb{P}\{\text{SINR}^{3}_{S,SL} > T_{SL} \cap \text{SINR}^{3}_{S,PL} > T_{PL}\}$ can be written as in (8.48), and $\mathbb{P}\{\text{SINR}^{3}_{S,SL} > T_{SL}\}$ can be expressed as in (8.49). Therefore, combining (8.48), (8.49), (8.58), and (8.62), $P^{3}_{c,PSL}(T_{PL}, T_{SL}, \alpha_p)$ can be finally expressed as

$$\begin{aligned} P^{3}_{c,PSL}(T_{PL}, T_{SL}, \alpha_p) =& P^{2}_{c,S,PSL}(T_{PL}, T_{SL}, \alpha_p)(1 - P^{3}_{c,M,PL}(T_{PL})) \\ &+ P^{2}_{c,S,SL}(T_{SL}, \alpha_p)P^{3}_{c,M,PL}(T_{PL}). \end{aligned} \tag{8.63}$$

8.5.2.2 Average Number of Served Users

For the mmWave multicast cluster, B_o, the average number of served users by the primary layer can be expressed as

$$\mathbb{E}^o[N^{3}_{PL}] \triangleq \mathbb{E}^o\left[\sum_{y\in\Phi_{U,B_o}} \mathbb{I}(E^{3}_{PL}(y))\right], \tag{8.64}$$

where $E^{3}_{PL}(y) = \{\text{SINR}^{3}_{M,PL} \geq 2^{R_{PL}} - 1 \cup \text{SINR}^{3}_{S,PL} \geq 2^{R_{PL}} - 1\}$. Given the user density λ_U, the average number of users covered by a beam with width θ_S is

$\lambda_U \theta_S (2\pi\lambda_S)^{-1}$. Further, considering the average coverage probability, $\mathbb{E}^o[N_{PL}^3]$ can be finally expressed as

$$\mathbb{E}^o[N_{PL}^3] = \lambda_U P_{c,PL}^3(T_{PL})\theta_S(2\pi\lambda_S)^{-1}. \tag{8.65}$$

The average number of served users by the secondary layer can be expressed as

$$\mathbb{E}^o[N_{PSL}^3] \triangleq \mathbb{E}^o\left[\sum_{y\in\Phi_{U,B_o}} \mathbb{I}(E_{PSL}^3(y))\right], \tag{8.66}$$

where $E_{PSL}^3(y) = \{\{\text{SINR}_{M,PL}^3 \geq 2^{R_{PL}} - 1 \cup \text{SINR}_{S,PL}^3 \geq 2^{R_{PL}} - 1\} \cap \{\text{SINR}_{S,SL}^3 \geq 2^{R_{SL}} - 1\}\}$. Similarly, $\mathbb{E}^o[N_{PSL}^3]$ can be finally expressed as

$$\mathbb{E}^o[N_{PSL}^3] = \lambda_U P_{c,PSL}^3(T_{PL}, T_{SL}, \alpha_p)\theta_S(2\pi\lambda_S)^{-1}. \tag{8.67}$$

8.5.2.3 Sum Rate

The sum rate for cooperative NOMA multicast is defined as the mean of the sum rate for all users in coverage of the multicast cluster and is equal to

$$\begin{aligned}\bar{R}_{\text{sum}}^3 &= R_{PL}(\mathbb{E}^o[N_{PL}^3] - \mathbb{E}^o[N_{PSL}^3]) + (R_{PL} + R_{SL})\mathbb{E}^o[N_{PSL}^3] \\ &= R_{PL}\mathbb{E}^o[N_{PL}^3] + R_{SL}\mathbb{E}^o[N_{PSL}^3].\end{aligned} \tag{8.68}$$

Combining (8.65), (8.67), and (8.68), the sum rate can be finally expressed as

$$\bar{R}_{\text{sum}}^3 = (R_{PL}P_{c,PL}^3(T_{PL}) + R_{SL}P_{c,PSL}^3(T_{PL}, T_{SL}, \alpha_p))\lambda_U\theta_S(2\pi\lambda_S)^{-1}. \tag{8.69}$$

8.5.3 Numerical Results

Figure 8.10 plots the coverage probabilities of cooperative multicast transmissions for mmWave-NOMA HetNets with fixed power ratio (0.8, 0.2). The results show that compared with multicast and NOMA multicast, cooperative NOMA multicast can further improve the coverage probabilities of the primary and secondary layers. More specifically, compared with multicast, cooperative NOMA multicast can achieve superior coverage in the low SINR threshold region, while compared with NOMA multicast, it can achieve better primary coverage, especially in the low and high SINR threshold regions, while it provides better secondary coverage, especially in the high SINR threshold region. This is because cooperative NOMA multicast enables the MBS tier to transmit a copy of the primary layer without power split as well. As a result, users can receive two copies of the primary layer, which increases the success probability of decoding the primary layer. This also increases the success probability of decoding the secondary layer, as users can try to further decode the secondary layer when they fail to decode the primary layer from a mmWave base station but succeed to decode the primary layer from a MBS.

Figure 8.11 plots the sum rates for cooperative multicast transmission for mmWave-NOMA HetNets with fixed power ratios (0.8, 0.2) and (0.95, 0.05). The results show that cooperative NOMA multicast can achieve higher sum rate than NOMA multicast, especially in the medium multicast rate region. This is because in cooperative NOMA multicast scheme, MBS transmits a replica of the primary layer as well, which overcomes the maximum SINR limit, $\frac{\alpha_P}{1-\alpha_P}$, of decoding the primary layer caused by NOMA transmission. As a result, this increases the success probability of decoding the primary and secondary layers. Comparing Fig. 8.11a, b, more power is allocated to the primary layer, a higher multicast rate can be provided, yet a lower sum multicast rate is achieved. This means that cooperative NOMA multicast gradually degrades to conventional multicast, with the increase of power allocated to the primary layer.

8.6 Summary

This chapter applied NOMA to mmWave networks and discussed unicast, multicast and cooperative multicast transmissions for mmWave-NOMA networks. An analytical framework for performance analysis of large-scale mmWave-NOMA networks by using stochastic geometry was also given. Based on this framework, analytical

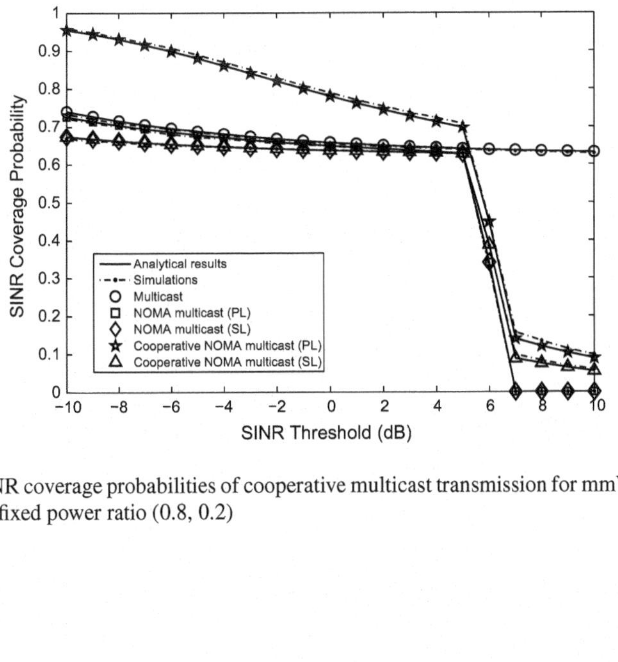

Fig. 8.10 SINR coverage probabilities of cooperative multicast transmission for mmWave-NOMA HetNets with fixed power ratio (0.8, 0.2)

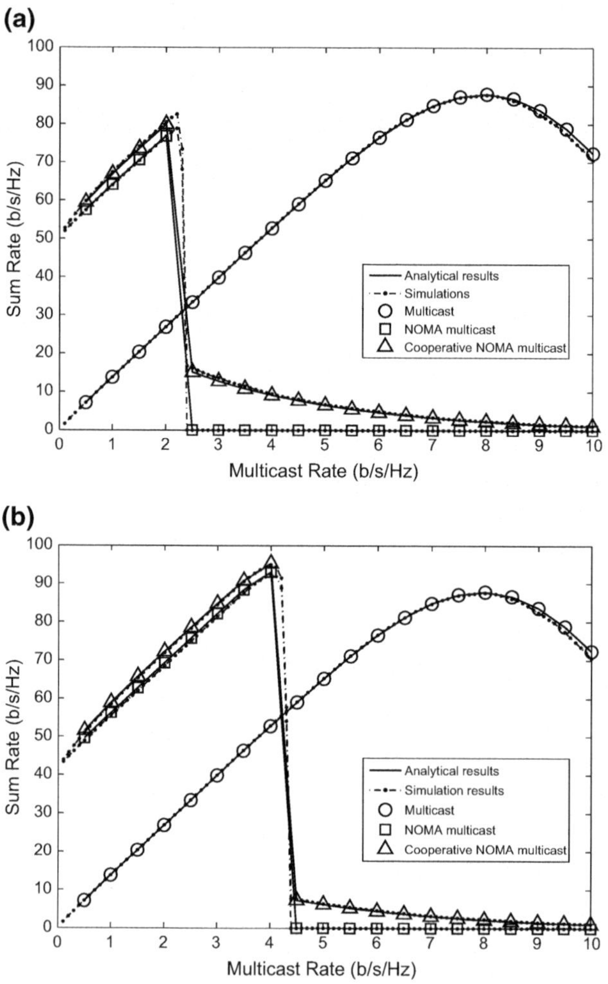

Fig. 8.11 Sum rates for cooperative multicast transmission for mmWave-NOMA HetNets with fixed power ratios: **a** (0.8, 0.2); **b** (0.95, 0.05)

expressions for SINR coverage probability, outage probability, and sum rate were provided to evaluate the performance of the presented schemes. It can be concluded that: (1) NOMA can achieve better performance than OMA, by multiplexing multiple users in the power domain; (2) compared with multicast, NOMA multicast multiplexes multiple data streams with different multicast rates in the power domain such that users can decode data streams according to their channel conditions, which significantly increases the sum rate; (3) the cooperative NOMA multicast increases the success probability of decoding data streams from the superposed signal with the help of cooperation of MBSs such that it further improves the performance of NOMA multicast. Some further research directions on NOMA for mmWave networks are pointed out as follows:

- Power Allocation for mmWave-NOMA Networks: the power ratio is a key factor that NOMA achieves better performance than the orthogonal one. The fixed power ratio is a simple way for NOMA, but it cannot always achieve optimal performance. This is because it does not utilize channel state information (CSI) in real time. The optimization of power ratio to further improve the system performance according to the instantaneous CSI is a great challenge, especially for multicast transmissions.
- Cooperation for mmWave-NOMA Networks: HetNet is an important deployment form for mmWave communications. This chapter only discusses cooperative multicast for mmWave-NOMA networks that MBSs cooperate transmission for the primary layer. It is important to exploit other cooperation schemes to further improve the performance of unicast and multicast transmissions for mmWave-NOMA networks.

References

1. T.S. Rappaport et al., Millimeter wave mobile communications for 5G cellular: it will work! IEEE Access **1**, 335–349 (2013)
2. W. Roh et al., Millimeter-wave beamforming as an enabling technology for 5G cellular communications: theoretical feasibility and prototype results. IEEE Commun. Mag. **52**, 106–113 (2014)
3. M. Xiao et al., Millimeter wave communications for future mobile networks. IEEE J. Sel. Areas Commun. **35**, 1909–1935 (2017)
4. L. Kong, M.K. Khan, F. Wu, G. Chen, P. Zeng, Millimeter-wave wireless communications for IoT-cloud supported autonomous vehicles: overview, design, and challenges. IEEE Commun. Mag. **55**, 62–68 (2017)
5. Z. Ding, Z. Yang, Z. Fan, H.V. Poor, On the performance of non-orthogonal multiple access in 5G systems with randomly deployed users. IEEE Signal Process. Lett. **21**, 1501–1505 (2014)
6. Z. Ding, X. Lei, G.K. Karagiannidis, R. Schober, J. Yuan, V.K. Bhargava, A survey on non-orthogonal multiple access for 5G networks: research challenges and future trends. IEEE J. Sel. Areas Commun. **35**, 2181–2195 (2017)
7. Z. Zhang, H. Sun, R.Q. Hu, Downlink and uplink non-orthogonal multiple access in a dense wireless network. IEEE J. Sel. Areas Commun. **35**, 2771–2784 (2017)
8. G. Araniti, M. Condoluci, P. Scopelliti, A. Molinaro, A. Iera, Multicasting over emerging 5G networks: challenges and perspectives. IEEE Netw. **31**, 80–89 (2017)
9. Z. Ding, P. Fan, H.V. Poor, Random beamforming in millimeter-wave NOMA networks. IEEE Access **5**, 7667–7681 (2017)
10. D. Zhang, Z. Zhou, C. Xu, Y. Zhang, J. Rodriguez, T. Sato, Capacity analysis of NOMA with mmWave massive MIMO systems. IEEE J. Sel. Areas Commun. **35**, 1606–1618 (2017)
11. B. Wang, L. Dai, Z. Wang, N. Ge, S. Zhou, Spectrum and energy-efficient beamspace MIMO-NOMA for millimeter-wave communications using lens antenna array. IEEE J. Sel. Areas Commun. **35**, 2370–2382 (2017)
12. Z. Ding, L. Dai, R. Schober, H.V. Poor, NOMA meets finite resolution analog beamforming in massive MIMO and millimeter-wave networks. IEEE Commun. Lett. **21**, 1879–1882 (2017)
13. A.J. Morgado, K.M.S. Huq, J. Rodriguez, C. Politis, H. Gacanin, Hybrid resource allocation for millimeter-wave NOMA. IEEE Wirel. Commun. **24**, 23–29 (2017)

14. Z. Zhang, Z. Ma, Y. Xiao, M. Xiao, G.K. Karagiannidis, P. Fan, Non-orthogonal multiple access for cooperative multicast millimeter wave wireless networks. IEEE J. Sel. Areas Commun. **35**, 1794–1808 (2017)
15. S. Naribole, E. Knightly, Scalable multicast in highly-directional 60-GHz WLANs. IEEE Trans. Netw. **25**, 2844–2857 (2017)
16. T. Bai, R.W. Heath, Coverage and rate analysis for millimeter-wave cellular networks. IEEE Trans. Wirel. Commun. **14**, 1100–1114 (2015)
17. J.G. Andrews, T. Bai, M. Kulkarni, A. Alkhateeb, A. Gupta, R.W. Heath, Modeling and analyzing millimeter wave cellular systems. IEEE Trans. Commun. **65**, 403–430 (2017)
18. T.S. Rappaport, G.R. MacCartney, M.K. Samimi, S. Sun, Wideband millimeter-wave propagation measurements and channel models for future wireless communication system design. IEEE Trans. Commun. **63**, 3029–3056 (2015)
19. H.S. Ghadikolaei, C. Fischione, G. Fodor, P. Popovski, M. Zorzi, Millimeter wave cellular networks: a MAC layer perspective. IEEE Trans. Commun. **63**, 3437–3458 (2015)
20. J.G. Andrews, F. Baccelli, R.K. Ganti, A tractable approach to coverage and rate in cellular networks. IEEE Trans. Commun. **59**, 3122–3134 (2011)
21. Andrews, J.G., Gupta, A.K., Dhillon, H.S.: *A Primer on Cellular Network Analysis Using Stochastic Geometry*, http://arxiv.org/abs/1604.03183
22. H.A. David, H.N. Nagaraja, *Order Statistics*, 3rd edn. (Wiley, Hoboken, New Jersey, 2003)

Chapter 9
Full-Duplex Non-Orthogonal Multiple Access Networks

Mohammed S. Elbamby, Mehdi Bennis, Walid Saad, Mérouane Debbah and Matti Latva-aho

9.1 Introduction

In conventional wireless networks, resources are assumed to be allocated exclusively to users by considering half-duplex (HD) transmissions in serving uplink (UL) and downlink (DL) requests. Operating simultaneously in UL and DL over the same frequency band, known as in-band full-duplex (IBFD) or more commonly as full-duplex (FD), has been avoided for a long time, due to the inability to suppress the self-interference in the full-duplex radio to a feasible operating point. Furthermore, orthogonal multiple access techniques, such as orthogonal frequency-division multiple access (OFDMA), is used in multi-carrier settings to allocate subcarriers to users in an exclusive manner. This restriction is imposed to avoid the multi-user interference (MUI) resulting from scheduling multiple users/messages over the same subcarrier. The unprecedented growth in data rate requirement and the number of connected devices mandates going beyond the traditional ways of handling the scarcity of bandwidth in future wireless networks. A fundamental shift in the in the way wireless resources are allocated and managed is thus necessary.

M. S. Elbamby (✉) · M. Bennis · M. Latva-aho
Centre for Wireless Communications, University of Oulu, Oulu, Finland
e-mail: mohammed.elbamby@oulu.fi

M. Bennis
e-mail: mehdi.bennis@oulu.fi

M. Latva-aho
e-mail: matti.latva-aho@oulu.fi

W. Saad
Wireless@VT, Bradley Department of Electrical and Computer Engineering, Virginia Tech, Blacksburg, VA, USA
e-mail: walids@vt.edu

M. Debbah
Mathematical and Algorithmic Sciences Lab, Huawei France, 92100 Paris, France
e-mail: merouane.debbah@huawei.com

M. Vaezi et al. (eds.), *Multiple Access Techniques for 5G Wireless Networks and Beyond*, https://doi.org/10.1007/978-3-319-92090-0_9

Full-duplex (FD) is ideally a spectrum efficiency doubler. By relaxing the constraint of orthogonal UL and DL transmissions, transceivers in the user and base station nodes can exploit the non-orthogonality to boost the spectral efficiency [21]. However, HD was adopted as the default setting in wireless networks for multiple reasons. First, FD radios can experience self-interference leaked from the UL transmitter to the DL receiver when operating in the same time–frequency resource. Recent advances in self-interference cancellation techniques have challenged this assumption. By using a combination of analogue and digital cancellation techniques, self-interference can be reduced to a level close to the receiver noise floor. Second, operating in FD results in increased inter-user interference due to the larger number of transmissions per resource. In particular, DL base station to UL base station and UL user to DL user interference will occur when operating simultaneously in UL and DL in the same frequency resource. Furthermore, this interference can occur at the intra-cell level, unlike the orthogonal resources per cell associated with OFDMA-based HD systems.

Another way of boosting the cellular network bandwidth utilization is the use of non-orthogonal multiple access (NOMA) technique to schedule users with potentially overlapping resources [11]. NOMA has been recently discussed as a promising way to boost the network spectral efficiency as compared to orthogonal multiple access (OMA) techniques. NOMA provides diversity in the power domain by transmitting different messages to/from different users using the same time–frequency resource. If the different signals are received with enough power disparity, the signals can be decoded for example using successive interference cancellation (SIC) techniques.

Hence, incorporating both FD and NOMA into current wireless networks will pose different challenges in terms of interference and network management. Different solutions are needed to address these challenges, ranging from smart resource scheduling, power allocation, duplex and multiple access mode switching. This chapter sheds light on both FD and NOMA network operation and discusses potential future research directions for this topic. First, preliminaries on FD, self-interference, and the interplay between FD and NOMA are discussed in Sect. 9.2. The objectives and tools used for FD and NOMA networks are discussed and surveyed in Sect. 9.3. Numerical results on the performance of FD-NOMA networks are presented in Sect. 9.4. Finally, Sect. 9.5 concludes the chapter and discusses some open problems.

9.2 Full-Duplex NOMA Networks

9.2.1 Preliminaries

Here, we provide the basics of FD and NOMA that can be helpful to the reader. We particularly define the basic concepts of FD and self-interference cancellation and then briefly introduce FD and NOMA network operation challenges.

9.2.1.1 Full-Duplex

Conventional wireless networks operate in HD mode, meaning that one direction of transmission is allowed at any given time and frequency resource. Different duplexing techniques have been considered in HD networks to duplex the UL and DL transmissions. In particular, time division duplex (TDD) and frequency division duplex (FDD) are both used commonly in today's networks. FDD dedicates frequency resources to UL and DL where communication in both directions is orthogonal in the frequency domain. TDD splits the time resources between UL and DL transmissions either in a static manner (fixed UL and DL duty cycles on each time frame) or a dynamic manner (where UL and DL duty cycles can change to match the network UL and DL load conditions [14].

Alternatively, bidirectional transmissions can be used simultaneously, namely full-duplex communication. FD can ideally double the spectrum efficiency by relaxing the constraint on UL and DL orthogonality [21]. This has been an ideal assumption for a long time. However, taking it into practice was hindered by the complexity of removing the self-interference leaked from the transmitter to the receiver of the same device. The transmitted signal is much stronger than the received one, as the latter is significantly weakened by the path and propagation losses. Hence, the transmitted signal saturates the receiver radio chains and prevents signal reception. Self-interference has been seen as the major impairment to FD operation. Ideally, self-interference should be cancelled to the same level as the receiver noise floor such that the received signal is decoded in the same level as in HD. Otherwise, the residual interference is added to the received signal and hence decreases the receive SNR and throughput.

9.2.1.2 Self-Interference Cancellation

Cancelling the self-interference from the transmitted signal of a FD radio is not as easy as it might sound. Although the transmitter knows what it is transmitting, this knowledge is of the signal in the baseband level [12, 20]. The baseband signal experiences several linear and nonlinear in the analogue radio chain before it is converted into a transmitted signal. Hence, subtracting the baseband signal does not help in removing the self-interference.

Recently, the self-interference cancellation capability has significantly evolved. The study in [20] has shown simultaneous transmission and reception (with a single antenna) is achievable using analogue and digital cancellation techniques to cancel the self-interference. The leaked self-interference is reduced to the noise floor such that the received signal is not degraded. This breakthrough in the self-interference cancellation capability motivated the consideration of full-duplex in future networks for user scheduling [17] as well as relaying [38]. Furthermore, the shift from traditional macrocells to low-power small-cell networks eases the process of integrating FD into future networks. As compared to 46 dBm transmit power of macrocells, femtocells operate in power levels as low as 20 dBm, which makes the self-interference process feasible.

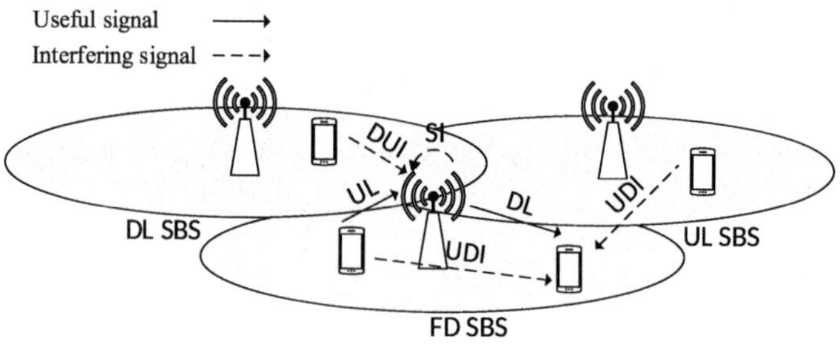

Fig. 9.1 An illustration of the different types of interference in multi-cell FD operation (UDI: UL-to-DL interference, DUI: DL-to-UL interference and SI: self-interference.)

9.2.1.3 FD Network Operation

Once the self-interference is cancelled, the link throughput of FD can potentially double that of an HD link. However, as FD operates in a network level, another issue arises which is the inter-link interference due to having simultaneous transmission and reception, as illustrated in Fig. 9.1. This interference situation is similar to that of dynamic TDD networks, where base stations operate with different TDD configurations to satisfy their individual UL and DL traffic requirements, resulting in additional inter-link inter-cell interference. In addition to that, FD networks will suffer intra-cell interference between the user pairs operating in UL and DL. Therefore, intelligent user pairing methods are needed to select the appropriate pairs of users to be scheduled in each time–frequency resource such that this interference is avoided.

9.2.1.4 FD-NOMA Network Operation

FD-NOMA refers to the concept of operating in IBFD mode and using NOMA to serve multiple requests. It is shown in the literature that NOMA can operate in both UL and DL. To reap the benefits promised by NOMA over OMA, power disparity has to be guaranteed. In UL, decoding the signals of multiple users sharing the time–frequency resource will occur in the base station level. It is intuitive therefore to select users with different receive power levels to be scheduled simultaneously. For example, scheduling a pair of cell-centre and cell-edge users is a convenient approach. In DL, different power levels should be allocated to the messages of different users, according to the decoding order they are going to select. This is essential to ensure a successful SIC process to cancel the intra-cell interference.

When FD is combined with NOMA, additional levels of interference are introduced generated from the opposite direction [3]. Different approaches can be considered to mitigate the effect of this unwanted interference. One approach [13] is to select between operating in FD while using an OMA technique or operating in HD and using NOMA. The scheduling and power allocation schemes play a key role in finding the optimal operating methods given the network conditions and optimization objectives. Prioritizing the serving of users in certain link direction means that NOMA should be selected to operate. On the other hand, FD can be selected to simultaneously serve UL and DL requests. While this approach restricts operation to FD-OMA or HD-NOMA at a certain time instant, it avoids the additional intra-cell interference of FD on NOMA operation and hence allows for more users to be served by NOMA in a certain link direction.

Another way to blend the use of FD and NOMA together is to allow the network to operate in FD and NOMA in a given time instant. In this regard, the inter-link interference can be treated as noise [35] and will affect the NOMA performance. In particular, DL users will experience high UL-to-DL interference from multiple users operating in UL-NOMA. This interference can be handled by limiting the maximum number of users allowed to operate in UL-NOMA if the performance of DL users is degraded.

9.2.2 Challenges of FD-NOMA Resource Optimization

As discussed above, resource optimization is a key role in reaping the benefits of FD-NOMA operation. Since the challenges of implementing them in practice are similar, there are common tools that can be used to optimize the resource and power allocation when either or both of them are considered. In this chapter, we focus on two main optimization problems that are fundamental for enabling NOMA, which are user pairing/scheduling and power allocation. Furthermore, we discuss the impact of the different objective functions on the resource optimization.

9.2.3 User Pairing and Power Optimization

To illustrate the problem of user scheduling FD-NOMA networks, we assume a general multi-cell wireless network that comprises a set $\mathscr{B}$ of B FD-capable base stations, with self-interference cancellation capability of ζ, and a set $\mathscr{U}$ of U users requesting either UL or DL service. Furthermore, we assume that both base stations and users can operate in NOMA scheme, where the resulting multi-user interference can be cancelled by performing SIC at the receiver side. The user scheduling parameters in UL and DL at time instant t are defined using the binary parameters $x_{bu}^{\mathrm{UL}}(t) \in X^{\mathrm{UL}}$, $x_{bu}^{\mathrm{DL}}(t) \in X^{\mathrm{DL}} \forall b \in \mathscr{B}, u \in \mathscr{U}$. Let f_b be the allocated bandwidth, $v_b^{\mathrm{UL}}(t)$ and $v_b^{\mathrm{DL}}(t)$ be the vectors of the UL and DL successive ordering index of all the users in base station b, where $v_{bu}^{(.)}(t) \in \mathbf{v}_b^{(.)}(t)$ is the decoding order of user u when it is scheduled by base station b, and R_{bu}^{UL} and R_{bu}^{DL} be the UL and DL data rates between base station b and user u, then for a general network with open-access policy, the UL and DL service rates of user u can be expressed using the Shannon formula as:

$$\begin{aligned} r_u^{\mathrm{UL}}(t) &= \sum_{b\in\mathscr{B}} x_{bu}^{\mathrm{UL}}(t) R_{bu}^{\mathrm{UL}}(t), \\ &= \sum_{b\in\mathscr{B}} x_{bu}^{\mathrm{UL}}(t) f_b \log_2(1+\Gamma_{bu}^{\mathrm{UL}}(t)), \end{aligned} \tag{9.1}$$

$$\begin{aligned} r_u^{\mathrm{DL}}(t) &= \sum_{b\in\mathscr{B}} x_{bu}^{\mathrm{DL}}(t) R_u^{\mathrm{DL}}(t), \\ &= \sum_{b\in\mathscr{B}} x_{bu}^{\mathrm{DL}}(t) f_b \log_2(1+\Gamma_{bu}^{\mathrm{DL}}(t)). \end{aligned} \tag{9.2}$$

Assuming that the information to be transmitted is encoded based on a Gaussian distribution and a zero-mean additive white Gaussian noise with variance N_0, the UL and DL signal-to-interference-plus-noise ratios (SINRs) between base station b and user u at time instant t are given by:

$$\Gamma_{bu}^{\mathrm{UL}}(t) = \frac{p_u^{\mathrm{UL}}(t) h_{bu}(t)}{N_0 + I_b^{\mathrm{UL\text{-}UL}}(t) + I_b^{\mathrm{DL\text{-}UL}}(t) + I_{bu}^{\mathrm{NOMA\text{-}UL}}(t) + p_b^{\mathrm{DL}}(t)/\zeta}, \tag{9.3}$$

$$\Gamma_{bu}^{\mathrm{DL}}(t) = \frac{p_{bu}^{\mathrm{DL}}(t) h_{bu}(t)}{N_0 + I_u^{\mathrm{DL\text{-}DL}}(t) + I_u^{\mathrm{UL\text{-}DL}}(t) + I_{bu}^{\mathrm{NOMA\text{-}DL}}(t)}, \tag{9.4}$$

where p_u^{UL} is the UL transmit power of user u, and $p_b^{\mathrm{DL}} = \sum_{u\in\mathscr{U}} p_{bu}^{\mathrm{DL}}$ is the total DL transmit power of base station b, $h_{x,y}(t) = |g_{x,y}(t)|^2$ is the channel gain between the two nodes x and y, $g_{x,y}(t)$ is the propagation channel between nodes x and y, and the interference terms in (9.3) and (9.4) are expressed as follows[1]:

$$\begin{aligned} I_b^{\mathrm{UL\text{-}UL}}(t) &= \sum_{u'\in\mathscr{U}\setminus\{u\}} p_{u'}^{\mathrm{UL}}(t) h_{bu'}(t), \\ I_b^{\mathrm{DL\text{-}UL}}(t) &= \sum_{b'\in\mathscr{B}\setminus\{b\}} p_{b'}^{\mathrm{DL}}(t) h_{b'b}(t), \\ I_u^{\mathrm{DL\text{-}DL}}(t) &= \sum_{b'\in\mathscr{B}\setminus\{b\}} p_{b'}^{\mathrm{DL}}(t) h_{b'u}(t), \\ I_u^{\mathrm{UL\text{-}DL}}(t) &= \sum_{u'\in\mathscr{U}} p_{u'}^{\mathrm{UL}}(t) h_{u',u}(t), \end{aligned}$$

[1]Note that the term $I_{\mathrm{UL\text{-}DL}}$ includes the intra-cell interference, to account for the interference due to FD operation.

$$I_{bu}^{\text{NOMA-UL}}(t) = \sum_{\substack{u' \in \mathscr{U} \setminus \{u\} | x_{bu'}^{\text{UL}}=1, \\ v_{bu}^{\text{UL}}(t) < v_{bu'}^{\text{UL}}(t)}} p_{u'}^{\text{UL}}(t) h_{bu'}(t),$$

$$I_{bu}^{\text{NOMA-DL}}(t) = \sum_{\substack{u' \in \mathscr{U} \setminus \{u\} | x_{bu'}^{\text{DL}}=1, \\ v_{bu}^{\text{DL}}(t) < v_{bu'}^{\text{DL}}(t)}} p_{bu'}^{\text{DL}}(t) h_{bu}(t),$$

$p_b^{\text{DL}}(t)/\zeta$ is the leaked self-interference.

Remark 1 To guarantee a successful NOMA operation in the DL, a user u' should decode the data of user u with an SINR level $\Gamma_{bu}^{u'\text{DL}}(t)$ that is at least equal to the user u received SINR $\Gamma_{bu}^{\text{DL}}(t)$. Otherwise, the data rate of user u is higher than what user u' can decode. Accordingly, the inequality $\Gamma_{bu}^{u'\text{DL}}(t) \geq \Gamma_{bu}^{\text{DL}}(t)$ must hold, where:

$$\Gamma_{bu}^{u'\text{DL}}(t) = \frac{p_{bu}^{\text{DL}}(t) h_{bu'}(t)}{p_{bu'}^{\text{DL}}(t) h_{bu'}(t) + N_0 + I_{u'}^{\text{DL-DL}}(t) + I_{u'}^{\text{UL-DL}}(t) + I_{bu'}^{\text{NOMA-DL}}(t)}.$$

This condition is met if the following metric is greater or equal to zero:

$$Y_{uu'} = \mathbb{1}_{v_{bu}^{\text{DL}}(t) < v_{bu'}^{\text{DL}}(t)} x_{bu}^{\text{DL}} x_{bu'}^{\text{DL}} (\Gamma_{bu}^{u'\text{DL}} - \Gamma_{bu}^{\text{DL}}).$$

Note that the above condition is satisfied by default in the UL, since all users' signals are decoded in the base station's receiver.

A general FD-NOMA resource optimization problem can be defined to be optimizing user scheduling, NOMA decoding ordering, and UL and DL power allocations, i.e., the scheduling matrices $X^{\text{UL}} = [x_{bu}^{\text{UL}}]$ and $X^{\text{DL}} = [x_{bu}^{\text{DL}}]$, the decoding ordering vectors $\boldsymbol{v}_b^{\text{UL}}$ and $\boldsymbol{v}_b^{\text{DL}}$, and the power allocation vectors, $\boldsymbol{p}^{\text{UL}} = [p_1^{\text{UL}}, ..., p_U^{\text{UL}}]$ and $\boldsymbol{p}_b^{\text{DL}} = [p_{b1}^{\text{DL}}, ..., p_{bU}^{\text{DL}}]$. Note that the selection of the scheduling parameters essentially includes the problems of user association, UL/DL mode selection, and OMA/NOMA mode selection. Accordingly, a general optimization problem can be cast as follows:

$$\textbf{P1:} \quad \max_{\substack{X^{\text{UL}}, X^{\text{DL}}, \\ \{\boldsymbol{v}_b^{\text{UL}}\}, \{\boldsymbol{v}_b^{\text{DL}}\} \\ \boldsymbol{p}^{\text{UL}}, \{\boldsymbol{p}_b^{\text{DL}}\}}} \quad U\left(\{r_u^{\text{UL}}\}, \{r_u^{\text{DL}}\}\right) \tag{9.5a}$$

$$\text{subject to} \quad p_u^{\text{UL}} \leq P_{\max}^{\text{UL}}, \ \forall u \in \mathscr{U}, \tag{9.5b}$$

$$p_u^{\text{DL}} \leq P_{\max}^{\text{DL}}, \ \forall b \in \mathscr{B}, \tag{9.5c}$$

$$\sum_{b \in \mathscr{B}} x_{bu}^{\text{UL}} + x_{bu}^{\text{DL}} \leq 1, \ \forall u \in \mathscr{U}, \tag{9.5d}$$

$$\sum_{u \in \mathscr{U}} x_{bu}^{\text{UL}} \leq q^{\text{UL}}, \sum_{u \in \mathscr{U}} x_{bu}^{\text{DL}} \leq q^{\text{DL}} \ \forall b \in \mathscr{B}. \tag{9.5e}$$

$$Y_{uu'} \geq 0 \ \forall u, u' \in \mathscr{U}. \tag{9.5f}$$

Constraints (9.5b) (9.5c) limit the UL and DL transmit powers to their maximum values $P_{\max}^{\mathrm{UL}}$ and $P_{\max}^{\mathrm{DL}}$, respectively. Constraint (9.5d) restricts the connection of a user to one base station in either UL or DL. The number of users a base station can simultaneously serve in UL or DL-NOMA is limited to a quota of q^{UL} and q^{DL} users, respectively, by constraint (9.5e).[2] Constraint (9.5f) guarantees a successful SIC in the DL.

The utility in (9.5) should be selected to reflect the objective of the network optimization problem. The work in [35] considers a utility that maximizes the sum rate in a multi-carrier setting. In this regard, the resource allocation includes the subcarrier allocation. In [13], a weighted rate maximization utility was developed from a stochastic optimization problem, where the user weights are derived from the user traffic queue and virtual queue backlogs.[3]

9.2.4 *Optimization Tools*

Rate-based utility maximizations are often combinatorial and non-convex optimization problems that are computationally complex to solve optimally. In particular, the user scheduling problem is a combinatorial problem whose complexity will increase exponentially as the number of base stations and users increase.

In many cases, the optimization problem has to be solved dynamically. For example, if the weighted rate maximization based on user's queue state is considered as the objective, both the user pairing and power allocation will need to be dynamically adjusted. Efficient and local solutions tools are therefore needed. Matching is a powerful tool that can solve the user scheduling problem [26]. Matching theory is a framework that provides solutions for the combinatorial problem of matching members of two disjoint sets of players in which each player is interested to associate with one or more player the other set. Matching is performed on the basis of preference profiles defined by the players of each side, providing a low-complexity stable matching. Although matching does not necessarily guarantee finding the optimal solution, its suitability for dynamic systems and local implementations made it a popular solution to reduce the complexity of the combinatorial problems [18, 19, 31]. Matching can also be used with other game-theoretic tools, such as cooperative game theory [28, 29], to further solve user grouping problems. Matching should be performed assuming an initial feasible power allocation policy. Subsequently, the power allocation is performed for the selected matching setting. The decoupling of the user scheduling and power allocation problems simplifies both problems, since power allocation is performed to the reduced set of scheduled users [13].

[2] Note that, in theory, the number of users that can be served simultaneously using NOMA is unrestricted. However, we impose a quota to avoid high-complexity SIC in the receiver side if a high number of users are scheduled.

[3] Virtual queues result from applying the Lyapunov framework to convert the time-averaged constraints into virtual queues such that the constraints are met as the virtual queues are stabilized.

In addition to the complexity of the user scheduling problem, the power allocation problem is non-convex, due to having interference terms in the denominator of the SINR in the rate expression. Several approaches have been proposed in the literature to deal with the non-convexity of the problem. In [35], the global optimal solution of the joint problem of resource and power allocation is solved using monotonic optimization. The solutions are, however, with high computational complexity. The authors provide a lower complexity solution to solve the problem using successive convex approximation (SCA), which is shown to achieve a close to global optimal solution. SCA and similar tools to convexify the non-convex terms in the optimization problem are used in several works [13, 15, 24, 35, 36]. In this case, the convexified problem is solved iteratively using convex optimization tools until some convergence criterion is met.

The NOMA decoding order significantly affects the user performance in both UL and DL. In DL-NOMA, decoding users with the lower channel strength first is optimal in the single-cell scenario [11]. In UL-NOMA, the opposite decoding order based on the user channel strength is shown in [6] to result in a gain over OMA based on users channel gain disparity. The decoding ordering in a multi-cell scenario is a challenging task, especially in the DL where different users might experience different inter-cell interference levels.

9.3 State of the Art in FD and NOMA Resource Optimization

This section surveys the recent works in the problem of wireless resource optimization in FD and NOMA networks. First, with the promises of doubling the link throughput, the study of the when and how the potential gains can be achieved in a network level is surveyed. Following that, we overview the studies of resource optimization in NOMA networks in UL and DL. Finally, we highlight the works that combine both FD and NOMA schemes.

9.3.1 FD Resource Optimization

In [16], a hybrid HD/FD scheduler for a single-cell network is proposed. The scheduler assigns FD resources only when it is advantageous over HD resource assignment. The joint problem of subcarrier and power allocation are optimized in [36] using an SCA algorithm. An auction-based algorithm to pair UL and DL users in a single-cell IBFD network is proposed in [33]. Subcarrier and power allocation is optimized using a heuristic algorithm with polynomial complexity for a single-cell IBFD network in [37]. The study also considers the case of imperfect self-interference cancellation on the performance of FD as compared to HD. It concludes that a higher number of users can be served using FD as the cancellation capability increases. In [17], the authors extended the work in [16] to multi-cell networks. To reduce the complexity of a centralized solution, a distributed resource allocation scheme is developed where each base station selects locally which user to serve, and then, it coordinates with the neighbouring base stations to coordinate their transmission powers such that the inter-cell interference is minimized. The study shows that FD can achieve up to double the throughput of HD in indoor scenarios and 65% throughput gain in outdoor scenarios. The work in [34] also considered the FD resource and power allocation in multi-cell FD networks but with frequency reuse allowed only once among different cells. Matching theory is leveraged in [7] to develop a resource allocation algorithm. A matching algorithm is proposed to assign subcarriers to UL and DL user pairs. In [5], the user scheduling in FD ultra-dense-networks is optimized using different schemes with and without power optimization. The scheduling is carried out locally assuming no knowledge of the inter-cell interference. Table 9.1-a summarizes the contributions on the FD resource allocation.

Remark 2 A common assumption in FD scheduling is that the base station knows the individual channel between its own users. This assumption is necessary to select FD pairs that do not significantly interfere on each other [16]. The assumption should be practical as FD is expected to be feasible in low-power small-cell networks where a low number of users are served by each base station.

9.3.2 NOMA Resource Optimization

Recently, several works have looked into the resource optimization in UL and/or DL-NOMA networks. Here, we shed light on the papers focusing on the scheduling and power optimization in NOMA networks. The authors in [8] propose a many-to-many algorithm to assign subcarriers to users in a single DL-NOMA network. Many-to-many matching is used in [9] for a multi-cell DL-NOMA scenario. Matching is also used for DL-NOMA resource allocation in [15] with a focus on energy efficiency. The power allocation problems are convexified and solved using difference of convex functions (DC) programming. DC programming is also used in [24] to optimize the power allocation in OFDM-based DL-NOMA networks, whereas a greedy algorithm is proposed for the user selection problem. A greedy approach is also used for the user scheduling in a multiple-input multiple-output (MIMO) DL-NOMA network in [32], and a minimum mean squared error (MMSE)-based power allocation is considered.

Several works have looked into NOMA in the UL direction. The performance of NOMA in the UL is investigated in [2] using an iterative channel allocation algorithm. In [30], the problem of user pairing in UL-NOMA for users with single and multi-antennas is optimized. User grouping and power optimization in UL-NOMA is studied in [6] where the impact of user ordering and imperfect SIC is

Table 9.1 Summary of existing literature in FD- and NOMA-based resource allocation problems

(a) FD

References	Network scenario	Implementation	FD scheduling	Power allocation
[16]	Single-cell	Local	HD/FD mode selection	×
[36]	Single-cell	Local	Joint subchannel and power allocation	✓
[33]	Single-cell	Local	FD user pairing and channel allocation	✓
[37]	Single-cell	Local	OFDMA channel allocation	✓
[17]	Multi-cell	Local	Suboptimal HD/FD user selection	✓
[34]	Multi-cell (subcarrier is reused once)	Central	Mode selection and subcarrier allocation	✓
[7]	Multi-cell	Central	Matching subcarriers to user pairs	✓
[5]	Multi-cell	Local	Local scheduling	✓

(b) NOMA

References	Link scenario	Network scenario	Scheduling and power allocation scheme
[8]	DL	Single-cell	Matching algorithm
[15]	DL	Single-cell	Subchannel assignment and power allocation
[24]	DL	Single-cell	User selection and power optimization
[32]	DL	Single-cell	User pairing and power allocation
[9]	DL	Multi-cell	Matching algorithm and power allocation
[30]	UL	Single-cell	User pairing for multi-antenna systems
[2]	UL	Single-cell	Iterative subcarrier and power allocation
[6]	UL	Heterogeneous	User clustering and power allocation
[4]	UL + DL	Multi-cell	Optimal power allocation for a limited number of users

(c) FD-NOMA

References	Network scenario	Scheduling and power allocation scheme
[35]	Single-cell	Joint subchannel and powerallocation
[13]	Multi-cell	Joint user scheduling and power allocation

investigated. In [4], the authors consider a multi-cell UL and DL-NOMA system where a user grouping and power optimization scheme are proposed. The optimal power allocation is derived for a single macro-cell and a limited number of users.

9.3.3 FD-NOMA Resource Optimization

Two recent studies have looked into the incorporation of FD into NOMA networks and the impact on the scheduling and power allocation. In [35], the authors proposed an FD-NOMA approach in which users can be scheduled simultaneously in UL and DL in the same time–frequency resource and NOMA can be used in both directions. SIC is used in UL and DL to decode the messages of different users, whereas the inter-link interference due to FD is treated as noise. The joint subcarrier and power optimization problem is formulated, and the global optimal solution is found using monotonic optimization. Due to the high complexity of finding the global optimal solution, a low-complexity solution based on SCA is found. The results have shown that FD-NOMA improves the spectral efficiency as compared to HD-NOMA. Moreover, the effect of imperfect SIC is shown to impact the performance of the FD scheme.

In [13], FD-NOMA is investigated in a dynamic multi-cell scenario where a stochastic optimization problem based on the Lyapunov framework is considered. The benefits of operating in HD or FD, as well as in OMA or NOMA modes, depending on traffic conditions, network density, and self-interference cancellation capabilities are investigated. The optimization problem is decomposed into two subproblems that are solved independently per small-cell base station. User association and mode selection are formulated as a many-to-one matching problem. A distributed matching algorithm aided by an inter-cell interference learning mechanism is proposed which is shown to converge to a pairwise stable matching. The matching algorithm allows small-cells to select between HD and FD and to operate either in OMA or NOMA schemes to serve their users. Subsequently, the UL/DL power optimization problem is formulated as a sequence of convex problems, and an iterative algorithm to allocate the optimal power levels for the matched users and their base stations is proposed. It was shown that using matching theory, the network can dynamically select when to operate in HD or FD and when to use OMA or NOMA to serve different users, which yields significant gains in UL and DL user throughput and packet throughput, as compared to HD-NOMA, FD-OMA, and HD-OMA schemes.

9.4 Numerical Results

In this section, we present some numerical results to assess the performance of the queue-aware FD and NOMA resource optimization. We consider a continuous utility function of time-averaged UL and DL service rates. The problem can be decomposed using the Lyapunov framework into an instantaneous weighted rate maximization in which the user weights are their queue backlogs. The network consists of small-cell base stations with a varying self-interference cancellation capability, serving multiple users in an open-access manner. Scheduling can be in HD or FD modes, and in HD mode, users can be scheduled in OMA or NOMA. To cancel the resulting multi-user interference, base stations or users operating in NOMA can perform SIC at the receiver side. The decoding ordering is assumed to be done in a descending order of channel strength in DL-NOMA, and an ascending order of channel strength in UL-NOMA. We assume that the user's mean packet arrival rate and mean packet size follow Poisson and exponential distributions, respectively. To satisfy queue stability requirements, base stations need to ensure that user's traffic queues are mean rate stable. Equivalently, constraints are imposed to ensure that the average service rate is higher or equal to the average arrival rate. The resource optimization problem consists of the scheduling problem (which includes the mode selection) and the power allocation problem. To reduce the complexity of the combinatorial scheduling problem, a many-to-one matching algorithm based on the *deferred acceptance (DA)* matching [13] is considered to dynamically schedule one or more users to each base station at each time instant. Preference profiles for users and base stations are selected as to maximize their individual weighted rates. The matching algorithm can be performed locally at each base station, which significantly reduce the amount of signalling exchange. After the matching is performed, each base station coordinates with its neighbours to optimize the allocated power, such that a feasible policy is achieved and inter-cell interference is minimized. The multi-cell power optimization is non-convex. Hence, the DC programming is used to convexify the problem, which is guaranteed to converge to a local optimal solution. System level simulations are carried out to show the gains brought by FD and NOMA, as well as to investigate the impact of queue stability constraints on the network performance. Simulation parameters are presented in Table 9.2. For the sake of comparison, the following schemes are considered in the simulation:

1. *HD-OMA scheme*: users are associated to the nearest base station and are allocated orthogonal resources for UL and DL. Requests are served using a round robin (RR) scheduler.
2. *HD-NOMA only scheme*: users are associated to the nearest base station, and RR scheduler is used to serve UL and DL queues. If there are multiple users in a scheduling queue, they are ranked according to their channel gains and are served using NOMA if the ratio between their channel gains is at least 2; otherwise, OMA is used. Power is allocated to NOMA users based on their channel ranking, in a uniform descending order for UL-NOMA and a uniform ascending order for DL-NOMA. Base stations operate either in UL or in DL depending on the queue length on each link direction.
3. *FD-OMA scheme*: users are associated to the nearest base station, and a pair of users is served in FD mode if the channel gain between them is greater than a certain threshold; otherwise, users are served in HD mode using RR scheduler.
4. *Uncoordinated scheme*: in this scheme, users can be served in either HD or FD and in OMA or NOMA modes. The many-to-one matching algorithm is used for mode selection and user scheduling, with the user queues taken into account in the weighted rate maximization. Power is assumed to be fixed for OMA and is similar to that of scheme 2 for NOMA. No inter-cell interference coordination is considered.

Table 9.2 Simulation parameters

Parameter	Value
System bandwidth	10 MHz
Duplex modes	TDD HD/ FD
Multiplexing mode	OMA/NOMA
Subframe duration	1 ms
Network size	500×500 m^2
Number of base stations	10
Avg. number of users per base station	10
Small-cell radius	40 m
Max. base station transmit power	22 dBm
Max. user transmit power	20 dBm
Path loss model	Multi-cell pico scenario [1]
Shadowing standard deviation	4 dB
Penetration loss	0 dB
Self-interference cancellation	110 dB
Packet arrival rate per user	10 packet/s
Max. quota of NOMA users q^{UL}, q^{DL}	5

5. *Proposed scheme*: In this scheme, the many-to-one matching algorithm is used for mode selection and user scheduling as in scheme 4. In addition, inter-cell interference coordination is considered by optimizing the UL and DL power allocation using DC programming.

We begin by evaluating the performance of the proposed scheme under different traffic intensity conditions. Traffic intensity is varied by changing the mean packet size between 50 and 400 kb. In Fig. 9.2, we show the impact of traffic intensity on the normalized UL and DL user throughput. The normalized user throughput is defined as the user service rate divided by its data arrival rate. Figure 9.2a shows that in the UL, all schemes but the proposed one achieve lower UL throughput as compared to the HD-OMA scheme. The performance drop is due to the DL-to-UL interference that has a significant impact on the uncoordinated schemes due to the higher transmitting power of base stations and the lower path loss between the base station and user. The proposed scheme outperforms the different schemes by mitigating the DL-to-UL interference through power optimization.

In Fig. 9.2b, we can see that, in the DL case, the effect of UL-to-DL interference is less significant, as it can be avoided within each cell through the pairing process. By leveraging both matching and the UL and DL power optimization, the proposed scheme outperforms the baseline schemes, with UL and DL gains of 10 and 20% over the HD-OMA scheme. The figure also shows that the coordination gain (over the uncoordinated scheme) is even more evident in the UL as in the DL due to the dominance of the DL-to-UL interference in the uncoordinated scenario.

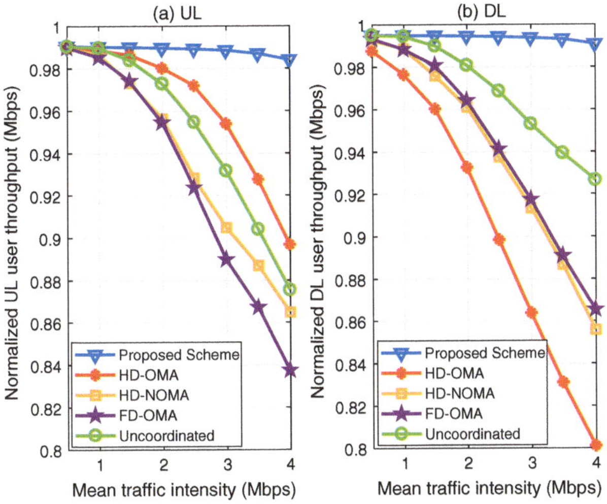

Fig. 9.2 Normalized **a** UL and **b** DL user throughput performance for different schemes as the user traffic intensity increases, for a network of ten base station and an average of ten users per base station

Fig. 9.3 Average user queue length performance for different schemes as the user traffic intensity increases, for a user of ten base station and an average of ten users per base station

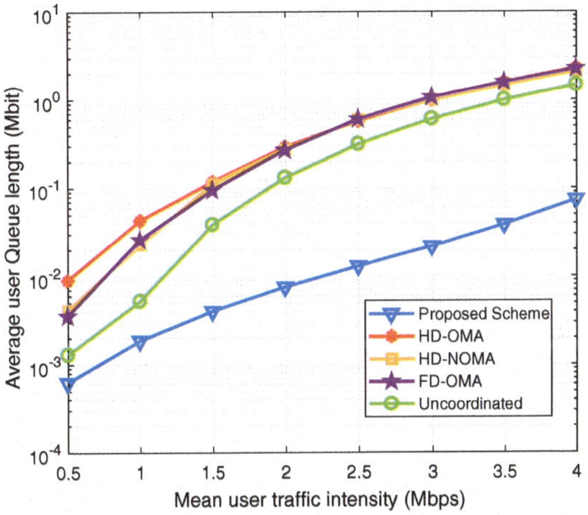

Next, we investigate the queue behaviour of the different schemes as the traffic intensity varies. In Fig. 9.3, the average queue length is shown as function of the traffic intensity. Figure 9.3 shows that the average queue length grows as the traffic intensifies. In low traffic conditions, FD-OMA and HD-NOMA have smaller queue lengths as compared to the HD-OMA scheme. Also, the uncoordinated FD-NOMA scheme maintains a smaller queue length, since coordination is not crucial in low traffic intensity conditions. As the traffic intensity increases, we can see that the queue length grows significantly for the uncoordinated schemes. The proposed scheme

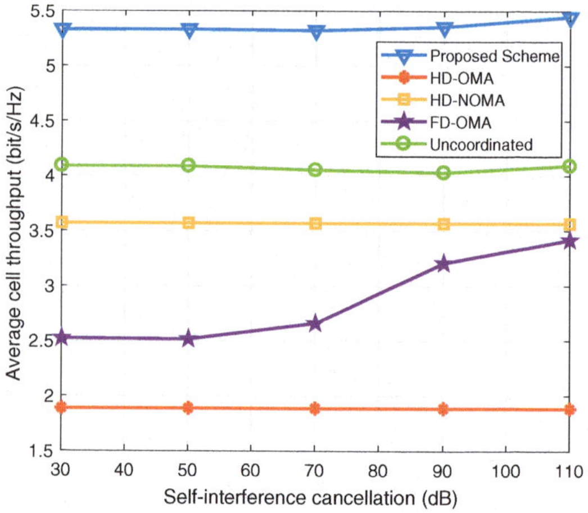

Fig. 9.4 Average cell throughput (in bit/s/Hz) performance for different schemes as the base station self-interference cancellation capability varies, for a network of ten base stations, an average of ten users per base station, and a mean traffic intensity rate of 3 Mbps per user

maintains small queue lengths as it seeks to stabilize the user queues through the weighted rate maximization.

Finally, we show the impact of the self-interference cancellation capability on the performance of the FD schemes. Figure 9.4 compares the average cell throughput (in bit/s/Hz) of the different schemes as the self-interference cancellation capability varies from 30 to 110 dB, which is the highest reported value [20]. As shown in the figure, the throughput of the FD schemes degrades with lower self-interference cancellation levels due to the interference leakage on the UL signal. It is also shown that only a slight degradation in the throughput of the proposed FD-NOMA is observed. As the proposed scheme optimizes the mode selection between HD/FD and OMA/NOMA, it can select more frequently the UL-NOMA instead of FD to serve UL users, such that it avoids high interference from the base station's DL. This shows that enabling both FD and NOMA has the potential to enable higher network spectral efficiency in different network conditions.

9.5 Conclusions and Open Problems

This chapter has provided an overview of the topic of full-duplex (FD) non-orthogonal multiple access (NOMA) from a network optimization point of view. Different challenges on the optimization of FD-NOMA networks are discussed. It has highlighted the particular importance of the user pairing and scheduling in both FD and NOMA networks. Several directions are still open for research. As was mentioned throughout the chapter, the decoding ordering is a key factor in the performance of the NOMA systems. Some studies are carried out on the optimal ordering of users, most of which are assuming single-cell operation and focus on the optimal solution from the point of sum rate maximization. Finding the optimal decoding ordering is a challenging task in a multi-cell scenario and is even challenging in FD networks in which both intra-cell and inter-cell interference impact the network performance. The objective of the decoding order optimization should also take into account the notion of fairness between the users with different channel and queue state conditions. Moreover, enabling NOMA for emerging 5G systems, such as vehicle-to-everything (V2X) networks [10, 25] and networks with unmanned aerial vehicles [22, 23, 27], poses a wide range of open problems. Finally, looking into different objectives beyond the average rate maximization problems is necessary, especially in the context of ultra-reliable and low latency communication (URLLC), which brings further challenges to the system design.

Acknowledgements This research was supported in part by the Academy of Finland CARMA Project, in part by the U.S. National Science Foundation under Grant CNS-1513697, and Grant CNS-1617896, and in part by the ERC Starting Grant MORE (Advanced Mathematical Tools for Complex Network Engineering) under Grant 305123.

References

1. 3GPP, Further enhancements to LTE TDD for DL-UL interference management and traffic adaptation. TR 36.828, 3rd Generation Partnership Project (3GPP) (2012), http://www.3gpp.org/DynaReport/36828.htm
2. M. Al-Imari, P. Xiao, M.A. Imran, R. Tafazolli, Uplink non-orthogonal multiple access for 5G wireless networks, in *2014 11th International Symposium on Wireless Communications Systems (ISWCS)* (2014), pp. 781–785
3. K.S. Ali, H. Elsawy, A. Chaaban, M.S. Alouini, Non-orthogonal multiple access for large-scale 5G networks: interference aware design. IEEE Access **5**, 21204–21216 (2017)
4. M.S. Ali, H. Tabassum, E. Hossain, Dynamic user clustering and power allocation for uplink and downlink non-orthogonal multiple access (noma) systems. IEEE Access **4**, 6325–6343 (2016)
5. I. Atzeni, M. Kountouris, G.C. Alexandropoulos, Performance evaluation of user scheduling for full-duplex small cells in ultra-dense networks, in *22th European Wireless Conference on European Wireless 2016* (2016), pp. 1–6
6. A. Celik, R.M. Radaydeh, F.S. Al-Qahtani, A.H.A. El-Malek, M.S. Alouini, Resource allocation and cluster formation for imperfect NOMA in DL/UL decoupled hetnets, in *IEEE Global Communications Conference (GLOBECOM)* (2017), pp. 1–7
7. B. Di, S. Bayat, L. Song, Y. Li, Radio resource allocation for full-duplex OFDMA networks using matching theory, in *2014 IEEE Conference on Computer Communications Workshops (INFOCOM WKSHPS)* (2014), pp. 197–198
8. Di, B., Bayat, S., Song, L., Li, Y.: Radio resource allocation for downlink non-orthogonal multiple access (NOMA) networks using matching theory. In: 2015 IEEE Global Communications Conference (GLOBECOM), pp. 1–6 (2015)
9. B. Di, L. Song, Y. Li, Sub-channel assignment, power allocation, and user scheduling for non-orthogonal multiple access networks. IEEE Trans. Wirel. Commun. **15**(11), 7686–7698 (2016)
10. B. Di, L. Song, Y. Li, G.Y. Li, Noma-based low-latency and high-reliable broadcast communications for 5G v2x services, in *2017 IEEE Global Communications Conference on GLOBECOM*, pp. 1–6 (2017)

11. Z. Ding, X. Lei, G.K. Karagiannidis, R. Schober, J. Yuan, V.K. Bhargava, A survey on non-orthogonal multiple access for 5G networks: research challenges and future trends. IEEE J. Sel. Areas Commun. **35**(10), 2181–2195 (2017)
12. M. Duarte, A. Sabharwal, Full-duplex wireless communications using off-the-shelf radios: feasibility and first results, in *2010 Conference Record of the Forty Fourth Asilomar Conference on Signals, Systems and Computers* (2010), pp. 1558–1562
13. M.S. Elbamby, M. Bennis, W. Saad, M. Debbah, M. Latva-aho, Resource optimization and power allocation in in-band full duplex-enabled non-orthogonal multiple access networks. IEEE J. Sel. Areas Commun. **35**(12), 2860–2873 (2017)
14. M.S. ElBamby, M. Bennis, W. Saad, M. Latva-aho, Dynamic uplink-downlink optimization in TDD-based small cell networks, in *2014 11th International Symposium on Wireless Communications Systems (ISWCS)* (2014), pp. 939–944
15. F. Fang, H. Zhang, J. Cheng, V.C.M. Leung, Energy-efficient resource allocation for downlink non-orthogonal multiple access network. IEEE Trans. Commun. **64**(9), 3722–3732 (2016)
16. Goyal, S., Liu, P., Panwar, S., DiFazio, R.A., Yang, R., Li, J., Bala, E.: Improving small cell capacity with common-carrier full duplex radios, in *2014 IEEE International Conference on Communications (ICC)* (2014), pp. 4987–4993
17. S. Goyal, P. Liu, S.S. Panwar, User selection and power allocation in full-duplex multicell networks. IEEE Trans. Veh. Technol. **66**(3), 2408–2422 (2017)
18. Y. Gu, W. Saad, M. Bennis, M. Debbah, Z. Han, Matching theory for future wireless networks: fundamentals and applications **53**(5), 52–59 (2015)
19. Y. Gu, W. Saad, M. Bennis, M. Debbah, Z. Han, Matching theory for future wireless networks: fundamentals and applications. IEEE Commun. Mag. **53**(5), 52–59 (2015)
20. M. Jain, J.I. Choi, T. Kim, D. Bharadia, S. Seth, K. Srinivasan, P. Levis, S. Katti, P. Sinha, Practical, real-time, full duplex wireless, in *Proceedings of the 17th Annual International Conference on Mobile Computing and Networking, MobiCom'11*, New York, NY, USA (2011), pp. 301–312, http://doi.acm.org/10.1145/2030613.2030647
21. R. Li, Y. Chen, G.Y. Li, G. Liu, Full-duplex cellular networks. IEEE Commun. Mag. **55**(4), 184–191 (2017)
22. M. Mozaffari, W. Saad, M. Bennis, M. Debbah, Unmanned aerial vehicle with underlaid device-to-device communications: performance and tradeoffs. IEEE Trans. Wirel. Commun. **15**(6), 3949–3963 (2016)
23. M. Mozaffari, W. Saad, M. Bennis, M. Debbah, Wireless communication using unmanned aerial vehicles (UAVS): optimal transport theory for hover time optimization. IEEE Trans. Wirel. Commun. **16**(12), 8052–8066 (2017)
24. P. Parida, S.S. Das, Power allocation in OFDM based NOMA systems: a DC programming approach, in *2014 IEEE Globecom Workshops (GC Wkshps)* (2014), pp. 1026–1031
25. L.P. Qian, Y. Wu, H. Zhou, X. Shen, Dynamic cell association for non-orthogonal multiple-access v2s networks. IEEE J. Sel. Areas Commun. **35**(10), 2342–2356 (2017)
26. A. Roth, M. Sotomayor, *Two-Sided Matching: A Study in Game-theoretic Modeling and Analysis* (Cambridge University Press, Cambridge, 1992)
27. N. Rupasinghe, Y. Yapici, I. Guvenc, Y. Kakishima, Non-Orthogonal Multiple Access for mmWave Drones with Multi-Antenna Transmission (2017), http://arxiv.org/abs/1711.10050
28. W. Saad, Z. Han, M. Debbah, A. Hjorungnes, A distributed coalition formation framework for fair user cooperation in wireless networks. IEEE Trans. Wirel. Commun. **8**(9), 4580–4593 (2009)
29. W. Saad, Z. Han, M. Debbah, A. Hjorungnes, T. Basar, Coalitional game theory for communication networks. IEEE Signal Process. Mag. **26**(5), 77–97 (2009)
30. M.A. Sedaghat, R.R. Müller, On user pairing in NOMA uplink (2017), http://arxiv.org/abs/1707.01846
31. O. Semiari, W. Saad, M. Bennis, Joint millimeter wave and microwave resources allocation in cellular networks with dual-mode base stations. IEEE Trans. Wirel. Commun. **16**(7), 4802–4816 (2017)

32. L. Shi, B. Li, H. Chen, Pairing and power allocation for downlink nonorthogonal multiple access systems. IEEE Trans. Veh. Technol. **66**(11), 10084–10091 (2017)
33. J.M.B. da Silva, Y. Xu, G. Fodor, C. Fischione, Distributed spectral efficiency maximization in full-duplex cellular networks, in *2016 IEEE International Conference on Communications Workshops (ICC)* (2016), pp. 80–86
34. R. Sultan, L. Song, K.G. Seddik, Y. Li, Z. Han, Mode selection, user pairing, subcarrier allocation and power control in full-duplex OFDMA HetNets, in *2015 IEEE International Conference on Communication Workshop (ICCW)* (2015), pp. 210–215
35. Y. Sun, D.W.K. Ng, Z. Ding, R. Schober, Optimal joint power and subcarrier allocation for full-duplex multicarrier non-orthogonal multiple access systems. IEEE Trans. Commun. **65**(3), 1077–1091 (2017)
36. Y. Sun, D.W.K. Ng, R. Schober, Joint power and subcarrier allocation for multicarrier full-duplex systems, in *2017 IEEE International Conference on Acoustics, Speech and Signal Processing (ICASSP)* (2017), pp. 6548–6552
37. P. Tehrani, F. Lahouti, M. Zorzi, Resource allocation in OFDMA networks with half-duplex and imperfect full-duplex users, in *2016 IEEE International Conference on Communications (ICC)* (2016), pp. 1–6
38. T.K. Vu, M. Bennis, S. Samarakoon, M. Debbah, M. Latva-aho, Joint load balancing and interference mitigation in 5G heterogeneous networks. IEEE Trans. Wirel. Commun. **16**(9), 6032–6046 (2017)

Chapter 10
Heterogeneous NOMA with Energy Cooperation

Bingyu Xu, Yue Chen and Yuanwei Liu

10.1 Background

10.1.1 Resource Allocation in NOMA HetNets

Heterogeneous networks (HetNets) are one of the "big three" technologies which can achieve the system performance requirement of fifth-generation (5G) networks. In conventional one-tier homogeneous networks, there is only one macro-BS in each cell. In HetNets, which are also called multi-tier networks, small BSs such as pico BSs and micro-BSs which have lower transmit powers and smaller coverages are deployed within the coverage of the macrocell. These smaller BSs obtain higher spectrum efficiency and throughput through reusing the spectrum. They are typically deployed in the hotpot zones (e.g., areas with high traffic demand) and dead zones (e.g., cell edge and areas cannot receive signal).

Recently, NOMA-enabled HetNets have attracted significant research interests. It is worth introducing NOMA in HetNets due to the following key advantages: (1) in HetNets, users are closer to their associated BSs, which can reduce the interference between users and increase the accuracy of successive interference cancelation (SIC) in NOMA systems. (2) NOMA is capable to deal with the fairness issue among users, which is one of the main challenges of HetNets.

Although intensive research contributions have been conducted on the design of NOMA transmission, resource allocation in NOMA-enabled HetNets remains an open problem. Considering the fact that user association (UA) determines that

B. Xu (✉) · Y. Chen · Y. Liu
Queen Mary University of London, Mile End Road, London E1 4NS, UK
e-mail: bingyu.xu@qmul.ac.uk

Y. Chen
e-mail: yue.chen@qmul.ac.uk

Y. Liu
e-mail: yuanwei.liu@qmul.ac.uk

M. Vaezi et al. (eds.), *Multiple Access Techniques for 5G Wireless Networks and Beyond*, https://doi.org/10.1007/978-3-319-92090-0_10

users should be connected to a specified BS to form a user group for superposition transmission [1], the number of users associated with a BS has a significant effect on the spectral and energy efficiency in NOMA multi-cell networks [2]. In addition, power control is of great importance in such networks, since the intra-cell interference and inter-cell interference need to be coordinated. Otherwise, the performance of cell edge users will be significantly degraded [3].

A cooperative NOMA scheme in HetNets was proposed in [4] where each user was served by a macro-BS and a pico BS simultaneously. Zhao et al. [5] and [6] focused on downlink joint spectrum and power allocation problem which aims at maximizing the sum rate of small cell users. The macro-BS used conventional OMA protocol while small cells use NOMA protocol. Meanwhile, many works combined the NOMA enabled HetNets with other 5G techniques, such as massive MIMO and cloud radio access. A user association scheme was proposed in [7] for massive MIMO enabled NOMA HetNets to maximize the biased average received power of users. Meanwhile, the downlink power allocation problem in NOMA HetNets with cloud radio access network was investigated in [8]. It analyzed the energy efficiency of the network by finding the optimal number of BSs.

10.1.2 Energy Cooperation

In 5G mobile systems, one main goal is to improve energy efficiency significantly compared to today's networks [9]. Indeed, such large level of connectivity will inevitably give rise to an unprecedented surge in global energy consumption, especially in networks with HetNets, which has a vast number of small cells. The latest analysis shows that the energy demand for information and communications technology already accounts for almost 10% of the world's total energy consumption [10]. In addition, critical environmental issues such as high carbon emissions are a big concern. Hence, "greener" solutions need to be developed to enhance the network energy efficiency. Among the emerging technologies, energy harvesting is regarded as one viable solution [11]. By allowing base stations (BSs) to harvest energy from renewable energy sources such as solar and wind, the conventional grid energy consumption of wireless networks can be greatly reduced.

Although renewable energy harvesting is a viable solution for cutting the conventional grid energy consumption in cellular networks, there are many challenges for integrating energy harvesting capabilities into BSs [12]. In renewable energy harvesting-enabled networks, BSs harvest variable amounts of renewable energy, due to the fluctuating nature of renewable energy sources. When the renewable energy harvested by BSs is insufficient to meet their load conditions, some user equipments (UEs) have to be offloaded to distant BSs with abundant energy and may suffer more from signal degradation. Moreover, some BSs may always have excessive harvested energy (e.g., because of more favorable weather conditions) that will eventually be wasted. Since the deployment of BSs with large energy storage capabilities brings high expenditure of networks [13], the energy fluctuation problem cannot be solely solved by using storage.

To improve the utilization rate of renewable energy, with the development of smart grid, the definition of energy cooperation is proposed which allows energy transferred through grids [14]. By this way, energy can be shared between BSs with acceptable energy loss during the energy transmission process.

Energy cooperation in the point-to-point transmission scenario has been studied in [14–16]. In [15], one-way energy transfer in the Gaussian two-way channel and multiple access channel were considered respectively. This line of work was extended to the two-way case in [16]. The implementation of energy cooperation in multiple access channels and multiple access relay networks were studied in [17] and [18], respectively. In [19], an energy cooperation scheme in cognitive radio networks was proposed to improve both the spectral and energy efficiency.

Recently, the potential of energy cooperation in renewable energy-enabled cellular networks has been explored, and various energy-cooperation optimization problems have been studied. In [13], a joint energy and spectrum allocation problem between two neighboring cellular systems was formulated, which aimed to minimize the cost of energy and bandwidth, and the problem was solved by convex optimization. The power control problem between two BSs was considered in [20] under the assumption that the harvested energy, and the energy demand of BSs were deterministic. In [21], the energy cost of cellular networks was minimized with the assumption that BSs traded energy via the smart grid with different prices. The work of [22] aimed to maximize the sum rate through optimizing the transmit powers of BSs in a coordinated multipoint cluster. In [23], the energy trading problem was formulated to minimize the average cost of energy exchange between BSs, and a dynamic algorithm was proposed based on the Lyapunov optimization technique, which did not require the statistical knowledge of the channel and energy.

Due to the energy-saving feature of energy cooperation, it is worth to deploy it in NOMA networks. Meanwhile, most of existing NOMA works such as [24–29] only consider the case consisting of one BS and a group of users. Besides, the practical multi-cell scenario is evaluated in this chapter considering the effect of inter-cell interference which has a substantial impact on the system performance. In this chapter, we study the power control and UA problem in energy-cooperation-enabled two-tier NOMA HetNets. The content of this chapter is mainly based on our previous published journal [30].

10.2 Network Model and Problem Formulation

In this section, the system model for energy cooperation in two-tier NOMA HetNets is presented, and the corresponding joint UA and power control problem is formulated.

10.2.1 Downlink NOMA Transmission

As shown in Fig. 10.1, a two-tier energy-cooperation-enabled HetNet consisting of one macro BS (MBS) and M pico BSs (PBSs) is considered, where the NOMA-based downlink transmission is utilized, and all BSs are assumed to share the same frequency band. In such a network, BSs are powered by both the conventional power grid and renewable energy sources, and energy can be shared between BSs through the smart grid. Let $m \in \{1, 2, 3, ..., M + 1\}$ be the mth BS, in which $m = 1$ denotes the MBS, and the other values denote PBSs. There are N randomly located user equipments (UEs) in this network, and each UE is associated with only one BS. All BSs and UEs are single-antenna nodes. It is assumed that the global perfect channel state information (CSI) is available. Let $j \in \{1, 2, 3, ..., N\}$ index the jth UE. According to the NOMA scheme [31, 32], the superimposed signal transmitted by the BS m is $s_m = \sum_{j=1}^{N} x_{jm}\sqrt{P_{jm}}s_{jm}$ with $\mathrm{E}\left[s_{jm}\left(s_{jm}\right)^H\right]=1, \forall m, j$, where $x_{jm} \in \{0, 1\}$ is the binary UA indicator, i.e., $x_{jm} = 1$ when the jth UE is associated with the mth BS and otherwise it is zero, s_{jm} is the jth user-stream and P_{jm} is the corresponding allocated transmit power. When the jth UE is associated with the mth BS, its received signal can be expressed as

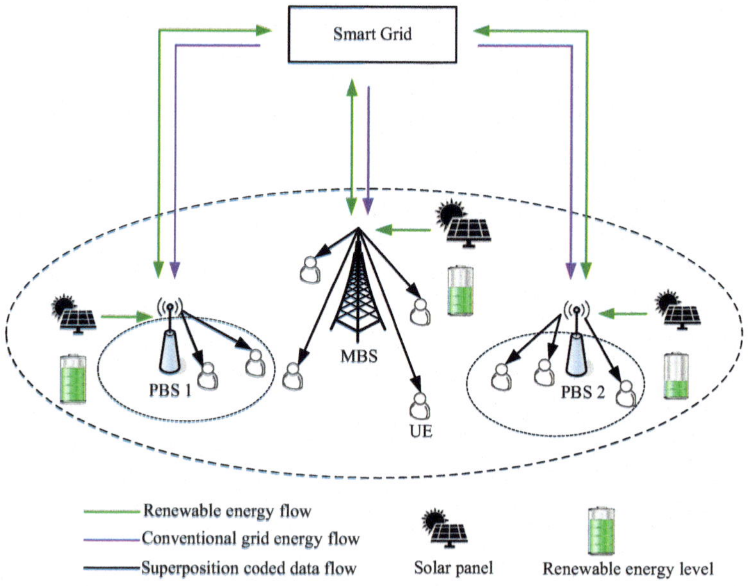

Fig. 10.1 An example of an energy-cooperation enabled two-tier NOMA HetNet powered by both solar panels and the conventional grid

$$y_{jm} = h_{jm}\sqrt{P_{jm}}s_{jm} + \underbrace{h_{jm}\sum_{j'=1,j'\neq j}^{N} x_{j'm}\sqrt{P_{j'm}}s_{j'm}}_{\text{Intra-cell interference}} + \underbrace{\sum_{m'=1,m'\neq m}^{M+1} h_{jm}^{m'}\left(\sum_{j'=1}^{N} x_{j'm'}\sqrt{P_{j'm'}}s_{j'm'}\right)}_{\text{Inter-cell interference}} + \varpi_m, \tag{10.1}$$

where $x_{j'm}, x_{j'm'} \in \{0, 1\}$, h_{jm} is the channel coefficient from the associated BS m, $h_{jm}^{m'}$ is the interfering channel coefficient from the BS m', and ϖ_m is the additive white Gaussian noise. The power density of ϖ_m is σ^2. In NOMA systems, SIC is employed at UEs, to cancel the intra-cell interference from the stronger UEs' data signals. Without loss of generality, assuming that there are k_m ($k_m \leq N$) UEs constituting a group that is served by the mth BS at the same time and frequency band, the corresponding channel to inter-cell interference plus noise ratios (CINRs) are ordered as

$$\frac{|h_{1m}|^2}{I_{1m}^{(2)} + \sigma^2} \geq \cdots \geq \frac{|h_{jm}|^2}{I_{jm}^{(2)} + \sigma^2} \geq \cdots \geq \frac{|h_{k_m m}|^2}{I_{k_m m}^{(2)} + \sigma^2}, \tag{10.2}$$

where $I_{jm}^{(2)}$ is the inter-cell interference power at the jth UE and σ^2 is the noise power. Based on the principle of multi-cell NOMA [31], the power allocation of the users' data signals in the mth cell needs to satisfy

$$0 < P_{1m} \leq \cdots \leq P_{jm} \leq \cdots \leq P_{k_m m}, \sum_{j=1}^{k_m} P_{jm} = P_m, \tag{10.3}$$

where P_m is the total transmit power of the mth BS. Such order is optimal for decoding and guarantees the user fairness [31], namely the data signals of users with weaker downlink channels and larger interference need to be allocated more transmit power to achieve the desired QoS. For the special case of single-cell, i.e., $I_{jm}^{(2)} = 0$, (10.3) reduces to the order based on the channel power gains, as seen in [32]. Therefore, based on (10.1), the data rate after SIC at the jth UE is given by

$$\tau_{jm} = W \log_2\left(1 + \gamma_{jm}\right), \tag{10.4}$$

where W is the system bandwidth, and γ_{jm} is the signal-to-interference-plus-noise ratio (SINR) given by

$$\gamma_{jm} = \frac{P_{jm}|h_{jm}|^2}{\underbrace{|h_{jm}|^2 \sum_{j'=1}^{j-1} P_{j'm}}_{I_{jm}^{(1)}} + \underbrace{\sum_{m'=1,m'\neq m}^{M+1} \left|h_{jm}^{m'}\right|^2 P_{m'}}_{I_{jm}^{(2)}} + \sigma^2}$$

$$= \frac{P_{jm}}{\sum_{j'=1}^{j-1} P_{j'm} + (I_{jm}^{(2)} + \sigma^2)/|h_{jm}|^2}, \quad j \leq k_m \tag{10.5}$$

in which $I_{jm}^{(1)}$ is the remaining intra-cell interference after SIC, and $P_{m'} = \sum_{j'=1}^{N} x_{j'm'} P_{j'm'}$ is the total transmit power of the m'th BS. Although this chapter focuses on the single-carrier system, it can be straightforwardly extended to the multi-carrier system by letting W be the subcarrier bandwidth and τ_{jm} multiply the subcarrier indicator to be the data rate of a subcarrier. Thus, the optimal solution over all subcarriers in the multi-carrier case can be iteratively obtained by following the decomposition approach of this chapter.

10.2.2 Energy Model

Each BS is powered by both the conventional grid and renewable energy sources. The energy drawn by the mth BS from the conventional grid is denoted as G_m. The energy harvested by the mth BS from renewable energy sources is denoted by E_m. The energy transferred from BS m to BS m' is denoted as $\mathcal{E}_{mm'}$, and the energy transfer efficiency factor between two BSs is denoted as $\beta \in [0, 1]$. Hence $(1 - \beta)$ specifies the level of energy loss during the energy transmission process. In addition, it is assume that there is no battery to avoid the time-consuming and expensive energy waste during the charging/discharging process, and the energy-cooperation problem in each time slot is independent. The time slot length is normalized as one to simplify the power-to-energy conversion. Therefore, the transmit energy consumption at the mth BS should satisfy.

$$P_m \leq G_m + E_m + \underbrace{\beta \sum_{m'=1,m'\neq m}^{M+1} \mathcal{E}_{m'm}}_{\text{Energy received from other BSs}} - \underbrace{\sum_{m'=1,m'\neq m}^{M+1} \mathcal{E}_{mm'}}_{\text{Energy transferred to other BSs}}, \tag{10.6}$$

where $P_m = \sum_{j=1}^{N} x_{jm} P_{jm}$ is the total transmit power of the mth BS.

From (10.6), it is seen that in energy-cooperation-enabled networks, the grid energy consumption of a BS depends on its harvested renewable energy, transferred energy and transmit power. Given a BS's transmit power, its grid energy consumption needs to be formulated as a random variable, since the amount of harvested renewable energy and transferred energy is uncertain, which is different from the conventional network without energy cooperation.

10.2.3 Problem Formulation

Our aim is to maximize the energy efficiency of such networks. The energy efficiency (bits/Joule) is defined as the ratio of the overall network data rate to the overall grid energy consumption; i.e., the network utility is

$$\mathcal{U}\left(\mathbf{x},\mathbf{P},\mathcal{E},\mathbf{G}\right)=\Big(\sum_{m=1}^{M+1}\sum_{j=1}^{N}x_{jm}\tau_{jm}\Big)/\sum_{m=1}^{M+1}G_m. \tag{10.7}$$

In this way, the harvested renewable energy can be maximally utilized to reduce the grid energy consumption [11]. Therefore, our problem can be formulated as follows:

$$\mathbf{P1}: \quad \max_{\mathbf{x},\mathbf{P},\mathcal{E},\mathbf{G}} \quad \mathcal{U}\left(\mathbf{x},\mathbf{P},\mathcal{E},\mathbf{G}\right) \tag{10.8}$$

$$\begin{aligned}
\text{s.t. } & \text{C1}: \sum_{m=1}^{M+1} x_{jm}\tau_{jm} \geq \overline{\tau}_{\min},\ \forall j,\\
& \text{C2}: \sum_{m=1}^{M+1} x_{jm} = 1,\ \forall j,\\
& \text{C3}: P_m + \sum_{m'=1,m'\neq m}^{M+1} \mathcal{E}_{mm'} \leq G_m + E_m + \beta \sum_{m'=1,m'\neq m}^{M+1} \mathcal{E}_{m'm},\ \forall m,\\
& \text{C4}: \sum_{j=1}^{N} x_{jm}P_{jm} = P_m,\ \forall m,\\
& \text{C5}: x_{jm} \in \{0,1\},\forall j,\forall m,\\
& \text{C6}: G_m \geq 0, \mathcal{E}_{mm'} \geq 0, \forall j, \forall m,\\
& \text{C7}: 0 \leq P_m \leq P_{\max}^m, P_{jm} \geq 0, \forall j, \forall m,
\end{aligned}$$

where $\mathbf{x}=\left[x_{jm}\right]$, $\mathbf{P}=\left[P_{jm}\right]$, $\mathcal{E}=\left[\mathcal{E}_{mm'}\right]$, $\mathbf{G}=\left[G_m\right]$, $\overline{\tau}_{\min}$ denotes the required minimum data rate for a UE, $P_{\max}^m$ is the maximum transmit power of the BS m. Constraint C1 guarantees the QoS. C2 and C5 ensure that each UE cannot be associated with multiple BSs. C3 is the energy consumption constraint, and C4 is the power allocation under NOMA principle in a cell. C6 indicates that the consumed grid energy and transferred energy are nonnegative values, and C7 is the maximum transmit power constraint.

From the objective of **P1** and its constraint C3, it can be found that when more renewable energy is harvested and shared between BSs, the total grid energy consumption of the network can be reduced, which boosts the energy efficiency.

10.3 Proposed Resource Allocation Scheme

10.3.1 Resource Allocation Under Fixed Transmit Power

P1 is a mixed integer nonlinear programming (MINLP) problem, and constitutes a challenging problem. In this section, it is assumed that the transmit power is fixed, and accordingly the original problem **P1** can be simplified as

$$\begin{aligned}\mathbf{P2}: \quad & \max_{\mathbf{x},\mathscr{E},\mathbf{G}} \quad \mathscr{U}\left(\mathbf{x},\mathscr{E},\mathbf{G}\right) \\ \text{s.t.} \quad & \text{C1, C2, C3, C4, C5, C6.}\end{aligned} \tag{10.9}$$

The problem **P2** is still a combinatorial problem due to its discrete nature. To efficiently solve it, we adopt a decomposition approach. For a given **G** and $\mathscr{E}$, the above problem can be rewritten as

$$\begin{aligned}\mathbf{P2.1}: \quad & \max_{\mathbf{x}} \quad \mathscr{U}\left(\mathbf{x}\right) \\ \text{s.t.} \quad & \text{C1, C2, C4, C5.}\end{aligned} \tag{10.10}$$

10.3.1.1 Lagrangian Dual Analysis

Based on **P2.1**, the Lagrangian function can be written as

$$L\left(\mathbf{x},\boldsymbol{\lambda},\boldsymbol{\theta}\right) = \mathscr{U}\left(\mathbf{x}\right) - \sum_{j=1}^{N} \lambda_j \left(\overline{\tau}_{\min} - \sum_{m=1}^{M+1} x_{jm}\tau_{jm}\right) - \sum_{m=1}^{M+1} \theta_m \left(\sum_{j=1}^{N} x_{jm}P_{jm} - P_m\right), \tag{10.11}$$

where λ_j and θ_m are the nonnegative Lagrange multipliers. Then, the dual function is given by

$$g(\boldsymbol{\lambda},\boldsymbol{\theta}) = \begin{cases} \max\limits_{x} L(x,\lambda,\theta) \\ s.t. \ C2, C5 \end{cases}, \tag{10.12}$$

and the dual problem of **P2.1** is expressed as

$$\min_{\boldsymbol{\lambda},\boldsymbol{\theta}} g\left(\boldsymbol{\lambda},\boldsymbol{\theta}\right). \tag{10.13}$$

Given the dual variables λ_j and θ_m, the optimal solution for maximizing the Lagrangian w.r.t. $\mathbf{x}$ is

$$x_{jm}^{*} = \begin{cases} 1, & \text{if} \quad m = m^{*} \\ 0, & \text{otherwise} \end{cases}, \tag{10.14}$$

where $m^{*} = \underset{m}{\operatorname{argmax}} \left(\mu_{jm}\right)$ with

$$\mu_{jm} = \tau_{jm} / \sum_{m=1}^{M+1} G_m + \lambda_j \tau_{jm} - \theta_m P_{jm}. \tag{10.15}$$

The solution of (10.14) can be intuitively interpreted based on the fact that given the grid energy consumption, users select BSs which provide the maximum data rates. Since the objective of the dual problem is not differentiable, we utilize the subgradient method to obtain the optimal solution $\left(\boldsymbol{\lambda}^{*}, \boldsymbol{\theta}^{*}\right)$ of the dual problem, which is given by

$$\lambda_j\left(t+1\right) = \left[\lambda_j\left(t\right) - \delta\left(t\right)\left(\sum_{m=1}^{M+1} x_{jm}\tau_{jm} - \overline{\tau}_{\min}\right)\right]^{+}, \tag{10.16}$$

$$\theta_m\left(t+1\right) = \left[\theta_m\left(t\right) - \delta\left(t\right)\left(P_m - \sum_{j=1}^{N} x_{jm}P_{jm}\right)\right]^{+}, \tag{10.17}$$

where $[a]^{+} = \max\{a, 0\}$, t is the iteration index, and $\delta\left(t\right)$ is the step size. Note that there exist several step size selections such as constant step size and diminishing step size. Here, the nonsummable diminishing step length is used [33].

After obtaining the optimal $\left(\boldsymbol{\lambda}^{*}, \boldsymbol{\theta}^{*}\right)$ based on (10.16) and (10.17), the corresponding $\mathbf{x}$ is the solution of the primal problem **P2.1**. Therefore, based on the Lagrangian dual analysis, UA can be determined in a centralized or distributed way. The centralized UA is intuitive and requires a central controller, which has the global CSI and determines which user is connected to a BS in this network. In this chapter, a distributed UA algorithm which does not require any centralized coordination is proposed, as summarized in Algorithm 1. Since our problem satisfies the conditions of the convergence proof in [33], the convergence of the proposed algorithm is guaranteed. The complexity of the proposed algorithm is $\mathcal{O}\left((M+1)N\right)$ for each iteration and the convergence is fast (less than 40 iterations in the simulation), which is much lower than the brute force algorithm $\mathcal{O}\left((M+1)^{N}\right)$. Note that the broadcast operations have a negligible effect on computational complexity (Table 10.1).

Table 10.1 Algorithm 1 distributed user association

Step 1: At user side

if $t = 0$
  Initialize $\lambda_j(t)$, $\forall j$. Each UE measures its received inter-cell interference via pilot signal from all BSs, and feedbacks the CINR values to the corresponding BSs. Meanwhile, each UE selects the BS with the largest CINR value.
else
  User j receives the values of μ_{jm} and τ_{jm} from BSs.
  Determines the serving BS m according to $m^* = \underset{m}{\text{argmax}} \left(\mu_{jm}\right)$.
  Update $\lambda_j(t)$ according to (10.16).
end if
$t \leftarrow t + 1$.
Each user feedbacks the UA request to the chosen BS, and broadcasts the value of $\lambda_j(t)$.

Step 2: At BS side

if $t = 0$
  Initialize $\theta_m(t)$, $\forall m$.
else
  Receives the updated user association matrix $\mathbf{x}$.
  Updates $\theta_m(t)$ according to (10.17), respectively.
  Each BS calculates μ_{jm} and τ_{jm} under NOMA principle.
end if
$t \leftarrow t + 1$.
Each BS broadcasts the values of μ_{jm} and τ_{jm}.

10.3.1.2 Genetic Algorithm

In this subsection, a genetic algorithm (GA)-based UA is proposed to solve the problem **P2.1**. Such algorithm will be compared with the proposed Algorithm 1. GA can achieve good performance when the population of candidate solutions is sufficient [34]. Specifically, each feasible chromosome represents a possible solution that satisfies the constraints of problem **P2.1**, which is defined as

$$\mathscr{D}_i = \{[m_{1i}], [m_{2i}], \ldots, [m_{Ni}]\}, \ i \in \{1, \ldots, K\}, \tag{10.18}$$

where m_{ji} is the gene representing the index of the BS that the jth UE is associated with, and it has an integer value varying from 1 to $M + 1$, and K is the population size. During each generation, the fitness of each chromosome is evaluated, to select high fitness chromosomes and produce higher fitness offsprings. Based on the objective of problem **P2.1**, the fitness value of the chromosome $\mathscr{D}_i$ is calculated as

Table 10.2 Algorithm 2 genetic algorithm-based user association algorithm

if $t = 0$
  Initialize a set of feasible chromosomes $\{\mathscr{D}_i\}$ with population size K, and the maximum number of generations $t_{\max}$.
else
  Rank $\{\mathscr{D}_i\}$ based on the fitness values given by (10.19).
  Based on the selection probability $\rho_s(r)$, chromosomes are selected to produce offspring via uniform crossover and mutation operations.
  if exceed the maximum number of generations
    $x^*_{jm} := \{\mathscr{D}^*_i\}$, where $\{\mathscr{D}^*_i\}$ is the feasible chromosome with the highest fitness value.
    break
  else
    $t \leftarrow t + 1$.
  end if
end if

$$\Phi_i(\mathscr{D}_i) = \mathscr{U}(\mathscr{D}_i). \tag{10.19}$$

Then, all chromosomes are ranked from the best to the worst with ranking r, based on their fitness values. The probability that a chromosome is selected as a parent to produce offspring is given by $\rho_s(r) = \frac{q(1-q)^{r-1}}{1-(1-q)^K}$ with a predefined value q [34]. In each generation process, a uniform crossover operation with the probability ρ_c is utilized to produce offspring by swapping and recombining genes based on the parental chromosomes. In addition, a uniform mutation operation with the probability ρ_m is employed. Such generation procedure is repeated until reaching the maximum number of generations, and is summarized in Algorithm 2. Given the maximum number of generations Ω and fixed population size K, the complexity of the proposed algorithm is $\mathscr{O}(\Omega K \log(K))$ [35]. The performance of the GA-based UA algorithm heavily depends on the population size and number of generations, due to the inherent nature of GA [34]. In the simulation results of Sect. 10.5, we will demonstrate that overall, the proposed Algorithm 1 outperforms GA-based Algorithm 2 which is shown in Table 10.2 when the population size of GA is not very large, and thus has lower complexity.

The aforementioned approach provides UA solutions for problem **P2.1**. After obtaining the UA solution $\mathbf{x} = \left[x^*_{jm}\right]$, the corresponding pair $(\mathbf{G}, \mathscr{E})$ is obtained by solving the following simple linear programming (LP):

Table 10.3 Algorithm 3 resource allocation algorithm under fixed transmit power

if $t = 0$
  For a fixed **P**, initialize $G_m, \forall j, m$.
else
  Determine $x_{jm}(t)$ under fixed $(\mathcal{E}, \mathbf{G})$ by selecting the user association algorithm from Algorithm 1 or Algorithm 2.
  Given $x_{jm}(t)$, update the energy allocation policy $(\mathcal{E}, \mathbf{G})$ by solving the LP **P2.2** via CVX.
  if convergence
    Obtain optimal resource allocation policy $(\mathbf{x}^*, \mathcal{E}^*, \mathbf{G}^*)$.
    break
  else
    $t \leftarrow t + 1$.
  end if
end if

Table 10.4 Algorithm 4 one-dimensional search algorithm

if $t = 0$
  Given χ_j, initialize $\nu_m^l = 0$, $\nu_m^h = \nu_m^{\max}, \forall m$, calculate $F_l = \sum_{j=1}^{N} x_{jm} P_{jm}^{*(l)}$ and $F_h = \sum_{j=1}^{N} x_{jm} P_{jm}^{*(h)}$, where $\left\{P_{jm}^{*(l)}\right\}$ and $\left\{P_{jm}^{*(h)}\right\}$ are the allocated transmit powers of the j-th UE's data stream for the cases of ν_m^l and ν_m^h respectively, which are calculated by using (10.28).
else
  while $F_l \neq \varphi_m$ and $F_h \neq \varphi_m$
    Let $\nu_m = \frac{\nu_m^l + \nu_m^h}{2}$, and compute F_m.
    if $F_m = \nu_m$
      The optimal dual variable ν_m^* is obtained.
      break
    elseif $F_m < \varphi_m$
      $\nu_m^h = \nu_m$.
    else $F_m > \varphi_m$
      $\nu_m^l = \nu_m$.
    end if
  end while
end if

Table 10.5 Algorithm 5 Joint User Association and Power Control

---

if $t = 0$

  Initialize $P_m, G_m, E_m, \forall m$

else

  Determine $x_{jm}(t)$ under $(\mathbf{P}, \mathbf{G}, \mathscr{E})$ by selecting the user association algorithm from Algorithm 1 or Algorithm 2.

  Given $x_{jm}(t)$ and the corresponding $(\mathbf{G}, \mathscr{E})$, update the transmit power $\mathbf{P}$ based on the following rule:

   Loop:

   a) Given $\Theta_{jm}^{(2)}$, loop over UE j:

    i): Obtain $\{\nu_m^*\}$ using Algorithm 4 given $\{\chi_j\}$

    ii): Obtain P_{jm} according to (10.28) with $\{\nu_m^*, \chi_j\}$.

    iii): Update $\{\chi_j\}$ using subgradient method.

    iv): Update P_{jm} using (10.29).

    Until convergence.

   b) Update $\Theta_{jm}^{(2)}$ using (10.27).

   Until convergence.

  Based on the updated $\mathbf{P}$, update G_m and $\mathscr{E}_{mm'}$ by solving LP problem **P2.2** via CVX.

  if convergence

   Obtain optimal resource allocation policy $(\mathbf{x}^*, \mathbf{P}^*, \mathscr{E}^*, \mathbf{G}^*)$.

   break

  else

   $t \leftarrow t + 1$.

  end if

end if

$$\mathbf{P2.2}: \min_{\mathscr{E}, \mathbf{G}} \sum_{m=1}^{M+1} G_m \quad \text{s.t. C3, C6.} \tag{10.20}$$

The problem **P2.2** can be efficiently solved by using existing software, e.g. CVX [36].

When no energy cooperation is allowed, i.e., $\mathscr{E}_{mm'} = 0, \forall j, \forall m$, the optimal grid energy consumption $\mathbf{G}$ of problem **P2.2** under the UA solution $\mathbf{x} = \left[x_{jm}^*\right]$ is directly obtained as

$$G_m^* = [P_m - E_m]^+, \tag{10.21}$$

where $P_m = \sum_{j=1}^{N} x_{jm}^* P_{jm}$.

Based on the solutions of subproblems **P2.1** and **P2.2**, we propose an iterative algorithm to solve the problem **P2**, which is summarized in Algorithm 3 (Table 10.3).

10.3.2 Resource Allocation Under Power Control

In this subsection, we consider the joint resource allocation and power control design. Specifically, we develop an algorithm to solve the MINLP problem **P1** through the decomposition approach. As discussed in the previous section, we first determine the UA indicators given the resource allocation policy (**P**, $\mathscr{E}$, **G**), which can be obtained by solving problem **P2.1** via Algorithm 1 or Algorithm 2. Then, under a fixed UA $\{x_{jm}\}$, the problem for optimizing (**P**, $\mathscr{E}$, **G**) is written as

$$\begin{aligned} \mathbf{P3}: \quad & \max_{\mathbf{P},\mathscr{E},\mathbf{G}} \quad \mathscr{U}(\mathbf{P},\mathscr{E},\mathbf{G}) \\ & \text{s.t.} \quad \text{C1, C3, C4, C6, C7.} \end{aligned} \tag{10.22}$$

From the utility function, we find that the power allocation vectors **P** and **G** are coupled within the objective of problem **P3**. Thus, given **G** and $\mathscr{E}$, the above problem can be decomposed into

$$\begin{aligned} \mathbf{P3.1}: \quad & \max_{\mathbf{P}} \quad \sum_{m=1}^{M+1}\sum_{j=1}^{N} x_{jm}\tau_{jm} \\ & \text{s.t.} \quad \text{C1, C3, C4, C7.} \end{aligned} \tag{10.23}$$

Problem **P3.1** is non-convex. Hence, we provide a tractable suboptimal solution based on the Karush–Kuhn–Tucker (KKT) conditions. The Lagrangian function of problem **P3.1** is

$$\begin{aligned} L(\mathbf{P},\boldsymbol{\nu},\boldsymbol{\chi}) = & \sum_{m=1}^{M+1}\sum_{j=1}^{N} x_{jm}\tau_{jm} - \sum_{j=1}^{N+1}\chi_j\left(\tau_{\min} - \sum_{m=1}^{M+1} x_{jm}\tau_{jm}\right) \\ & - \sum_{m=1}^{M+1}\nu_m\left(\sum_{j=1}^{N} x_{jm}P_{jm} - \varphi_m\right), \end{aligned} \tag{10.24}$$

where $\varphi_m = \min\left\{G_m + E_m + \beta\sum_{m'=1,m'\neq m}^{M+1}\mathscr{E}_{m'm} - \sum_{m'=1,m'\neq m}^{M+1}\mathscr{E}_{mm'}, P_{\max}^m\right\}$ according to constraints C3 and C7, and χ_j and ν_m are the nonnegative Lagrange multipliers.

Without loss of generality, assuming that the jth UE is associated with the BS m, i.e., $x_{jm} = 1$, based on the KKT conditions, we have

$$\begin{aligned} \frac{\partial L}{\partial P_{jm}} &= \left(1+\chi_j\right)\left(\frac{W\Lambda_{jm}}{1+P_{jm}\Lambda_{jm}}\right) - \Theta_{jm}^{(1)} - \Theta_{jm}^{(2)} - \nu_m\log(2) \\ &= 0, \end{aligned} \tag{10.25}$$

where $\Lambda_{jm} = \frac{|h_{jm}|^2}{I_{jm}^{(1)}+I_{jm}^{(2)}+\sigma^2}$ is referred to as the channel-to-interference-plus-noise ratio at the jth UE. Based on (10.3) and (10.5), $\Theta_{jm}^{(1)}$ resulting from the intra-cell interference is given by

$$\Theta_{jm}^{(1)} = \sum_{\ell>j}^{k_m} (1+\chi_\ell) \frac{W\gamma_{\ell m}}{1+\gamma_{\ell m}} \Lambda_{lm}, \tag{10.26}$$

and $\Theta_{jm}^{(2)}$ resulting from the inter-cell interference is given by

$$\Theta_{jm}^{(2)} = \sum_{m'=1,m'\neq m}^{M+1} \sum_{j'=1}^{N} \frac{\left(1+\chi_{j'}\right) x_{j'm'} W\gamma_{j'm'} \left|h_{j'm'}^{m}\right|^2}{\left(1+\gamma_{j'm'}\right)\left(I_{j'm'}^{(1)} + I_{j'm'}^{(2)} + \sigma^2\right)}. \tag{10.27}$$

Based on (10.25), the transmit power allocated to the jth user-stream in the mth cell is obtained as

$$P_{jm}^{*} = \left[\frac{(1+\chi_j)W}{\Theta_{jm}^{(1)} + \Theta_{jm}^{(2)} + \nu_m \log(2)} - \frac{1}{\Lambda_{jm}}\right]^{+}. \tag{10.28}$$

In (10.28), the allocated transmit power is a monotonic function of ν_m. As such, given $\{\chi_j\}$, we adopt a one-dimensional search over the Lagrange multipliers $\{\nu_m\}$, which can efficiently obtain the optimal $\boldsymbol{\nu}^*$ that satisfies constraints C3 and C7. According to (10.28), we can easily find that ν_m^* needs to satisfy $0 \leq \nu_m^* \leq \nu_m^{\max}$, where $\nu_m^{\max} = \max_j \left\{\left((1+\chi_j)W\Lambda_{jm} - \Theta_{jm}^{(1)} - \Theta_{jm}^{(2)}\right)/\log(2)\right\}$. Here, $\nu_m^* = 0$ represents that there is no limitation on the transmit power of the jth user-stream and $\nu_m^* = \nu_m^{\max}$ corresponds to the case that no transmit power is allocated to the jth user-stream. Thus, by fixing $\{\chi_j\}$, $\boldsymbol{\nu}^*$ can be obtained by using Algorithm 4 (Table 10.4). For achieving a specific accuracy ς, the complexity of Algorithm 4 is $\mathcal{O}(\log(1/\varsigma))$. After obtaining $\boldsymbol{\nu}^*$, the Lagrange multiplier χ_j can be updated by using the subgradient method, which is similar to (10.16).

To ensure the system stability, we utilize the Mann iterative method to update the transmit power in each iteration [37], which is given by

$$P_{jm}^{(\ell+1)} = (1-\eta(\ell))P_{jm}^{(\ell)} + \eta(\ell)P_{jm}^{*}, \tag{10.29}$$

where ℓ is the iteration index, $0 < \eta(\ell) < 1$ is the step size, which is usually chosen as $\eta(\ell) = \frac{\ell}{2\ell+1}$. After obtaining the optimal solution of problem **P3.1**, the corresponding $(\mathbf{G}, \mathcal{E})$ can be updated by solving the LP problem **P2.2** via CVX. As such, the solution of problem **P3** can be iteratively obtained. Note that the convergence of the KKT-based algorithm is usually faster than the gradient-based designs [38].

Based on the previous analysis, the proposed joint UA and power control scheme in energy-cooperation enabled NOMA HetNets is summarized in Algorithm 5 (Table 10.5).

10.3.3 Comparison with FTPA

In 4G networks, fractional transmission power allocation (FTPA) scheme is adopted [31]. The rule of FTPA is that the transmit power will be allocated based on the UEs' channel conditions, i.e., the data signals of UEs with weaker downlink channels will own more transmit power. Based on the CINR order in (10.30), the transmit power allocated to the jth UE's data stream in the mth cell under FTPA protocol is expressed as [31]

$$P_{jm} = P_m \left(\frac{|h_{jm}|^2}{I_{jm}^{(2)} + \sigma^2} \right)^{-\alpha} / \sum_{l=1}^{N} x_{lm} \left(\frac{|h_{lm}|^2}{I_{lm}^{(2)} + \sigma^2} \right)^{-\alpha}, \tag{10.30}$$

where $0 \leq \alpha \leq 1$ is the decay factor. Here, $\alpha = 0$ represents equal power allocation. For larger α, the transmit power allocated to the data-stream of the user with largest CINR becomes lower, and more power will be allocated to the data-stream of the user with the lowest CINR, in order to achieve the user fairness and the optimal decoding. However, the detrimental effect of using such simple power allocation scheme is that distant users may receive severer inter-cell interference without power control among BSs, due to the fact that each BS has to assign larger transmit power to the far-away users. Therefore, compared to the single-cell NOMA case [32], the inter-cell interference has a significant impact on the power allocation of multi-tier NOMA HetNets.

10.3.4 Comparison with No Renewable Energy

When there is no renewable energy harvesting (i.e., $E_m = 0, \ \forall m$), no renewable energy can be shared between BSs (i.e., $\mathscr{E}_{mm'} = \mathscr{E}_{m'm} = 0, \ \forall m, m'$), and thus, the required energy can only be supplied by the conventional grid. In this case, $P_m = G_m, \ \forall m$, and the original problem **P1** reduces to

$$\begin{aligned} \mathbf{P4}: \ & \max_{\mathbf{x},\mathbf{P}} \ \frac{\sum_{m=1}^{M+1} \sum_{j=1}^{N} x_{jm} \tau_{jm}}{\sum_{m=1}^{M+1} \sum_{j=1}^{N} x_{jm} P_{jm}} \\ & \text{s.t.} \ \ C1, C2, C4, C5, C7. \end{aligned} \tag{10.31}$$

The above problem is nonlinear fractional programming and NP-hard, which can be solved by using the proposed Algorithm 5 with $E_m = 0$ and $\mathscr{E}_{mm'} = \mathscr{E}_{m'm} = 0$.

10.3.5 Comparison with No Energy Cooperation

In this case, the energy transfer efficiency β is set to 0, which means that the harvested renewable energy cannot be transferred between BSs. Each BS is powered by the conventional grid and its harvested renewable energy; i.e., the transmit energy consumption at a BS needs to satisfy $P_m \leq G_m + E_m, \ \forall m$. Then, the proposed Algorithm 5 can still be applied to solve this problem, and during each iteration, the grid energy consumption is updated as $G_m = [P_m - E_m]^+$ based on the updated P_m.

10.4 Simulation Results

In this section, we present numerical results to demonstrate the effectiveness of the proposed algorithm compared with other schemes as well as the conventional counterpart. Since the renewable energy arrival rate changes slowly in practice and is stationary at each information transmission time slot [39], we consider the amounts of harvested energy at the MBS and PBSs to be constant and each PBS has the same level of renewable energy during each transmission time slot for the sake of simplicity. For simplicity, the amount of harvested energy E_m of BS m is modeled as a uniform distribution $U_m[a_m, b_m]$, and varies across different transmission blocks, where a_m and b_m are the minimum and maximum harvested energy values of BS m follows uniform distributions as shown in Table 10.6, respectively [40]. Our analysis and proposed algorithm are independent of the specific renewable energy distribution. For the channel h_{jm}, we focus on the large-scale channel fading condition in low mobility environment, due to the fact that UA is carried out in a large time scale and the small-scale fading can be averaged out [41, 42]. In addition, PBSs and UEs are uniformly distributed in a macrocell geographical area. The basic simulation parameters are shown in Table 10.6.

10.4.1 User Association Under Fixed Transmit Power

In this subsection, we study different UA algorithms under fixed transmit power, i.e., power control is unavailable at BSs. Based on the NOMA power allocation condition in (10.3), we consider that the total transmit power at each BS is $P_m = P_{\max}^m$, and adopt an arithmetic progression power allocation approach for the sake of simplicity, namely the transmit power of the jth user's data signal is $P_{jm} = \frac{2j}{k_m(1+k_m)} P_m$, $j \in \{1, 2, 3, ..., k_m\}$ when k_m users are multiplexed in the power domain of the mth cell. We also provide the comparison with the conventional reference signal received power (RSRP)-based UA. The aim of this subsection is to show the performance of different UA algorithms under the same fixed power allocation condition.

Table 10.6 Simulation parameters

Parameter	Value
System bandwidth	10 MHz
Noise power density	−174 dBm/Hz
Cell radius	500 m
Path loss of MBS	$128.1 + 37.6\log_{10}d$ (km)
Path loss of PBS	$140.7 + 36.7\log_{10}d$ (km)
Min harvested energy of MBS	575 W
Max harvested energy of MBS	660 W
Min harvested energy of PBS	15 W
Max harvested energy of PBS	25 W
Max transmit power of MBS	46 dBm [43]
Max transmit power of PBS	30 dBm [43]

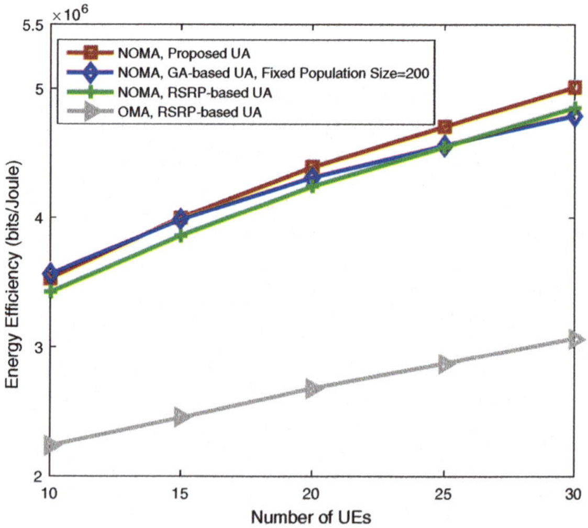

Fig. 10.2 Energy efficiency versus the number of UEs for different UA algorithms

Figure 10.2 shows the energy efficiency versus the number of UEs with the number of PBSs $M = 6$ and the energy transfer efficiency factor $\beta = 0.9$. We set the minimum QoS as $\overline{\tau}_{\min} = 0.1$ bits/s/Hz and the amount of energy harvested by MBS and PBS as 37 dBm and 27 dBm, respectively.[1] The maximum number of generations for the GA-based UA is 10, $q = 0.1$, and $\rho_c = \rho_m = 0.4$. The proposed UA scheme with NOMA achieves better energy efficiency than the other cases. The energy efficiency

[1]In real networks, the renewable energy generation rate is constant during a certain period, and the time scale of the UA and power control process is much shorter, typically less than several minutes [41, 42]. In addition, the amount of energy harvested by an MBS is usually larger than that at a PBS, since MBS can fit larger solar panel [42, 44].

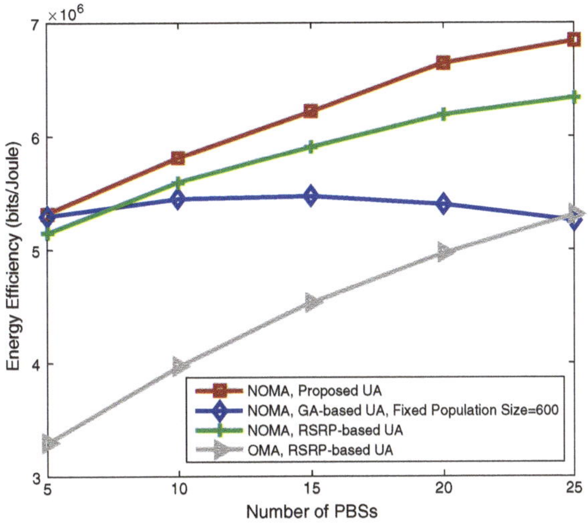

Fig. 10.3 Energy efficiency versus the number of PBSs for different UA algorithms

increases with the number of UEs because of the multiuser diversity gain (i.e., different users experience different path loss, and more users with lower path loss help enhance the overall energy efficiency.) [45]. The use of NOMA outperforms OMA. By using the GA-based UA, the energy efficiency slowly increases with the number of UEs, due to the fact that the efficiency of the GA-based algorithm depends on the population size [34]. In other words, given the population size (e.g., $K = 200$ in this figure), the GA algorithm may not obtain good solutions when the number of UEs grows large, which indicates that larger populations of candidate solutions are needed [34].

Figure 10.3 shows the energy efficiency versus the number of PBSs with the number of UEs $N = 40$ and the energy transfer efficiency factor $\beta = 0.9$. We set the minimum QoS as $\overline{\tau}_{\min} = 0.1$ bits/s/Hz and the amount of harvested energy at MBS and PBS as 37 dBm and 27 dBm, respectively. The maximum number of generations for GA is 10, $q = 0.1$, and $\rho_c = \rho_m = 0.4$. NOMA achieves higher energy efficiency than OMA, since NOMA can achieve higher spectral efficiency. The proposed UA algorithm outperforms the other cases and the performance gap between the proposed UA and the conventional RSRP-based UA is larger when deploying more PBSs, due to the fact that the proposed UA can achieve more BS densification gains [9]. For the GA-based UA algorithm with the population size $K = 600$, solutions are inferior when the number of PBSs is large, as larger populations of candidate solutions are needed [34].

10.4.2 Power Control Under Fixed User Association

In, this subsection, we consider three power allocation schemes, namely the power control method proposed in Sect. 10.4, fractional transmission power allocation (FTPA) and the conventional fixed transmit power, to confirm the advantages of our proposal. We adopt the conventional RSRP-based UA in the simulation, and all the considered cases experience the same UA condition. In addition, BSs use their maximum transmit powers in the OMA scenario, and the total transmit power of each BS for FTPA is set as $P_m = P_{\max}^m$, $m \in \{1, 2, 3, ..., M+1\}$.

Figure 10.4 shows the energy efficiency versus the number of PBSs with the number of UEs $N = 50$ and the energy transfer efficiency factor $\beta = 0.9$. We set the minimum QoS as $\overline{\tau}_{\min} = 1$ bits/s/Hz and the amount of energy harvested by MBS and PBS as 37 dBm and 27 dBm, respectively. We see that by using NOMA with the proposed power control, energy efficiency rapidly increases with the number of PBS. The proposed algorithm achieves better performance than the other cases. When deploying more PBSs, the performance gap between the proposed solution and the other cases is larger, which indicates that the proposed power control algorithm can achieve more BS densification gains and efficiently coordinate the inter-cell interference. When the number of PBSs is not large, NOMA with FTPA can outperform the conventional OMA case, since NOMA can achieve better spectral efficiency than OMA [32]. However, when adding more PBSs, NOMA with FTPA may not provide higher energy efficiency. The reason is that more UEs will be offloaded to picocells, and the inter-cell interference will become severer, which means that the transmit power of each user-stream needs to be larger to combat the inter-cell interference. As suggested in Sect. 10.3.3, FTPA with $\alpha = 0$ achieves a higher energy efficiency of the network than the $\alpha = 0.7$ case, since the data-streams for UEs with poorer

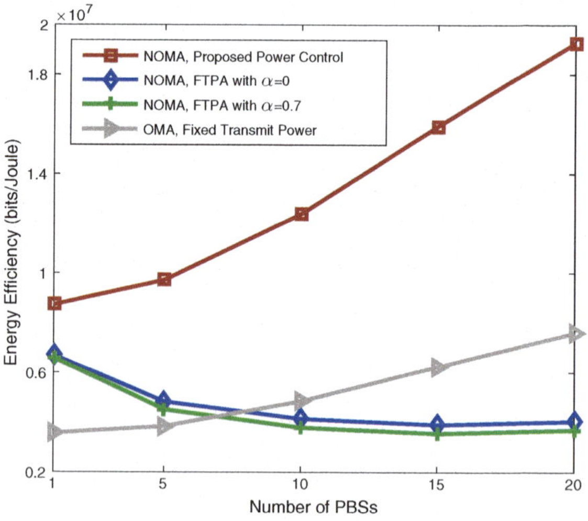

Fig. 10.4 Energy efficiency versus the number of PBSs for different power allocation policies

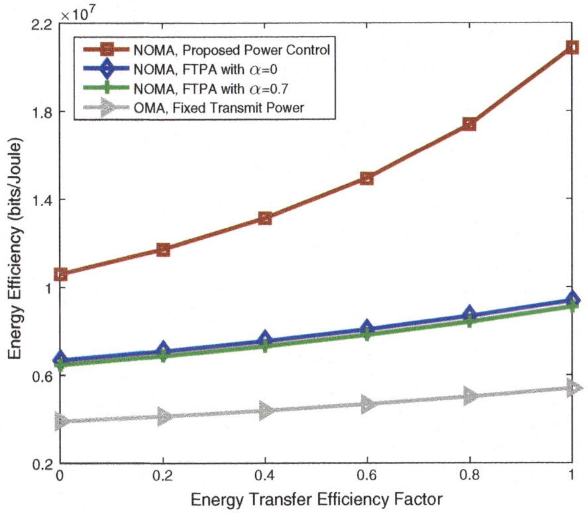

Fig. 10.5 Energy efficiency versus energy transfer efficiency factor for different power allocation policies

channel condition (i.e., lower CINR) have to be allocated more power in the case of FTPA with $\alpha = 0.7$, which reduces the total throughput of the network under the same energy consumption.

Figure 10.5 shows the energy efficiency versus the energy transfer efficiency factor β with the number of PBSs $M = 3$ and the number of UEs $N = 40$. We set the minimum QoS to $\overline{\tau}_{\min} = 1$ bits/s/Hz and the amount of harvested energy at MBS and PBS to 40 dBm and 35 dBm, respectively. Compared to the no energy-cooperation case (i.e., $\beta = 0$), the use of energy cooperation can enhance the energy efficiency, particularly when the energy transfer efficiency factor is large. The implementation of NOMA can achieve higher energy efficiency than the conventional OMA system because of higher spectral efficiency, and the proposed power control algorithm outperforms the other cases. Moreover, the energy efficiency grows at a much higher speed when applying the proposed algorithm. For a specified β, FTPA with $\alpha = 0$ achieves higher energy efficiency of the network than the $\alpha = 0.7$ case, as suggested in Fig. 10.4.

Figure 10.6 shows the trade-off between the energy efficiency and the minimum QoS with the number of PBSs $M = 3$ and the number of UEs $N = 30$. We set the energy transfer efficiency factor to $\beta = 0.9$ and the amount of energy harvested by MBS and PBS to 37 dBm and 27 dBm, respectively. For a given minimum QoS, the proposed power control under NOMA achieves higher energy efficiency than conventional OMA. When better QoS is required by the UE, energy efficiency of both NOMA and OMA cases decreases. The reason is that for the proposed solution, more transmit power will be allocated to the UEs with lower CINRs to achieve such minimum QoS, which results in more energy consumption; for conventional OMA, it means that more users cannot obtain the desired QoS and have to experience an outage. We see that energy efficiency decreases significantly in the low minimum QoS regime, because many UEs receive low QoS and increasing the level of the

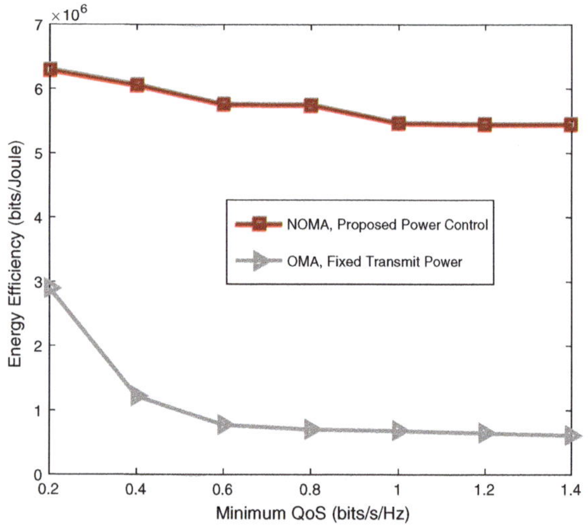

Fig. 10.6 Trade-off between the energy efficiency and the minimum QoS for NOMA and OMA

minimum QoS means that these UEs cannot be served. In practice, the minimum QoS can be found in an off-line manner [46].

10.4.3 Joint User Association and Power Control

In this subsection, we examine the benefits of joint UA and power control design in energy-cooperation-enabled NOMA HetNets. We also present comparisons by considering different power allocation schemes with the conventional RSRP-based UA. In the OMA scenario, transmit power at the BS is set to $P_m = P_{\max}^m$ in the OMA scenario.

Figure 10.7 shows the energy efficiency versus the number of UEs with the number of PBSs $M = 5$ and the energy transfer efficiency factor $\beta = 0.9$. We set the minimum QoS as $\overline{\tau}_{\min} = 0.5$ bits/s/Hz and the amount of harvested energy at MBS and PBS as 32 dBm and 22 dBm, respectively. We see that the proposed joint UA and power control algorithm achieves higher energy efficiency than the other cases, and significantly improves the performance when more UEs are served in the network. The reason is that the proposed algorithm is capable of obtaining larger multiuser diversity gains. The use of NOMA can obtain higher energy efficiency than the OMA case, due to NOMA's capability of achieving higher spectral efficiency. Additionally, when equal power allocation is adopted in NOMA HetNets with the conventional RSRP-based UA, energy efficiency decreases with increasing the number of UEs of the network, which can be explained by the fact that given the total transmit power of a BS, the transmit power allocated to the data-streams of the UEs with better channel condition reduces when more UEs are served simultaneously.

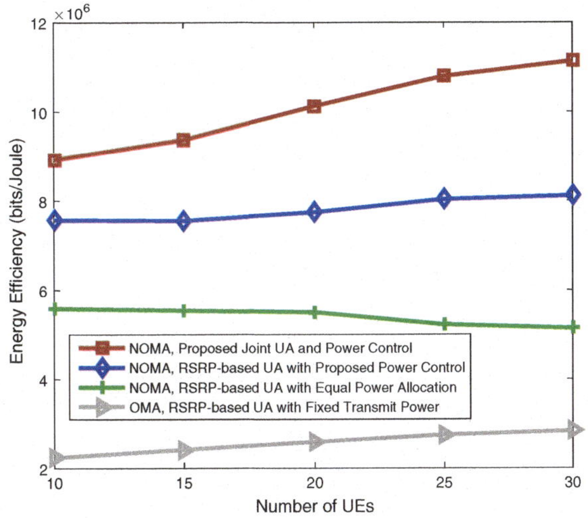

Fig. 10.7 Energy efficiency versus the number of UEs for different joint UA and power allocation designs

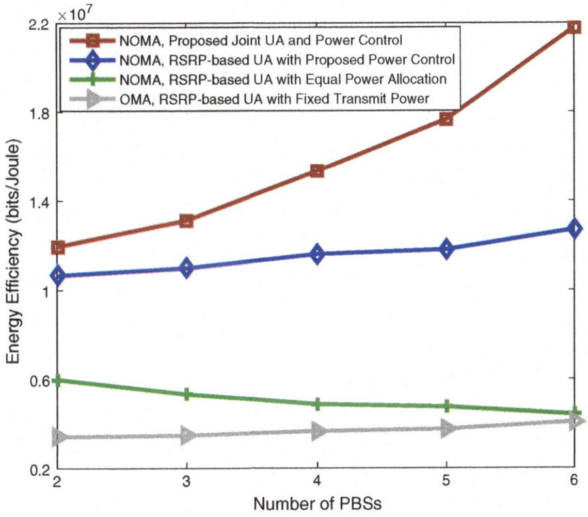

Fig. 10.8 Energy efficiency versus the number of PBSs for different joint UA and power allocation designs

Figure 10.8 shows the energy efficiency versus the number of PBSs with the number of UEs $N = 50$ and the energy transfer efficiency factor $\beta = 0.9$. We set the minimum QoS as $\overline{\tau}_{\min} = 0.1$ bits/s/Hz and the amount of energy harvested by MBS and PBS as 37 dBm and 27 dBm, respectively. The proposed design outperforms the other cases. By using the proposed joint UA and power control with NOMA, the energy efficiency significantly increases with the PBS number, since the proposed design can obtain more BS densification gains. Again, the use of NOMA achieves better performance than OMA. For the case of RSRP-based UA with NOMA and equal power allocation, energy efficiency decreases with increasing the number of

PBSs, because the inter-cell interference has a big adverse effect on the NOMA transmission [3].

10.5 Conclusion and Future Work

This chapter studied UA and power control in energy-cooperation- aided two-tier HetNets with NOMA. A distributed UA algorithm was proposed based on the Lagrangian dual analysis, which does not require a central controller. Then, we proposed a joint UA and power control algorithm which achieves higher energy efficiency performance than the existing schemes. The proposed power control algorithm satisfies the KKT optimality conditions. Simulation results demonstrate the effectiveness of the proposed algorithms. The results showed that the proposed algorithm can efficiently coordinate the intra-cell and inter-cell interference and has the capability of exploiting the multiuser diversity and BS densification. The application of NOMA can achieve larger energy efficiency than OMA due to the higher spectral efficiency of NOMA.

To further extend this line of work, other UA optimization designs in multi-cell NOMA networks such as proportional fairness or max-min fairness would be of interest, and they are not trivial extensions since the optimization problems involved will be distinct. Moreover, imperfect CSI can have a substantial effect on outage probability and average data rate in NOMA networks, as analyzed in [47]. One of the challenges for optimization designs under imperfect CSI is that error propagation occurs since intra-cell interference cannot be perfectly canceled. Therefore, robust optimization designs need to be developed in multi-cell NOMA networks. In addition, the application of MIMO technology in NOMA networks is another important research area, which can significantly improve the performance gain [32]. In MIMO-NOMA networks, inter-user pair/group interference can deteriorate the performance, as analyzed in [32, 48]. Therefore, how to mitigate the inter-user pair/group interference is crucial. Currently, UA and power control solutions in multi-cell MIMO-NOMA networks are not available, and more research efforts need to be made in this area.

References

1. D. Liu, L. Wang, Y. Chen, M. Elkashlan, K.K. Wong, R. Schober, L. Hanzo. User association in 5G networks: a survey and an outlook. IEEE Commun. Surv. Tutor. **18**(2), 1018–1044 (2016)
2. Y. Liu, Z. Qin, M. Elkashlan, Z. Ding, A. Nallanathan, L. Hanzo, Non-orthogonal multiple access for 5g and beyond. Proc. IEEE **105**(12), 2347–2381 (2017)
3. W. Shin, M. Vaezi, B. Lee, D.J. Love, J. Lee, H.V. Poor, Non-orthogonal multiple access in multi-cell networks: Theory, performance, and practical challenges. IEEE Commun. Mag. **55**(10), 176–183 (2017)

4. Y. Xu, H. Sun, R. Q. Hu, Y. Qian, Cooperative non-orthogonal multiple access in heterogeneous networks, in *Proceedings of IEEE GLOBECOM* (San Diego, CA, USA, Dec 2015), pp. 1–6
5. J. Zhao, Y. Liu, K.K. Chai, A. Nallanathan, Y. Chen, Z. Han, Resource allocation for non-orthogonal multiple access in heterogeneous networks. In *Proceedings of IEEE ICC* (Paris, France, May 2017), pp. 1–6
6. J. Zhao, Y. Liu, K.K. Chai, A. Nallanathan, Y. Chen, Z. Han, Spectrum allocation and power control for non-orthogonal multiple access in hetnets. IEEE Trans. Wirel. Commun. **16**(9), 5825–5837 (2017)
7. Y. Liu, Z. Qin, M. Elkashlan, A. Nallanathan, J.A. McCann, Non-orthogonal multiple access in large-scale heterogeneous networks. IEEE J. Sel. Areas Commun. **35**(12), 2667–2680 (2017)
8. Q.T. Vien, T.A. Le, B. Barn, C.V. Phan, Optimising energy efficiency of non-orthogonal multiple access for wireless backhaul in heterogeneous cloud radio access network. IET Commun. **10**(18), 2516–2524 (2016)
9. J.G. Andrews, S. Buzzi, W. Choi, S.V. Hanly, A. Lozano, A.C.K. Soong, J.C. Zhang, What will 5g be? IEEE J. Sel. Areas Commun. **32**(6), 1065–1082 (2014)
10. D. Jiang, P. Zhang, Z. Lv, H. Song, Energy-efficient multi-constraint routing algorithm with load balancing for smart city applications. IEEE Internet Things J. **3**(6), 1437–1447 (2016)
11. T. Han, N. Ansari, Powering mobile networks with green energy. IEEE Wirel. Commun. **21**(1), 90–96 (2014)
12. O. Ozel, K. Tutuncuoglu, J. Yang, S. Ulukus, A. Yener, Transmission with energy harvesting nodes in fading wireless channels: optimal policies. IEEE J. Sel. Areas Commun. **29**(8), 1732–1743 (2011)
13. Y. Guo, J. Xu, L. Duan, R. Zhang, Joint energy and spectrum cooperation for cellular communication systems. IEEE Trans. Commun. **62**(10), 3678–3691 (2014)
14. B. Gurakan, O. Ozel, J. Yang, S. Ulukus, Energy cooperation in energy harvesting communications. IEEE Trans. Commun. **61**(12), 4884–4898 (2013)
15. B. Gurakan, O. Ozel, J. Yang, S. Ulukus, Two-way and multiple-access energy harvesting systems with energy cooperation, in *Proceedings of IEEE ASILOMAR* (Pacific Grove, CA, USA, Nov. 2012), pp. 58–62
16. B. Gurakan, O. Ozel, J. Yang, S. Ulukus, Energy cooperation in energy harvesting two-way communications. in *Proceedings of IEEE ICC* (Budapest, Hungary, June 2013), pp. 3126–3130
17. K. Tutuncuoglu, A. Yener, Multiple access and two-way channels with energy harvesting and bi-directional energy cooperation. in *Proceedings of IEEE ITA* (San Diego, CA, USA, Feb 2013), pp. 1–8
18. K. Tutuncuoglu, A. Yener, Cooperative energy harvesting communications with relaying and energy sharing. in *Proceedings of IEEE ITW* (Sevilla, Spain, Sept 2013), pp. 1–5
19. D. Wang, P. Ren, Y. Wang, Q. Du, L. Sun, Energy cooperation for reciprocally-benefited spectrum access in cognitive radio networks, in *Proceedings of IEEE GlobalSIP* (Atlanta, GA, USA, Dec 2014), pp. 1320–1324
20. Y.K. Chia, S. Sun, R. Zhang, Energy cooperation in cellular networks with renewable powered base stations. IEEE Trans. Wirel. Commun. **13**(12), 6996–7010 (2014)
21. J. Xu, L. Duan, R. Zhang, Cost-aware green cellular networks with energy and communication cooperation. IEEE Commun. Mag. **53**(5), 257–263 (2015)
22. J. Xu, R. Zhang, Comp meets smart grid: a new communication and energy cooperation paradigm. IEEE Trans. Veh. Technol. **64**(6), 2476–2488 (2015)
23. S. Lakshminarayana, T.Q.S. Quek, H.V. Poor, Cooperation and storage tradeoffs in power grids with renewable energy resources. IEEE J. Sel. Areas Commun. **32**(7), 1386–1397 (2014)
24. Y. Zhang, H.M. Wang, T.X. Zheng, Q. Yang, Energy-efficient transmission design in non-orthogonal multiple access. IEEE Trans. Veh. Technol. **66**(3), 2852–2857 (2017)

25. Z. Yang, Z. Ding, P. Fan, N. Al-Dhahir, A general power allocation scheme to guarantee quality of service in downlink and uplink NOMA systems. IEEE Trans. Wireless Commun. **15**(11), 7244–7257 (2016)
26. J. Cui, Z. Ding, P. Fan, A novel power allocation scheme under outage constraints in NOMA systems. IEEE Signal Process. Lett. **23**(9), 1226–1230 (2016)
27. F. Fang, H. Zhang, J. Cheng, V.C.M. Leung, Energy-efficient resource allocation for downlink non-orthogonal multiple access network. IEEE Trans. Commun. **64**(9), 3722–3732 (2016)
28. Z. Wei, D.W.K. Ng, J. Yuan, Power-efficient resource allocation for MC-NOMA with statistical channel state information, in *Proceedings of IEEE GLOBECOM* (Washington, DC, USA, Dec 2016), pp. 1–7
29. Y. Sun, D.W.K. Ng, Z. Ding, R. Schober, Optimal joint power and subcarrier allocation for MC-NOMA systems, in *Proceedings of IEEE GLOBECOM* (Washington, DC, USA, Dec 2016), pp. 1–6
30. B. Xu, Y. Chen, J.R. Carrión, T. Zhang, Resource allocation in energy-cooperation enabled two-tier noma hetnets toward green 5g. IEEE J. Sel. Areas Commun. **35**(12), 2758–2770 (2017)
31. Y. Saito, Y. Kishiyama, A. Benjebbour, T. Nakamura, A. Li, K. Higuchi, Non-orthogonal multiple access (NOMA) for cellular future radio access, in *Proceedings of IEEE VTC (Spring)* (Dresden, Germany, June 2013) pp. 1–5,
32. Z. Ding, F. Adachi, H.V. Poor, The application of MIMO to non-orthogonal multiple access. IEEE Trans. Wirel. Commun. **15**(1), 537–552 (2016)
33. S. Boyd, A. Mutapcic. *Subgradient Methods*. Stanford University (2008)
34. K. Yasuda, L. Hu, Y. Yin, A grouping genetic algorithm for the multi-objective cell formation problem. Int. J. Prod. Res. **43**(4), 829–853 (2005)
35. D.E. Goldberg, K. Deb, *A Comparative Analysis of Selection Schemes Used in Genetic Algorithms* (Morgan Kaufmann Publishers Inc., San Mateo, CA, 1991)
36. M. Grant, S. Boyd. Cvx: Matlab software for disciplined convex programming
37. Z. Han, D. Niyato, W. Saad, T. Baar, A. Hjrungnes, *Game Theory in Wireless and Communication Networks: Theory, Models, and Applications* (Cambridge University, Press, 2012)
38. M. Kobayashi, G. Caire, An iterative water-filling algorithm for maximum weighted sum-rate of Gaussian MIMO-BC. IEEE J. Sel. Areas Commun. **24**(8), 1640–1646 (2006)
39. S. Zhang, N. Zhang, S. Zhou, J. Gong, Z. Niu, X.S. Shen, Energy-aware traffic offloading for green heterogeneous networks. IEEE J. Sel. Areas Commun. **34**(5), 1116–1129 (2016)
40. M. Zheng, P. Pawelczak, S. Stanczak, H. Yu, Planning of cellular networks enhanced by energy harvesting. IEEE Commun. Lett. **17**(6), 1092–1095 (2013)
41. K. Son, H. Kim, Y. Yi, B. Krishnamachari, Base station operation and user association mechanisms for energy-delay tradeoffs in green cellular networks. IEEE J. Sel. Areas Commun. **29**(8), 1525–1536 (2011)
42. D. Liu, Y. Chen, K.K. Chai, T. Zhang, M. Elkashlan, Two-dimensional optimization on user association and green energy allocation for HetNets with hybrid energy sources. IEEE Trans. Commun. **63**(11), 4111–4124 (2015)
43. A. Ghosh, N. Mangalvedhe, R. Ratasuk, B. Mondal, M. Cudak, E. Visotsky, T.A. Thomas, J.G. Andrews, P. Xia, H.S. Jo, H.S. Dhillon, T.D. Novlan, Heterogeneous cellular networks: from theory to practice. IEEE Commun. Mag. **50**(6), 54–64 (2012)
44. T. Han, N. Ansari, Green-energy aware and latency aware user associations in heterogeneous cellular networks, in *Proceedings of IEEE GLOBECOM* (Atlanta, GA, USA, Dec. 2013), pp. 4946–4951
45. D. Tse, P. Viswanath, *Fundamentals of Wireless Communication* (Cambridge University Press, Cambridge, U.K., 2005)

46. D.W.K. Ng, E.S. Lo, R. Schober, Energy-efficient resource allocation in ofdma systems with large numbers of base station antennas. IEEE Trans. Wirel. Commun. **11**(9), 3292–3304 (2012)
47. Z. Yang, Z. Ding, P. Fan, G.K. Karagiannidis, On the performance of non-orthogonal multiple access systems with partial channel information. IEEE Trans. Commun. **64**(2), 654–667 (2016)
48. Z. Ding, R. Schober, H.V. Poor, A general MIMO framework for NOMA downlink and uplink transmission based on signal alignment. IEEE Trans. Wirel. Commun. **15**(6), 4438–4454 (2016)

Chapter 11
NOMA in Vehicular Communications

Yingyang Chen, Li Wang, Yutong Ai, Bingli Jiao and Lajos Hanzo

11.1 Background and Motivation

With the rapid development of intelligent transportation systems (ITS), the broad objective of vehicular communications is to improve the travel experience of users. To support a variety ITS applications, the integrated vehicular networking concept termed as 'vehicle-to-everything' (V2X) has been proposed. To elaborate a little further, this includes four main types of communications, namely vehicle-to-vehicle (V2V), vehicle-to-pedestrian (V2P), vehicle-to-infrastructure (V2I), and vehicle-to-network (V2N) scenarios, where the ultimate objective is that of offering improved road safety, traffic efficiency, and infotainment services [1].

The IEEE 802.11p standard was conceived in support of wireless access for vehicular environments (WAVE), specifically dedicated to vehicular safety applications and to the provision of data rates ranging from 6 to 27 Mbps for short-transmission distances [2]. However, 802.11p fails to support flexible scalability and to provide quality of service (QoS) guarantees. Furthermore, it has a potentially unbounded delay. These characteristics of the IEEE 802.11p standard prevent its employment in demanding V2X services requiring low latency and high reliability [3, 4]. Moreover,

Y. Chen · B. Jiao
School of Electronics Engineering and Computer Science, Peking University, Beijing 100871, China
e-mail: chenyingyang@pku.edu.cn

B. Jiao
e-mail: jiaobl@pku.edu.cn

L. Wang (✉) · Y. Ai
School of Electronic Engineering, Beijing University of Posts and Telecommunications, Beijing 100876, China
e-mail: liwang@bupt.edu.cn

Y. Ai
e-mail: ytailiwang@bupt.edu.cn

L. Hanzo
School of Electronics and Computer Science, University of Southampton, Southampton SO17 1BJ, U.K.
e-mail: lh@ecs.soton.ac.uk

M. Vaezi et al. (eds.), *Multiple Access Techniques for 5G Wireless Networks and Beyond*, https://doi.org/10.1007/978-3-319-92090-0_11

due to the asynchronous nature of the IEEE 802.11p transmission, its performance is inevitably degraded by the packet collisions imposed by the hidden node problem encountered in carrier sense multiple access with collision avoidance (CSMA/CA) solutions. Finally, the current status of IEEE 802.11p does not support an evolutionary path for improving its reliability, robustness, and coverage [5].

As an alternative to the IEEE 802.11p-based vehicular ad hoc network (VANET) concept, the long-term evolution (LTE)-based V2X solution has been actively developed by the 3rd Generation Partnership Project (3GPP), and it was redefined as LTE V2X within the 3GPP standardization framework so as to provide a beneficial solution for V2X communications [6]. As a benefit of the global deployment and commercialization of LTE systems, LTE V2X serves as an integrated solution for vehicular communications. To be more specific, given an LTE network, both V2I and V2N services can be supported at a high data rate, whilst maintaining an excellent QoS with the aid of the so-called eNodeBs. Meanwhile, LTE can be extended to support V2V and V2P services by invoking direct device-to-device (D2D) communications for satisfying the QoS requirements even in the case of high vehicular densities [7]. Finally, in contrast to 802.11p, the hidden node problem can be avoided in LTE V2X scenario due to its synchronous nature, whilst both the reliability and latency can also be improved.

The rest of this section is organized as follows. First, we provide an overview of existing LTE-based V2X systems. Then, we describe the applicability of a range of popular transmission techniques to vehicular communications. A non-orthogonal multiple access (NOMA) and spatial modulation (SM)-based transmission scheme are proposed for supporting the high data rate and high-reliability demands of V2X systems. Finally, we detail the outline of this chapter.

11.1.1 Overview of LTE-Based V2X

11.1.1.1 Typical V2X Services

The operational LTE system is already capable of supporting ITS applications. The V2X services can be broadly classified into three typical types, namely *road safety*, *traffic efficiency*, and *infotainment* enhancement applications [1, 8]. *Road safety* enhancements aim for reducing the risk of accidents and hence have to satisfy stringent reliability and latency specifications. Basically, the road safety services are short messages and periodically broadcast from each vehicle to its neighbours within a particular geographic region.

As the second category of vehicular applications, *traffic efficiency* enhancements aim for optimizing the platooning of vehicles by reducing traffic congestion. The vehicles are required to collect sensed data and send them to the remote management servers for route planning. Although the reliability and delay requirements related to traffic efficiency enhancements are less strict than those of the safety enhancements, it is still necessary to keep the packet loss and latency low in high-velocity environments.

In contrast to the previous two categories, *infotainment* services include a range of traditional and emerging Internet applications, such as popular content download and dissemination, social networking, and Web browsing, with the goal of providing an improved driving experience.

11.1.1.2 LTE-Based V2X Communication Modes

To support key applications in vehicular communications, the LTE system offers a pair of communication modes.

LTE D2D for V2V/V2P: Although the original system does not support V2V/V2P communications natively, LTE has been extended to support V2V/V2P direct communications based on a device-to-device (D2D) sidelink design through the so-called PC5 interface [7]. The LTE D2D mode allows the terminals in close proximity of each other to communicate directly without involving the base stations. As a benefit, the end-to-end latency can be reduced for satisfying the associated QoS requirements. At the time of writing, the LTE D2D mechanism is considered as the baseline for PC5-based V2V/V2P communications [9]. However, in high user density scenarios, the attributes of V2X services are different from those of the legacy LTE D2D communications, since V2X services are periodic or event-triggered. Hence, efficient resource allocation has to be conceived for dense, high-mobility scenarios.

Cellular LTE for V2I/V2N: Cellular LTE refers to the common communication mode between vehicles and infrastructure/network units. Specifically, there are two main cellular LTE mechanisms, namely *unicast* and *multicast*. In the case of *unicast*, the vehicles are addressed individually. By contrast, in the *multicast* case, all vehicles in the relevant area are collectively addressed. LTE supports high-quality multicast and transmissions through the evolved multimedia broadcast multicast service (eMBMS) capabilities in the radio access network [10]. Compared to unicast services, multicast offers the capability of geocasting the data to a set of users more resource efficiently, although at the cost of longer delays due to the cumbersome eMBMS session set-up, especially in the face of a heavy traffic load.

11.1.1.3 LTE-Challenges in Vehicular Scenarios

The LTE system is capable of providing a round-trip delay below 10 ms and a radio access latency of less than 100 ms. It is based on orthogonal frequency-division multiple access (OFDMA) in the downlink and single-carrier frequency-division multiple access (SC-FDMA) in the uplink. It exhibits flexible resource allocation and scheduling. The LTE system also relies on multiple-input multiple-output (MIMO) techniques for improving the diversity and/or multiplexing gain of the previous generations, making LTE attractive in dynamic vehicular wireless propagation environments.

However, the ever-growing demands for vehicular communications increase the tele-traffic congestion. Hence, it is desirable to achieve a **high bandwidth efficiency, massive connectivity, high reliability, and low latency** in V2X communications. One of the limitations of the LTE systems arises from the fact that LTE was designed for supporting the user terminals sharing the wireless resources using orthogonal multiple access (OMA), which can be potentially improved by NOMA schemes in V2X communications.

In order to support advanced V2X services, given their stringent reliability and latency requirements, multiple-input multiple-output (MIMO) techniques may be invoked. Traditionally, MIMO schemes have been designed either for enhancing the diversity gain by combating the channel fading (e.g. Alamouti code), or for spatial multiplexing (e.g. Vertical Bell Laboratories Layered Space-Time, termed VBLAST), albeit they are amalgamated by the multi-functional MIMO concept of [11, 12]. To accommodate the ever-increasing demands of multimedia services and applications, the massive MIMO concept emerged [13, 14]. Theoretically, massive MIMO is able to reap all the benefits of conventional MIMO and offers abundant degrees of freedom (DoFs). By exploiting the knowledge of the channel state information at the transmitter (CSIT), a massive antenna array becomes capable of simultaneously serving a large number of users by sharing its multiplexing gain among them, while providing higher data rates and transmission reliability. Furthermore, in contrast to shirt-pocket-sized handsets, the employment of large-scale MIMO schemes becomes realistic in V2X scenarios, since multiple antennas can be realistically accommodated [15, 16].

However, massive MIMOs suffer from various problems, including the inter-antenna interference (IAI) and the high complexity of the receivers. It would be a particularly costly process to acquire CSIT in frequency-division duplexing (FDD) systems. Moreover, the hardware cost (e.g. a dedicated radio frequency (RF) chain associated with each antenna) becomes excessive for large antenna arrays. In vehicular wireless communications, the gravest challenge is the hostile high-Doppler propagation imposed. For example, the dominant Doppler effect aggravates the inter-subcarrier interference of orthogonal frequency-division multiplexing (OFDM), and the strong line of sight (LoS) component of V2V channels would aggravate the spatial correlation between antennas. Therefore, the direct applications of massive MIMO in vehicular transmissions are deemed to be problematic, and another version of massive antenna technology is required to be fit for LTE V2X communications.

11.1.2 The Applicability of NOMA to V2X Communications

To mitigate the probability of access collision in V2X environments, a range of novel multiple access techniques has been proposed, such as sparse code multiple access (SCMA), pattern division multiple access (PDMA), and non-orthogonal multiple access (NOMA) to support higher bandwidth efficiency and massive connectivity [17, 18]. Among these techniques, NOMA exhibits an appealing low receiver complexity, high bandwidth efficiency, and massive connectivity by allowing multiple users to share the same channel resource via power domain multiplexing. Thus, NOMA is considered to be a promising candidate for future wireless access [19]. To mitigate the multiple access interference (MAI), multi-user detection (MUD) techniques such as successive interference cancellation (SIC) [20] can be applied to the end-user receivers for detecting the desired signals. Through power domain multiplexing at the transmitter and SIC at the receivers, NOMA becomes capable of fully exploiting its capacity region hence outperforming the OMA schemes [21].

The specific design aspects of NOMA schemes in cellular environments have been discussed in [22–24]. Explicitly, in [22], the concept of basic NOMA with SIC was introduced and its performance was compared to that of the traditional orthogonal frequency-division multiple access (OFDMA) scheme through a system-level evaluation. A beneficial power allocation scheme was designed in [23] for striking compelling tradeoffs between the user fairness and system throughput. Lv et al. [24] studied a new cooperative NOMA transmission scheme and derived the outage probability associated with fixed power allocation.

In vehicular environments, NOMA provides a new dimension for V2X services to alleviate the access collisions, thereby improving the bandwidth efficiency as well as supporting massive connectivity. The authors of [25] proposed an contention-based uplink NOMA solution in order to reduce the control signalling overhead. In [26], the NOMA concept was exploited to enhance the transmission of safety information, which required low latency and high reliability within a dense vehicular communication network. The authors of [27] invoked the NOMA principle for boosting the bandwidth efficiency of the infotainment applications in V2X services. In conclusion, NOMA is eminently applicable for supporting V2X services with enhanced bandwidth efficiency and QoS support.

11.1.3 The Applicability of SM to V2X Communications

In recent years, spatial modulation (SM) [28] has grown in popularity, because in contrast to the traditional MIMO configurations, it only activates a single transmit antenna (Tx) at every transmission instance. Hence, it only requires a single-RF chain. As a benefit, the inter-antenna interference (IAI) can be completely eliminated. Thus, a reduced implementational cost and complexity are achieved [29, 30].

The basic idea of SM was initially derived from Chau and Yu's work dating back to 2001 [31], where the receiver decodes the signals transmitted from different antennas. Then, a compelling SM-MIMO solution was proposed by Mesleh et al. in [32]. Since then, SM has been extensively studied in the scenario of point-to-point communications. In [28], the authors studied the channel capacity of the SM system under the parlance of information-guided channel hopping (IGCH). It was shown that IGCH provides better spectral efficiency than orthogonal space-time block coding (OSTBC). In [33], the SM concept was studied by using a low-complexity two-stage demodulator, and the potential advantages of SM-MIMO compared to the existing spatial-multiplexing and Alamouti schemes were shown. In [34], Jeganathan et al. developed the maximum likelihood (ML)-optimum demodulator for SM-MIMO and a range of performance improvements was shown compared to the suboptimal demodulator introduced in [33].

The SM philosophy is that not only the classic quadrature amplitude modulation (QAM) symbols but also the index of the active Tx (spatial constellation) convey information for the sake of achieving bandwidth efficiency enhancements without sacrificing the advantages of a single-RF stage. Consequently, SM was proposed to be combined with massive MIMOs, yielding the novel concept of massive SM-MIMO, where each UE still has one RF chain combined with a massive Tx configuration [35, 36, 38]. Due to the single-RF structure of SM, both the cost and the design complexity of each user terminal remain similar to those of SM-MIMOs, while the data rates can be boosted by conveying more information bits via employing a large Tx array. More specifically, a large-scale multi-user SM-MIMO system was proposed in [35] along with multi-user detection (MUD) schemes. In [36], Wang et al. proposed an uplink transceiver scheme for massive SM-MIMO within frequency-selective fading environments. The authors of [37] investigated the achievable uplink spectral efficiency in a multi-cell massive SM-MIMO scenario, and [38] further investigated the optimal number of Txs at the user equipment.

Indeed, a recent survey of SM can be found in [39]. In [40–43], SM and its extensions were considered in vehicular environments. A differential SM scheme was proposed for vehicle communications in [40], exhibiting robustness against time-selective fading and Doppler effects. Fu et al. [41] studied the bit error rate (BER) performance of SM under a three-dimensional V2V channel model. Peppas et al. [42] applied space shift keying (SSK) in inter-vehicular communications and derived a closed-form expression for the pairwise error probability. In [43], the performance of massive SM-MIMO over a spatio-temporally correlated Rician channel was analysed under a high-speed railway scenario. Moreover, Cui and Fang have demonstrated that by activating a single Tx, SM is capable of alleviating the channel correlation. In conclusion, SM has become increasingly appealing for V2V systems.

11.1.4 NOMA-SM Tailored for Vehicular Communications

Let us continue by conceiving a novel transmission scheme, termed NOMA-SM, by intrinsically amalgamating NOMA and SM in support of vehicular communications [27]. Specifically, in synergy with the inherent requirement of high bandwidth efficiency, NOMA is invoked for non-orthogonally accessing all the resources combined with the single-RF benefits of SM. The bandwidth efficiency of the proposed NOMA-SM scheme is further boosted by a massive Tx configuration.

Against this background, the key points of the proposed scheme are threefold: firstly, the novel NOMA-SM concept is proposed and its link reliability is quantified. Secondly, the capacity of NOMA-SM is derived and verified by Monte Carlo simulations. Thirdly, a pair of upper bounds on the capacity of NOMA-SM is formulated in closed form and a power allocation optimization is considered.

Explicitly, instead of simply combining a pair of popular techniques, their benefits are intrinsically amalgamated. By investigating the BER performance of NOMA in comparison to different MIMO techniques and the bandwidth efficiency of SM combined with distinct multiple access methods, NOMA and SM are shown to cooperatively improve V2V transmissions.

11.1.5 Outline of the Chapter

The rest of this chapter is organized as follows. In Sect. 11.2, the system model of NOMA-SM is presented, while Sect. 11.3 provides the capacity analysis and mutual information (MI) evaluation of NOMA-SM. Our capacity upper bound derivations and power allocation problem are considered in Sect. 11.4. Simulation results and discussions of the BER performance are provided in Sect. 11.5, together with the numerical capacity analysis and power allocation optimization. In the final section, we offer the main conclusions of this chapter and discuss some open problems as well as a range of promising potential research directions. For convenience, we list the most important notations here.

Notation: Uppercase and lowercase bold-faced letters indicate matrices and vectors, respectively. $(\cdot)^{-1}$, $(\cdot)^{H}$, $\det(\cdot)$, and $[\cdot]_{p,q}$ represent inverse, conjugate-transpose, determinant, and the entry in the pth row and q-column of a matrix, respectively. $\mathbb{E}_X\{\cdot\}$ denotes the expectation on the random variable X. $\mathbf{A} \in \mathbb{C}^{M\times N}$ is a complex-element matrix with dimensions $M \times N$, and $\mathbf{I}_N$ is an $N \times N$ identity matrix. $|\cdot|$ and $(\cdot)^*$ imply the absolute value and the conjugate of a complex scalar, while $\|\cdot\|$ denotes the Euclidean norm of a vector. Finally, $x \sim \mathscr{CN}\left(\mu, \sigma^2\right)$ indicates that the random variable x obeys a complex Gaussian distribution with mean μ and variance σ^2.

11.2 System Model

We consider a generic vehicular communication system, where the vehicle-to-infrastructure (V2I), V2V, and intra-vehicle transmissions are all included. As shown in Fig. 11.1, a base station (BS) is located at the roadside while the vehicles V_1 and V_2 are in motion. There is a mobile user U in V_1 who requests to download a file locally cached at the BS. Vehicle V_2 also requests to download its own intended signal from BS. We assume that V_1 has also acquired the signal of V_2, as a result of the first transmission phase, during which the messages of V_1 and V_2 are transmitted simultaneously from the BS. For example, BS employs a NOMA technique to multiplex signals of V_1 and V_2 in the power domain. By involving the classical SIC, V_1 extracts the signal of V_2 in the spirit of cooperation. Another appropriate interpre-

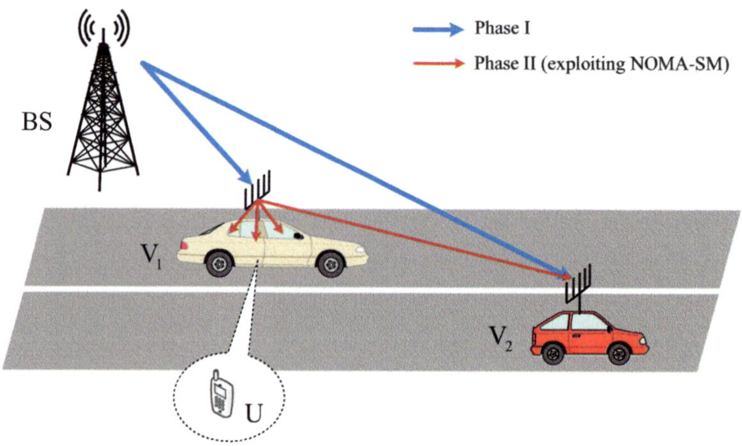

Fig. 11.1 An illustration of the considered vehicular communication system, where NOMA-SM is applied in Phase II

tation is related to the distribution of popular multimedia contents in VANET [44], using peer-to-peer protocols for exchanging popular packets through V2V channels.

Therefore, as shown in Fig. 11.1, cooperative inter-vehicle transmission is constructed during the second phase to enhance the reception reliability. Specifically, V_1 forwards the desired signal to V_2 for cooperatively enhancing the reception at V_2. Furthermore, the second phase scenario can be generalized to various situations. For example, user U can be a roadside unit, aiming for exchanging information with the onboard unit of the vehicle V_1. While U may be a vehicle which is much closer to V_1 than V_2. Similar to the concept in [45], a VANET is formed among these vehicles for exchanging safety information, or for cooperatively distributing popular multimedia contents within a geographical area of interest. In general, our model is valid in a wide range of vehicular scenarios.

In the light of bandwidth scarcity, cognitive radio techniques can be exploited in the second stage to opportunistically exploit the spectrum holes in the licensed spectrum. For example, V_1 may be permitted to share the cellular uplink, for which the data traffic is typically lighter than for the downlink, hence resulting in potential spectrum wastage [46]. Basically, underlay cognitive transmission is feasible without traversing through the primary network. However, the interference imposed by V_1 on the BS in the second stage should be carefully managed, albeit this is beyond the scope of this article. Our main focus is on the second stage of the cooperative transmission in Fig. 11.1, since the performance in the first phase can be analysed similarly. Particularly, the NOMA-SM strategy is employed in the second stage for both V_1–V_2 and V_1–U links.

The schematic diagram of NOMA-SM operated in the second stage is presented in Fig. 11.2, where V_1 assigns distinct transmit power to V_2 and U. The user access is based on NOMA, combined with SM. Although there is the literature proposing

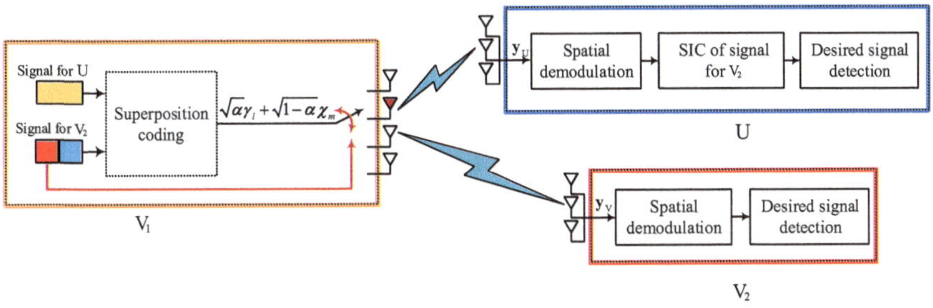

Fig. 11.2 The schematic diagram of the proposed NOMA-SM strategy

multi-user SM schemes [47, 48], we use a classical SM designed for point-to-point transmission [28, 49] in vehicular environments. In what follows, we first elaborate on the principles of the proposed NOMA-SM scheme. Then, our V2V channel model is detailed.

11.2.1 The Principles of NOMA-SM

Let us assume that N_t, N_r, and N_u omnidirectional antennas are employed at V_1, V_2, and U, respectively. As illustrated in Fig. 11.2, the proposed NOMA-SM strategy is applied both for the V_1–V_2 and V_1–U links. At the transmitter V_1, two independent bit streams are prepared for transmission. The bit stream for V_2 is partitioned into two parts: the first $\log_2(N_t)$ bits are used for Tx activation, activating a specific Tx index n_t ($n_t \in \{1, \ldots, N_t\}$). The other $\log_2(M)$ bits destined for V_2 are combined with $\log_2(L)$ bits for U, employing superposition coding.

Subsequently, the modulated symbol $\sqrt{\alpha}\gamma_l + \sqrt{1-\alpha}\chi_m$ is radiated from the activated Tx n_t, where γ_l and χ_m are intended for the in-car user U of V_1 and for V_2, respectively, satisfying $\mathbb{E}\{|\gamma_l|^2\} = \mathbb{E}\{|\chi_m|^2\} = E_s$, where E_s is the average energy per transmission at V_1, while α is the power allocation factor. According to the NOMA principle [23], the transmit power of the distant user in Fig. 11.2 must be higher than that of the close-by user, that is $(1-\alpha)E_s > \alpha E_s$. With this, $0 < \alpha < \frac{1}{2}$ should be guaranteed since the in-car user has a good channel. As a result, a block of $\log_2(N_t ML)$ bits unambiguously identifies the active Tx n_t and the superimposed complex symbol $\sqrt{\alpha}\gamma_l + \sqrt{1-\alpha}\chi_m$ transmitted from it. Hence, a NOMA-SM super symbol can be expressed as

$$\mathbf{x} = \mathbf{e}_{n_t}\left(\sqrt{\alpha}\gamma_l + \sqrt{1-\alpha}\chi_m\right),$$

where $\mathbf{e}_{n_t}$ is the n_tth column of the identity matrix $\mathbf{I}_{N_t}$, indicating that the n_tth Tx of V_1 is activated, while the other $(N_t - 1)$ Txs are deactivated. Furthermore, χ_m is the mth symbol in the M-ary amplitude-phase modulation (APM) used for V_1–V_2 transmission, while γ_l is the lth symbol in the L-ary APM for V_1–U transmission.

Considering the propagation inside the vehicle V_1, we assume that the in-car user U experiences a frequency-flat Rayleigh channel. For example, the Txs of V_1 are installed on the central column of the vehicular dashboard, while the receive antennas (RAs) of U are placed behind the passenger front seat, without LoS from V_1. In [50], this scenario has been shown to be well suited to characterize diffuse scattering. Thus, we let $\mathbf{G} \in \mathbb{C}^{N_r \times N_t}$ denote the channel matrix between V_1 and U, and assume that all entries of $\mathbf{G}$ are independent identically distributed (i.i.d), obeying the distribution $\mathcal{CN}(0, 1)$. The signal vector received at U and V_2 can be written as

$$\mathbf{y}_{\mathrm{U}} = \mathbf{g}_{n_t}\left(\sqrt{\alpha}\gamma_l + \sqrt{1-\alpha}\chi_m\right) + \mathbf{w}_{\mathrm{U}}, \tag{11.1}$$

$$\mathbf{y}_{\mathrm{V}} = \sqrt{p_0}\mathbf{h}_{n_t}\left(\sqrt{\alpha}\gamma_l + \sqrt{1-\alpha}\chi_m\right) + \mathbf{w}_{\mathrm{V}}, \tag{11.2}$$

respectively, where p_0 represents the average power drop between V_1 and V_2 due to the large-scale fading. Furthermore, $\mathbf{g}_{n_t} \in \mathbb{C}^{N_u \times 1}$ is the n_tth column of $\mathbf{G}$, representing the channel vector between U and the n_tth Tx of V_1, while $\mathbf{h}_{n_t} \in \mathbb{C}^{N_r \times 1}$ is the n_tth column of the V2V channel matrix $\mathbf{H} \in \mathbb{C}^{N_r \times N_t}$, indicating the complex fading envelope between V_2 and the n_tth Tx of V_1. Finally, $\mathbf{w}_{(\cdot)}$ denotes a complex additive white Gaussian noise (AWGN) vector with a power spectrum density of σ_0^2 per entry. For the inter-vehicle channel, the path loss is considerable in (11.2), while it is neglected between the in-car user and the antenna array of V_1.

In our system, the transmitter and both receivers are assumed to have perfect synchronization in both time and frequency. Full channel state information is assumed to be available at receivers (i.e. CSIR). In principle, both V_2 and U first have to detect the signal destined for V_2, i.e. the activated Tx index $\hat{n}_t$ and the APM symbol $\chi_{\hat{m}}$ at each particular time instant. The corresponding optimum maximum likelihood (ML) detector is invoked at U and V_2 according to

$$\left(\hat{n}_t, \chi_{\hat{m}}\right) = \arg\min_{n_t, m} \left\|\mathbf{y}_U - \sqrt{1-\alpha}\mathbf{g}_{n_t}\chi_m\right\|^2, \tag{11.3}$$

$$\left(\hat{n}_t, \chi_{\hat{m}}\right) = \arg\min_{n_t, m} \left\|\mathbf{y}_V - \sqrt{p_0(1-\alpha)}\mathbf{h}_{n_t}\chi_m\right\|^2. \tag{11.4}$$

After eliminating the interference imposed by $\left(\hat{n}_t, \chi_{\hat{m}}\right)$ on $\mathbf{y}_{\mathrm{U}}$, U becomes capable of performing another ML detection to acquire the desired signal $\gamma_{\hat{l}}$.

11.2.2 V2V Massive MIMO Channel Model

In contrast to the conventional fixed-to-mobile cellular radio systems, in V2V systems, both the transmitter and receiver are in motion and both are equipped with low-elevation antennas, which will result in quite different propagation conditions. Hence, a non-isotropic scattering V2V stochastic model was proposed in [51] for characterizing a wide variety of V2V scenarios by adjusting relevant model parameters. In [41], a novel three-dimensional V2V geometry-based stochastic channel was proposed for accurately capturing the effect of vehicular traffic density on the channel.

In this article, we consider a spatio-temporally correlated Rician channel model for characterizing our narrowband V2V massive MIMO channel, which has also been exploited in [43] and [52]. We describe the underlying V2V channel as a matrix of complex fading envelopes, i.e. $\mathbf{H} \in \mathbb{C}^{N_t \times N_r}$, which can be expressed as

$$\mathbf{H} = \sqrt{\frac{K}{K+1}}\bar{\mathbf{H}} + \sqrt{\frac{1}{K+1}}\tilde{\mathbf{H}},$$

where K is the Rician factor, while $\bar{\mathbf{H}}$ is the fixed part related to the LoS component. Furthermore, $\tilde{\mathbf{H}}$ represents the variable part, whose entries are correlated complex Gaussian variables. Given $\left[\tilde{\mathbf{H}}\right]_{p,q} = \tilde{h}_{p,q}$, we assume that

$$\mathbb{E}\left\{\tilde{h}^R_{p,q}\tilde{h}^R_{\hat{p},\hat{q}}\right\} = \mathbb{E}\left\{\tilde{h}^I_{p,q}\tilde{h}^I_{\hat{p},\hat{q}}\right\},$$
$$\mathbb{E}\left\{\tilde{h}^R_{p,q}\tilde{h}^I_{\hat{p},\hat{q}}\right\} = \mathbb{E}\left\{\tilde{h}^I_{p,q}\tilde{h}^R_{\hat{p},\hat{q}}\right\} = 0,$$

where $p, \hat{p} \in \{1, \ldots, N_r\}$ and $q, \hat{q} \in \{1, \ldots, N_t\}$. Explicitly, for each $\tilde{h}_{p,q}$, the auto-correlations of the real and imaginary parts are identical and the cross-correlations between real and imaginary parts are equal to zero. Hence, the correlated channel matrix can be described by the widely used Kronecker correlation model [53], which is expressed as

$$\tilde{\mathbf{H}} = \mathbf{\Sigma}_r^{\frac{1}{2}}\hat{\mathbf{H}}\mathbf{\Sigma}_t^{\frac{1}{2}}.$$

Here, $\mathbf{\Sigma}_t \in \mathbb{C}^{N_t \times N_t}$ and $\mathbf{\Sigma}_r \in \mathbb{C}^{N_r \times N_r}$ are the correlation matrices at V_1 and V_2, respectively, with the elements defined as $[\mathbf{\Sigma}_t]_{q,\hat{q}} = \sigma^t_{q,\hat{q}}$ for $q, \hat{q} \in \{1, \ldots, N_t\}$, and $[\mathbf{\Sigma}_r]_{p,\hat{p}} = \sigma^r_{p,\hat{p}}$ for $p, \hat{p} \in \{1, \ldots, N_r\}$. Furthermore, $\hat{\mathbf{H}}$ is the independent Rayleigh channel matrix whose entries are i.i.d complex Gaussian random variables, i.e. $[\hat{\mathbf{H}}]_{p,q} = \hat{h}_{p,q} \sim \mathcal{CN}(0, 1)$. Specifically, the correlation matrices $\mathbf{\Sigma}_t$ and $\mathbf{\Sigma}_r$ can be determined according to a concrete model. Here, the exponential model of Loyka [54] is adopted, and the correlation matrix entries are formed as $\sigma^t_{q,\hat{q}} = \kappa_t^{|q-\hat{q}|}$ and $\sigma^r_{p,\hat{p}} = \kappa_r^{|p-\hat{p}|}$, where κ_t and κ_r are the adjacent antenna correlation coefficients at V_1 and V_2, respectively.

In order to mimic the influence of the V2V channel's time-varying effects, we take the temporal correlation into consideration, which is defined as

$$\delta(\tau) = \mathbb{E}\left\{\hat{\mathbf{H}}(t)\,\hat{\mathbf{H}}(t+\tau)\right\},$$

where τ is the sampling time. In [43], Jakes' model is used for describing the temporal correlation expressed as $\delta(\tau) = J_0(2\pi f_D \tau)$, where f_D is the maximum Doppler frequency related to both the carrier frequency and the velocity of the terminal. For simplicity of analysis, in the following, we omit the index τ. Observe that $\delta = 1$ indicates that the underlying V2V channel is quasi-static, while $\delta < 1$ is related to a time-varying channel due to mobility. Naturally, both the spatial and temporal correlations would affect the performance of the receivers.

11.3 Capacity Analysis of the NOMA-SM System

Recall that the proposed NOMA-SM transmission scheme relies on a pair of independent spaces: the classical signal-domain, pertaining to the radiated superimposed symbol $\sqrt{\alpha}\gamma_l + \sqrt{1-\alpha}\,\chi_m$, and the Tx-domain, pertaining to the activated Tx index n_t. More specifically, the message intended for V_2 is conveyed by both of the two streams. While the message destined for U is only mapped to the classical signal-domain, superimposed with part of V_2's signal in the power domain. In what follows, we investigate the capacity of the collaboration-aided vehicle V_2 and the in-car user U. Monte Carlo estimates are also provided for MI evaluation, followed by an illustrative example to augment the theoretical analysis.

11.3.1 Capacity Analysis of the Collaboration-Aided Vehicle

In the NOMA protocol, the transmit power assigned by V_1 to the distant user V_2 has to be higher than that to the close-by user U. Then, the distant user directly detects its signal, since the interference induced by the close-by user is lower and can thus be treated as background noise. Considering that all Txs of V_1 are activated with the same probability for NOMA-SM, the instantaneous capacity pertaining to the classical signal-domain of V2V transmission is given by

$$\begin{aligned} C_V^{sig} &= \max_{f_\chi} I\left(\chi; \mathbf{y}_V \,|n_t\right) \\ &= \frac{1}{N_t}\sum_{i=1}^{N_t} \log_2\left(\frac{E_s p_0 \|\mathbf{h}_i\|^2+\sigma_0^2}{\alpha E_s p_0 \|\mathbf{h}_i\|^2+\sigma_0^2}\right). \end{aligned} \tag{11.5}$$

Observe that no practical modulation constellation is assumed, when performing these capacity calculations. Since the channel capacity relates to the highest rate in bits per channel use at which information can be sent with arbitrarily low probability of error, in (11.5), we substitute χ_m by χ, which denotes a random input signal alphabet with a distribution of f_χ. On the other hand, the MI conveyed by the spatial-domain Tx-constellations can be written as

$$I\left(n_t; \mathbf{y}_V\right) = \frac{1}{N_t} \sum_{i=1}^{N_t} \int \Pr\left(\mathbf{y}_V \,|\, \mathbf{h}_i\right) \log_2 \frac{\Pr\left(\mathbf{y}_V \,|\, \mathbf{h}_i\right)}{\Pr\left(\mathbf{y}_V\right)} d\mathbf{y}_V, \tag{11.6}$$

where $\Pr\left(\mathbf{y}_V \,|\mathbf{h}_i\right)$ denotes the probability density function (PDF) of the channel output $\mathbf{y}_V$ received over the ith channel vector of $\mathbf{H}$, given by

$$\Pr\left(\mathbf{y}_V \,|\mathbf{h}_i\right) = \frac{1}{\pi^{N_r} \det\left(\boldsymbol{\Sigma}_i\right)} \exp\left\{-\mathbf{y}_V^H \boldsymbol{\Sigma}_i^{-1} \mathbf{y}_V\right\},$$

where $\boldsymbol{\Sigma}_i = \sigma_0^2 \mathbf{I} + pE_s \mathbf{h}_i \mathbf{h}_i{}^H$. As a result, the instantaneous capacity of V_2 in the NOMA-SM system is formulated as

$$C_V = C_V^{sig} + I\left(n_t; \mathbf{y}_V\right). \tag{11.7}$$

Remark It is worth noting that in (11.5), C_V^{sig} is achievable where the optimum input distribution for χ is Gaussian. In fact, this optimum input distribution is also regarded as the optimum input distribution for a conventional SM system. This is a common assumption in the majority of SM capacity-related contributions [28, 55–57], effectively simplifying the analysis. Nevertheless, a fundamental weakness of the Gaussian input assumption is that f_χ affects both $I\left(\chi; \mathbf{y}_V \,|n_t\right)$ and $I\left(n_t; \mathbf{y}_V\right)$. Clearly, the Gaussian input distribution maximizes $I\left(\chi; \mathbf{y}_V \,|n_t\right)$, but it is unclear whether it maximizes $I\left(n_t; \mathbf{y}_V\right)$. In addition, the equiprobable activation of antennas is a widely accepted assumption for SM-enabled systems, albeit this activation regime cannot guarantee the optimal spatial design capable of achieving the capacity in the Tx-domain. Actually, Liu et al. in [55] studied the optimal antenna activation required for Tx-domain capacity maximization. Moreover, Basnayaka et al. [30] have demonstrated that the Gaussian input does not achieve the upper limit of the MI provided by an SM-aided system. As a further insight, although the MI conveyed by the Tx-domain cannot be formulated as an analytical expression, we are inspired to derive the capacity upper bound and to conceive the associated power allocation optimization schemes, which will be addressed in Sect. 11.5.

11.3.2 Capacity Analysis of the In-Car User

In contrast to the receiver of V_2, the receiver of U can detect its own signal after removing the interference imposed by V_2, as seen in Fig. 11.2. To demonstrate the feasibility of this SIC procedure, we first deduce the maximum rate of which U can detect the message of V_2. Specifically, the maximum rate for U detecting the message related to the classical signal-domain of V_2 is given by

$$C_U^{V,sig} = \frac{1}{N_t}\sum_{i=1}^{N_t} \log_2\left(\frac{E_s\|\mathbf{g}_i\|^2 + \sigma_0^2}{\alpha E_s\|\mathbf{g}_i\|^2 + \sigma_0^2}\right). \tag{11.8}$$

The MI associated with U detecting the information embedded in the Tx-constellation of V_2 can be written as

$$I\left(n_t; \mathbf{y}_U\right) = \frac{1}{N_t}\sum_{i=1}^{N_t}\int \Pr\left(\mathbf{y}_U | \mathbf{g}_i\right) \log_2 \frac{\Pr\left(\mathbf{y}_U | \mathbf{g}_i\right)}{\Pr\left(\mathbf{y}_U\right)} \mathbf{d}\mathbf{y}_U, \tag{11.9}$$

where $\Pr\left(\mathbf{y}_U \left| \mathbf{g}_i\right.\right)$ denotes the PDF of the channel output $\mathbf{y}_U$ received over the ith channel vector of $\mathbf{G}$ given by

$$\Pr\left(\mathbf{y}_U \left| \mathbf{g}_i\right.\right) = \frac{1}{\pi^{N_u}\det\left(\boldsymbol{\Omega}_i\right)}\exp\left\{-\mathbf{y}_U^H \boldsymbol{\Omega}_i^{-1}\mathbf{y}_U\right\},$$

where $\boldsymbol{\Omega}_i = \sigma_0^2\mathbf{I} + E_s\mathbf{g}_i\mathbf{g}_i{}^H$. As a result, the instantaneous capacity for U detecting the signal of V_2 can be expressed as

$$C_U^V = C_U^{V,sig} + I\left(n_t; \mathbf{y}_U\right). \tag{11.10}$$

It may be readily seen that $C_U^V > C_V$ is always satisfied, since $\|\mathbf{g}_i\|^2 > p_0\|\mathbf{h}_i\|^2$, guaranteeing the success of SIC. Hence, the capacity of U detecting its own desired signal is written as

$$\begin{aligned} C_U &= \max_{f_\gamma} I\left(\gamma; \mathbf{y}_U | n_t, \chi, \mathbf{G}\right) \\ &= \tfrac{1}{N_t}\sum_{i=1}^{N_t}\log_2\left(1 + \tfrac{\alpha E_s}{\sigma_0^2}\|\mathbf{g}_i\|^2\right), \end{aligned} \tag{11.11}$$

where γ denotes the random input signal variable related to the desired message of U, with a distribution of f_γ. The capacity for U detecting γ indeed becomes achievable when the channel's input distribution f_γ is Gaussian.

11.3.3 Mutual Information

To appreciate the above theoretical analysis in terms of its relevance, next, we characterize the bandwidth efficiency of the proposed NOMA-SM. Assuming perfect knowledge of the instantaneous channel state information at both receivers, the MI achieved by V_2 and U with the aid of practical APM constellations is evaluated by the classical Monte Carlo method. For the collaboration-aided vehicle V_2, the MI between a discrete signal input (n_t, χ_m) and the received signal $\mathbf{y}_V$ can be formulated as

$$\begin{aligned} I(n_t, \chi_m; \mathbf{y}_V | \mathbf{H}) &= \mathbb{E}_{n_t, \chi_m, \mathbf{y}_V} \left\{ \log_2 \frac{\Pr(\mathbf{y}_V | n_t, \chi_m, \mathbf{H})}{\Pr(\mathbf{y}_V | \mathbf{H})} \right\} \\ &= \frac{1}{N_t M} \times \int \Pr(\mathbf{y}_V | \chi_m, \mathbf{h}_i) \log_2 \frac{\Pr(\mathbf{y}_V | \chi_m, \mathbf{h}_i)}{\Pr(\mathbf{y}_V | \mathbf{H})} d\mathbf{y}_V, \end{aligned} \tag{11.12}$$

where the conditional probability $\Pr(\mathbf{y}_V | \chi_m, \mathbf{h}_i)$ is expressed as

$$\begin{aligned} \Pr(\mathbf{y}_V | \chi_m, \mathbf{h}_i) = \frac{1}{\pi^{N_r} \det(\Psi_i)} \exp\Big\{ -\left(\mathbf{y}_V - \sqrt{p_0(1-\alpha)} \mathbf{h}_i \chi_m\right)^H \\ \times \boldsymbol{\Psi}_i^{-1} \left(\mathbf{y}_V - \sqrt{p_0(1-\alpha)} \mathbf{h}_i \chi_m\right) \Big\}, \end{aligned}$$

with $\boldsymbol{\Psi}_i = \sigma_0^2 \mathbf{I} + \alpha p_0 E_s \mathbf{h}_i \mathbf{h}_i^H$. With regard to the in-car user U performing SIC first, the MI between the information input (n_t, χ_m) and the received signal $\mathbf{y}_U$ is given by

$$\begin{aligned} I(n_t, \chi_m; \mathbf{y}_U | \mathbf{G}) = \frac{1}{N_t M} \times \\ \int \Pr(\mathbf{y}_U | \chi_m, \mathbf{g}_i) \log_2 \frac{\Pr(\mathbf{y}_U | \chi_m, \mathbf{g}_i)}{\Pr(\mathbf{y}_U | \mathbf{G})} d\mathbf{y}_U, \end{aligned} \tag{11.13}$$

where the conditional probability $\Pr(\mathbf{y}_U | \chi_m, \mathbf{g}_i)$ is expressed as

$$\begin{aligned} \Pr(\mathbf{y}_U | \chi_m, \mathbf{g}_i) = \frac{1}{\pi^{N_u} \det(\boldsymbol{\Phi}_i)} \times \\ \exp\Big\{ -\left(\mathbf{y}_U - \sqrt{1-\alpha} \mathbf{g}_i \chi_m\right)^H \boldsymbol{\Phi}_i^{-1} \left(\mathbf{y}_U - \sqrt{1-\alpha} \mathbf{g}_i \chi_m\right) \Big\}, \end{aligned}$$

with $\boldsymbol{\Phi}_i = \sigma_0^2 \mathbf{I} + \alpha E_s \mathbf{g}_i \mathbf{g}_i^H$.

Subsequently, the MI between the information input γ_l and the received signal $\mathbf{y}_U$ after perfect SIC is expressed as

$$I\left(\gamma_l; \tilde{\mathbf{y}}_U | \mathbf{g}_{n_t}\right) = \frac{1}{N_t L} \int \Pr(\tilde{\mathbf{y}}_U | \gamma_l, \mathbf{g}_i) \log_2 \frac{\Pr(\tilde{\mathbf{y}}_U | \gamma_l, \mathbf{g}_i)}{\frac{1}{N_t L} \sum_{k,j} \Pr(\tilde{\mathbf{y}}_U | \gamma_k, \mathbf{g}_j)} d\tilde{\mathbf{y}}_U, \tag{11.14}$$

where $\tilde{\mathbf{y}}_U = \mathbf{y}_U - \sqrt{1-\alpha} \mathbf{g}_i \chi_m$ with $i \in \{1, \ldots, N_t\}$ and $m \in \{1, \ldots, M\}$ denotes the received vector after SIC. The conditional probability $\Pr(\tilde{\mathbf{y}}_U | \gamma_l, \mathbf{g}_i)$ is given by

$$\Pr(\tilde{\mathbf{y}}_U | \gamma_l, \mathbf{g}_i) = \frac{1}{\left(\pi \sigma_0^2\right)^{N_u}} \exp\left\{ -\frac{\left\| \tilde{\mathbf{y}}_U - \sqrt{\alpha} \mathbf{g}_i \gamma_l \right\|^2}{\sigma_0^2} \right\}.$$

11.3.4 An Illustration

In this part, a simulation-based study of our theoretical expressions is provided with the aid of the MI attained by practical APM constellations. We set $N_t = 64$, $N_r = N_u = 2$ for our MIMO configurations in conjunction with $\alpha = 0.1$, $E_s = 1$ and $p_0 = 10^{-3}$ are given. The channel matrix $\mathbf{H}$ is generated according to Sect. 11.2.2, where $K = 0.2$, $\kappa_t = \kappa_r = 0.5$, and $\delta = 1$ are used. Each entry of $\mathbf{G}$ is identically and independently generated according to a complex Gaussian distribution $\mathscr{CN}(0, 1)$. In our Monte Carlo evaluations, the 16PSK signal constellation is chosen as the APM for χ_m and γ_l; hence, we have $M = L = 16$. The effective transmit signal-to-noise ratio (SNR) at V_1 is given by $p_0 E_s/\sigma_0^2$ as the horizontal axis of Fig. 11.3. Notice that the transmit-SNR cannot be readily interpreted physically, because it relates the transmitter power to the noise power at the receiver, but its notion is convenient to use in NOMA-aided scenarios. Given the effective transmit-SNR at V_1 as $\text{SNR} = p_0 E_s/\sigma_0^2$, the average receive-SNR at V_2 can be computed as

$$\text{SNR}_r^{V_2} = \frac{(1-\alpha)\,\text{SNR}}{1+\alpha\text{SNR}}.$$

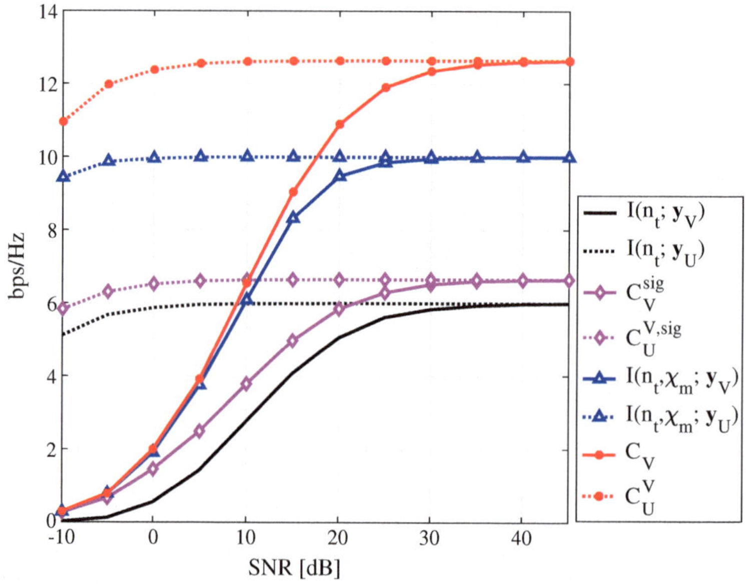

Fig. 11.3 Capacity and MI performance for $N_t = 64$, $N_r = N_u = 2$, $M = L = 16$, and $\alpha = 0.01$. Specifically, C_V^{sig}, $C_U^{V,sig}$, $I(n_t; \mathbf{y}_V)$, and $I(n_t; \mathbf{y}_U)$ are obtained from (11.5), (11.8), (11.6), and (11.9), respectively. While $I(n_t, \chi_m; \mathbf{y}_V)$ and $I(n_t, \chi_m; \mathbf{y}_U)$ are generated from (11.12) and (11.13), respectively, after averaging over multiple channel realizations. Finally, C_V and C_U^V are evaluated from (11.7) and (11.10), respectively

Similarly, the average receive-SNR at U for detecting the signal of V_2 and that of itself is respectively expressed as

$$\begin{aligned} \mathrm{SNR}_r^{U,V_2} &= \tfrac{(1-\alpha)\mathrm{SNR}}{p_0+\alpha \mathrm{SNR}}, \\ \mathrm{SNR}_r^{U} &= \tfrac{\alpha \mathrm{SNR}}{p_0}. \end{aligned}$$

Hence, the effective transmit-SNR at V_1 is unambiguously related to the SNRs at each receiver. Furthermore, we use $\mathrm{SNR} = p_0 E_s/\sigma_0^2$ in all of the subsequent performance analyses.

The relevant results of Fig. 11.3 are discussed as follows.

- The capacity of V_2 gleaned from the signal-domain, that is C_V^{sig} obtained from (11.5), increases steadily up to a saturation point as the SNR increases. By contrast, the capacity for U detecting the signal-domain destined for V_2, i.e. $C_U^{V,sig}$ obtained from (11.8), is higher than C_V^{sig} in the low and moderate SNR domain. Clearly, a successful detection of the signal-domain of V_2 can be performed by U.
- The MI $I(n_t; \mathbf{y}_V)$ generated using (11.6) increases with the SNR and saturates at 6 bps/Hz, since the input entropy of the Tx-domain space is $\log_2(N_t)$. By contrast, the MI $I(n_t; \mathbf{y}_U)$ attained by (11.9) is as high as 6 bps/Hz across almost the entire SNR range, since the channel quality of U is much higher than that of V_2, implying that U can successfully detect the signal of V_2 embedded in the Tx-domain.
- The capacity of V_2, i.e. C_V, grows steadily as the SNR increases up to its saturation at high SNRs, but it remains lower than C_U^V. Since C_V is obtained by the summation of C_V^{sig} and $I(n_t; \mathbf{y}_V)$, and C_U^V equals to the sum of $C_U^{V,sig}$ and $I(n_t; \mathbf{y}_U)$. Naturally, $C_U^V > C_V$ is satisfied, as $C_U^{V,sig}$ and $I(n_t; \mathbf{y}_U)$ are higher than C_V^{sig} and $I(n_t; \mathbf{y}_V)$, respectively. Therefore, U can always perform successful SIC.
- The MI curves $I(n_t, \chi_m; \mathbf{y}_V)$ and $I(n_t, \chi_m; \mathbf{y}_U)$ are generated from (11.12) and (11.13), respectively, after averaging over multiple channel realizations. It may be observed that the simulated curve $I(n_t, \chi_m; \mathbf{y}_V)$ matches the analytical capacity C_V quite closely upto an SNR of 5 dB, but beyond that $I(n_t, \chi_m; \mathbf{y}_V)$ starts to drift away from C_V. By contrast, the drift of $I(n_t, \chi_m; \mathbf{y}_U)$ from C_U^V remains nearly unchanged. Both drifts are due to the fact that the MI attained with the aid of practical APM modulation is upper bounded by the capacity, namely by the maximum data rate related to the optimal input distribution.

11.4 Power Allocation Algorithms

It has been demonstrated that the MI conveyed by the Tx-domain cannot be readily formulated as a closed-form expression, only by resorting to simulations. Thus, it is very hard to perform an optimal power allocation for NOMA-SM. To circumvent this problem, we first derive an upper bound of the NOMA-SM capacity. Then the power allocation, which is capable of maximizing the capacity bound is considered, leading to the optimal solution.

11.4.1 Problem Formulation

Theoretically, the instantaneous capacity of V_2 in the NOMA-SM system can be expressed as

$$C_V = \max_{f_\chi} I\left(n_t, \chi; \mathbf{y}_V\right) = \max_{f_\chi} h\left(\mathbf{y}_V\right) - h\left(\mathbf{y}_V \left| n_t, \chi\right.\right), \tag{11.15}$$

where $h(\cdot)$ denotes the differential entropy. The conditional differential entropy $h\left(\mathbf{y}_V \left| n_t, \chi\right.\right)$ in (11.15) is explicitly given by

$$h\left(\mathbf{y}_V \left| n_t, \chi\right.\right) = \frac{1}{N_t}\sum_{i=1}^{N_t} \log_2 \det\left[\pi e\left(p_0 \alpha E_s \mathbf{h}_i \mathbf{h}_i^H + \sigma_0^2 \mathbf{I}\right)\right].$$

To determine C_V, we have to evaluate $h\left(\mathbf{y}_V\right)$, which requires the knowledge of the distribution of $\mathbf{y}_V$. It may be readily seen that the MI $I\left(n_t, \chi; \mathbf{y}_V\right)$ is maximized if the vector variable $\mathbf{y}_V$ has a Gaussian distribution. Thus, we assume that the received vector $\mathbf{y}_V$ has a Gaussian distribution, which is a zero-mean vector having a covariance matrix presented as

$$\begin{aligned}
\mathbb{E}\left\{\mathbf{y}_V \mathbf{y}_V^H\right\} &= \mathbf{H}\mathbb{E}_{n_t}\left\{\mathbf{e}_{n_t}\mathbb{E}_\chi\left\{p_0\left(1-\alpha\right)\chi\chi^*\right\}\mathbf{e}_{n_t}^H\right\}\mathbf{H}^H \\
&\quad + \mathbf{H}\mathbb{E}_{n_t}\left\{\mathbf{e}_{n_t}\mathbb{E}_\gamma\left\{p_0 \alpha \gamma\gamma^*\right\}\mathbf{e}_{n_t}^H\right\}\mathbf{H}^H + \sigma_0^2\mathbf{I} \\
&= \mathbf{H}\left\{\frac{1}{N_t}\sum_{i=1}^{N_t}\mathbf{e}_i\mathbf{e}_i^H p_0\left(1-\alpha\right)E_s\right\}\mathbf{H}^H \\
&\quad + \mathbf{H}\left\{\frac{1}{N_t}\sum_{n_t=1}^{N_t}\mathbf{e}_i\mathbf{e}_i^H p_0 \alpha E_s\right\}\mathbf{H}^H + \sigma_0^2\mathbf{I} \\
&= \frac{p_0 E_s}{N_t}\mathbf{H}\mathbf{H}^H + \sigma_0^2\mathbf{I}.
\end{aligned}$$

An upper bound of $h\left(\mathbf{y}_V\right)$ can be formulated as

$$h\left(\mathbf{y}_V\right) \le \log_2 \det\left(\pi e\left(\frac{p_0 E_s}{N_t}\mathbf{H}\mathbf{H}^H + \sigma_0^2\mathbf{I}\right)\right).$$

Hence, we obtain an upper bound of C_V which is written as

$$
\begin{aligned}
C_V &\leq \log_2 \det\left(\pi e \left(\tfrac{p_0 E_s}{N_t} \mathbf{H}\mathbf{H}^H + \sigma_0^2 \mathbf{I}\right)\right) \\
&\quad - \tfrac{1}{N_t} \sum_{i=1}^{N_t} \log_2 \det\left(\pi e \left(p_0 \alpha E_s \mathbf{h}_i \mathbf{h}_i^H + \sigma_0^2 \mathbf{I}\right)\right) \\
&= \sum_{j=1}^{N_r} \log_2 \left(\tfrac{p_0 E_s}{N_t} \lambda_j^2 + \sigma_0^2\right) \\
&\quad - \tfrac{1}{N_t} \sum_{i=1}^{N_t} \log_2 \left(p_0 \alpha E_s \|\mathbf{h}_i\|^2 + \sigma_0^2\right) \triangleq C_V^{B_1},
\end{aligned}
\tag{11.16}
$$

where λ_j is the jth singular value of $\mathbf{H}$ with $j \in \{1, \ldots, N_r\}$. Clearly, $C_V^{B_1}$ has N_r DoFs, and it is the same as the capacity of an $(N_t \times N_r)$-element spatially multiplexed MIMO system, subject to inter-user interference.

On the other hand, the MI of the Tx-domain has a natural upper bound written as

$$
I\left(n_t; \mathbf{y}_V\right) \leq \log_2\left(N_t\right),
$$

which corresponds to the maximum MI that can be conveyed by the Tx-domain of the V2V transmission link. Now, another upper bound of C_V may also be formulated as

$$
\begin{aligned}
C_V &\leq C_V^{sig} + \log_2\left(N_t\right) \\
&= \tfrac{1}{N_t} \sum_{i=1}^{N_t} \log_2 \left(\tfrac{E_s p_0 \|\mathbf{h}_i\|^2 + \sigma_0^2}{\alpha E_s p_0 \|\mathbf{h}_i\|^2 + \sigma_0^2}\right) + \log_2\left(N_t\right) \triangleq C_V^{B_2}.
\end{aligned}
\tag{11.17}
$$

Before proceeding, we provide a numerical illustration in order to evaluate both of the upper bounds on the capacity of V_2. Figure 11.4 depicts C_V and both upper bounds of the NOMA-SM system in conjunction with $N_t = 64$, $N_r = 2$, $M = 16$, and $\alpha = 0.1$, which exhibit distinct approximations of C_V within certain SNR regions. The upper bound $C_V^{B_1}$ gives a tight bound of C_V at low SNRs, indicating that the NOMA-SM capacity at V_2 is almost the same as that of a spatially multiplexed MIMO system of the same configuration in the presence of inter-user interference. However, the MI embedded in the Tx-domain saturates as the SNR increases, which is due to the fact that N_t is finite. Hence, at high SNRs, $C_V^{B_2}$ is much tighter.

Based on the above observations, a refined upper bound on the capacity of V_2 in the NOMA-SM system is represented as

$$
C_V^B \triangleq \min\left(C_V^{B_1}, C_V^{B_2}\right).
\tag{11.18}
$$

Considering the QoS of the two receivers from a practical perspective, we define the minimum rate requirement of V_2 and U as $\tilde{C}_V$ and $\tilde{C}_U$, respectively. The optimization problem constructed for maximizing the sum capacity with a power allocation factor of α can be formulated as

$$
\mathscr{P}: \max_{\alpha} C_U + C_V^B
$$

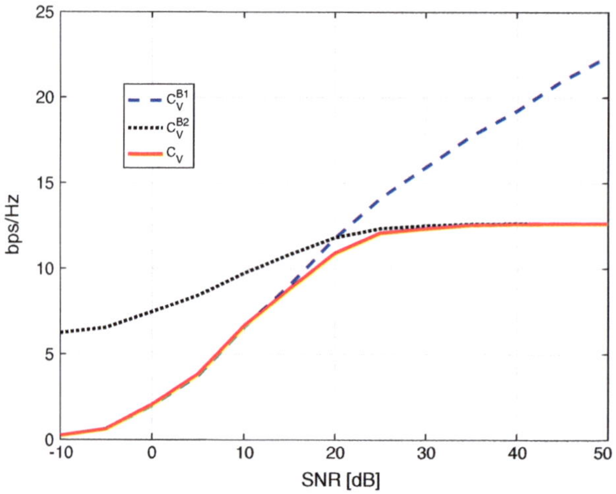

Fig. 11.4 Capacity and two upper bounds of the V2V transmission link with $N_t = 64$, $N_r = 2$, $M = 16$, and $\alpha = 0.01$. Specifically, C_V, $C_V^{B_1}$, and $C_V^{B_2}$ are evaluated from (11.7), (11.16), and (11.17), respectively

$$s.t. \begin{cases} C_U \geq \tilde{C}_U, & (a) \\ C_V^B \geq \tilde{C}_V, & (b) \\ 0 < \alpha < \frac{1}{2}. & (c) \end{cases} \tag{11.19}$$

11.4.2 The Proposed Power Allocation Algorithm

To solve the proposed optimization problem, we first express the derivatives of C_U, $C_V^{B_1}$, and $C_V^{B_2}$ with respect to α as

$$\begin{aligned} \frac{\mathrm{d}C_U}{\mathrm{d}\alpha} &= \frac{1}{N_t}\sum_{i=1}^{N_t}\frac{E_s\|\mathbf{g}_i\|^2}{\alpha E_s\|\mathbf{g}_i\|^2+\sigma_0^2}, \\ \frac{\mathrm{d}C_V^{B_1}}{\mathrm{d}\alpha} &= -\frac{1}{N_t}\sum_{n_t=1}^{N_t}\frac{E_s p_0\|\mathbf{h}_i\|^2}{\alpha E_s p_0\|\mathbf{h}_i\|^2+\sigma_0^2}, \\ \frac{\mathrm{d}C_V^{B_2}}{\mathrm{d}\alpha} &= -\frac{1}{N_t}\sum_{n_t=1}^{N_t}\frac{E_s p_0\|\mathbf{h}_i\|^2}{\alpha E_s p_0\|\mathbf{h}_i\|^2+\sigma_0^2}, \end{aligned} \tag{11.20}$$

respectively. Observe from (11.20) that C_U is a monotonically increasing function of α, given its positive derivative, while both $C_V^{B_1}$ and $C_V^{B_2}$ are decreasing ones. Thus, when the constraint (c) of (11.19) is taken into account, there exist both minimum and maximum capacities that V_2 and U can achieve. Furthermore, to satisfy the constraint (a) and (b), we have the following conditions for $\tilde{C}_U$ and $\tilde{C}_V$, respectively

$$0 < \tilde{C}_U < C_U\left(\alpha = \frac{1}{2}\right),$$

$$C_V^B\left(\alpha = \frac{1}{2}\right) < \tilde{C}_V < C_V^B\left(\alpha = 0\right).$$

Given the above conditions, we can rewrite the constraints of problem $\mathscr{P}$ in a compact form as

$$g^{-1}\left(\tilde{C}_U\right) < \alpha < f^{-1}\left(\tilde{C}_V\right),$$

where $g^{-1}(\cdot)$ and $f^{-1}(\cdot)$ indicate the inverse function of C_U and C_V^B, respectively. To guarantee that the feasible set of problem $\mathscr{P}$ is non-empty, a further refined condition for setting $\tilde{C}_V$ is given by

$$C_V^B\left(\alpha = \frac{1}{2}\right) < \tilde{C}_V < C_V^B\left[\alpha = g^{-1}\left(\tilde{C}_U\right)\right].$$

Moreover, since $\left\|\mathbf{g}_{n_t}\right\|^2 > p_0\left\|\mathbf{h}_{n_t}\right\|^2$ is always satisfied, the derivative of $\left(C_U + C_V^B\right)$ can be guaranteed to have a positive value. Accordingly, the objective function of problem $\mathscr{P}$ is a monotonically increasing function and can be maximized, when α reaches the upper bound of its feasible set. With $\tilde{C}_U$ and $\tilde{C}_V$ being appropriately set, we find that the upper bound of α's feasible set is related to the constraint (b) of (11.19), and the lower bound corresponds to the constraint (a) of (11.19). Thus, the optimal solution of problem $\mathscr{P}$ is

$$\alpha_{opt}^{\mathscr{P}} = f^{-1}\left(\tilde{C}_V\right). \tag{11.21}$$

This optimal solution implies that the amount of power allocated to V_2 is 'just' sufficient to meet the minimum rate requirement $\tilde{C}_V$, while the remaining power is used for U, aiming for maximizing its capacity. Nevertheless, we should notice that there may exist some practical considerations, which require us to give high priority to the V2V transmission link, such as those of safety applications, which have to be served reliably. By contrast, the transmissions for in-car users are typically related to infotainment applications, for example peer-to-peer video sharing and multimedia advertisements [58]. Hence, it may be desirable to maximize the data rate of the V2V link, while guaranteeing the minimum rate requirement of the in-car user. To this end, we develop an alternative optimization problem formulated as

$$\mathscr{O}: \max_{\alpha} C_V^B$$

$$s.t. \begin{cases} C_U \geq \tilde{C}_U, & (a) \\ C_V^B \geq \tilde{C}_V, & (b) \\ 0 < \alpha < \frac{1}{2}. & (c) \end{cases} \tag{11.22}$$

Clearly, the objective function of (11.22) is a monotonically decreasing function of α, and it is maximized, when the constraint (a) is inactive. Therefore, the optimal solution of problem $\mathscr{O}$ can be written as

$$\alpha_{opt}^{\mathscr{O}} = g^{-1}\left(\tilde{C}_U\right). \tag{11.23}$$

So far, we have proposed a pair of power allocation schemes and analysed the solvability of the optimization problems considered. Explicitly, we provided an algorithm for finding the optimal solution of each problem, which are summarized in Table 11.1. The proposed algorithm essentially performs bounding through with the aid of a bisection procedure, yielding globally optimal solutions at linearly increasing computational complexity [59]. In specific, the minimum rate requirements of V_2 and U are respectively set as

$$\begin{aligned} \tilde{C}_U &= \frac{C_U\left(\alpha=\frac{1}{2}\right)}{2}, \\ \tilde{C}_V &= \frac{C_V^B\left(\alpha=\frac{1}{2}\right)+C_V^B\left[\alpha=g^{-1}\left(\tilde{C}_U\right)\right]}{2}, \end{aligned} \tag{11.24}$$

for simplicity. Basically, both of the two power allocation optimization problems satisfy realistic practical considerations, and the suitable one can be flexibly selected based on the specific data priority of the distinct transmission links.

11.5 Simulations and Discussions

In this section, simulation results are provided for evaluating the performance of the proposed NOMA-SM scheme. The system parameters are summarized as follows. The MIMO configurations for the NOMA-SM system are set as $N_t = 64$, $N_r = N_u = 2$. We fix $p_0 = 10^{-3}$, or, equivalently, the path loss exponential is set to 3, and the distance between V_1 and V_2 is assumed to be 10 m, which is typical for urban environments, especially during rush hours.

Table 11.1 Power Allocation Algorithm

Power Allocation Algorithm for Problem $\mathscr{P}$ and Problem $\mathscr{O}$

1. Initialization

Set tolerance $0 < \varepsilon \ll 1$. Calculate $C_U\left(\alpha = \frac{1}{2}\right)$ and set $\tilde{C}_U = C_U\left(\alpha = \frac{1}{2}\right)/2$.

2. Determine the lower bound of α and find the optimal solution of problem $\mathscr{O}$

Set $\alpha_L = 0$ and $\alpha_U = \frac{1}{2}$.
While $\alpha_L - \alpha_U > \varepsilon$
  Set $\alpha = \frac{\alpha_L + \alpha_U}{2}$. Calculate $C_U(\alpha)$.
  If $C_U(\alpha) - \tilde{C}_U > 0$
    $\alpha_U = \alpha$
  Else
    $\alpha_L = \alpha$.
  End
End
Set $\tilde{C}_V = \left[C_V^B\left(\alpha = \frac{1}{2}\right) + C_V^B\left(\alpha = \frac{\alpha_L + \alpha_U}{2}\right)\right]/2$.
The optimal solution to the problem $\mathscr{O}$ is obtained as $\alpha_{opt}^{\mathscr{O}} = \frac{\alpha_L + \alpha_U}{2}$. Calculate $C_U\left(\alpha_{opt}^{\mathscr{O}}\right)$ and $C_V^B\left(\alpha_{opt}^{\mathscr{O}}\right)$.

3. Determine the upper bound of α and find the optimal solution of problem $\mathscr{P}$

Set $\alpha_{\min} = \frac{\alpha_L + \alpha_U}{2}$ and $\alpha_{\max} = \frac{1}{2}$.
While $\alpha_{\max} - \alpha_{\min} > \varepsilon$
  Set $\alpha = \frac{\alpha_{\min} + \alpha_{\max}}{2}$. Calculate $C_V^B(\alpha)$.
  If $C_V^B(\alpha) - \tilde{C}_V > 0$
    $\alpha_{\min} = \alpha$
  Else
    $\alpha_{\max} = \alpha$.
  End
End
The optimal solution of the problem $\mathscr{P}$ is obtained as $\alpha_{opt}^{\mathscr{P}} = \frac{\alpha_{\min} + \alpha_{\max}}{2}$. Calculate $C_U\left(\alpha_{opt}^{\mathscr{P}}\right)$ and $C_V^B\left(\alpha_{opt}^{\mathscr{P}}\right)$.

11.5.1 BER Results and Discussions

In this subsection, the BER performance of the NOMA-SM scheme is compared to NOMA relying on the popular VBLAST technique, termed NOMA-VBLAST. Specifically, we focus on the receiver performance of V_2. The effects of the Rician K-factor, adjacent antenna correlation coefficient, temporal correlation, and power allocation factor are all taken into consideration. The Rician K-factors are configured as $K = 2.186$ and $K = 0.2$ for low and high vehicular traffic density, respectively (see [51] for more details). More specifically, we consider a pair of references: NOMA-VBLAST applied with 16QAM and $N_t = 2$, and NOMA-VBLAST adopted QPSK and $N_t = 4$. The MIMO configuration of the references is the same as that of NOMA-SM except for N_t. Besides, QPSK is applied for NOMA-SM. Thus, the following BER comparisons are carried out for the same bandwidth efficiency of 8 bits per channel use (bpcu). The optimum ML detector described in (11.4) is

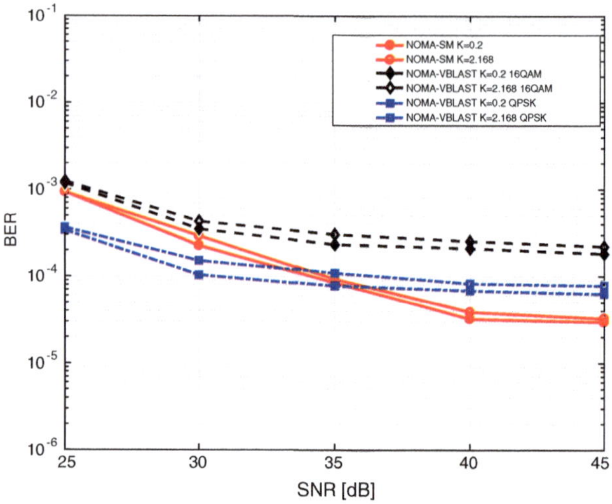

Fig. 11.5 BER comparisons with different Rician K-factor when $\kappa_t = \kappa_r = 0.2$ and $\delta = 1$ are given, and the power allocation factor is fixed at $\alpha = 0.001$, as evaluated by the Monte Carlo simulation with 10^6 channel realizations

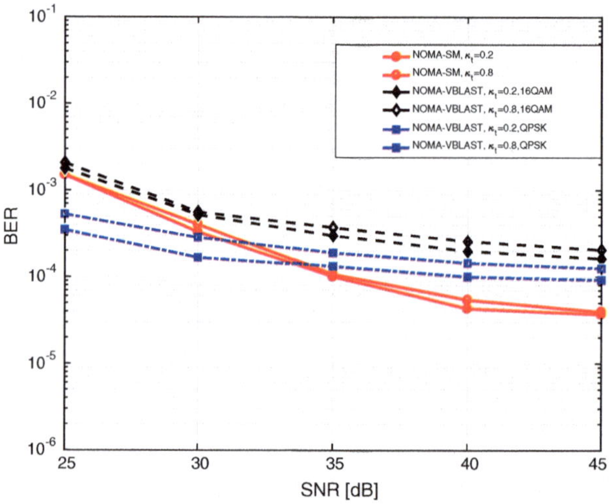

Fig. 11.6 BER comparisons with different adjacent antenna correlation coefficient at V_1, i.e. κ_t, when $K = 0.2, \kappa_r = 0.5$, and $\delta = 1$ are given, and the power allocation factor is fixed at $\alpha = 0.001$, as evaluated by the Monte Carlo simulation with 10^6 channel realizations

employed at V_2 in both schemes. All simulation results of this subsection are obtained through a Monte Carlo method.

In Fig. 11.5, we show the BER performance for different Rician K-factor. It is observed that NOMA-SM outperforms the benchmark especially in the high SNR regime. Additionally, the increase of K imposes a more dominant degradation on

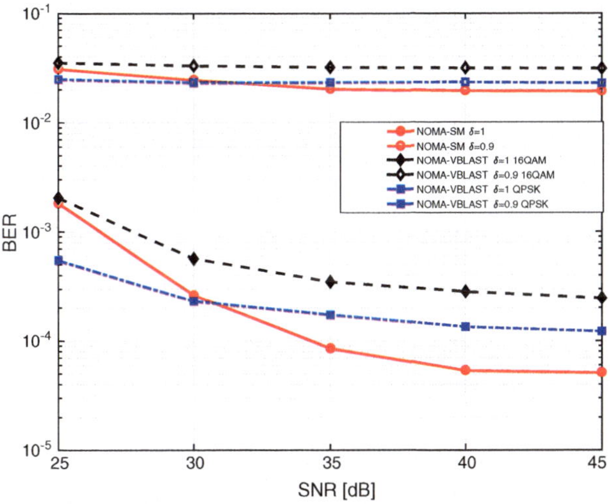

Fig. 11.7 BER comparisons with different temporal correlation coefficient δ when $K = 0.2$ and $\kappa_t = \kappa_r = 0.5$ are given, and the power allocation factor is fixed at $\alpha = 0.001$, as evaluated by the Monte Carlo simulation with 10^6 channel realizations

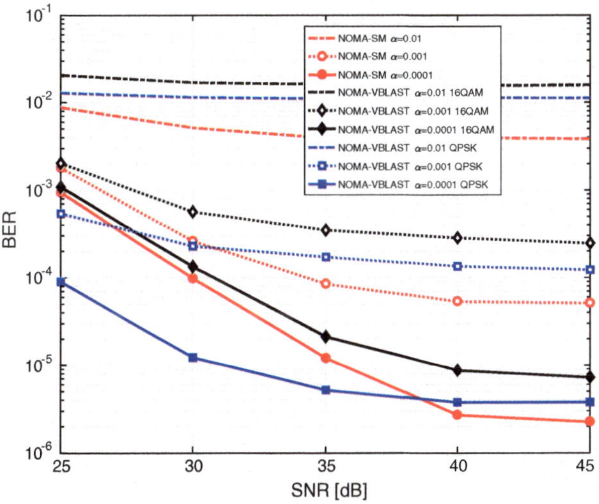

Fig. 11.8 BER comparisons with different power allocation factor α when $K = 0.2, \kappa_t = \kappa_r = 0.5$, and $\delta = 1$ are given, as evaluated by the Monte Carlo simulation with 10^6 channel realizations

both of the NOMA-VBLAST schemes, which rely more vitally on the presence of rich non-LoS scattering. This phenomenon can be explained as follows. The higher Rician factor K represents a stronger LoS component, which increases the spatial correlation among the adjacent channel paths. For NOMA-VBLAST schemes, the multiple-stream information is conveyed with the aid of multiple DoFs. By contrast, for NOMA-SM, although the more severe spatial correlation of the LoS scenario makes it difficult to determine the index of the activated Tx, the remaining information related to the APM signal-domain is transmitted over a single DoF; hence, it is less susceptible to spatial correlation.

Figure 11.6 investigates the BER results associated with different adjacent Tx-correlation coefficients at V_1. Compared to $\kappa_t = 0.8$, $\kappa_t = 0.2$ represents an insignificant spatial correlation. Again, observe from Fig. 11.6 that NOMA-SM is less susceptible to spatial correlation. This phenomenon can be interpreted similarly to the trend of Fig. 11.5. Besides, for these two figures, we notice that NOMA-SM temporarily loses its advantage over NOMA-VBLAST adopted QPSK and $N_t = 4$ within moderate SNR regime. This observation results from the superiority that QPSK brings to NOMA-VBLAST compared to 16QAM. However, as the increment of SNR, NOMA-SM achieves its dominance in terms of higher diversity gain.

Below, we investigate the impact of the V2V channel's time-varying nature. Observe from Fig. 11.7 that compared to the performance of no time-varying effect associated with $\delta = 1$, the BER has been substantially degraded in all schemes for $\delta = 0.9$. Although a perfect channel estimation procedure is assumed for the receivers, the estimated channel coefficients used for ML detection becomes partially outdated due to the channel's time-varying nature, hence resulting in a degraded BER performance. Nevertheless, the proposed NOMA-SM scheme maintains its advantage over the reference within the medium and high SNR regime, regardless of the grade of temporal correlation.

Figure 11.8 shows the BER performance associated with different α values. For all schemes, the lower α values exhibit a better detection performance, since less power is allocated to U and hence V_2 suffers from a lower inter-user interference. More importantly, we observe that NOMA-SM consistently outperforms the NOMA-VBLAST scheme applied with 16PSK and $N_t = 2$. For the cases of $\alpha = 0.001$ and $\alpha = 0.0001$, though NOMA-VBLAST adopted QPSK and $N_t = 4$ holds a dominance within the moderate SNR regime, NOMA-SM keeps improving its performance with the increase of SNR and outperform the references in terms of higher diversity gain. By jointly considering the above observations, we conclude that NOMA-SM constitutes a potent amalgam.

11.5.2 Capacity Results and Discussions

Below, we evaluate the capacity of the NOMA-SM system associated with different power allocation strategies. All results presented in this subsection are obtained by averaging the instantaneous capacities over multiple channel realizations. In particular, we fix $K = 0.2$, $\kappa_t = \kappa_r = 0.5$, and $\delta = 1$ unless otherwise stated. For benchmarking, we use an OMA-SM system, where V_1 transmits messages to V_2 using SM in the first slot. Then, V_1 sends messages through the previously activated antenna to U, without activating another antenna. This OMA-SM model constitutes a fair reference for the NOMA-SM system, since the signal intended for V_2 is conveyed by both the APM signal- and Tx-domain, whereas the signal destined for U is only embedded in the classical signal-domain. The distinctive feature of OMA-SM is that data transmissions destined for V_1–V_2 and V_1–U are operated in an orthogonal time division way within the classical APM signal-domain. Accordingly, the capacity upper bound for V_2 and the capacity for U in the OMA-SM system are expressed as

$$\begin{aligned} \mathscr{C}_V^B &= \min\left\{\mathscr{C}_V^{B_1}, \mathscr{C}_V^{B_2}\right\}, \\ \mathscr{C}_U &= \frac{1}{2N_t}\sum_{i=1}^{N_t}\log_2\left(1+\frac{\alpha E_s}{\sigma_0^2}\|\mathbf{g}_i\|^2\right), \end{aligned} \tag{11.25}$$

respectively, where

$$\begin{aligned} \mathscr{C}_V^{B_1} &= \frac{1}{2N_t}\sum_{i=1}^{N_t}\log_2\left(1+\frac{(1-\alpha)E_s p_0}{\sigma_0^2}\|\mathbf{h}_i\|^2\right)+\frac{1}{2}\log_2\left(N_t\right), \\ \mathscr{C}_V^{B_2} &= \frac{1}{2}\log_2\det\left[\mathbf{I}+\frac{(1-\alpha)E_s p_0}{\sigma_0^2 N_t}\mathbf{H}\mathbf{H}^H\right]. \end{aligned} \tag{11.26}$$

Let us first check the capacity associated with a fixed power allocation, that is $\alpha = 0.01$. Figure 11.9 depicts the capacity of V_2 and U, as well as the sum capacity versus SNR for both NOMA-SM and OMA-SM. Compared to OMA-SM, NOMA-SM provides substantial capacity gains both for the collaboration-aided vehicle V_2 and for the in-car user U and accordingly obtains a significant sum capacity enhancement. Specifically, the capacity C_U has been beneficially boosted by the proposed scheme, about twice as high as that of OMA-SM. Since the APM signal-domain of the proposed scheme is combined with a NOMA strategy, each user accesses the channel resources via power domain multiplexing.

Subsequently, we investigate the efficiency of the proposed power allocation optimization. Specifically, the power allocation optimization denoted by $\mathscr{Q}$ is considered for OMA-SM, which is formulated as

$$\begin{aligned} \mathscr{Q}: \max_{\alpha} \quad & \mathscr{C}_U + \mathscr{C}_V^B \\ s.t. \quad & \begin{cases} \mathscr{C}_U \geq \tilde{\mathscr{C}}_U, \\ \mathscr{C}_V^B \geq \tilde{\mathscr{C}}_V^B, \\ 0 < \alpha < 1. \end{cases} \end{aligned} \tag{11.27}$$

For simplicity, the minimum rate requirements of V_2 and U are set to $\tilde{\mathscr{C}}_U = \frac{\mathscr{C}_U(\alpha=1)}{2}$ and $\tilde{\mathscr{C}}_V^B = \frac{\mathscr{C}_V^B(\alpha=0)}{2}$, which respectively correspond to the lower bound and upper

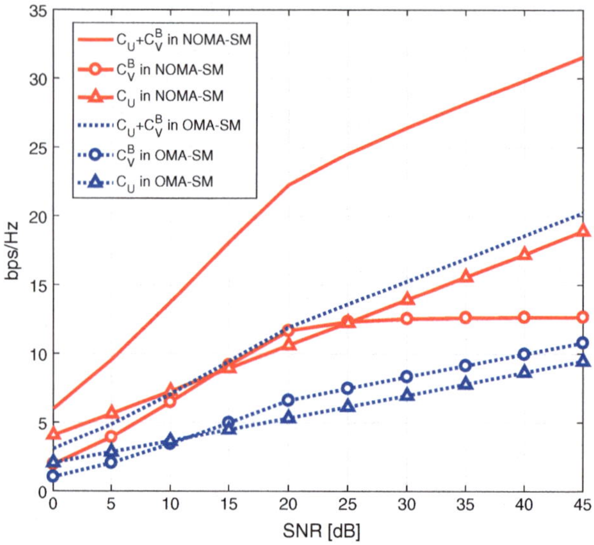

Fig. 11.9 Capacity of V_2 and U, or the sum capacity versus SNR for the NOMA-SM and OMA-SM scheme with a fixed power allocation factor, i.e. $\alpha = 0.01$. Specifically, C_V^B and C_U in NOMA-SM are evaluated from (11.18) and (11.11), while C_V^B and C_U in OMA-SM are obtained from (11.25)

bound of α's feasible set. Then, a full-search algorithm is applied for OMA-SM within the feasible set.

Figure 11.10 illustrates the capacity of V_2 and U for NOMA-SM with optimization $\mathscr{P}$ or $\mathscr{O}$, where the QoS of the collaboration-aided vehicle V_2 and the in-car user U, i.e. $\tilde{C}_V^B$ and $\tilde{C}_U$, are also plotted for reference. It can be observed that C_V^B always meets the requirement of $\tilde{C}_V^B$ with the aid of the optimization $\mathscr{P}$, and C_U associated with the optimization $\mathscr{O}$ exactly meets the QoS $\tilde{C}_U$. This observation is in accordance with the foregoing analysis, which indicates that the optimization $\mathscr{P}$ intends to maximize C_U, while maintaining the QoS $\tilde{C}_V^B$ for V2V transmission, whereas the optimization $\mathscr{O}$ aims for maximizing C_V^B while guaranteeing the minimum rate requirement $\tilde{C}_U$ for the in-car user. Thus, we find that the optimized C_U of $\mathscr{P}$ is higher than that of $\mathscr{O}$, whereas the optimized C_V^B of $\mathscr{O}$ outperforms that of $\mathscr{P}$. Accordingly, the more appropriate optimization scheme can be readily selected based on the data priority of distinct transmission links.

Figure 11.11 compares the results of the optimization $\mathscr{Q}$ to that of $\mathscr{P}$ and $\mathscr{O}$. Let us contrast $\mathscr{P}$ and $\mathscr{Q}$ first. Clearly, both C_V^B and C_U in NOMA-SM with optimization $\mathscr{P}$ have been remarkably improved, demonstrating that the NOMA strategy offers a bandwidth efficiency improvement. By considering the results of $\mathscr{O}$ and $\mathscr{Q}$ in Fig. 11.11, we find that C_U of NOMA-SM associated with optimization $\mathscr{O}$ is tightly lower bounded by that of OMA-SM associated with optimization $\mathscr{Q}$, and C_V^B with $\mathscr{O}$ provides a substantial gain, achieving nearly twice that of $\mathscr{Q}$.

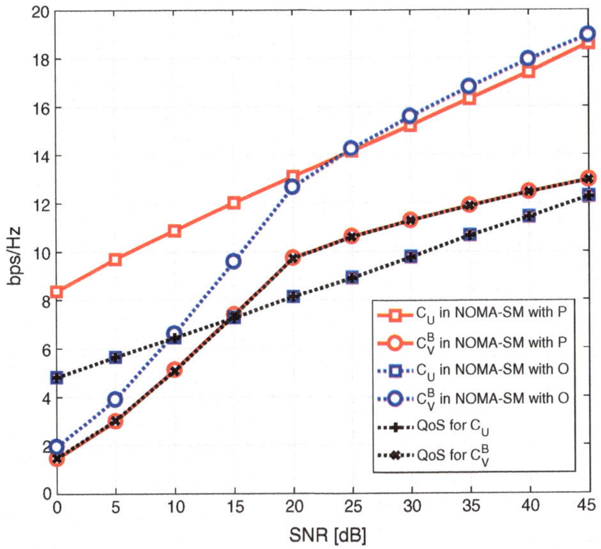

Fig. 11.10 Capacity of V_2 and U, or the respective QoS versus SNR for NOMA-SM with power allocation optimization $\mathscr{P}$ or $\mathscr{O}$. Specifically, C_V^B and C_U in NOMA-SM with $\mathscr{P}$ or $\mathscr{O}$ are evaluated with the aid of the algorithm in Table 11.1. The QoS for C_V^B and C_U, i.e. $\tilde{C}_V$ and $\tilde{C}_U$, are set according to (11.24)

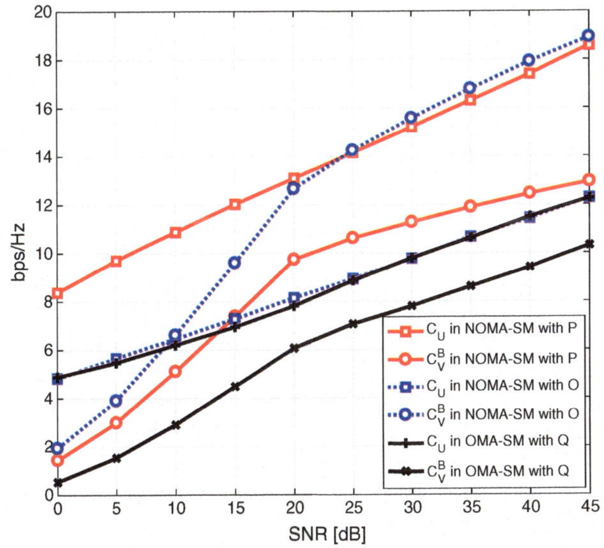

Fig. 11.11 Capacity of V_2 and U versus SNR for NOMA-SM with power allocation optimization $\mathscr{P}$ or $\mathscr{O}$, and OMA-SM with power allocation optimization $\mathscr{Q}$, respectively. Specifically, C_V^B and C_U in NOMA-SM with $\mathscr{P}$ or $\mathscr{O}$ are evaluated with the aid of the algorithm in Table 11.1. While C_V^B and C_U in OMA-SM with $\mathscr{Q}$ are obtained from a full-search algorithm

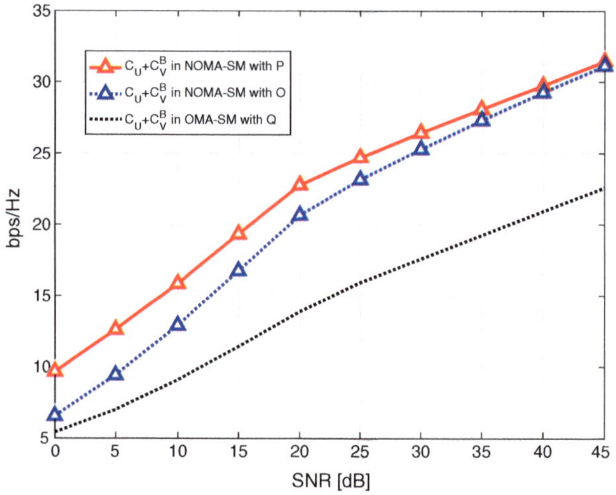

Fig. 11.12 Sum capacity versus SNR for NOMA-SM with power allocation optimization $\mathscr{P}$ or $\mathscr{O}$, and OMA-SM with power allocation optimization $\mathscr{Q}$, respectively. Specifically, C_V^B and C_U in NOMA-SM with $\mathscr{P}$ or $\mathscr{O}$ are evaluated with the aid of the algorithm in Table 11.1. While C_V^B and C_U in OMA-SM with $\mathscr{Q}$ are obtained from a full-search algorithm

Furthermore, it can be observed from Fig. 11.12 that the NOMA-SM systems achieve higher sum capacity than OMA-SM. Specifically, optimization $\mathscr{P}$ provides higher capacity gain than $\mathscr{O}$, since $\mathscr{P}$ aims for maximizing the data rate of the in-car user U, which experiences a much better channel than the collaboration-aided vehicle V_2.

11.6 Chapter Summary and Future Outlook

In this chapter, we introduce NOMA and SM techniques into V2X scenarios in order to support high bandwidth efficiency and enhanced link reliability. The BER performance of the new NOMA-SM transmission strategy has been investigated with the impact of the Rician K-factor, spatial correlation of antenna array, time-varying effect of the V2V channel, and the power allocation factor being discussed. Compared to NOMA relying on VBLAST, NOMA-SM has been demonstrated to exhibit improved robustness against the spatial and temporal effects of the V2V channel. By analysing the capacity and deriving analytical upper bounds in closed form, a pair of power allocation optimization schemes have been formulated for NOMA-SM. The optimal solutions have also been shown to be achievable with the aid of the proposed power allocation algorithm. Our numerical results have verified that with the aid of an appropriate power allocation, NOMA-SM is capable of satisfying the QoS support of a low priority flow, whilst maximizing the throughput of the high priority flow. In summary, NOMA-SM has been demonstrated to cooperatively improve the link reliability and bandwidth efficiency of V2V transmissions.

Nonetheless, several open issues still need to be carefully addressed before NOMA can be practically exploited in vehicular environments. Here, we discuss two potential research topics in this field as examples.

Parallel Interference Cancellation-Aided NOMA: There is a much broader range of V2X applications to be considered in VANETs, especially within the automated driving field, whose characteristics are more stringent, as captured by the ultra-reliable low-latency constraints. The traditional NOMA schemes use the classic SIC technique, where a high received signal power difference is preferred. However, this condition cannot always be guaranteed, especially in a traffic jam, where all cars tend to have similar channel conditions. Furthermore, SIC receivers were reported to exhibit an error floor in high-order modulation modes due to error propagation across the cancellation stages [60]. By contrast, parallel interference cancellation (PIC) does not require any specific detection order and all users reconstruct the signals of all the other users in parallel. Then, they subtract the reconstructed signals from the composite signal. Hence, PIC outperforms SIC when the received signal powers for all users are similar. With the advent of a PIC receiver, NOMA is expected to possess enhanced transmission reliability as well as better applicability to highly-loaded vehicular scenarios. In a nutshell, the performance analysis of NOMA with PIC should be addressed in our further research.

NOMA for Cognitive V2X: The large amount of data generated by the vehicles might impose excessive traffic demands on traditional cellular traffic. Hence, cognitive NOMA principles can be conceived for delay-tolerant vehicular communications services to opportunistically access the channels originally occupied by the cellular users. Given a dedicated spectral band, cellular users and vehicles can be regarded as primary users and secondary users, respectively. The vehicles would only be permitted to access the channel when the services of cellular users are not affected. In contrast to the traditional cognitive radio scheme, both of the power control and resource allocation need to be designed elaborately, and efficient transmission schemes accommodating both primary and secondary users should be proposed and analysed carefully.

References

1. 3GPP TS 22.185, Service requirements for V2X services, Feb 2016
2. D. Jiang, L. Delgrossi, IEEE 802.11p: towards an international standard for wireless access in vehicular environments, in *Proceedings VTC Spring 2008—IEEE Vehicular Technology Conference* (Singapore, May 2008), pp. 2036–2040
3. S.H. Sun, J.L. Hu, Y. Peng, X.M. Pan, L. Zhao, J.Y. Fang, Support for vehicle-to-everything services based on LTE. IEEE Wirel. Commun. **23**(3), 4–8, June 2016
4. M. Amadeo, C. Campolo, A. Molinaro, Enhancing IEEE 802.11p/WAVE to provide infotainment applications in VANETs. Elsevier Ad Hoc Netw. **10**(2), 253–269 (2012)
5. S. Chen et al., Vehicle-to-everything (v2x) services supported by LTE-based systems and 5G. IEEE Commun. Stand. Mag. **1**(2), 70–76 (2017)

6. S. Chen, J. Hu, Y. Shi, L. Zhao, LTE-V: a TD-LTE-based V2X solution for future vehicular network. IEEE Internet Things J. **3**(6), 997–1005 (2016)
7. 3GPP RP-152293, New WI proposal: support for V2V services based on LTE sidelink, Dec 2015
8. G. Araniti, C. Campolo, M. Condoluci, A. Iera, A. Molinaro, LTE for vehicular networking: a survey. IEEE Commun. Mag. **51**(5), 148–157 (2013)
9. 3GPP TS 23.285, v.14.1.0, Architecture enhancements for V2X services, Dec 2016
10. 3GPP, TS 36.440, General aspects and principles for interfaces supporting multimedia broadcast multicast service (MBMS) within E-UTRAN, Rel. 11, Sept 2012
11. L. Hanzo, O. Alamri, M. El-Hajjar, N. Wu, Near-capacity multi-functional MIMO systems: sphere-packing, iterative detection and cooperation (Wiley, New York, NY, USA, 2009)
12. L. Wang, R. Li, C. Cao, G.L. Stüber, SNR analysis of time reversal signaling on target and unintended receivers in distributed transmission. IEEE Trans. Commun. **64**(5), 2176–2191 (2016)
13. E.G. Larsson, O. Edfors, F. Tufvesson, T.L. Marzetta, Massive MIMO for next generation wireless systems. IEEE Commun. Mag. **52**(2), 186–195 (2014)
14. Y. Wu, R. Schober, D.W.K. Ng, C. Xiao, G. Caire, Secure massive MIMO transmission with an active eavesdropper. IEEE Trans. Inf. Theory **62**(7), 3880–3900 (2016)
15. R. Zhang, Z. Zhong, J. Zhao, B. Li, K. Wang, Channel measurement and packet-level modeling for V2I spatial multiplexing uplinks using massive MIMO. IEEE Trans. Veh. Technol. **65**(10), 7831–7843 (2016)
16. P. Harris et al., Performance characterization of a real-time massive MIMO system with LOS mobile channels. IEEE J. Select. Areas Commun. **35**(6), 1244–1253 (2017)
17. Y. Saito, Y. Kishiyama, A. Benjebbour, T. Nakamura, A. Li, K. Higuchi, Non-orthogonal multiple access (NOMA) for cellular future radio access, in *Proceedings IEEE 77th Vehicular Technology Conference* (Dresden, Germany, June 2013), pp. 1–5
18. Y. Chen, L. Wang, B. Jiao, Cooperative multicast non-orthogonal multiple access in cognitive radio, in *Proceedings 2017 IEEE ICC* (Paris, France, May 2017), pp. 1–6
19. L. Dai, B. Wang, Y. Yuan, S. Han, I. Chih-Lin, Z. Wang, Non-orthogonal multiple access for 5G: solutions, challenges, opportunities, and future research trends. IEEE Commun. Mag. **53**(9), 74–81 (2015)
20. J. Liberti, S. Moshavi, P. Zablocky, Successive interference cancellation. U.S. Patent 8670418 B2, 11 Mar 2014
21. Q. Sun, S. Han, I. Chih-Lin, Z. Pan, On the ergodic capacity of MIMO NOMA systems. IEEE Wirel. Commun. Lett. **4**(4), 405–408, Aug 2015
22. Y. Saito, A. Benjebbour, Y. Kishiyama, T. Nakamura, System-level performance evaluation of downlink non-orthogonal multiple access (NOMA), in *Proceedings IEEE 24th International Symposium Personal Indoor and Mobile Radio Communications (PIMRC)* (London, UK, Sept 2013), pp. 611–615
23. Z. Yang, Z. Ding, P. Fan, N. Al-Dhahir, A general power allocation scheme to guarantee quality of service in downlink and uplink NOMA systems. IEEE Trans. Wirel. Commun. **15**(11), 7244–7257 (2016)
24. L. Lv, J. Chen, Q. Ni, Cooperative non-orthogonal multiple access in cognitive radio. IEEE Commun. Lett. **20**(10), 2059–2062 (2016)
25. K. Au et al., Uplink contention based SCMA for 5G radio access, in *Proceedings IEEE GLOBECOM Workshop* (Austin, TX, Dec, 2014), pp. 900–905
26. B. Di, L. Song, Y. Li, G.Y. Li, Non-orthogonal multiple access for high-reliable and low-latency V2X communications in 5G systems. IEEE J. Select. Areas Commun. **35**(10), 2383–2397 (2017)
27. Y. Chen, L. Wang, Y. Ai, B. Jiao, L. Hanzo, Performance analysis of NOMA-SM in vehicle-to-vehicle massive MIMO. IEEE J. Select. Areas Commun. **35**(12), 2653–2666 (2017)
28. Y. Yang, B. Jiao, Information-guided channel-hopping for high data rate wireless communication. IEEE Commun. Lett. **12**(4), 225–227 (2008)

29. L. He, J. Wang, J. Song, L. Hanzo, On the multi-user, multi-cell massive spatial modulation uplink: how many antennas for each user? IEEE Trans. Wirel. Commun. **16**(3), 1437–1451 (2017)
30. D.A. Basnayaka, M. Di Renzo, H. Haas, Massive but few active MIMO. IEEE Trans. Veh. Technol. **65**(9), 6861–6877 (2016)
31. Y. Chau, S.-H. Yu, Space modulation on wireless fading channels, in *Proceedings IEEE Vehicular Technology Conference—Fall*, vol. 3 (Atlantic City, NJ, Oct 2001), pp. 1668–11671
32. R. Mesleh, H. Haas, C. W. Ahn, S. Yun, Spatial modulation—a new low complexity spectral efficiency enhancing technique, in *Proceedings 2006 First International Conference on Communications and Networking in China* (Beijing, 2006), pp. 1–5
33. R.Y. Mesleh, H. Haas, S. Sinanovic, C.W. Ahn, S. Yun, Spatial modulation. IEEE Trans. Veh. Technol. **57**(4), 2228–2241 (2008)
34. J. Jeganathan, A. Ghrayeb, L. Szczecinski, Spatial modulation: optimal detection and performance analysis. IEEE Commun. Lett. **12**(8), 545–547 (2008)
35. T.L. Narasimhan, P. Raviteja, A. Chockalingam, Large-scale multiuser SM-MIMO versus massive MIMO, in *Proceedings, Information Theory and Applications Workshop (ITA)* (San Diego, CA, 2014), pp. 1–9
36. S. Wang, Y. Li, M. Zhao, J. Wang, Energy-efficient and low-complexity uplink transceiver for massive spatial modulation MIMO. IEEE Trans. Veh. Technol. **64**(10), 4617–4632 (2015)
37. P. Patcharamaneepakorn et al., Spectral, energy, and economic efficiency of 5G multicell massive MIMO systems with generalized spatial modulation. IEEE Trans. Veh. Technol. **65**(12), 9715–9731 (2016)
38. L. He, J. Wang, J. Song, On massive spatial modulation MIMO: spectral efficiency analysis and optimal system design, in *Proceedings IEEE GLOBECOM* (2016), pp. 1–6
39. M. Di Renzo, H. Haas, A. Ghrayeb, S. Sugiura, L. Hanzo, Spatial modulation for generalized MIMO: challenges, opportunities, and implementation. Proc. IEEE **102**(1), 56–103 (2014)
40. M. Zhang, X. Cheng, L.Q. Yang, Differential spatial modulation in V2X, in *Proceedings 2015 9th European Conference on Antennas and Propagation (EuCAP)* (Lisbon, Apr 2015), pp. 1–5
41. Y. Fu et al., BER performance of spatial modulation systems under 3-D V2V MIMO channel models. IEEE Trans. Veh. Technol. **65**(7), 5725–5730 (2016)
42. K.P. Peppas, P.S. Bithas, G.P. Efthymoglou, A.G. Kanatas, Space shift keying transmission for intervehicular communications. IEEE Trans. Intell. Transp. Syst. **17**(12), 3635–3640 (2016)
43. Y. Cui, X. Fang, Performance analysis of massive spatial modulation MIMO in high-speed railway. IEEE Trans. Veh. Technol. **65**(11), 8925–8932 (2016)
44. T. Wang, L. Song, Z. Han, Coalitional graph games for popular content distribution in cognitive radio VANETs. IEEE Trans. Veh. Technol. **62**(8), 4010–4019 (2013)
45. H. Ilhan, I. Altunbas, M. Uysal, Optimized amplify-and-forward relaying for vehicular ad-hoc networks, in *Proceedings IEEE Vehicular Technology Conference* (Calgary, BC, Sept 2008), pp. 1–5
46. K. Huang, V.K.N. Lau, Y. Chen, Spectrum sharing between cellular and mobile ad hoc networks: transmission-capacity trade-off. IEEE J. Select. Areas Commun. **27**(7), 1256–1267 (2009)
47. T. Lakshmi Narasimhan, P. Raviteja, A. Chockalingam, Generalized spatial modulation in large-scale multiuser MIMO systems. IEEE Trans. Wirel. Commun. **14**(7), 3764–3779 (2015)
48. Y. Chen, L. Wang, Z. Zhao, M. Ma, B. Jiao, Secure multiuser MIMO downlink transmission via precoding-aided spatial modulation. IEEE Commun. Lett. **20**(6), 1116–1119 (2016)
49. L. Wang, S. Bashar, Y. Wei, R. Li, Secrecy enhancement analysis against unknown eavesdropping in spatial modulation. IEEE Commun. Lett. **19**(8), 1351–1354 (2015)
50. O. Delangre, S. Van Roy, P. De Doncker, M. Lienard, P. Degauque, Modeling in-vehicle wideband wireless channels using reverberation chamber theory, in *Proceedings 2007 IEEE 66th Vehicular Technology Conference* (Baltimore, MD, Oct 2007), pp. 2149–2153
51. X. Cheng, C.X. Wang, D.I. Laurenson, S. Salous, A.V. Vasilakos, An adaptive geometry-based stochastic model for non-isotropic MIMO mobile-to-mobile channels. IEEE Trans. Wirel. Commun. **8**(9), 4824–4835 (2009)

52. M. Koca, H. Sari, Performance analysis of spatial modulation over correlated fading channels, in *Proceedings, IEEE Vehicular Technology Conference (VTC Fall)* (Quebec City, QC, Sept 2012), pp. 1–5
53. J.P. Kermoal, L. Schumacher, K.I. Pedersen, P.E. Mogensen, F. Frederiksen, A stochastic MIMO radio channel model with experimental validation. IEEE J. Select. Areas Commun. **20**(6), 1211–1226 (2002)
54. S.L. Loyka, Channel capacity of MIMO architecture using the exponential correlation matrix. IEEE Commun. Lett. **5**(9), 369–371 (2001)
55. C. Liu, M. Ma, Y. Yang, B. Jiao, Optimal spatial-domain design for spatial modulation capacity maximization. IEEE Commun. Lett. **20**(6), 1092–1095 (2016)
56. X. Guan, Y. Cai, W. Yang, On the mutual information and precoding for spatial modulation with finite alphabet. IEEE Wirel. Commun. Lett. **2**(4), 383–386 (2013)
57. Z. An, J. Wang, J. Wang, S. Huang, J. Song, Mutual information analysis on spatial modulation multiple antenna system. IEEE Trans. Commun. **63**(3), 826–843 (2015)
58. Y. Toor, P. Muhlethaler, A. Laouiti, A.D. La Fortelle, Vehicle Ad Hoc networks: applications and related technical issues, IEEE Commun. Surv. Tutor. **10**(3), 74–88, Third Quarter (2008)
59. S. Boyd, L. Vandenberghe, *Convex Optimization* (Cambridge Univ. Press, Cambridge, U.K., 2004)
60. L. Hanzo, L.-L. Yang, E.-L. Kuan, K. Yen. Single-and multi-carrier DS-CDMA: multi-user detection, space-time spreading, synchronisation, standards and networking (Wiley, 2003)

Part III
NOMA in Code and Other Domains

Chapter 12
Sparse Code Multiple Access (SCMA)

Zheng Ma and Jinchen Bao

12.1 General Description

Overloaded systems, in which the number of users is greater than the dimension of signal-space, are of practical interest in bandwidth-efficient multi-user communications. One kind of such systems is sparse code multiple access (SCMA), which is a promising code-domain non-orthogonal multiple access technique to address the challenges for the fifth-generation (5G) mobile networks [1–4]. Non-orthogonal multiple access has the potential to accommodate more users with limited resources, which provides many advantages over orthogonal multiple access including multi-user capacity, supporting overloaded transmission, enabling reliable and low latency grant-free transmission, enabling flexible service multiplexing, etc. Applications of non-orthogonal signaling for multi-user communications have been investigated several years ago, significant efforts were paid to the optimal signaling design and intensive multi-user detection techniques, to suppress the multiple access interference (MAI) for lowering probability of error or increasing capacity. Hoshyar and Guo suggest the low-density signature (LDS)-based multiple access [5], or sparsely spread code-division multiple access (CDMA) [6], which intentionally arranges each user to spread its data over a fraction of the chips, instead of all chips, to reduce both the MAI and the complexity of multi-user detection. Inspired by the overloading capability and the low-complexity feature of LDS, SCMA is developed by inheriting from LDS the sparse sequence structure, such that the message-passing algorithm (MPA) is available in multi-user detection to achieve near-optimal performance. In contrast to the LDS scheme, multi-dimensional signal constellations, instead of the spreading, are utilized in SCMA to combat the channel fading and MAI. As a result,

Z. Ma (✉) · J. Bao
Southwest Jiaotong University, West Section, High-tech Zone, Chengdu, Sichuan, China
e-mail: zma@swjtu.edu.cn

J. Bao
e-mail: jinchen_bao@my.swjtu.edu.cn

M. Vaezi et al. (eds.), *Multiple Access Techniques for 5G Wireless Networks and Beyond*, https://doi.org/10.1007/978-3-319-92090-0_12

the larger coding gain and better spectrum efficiency are achievable for SCMA due to the improved codebooks, compared to LDS.

As one of NOMA family, SCMA is capable of supporting overloaded access over the coding domain, hence increasing the overall rate and connectivity. By carefully designing the codebook and multi-dimensional modulation constellations, the coding and shaping gain can be obtained simultaneously. In an SCMA system, users occupy the same resource blocks in a low-density way, which allows affordable low multi-user joint detection complexity at receiver. The sparsity of signal guarantees a small collision even for a large number of concurrent users, and the spread-coding like codes design brings good coverage and anti-interference capability due to spreading gain as well.

12.1.1 System Model

12.1.1.1 Multiple Access Procedure

An SCMA transmission system can be simply illuminated in Fig. 12.1. Suppose that there are J synchronous users multiplexing over K shared orthogonal resources, e.g., K time slots or orthogonal frequency division multiplexing (OFDM) tones, and each user employs one SCMA layer.[1] The forward error control (FEC) coding scheme can be low-density parity-check (LDPC) codes or polar codes which have been adopted for 5G recently. Each SCMA modulator/encoder maps the coded bits to a K-dimensional complex codeword, and the resulted J codewords constitute an SCMA block, as is shown in Fig. 12.1 ($J = 6$, $K = 4$ in the figure). The multi-user codewords in each SCMA block are multiplexed over the air transmissions in uplink multiple access channel (MAC), or they are superimposed at the transmitter of the downlink broadcast channel (BC). Since each SCMA block occupies K resources for codeword transmitting, the resulted *overloading factor* is J/K. This multiple access process is similar to that of CDMA, where the spread signals in CDMA are replaced with the SCMA codewords. Multi-user detection is carried out at the receiver to recover the colliding codewords.

For the uplink MAC, the received signal vector after the synchronous user multiplexing is expressed as

$$\mathbf{y} = \sum_{j=1}^{J} \mathrm{diag}(\mathbf{h}_j)\mathbf{x}_j + \mathbf{n} \tag{12.1}$$

where $\mathbf{x}_j = \left[x_j[1], \ldots, x_j[K]\right]^t$ and $\mathbf{h}_j = [h_j[1], \ldots, h_j[K]]^t$, are the K-dimensional codeword and the corresponding channel gain for the jth user, respectively, and $\mathrm{diag}(\mathbf{h}_j)$ denotes the diagonal matrix with $h_j[k]$ being the kth diagonal

[1] In practical scenarios, each user employs one or multiple layers.

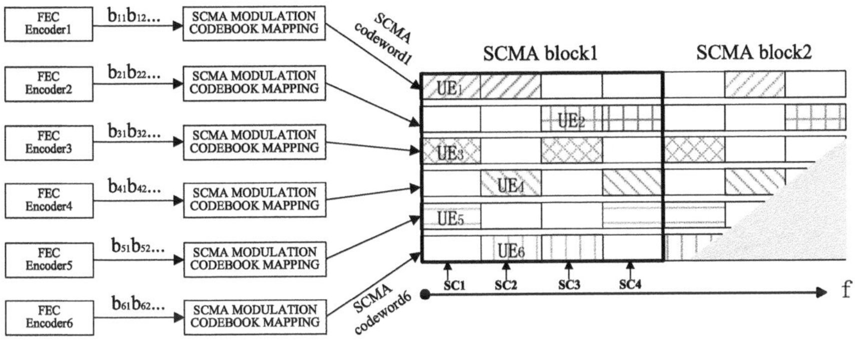

Fig. 12.1 The system model for SCMA

element. The K-vector $\mathbf{n}$ is the additive white Gaussian noise (AWGN) with zero mean and variance N_0 per dimension. It is convenient to view the MAC model as an equivalent "MIMO" communication system, and the received vector in (12.1) becomes

$$\mathbf{y} = \mathbf{HX} + \mathbf{n} \tag{12.2}$$

where $\mathbf{H} = \left[\mathrm{diag}(\mathbf{h}_1), \mathrm{diag}(\mathbf{h}_2), \ldots, \mathrm{diag}(\mathbf{h}_J)\right]$, is the equivalent "MIMO" channel matrix, and $\mathbf{X} = \left[\mathbf{x}_1^t, \mathbf{x}_2^t, \ldots, \mathbf{x}_J^t\right]^t$, is the combined multi-user codeword representing an SCMA block.

For the downlink BC, the codewords from multiple users are superimposed before the transmission, so that they experience the same fading. In the case of absence of interference between K resources, the received signal vector is given by

$$\mathbf{y} = \mathrm{diag}(\mathbf{h}) \sum_{j=1}^{J} \mathbf{x}_j + \mathbf{n} = \mathrm{diag}(\mathbf{h})\mathbf{X} + \mathbf{n} \tag{12.3}$$

where a single receiver is considered here for simplicity, and $\mathbf{X} = \sum_{j=1}^{J} \mathbf{x}_j$, is the superimposed codeword of J users at the input of a BC, which also represents an SCMA codeword block.

In the following, the upper case $\mathbf{X}$ always denotes the combined multi-user codeword of J users in the MAC model, or the superimposed codeword in the BC model.

12.1.1.2 SCMA Codebook Mapping

Unlike the modulation used for 3G and 4G, the modulation and codebook mapping in SCMA are designed jointly in a multi-dimensional and sparsely spread way. An SCMA modulator/encoder maps the input bits to a K-dimensional sparse codeword,

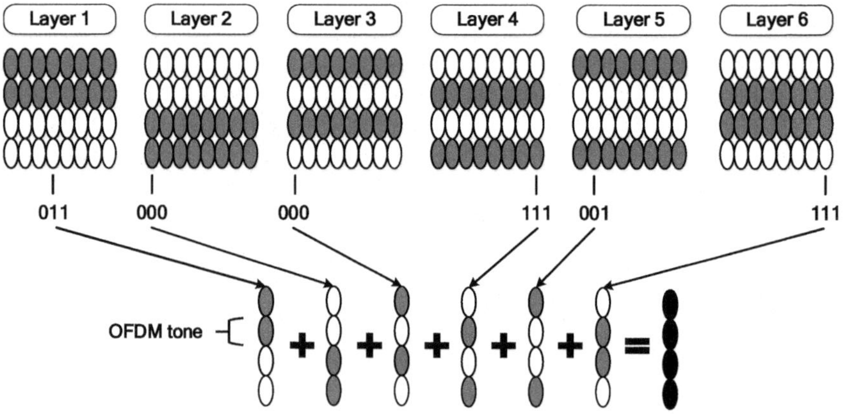

Fig. 12.2 Illustration of SCMA codebooks and bits to codeword mapping

which is selected from a layer-specific codebook of size M. The K-dimensional complex codewords of the codebook are sparse vectors with $N < K$ nonzero entries, and all the codewords contain 0 in the same dimensions. Then, the codebook is sparse, and this is where the "sparse code multiple access" is named from.

The codebooks are constructed by a mapping from an N-dimensional complex constellation with a mapping matrix. Denote the constellation for the jth layer/user with $\mathbb{C}_j$, which contains M_j constellation points of length N_j. The mapping matrix $\mathbf{V}_j$ maps the N_j-dimensional constellation points to SCMA codewords to form the codebook $\mathbb{X}_j$. To simplify our illumination and analysis, we assume that all layers have the same constellation size and length, i.e., $M_j = M$, $N_j = N$, $\forall j$. In summary, the resulting codebook for the jth user contains M codewords, each codeword consists of K complex values from which only N are nonzero specified by the mapping matrix $\mathbf{V}_j$.

An example of the codebook mapping is shown in Fig. 12.2, where a codebook set containing 6 codebooks for transmitting 6 SCMA layers is illustrated ($J = 6$). Each codebook contains 8 four-dimensional codewords ($M = 8$, $K = 4$), and two of the four entries in the codewords are nonzero ($N = 2$). Upon transmission, the codeword of each layer is selected based on the labeling of the bit sequence.

12.1.1.3 Factor Graph Representation

The low-density structure of SCMA codewords can be efficiently characterized by a factor graph, which is analogous to that for LDPC codes. A binary column vector $\mathbf{f}_j$ of length K is used to indicate the positions of zero (with digit 0) and nonzero (with digit 1) entries of the jth codebook. Then, a $K \times J$ sparse matrix $\mathbf{F} = [\mathbf{f}_1, \ldots, \mathbf{f}_J]$, called *factor graph matrix*, can be used to indicate the relationships between the layers and resources. The rows of $\mathbf{F}$ indicate the resources and the columns indicate

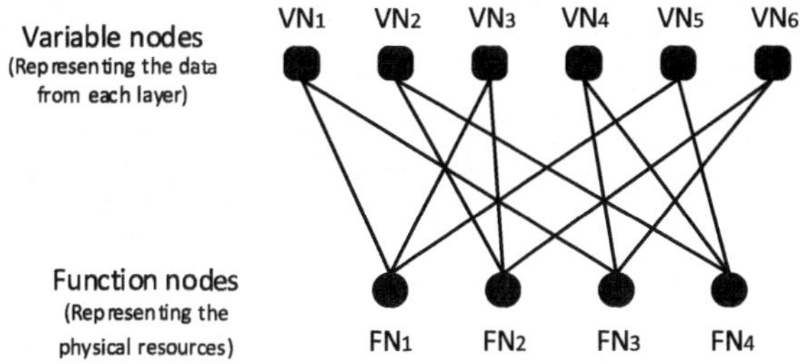

Fig. 12.3 Factor graph representation for SCMA

the layers. The (k, j)th element of $\mathbf{F}$, denoted as $f_{k,j}$, is 1 if the jth layer contributes its data to the kth resource.

Correspondingly, let the J variable nodes (VNs) and K function nodes (FNs) in the factor graph represent the layers and resources, respectively, and the jth VN is connected to the kth FN if and only if $f_{k,j} = 1$. In the following, we denote

$$\begin{aligned} \phi_k &= \{j : 1 \leq j \leq J, f_{k,j} = 1\}, \\ \varphi_j &= \{k : 1 \leq k \leq K, f_{k,j} = 1\} \end{aligned} \tag{12.4}$$

the index set of layers contributing to the kth resource, and the index set of resources occupied by the jth layer, respectively. For a regular factor graph matrix, $|\phi_1| = \cdots = |\phi_K|$ and $|\varphi_1| = \cdots = |\varphi_J|$, and let $d_f = |\phi_k|$ and $d_v = |\varphi_j|$.

Example 1 Consider a 6-user SCMA transmission system with $J = 6$, $K = 4$, such a system permits a transmission overloading 150%, and the system model is depicted in Fig. 12.1. If we carefully design the factor graph matrix $\mathbf{F}$ to allow the users to collide over only one nonzero element, then a choice of $\mathbf{F}$ is given by

$$\mathbf{F} = \begin{bmatrix} 1\,0\,1\,0\,1\,0 \\ 0\,1\,1\,0\,0\,1 \\ 1\,0\,0\,1\,0\,1 \\ 0\,1\,0\,1\,1\,0 \end{bmatrix} \tag{12.5}$$

In the sparse matrix settings, matrix (12.5) has $d_f = 3$ and $d_v = 2$, which means that each FN is connected to three VNs and each VN is connected to two FNs. The corresponding factor graph is shown in Fig. 12.3, and an example of a codebook (with size $M = 4$) is listed in Table 12.1.

In summary, the main features of SCMA lie in:

- Code-domain non-orthogonal multiplexing: SCMA allows superposition of multiple codewords from different users over several resources, which supports overloading. The superposition pattern on each resource is defined in codebooks.
- Sparse spreading: SCMA uses sparse spreading to reduce inter-layer interference, so that more codewords collisions can be tolerated with low receiver complexity.

Table 12.1 An Example of SCMA Codebook ($K = M = 4$, $N = 2$, $J = 6$)

SCMA codebook index	SCMA Codebook for each layer
Codebook 1	$\begin{bmatrix} 0.7851 & -0.2243 & 0.2243 & -0.7851 \\ 0 & 0 & 0 & 0 \\ -0.1815-0.1318i & -0.6351-0.4615i & 0.6351+0.4615i & 0.1815+0.1318i \\ 0 & 0 & 0 & 0 \end{bmatrix}$
Codebook 2	$\begin{bmatrix} 0 & 0 & 0 & 0 \\ -0.1815-0.1318i & -0.6351-0.4615i & 0.6351+0.4615i & 0.1815+0.1318i \\ 0 & 0 & 0 & 0 \\ 0.7851 & -0.2243 & 0.2243 & -0.7851 \end{bmatrix}$
Codebook 3	$\begin{bmatrix} -0.6351+0.4615i & 0.1815-0.1318i & -0.1815+0.1318i & 0.6351-0.4615i \\ 0.1392-0.1759i & 0.4873-0.6156i & -0.4873+0.6156i & -0.1392+0.1759i \\ 0 & 0 & 0 & 0 \\ 0 & 0 & 0 & 0 \end{bmatrix}$
Codebook 4	$\begin{bmatrix} 0 & 0 & 0 & 0 \\ 0 & 0 & 0 & 0 \\ 0.7851 & -0.2243 & 0.2243 & -0.7851 \\ -0.0055-0.2242i & -0.0193-0.7848i & 0.0193+0.7848 & 0.0055+0.2242i \end{bmatrix}$
Codebook 5	$\begin{bmatrix} -0.0055-0.2242i & -0.0193-0.7848i & 0.0193+0.7848 & 0.0055+0.2242i \\ 0 & 0 & 0 & 0 \\ 0 & 0 & 0 & 0 \\ -0.6351+0.4615i & 0.1815-0.1318i & -0.1815+0.1318i & 0.6351-0.4615i \end{bmatrix}$
Codebook 6	$\begin{bmatrix} 0 & 0 & 0 & 0 \\ 0.7851 & -0.2243 & 0.2243 & -0.7851 \\ 0.1392-0.1759i & 0.4873-0.6156i & -0.4873+0.6156i & -0.1392+0.1759i \\ 0 & 0 & 0 & 0 \end{bmatrix}$

- Multi-dimensional modulation: SCMA employs multi-dimensional constellations for better spectral efficiency.

12.1.2 Multi-user Detection

This subsection discusses multi-user detection schemes for SCMA, including the optimal detection, the MPA receiver and other advanced receivers.

12.1.2.1 Optimal/Quasi-optimal Multi-user Detection

A. Optimal Multi-user Detection

Assume that channel state is perfectly estimated at the receiver, given the received signal vector $\mathbf{y}$, the joint optimum maximum a posteriori probability (MAP) detection, for multi-user codeword $\mathbf{X}$ and for the jth user's codeword $\mathbf{x}_j$, can be written as

$$\hat{\mathbf{X}} = \arg \max_{\mathbf{x}_j \in \mathbb{X}_j, \forall j} p(\mathbf{X}|\mathbf{y}), \quad \text{and} \quad \hat{\mathbf{x}}_j = \arg \max_{\mathbf{x}_j \in \mathbb{X}_j} \sum_{\mathbf{x}_i \in \mathbb{X}_i, \forall i \neq j} p(\mathbf{X}|\mathbf{y}) \tag{12.6}$$

respectively. Using Bayes' rule

$$p(\mathbf{X}|\mathbf{y}) = \frac{p(\mathbf{y}|\mathbf{X})p(\mathbf{X})}{p(\mathbf{y})} \propto p(\mathbf{y}|\mathbf{X})P(\mathbf{X})$$

where

$$p(\mathbf{X}) = \prod_{j=1}^{J} p(\mathbf{x}_j), \quad \text{and} \quad p(\mathbf{y}) = \sum_{\mathbf{x}_j \in \mathbb{X}_j, \forall j} p(\mathbf{y}|\mathbf{X})p(\mathbf{X})$$

are the joint a prior probability[2] for each user's codeword, and the probability of the received signal vector, respectively.

By assuming that the noise components over the K resources are identically independently distributed (i.i.d.), it holds

$$p(\mathbf{y}|\mathbf{X}) = \prod_{k=1}^{K} p(y[k]|\mathbf{X})$$

and considering only d_f users actually collided over the kth resource, we have

$$p(y[k]|\mathbf{X}) = \frac{1}{\pi N_0} \exp\left(-\frac{1}{N_0}\left|y[k] - \sum_{j \in \phi_k} h_j[k]x_j[k]\right|^2\right) \tag{12.7}$$

Thus, the MAP decision for the jth user's codeword is given by

$$\hat{\mathbf{x}}_j = \arg \max_{\mathbf{x}_j \in \mathbb{X}_j} \sum_{\mathbf{x}_i \in \mathbb{X}_i, \forall i \neq j} P(\mathbf{X}) \prod_{k=1}^{K} p(y[k]|\mathbf{X}), \quad \forall j \tag{12.8}$$

With the codeword probability for each user, it is straightforward to calculate the log-likelihood rate (LLR) for each coded bit, so that they can serve as the input for

[2] Without feedback from the FEC decoder, $p(\mathbf{x}_j) = \frac{1}{M}$ for all the users.

the FEC decoder. For the jth user, the LLR considering the mth bit $b_{j,m}$ is calculated by

$$\begin{aligned}\Lambda(b_{j,m}) &= \log \frac{\Pr\{b_{j,m} = 1|\mathbf{y}\}}{\Pr\{b_{j,m} = 0|\mathbf{y}\}} \\ &= \log \frac{\sum_{\mathbf{x}_j \in \mathbb{X}_{j,m}^1} \sum_{\mathbf{x}_i \in \mathbb{X}_i, \forall i \neq j} P(\mathbf{X}) \prod_{k=1}^{K} p(y[k]|\mathbf{X})}{\sum_{\mathbf{x}_j \in \mathbb{X}_{j,m}^0} \sum_{\mathbf{x}_i \in \mathbb{X}_i, \forall i \neq j} P(\mathbf{X}) \prod_{k=1}^{K} p(y[k]|\mathbf{X})}\end{aligned} \tag{12.9}$$

where $\mathbb{X}_{j,m}^1$ and $\mathbb{X}_{j,m}^0$ are subsets of $\mathbb{X}_j$ for which the mth bit of the jth user $b_{j,m} = 1$ and $b_{j,m} = 0$, respectively. Note that solving (12.8) is equivalent to solve the marginal product of functions (MPF) problem, which is of exponential complexity with *brute-force* searching, and is prohibitive to employ when the number of users increases.

B. MPA Detection

As the SCMA encoding can be represented by a factor graph with sparse property, the low-complexity MPA can be used to solve the MPF problem with near-optimum performance.

Let $I_{k\to j}^{(t)}$ be the extrinsic information to be passed from FN k to VN j at the tth iteration, and $I_{j\to k}^{(t)}$ be the extrinsic information to be passed from VN j to FN k. Given the *a prior* probability $p(\mathbf{x}_j)$, the probability that $\mathbf{x}_j$ is transmitted by the jth user given the channel sample is updated as

$$I_{j\to k}^{(t)}(\mathbf{x}_j) = p(\mathbf{x}_j) \prod_{l \in \varphi_j \backslash k} I_{l\to j}^{(t)}(\mathbf{x}_j)$$

Then, for any $\mathbf{x}_j \in \mathbb{X}_j$, the probability of the received signal $y[k]$ given that $\mathbf{x}_j$ is transmitted by the jth user, marginalized over all possible codewords of the other users, is given by

$$I_{k\to j}^{(t)}(\mathbf{x}_j) = \sum_{\mathbf{x}_i \in \mathbb{X}_i, \forall i \in \phi_k \backslash j} p\left(y[k]|\mathbf{X}\right) \prod_{i \in \phi_k \backslash j} I_{i\to k}^{(t-1)}(\mathbf{x}_i) \tag{12.10}$$

After a number of iterations, the posterior probability of $\mathbf{x}_j$ produced by the MPA is proportional to

$$I_j(\mathbf{x}_j) = p(\mathbf{x}_j) \prod_{k \in \varphi_j} I_{k\to j}^{(T)}(\mathbf{x}_j), \quad \mathbf{x}_j \in \mathbb{X}_j, j = 1, \ldots, J \tag{12.11}$$

where T is the number of iterations at the termination.

Similar to that for MAP detection, the LLR of the mth bit of the jth user $b_{j,m}$ is calculated by

$$\Lambda(b_{j,m}) = \log \frac{\sum_{\mathbf{x}_j \in \mathbb{X}_{j,m}^1} I_j(\mathbf{x}_j)}{\sum_{\mathbf{x}_j \in \mathbb{X}_{j,m}^0} I_j(\mathbf{x}_j)} \tag{12.12}$$

where $\mathbb{X}_{j,m}^1$ and $\mathbb{X}_{j,m}^0$ are the same as that in (12.9).

The main complexity of MPA comes from the calculation of (12.10), the summation over $\mathbf{x}_i$ adds up $M^{|\phi_k|-1}$ terms while M probabilities should be calculated in each iteration, which leads to a complexity order $O(TKM^{d_f})$, and is far below that of the optimal MAP detection. In practical implementations, the exponential function in MPA algorithm may cause large dynamic ranges and very high storage burden, then the logarithmic domain MPA is preferred to avoid the exponential operations. For the log-MPA operation, the information exchanged between the FNs and VNs can be expressed as

$$I_{j\to k}^{(t)}(\mathbf{x}_j) = \log p(\mathbf{x}_j) + \sum_{l\in\varphi_j\backslash k} I_{l\to j}^{(t)}(\mathbf{x}_j)$$

$$I_{k\to j}^{(t)}(\mathbf{x}_j) = \max_{\mathbf{x}_i\in\mathbb{X}_i,\forall i\in\phi_k\backslash j}\left\{\log p\left(y[k]|\mathbf{X}\right) + \sum_{i\in\phi_k\backslash j} I_{i\to k}^{(t-1)}(\mathbf{x}_i)\right\}$$

where Jacobi's logarithm formula $\log\left(\sum_i e^{p_i}\right) \approx \max_i p_i$ is applied for a complexity reduction to a certain degree, which results in the max-log-MPA detection. The output LLR of the MPA detector is given by

$$\Lambda(b_{j,m}) = \max_{\mathbf{x}_j\in\mathbb{X}_{j,m}^1} I_j(\mathbf{x}_j) - \max_{\mathbf{x}_j\in\mathbb{X}_{j,m}^0} I_j(\mathbf{x}_j)$$

where

$$I_j(\mathbf{x}_j) = \log p(\mathbf{x}_j) + \sum_{k\in\varphi_j} I_{k\to j}^{(T)}(\mathbf{x}_j)$$

12.1.2.2 Other Advanced Detectors

The MPA detector still has exponential complexity with respect to the codebook size (M) and the number of accessed users at each resource (d_f), which may become impractical for the implementation of very large codebook size (e.g., $M \geq 64$) and very high overload (e.g., $d_f \geq 8$). Some other advanced detectors can harness the potential gain of SCMA while provide sufficient flexibility for a good trade-off between the performance and detection complexity [7, 8].

A. EPA Detector

Expectation propagation algorithm (EPA) is an approximate Bayesian inference method in machine learning for estimating sophisticated posterior distributions with simple distributions through distribution projection, and it turns out to be an efficient iterative multi-user detector for SCMA as well as some other multiple access schemes [8]. It approximates the discrete message in MPA as continuous Gaussian message using the minimum Kullback–Leibler (KL) divergence criterion, and use the a posteriori probabilities fed back from the FEC decoder to compute the approximate symbol belief and the approximate message, such that the message passing reduces to mean and variance parameters update. The detailed algorithm is given in [8]. With EPA, the complexity order of SCMA detection is reduced to linear complexity, i.e., it only scales linearly with the codebook size M and the average degree of the factor nodes d_f, while simulation results show that the EPA detector shows nearly the same error performance as MPA for SCMA with receiver diversity. As a result, the computation burden of the SCMA receiver is significantly alleviated and is no longer a problem for implementation in real systems.

B. SIC-MPA Detector

Successive interference cancelation (SIC) receiver is a kind of multi-user receiver that treats all the other undecoded users as interference when decoding a target user, and can be implemented as either symbol level or codeword level. It works well when the received SNR among users are quite different from each other. However, the detecting performance deteriorates when the SNR difference is not obvious between users, in which case error propagation happens.

To strike a good balance between link performance and implementation complexity, it is reasonable to combine SIC with an MPA (SIC-MPA) receiver. More specifically, MPA is applied to a limited number of users firstly, so that the number of colliding users over each resource does not exceed a given threshold value (e.g., d_s users). Then, the successfully decoded users are removed by SIC and the procedure continues until all users are successfully decoded or no new user gets successfully decoded in MPA. In the case of $d_s = d_f$, full MPA is realized, and when $d_s = 1$, it becomes a pure SIC receiver. Due to the fact that MPA is used for a very limited number of users instead of all the users, the decoding complexity is greatly reduced, which is of the order $O(TKM^{d_s})$.

12.2 Performance Evaluation

The error performance and capacity are excellent measures that indicating the goodness of a system, and more importantly, they serve as powerful tools for the practical system design. For SCMA, the multi-user codebook plays a key role in the system performance improvement, and it is necessary to establish performance criterion to guide the codebooks design. In this section, error performance and capacity analysis for uplink and downlink SCMA systems are provided, and independent Rayleigh fadings are assumed.

12.2.1 Average Error Probability

The error probability, e.g., the average codeword error probability (ACEP), is one of the most important performance criteria, since it is most revealing about the nature of a system behavior. However, it is quite difficult to evaluate the exact ACEP for SCMA systems, since one needs to average over several fading statistics due to the multi-channel transmissions, and the integration involves a decision cell of a multi-dimensional signal point. As an alternative approach, it is convenient to resort to an upper bound or approximation on the ACEP [9]. In this subsection, we use union bound to evaluate the error performance of uplink and downlink SCMA under joint maximum likelihood (ML) multi-user detection.

12.2.1.1 PEP over Uplink MACs

Consider the equivalent "MIMO" channel (12.2). Under the assumption of perfect channel estimation at the receiver, the joint ML detection of multi-user codewords is equivalent to the joint minimum distance decoding

$$\hat{\mathbf{X}} = \arg\min_{\mathbf{X}} \|\mathbf{y} - \mathbf{H}\mathbf{X}\|$$

The pairwise error probability (PEP), defined as the probability that received signal vector $\mathbf{y}$ is detected into $\mathbf{X}_b$ given that $\mathbf{X}_a$ is transmitted, is given by [10]

$$P\{\mathbf{X}_a \to \mathbf{X}_b\} = \mathbb{E}_{\mathbf{H}}\left[Q\left(\sqrt{\frac{\|\mathbf{H}(\mathbf{X}_a - \mathbf{X}_b)\|^2}{2N_0}} \right) \right] \tag{12.13}$$

where, $Q(x) = \frac{1}{\sqrt{2\pi}} \int_x^{\infty} e^{-t^2/2} \mathrm{d}t$, is the well-known Gaussian function [11], and $\mathbb{E}_{\mathbf{H}}[\cdot]$ denotes the mean.

Let $x_{j,a}[k]$ and $x_{j,b}[k]$ be the kth entries of the jth user's codewords $\mathbf{x}_{j,a}$ and $\mathbf{x}_{j,b}$, corresponding to $\mathbf{X}_a$ and $\mathbf{X}_b$, respectively. Due to the sparseness of the codewords, $x_{j,a}[k] = x_{j,b}[k] = 0$ whenever $j \notin \phi_k$. Now we define a distance for the MAC.

Definition 1 The kth dimension-wise distance, between the multi-user combined codewords $\mathbf{X}_a$ and $\mathbf{X}_b$, for the uplink MAC is defined as

$$\lambda_k^2 = \sum_{j=1}^{J} |x_{j,a}[k] - x_{j,b}[k]|^2 = \sum_{j \in \phi_k} |\delta_j[k]|^2, \quad \forall k \tag{12.14}$$

where $\delta_j[k] = x_{j,a}[k] - x_{j,b}[k]$.

Assume that there are repeated values among the set $\{\lambda_1^2, \ldots, \lambda_K^2\}$, such that they can be divided into V $(1 \le V \le K)$ groups, and each group contains the collection of a certain value $\hat{\lambda}_v^2$. Let $\hat{\boldsymbol{\lambda}} = [\hat{\lambda}_1^2, \ldots, \hat{\lambda}_V^2]^t$, be the vector of V distinct elements among $\{\lambda_1^2, \ldots, \lambda_K^2\}$, and $\mathbf{r} = [r_1, \ldots, r_V]^t$, where r_v is the number of elements in $\{\lambda_1^2, \ldots, \lambda_K^2\}$ that equals to $\hat{\lambda}_v^2$, such that $\sum_{v=1}^{V} r_v = K$.

Definition 2 Define the parameter

$$A_{\mathbf{r},\lambda} = \prod_{k=1}^{K} \lambda_k^{-2} = \prod_{v=1}^{V} \hat{\lambda}_v^{-2r_v} \tag{12.15}$$

as the reciprocal of the product of the dimension-wise distances.

Definition 3 For positive integers l, v and vectors $\mathbf{r}$ and $\hat{\boldsymbol{\lambda}}$, define the parameter

$$B_{v,l,\mathbf{r},\hat{\lambda}} = (-1)^{l+1} \sum_{\eta \in \Omega_{v,l}} \prod_{j=1, j\neq v}^{V} \binom{\eta_j + r_j - 1}{\eta_j} \left(\frac{1}{\hat{\lambda}_j^2} - \frac{1}{\hat{\lambda}_v^2} \right)^{-(r_j+\eta_j)} \tag{12.16}$$

where the vector $\boldsymbol{\eta} = [\eta_1, \ldots, \eta_V]^t$ is created from the set $\Omega_{v,l}$ of all nonnegative integer partitions of $l-1$ (with $\eta_v = 0$). The set $\Omega_{v,l}$ is defined as

$$\Omega_{v,l} = \left\{ \boldsymbol{\eta} = [\eta_1, \ldots, \eta_V]^t \in \mathbb{Z}^V; \sum_{j=1}^{V} \eta_j = l-1, \eta_v = 0, \eta_j \geq 0 \, \forall j \right\}$$

Next, we provide the main result regarding the PEP.

Theorem 1 *For J users using K-dimensional codebooks in the uplink SCMA systems, the PEP between $\mathbf{X}_a$ and $\mathbf{X}_b$ is given by*

$$\begin{aligned} P\{\mathbf{X}_a \to \mathbf{X}_b\} &= A_{\mathbf{r},\hat{\lambda}} \sum_{v=1}^{V} \sum_{L=1}^{r_v} \hat{\lambda}_v^{2L} B_{v,r_v-L+1,\mathbf{r},\hat{\lambda}} \\ &\times \left(\frac{1-\mu_v}{2} \right)^L \sum_{k=0}^{L-1} \binom{L-1+k}{k} \left(\frac{1+\mu_v}{2} \right)^k \end{aligned} \tag{12.17}$$

where

$$\mu_v = \sqrt{\frac{\hat{\lambda}_v^2}{4N_0 + \hat{\lambda}_v^2}}$$

Proof Consider the metric in (12.13),

$$\|\mathbf{H}(\mathbf{X}_a - \mathbf{X}_b)\|^2 = \sum_{k=1}^{K} \|\mathbf{h}[k]^\dagger (\mathbf{x}_a[k] - \mathbf{x}_b[k])\|^2$$

where $\mathbf{x}_a[k] = [x_{1,a}[k], \ldots, x_{J,a}[k]]^t$, is the vector of the kth component for J users, and $\mathbf{h}[k]^\dagger = [h_1[k], \ldots, h_J[k]]$, are the corresponding channel gains, and $[\cdot]^\dagger$ denotes conjugate transpose. Using the matrix decomposition, it holds that

$$(\mathbf{x}_a[k] - \mathbf{x}_b[k])(\mathbf{x}_a[k] - \mathbf{x}_b[k])^\dagger = \mathbf{U}_k \boldsymbol{\Lambda}_k \mathbf{U}_k^\dagger$$

where $\mathbf{U}_k$ is unitary and $\boldsymbol{\Lambda}_k$ is a diagonal matrix, i.e., $\boldsymbol{\Lambda}_k = \text{diag}(\tilde{\lambda}_{k,1}^2, \ldots, \tilde{\lambda}_{k,J}^2)$, with $\tilde{\lambda}_{k,j}^2$ being the ordered singular values of the matrix $(\mathbf{x}_a[k] - \mathbf{x}_b[k])(\mathbf{x}_a[k] - \mathbf{x}_b[k])^\dagger$. Note that the matrix $(\mathbf{x}_a[k] - \mathbf{x}_b[k])(\mathbf{x}_a[k] - \mathbf{x}_b[k])^\dagger$ is of rank 1 and the unique nonzero singular value in $\boldsymbol{\Lambda}_k$ is

$$\tilde{\lambda}_{k,1}^2 = \|\mathbf{x}_a[k] - \mathbf{x}_b[k]\|^2 \overset{(a)}{=} \sum_{j \in \phi_k} |x_{j,a}[k] - x_{j,b}[k]|^2$$

where (a) is due to the sparseness of the codebooks. Obviously, the nonzero eigenvalue is equal to the dimension-wise distances defined in Definition 1, namely $\tilde{\lambda}_{k,1}^2 = \lambda_k^2$. Hence,

$$\begin{aligned}\|\mathbf{h}[k]^\dagger(\mathbf{x}_a[k] - \mathbf{x}_b[k])\|^2 &= \mathbf{h}[k]^\dagger \mathbf{U}_k \boldsymbol{\Lambda}_k \mathbf{U}_k^\dagger \mathbf{h}[k] \\ &= \tilde{\mathbf{h}}[k]^\dagger \boldsymbol{\Lambda}_k \tilde{\mathbf{h}}[k] = \lambda_k^2 |\tilde{h}_1[k]|^2\end{aligned}$$

where we define, $\tilde{\mathbf{h}}[k]^\dagger = \mathbf{h}[k]^\dagger \mathbf{U}_k = [\tilde{h}_1[k], \ldots, \tilde{h}_J[k]]$. Thus, $\tilde{\mathbf{h}}[k]$ has the same distribution as $\mathbf{h}[k]$, since multiplying with unitary matrix $\mathbf{U}_k$ doesn't change the amplitudes. Thus, the average PEP in (12.13) is equal to

$$P\{\mathbf{X}_a \to \mathbf{X}_b\} = \mathbb{E}_{\tilde{h}[1],\ldots,\tilde{h}[K]}\left[Q\left(\sqrt{\frac{\sum_{k=1}^K \lambda_k^2 |\tilde{h}[k]|^2}{2N_0}}\right)\right] \tag{12.18}$$

where the index 1 is dropped here for $\tilde{h}_1[k]^2$.

For i.i.d. Rayleigh fading, $\tilde{h}[1], \ldots, \tilde{h}[K]$ are i.i.d. complex Gaussian random variables with zero mean and unit variance. Thus, $\sum_{k=1}^K \lambda_k^2 |\tilde{h}[k]|^2$ is the sum of K exponential random variables with different means, or a linear combination of V independent χ^2-distributed random variables with $2r_1, \ldots, 2r_V$ degrees of freedom, which follows Gamma or Generalized chi-squared distribution [12], with PDF given by

$$f(x; \mathbf{r}, \hat{\boldsymbol{\lambda}}) = A_{\mathbf{r},\hat{\boldsymbol{\lambda}}} \sum_{v=1}^{V} \sum_{l=1}^{r_v} \frac{B_{v,l,\mathbf{r},\hat{\boldsymbol{\lambda}}}}{(r_v - l)!} x^{r_v - l} e^{-\frac{x}{\hat{\lambda}_v^2}}$$

Then, the PEP can be obtained as

$$
\begin{aligned}
P\{\mathbf{X}_a \to \mathbf{X}_b\} &= \int_0^{\infty} Q\left(\sqrt{\frac{x}{2N_0}}\right) f(x;\mathbf{r},\hat{\boldsymbol{\lambda}})\mathrm{d}x \\
&= A_{\mathbf{r},\hat{\boldsymbol{\lambda}}} \sum_{v=1}^{V}\sum_{l=1}^{r_v} \frac{B_{v,l,\mathbf{r},\hat{\boldsymbol{\lambda}}}}{(r_v-l)!} \int_0^{\infty} Q\left(\sqrt{\frac{x}{2N_0}}\right) x^{r_v-l} e^{-\frac{x}{\hat{\lambda}_v^2}} \mathrm{d}x \\
&= A_{\mathbf{r},\hat{\boldsymbol{\lambda}}} \sum_{v=1}^{V}\sum_{l=1}^{r_v} B_{v,l,\mathbf{r},\hat{\boldsymbol{\lambda}}} \left(\frac{(1-\mu_v)\hat{\lambda}_v^2}{2}\right)^{r_v-l+1} \\
&\quad \times \sum_{k=0}^{r_v-l} \binom{r_v-l+k}{k} \left(\frac{1+\mu_v}{2}\right)^k
\end{aligned} \tag{12.19}
$$

where the last step follows from (13.4-15) in [11]. Substituting l with $r_v - L + 1$, (12.17) is proved. This concludes the proof.

The PEP is uniquely determined by the set of all λ_k^2s and the SNR, which is valid for any multi-dimensional codebooks and an arbitrary number of users. By reducing the number of users to 1, $P\{\mathbf{X}_a \to \mathbf{X}_b\}$ becomes the PEP between two multi-dimensional constellation points for a single-user transmission system. Therefore, the PEP of a joint multi-user detector is actually identical to that of the PEP of a single-user transmitting over a fading channel, where an equivalent K-dimensional constellation is employed such that the dimension-wise distances between the two constellation points are $\lambda_1^2, \ldots, \lambda_K^2$.

12.2.1.2 PEP over Downlink BCs

Consider the received signal vector of the downlink BC in (12.3), where $\mathbf{X} = \sum_{j=1}^{J} \mathbf{x}_j$ is the superimposed codeword of multiple users at the transmitter. Obviously, the model is exactly the same with that in the single-user communications, where $\mathbf{X}$ is used as the K-dimensional transmitted codeword. The ML multi-user detection for the superimposed codeword $\mathbf{X}$ becomes

$$\hat{\mathbf{X}} = \arg\min_{\mathbf{X}} \|\mathbf{y} - \mathrm{diag}(\mathbf{h})\mathbf{X}\|$$

Similar to that in uplink SCMA, we define the distances for downlink BC model.

Definition 4 Let $\mathbf{X}_a$ and $\mathbf{X}_b$ be two superimposed codewords, and $x_{j,a}[k]$ and $x_{j,b}[k]$ are the kth entries of the jth user's codeword corresponding to $\mathbf{X}_a$ and $\mathbf{X}_b$, respectively. The kth dimension-wise distance between $\mathbf{X}_a$ and $\mathbf{X}_b$, for the downlink broadcast channel, is defined as

$$\tau_k^2 = \left|\sum_{j=1}^{J}\left(x_{j,a}[k] - x_{j,b}[k]\right)\right|^2 = \left|\sum_{j\in\phi_k}\delta_j[k]\right|^2, \quad \forall k \tag{12.20}$$

where $\delta_j[k] = x_{j,a}[k] - x_{j,b}[k]$.

Theorem 2 *The PEP of a Rayleigh broadcast channel is the same as that in (12.17), after the substitution of λ_k^2 with τ_k^2.*

Proof Similar to that of the uplink case, the average PEP between $\mathbf{X}_a$ and $\mathbf{X}_b$ is equal to

$$\begin{aligned} P\{\mathbf{X}_a \to \mathbf{X}_b\} &= \mathbb{E}_{\mathbf{h}}\left[Q\left(\sqrt{\frac{\|\mathrm{diag}(\mathbf{h})\left(\mathbf{X}_a - \mathbf{X}_b\right)\|^2}{2N_0}}\right)\right] \\ &= \mathbb{E}_{\mathbf{h}}\left[Q\left(\sqrt{\frac{\sum_{k=1}^{K}\tau_k^2|h[k]|^2}{2N_0}}\right)\right]. \end{aligned} \tag{12.21}$$

As $h[1], \ldots, h[K]$ are independent Rayleigh distributed random variables, the integral has been solved in (12.18), and the PEP has the similar expression as that in the MAC case, after the substitution of λ_k^2 with τ_k^2. This completes the proof.

It should be noted that, while the PEP of a BC can be evaluated through the same expression as that in the MAC case, τ_k^2 is different from the dimension-wise distance λ_k^2 in MAC, due to the absence of cross components $\delta_j[k] \times \delta_i[k]$, $j \neq i$, between different users. This is because in the MAC, the receiver distinguishes the multi-user signals by exploiting the differences among the channel coefficients, and only the amplitude of $\delta_j[k]$ contributes to the PEP. However, in the broadcast channel case, since the receiver exploits the differences among the multiple users' signals to perform the joint detection, both the amplitude and signs of $\delta_j[k]$ will influence the result of PEP.

12.2.1.3 PEP over the AWGN Channel

For the AWGN channel, where $h_j[k]$ is a constant for all j and k (assume that $|h_j[k]| = c$), the expressions of the received signal vector in (12.1) for uplink channels and (12.3) for downlink channels are the same. Then, according to (12.21), it is straightforward to derive the PEP as

$$\begin{aligned} P\{\mathbf{X}_a \to \mathbf{X}_b\} &= Q\left(\sqrt{\frac{c^2\sum_{k=1}^{K}\tau_k^2}{2N_0}}\right) \\ &= Q\left(\sqrt{\frac{c^2\sum_{k=1}^{K}\left|\sum_{j\in\phi_k}\delta_j[k]\right|^2}{2N_0}}\right) \end{aligned} \tag{12.22}$$

where τ_k^2 is the dimension-wise distance defined in (12.20), and $\delta_j[k] = x_{j,a}[k] - x_{j,b}[k]$.

12.2.1.4 Upper Bounds on PEP

In the codebook design, sometimes it is sufficient and easier to optimize the performance through a bound or an approximation of PEP. The exact PEP in (12.17) is a little complicated for large K, due to the large number of enumerations in $\Omega_{v,l}$, when calculating $B_{v,r_v-L+1,\mathbf{r},\boldsymbol{\lambda}}$. An alternative way to evaluate (12.17) is to use an upper bound for the Q-function as [13]

$$Q(x) \leq \sum_{i=1}^{N} a_i e^{-b_i x^2}, \quad \text{for } x > 0,$$

where N, a_i, b_i are constants. Note that the upper bound in Sect. 12.2.1.4 tends to the exact value as N increases.

For the multiple access and broadcast channels, since $X = |\hat{h}[k]|^2$ is an exponential random variable with unit mean, holds that $\mathbb{E}_X[e^{tX}] = \int_0^\infty e^{tx}e^{-x}dx = \frac{1}{1-t}$, for $t \leq 1$, and

$$\begin{aligned} P\{\mathbf{X}_a \to \mathbf{X}_b\} &\leq \mathbb{E}_{\tilde{h}[1],\ldots,\tilde{h}[K]}\left[\sum_{i=1}^{N} a_i \exp\left(-\frac{b_i \sum_{k=1}^{K} \lambda_k^2 |\tilde{h}[k]|^2}{2N_0}\right)\right] \\ &= \sum_{i=1}^{N} a_i \prod_{k=1}^{K} \mathbb{E}_{\tilde{h}[k]}\left[\exp\left(-\frac{b_i \lambda_k^2 |\tilde{h}[k]|^2}{2N_0}\right)\right] \\ &= \sum_{i=1}^{N} a_i \prod_{k=1}^{K} \frac{2N_0}{2N_0 + b_i \lambda_k^2} \end{aligned}$$

By choosing $N = 1, a_1 = b_1 = \frac{1}{2}$, we get the Chernoff bound with a scaling factor of 0.5 as

$$P_{\text{ch}}\{\mathbf{X}_a \to \mathbf{X}_b\} \leq \frac{1}{2} \prod_{k=1}^{K} \frac{4N_0}{4N_0 + \lambda_k^2} \tag{12.23}$$

In general, the Chernoff bound may be a little loose, but this does not affect the optimization criteria in the constellation design. It is obvious from (12.23) that a good direction is to design multi-dimensional multi-user codebooks, such that λ_k^2 to span in as many dimensions as possible (maximizing the diversity) and to make the maximum PEP or maximum of $P_{\text{ch}}\{\mathbf{X}_a \to \mathbf{X}_b\}$ as small as possible. If for any codeword pair $\mathbf{X}_a$ and $\mathbf{X}_b$, all the λ_k^2 are positive, then the maximal diversity order of K can be achieved. Due to the sparseness of the codebooks, the diversity is always less than K. A tight and simple bound (or approximation) is to choose $N = 2$, $a_1 = \frac{1}{12}$, $a_2 = \frac{1}{4}$, $b_1 = \frac{1}{2}$, $b_2 = \frac{2}{3}$, which is denoted as $P_{\text{ub}}\{\mathbf{X}_a \to \mathbf{X}_b\}$.

12.2.1.5 A Universal Bound of ACEP for Joint ML Detection of Multiple Signals

A commonly used approach for the error performance analysis is the evaluation of the ACEP by using a union bound, assuming that the codewords are equiprobable transmitted. In general, the ACEP is dominated by the nearest neighbors of codewords, which result in a tight upper bound. However, it is quite difficult (if not impossible) to find the nearest neighbors in multi-user scenarios. To deal with this, we take into account all possible codewords that contribute to the ACEP.

Let $M_1, \ldots, M_J$ be the codebook size for J users, respectively. We define $\{\mathbb{X}_j\}_{j=1}^J$ as the set of all $\prod_{j=1}^J M_j$ possible combined codewords of J users, and let $\mathbf{X}_a, \mathbf{X}_b \in \{\mathbb{X}_j\}_{j=1}^J$ be two different elements of $\{\mathbb{X}_j\}_{j=1}^J$. Here, the combined codeword $\mathbf{X}_a$ and $\mathbf{X}_b$ are a JK-dimensional vector for the MAC, or the sum of J K-dimensional codewords for the BC. Denote $\mathbf{x}_{j,a}$ and $\mathbf{x}_{j,b}$ the transmitted codewords of the jth user corresponding to $\mathbf{X}_a$ and $\mathbf{X}_b$. Then, there are M_j possible values for $\mathbf{x}_{j,a}$ and $\mathbf{x}_{j,b}$. Note that $\mathbf{x}_{j,a}$ and $\mathbf{x}_{j,b}$ are K-dimensional vector with complex entries, i.e., SCMA codeword. Following the approach in [14] for multiple signals and [15] for MIMO channels, the ACEP for the jth user with joint ML detection of J users' signals is upper bounded by

$$P_j(e) \leq \frac{1}{\prod_{j=1}^J M_j} \sum_{\mathbf{X}_a} \left(\sum_{\mathbf{X}_b, \mathbf{x}_{j,b} \neq \mathbf{x}_{j,a}} P\left\{\mathbf{X}_a \to \mathbf{X}_b\right\} \right) \tag{12.24}$$

The ACEP of the system can be obtained by taking the mean of all the single-user ACEPs, namely $P(e) = \frac{1}{J}\sum_{j=1}^J P_j(e)$.

On the right-hand side of (12.24), the summation over $\mathbf{X}_a$ will add up $\prod_{j=1}^J M_j$ terms and the summation over $\mathbf{X}_b$ will add up $(M_1 - 1)\prod_{j=2}^J M_j$ terms. Thus, there will be up to $(M_1^2 - M_1)\prod_{j=2}^J M_j^2$ PEPs in (12.24), which is intractable for a large constellation size and number of users. However, we can simplify it by using the symmetry of the dimension-wise distances. For example, consider the ACEP for the first user $P_1(e)$ here. The upper bound of $P_1(e)$ can be decomposed into the summation of two parts as

$$
\begin{aligned}
&\frac{1}{\prod_{j=1}^{J} M_j} \sum_{\mathbf{X}_a} \sum_{\substack{\mathbf{X}_b, \mathbf{x}_{1,b} \neq \mathbf{x}_{1,a}, \\ [\mathbf{x}_{2,b}, \ldots, \mathbf{x}_{j,b}] = [\mathbf{x}_{2,a}, \ldots, \mathbf{x}_{j,a}]}} P\{\mathbf{X}_a \rightarrow \mathbf{X}_b\} \\
&+ \frac{1}{\prod_{j=1}^{J} M_j} \sum_{\mathbf{X}_a} \sum_{\substack{\mathbf{X}_b, \mathbf{x}_{1,b} \neq \mathbf{x}_{1,a}, \\ [\mathbf{x}_{2,b}, \ldots, \mathbf{x}_{j,b}] \neq [\mathbf{x}_{2,a}, \ldots, \mathbf{x}_{j,a}]}} P\{\mathbf{X}_a \rightarrow \mathbf{X}_b\}.
\end{aligned}
\tag{12.25}
$$

The first part in (12.25) is the union bound of the probability of the event that all users' signals are correctly detected except for the first user, namely the ACEP for the first user with single-user detection in the absence of interference. This part is a summation of $(M_1 - 1)\prod_{j=1}^{J} M_j$ PEPs, while only $\frac{1}{2}M_1(M_1 - 1)$ different PEP values should be calculated, due to the symmetry of the dimension-wise distance for the first user. The second part is the probability of the event that the errors happen for the first user and for at least one user among $\{2, \ldots, J\}$, which is the summation of $(M_1 - 1)(\prod_{j=2}^{J} M_j - 1)\prod_{j=1}^{J} M_j$ PEPs, but only one-fourth of them should be considered. A further simplification can be achieved for the MAC by considering more decompositions of the second part.

In general, SCMA codebooks of all users are constructed from a common mother constellation [16], with some layer-specific operations over this constellation to get their own layer's codebook. These layer-specific operations do not change the fundamental properties of the mother constellation, such as the Euclidean distance. The layer operation losses their efficiency in the uplink multiple access fading channels, due to the distinctness of each user's channel gain. If the factor graph matrix is regular as that in (12.5), every user will suffer from the same interference from other users. Then, the system results in the same performance for all users, while for other cases, the ACEP is asymmetric for each user.

The MPA detection is believed to be an efficient approach for SCMA systems. Theoretically, the MPA detector is asymptotically equivalent to the optimal MAP detector [17, 18] (or ML conditioned on equal probably transmissions) for a sparsely spread system with long signatures. The analytical bounds, proposed in this subsection, work for ML detector as well as for the MPA detector.

12.2.1.6 Numerical Results and Simulations

We consider an SCMA system illustrated in Example 1, and the four-dimensional four-ary codebooks are listed in Table 12.1. The ACEP of SCMA over AWGN and uplink Rayleigh fading channels for 2, 4, and 6 user cases are evaluated. For the Rayleigh fading channel, we give analytical results of the union bound on the ACEP, corresponding to exact PEP (denoted as P_{jml}), the upper bound on PEP P_{ub}, and the scaled Chernov bound P_{ch}, respectively.

Results for AWGN channels are shown in Fig. 12.4a. The analytical bound of a joint ML detector closely coincides with the simulation curves for large SNR. The bound is quite tight for values of ACEP below 10^{-3}, even for six users. Thus, this

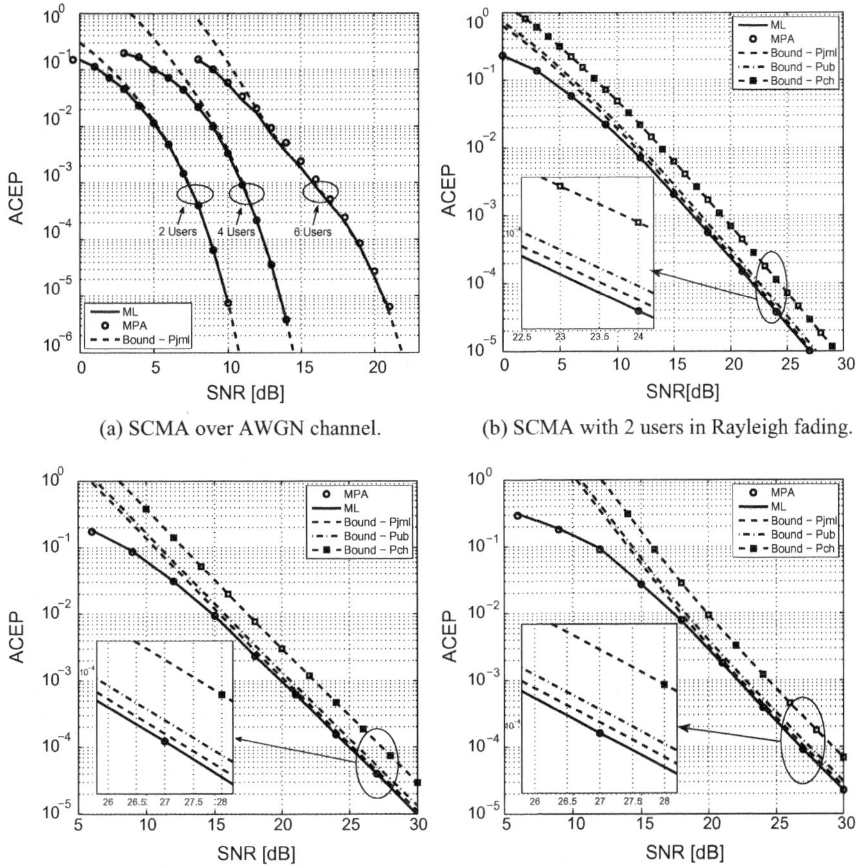

(a) SCMA over AWGN channel.

(b) SCMA with 2 users in Rayleigh fading.

(c) SCMA with 4 users in Rayleigh fading.

(d) SCMA with 6 users in Rayleigh fading.

Fig. 12.4 ACEP of uplink SCMA over AWGN and Rayleigh fading channels

bound is sufficient for the analysis and design of a signal constellation in AWGN. Surprisingly, there are bends for the ACEP curves of an ML detector and analytical bound for the six user case. The performance turns better than expected within the SNR region from 12 to 18 dB, which is due to the sparse codebooks. This phenomenon happens if the distance profile of the multi-user codebooks is uneven. For example, a quite small distance exists while the others are very large. In general, the ACEP of a constellation in AWGN channels is proportional to the summation of Q-function of the distances d and SNR, i.e., $P(\text{SNR}) \propto \sum_d Q(\sqrt{d\text{SNR}})$, where d is the set of distances among the constellation points. If there is a large difference between two distance components, $P(\text{SNR})$ is not a convex function and a bend appears in the $P(\text{SNR})$ vs SNR curves in log–log scale at low SNR. The theoretical bound is still quite close to the actual ACEP within the bend region. It can be seen from Fig. 12.4a that there is nearly 0.4 dB gap between the performance of the MPA detector and the

ML detector at the SNR of 14 dB. The performance of the MPA detector is improved asymptotically and approaches that of ML detector at high SNRs.

Figure 12.4b–d present the performance for 2, 4, and 6 users over Rayleigh fading channels, respectively. All the bounds are asymptotically tight as SNR increases. The analytical bound P_{jml} is quite tight for values of ACEP below 10^{-3} for all numbers of users, and the gap between P_{jml} and the exact ACEP is almost constant at high SNRs, when the number of users increases. Moreover, the bounds become looser at low SNRs as the number of users increases. The upper bound P_{ub} shows superiority over all the other bounds, since it is much easier to calculate than P_{jml} while it has only a little difference. It should be noted that the scaled P_{ch} is much looser compared to P_{ub}. As expected, the MPA detector shows exactly the same performance as the ML detector for any number of users and any values of SNR over Rayleigh fading channels.

12.2.2 Capacity and Cutoff Rate

This subsection discusses the sum rate analysis of SCMA systems. The channel capacity characterizes the limit information rate that can be reliably transmitted over a channel. It is well known that the sum rate of multi-channel transmissions is simply the sum of per channel rate, and in the uplink SCMA, the communications over each SCMA resource constitutes a multiple access process, then, the sum rate of uplink SCMA is

$$\begin{aligned} C &= \sum_{k=1}^{K} \mathbb{E}_{h_1[1],\ldots,h_J[K]} \left[\log_2 \left(1 + \rho \sum_{j\in\phi_k} |h_j[k]|^2 \right) \right] \\ &= \frac{K e^{1/\rho}}{\rho^{d_f} \ln 2} \sum_{i=1}^{d_f} \sum_{j=0}^{d_f - i} \frac{(-1)^{d_f - j - i} \rho^{i+j}}{j!(d_f - i - j)!} \Gamma(j, 1/\rho) \end{aligned} \tag{12.26}$$

where ρ is the SNR, and $\Gamma(a, x) = \int_x^{\infty} t^{a-1} e^{-t} \mathrm{d}t$, is the incomplete Gamma function. In the above sum rate evaluation, we assume that the users have the same transmitting power, and each SCMA resource carries the same number of users, i.e., $d_f = |\phi_k|$. To achieve any point on the sum rate curve, codebooks with Gaussian distributions and successive interference cancelation (SIC) receivers are generally required.

In practical cases, it is more valuable to investigate the capacity restricted by specific codebooks, i.e., the discrete codebook-constrained capacity (DCCC). Consider the equivalent linear system of uplink SCMA in (12.2). Assuming that perfect channel knowledge is available at the receiver. The conditional probability density function (PDF) of the received signal vector is

$$f(\mathbf{y}|\mathbf{X},\mathbf{H}) = \frac{1}{(\pi N_0)^K} \exp\left(-\frac{\|\mathbf{y}-\mathbf{H}\mathbf{X}\|^2}{N_0}\right) \tag{12.27}$$

The mutual information $I(\mathbf{X};\mathbf{y})$ between the discrete input $\mathbf{X}$ and the continuous output $\mathbf{y}$, or the DCCC, is given by [11]

$$\begin{aligned} I(\mathbf{X};\mathbf{y}) &= \log_2 M^J - \mathbb{E}_{\mathbf{y},\mathbf{X}_a,\mathbf{H}}\left[\log_2 \frac{\sum_{\mathbf{X}_b} f(\mathbf{y}|\mathbf{X}_b,\mathbf{H})}{f(\mathbf{y}|\mathbf{X}_a,\mathbf{H})}\right] \\ &= \log_2 M^J - \mathbb{E}_{\mathbf{H}}\left[\frac{1}{M^J}\int_{\mathbf{y}}\sum_{\mathbf{X}_a} f(\mathbf{y}|\mathbf{X}_a,\mathbf{H}) \log \frac{\sum_{\mathbf{X}_b} f(\mathbf{y}|\mathbf{X}_b,\mathbf{H})}{f(\mathbf{y}|\mathbf{X}_a,\mathbf{H})} \mathrm{d}\mathbf{y}\right] \end{aligned} \tag{12.28}$$

where $\mathbf{X}_a, \mathbf{X}_b \in \{\mathbb{X}_j\}_{j=1}^J$ are two combined codewords for J users. Obviously, it is quite difficult—if not impossible—to deal with the expression for the mutual information, and a closed-form solution is unattainable. In the following, we resort to the cutoff rate analysis.

12.2.2.1 Cutoff Rate of Uplink MAC

The channel cutoff rate R_0, which is a lower bound on the channel capacity, is another commonly used metric characterizing the channel rate. The cutoff rate is more informative than the DCCC, since it provides a good estimate of the capacity as well as a tight upper bound on the error probability of an optimal detector.

The cutoff rate can be defined by [11]

$$R_0 = -\log_2\left[\sum_{\mathbf{X}_a}\sum_{\mathbf{X}_b} p(\mathbf{X}_a)p(\mathbf{X}_b)\boldsymbol{\Delta}_{\mathbf{X}_a,\mathbf{X}_b}\right] \tag{12.29}$$

where $p(\mathbf{X}_a) = p(\mathbf{X}_b) = \frac{1}{M^J}$, and $\boldsymbol{\Delta}_{\mathbf{X}_a,\mathbf{X}_b}$ is the Bhatacharyya bound on the PEP between $\mathbf{X}_a$ and $\mathbf{X}_b$, which is given by [11]

$$\begin{aligned} \boldsymbol{\Delta}_{\mathbf{X}_a,\mathbf{X}_b} &= \mathbb{E}_{\mathbf{H}}\left[\int \sqrt{p\left(\mathbf{y} \mid \mathbf{X}_a,\mathbf{H}\right) p\left(\mathbf{y} \mid \mathbf{X}_b,\mathbf{H}\right)} \,\mathrm{d}\mathbf{y}\right] \\ &= \mathbb{E}_{\mathbf{H}}\left[e^{-\frac{1}{4N_0}\|\mathbf{H}(\mathbf{X}_a-\mathbf{X}_b)\|^2} \int \frac{1}{(\pi N_0)^K} e^{-\frac{1}{N_0}\left\|\mathbf{y}-\frac{\mathbf{H}(\mathbf{X}_a+\mathbf{X}_b)}{2}\right\|^2} \mathrm{d}\mathbf{y}\right] \\ &= \mathbb{E}_{\mathbf{H}}\left[e^{-\frac{1}{4N_0}\|\mathbf{H}(\mathbf{X}_a-\mathbf{X}_b)\|^2}\right] \end{aligned} \tag{12.30}$$

Note that $\boldsymbol{\Delta}_{\mathbf{X}_a,\mathbf{X}_a} = 1$, then the cutoff rate can be written as

$$R_0 = \log_2 M^J - \log_2\left(1 + \frac{1}{M^J}\sum_{\mathbf{X}_a}\sum_{\mathbf{X}_b \neq \mathbf{X}_a} \boldsymbol{\Delta}_{\mathbf{X}_a,\mathbf{X}_b}\right) \quad (12.31)$$

It is observed that, the term $\frac{1}{M^J}\sum_{\mathbf{X}_a}\sum_{\mathbf{X}_b\neq\mathbf{X}_a}\boldsymbol{\Delta}_{\mathbf{X}_a,\mathbf{X}_b}$, inside the bracket of (12.31), is the union-Bhatacharyya bound on the joint codeword error probability for multiple users. Therefore, optimizing the mean cutoff rate is equivalent to the optimization of the error probability, and cutoff rate can be used as a good performance criterion for the system design.

For the uplink MAC, according to the analysis in Theorem 1,

$$\|\mathbf{H}(\mathbf{X}_a - \mathbf{X}_b)\|^2 = \sum_{k=1}^{K} \lambda_k^2 |\tilde{h}[k]|^2$$

where λ_k^2 is the kth dimension-wise distance in the MAC defined in Definition 1, and $|\tilde{h}[1]|, \ldots, |\tilde{h}[K]|$ are independent Rayleigh distributed random variables. Therefore,

$$\boldsymbol{\Delta}_{\mathbf{X}_a,\mathbf{X}_b} = \prod_{k=1}^{K} \mathbb{E}_{\tilde{h}[k]}\left[e^{-\frac{1}{4N_0}\lambda_k^2|\tilde{h}[k]|^2}\right] = \prod_{k=1}^{K}\left(1 + \frac{\lambda_k^2}{4N_0}\right)^{-1}$$

and thus the cutoff rate for uplink MAC is given by

$$R_0 = \log M^J - \log\left[1 + \frac{1}{M^J}\sum_{\mathbf{X}_a}\sum_{\mathbf{X}_b, b\neq a}\prod_{k=1}^{K}\left(1 + \frac{\lambda_k^2}{4N_0}\right)^{-1}\right] \quad (12.32)$$

The average sum rate and cutoff rate for uplink SCMA in Rayleigh fading are depicted in Fig. 12.5, where the 4-ary codebook in Table 12.1 is adopted. The sum rates of SCMA with 2, 4 and 6 users are represented by the uppermost curves, which increase almost linearly with the SNR when SNR becomes very large. Due to the discrete codebooks, the DCCC and the cutoff rate are upper bounded by $\frac{J}{K}\log_2(M)$. However, significant rate improvement can be achieved by overloaded access for moderate to large SNRs. As it is observed, the cutoff rate establish a lower bound to the DCCC, and it asymptotically approaches the DCCC with increasing SNRs.

12.2.2.2 Cutoff Rate of Downlink BC

Consider the downlink BC model in (12.3), for the jth user, the cutoff rate corresponding to the mutual information $I(\mathbf{x}_j; \mathbf{y})$ is given by [19]

$$R_0 = \log_2 M - \log_2\left(1 + \frac{1}{M}\sum_{\mathbf{x}_{j,a}}\sum_{\mathbf{x}_{j,b}\neq\mathbf{x}_{j,a}} \boldsymbol{\Delta}_{\mathbf{x}_{j,a},\mathbf{x}_{j,b}}\right)$$

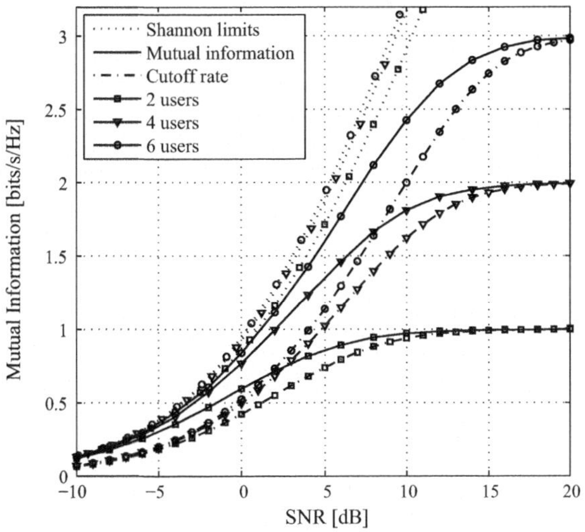

Fig. 12.5 Capacity and cutoff rate of uplink SCMA in Rayleigh fading

and the Bhatacharyya parameter is

$$\begin{aligned}\boldsymbol{\Delta}_{\mathbf{x}_{j,a},\mathbf{x}_{j,b}} &= \mathbb{E}_{\mathbf{h}}\left[\int \sqrt{p\left(\mathbf{y} \mid \mathbf{x}_{j,a},\mathbf{h}\right) p\left(\mathbf{y} \mid \mathbf{x}_{j,b},\mathbf{h}\right)}\mathrm{d}\mathbf{y}\right] \\ &= \mathbb{E}_{\mathbf{h}}\left[\frac{1}{M^{J-1}}\int\sqrt{\sum_{\mathbf{x}_{i,a}\in\mathbb{X}_i,\forall i\neq j} p\left(\mathbf{y}|\mathbf{X}_a,\mathbf{h}\right)\sum_{\mathbf{x}_{i,b}\in\mathbb{X}_i,\forall i\neq j} p\left(\mathbf{y}|\mathbf{X}_b,\mathbf{h}\right)}\mathrm{d}\mathbf{y}\right]\end{aligned} \tag{12.33}$$

In the integral of (12.33) a square root of the double sum of the products "$p\left(\mathbf{y}|\mathbf{X}_a,\mathbf{h}\right)\ p\left(\mathbf{y}|\mathbf{X}_b,\mathbf{h}\right)$" is involved, which makes it excessively complex for a large number of users. Hence, we will attempt to obtain reasonable bounds for the cutoff rate. To deal with the expression, we first calculate $\boldsymbol{\Delta}_{\mathbf{X}_a,\mathbf{X}_b}$. According to the PEP analysis in (12.21) for downlink SCMA, the channel-dependent metric is equal to

$$\|\mathrm{diag}(\mathbf{h})(\mathbf{X}_a-\mathbf{X}_b)\|^2 = \sum_{k=1}^{K}\tau_k^2|h[k]|^2$$

where τ_k^2 is the dimension-wise distance defined in Definition 2, and $|h[k]|$ is the Rayleigh distributed random variables. Similar to that in the uplink case, the Bhatacharyya parameter considering the superimposed codewords $\mathbf{X}_a$ and $\mathbf{X}_b$ is given by

$$\begin{aligned}\boldsymbol{\Delta}_{\mathbf{X}_a,\mathbf{X}_b} &= \mathbb{E}_{\mathbf{h}}\left[e^{-\frac{1}{4N_0}\|\mathrm{diag}(\mathbf{h})(\mathbf{X}_a-\mathbf{X}_b)\|^2}\right] \\ &= \mathbb{E}_{\mathbf{h}}\left[e^{-\frac{1}{4N_0}\sum_{k=1}^{K}\tau_k^2|h[k]|^2}\right] = \prod_{k=1}^{K}\left(1+\frac{\tau_k^2}{4N_0}\right)^{-1}\end{aligned}$$

By applying Holder's inequality,

$$\sqrt{\sum_{\mathbf{x}_{i,a}\in\mathbb{X}_i,\forall i\neq j} p\left(\mathbf{y}|\mathbf{X}_a,\mathbf{h}\right) \sum_{\mathbf{x}_{i,b}\in\mathbb{X}_i,\forall i\neq j} p\left(\mathbf{y}|\mathbf{X}_b,\mathbf{h}\right)} \geq \sum_{(\mathbf{x}_{i,a},\mathbf{x}_{i,b})\in\mathscr{P}_i,\forall i\neq j} \sqrt{p\left(\mathbf{y}|\mathbf{X}_a,\mathbf{h}\right) p\left(\mathbf{y}|\mathbf{X}_b,\mathbf{h}\right)}$$

where $\mathscr{P}_i$ is the set of a point-to-point pairing of codewords $(\mathbf{x}_{i,a}, \mathbf{x}_{i,b})$ for all $i \neq j$, which contains M elements.[3] Then it holds that

$$\begin{aligned}
\boldsymbol{\Delta}_{\mathbf{x}_{j,a},\mathbf{x}_{j,b}} &\geq \mathbb{E}_{\mathbf{h}}\left[\frac{1}{M^{J-1}} \sum_{(\mathbf{x}_{i,a},\mathbf{x}_{i,b})\in\mathscr{P}_i,\forall i\neq j} \int \sqrt{p\left(\mathbf{y}|\mathbf{X}_a,\mathbf{h}\right) p\left(\mathbf{y}|\mathbf{X}_b,\mathbf{h}\right)} d\mathbf{y}\right] \\
&= \frac{1}{M^{J-1}} \sum_{(\mathbf{x}_{i,a},\mathbf{x}_{i,b})\in\mathscr{P}_i,\forall i\neq j} \boldsymbol{\Delta}_{\mathbf{X}_a,\mathbf{X}_b}
\end{aligned}$$

Thus, an upper bound on the cutoff rate of downlink SCMA is

$$R_0|^{\text{upper}} = \log_2 M - \log_2\left[1 + \frac{1}{M^J}\sum_{\mathbf{x}_{j,a}} \sum_{\mathbf{x}_{j,b}\neq\mathbf{x}_{j,a}} \sum_{(\mathbf{x}_{i,a},\mathbf{x}_{i,b})\in\mathscr{P}_i,\forall i\neq j} \prod_{k=1}^{K}\left(1+\frac{\tau_k^2}{4N_0}\right)^{-1}\right] \tag{12.34}$$

For the sake of deriving the lower bound of the cutoff rate, we may invoke the following simple inequality

$$\begin{aligned}
&\sqrt{\sum_{\mathbf{x}_{i,a}\in\mathbb{X}_i,\forall i\neq j} p\left(\mathbf{y}|\mathbf{X}_a,\mathbf{h}\right) \sum_{\mathbf{x}_{i,b}\in\mathbb{X}_i,\forall i\neq j} p\left(\mathbf{y}|\mathbf{X}_b,\mathbf{h}\right)} \\
&\leq \sum_{\mathbf{x}_{i,a}\in\mathbb{X}_i,\forall i\neq j} \sum_{\mathbf{x}_{i,b}\in\mathbb{X}_i,\forall i\neq j} \sqrt{p\left(\mathbf{y}|\mathbf{X}_a,\mathbf{h}\right) p\left(\mathbf{y}|\mathbf{X}_b,\mathbf{h}\right)}
\end{aligned}$$

we get that

$$\boldsymbol{\Delta}_{\mathbf{x}_{j,a},\mathbf{x}_{j,b}} \leq \frac{1}{M^{J-1}} \sum_{\mathbf{x}_{i,a}\in\mathbb{X}_i,\forall i\neq j} \sum_{\mathbf{x}_{i,b}\in\mathbb{X}_i,\forall i\neq j} \boldsymbol{\Delta}_{\mathbf{X}_a,\mathbf{X}_b}$$

and a lower bound on the cutoff rate is obtained

$$\begin{aligned}
R_0|^{\text{lower}} &= \log_2 M - \log_2\left[1 + \frac{1}{M^J}\sum_{\mathbf{x}_{j,a}} \sum_{\mathbf{x}_{j,b}\neq\mathbf{x}_{j,a}} \sum_{\mathbf{x}_{i,a}\in\mathbb{X}_i,\forall i\neq j} \sum_{\mathbf{x}_{i,b}\in\mathbb{X}_i,\forall i\neq j} \prod_{k=1}^{K}\left(1+\frac{\tau_k^2}{4N_0}\right)^{-1}\right] \\
&= \log_2 M - \log_2\left[1 + \frac{1}{M^J}\sum_{\mathbf{X}_a} \sum_{\mathbf{X}_b,\mathbf{x}_{j,b}\neq\mathbf{x}_{j,a}} \prod_{k=1}^{K}\left(1+\frac{\tau_k^2}{4N_0}\right)^{-1}\right]
\end{aligned} \tag{12.35}$$

[3] There are $M!$ possible pairing patterns for $(\mathbf{x}_{i,a}, \mathbf{x}_{i,b})$, hence $M!$ choices for $\mathscr{P}_i$. The tightness of the bound is determined by the specific selection of the pairing patterns. A detailed seek for the appropriate pairing pattern can be found in [19].

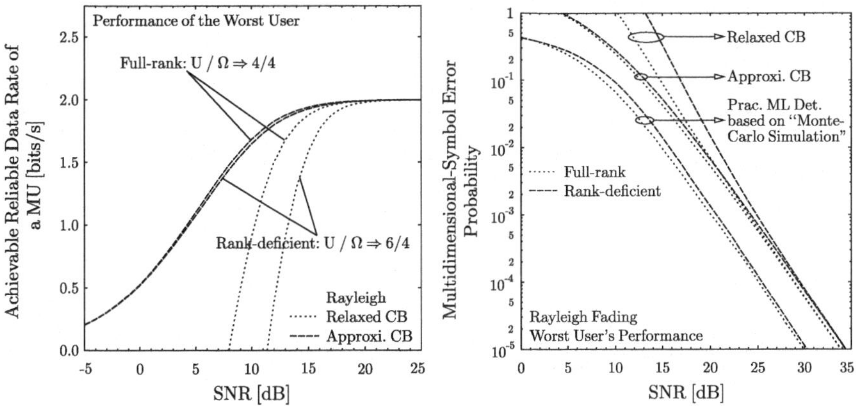

(a) Cutoff rate. (b) ACEP.

Fig. 12.6 Cutoff rate and ACEP of the worst user for downlink SCMA in Rayleigh fading

As discussed in (12.24), the expression inside the square bracket of (12.35) is the union-Bhatacharyya bound on the ACEP for the jth user.

With the derived upper and lower bound on R_0 or $\boldsymbol{\Delta}_{\mathbf{x}_{j,a},\mathbf{x}_{j,b}}$, the corresponding bounds to the union-Bhatacharyya bound on ACEPs, i.e., $\frac{1}{M}\sum_{\mathbf{x}_{j,a}}\sum_{\mathbf{x}_{j,b}\neq\mathbf{x}_{j,a}}\boldsymbol{\Delta}_{\mathbf{x}_{j,a},\mathbf{x}_{j,b}}$, can be obtained straightforwardly.

We will now verify the cutoff rate bounds and the corresponding bounds for ACEPs by simulations. If $R_0|^{\text{upper}}$ is sufficiently tight, it may be regarded as satisfactory approximation of R_0. In order to emphasize the primary common characteristic between $R_0|^{\text{upper}}$ and $R_0|^{\text{lower}}$, we can readily refer to $R_0|^{\text{upper}}$ as the approximated Chernov bound (Approxi. CB), and $R_0|^{\text{lower}}$ as relaxed Chernov bounds (Relaxed CB). The results for the cutoff rate of the worst user in downlink SCMA and the corresponding ACEPs are plotted in Fig. 12.6a and Fig. 12.6b, respectively. The curve of "Approxi. CB" and that of "Relaxed CB" merge with each other in the high-SNR region, while the "Approxi. CB" bound gets significantly close to the associated practical performance. Hence we may claim that $R_0|^{\text{upper}}$ indeed represents a satisfactory approximation of R_0. This conclusion may be verified by the associated simulation results shown in Fig. 12.6b, where the "Approxi. CB" over ACEP gives a better estimation of the practical ACEP than that of "Relaxed CB", for both full-rank (with 4 users) and rank-deficient (with 6 users) SCMA systems.

12.3 Codebook Design

As the performance of SCMA strongly depends on the multi-dimensional codebooks, codebook design constitutes one of the most important issues for SCMA, and it is what distinguishes SCMA from other non-orthogonal multiple access schemes.

Incorporating a sophisticated codebook design into SCMA has the potential of significantly improving the spectrum efficiency, and reducing the detection complexity.

12.3.1 General Design Rules

The design of SCMA codebook is a joint optimization of the sparse mapping matrix and the multi-dimensional constellations. Assume that all layers have the same constellation size and length. An SCMA codebook can be represented by structure $\mathfrak{S}(\mathcal{V}, \mathcal{C}; J, M, N, K)$, where $\mathcal{V} = \{\mathbf{V}_j\}_{j=1}^{J}$, is the set of mapping matrices, and $\mathcal{C} = \{\mathbb{C}_j\}_{j=1}^{J}$, is the set of signal constellations for J layers. Thus, the SCMA codebook designing is equivalent to solve the optimization problem [1]

$$\mathcal{V}^+, \mathcal{C}^+ = \arg\max_{\mathcal{V}, \mathcal{C}} \Upsilon(\mathfrak{S}(\mathcal{V}, \mathcal{C}; J, M, N, K)) \tag{12.36}$$

where the function $\Upsilon(\cdot)$ is somehow the design criterion.

Unfortunately, for a given criterion $\Upsilon(\cdot)$ and such a multi-dimensional problem, the optimum solution cannot be found. In practice, a suboptimal multi-stage optimization approach is adopted, by optimizing the mapping matrices and constellations separately. The set of mapping matrices $\mathcal{V}$ is generally selected in order to meet the maximum overloading, while the design of J multi-dimensional constellations is simplified to the design of a *mother constellation* and multiple layer-specific operators.

12.3.1.1 Mapping Matrices

The set of mapping matrices $\mathcal{V}$ should be pre-determined before the constellation design, since it determines the number of users/layers interfering at each resource node and complexity of the multi-user detection. As $\mathcal{V}$ can be characterized and uniquely determined by the factor graph matrix, the design of $\mathcal{V}$ can borrow the idea from the design of LDPC codes. However, here we introduce general rules for the designing:

- $\mathbf{V}_j \in \mathbb{B}^{K \times N}$, and $\mathbf{V}_i \neq \mathbf{V}_j, \forall i \neq j$
- $\mathbf{V}_j^{[\phi]} = \mathbf{I}_N$, where $\mathbf{V}_j^{[\phi]}$ is obtained by removing all-zero rows in $\mathbf{V}_j$

Thus we may insert $K - N$ all-zero row vectors into rows of $\mathbf{I}_N$ to obtain the unique solution $\mathcal{V}^+$ for problem (12.36).

If we take Example 1 as the illumination, we have following properties and relations for SCMA encoding parameters

- Choose the constellation length $N = 2$, and the codebook length $K = 4$
- The maximum number of layers $J = \binom{K}{N} = 6$

- The number of multiplexed layers over each resource $d_f = \frac{JN}{K} = 3$
- Overloading factor $\lambda = \frac{J}{K} = 1.5$
- $\max(0, 2N - K) \le l \le N - 1$, where l is the number of the overlapping elements of any two distinct $\mathbf{f}_j$ vectors. Thus $0 \le l \le 1$ if $K = 4$ means that the codeword are either orthogonal or collide at only one overlap nonzero element over any two rows.

The resulting factor graph matrix $\mathbf{F}$ is the same as (12.5) and the factor graph is shown in Fig. 12.3.

12.3.1.2 Multi-dimensional Constellations

Having the mapping set $\mathscr{V}^+$, the optimization problem of an SCMA is reduced to

$$\mathscr{C}^+ = \arg\max_{\mathscr{C}} \Upsilon\left(\mathfrak{S}(\mathscr{V}^+, \mathscr{C}; J, M, N, K)\right) \tag{12.37}$$

which is to find J different N-dimensional complex constellations, each contains M signal points. In general, the joint design of multiple multi-dimensional constellations is challenging, a further simplification of (12.37) can be conducted by dividing the problem into the design of a *mother constellation* and J layer-specific operators, and optimizing them separately. Without loss of generality, define $\mathbb{C}_j \equiv \boldsymbol{\Theta}_j(\mathbb{C}), \forall j$, where $\boldsymbol{\Theta}_j(\cdot)$ denotes a *constellation operator*. Thus the optimization problem in (12.37) becomes

$$\mathbb{C}^+, \left\{\boldsymbol{\Theta}_j^+\right\}_{j=1}^J = \arg\max_{\mathbb{C}, \{\boldsymbol{\Theta}_j\}_{j=1}^J} \Upsilon\left(\mathfrak{S}(\mathscr{V}^+, \mathscr{C} \equiv \left\{\boldsymbol{\Theta}_j(\mathbb{C})\right\}_{j=1}^J; J, M, N, K)\right) \tag{12.38}$$

A. Mother Constellation

In general, a constellation with large minimum Euclidean distance achieves good performance when no collisions occur among users/layers over a tone. With increasing number of users/layers, the collisions are unavoidable and the multi-user interference will be introduced. To mitigate such interference, it is required to induce dependency among the nonzero elements of the codewords, such that the receiver can recover colliding codewords from other tones. In general, the mother constellation can be any form of a multi-dimensional constellation with a maximized minimum Euclidean distance. To control dimensional dependency and power variation without destroying the Euclidean distance profile, a unitary rotation can be applied to the mother constellation. For transmission over fading channels, the performance is dominated by the product distance of a constellation at high-SNR region. Thus the goal of designing a good mother constellation for SCMA is trying to optimize both the minimum Euclidean distance and product distance. Fortunately, the optimization of the product distance could be realized by unitary rotation as well. Thus the two types of distances can be optimized separately. In [20], using the Chernoff bounding technique, it is shown that for Rayleigh fading channels, the error probability of a multi-dimensional signal set is essentially dominated by four factors. To improve performance is necessary to

- minimize the average energy per constellation point;
- maximize the modulation or signal-space diversity;
- maximize the minimum product distance

$$d_{p,\min} = \min_{\mathbf{x}_a,\mathbf{x}_b} \prod_{x_a[k] \neq x_b[k]} |x_a[k] - x_b[k]| \tag{12.39}$$

  between any two points $\mathbf{x}_a$ and $\mathbf{x}_b$ in the constellation;
- minimize the product *kissing number* for the minimum product distance, i.e., the total number of points at the minimum product distance.

For low rates, constellation design can be done by brute-force searching, however, this is not necessarily the case for higher rates and a larger number of users/layers due to the prohibitive searching complexity. Under this circumstance, the structured construction is required. Lattice constellation construction can be considered as a possible way to design good mother constellations. If we construct a constellation from the lattice $\mathbb{Z}^{2N}$ with gray labeling, the construction could be done effectively by forming orthogonal QAM constellations on different complex planes. To maximize the minimum product distance of rotated lattice, the unitary rotations of QAM lattice constellations might be optimized as in [20, 21].

B. Constellation Operators

After obtaining the mother constellation $\mathbb{C}^+$, layer-specific operators should be designed to guarantee the unique decodability of the multi-layer signals at the receiver, and also lower the multi-user interference. The optimization problem for the operators can be formulated as

$$\{\boldsymbol{\Theta}_j^+\}_{j=1}^J = \arg \max_{\{\boldsymbol{\Theta}_j\}_{j=1}^J} \Upsilon\left(\mathfrak{S}(\mathscr{V}^+, \mathscr{C} \equiv \{\boldsymbol{\Theta}_j(\mathbb{C}^+)\}_{j=1}^J; J, M, N, K)\right) \tag{12.40}$$

Note that here, the design criterion $\Upsilon(\cdot)$ are not necessarily the same as that in (12.38) for the joint design of mother constellation and constellation operators.

The constellations for different SCMA layers might be constructed with different operators $\boldsymbol{\Theta}_j(\cdot)$, and the constellation operators generally include complex conjugate, phase rotation and dimensional permutation. Generally speaking, if the different users have different power levels, the interfering codewords would be easily separated at receiver due to the power diversity. To do this, it is obliged to have a diverse average power level over the constellation dimensions when designing the mother constellations, which could be done by an appropriate rotation of the lattice constellation as discussed in [16]. Thus the task of optimization problem can be the permutation operators which enable the SCMA codebooks to capture as much power diversity as possible over the interfering users. The optimization for power variation over users can be designed to permute each codebook set to avoid interfering with the same dimensions of a mother constellation over a resource node.

As discussed in Sect. 12.2.1.2, the constellation operators is unnecessary for the uplink SCMA in fading channels. On the one hand, in MAC, the fading itself takes the role of constellation operations, and the receiver exploiting the differences among the channel fadings to separate the multi-user signals. On the other hand, the constellation operators like phase rotation and complex conjugate don't change λ_k^2 in Definition 1, hence don't change the error probability. However, it is important to design layer-specific operator for downlink SCMA, because all users experience the same channel condition and the destructive codeword collision can be avoided by careful design of $\boldsymbol{\Theta}_j(\cdot)$ in the downlink.

12.3.1.3 Constellations for Lower Receiver Complexity

This part introduces two kinds of multi-dimensional constellations for SCMA, that allow MPA receiving with reduced complexity.

A. Shuffled Multi-dimensional Constellation

The dependency among the complex dimensions of the mother constellation guarantees an efficient detection and diversity for fading channels. It is possible to construct a mother constellation such that the real part and imaginary part are independent with each other, while the complex dimensions are still dependent. One kind of approach is the *shuffling* [16], which enables the MPA to reduce the complexity from M^{d_f} to $M^{d_f/2}$. The shuffling method rotates two independent N-dimensional real constellations to maximize the minimum product distances, with the same or different unitary rotations, then generates an N-dimensional complex mother constellation by concatenation of the two N-dimensional rotated real constellations. One of the two N-dimensional real constellations corresponds to the real part of the points of

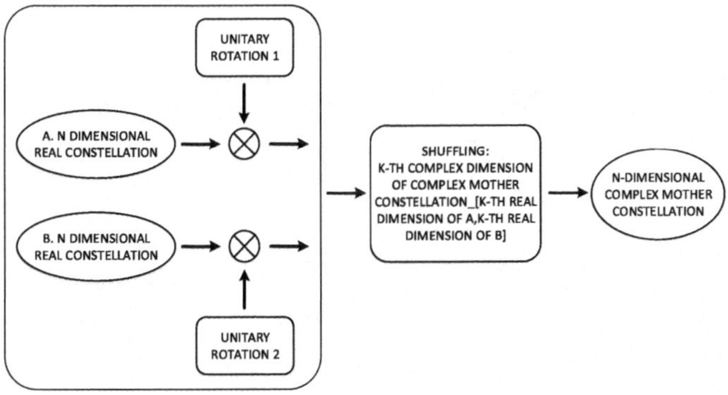

Fig. 12.7 Shuffling construction of the mother constellation [16]

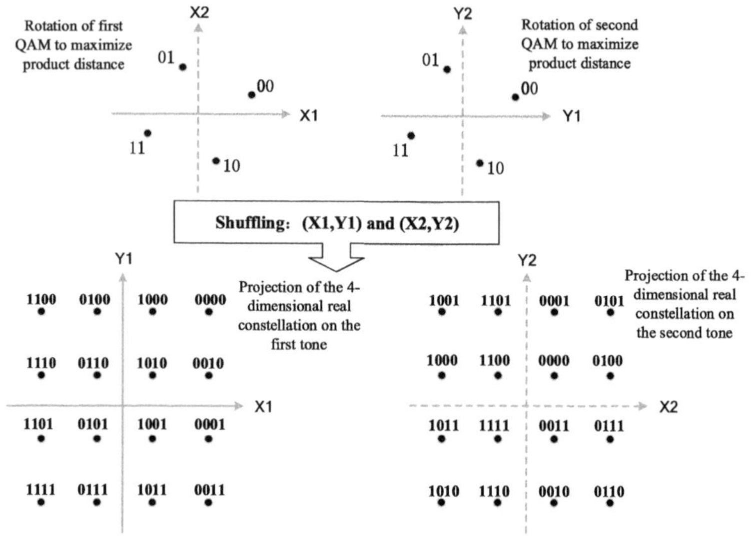

Fig. 12.8 An example of shuffling construction of two-dimensional 16-ary SCMA constellation [16]

the mother constellation, and the other one corresponds to the imaginary part. The construction is illustrated in Fig. 12.7.

Example 2 The construction of a 16-point SCMA mother constellation applicable to codebooks with two nonzero position ($N = 2$) by shuffling is illuminated in Fig. 12.8. Its optimum rotation angle is $\tan^{-1}\left(\frac{1+\sqrt{5}}{2}\right)$, which maximizes the minimum product distance.

B. Low-Projected Multi-dimensional Constellation

A key feature of SCMA codebooks is that the multi-dimensional constellation allows a few constellation points to collide over some of the dimensions, as they can still be separated through other components. An example is shown in Fig. 12.9, in which the constellation points corresponding to 01 and 10 collide over the first dimension, but are separated over the second tone, making the number of projection points equal to 3 instead of 4. By employing this low-projected constellation, the MPA receiver is able to reduce the number of probability calculations at the FNs during each iteration. As a result, the complexity is reduced to $O(M_p^{d_f})$, where $M_p \leq M$ is the number of projection points.

To do this, it is obliged to let the minimum "product distance"[4] be zero during the mother constellations design by rotation. However, the zero minimum product

[4]This is the relaxed product distance that takes the product of all the dimension-wise distance between two points into consideration.

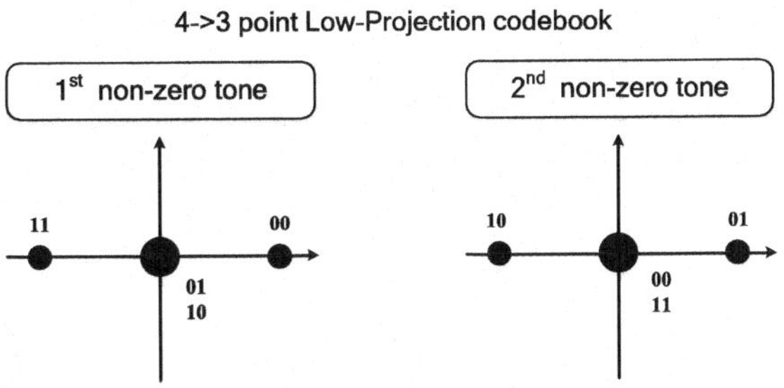

Fig. 12.9 An example of a 4-ary constellation with 3 projections per complex plane

distance would cause the performance degradation at high SNR, thus the trade-off between the performance and complexity should be considered for different scenarios.

12.3.2 Multi-user Codebooks Design for Uplink SCMA Systems

In this subsection, we introduce a practical codebook design approach for uplink SCMA systems over Rayleigh fading channels. Instead of optimizing the mother constellation and constellation operators separately, we address the joint design of multi-user constellations for small constellation size and number of users [22].

12.3.2.1 Design Criterion

To address the design of good codebooks, we need to establish appropriate performance criteria for a given system, i.e., determine the $\Upsilon(\cdot)$ in (12.37). It is straightforward to use the DCCC $I(\mathbf{X}, \mathbf{y})$ in (12.28), or the ACEP in (12.24), as the criterion, for increasing capacity or lowering probability of error. However, it is inefficient to use the DCCC or the ACEP as the cost function directly, since the evaluation of $I(\mathbf{X}, \mathbf{y})$ involves either Monte Carlo simulations or a large amount of numerical integration, and the calculation of the union bound on the ACEP is a little bit complicated.

As an alternative metric, the cutoff rate also gives an approximated evaluation for the capacity as well as the error probability and allows us to optimize the codebooks at a target value of SNR. Therefore, we can formulate the criterion for the multi-user codebooks design, by making the cutoff rate as large as possible, or equivalently the union-Bhatacharyya bound on the ACEP as small as possible. According to the cutoff rate analysis in (12.32) for MAC, maximizing R_0 is equivalent to choose the combined codewords such that

$$\begin{aligned} \{\mathbb{X}_j^+\}_{j=1}^J &= \arg \min_{\{\mathbb{X}_j\}_{j=1}^J} \sum_{\mathbf{X}_a} \sum_{\mathbf{X}_b, b \neq a} \prod_{k=1}^{K} \left(1 + \frac{\lambda_k^2}{4N_0}\right)^{-1} \\ \text{s.t.} \quad & \mathbb{E}\left[\|\mathbf{X}\|^2\right] = E_s \end{aligned} \tag{12.41}$$

where E_s is the total power constraint on the multi-user codebooks. It is expected that the criterion is optimal in the sense of designing codebooks with large DCCC or small ACEP, since it involves all pairs of possible codewords for multiple users.

12.3.2.2 Signal-Space Diversity Scheme

Recall that the problem of designing multi-dimensional constellations for fading channels has been solved by using signal-space diversity (SSD), which rotates the QAM constellations with a unitary matrix, constructed either from the algebraic number theory or by computer search [21, 23]. Therefore, here, we use the rotated constellation to build the multi-user codebooks. In particular, we obtain the rotation matrices through computer search over compact parameterizations of unitary matrices. Note, that the null dimensions of codebooks are discarded before the rotations. The $N \times N$ unitary matrix can be written as [23]

$$\mathbf{R} = \prod_{m=1}^{N-1} \prod_{n=m+1}^{N} \mathbf{T}_{m,n} \tag{12.42}$$

where $\mathbf{T}_{m,n}$ is a complex Givens matrix given by [24]

$$\mathbf{T}_{m,n} = \begin{bmatrix} \ddots & & & & \\ & \cos\theta_{m,n} & \ldots & e^{-i\eta_{m,n}} \sin\theta_{m,n} & \\ & \vdots & \ddots & \vdots & \\ & -e^{i\eta_{m,n}} \sin\theta_{m,n} & \ldots & \cos\theta_{m,n} & \\ & & & & \ddots \end{bmatrix}$$

which changes the identity matrix by replacing its (m, m)th, (n, n)th, (m, n)th and (n, m)th elements with $\cos\theta_{m,n}$, $\cos\theta_{m,n}$, $e^{-i\eta_{m,n}} \sin\theta_{m,n}$ and $-e^{i\eta_{m,n}} \sin\theta_{m,n}$, respectively. The angles satisfy $\theta_{m,n} \in [-\pi, \pi]$ and $\eta_{m,n} \in [-\pi/2, \pi/2]$. So, the search for $\mathbf{R}$ is reduced to the search for a sequence of phase of

$$\begin{aligned} \boldsymbol{\theta} &= \{\theta_{1,2}, \ldots, \theta_{1,N}, \theta_{2,3}, \ldots, \theta_{2,N}, \ldots, \theta_{N-1,N}\}, \\ \boldsymbol{\eta} &= \{\eta_{1,2}, \ldots, \eta_{1,N}, \eta_{2,3}, \ldots, \eta_{2,N}, \ldots, \eta_{N-1,N}\}. \end{aligned} \tag{12.43}$$

It seems that optimizing the constellations using the above criterion is intractable even for a moderate number of users and codebook size, since we need to search

$N(N-1)J$ angles, and the summation in the right-hand side of (12.41) will add up $M^J(M^J-1)$ terms. However, searching results for two-dimensional constellations with a small number of users show that the rotation matrices are the same for all codebooks, and are independent of the number of users. Therefore, we simplify the optimization process by searching over a single rotation matrix, and reducing the number of accessed users, even though this is suboptimal. Furthermore, we can use the approach developed in [25], where all the entries of the rotation matrix are equal in magnitude. Therefore, by expanding the product in (12.42), we get $\boldsymbol{\theta} = \left\{\frac{\pi}{4}\right\}$ and $\boldsymbol{\theta} = \left\{\frac{\pi}{4}, 0.6155, \frac{\pi}{4}\right\}$ for $N=2$ and $N=3$, respectively. Exhaustive search is computationally feasible, provided, that each user occupies a moderate number of resources such that $N \leq 3$.

In the signal-space diversity scheme, the constellations are restricted to lattice constellations such that the rotated QAMs are suggested. In practice, the rotation can be done over any multi-dimensional constellations to improve the cutoff rate, e.g., the rotated spherical codebook [26] and rotations over the product of other low-dimensional constellations.

12.3.2.3 Simulations and Discussions

Consider the SCMA system in Example 1, which supports an overloading factor 150%. Simulation results of packet error rate (PER) for uncoded SCMA and DCCC are provided, which are performed over i.i.d Rayleigh fading channels for 4-ary and 16-ary codebooks. Four kinds of codebooks including the codebooks through SSD scheme discussed in this subsection (named as $G_{4,4}/G_{16,16}$), the codebook from [16] (named as $C_{4,4}/C_{16,16}$), spherical codebook [26], star-QAM-based codebooks [27], are employed, and we also provide the results of the star-QAM-based codebooks after optimization using the criterion (12.41), for which we extend α to complex numbers and get $\beta = 1, \alpha = -i$ and $\alpha = 0.8 - 0.8i$ for 4-ary and 16-ary codebooks, respectively.[5]

Figure 12.10 plots the DCCC of 4-ary and 16-ary codebooks for four users, together with the theoretical limit rates of i.i.d Gaussian inputs. As it is evident, the SSD scheme outperforms all the other codebooks in the high rate region for both 4-ary and 16-ary codebooks, while the mutual information gain is more clear for 4-ary case. While the rate of the star-QAM scheme is small, a significant gain is achieved after optimization with the criterion in (12.41), and it becomes as good as the SSD scheme for 4-ary codebook.

Figure 12.11 compares the PER performance of different codebooks for uplink SCMA with six users, where two antennas are employed for receive diversity, and the MPA detector is used with six iterations all the time. As it is observed, the SSD scheme has a gain about 0.8 dB over $C_{4,4}$ and 0.6 dB over $C_{16,16}$, and a gain about 0.5 dB and 0.3 dB over the spherical codebook for 4-ary and 16-ary cases, respectively.

[5]The star-QAM-based codebook targets on downlink channels, while its performance deteriorates in the uplink and for large constellation size.

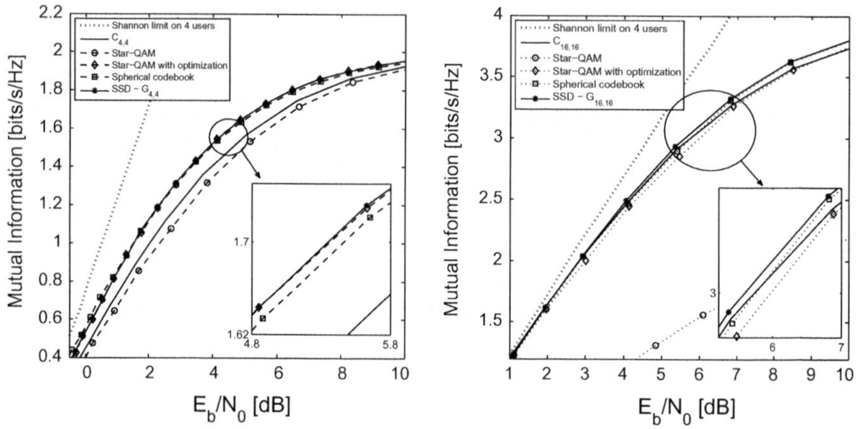

(a) 4-ary SCMA codebooks. (b) 16-ary SCMA codebooks.

Fig. 12.10 Mutual information of uplink SCMA with 4 users

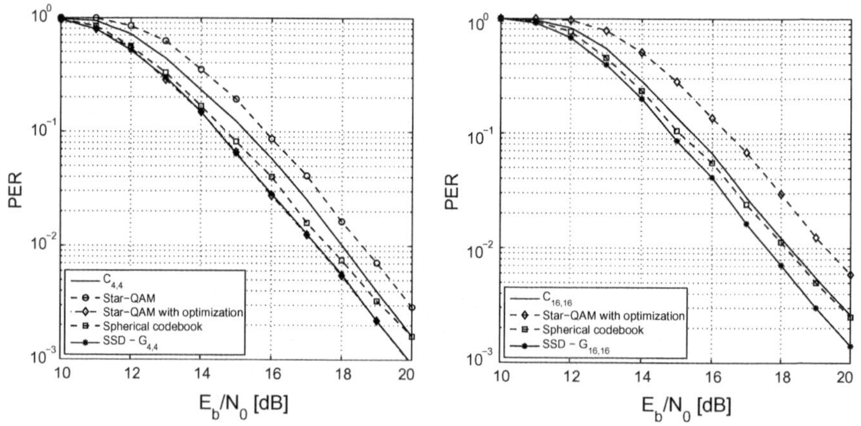

(a) 4-ary SCMA codebooks. (b) 16-ary SCMA codebooks.

Fig. 12.11 PER of uplink SCMA systems over Rayleigh fading channels

Without optimization, the star-QAM scheme yields the worst error performance. However, it performs much better after optimization, which coincides with the result of mutual information in Fig. 12.10.

12.3.3 Low-Projected Multi-dimensional Constellations Design

As is discussed above, by employing the multi-dimensional constellations with low projections, the MPA receiver is able to utilize the constellation structure to reduce the receiver complexity. This subsection introduces an approach of constructing low-projected multi-dimensional constellations for uplink coded SCMA. In particular, constellation optimization for bit-interleaved convolutional coded SCMA with iterative multi-user detection is considered.

12.3.3.1 Transfer Characteristics of Turbo-MPA Detector

Extrinsic information transfer (EXIT) characteristics are investigated to find the effect of multi-user constellations on the performance of the MPA detector, and give us insights on the constellation optimization criteria. For each user, the EXIT chart analysis computes the average mutual information (AMI) between the extrinsic LLR (L_e), or the a priori LLR (L_a), and each coded bit. Thus, the extrinsic AMI is calculated as [28]

$$I_{\text{det,e}} = 1 - \frac{1}{\sqrt{2\pi\sigma_e^2}} \int_{-\infty}^{+\infty} \exp\left[\frac{-\left(l - \sigma_e^2/2\right)^2}{2\sigma_e^2}\right] \log_2(1 + e^{-l})\mathrm{d}l$$

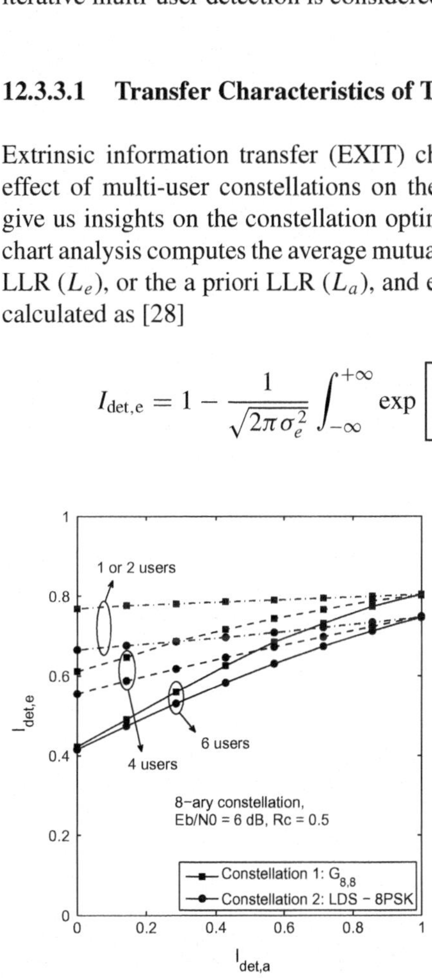

(a) EXIT charts for different constellations.

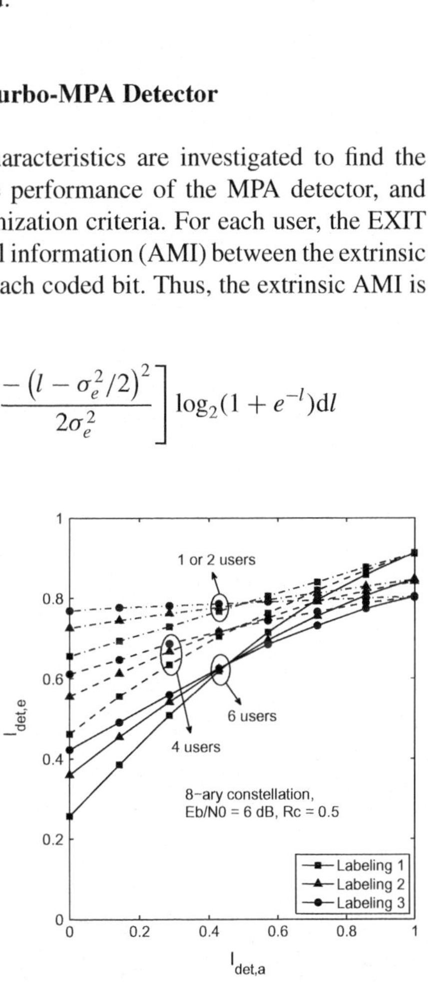

(b) EXIT charts for different labelings.

Fig. 12.12 Impacts of constellations and labelings on the detector's transfer characteristics ($E_b/N_0 = 6$ dB, MPA detector with 3 iterations) where σ_e^2 is the variance of the extrinsic LLR. It is worth noting that due to the multi-user interference, the *a priori* AMI ($I_{\text{det,a}}$) and extrinsic AMI ($I_{\text{det,e}}$) for each user will be influenced by the other users, and hence, a J-dimensional EXIT chart is necessary to characterize the transfer function. Here, the AMI is averaged over all the users such that the EXIT curves can be depicted on a one-dimensional complex plane.

Now, we investigate the transfer characteristics of the turbo-MPA detector for uplink SCMA over i.i.d. Rayleigh fading, and a factor graph matrix as in (12.5) is considered. Figure 12.12 presents the detector's transfer characteristics of two-dimensional 8-ary constellations for different number of users (J) at $E_b/N_0 = 6$ dB, where the MPA detector performs 3 iterations. Note that when $J = 2$, the signals from the two users are orthogonal with each other, so that they have the same AMI as that in the single-user case.

The impact of different constellations on the detector's transfer characteristics is shown in Fig. 12.12a, where the detector's EXIT curves of two different 8-ary constellations[6] with the same labelings are provided. Obviously, constellation 1 outperforms constellation 2, and the superiority of constellation 1 over constellation 2 is independent with the number of users. This implies that the effect of the constellation on the single-user system agrees with its multi-user counterpart, even though the EXIT curves become steeper as the number of users increases. In Fig. 12.12b, the results for the same 8-ary constellation with different mappings/labelings are demonstrated. It is observed that different labelings result in transfer characteristics curves of different slopes, for all the number of users cases, and the labeling with a steeper EXIT curve in the single-user case shows the larger slope in its multi-user counterpart.

The conclusion implies that the influence of constellations and labelings on the single-user system is consistent with that on the multi-user case. More precisely, a constellation or a labeling that is good for single-user systems will be beneficial to the multi-user systems. Therefore, we suggest to simplify the complicated multi-user constellations optimization in SCMA to the suboptimal single-user system design. It is expected that the constellation designing criteria for the single-user system is efficient for multi-user cases.

12.3.3.2 Design Criteria of Multi-dimensional Constellations

A. Links Between EXIT Charts and Constellation Design

The EXIT chart is a good tool to guide the system design. For iteratively decoded systems, given an outer convolutional code, the constellation should be designed to form a tunnel between the transfer curves of the detector and the decoder, and the

[6]The constellation 1 is constructed by rotation over the product of a binary phase-shift keying (BPSK) and a quadrature phase-shift keying (QPSK) constellation with Gray labelings, using the approach in Sect. 12.3.2.2 ($G_{8,8}$), and constellation 2 is the repetition over an 8PSK constellation with Gray labeling, i.e., the LDS scheme [29].

starting point of the detector curve and the intersection point between the detector curve and the decoder curve should be as high as possible, to guarantee a low threshold as well as a low error floor.

At a given value of SNR, the transfer characteristics of the detector are affected by both the constellation itself and the labeling, as shown in Fig. 12.12. In terms of the constellation, it is known that the area under the detector's EXIT curve is approximately equal to the DCCC per number of bits of a constellation point [30]. Based on this property, once a constellation with a larger DCCC is constructed, a larger area is obtained and then it has the potential of providing a wider EXIT tunnel, or equivalently, it would be easier to let the detector's curve to be above the decoder's curve. In terms of the labeling, for a given constellation, the detector curves corresponding to distinct labelings are rotations with each other, since the labeling does not change the DCCC and hence the area below the detector curve. A good labeling rotates the detector curve such that a large AMI is produced when $I_{\text{det,a}} = 1$, which provides an error floor that reaches the BER range of practical interest, and at the same time, to make the tunnel between the transfer curves of the detector and the decoder still open.

Based on these facts, we divide the constellation optimization framework into two steps. First, try to design the multi-dimensional constellation by maximizing the DCCC; Second, optimize the labeling by EXIT curve-fitting. In the following, we introduce two figures of merits for the constellation and the labeling.

B. Constellation Figure of Merit

As discussed above, the cutoff rate, corresponding to the DCCC, is a good criterion that allows us to optimize the constellation at a target value of SNR. Considering the received signal $\mathbf{y} = \text{diag}(\mathbf{h})\mathbf{x} + \mathbf{n}$, the cutoff rate constrained by an M-ary K-dimensional signal set $\mathbb{C}$ in i.i.d. Rayleigh fading, is given by [25]

$$\Psi_{\text{CFM}}(\mathbb{C}) = \log_2 M - \log\left[1 + \frac{1}{M}\sum_{\mathbf{x}_a \in \mathbb{C}}\ \sum_{\mathbf{x}_b \in \mathbb{C}, \mathbf{x}_b \neq \mathbf{x}_a}\ \prod_{k=1}^{K}\left(1 + \frac{\delta_k^2}{4N_0}\right)^{-1}\right] \tag{12.44}$$

where $\delta_k = |x_a[k] - x_b[k]|$, is the dimension-wise distance between any two distinct K-dimensional symbols $\mathbf{x}_a$ and $\mathbf{x}_b$. We take the quantity $\Psi_{\text{CFM}}(\mathbb{C})$ as the SNR-dependent constellation figure of merit, which is a function of SNR and the constellation $\mathbb{C}$, or the set of all pairwise distances between the constellation points. It involves all pairs of multi-dimensional symbols, and is independent of the labeling or any channel codes.

C. Labeling Figure of Merit

The constellation labeling is a crucial design parameter to achieve a high coding gain over the iterations for iteratively decoded bit-interleaved coded modulation (BICM) systems. To obtain an optimization criterion for the labelings, we resort to the error performance of multi-dimensional constellations under ideal interleaving. Let $\tilde{\mathbf{x}}_{(i)} = [\tilde{x}_{(i)}[1], \ldots, \tilde{x}_{(i)}[K]]^t$, be the symbol having the same label with that of $\mathbf{x}$

except at the ith bit position. The effect of labeling μ on the performance of BICM with iterative decoding (BICM-ID) systems employing multi-dimensional signal constellation can be characterized by [31]

$$\Psi_{\text{LFM}}(\mu) = \frac{1}{mM} \sum_{i=1}^{m} \sum_{b=0}^{1} \sum_{\mathbf{x} \in \mathbb{C}_i^b} \prod_{k=1}^{K} \left(1 + \frac{1}{4N_0} \delta_k^2\right)^{-1} \tag{12.45}$$

where $\delta_k = |x[k] - \tilde{x}_{(i)}[k]|$, and $\mathbb{C}_i^b$ is the subset of $\mathbb{C}$ that consisting of symbols whose label has the value b in the ith bit position. The SNR-dependent object function $\Psi_{\text{LFM}}(\mu)$ is able to characterize the influence of both the constellation $\mathbb{C}$ and the labeling μ to the bit error rate (BER) performance of BICM-ID systems. With this criterion, one can optimize the bit labeling when fixing the signal constellation, or optimize the constellation for a given labeling, or optimize them jointly. Since optimizing the labeling μ by decreasing $\Psi_{\text{LFM}}(\mu)$ improves the BER performance, we take $\frac{1}{\Psi_{\text{LFM}}(\mu)}$ as the labeling figure of merit to guide the labeling design for a given multi-dimensional constellation.

12.3.3.3 Design Multi-dimensional Constellations

The multi-dimensional constellation with the same projections over each dimension can be viewed as a multi-modulation scheme [32], where the data bits are modulated into multiple one-dimensional symbols that are chosen from a one-dimensional complex constellation $\mathbb{A}$, called subconstellations in the following. The difference among the modulations for each dimension is that they have different labelings. This implies that the multi-dimensional constellation can be constructed by permutations of the one-dimensional subconstellation $\mathbb{A}$, dimensionally. Therefore, the problem is to design an M-ary subconstellation $\mathbb{A}$ with M_p distinct signal points, and the specific mapping or permutation for each dimension. In the following, we propose a multi-stage optimization, and the K-dimensional constellation is constructed by three steps:

(a) Determine the desired number of projection points M_p such that $M_p \leq M$, choose a one-dimensional M-ary subconstellation $\mathbb{A}$ with M_p projections;
(b) Based on the one-dimensional subconstellation $\mathbb{A}$, construct a K-dimensional constellation $\mathbb{C}$ using permutations;
(c) Design a labeling for the K-dimensional constellation $\mathbb{C}$.

A. Design One-Dimensional Subconstellation $\mathbb{A}$

Different from the traditional constellation design, the M-ary constellation with M_p projections imply that there are $M - M_p$ signal points that overlap with others. We first choose an M_p-ary constellation $\mathbb{A}_p$ without overlappings, then allocate the M_p signal points with M labels to obtain $\mathbb{A}$. The choice of $\mathbb{A}_p$ is various, any one-dimensional complex constellation, e.g., quadrature amplitude modulation (QAM) or phase-shift keying (PSK), is available. Here, we construct $\mathbb{A}_p$ using the amplitude phase-shift keying (APSK) constellation, since it is able to provide good DCCC compared to other conventional modulations [33, 34].

An M_p-APSK constellation is composed of L concentric rings, each with uniformly spaced PSK points. The M_p-APSK constellation can be expressed as [33]

$$\mathbb{A}_p = \left\{r_1 e^{j\theta_1}\mathbb{P}(m_1), \ldots, r_L e^{j\theta_L}\mathbb{P}(m_L)\right\}$$

where $\mathbb{P}(m_l)$ is an m_l-ary PSK constellation with unit average energy, and r_l, θ_l are the radius and phase offset of the lth ring, respectively. Let $\mathbf{m} = [m_1, \ldots, m_L]^t$, be the vector of the number of points over each ring so that $M_p = \sum_{l=1}^{L} m_l$. To guarantee a good distance profile, it is preferred to locate fewer constellation points on the inner rings than that on the outer rings. Then, for a set of ordered radius $r_1 < \cdots < r_L$, it is suggested that $m_1 \leq \cdots \leq m_L$.

Following the general APSK design procedure proposed in [34], the M_p-APSK constellation can be constructed as

- Select the number of rings L and the number of constellation points on each ring m_l, such that $M_p = \sum_{l=1}^{L} m_l$;
- Determine the radius of each ring r_l,

$$r_l = \sqrt{-\ln\left[1 - \frac{1}{M_p}\left(\sum_{i=1}^{l-1} m_i + \frac{m_l}{2}\right)\right]};$$

- Set θ_l as 0 or π/m_l.

Given the designed M_p-APSK constellation $\mathbb{A}_p$, we allocate the M-ary constellation with the M_p signal points. The problem can be formulated as how to put M numbers, $0, 1, \ldots, M-1$, into M_p sets, where each set represents a signal point in $\mathbb{A}_p$. The allocation strategy is preferred to follow the rules:

- The numbers that are allocated to a set should be less than or equal to M_p, and greater than or equal to 1, such that the overlapped points can be separated through other dimensions;
- The numbers in each set should be as less as possible, such that the resulted multidimensional constellation has a good distance profile;
- Symmetry of the constellation $\mathbb{A}$ is preferred so that it has a zero mean;
- The sets with low power levels may be allocated with more numbers, such that the constellation has a small average energy.

Note that in some cases, the allocation yields a constellation $\mathbb{A}$ with nonzero mean, then we shift $\mathbb{A}$ toward the origin, such that the mean of all signals is zero and therefore more energy-efficient.

Now, we give an example to illustrate the allocations. For a given 9-APSK $\mathbb{A}_p$ that is constructed with 3 rings and $\mathbf{m} = [1, 3, 5]^t$, a 16-ary subconstellation $\mathbb{A}$ can be obtained by allocating 16 numbers into 9 sets. Some possible allocations are given in Fig. 12.13. Among the four strategies A, B, C, and D, while the strategy A is the most energy-efficient, it shows the worst performance when used to construct multi-dimensional constellations. This is because too many points overlap with each other, leading to a very poor distance profile for the multi-dimensional constellation. Numerical results show that the strategies B and C are equally efficient, and the strategy D is the best one, since the largest number of overlappings is only two.

B. Construct K-dimensional Constellation $\mathbb{C}$

Given the designed M-ary subconstellation $\mathbb{A}$, denote $\pi_k(\mathbb{A})$ a column vector of the kth permutation of the signals in $\mathbb{A}$, and let $\pi_1(\mathbb{A}) = \mathbb{A}$. The K-dimensional constellation through the permutation construction can be expressed with a $K \times M$ matrix as

$$\mathbb{C} = [\pi_1(\mathbb{A}), \ldots, \pi_K(\mathbb{A})]^t$$

where each column of the matrix corresponds to a K-dimensional symbol. Then, constructing a K-dimensional constellation requires to find $K-1$ permutations $\pi_2, \ldots, \pi_K$, such that the constellation figure of merit $\Psi_{\text{CFM}}(\mathbb{C})$ in (12.44) is maximized.

We focus on two-dimensional constellations ($K = 2$), by maximizing the constellation figure of merit, the unique permutation function π is selected as

$$\pi = \arg\min_{\hat{\pi}} \sum_{\mathbf{x}_a \in \mathbb{C}} \sum_{\mathbf{x}_b \in \mathbb{C}, \mathbf{x}_b \neq \mathbf{x}_a} \prod_{k=1}^{K} \left(1 + \frac{|x_a[k] - x_b[k]|^2}{4N_0}\right)^{-1}$$

There are $M!$ different choices for the permutations, for small constellation size where $M \leq 8$, the optimum solution can be solved by exhaustive search with a reasonable complexity. However, it becomes intractable for high order constellations. Note that this problem is similar to the labeling map of a constellation in bit-interleaved coded modulation with iterative decoding (BICM-ID) systems, which can be efficiently solved by using the binary switching algorithm (BSA) [35], or iteratively searching inside a randomly selected list, and a local optimum permutation can be found for a given cost function.

As for a larger dimension where $K > 2$, the search for $K-1$ permutations is challenging. A suboptimal solution can be used by successively optimizing the multi-dimensional constellation from lower dimensions to higher dimensions, such that only one permutation is needed to be checked in every round.

C. Labeling the K-dimensional Constellation

When a multi-dimensional constellation is found, we should choose an appropriate labeling for the constellation. In terms of EXIT chart, optimizing the labeling is to adjust the slope of the detector's curve. Our approach is to obtain a set of labelings with various slopes in their EXIT curves, firstly. Then, to choose a labeling from the set such that the detector EXIT curve matches with the decoder curve of a given convolutional code.

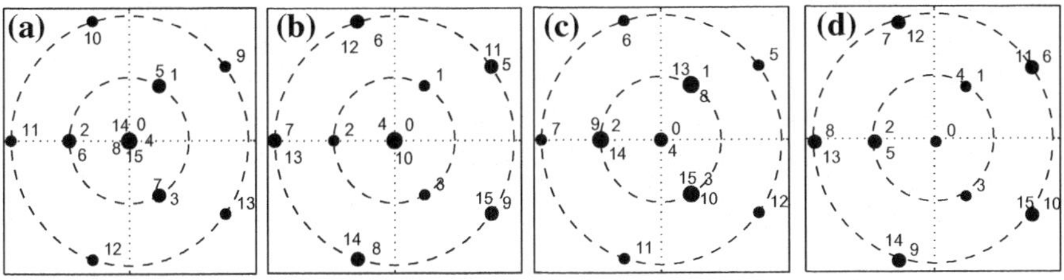

Fig. 12.13 Examples of 16-ary subconstellation $\mathbb{A}$ based on 9-APSK with $\mathbf{m} = [1, 3, 5]^t$

The labeling figure of merit $\frac{1}{\Psi_{\text{LFM}}(\mu)}$ in (12.45) represents the ultimate performance with perfectly known *a priori* AMI, and in EXIT charts, it corresponds to the maximum achievable value of $I_{\text{det,e}}$ with $I_{\text{det,a}} = 1$, denoted as I^*, and I^* becomes larger with decreasing $\Psi_{\text{LFM}}(\mu)$. It is observed in Fig. 12.12a, b that for single-user systems, the detector's EXIT curves corresponding to distinct labelings can be approximated to straight lines, with a common intersection around the point with $I_{\text{det,a}} = 0.5$. Then, the slope of the EXIT curve corresponding to a labeling can thus be determined by I^*, approximately. A labeling with a larger I^* can have a steeper transfer curve. Therefore, we can use $\Psi_{\text{LFM}}(\mu)$ to approximately control the slope of the EXIT curves, and the detector's curve becomes steeper with decreasing $\Psi_{\text{LFM}}(\mu)$.

Denote the set of labelings as Ω. For constellations with small sizes ($M \leq 8$), Ω is chosen to be the set of all possible labelings with distinct $\Psi_{\text{LFM}}(\mu)$. For higher order constellations, the BSA can be used once again to obtain Ω. Begin with a given original labeling, by minimizing $\Psi_{\text{LFM}}(\mu)$ using the BSA, new labelings with increasing slopes may be obtained during the search, we output these labelings and store them into Ω. Similarly, new labelings with decreasing slopes may be obtained by maximizing $\Psi_{\text{LFM}}(\mu)$ with the same original labeling. Then, the labelings in Ω are sorted with increasing values of $\Psi_{\text{LFM}}(\mu)$. The set Ω can also be obtained by iteratively searching inside a randomly selected list.

Now, we choose a labeling from Ω, with the aid of EXIT chart analysis. At an appropriate SNR, the following two conditions have to be fulfilled for the labeling:

(a) the slope of either the single-user or multi-user detector EXIT curve should be as steep as possible, to achieve a low BER error floor;
(b) the tunnel between the decoder and the multi-user detector curves should be open, or the intersection point between them should be as high as possible, to guarantee a low threshold.

As an illustrative example, Fig. 12.14 shows the choices of labelings for a two-dimensional 8-ary constellation (with 3 projections) and a 16-ary constellation (with 9 projections) for SCMA. The detector curves of several labelings, with distinct $\Psi_{\text{LFM}}(\mu)$, as well as the decoder curve are provided, where a half-rate four-state non-recursive convolutional code with generator $[5, 7]_8$ is employed as the outer channel code, and $I_{\text{dec,a}}(I_{\text{dec,e}})$ denotes the AMI between the *a priori* (extrinsic) LLR and the transmitted coded bit at the input (output) of the convolutional decoder.

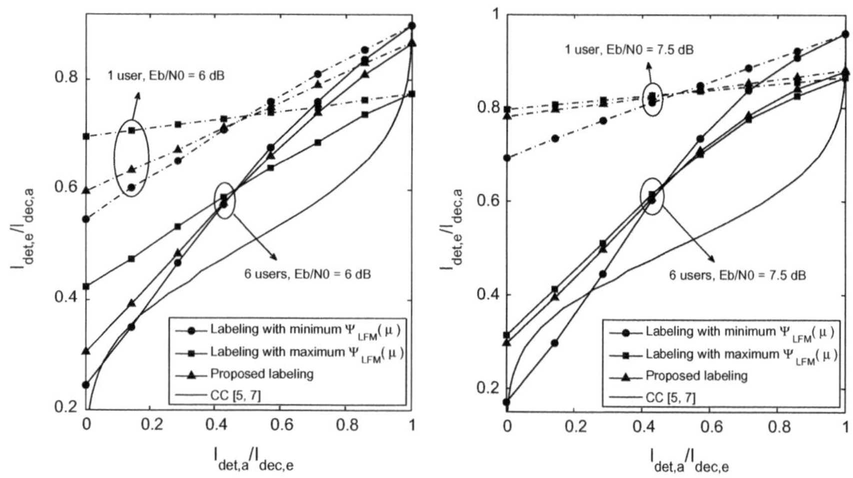

(a) Labeling for an 8-ary constellation with 3 projections.

(b) Labeling for a 16-ary constellation with 9 projections.

Fig. 12.14 Examples of labelings for two-dimensional constellations

As it is observed in Fig. 12.14, the labeling with the maximum $\Psi_{\text{LFM}}(\mu)$ yields a relatively flat slope in the 6 users case, and the one with the minimum $\Psi_{\text{LFM}}(\mu)$ closes the tunnel between the decoder and the multi-user detector curves. In contrast, the proposed labeling, which shows a very steep slope in the EXIT curve while still keeps the tunnel open, achieves a good trade-off between the threshold and the BER performance.

With the proposed approach, it is possible to construct constellations with any projections. Figure 12.15a and 12.15b show the examples of the designed two-dimensional 8-ary constellations with various projections, and a four-dimensional 64-ary constellation with 16 projections, respectively.

12.3.3.4 Simulations and Discussions

In the following, simulations are conducted to evaluate the performance of low-projected constellation for an uplink convolutional coded SCMA system. The detailed simulation configuration is given in Table 12.2. The SCMA system follows a factor graph matrix as (12.5), which supports an effective system loading of 75% (JR_c/K). For the sake of simplicity, let A_{M,M_p} denote the APSK-based M-ary constellation with M_p projections. The constellations in [16, 36, 37] are named as C_{M,M_p} (the SSD scheme in Sect. 12.3.2 is denoted as G_{M,M_p}), which are used as the benchmark.

The simulated BER performances of 4-ary, 8-ary and 16-ary codebooks are depicted in Figs. 12.16 and 12.17, respectively. It is obvious that the A_{M,M_p} code-

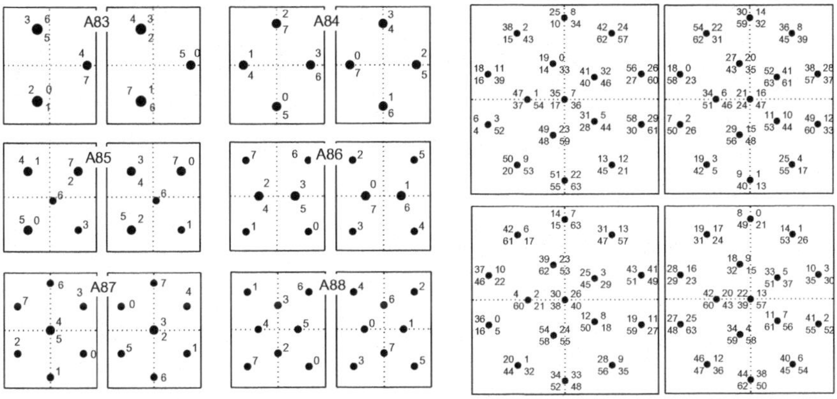

(a) Two-dimensiona l8-ary constellations with various projections.

(b) A four-dimensional 64-ary constellation with 16 projections.

Fig. 12.15 Examples of APSK-based low-projected SCMA constellations

Table 12.2 Simulation parameters

Parameters	Values
Channel model	Uplink Rayleigh fading channel
Target spectral efficiency	1.5, 2.25, 3 bits/resource
Number of users	6
FEC coding	1/2-rate convolutional code with generator $[5, 7]_8$
Interleaving	Random interleaver, interleave length: 1024 bits
Codebooks	C_{43}/A_{43}, $C_{44}/A_{44}/G_{44}$, C_{83}/A_{83}, C_{84}/A_{84}, C_{85}/A_{85}, A_{88}/G_{88} C_{169}/A_{169}, $C_{1616}/G_{1616}/A_{1616}$
Receiver	Turbo-MPA (3 MPA iterations + 6 BICM iterations)

books outperform others in the large SNR region, for almost all the simulation cases. In Fig. 12.16a, the BER floor of $A_{4,3}$ is lower than $C_{4,3}$ when SNR is less than 8 dB, and $A_{4,4}$ shows much better performance than $C_{4,4}$, and has a gain about 0.25 dB over $G_{4,4}$. Note that $A_{4,4}$ outperforms $G_{4,4}$ in the whole SNR region. This is because $A_{4,4}$ has the same labeling with $G_{4,4}$ but the larger DCCC. For the 8-ary codebooks shown in Fig. 12.16b, the error floors of the A_{M,M_p} codebooks happen at the BER level below 10^{-5}, which are much lower than the other codebooks. Thus, much smaller values of SNR are required to achieve a BER value of 10^{-6}. Similar results are also obtained for 16-ary codebooks, which are shown in Fig. 12.17. The gain of the A_{M,M_p} codebooks over C_{M,M_p} is smaller than that in the 8-ary codebooks case. The BER curve of $C_{16,16}$ degrades earlier than the others, but it arrives the error

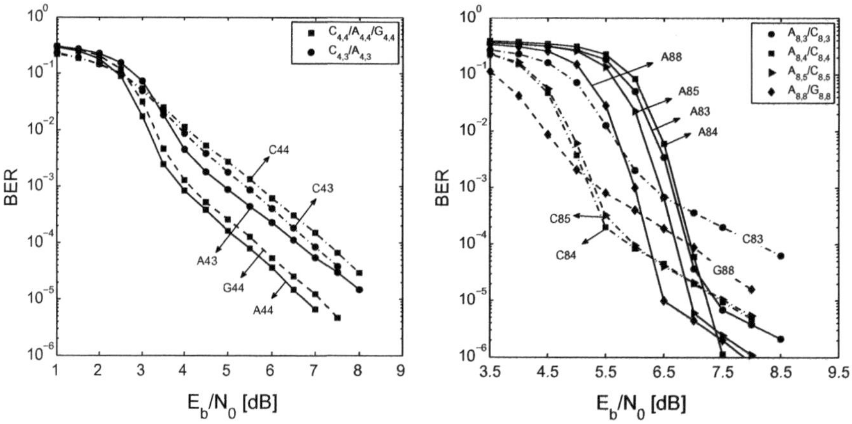

(a) 4-ary codebooks. (b) 8-ary codebooks.

Fig. 12.16 BER performance of uplink coded SCMA for 4-ary and 8-ary codebooks

Fig. 12.17 BER performance of uplink coded SCMA for 16-ary codebooks

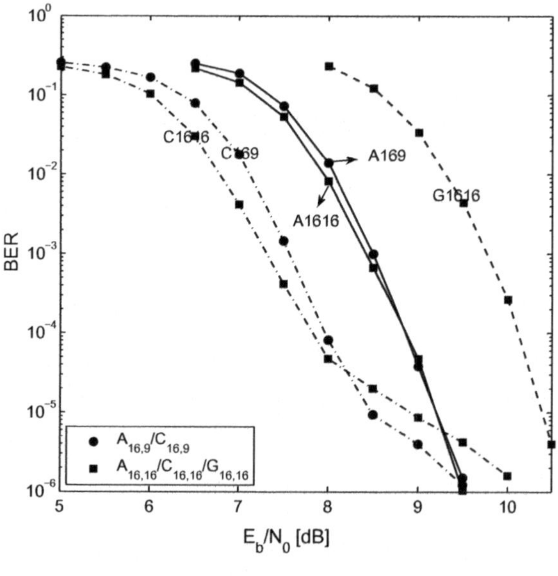

floor quickly. The codebook $G_{16,16}$ shows a very large threshold, and $A_{16,16}$ achieves a good trade-off. Even though the BER threshold of $A_{16,9}$ is larger than $C_{16,9}$, $A_{16,9}$ shows almost the same BER performance with $C_{16,9}$ at the SNR around 9.5 dB. To achieve the BER performance of 10^{-6}, equal or less SNR are required for the A_{M,M_p} codebooks.

12.4 SCMA for 5G Radio Transmission

12.4.1 Application Scenarios for 5G Networks

In addition to achieving higher transmission rates, faster access, supporting of larger user density and better user experience in enhanced mobile broadband (eMBB), the 5G air interface connects to new vertical industries and new devices, creating new application scenarios such as massive machine type communications (mMTC) and ultra reliable low latency communications (URLLC) services, by supporting massive number of devices and enabling mission-critical transmissions with ultra high reliability and ultra-low latency requirement, respectively. This presents new challenges and considerations for the radio multiple access to be fully scalable to support these diverse service requirements. The current orthogonal multiple access might not be able to fulfill some of the requirements, such as services for dense MTC devices deployments, and SCMA can be considered as a promising candidate to meet the 5G performance requirements. In particular, SCMA is proposed for 5G to achieve the following benefits:

- for eMBB: larger capacity region by non-orthogonal multiplexing; robustness to fading and interference with code-domain design; robust link adaptation with relaxed CSI accuracy.
- for URLLC: higher reliability through diversity gain achieved by multi-dimensional constellations, and robustness to collision by carefully design the codebooks; latency reduction and more transmission opportunities by enabling grant-free access; Non-Orthogonal Multiplexing of mixed traffic types.
- for mMTC: higher connection density with high overloading; reduction of signaling overhead and power consumption by enabling grant-free access.

Moreover, it is also possible to extend SCMA application to unlicensed spectrum and V2X systems, since the non-orthogonal transmissions can help to increase the system efficiency and deal with the interference.

The link-level performance evaluation for some uplink SCMA scenarios is provided in [38], which compares SCMA with orthogonal frequency division multiple access (OFDMA) in typical scenarios and investigate the robustness of SCMA to overloading and codebook collision. Results show that SCMA achieves significant gain over orthogonal multiple access with good codebook design, and the gain increases as the supported number of users and target spectrum efficiency increases. Moreover, high overloading with stable performance is feasible with SCMA design, which enables robust overloaded transmission, and the performance loss with codebook collision is negligible with SCMA design, which enables robust grant-free transmission.

12.4.2 Challenges and Future Works

While SCMA is able to greatly enhance the system capability for 5G networks, some further issues on design and implementation of SCMA remain to be resolved, which can be listed as follows:

- Reduced complexity receiver design: Even though MPA or EPA receiver is able to significantly reduce the complexity of SCMA, the complexity of MPA is still very high and iterative multi-user receiver is usually required, which brings several challenging issues for practical implementation:
  - It limits the capability of SCMA to support massive connectivity;
  - The iterative multi-user detection brings a large processing delay;
  - The complexity makes it difficult for SCMA to employ constellations with large sizes, hence limits the transmission rate.

  Sophisticated multi-user detection schemes should be developed to address the high complexity.
- Theoretical analysis: Further theoretical analysis of SCMA is needed to get more insights on the practical system design. For example, the capacity or error performance with randomly codebook allocations. Also, interference cancelation may be incorporated into the MPA detection for lower complexity, then it is desirable to determine the performance and capacity under practical detectors.
- Codebook design: The codebook design is complicated, especially for high-dimensional codebooks and that with large size. Advanced multi-dimensional constellation construction is necessary, and the joint design of factor graph matrix and constellations is to be developed, for further performance improvement. Moreover, the design for the scenario that all the overloaded users have different transmission rates (codebook sizes) is to be investigated, to improve the link adaptation.
- Other issues: System scalability of supporting various loading, SCMA in both uplink and downlink transmissions, supporting of other techniques such as MIMO, resource/codebook allocation, channel estimation for uplink SCMA, etc.

References

1. H. Nikopour, H. Baligh, Sparse code multiple access, in *IEEE 24th International Symposium on Personal Indoor and Mobile Radio Communications (PIMRC'13)* (2013), pp. 332–336
2. S. Zhang, X. Xu, L. Lu, Y. Wu, G. He, Y. Chen, Sparse code multiple access: an energy efficient uplink approach for 5G wireless systems, in *IEEE Global Communications Conference (GLOBECOM'14)* (2014), pp. 4782–4787
3. Y. Wu, S. Zhang, Y. Chen, Iterative multiuser receiver in sparse code multiple access systems, in *Proceedings of IEEE International Conference on Communications (ICC'15)* (2015), pp. 2918–2923
4. J. Zhang, L. Lu, Y. Sun et. al., PoC of SCMA-Based Uplink Grant-Free Transmission in UCNC for 5G. IEEE J. Sel. Areas Commun. **35**, 1353–1362 (2017)

5. R. Hoshyar, F.P. Wathan, R. Tafazolli, Novel low-density signature for synchronous CDMA systems over AWGN channel. IEEE Trans. Signal Process. **56**, 1616–1626 (2008)
6. D. Guo, C. Wang, Multiuser detection of sparsely spread CDMA. IEEE J. Sel. Areas Commun. **26**, 421–431 (2008)
7. R1-166098: Discussion on feasibility of advanced MU-detector. Huawei, HiSilicon, 3GPP TSG RAN WG1 Meeting #86 (2016)
8. X. Meng, Y. Wu, Y. Chen, M. Cheng M, Low complexity receiver for uplink SCMA system via expectation propagation, in *Proceedings of Wireless Communications and Networking Conference (WCNC' 17)*, (2017), pp. 1–5
9. J. Bao, Z. Ma, G.K. Karagiannidis, M. Xiao, Z. Zhu, Joint multiuser detection of multidimensional constellations over fading channels. IEEE Trans. Commun. **65**, 161–172 (2017)
10. D. Tse, P. Viswanath, *Fundamentals of Wireless Communications* (Cambridge University Press, 2005)
11. J.G. Proakis, M. Salehi, *Digital Communications* (McGraw-Hill, New York, 2008)
12. E. Björnson, D. Hammarwall, B. Ottersten, Exploiting quantized channel norm feedback through conditional statistics in arbitrarily correlated MIMO systems. IEEE Trans. Signal Process. **57**, 4027–4041 (2009)
13. M. Chiani, D. Dardari, M.K. Simon, New exponential bounds and approximations for the computation of error probability in fading channels. IEEE Trans. Wirel. Commun. **2**, 840–845 (2003)
14. S.J. Grant, J.K. Cavers, Performance enhancement through joint detection of cochannel signals using diversity arrays. IEEE Trans. Commun. **46**, 1038–1049 (1998)
15. X. Zhu, R.D. Murch, Performance analysis of maximum likelihood detection in a MIMO antenna system. IEEE Trans. Commun. **50**, 187–191 (2002)
16. M. Taherzadeh, H. Nikopour, A. Bayesteh, H. Baligh, SCMA codebook design, in *Proceedings of IEEE 80th Conference on Vehicular Technology (VTC Fall'14)* (2014), pp. 1–5
17. A. Montanari, D. Tse, Analysis of belief propagation for nonlinear problems: the example of CDMA (or: How to prove Tanaka's formula), in *Proceeding IEEE Information Theory Workshop (ITW)* (2006), pp. 122–126
18. C.C. Wang, D. Guo, Belief propagation is asymptotically equivalent to MAP estimation for sparse linear systems, in *Proceedings of 44th Annual Allerton Conference on Communication, Control, and Computing* (2006), pp. 926–935
19. L. Li, Z. Ma, L. Wang, P. Fan, L. Hanzo, Cutoff rate of sparse code multiple access in downlink broadcast channels. IEEE Trans. Commun. **65**, 3328–3342 (2017)
20. J. Boutros, E. Viterbo, C. Rastello, J.C. Belfiore, Good lattice constellations for both Rayleigh fading and Gaussian channels. IEEE Trans. Inf. Theory. **42**, 502–518 (1996)
21. J. Boutros, E. Viterbo, Signal space diversity: a power- and bandwidth-efficient diversity technique for the rayleigh fading channel. IEEE Trans. Inf. Theroy **44**, 1453–1467 (1998)
22. J. Bao, Z. Ma, Z. Ding, G.K. Karagiannidis, Z. Zhu, On the design of multiuser codebooks for uplink SCMA systems. IEEE Commun. Lett. **20**, 1920–1923 (2016)
23. Y. Xin, Z. Wang, G.B. Giannakis, Space-time diversity systems based on linear constellation precoding. IEEE Trans. Wireless Commun. **2**, 294–309 (2003)
24. G.H. Golub, C.F. Van Loan, *Matrix Computations* (Johns Hopkins University Press, 1996)
25. S.P. Herath, N.H. Tran, T. Le-Ngoc, Rotated multi-D constellations in rayleigh fading: mutual information improvement and pragmatic approach for near-capacity performance in high-rate regions. IEEE Trans. Commun. **60**, 3694–3704 (2012)
26. J. Bao, Z. Ma, M.A. Mahamadu, Z. Zhu, D. Chen, Spherical codes for SCMA codebook, in *Proceedings of IEEE 83th Conference on Vehicular Technology (VTC Spring'16)* (2016), pp. 1–5
27. L. Yu, X. Lei, P. Fan, D. Chen, An optimized design of SCMA codebook based on star-QAM signaling constellations, in *Proceedings of International Conference on Wireless Communications & Signal Processing (WCSP'15)* (2015), pp. 1–5
28. S.T. Brink, Convergence behavior of iteratively decoded parallel concatenated codes. IEEE Trans. Commun. **49**, 1727–1737 (2001)

29. J.V.D. Beek, B.M. Popović, Multiple access with low-density signatures, in *Proceedings of IEEE Conference on Global Communications (GLOBECOM)* (2009)
30. A. Ashikhmin, G. Kramer, S.T. Brink, Extrinsic information transfer functions: model and erasure channel properties. IEEE Trans. Inf. Theory. **50**, 2657–2673 (2004)
31. N.H. Tran, H.H. Nguyen, Design and performance of BICM-ID systems with hypercube constellations. IEEE Trans. Wirel. Commun. **5**, 1169–1179 (2006)
32. A. Seyedi, Multi-QAM modulation: a low-complexity full rate diversity scheme, in *Proceedings of IEEE International Conference on Communications (ICC)* (2006), pp. 1470–1475
33. C.M. Thomas, M.Y. Weidner, S.H. Durrani, Digital amplitude-phase keying with M-ary alphabets. IEEE Trans. Commun. **22**, 168–180 (1974)
34. Q. Xie, Z. Yang, J. Song, L. Hanzo, EXIT-chart-matching-aided near-capacity coded modulation design and a BICM-ID design example for both gaussian and rayleigh channels. IEEE Trans. Veh. Tech. **62**, 1216–1227 (2013)
35. F. Schreckenbach, N. Görtz, J. Hagenauer, G. Bauch, Optimization of symbol mappings for bit-interleaved coded modulation with iterative decoding. IEEE Commun. Lett. **7**, 593–595 (2003)
36. M.T. Boroujeni, A. Bayesteh, H. Nikopour, M. Baligh, System and method for generating codebooks with small projections per complex dimension and utilization thereof, U.S. Patent 0,049,999, 18 Feb 2016
37. A. Bayesteh, H. Nikopour, M. Taherzadeh, H. Baligh, J. Ma, Low complexity techniques for SCMA detection, in *Proceedings of IEEE Globecom Workshops* (2015), pp. 1–6
38. R1-164037: LLS results for uplink multiple access. Huawei, HiSilicon, 3GPP TSG RAN WG1 Meeting #85 (2016)

Chapter 13
Interleave Division Multiple Access (IDMA)

Yang Hu and Li Ping

13.1 Overview

The capacity of a multiple access channel was studied in [1, 2]. It can generally be achieved by random coding together with other techniques, e.g., power control, linear precoding, and dirty paper coding [3–6]. Random coding does not involve orthogonality among users so it is inherently non-orthogonal. The sub-optimality of orthogonal multiple access (OMA) was investigated in [7, 8]. The gain of non-orthogonal multiple access (NOMA) over OMA was assessed in [9, 10] for both single-input single-output (SISO) and multiple-input multiple-output (MIMO) systems. Recently, NOMA has been promoted for improving system fairness [11, 12].

However, many practical systems still belong to OMA category. This is mainly due to complexity concerns. OMA can work with low-cost single-user detection (SUD), while NOMA may require more complex multi-user detection (MUD). Thus, these two options entail different trade-off between cost and performance.

Historically, the third-generation (3G) direct-sequence code-division multiple access (DS-CDMA) system is non-orthogonal. Normally, only SUD is used in DS-CDMA to avoid high complexity, which is sub-optimal. The spreading operation in DS-CDMA reduces rate, so DS-CDMA is not convenient for high-rate applications. The fourth-generation (4G) orthogonal frequency division multiple access (OFDMA) system returns to OMA. OFDMA allows flexibility for resource allocation over time and frequency, which can bring about noticeable gain.

Y. Hu (✉) · L. Ping
Department of Electronic Engineering, City University of Hong Kong,
Hong Kong, SAR, China
e-mail: yhu228-c@my.cityu.edu.hk

L. Ping
e-mail: eeliping@cityu.edu.hk

M. Vaezi et al. (eds.), *Multiple Access Techniques for 5G Wireless Networks and Beyond*, https://doi.org/10.1007/978-3-319-92090-0_13

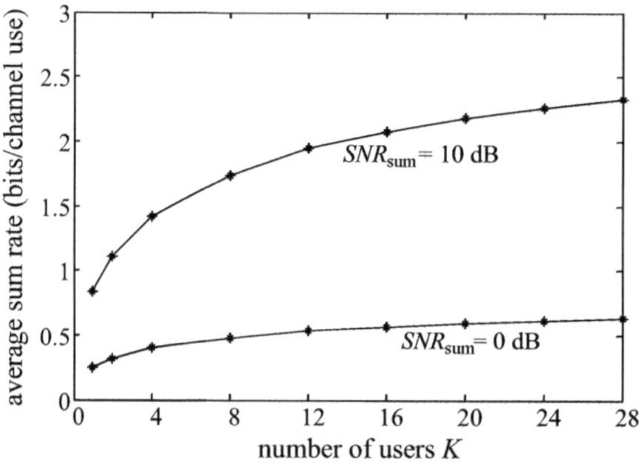

Fig. 13.1 Achievable sum-rate of NOMA and OMA under equal energy constraint. Note that the two strategies have exactly the same performance. SNR_{sum} is the sum signal-to-noise ratio (SNR) of all users. Complex channels with both slow- and fast-fading factors are considered. Path loss is based on a hexagon cell with a normalized side length = 1. The minimum normalized distance between users and the base station is 35/289, corresponding to an unnormalized distance of 35 m for an LTE cell with radius 289 m. Path loss factor = 3.76 and lognormal fading deviation = 8 dB. The channel samples are normalized such that the average power gain = 1

Recently, NOMA has been discussed widely for the fifth-generation (5G) [13–16]. A natural question is whether the possible gain of NOMA can justify its higher receiver cost. We examine this question using achievable sum-rate below.

In Fig. 13.1, we compare the achievable sum-rate of NOMA and OMA with equal energy constraint per frame per user in a multi-user SISO system. We can see NOMA improves sum-rate when the number of users K increases, but OMA can achieve the same gain through resource allocation. NOMA has no advantage here since OMA is capacity achieving in this case.

The situation is slightly different if the energy per frame per user can also be freely optimized under the sum energy constraint. For example, consider maximizing sum-log-rate under the proportional fairness criterion [3]. Figure 13.2 illustrates the related numerical results for $K = 2$. The sum-rate curves in Fig. 13.2 show that NOMA is only slightly better than OMA with resource allocation (about 8% gain at $SNR_{\text{sum}} = 10\,\text{dB}$).

The advantage of NOMA over OMA seen in Figs. 13.1 and 13.2 is disappointing compared with many results in the literature. This is mainly for two reasons. First, comparisons with OMA without resource allocation are not fair as resource allocation has already been widely used in LTE. Second, a practical signal-to-noise ratio (SNR) range should be used for comparison. A standard way for this purpose is using the following approximation of the signal-to-noise-plus-interference ratio (SINR) in a cellular system [3]:

$$SINR_{\text{sum}} = \frac{P_{\text{sum}}}{\beta P_{\text{sum}} + \sigma^2}, \tag{13.1}$$

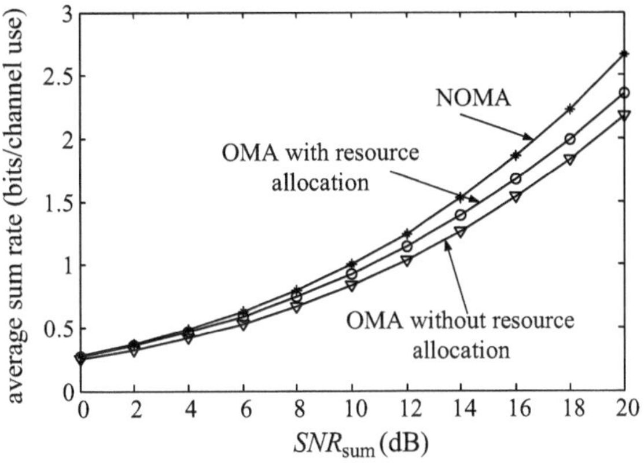

Fig. 13.2 Achievable sum-rate of NOMA and OMA under sum energy constraint with proportional fairness, $K = 2$. Other system settings are the same as those in Fig. 13.1

where P_{sum} is the sum received powers of all users in a cell, β a cross-cell interference factor and σ^2 the noise power. As a rule of thumb, a typical value is $\beta = 0.6$. Then, we have $SINR_{\text{sum}} \leq 2.2$ dB. Treating interference as noise and allowing a certain range of β, we may consider 0–10 dB as a typical range for SNR_{sum}, as used in Figs. 13.1 and 13.2. Clearly, the gain of NOMA is marginal over such a practical SNR range.

Furthermore, NOMA performance may deteriorate seriously if a practical forward error control (FEC) code is used. To see this, consider a successive interference cancelation (SIC) process with K users in a descendant order of user index k. We employ an FEC code that can achieve (almost) error-free decoding at SNR = Γ. Assume that, when we decode for user k, the signals of all users with indexes $k' > k$ have been successfully decoded and subtracted from the received signal. Let q_k be the received power of user k. Then, user k can achieve error-free decoding provided that

$$SNR_k = \frac{q_k}{\sum_{k'<k} q_{k'} + \sigma^2} \geq \Gamma. \tag{13.2a}$$

For sum-power minimization, q_k can be calculated using the following recursion (with $q_0 = 0$):

$$q_k = \Gamma \times \left(\sum_{k'<k} q_{k'} + \sigma^2 \right). \tag{13.2b}$$

Ideally, the minimum value for Γ can be found from Shannon capacity $R = \log_2(1 + \Gamma)$. For a practical code, however, a larger Γ is required to ensure (almost) error-free decoding. This overhead accumulates during SIC, which can amount to a considerable loss.

Specifically, we consider a practical rate-1/6 channel coding as an example. It achieves bit error rate (BER) $\approx 10^{-5}$ (approximately error-free) at about SNR = -3.5 dB with a relatively short block length. There is a 2.35 dB gap compared with SNR = -5.85 dB calculated from the Shannon formula for Gaussian signaling. Such

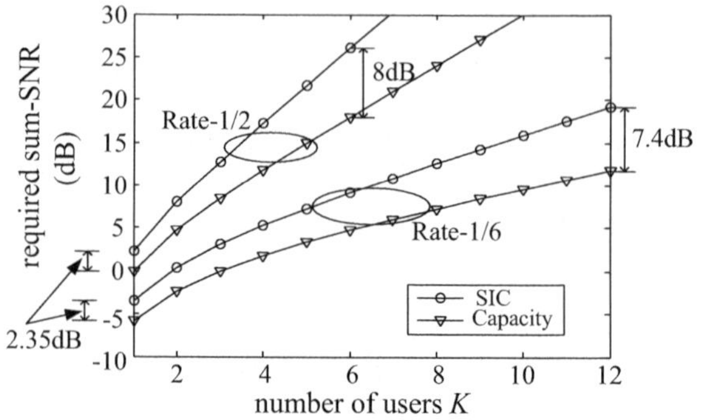

Fig. 13.3 Accumulated loss in the SIC process

loss accumulates in the SIC process as shown in Fig. 13.3. The accumulated loss is roughly 7.4 dB for 12 users.

The problem is more serious for higher rate, as seen from Fig. 13.3 for rate-1/2 coding. Assume the same initial single-user gap of 2.35 dB. The accumulated gap for six users is 8 dB.

Nevertheless, NOMA is still useful. In many situations, it is difficult to establish orthogonality due to the lack of centralized control or accurate channel state information (CSI). Then, we may have to resort to NOMA. In particular, as we will show below, NOMA based on interleave-division multiple access (IDMA) [17–22] offers robust and flexible solutions in such environments. IDMA can also recover a considerable portion of the accumulated loss suffered by SIC as shown in Fig. 13.3. We will use numerical results to verify these claims. Some of the software used in this chapter are available at: http://www.ee.cityu.edu.hk/%7Eliping/Research/Simulationpackage/.

13.2 Basic Principles of IDMA

Following the advent of turbo and low-density parity-check (LDPC) codes [23–25], iterative detection techniques were developed in late 1990s for equalization in multipath channels [26] and MUD in DS-CDMA systems [27–29]. It was shown in [30] that two independently interleaved code sequences can be separated by iterative detection. This inspired the IDMA scheme in which users are solely separated by interleavers [17]. Intuitively, a randomly interleaved code results in a different code. Thus we may also say that different users in IDMA are separated by different codes. This follows the basic principle of CDMA, except now the set of codes are generated by a master code followed by different interleavers. We therefore can regard IDMA as a special case of CDMA. However, IDMA is fundamentally different from DS-CDMA. The former does not rely on spreading for user separation and so can avoid the rate loss suffered by the latter.

In the following, we will start with a graphic model originally presented in [31]. We will show that the primary motivation behind IDMA is to break short cycles, since the latter is detrimental for message-passing detection. The same principles have been successfully used in turbo and LDPC codes.

13.2.1 IDMA Transmitter Principles

Throughout this chapter, we will assume an underlying OFDMA layer that removes inter-symbol interference (ISI). We will focus on uplink multiple access techniques built on this OFDMA layer.

Let K be the user number and $\boldsymbol{c}_k = \{c_k(j), j = 1, 2, \ldots, J\}$ a length-J codeword generated by users k. The transmitted symbols $\{x_k(j), j = 1, 2, \ldots, J\}$ are generated from $\{c_k(j)\}$ after certain operations, such as spreading, scrambling, interleaving, and modulations. At the receiver, the received signals $\{y(j)\}$ are given by

$$y(j) = \sum_{k=1}^{K} h_k \sqrt{e_k} x_k(j) + \eta(j), j = 1, 2, \ldots, J, \tag{13.3}$$

where h_k is the channel coefficient for user k, e_k the transmitted power of user k and $\eta(j)$ an additive white Gaussian noise (AWGN) sample with variance σ^2 per dimension.

Figure 13.4 illustrates the factor graph representation [32] for (13.3) with $K = 2$, $J = 8$, and the same LDPC coding for all users. A circle in Fig. 13.4 represents a variable and a square a constraint. Three types of constraints are involved, namely a square marked with "+" for linear additions in (13.3), a white square for LDPC coding, and a square marked with "×" for modulation.

Optimal detection for the system in (13.3) typically requires prohibitively high complexity. Low-cost message-passing detection, similar to that used for LDPC codes, can be applied instead, as Fig. 13.4 is sparse when $K \ll J$. However, short cylices constitute a problem. To see this, let us call a circle involving m coded

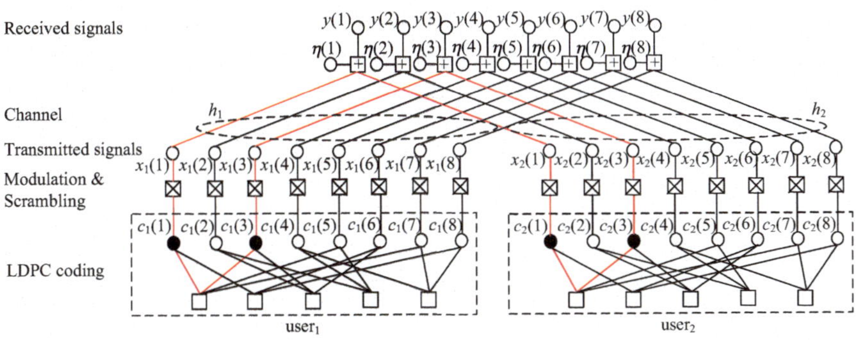

Fig. 13.4 Factor graph of a NOMA LDPC-coded system where $K = 2$ and $J = 8$

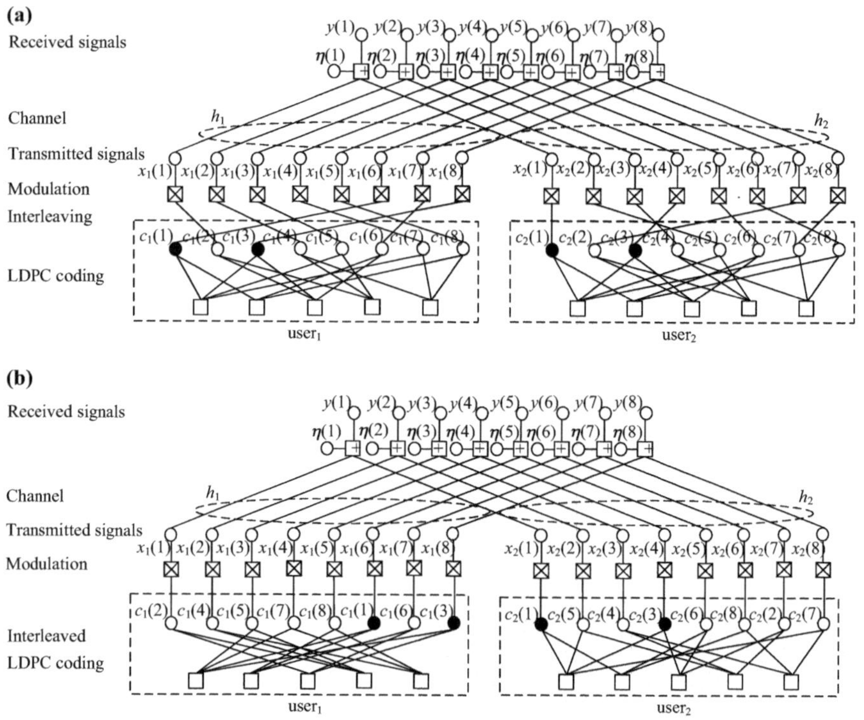

Fig. 13.5 **a** Factor graph of a two-user IDMA system. **b** An equivalence form of (**a**) with re-shuffled $\{c_k(j)\}$. Note that in (**a**) the interleavers for LDPC coding are the same for both users, while in (**b**) they are different

bits as a size-m cycle. An example of a size-4 cycle is shown by the black circles $\{c_1(1), c_1(3), c_2(1), c_2(3)\}$ in Fig. 13.4. There are a large number of such size-4 cycles in Fig. 13.4. Correlation may build up along these short cycles during message-passing detection, which is detrimental to performance [33].

Figure 13.4 can also be used to represent a DS-CDMA system. For example, a spreading operation involving binary sequence of ± 1 can be regarded as a repetition code plus a scrambling operation of sign changes. Repetition coding can be merged with FEC coding. Scrambling can be incorporated in function of the modulation nodes in Fig. 13.4. Note that scrambling does not change the topology of the graph. The problem of short cycles remains the same with or without scrambling. The same conclusion applies to other modulation techniques.

Inspired by the success of turbo and LDPC codes [23, 25], the IDMA scheme proposed in [31] employs user-specific interleaving to reduce short cycles in a statistical sense. This is illustrated by the shuffled edge connections between $\{c_k(j)\}$ and $\{x_k(j)\}$ in Fig. 13.5a. For example, compared with Fig. 13.4, $\{c_1(1), c_1(3), c_2(1), c_2(3)\}$ no longer form a size-4 cycle after interleaving in Fig. 13.5a. This is beneficial for

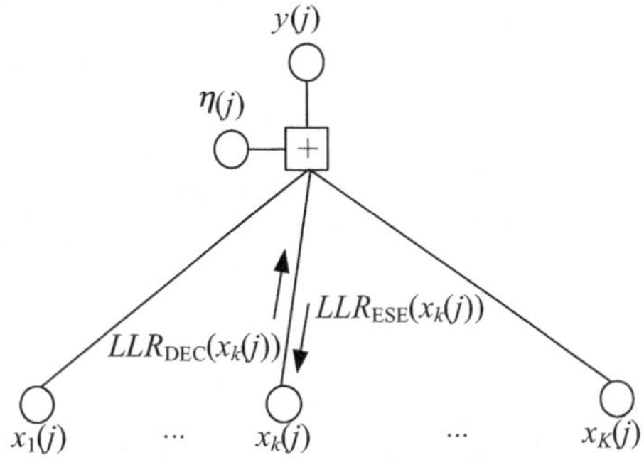

Fig. 13.6 An MA node. Here, an inbound message $LLR_{\mathrm{DEC}}(x_k(j))$ is from an LDPC decoder. An outbound message $LLR_{\mathrm{ESE}}(x_k(j))$ is generated from the MA node. ESE stands for "elementary signal estimation" and DEC for decoder

message-passing detection. Similar principles can also be applied to systems involving convolutional or turbo coding.

Two interleavers are used by each user in Fig. 13.5a, one for LDPC encoding and one for multiple access. The former is the same for all users, and the latter is user-specific. We can combine these two interleavers by re-shuffling $\{c_k(j)\}$ for each user, resulting in Fig. 13.5b. The latter interpretation of IDMA is based on [34].

The advantage of Fig. 13.5b is its simpler implementation. If an interleaver is random, its shifted version can be approximately regarded as another independent random interleaver. Thus, Fig. 13.5b can actually be realized by using a shifted version of an LDPC encoder involving a common underlying interleaver. The users, in this case, are separated by the amount of shift after the common encoder structure. Such shifting can be realized with very little cost. We may name such a scheme as code-shift division multiple access (CsDMA).[1] This concept was first discussed in [35].

Note that the use of interleavers in Fig. 13.5 does not incur any rate loss. This is a noticeable advantage of IDMA for high-rate applications.

13.2.2 Operations on a Multiple Access Node

We now consider message-passing detection based on Fig. 13.5. For convenience, we refer to a square marked with "+" in Fig. 13.5 as a multiple access (MA) node. We will only discuss the operations for such an MA node shown in Fig. 13.6. The operations for other nodes follow the standard treatments for an LDPC code [25].

Denote by DEC k the decoder for user k. We define two messages: an inbound message $LLR_{\mathrm{DEC}}(x_k(j))$ and an outbound message $LLR_{\mathrm{ESE}}(x_k(j))$ that are, respectively log-likelihood ratios (LLRs), generated by DEC k and elementary signal estimation

---

[1]The underlying code in CsDMA should be properly interleaved. An LDPC code naturally meets this requirement. Without interleaving, however, the correlation among the consecutive bits in a convolutional code may cause a problem in CsDMA. This problem can be easily avoided by shuffling the coded sequence.

(ESE) operations at the jth MA node in Fig. 13.6. The discussions for $LLR_{\mathrm{DEC}}(x_k(j))$ follow the standard LDPC decoding principles [25] and so will be omitted. We will focus on $LLR_{\mathrm{ESE}}(x_k(j))$ below since its computation is not part of a standard LDPC decoder.

For simplicity, let us first assume binary phase-shift keying (BPSK) modulation $x_k(j) = \pm 1$ for all k. We define $LLR_{\mathrm{ESE}}(x_k(j))$ by the following LLR for $x_k(j)$:

$$LLR_{\mathrm{ESE}}\,(x_k(j)) = \log \frac{\Pr\,(y(j)|x_k(j) = +1)}{\Pr\,(y(j)|x_k(j) = -1)}, \tag{13.4}$$

where $y(j)$ and $x_k(j)$ are defined in (13.3). Assume that $\eta(j)$ in (13.3) is Gaussian with mean $\mu(j) = \mathrm{E}(\eta(j))$ and variance $v = \mathrm{Var}(\eta(j))$. (For simplicity, we will assume that v is not a function of j. We will explain the rationale for this assumption later.) For a single-user system with $K = 1$ in (13.3), the conditional probabilities in (13.4) are given by

$$\Pr\,(y(j)|x_1(j) = \pm 1) = \frac{\exp\left(-\frac{\left(y(j)-(\mu(j)\pm h_1\sqrt{e_1})\right)^2}{2v}\right)}{\sqrt{2\pi v}}, \tag{13.5}$$

so

$$LLR_{\mathrm{ESE}}\,(x_1(j)) = 2h_1\sqrt{e_1}\frac{y(j) - \mu(j)}{v}. \tag{13.6}$$

For $K > 1$, the problem is much more complicated. We need to consider all possible combinations of $\{x_k(j)\}$. The exact result is the maximum likelihood (ML) estimator [36] below:

$$LLR_{\mathrm{ESE}}\,(x_k(j)) = \log \frac{\sum_i \Pr\left(y(j)|x_k(j) = +1, X^i_{\sim k}(j)\right) \Pr\left(X^i_{\sim k}(j)\right)}{\sum_i \Pr\left(y(j)|x_k(j) = -1, X^i_{\sim k}(j)\right) \Pr\left(X^i_{\sim k}(j)\right)}, \tag{13.7}$$

where $X^i_{\sim k}(j)$ is one among all 2^{K-1} possibilities of the set $\{x_1(j), x_2(j), \ldots, x_{k-1}(j), x_{k+1}(j), \ldots, x_K(j)\}$ (since $x_{k'}(j) \in \{-1, +1\}, \forall k'$). In (13.7), $\Pr\left(y(j)|x_k(j) = \pm 1, X^i_{\sim k}(j)\right)$ can be computed similarly to (13.5) and $\Pr\left(X^i_{\sim k}(j)\right)$ can be computed from messages $\{LLR_{\mathrm{DEC}}(x_k(j))\}$. Following [25], we define $LLR_{\mathrm{DEC}}(x_k(j))$ by an LLR:

$$LLR_{\mathrm{DEC}}\,(x_k(j)) = \log \frac{\Pr\,(x_k(j) = +1)}{\Pr\,(x_k(j) = -1)}. \tag{13.8}$$

We can obtain $\Pr(x_k(j) = \pm 1)$ by solving (13.8) together with $\Pr(x_k(j) = +1) + \Pr(x_k(j) = -1) = 1$. Then,

$$\Pr\left(X^i_{\sim k}(j)\right) = \prod_{k'=1, k' \neq k} \Pr\,(x_{k'}(j)). \tag{13.9}$$

The complexity of ML is $O(2^K)$ for BPSK, which increases exponentially with K. For a higher order modulation with an M-point constellation, the complexity of ML is $O(M^K)$. This can be a serious problem in practice.

Gaussian approximation (GA) is a low-cost alternative. We rewrite (13.3) as

$$y(j) = h_k\sqrt{e_k}x_k(j) + \zeta_k(j), \tag{13.10a}$$

where

$$\zeta_k(j) = y(j) - h_k\sqrt{e_k}x_k(j) = \sum_{k'=1,k'\neq k}^{K} h_{k'}\sqrt{e_{k'}}x_{k'}(j) + \eta(j) \tag{13.10b}$$

is the distortion (including interference-plus-noise) with respect to user k. From the central limit theorem, we apply GA to $\zeta_k(j)$ in (13.10b) and assume $\zeta_k(j) \sim N(\mu_k(j), \mathrm{Var}(\zeta_k(j)))$. Now we can treat (13.10a) as a single-user system. For simplicity, we assume a real channel. (We will discuss a complex channel in Sect. 13.5.2.) Then, we have

$$\Pr\left(y(j)|x_k(j) = \pm 1\right) = \frac{\exp\left(-\frac{\left(y(j)-(\mu_k(j)\pm h_k\sqrt{e_k})\right)^2}{2\mathrm{Var}(\zeta_k(j))}\right)}{\sqrt{2\pi\,\mathrm{Var}(\zeta_k(j))}}. \tag{13.11}$$

Substituting (13.11) into (13.4) and evaluating $\mu_k(j)$ via (13.10b), we have the following ESE operations for the jth MA node in Fig. 13.6. (We will discuss the generation of $\mathrm{Var}(\zeta_k(j))$ in (13.12c) later in (13.13).)

ESE operations

$$\text{(i) } \mathrm{E}(x_k(j)) = \Pr(x_k(j) = +1) - \Pr(x_k(j) = -1), \tag{13.12a}$$

$$\text{(ii) } \mu_k(j) = \sum_{k'=1}^{K} h_{k'}\sqrt{e_{k'}}\mathrm{E}(x_{k'}(j)) - h_k\sqrt{e_k}\mathrm{E}(x_k(j)), \tag{13.12b}$$

$$\text{(iii) } LLR_{\mathrm{ESE}}\left(x_k(j)\right) = 2h_k\sqrt{e_k}\frac{y(j) - \mu_k(j)}{\mathrm{Var}(\zeta_k(j))}. \tag{13.12c}$$

The following are some details related to the ESE operations.

- Initially, there is no decoder feedback, and we can set $\mathrm{E}(x_k(j)) = 0$ in (13.12a) for $\forall k, j$.
- GA is approximate. However, we observed a very good performance based on GA.
- The summation in (13.12b) can be shared by all users. The cost per information bit per user is independent of the number of users K.
- The principles of GA for higher-order modulations (such as quadrature phase-shift keying (QPSK)) and complex channels can be derived similarly. Related discussions will be shown in Sect. 13.5.2.

We now discuss the evaluation of $\mathrm{Var}(\zeta_k(j))$ involved in (13.12c). From (13.10b),

$$\mathrm{Var}(\zeta_k(j)) = \sum_{k'=1}^{K} |h_{k'}|^2 e_{k'} \mathrm{Var}(x_{k'}(j)) - |h_k|^2 e_k \mathrm{Var}(x_k(j)) + \sigma^2. \qquad (13.13a)$$

We observed that the system performance is not sensitive to $\mathrm{Var}(\zeta_k(j))$. Therefore, we take the following approximation

$$\mathrm{Var}(\zeta_k(j)) \approx \sum_{k'=1}^{K} |h_{k'}|^2 e_{k'} v_{k'} - |h_k|^2 e_k v_k + \sigma^2 \qquad (13.13b)$$

based on the following assumption

$$\mathrm{Var}(x_k(j)) = 1 - (\mathrm{E}(x_k(j)))^2 \approx v_k, \forall j. \qquad (13.13c)$$

To evaluate v_k in (13.13c), we can simply compute a few samples of $\mathrm{Var}(x_k(j))$ and take their average. In practice, computation for each $\mathrm{Var}(x_k(j))$ can be implemented using a look-up table. We observed that the required number of samples is small, so the related cost is negligible.

From the above discussions, the total cost for the ESE operations is (approximately) four additions and two multiplications per chip per iteration. (Some operations, such as $h_k\sqrt{e_k}$ and $2h_k\sqrt{e_k}/\mathrm{Var}(\zeta_k(j))$, can be precalculated and need not be repeated for every j. The related cost is negligible.)

13.2.3 Overall IDMA Receiver

Now return to an overall IDMA system in Fig. 13.5. We divide the receiver into an ESE module and K DEC modules. The operations in the ESE module is based on (13.12). Consider two types of schedules below.

- Serial schedule: In each iteration, operations are carried out as follows:
  ESE for user 1, DEC for user 1, ESE for user 2, DEC for user 2, ...
  As an example, ESE for user 1 means running (13.12) for every j with k fixed to 1. The LLRs generated in (13.12) are then fed to DEC for user 1. Then, $\{\mu_1(j)\}$ and v_1 are updated based on the DEC outputs and the process continues to user 2.
- Parallel schedule: In each iteration, operations are carried out as follows:
  ESE for all users in parallel, then
  DEC for all users.
  In the above, in each iteration, (13.12) is run through every pair of j and k. After the ESE operations, the outputs are fed to the K local DECs. Then, the K DECs are run in parallel. Afterward, $\{\mu_k(j)\}$ and v_k are updated simultaneously for all k.

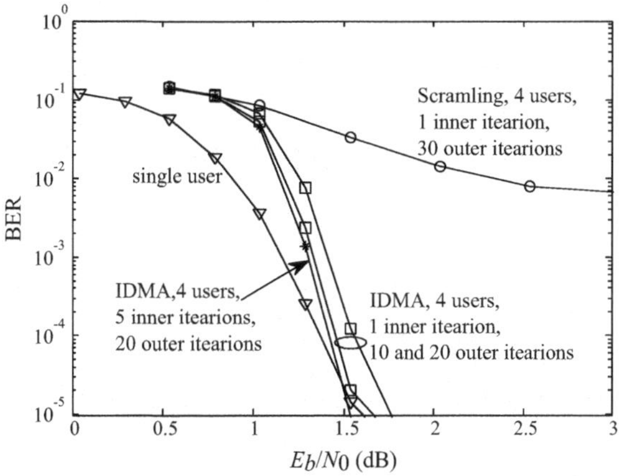

Fig. 13.7 Performance of IDMA with GA in AWGN channels

Note that in the parallel schedule, the value of the summation in (13.12b), that is generated using the results of the previous iteration, remains the same for all users in one iteration. In the serial schedule, this summation is updated user-by-user. We noticed that serial scheduling converges slightly faster than the parallel one.

A question arises whether it is helpful if multiple inner DEC iterations are carried out between two consecutive ESE iterations. For example, with the serial schedule, we can run multiple LDPC decoding iterations for one user before going to the next user. We observed that such inner iterations are generally unnecessary. For fixed overall cost, better performance is achieved without inner iterations. However, slipping some inner or outer iterations may lead to reduced complexity.

Intuitively, we can treat an IDMA system in Fig. 13.5 as a generalized code system on a graph (where an MA node is just for a special type of constraint). The inner-iteration method means more iterations on some parts of the graph for LDPC coding constraints. Such uneven message-passing process does not help in general.

For comparison of overall cost, let us consider a K-user TDMA system in which LDPC decoding is individually carried out for each user. The only difference between SUD for such a TDMA system and MUD for IDMA is the ESE operations. As we have seen above, the cost of the extra ESE operations is quite moderate. Thus, an IDMA receiver involving (13.12) has only moderately higher complexity than SUD for corresponding TDMA systems with the same number of users.

Figure 13.7 shows an example of a four-user IDMA system in AWGN channels. A rate-1/2 LTE turbo code with 1200 information bits per user is used, followed by a rate-1/8 repetition code and QPSK modulation. We can see that iterative detection nearly converges with about ten outer iterations between ESE and DECs with one inner-iteration. Here, one inner-iteration means running each component decoder once in a turbo decoder. We can also run multiple iterations in each turbo decoder within each outer iteration. The result of five inner iterations is shown in Fig. 13.7. We can see that multiple inner iterations can only offer marginal improvement on performance, even though at considerably higher overall cost.

Figure 13.7 also shows the result with user-specific scrambling by spreading each coded bit with a random binary sequence of +1 and −1 before modulation. No user-specific interleaving is used in this case. It is seen that interleaving offers better performance than scrambling. This is due to the short cycle problem as explained earlier.

13.2.4 Performance Evaluation Through SNR Evolution

We now outline an SNR evolution technique [17] for tracking IDMA performance. Similar techniques have been successfully applied to turbo and LDPC codes [25] and more recently to AMP algorithms [37]. This analysis method also provides the basis of power allocation for IDMA performance optimization in the next section.

Figure 13.8a is a so-called protograph [38] representation of Fig. 13.5, in which each circle represents a vector and each edge represents a vector connection. Relatively thick lines are used in Fig. 13.8a to distinguish it from Fig. 13.5. The messages $\boldsymbol{LLR}_{\mathrm{DEC},k}$ and $\boldsymbol{LLR}_{\mathrm{ESE},k}$ in Fig. 13.8a are LLR sequences generated by the coding and MA constraints, respectively.

We will use the following SNR-variance relationship to characterize the behavior of the system in Fig. 13.8a. Recall (13.10a): $y_k(j) = h_k\sqrt{e_k}x_k(j) + \zeta_k(j)$. We define the average SNR for user k as

$$SNR_k \equiv \frac{\mathrm{E}\left(|h_k\sqrt{e_k}x_k(j)|^2\right)}{\mathrm{E}(\mathrm{Var}(\zeta_k(j)))} = \frac{|h_k|^2 e_k}{\mathrm{E}(\mathrm{Var}(\zeta_k(j)))}. \tag{13.14}$$

From (13.13), we have

$$\mathrm{E}(\mathrm{Var}(\zeta_k(j))) = \sum_{k'=1,k'\neq k}^{K} |h_{k'}|^2 e_{k'} v_{k'} + \sigma^2. \tag{13.15}$$

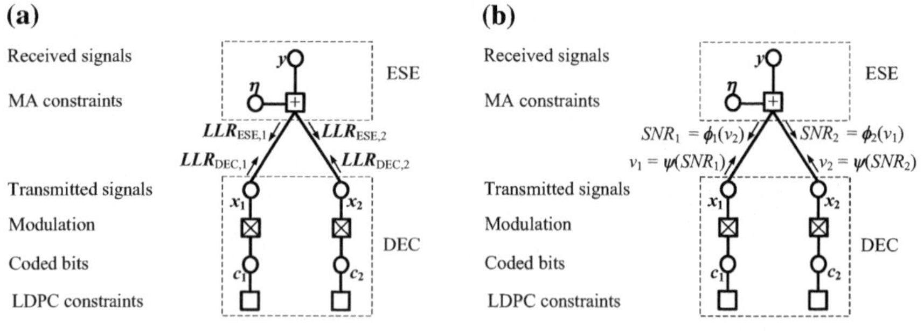

Fig. 13.8 **a** Protograph representation of Fig. 13.5. **b** Evolution characterization for (**a**)

Since $\boldsymbol{LLR}_{\text{DEC},k}$ is the output of DEC k with input $\boldsymbol{LLR}_{\text{ESE},k}$ characterized by SNR_k, we can write v_k as a function $v_k = \psi(SNR_k)$. Here for simplicity, we assume that all users employ the same LDPC code and therefore have the same $\psi(\cdot)$. (Note that the re-shuffling operation in Fig. 13.5b has no effect on $\psi(\cdot)$.) Combining this with (13.14) and (13.15), we have the following recursion to characterize an iterative IDMA detector:

$$SNR_k^{(t)} = \frac{|h_k|^2 e_k}{\sum_{k'=1,k'\neq k}^{K} |h_{k'}|^2 e_{k'} v_{k'}^{(t)} + \sigma^2} \equiv \phi_k\left(v_1^{(t)}, \ldots v_{k-1}^{(t)}, v_{k+1}^{(t)}, \ldots v_K^{(t)}\right), \tag{13.16a}$$

$$v_k^{(t)} = \psi\left(SNR_k^{(t-1)}\right), \tag{13.16b}$$

where t is an iteration index. The initialization is $v_k^{(0)} = 1, \forall k$, implying no information from DECs. In general, there is no closed form expression for $\psi(\cdot)$, but it can be obtained by simulating a single-user APP decoder in an AWGN channel with specified SNRs. Using $\{SNR_k^{(T)}\}$ in the final iteration in (13.16), we can estimate the BER by a function

$$BER_k = g\left(SNR_k^{(T)}\right), \tag{13.17}$$

where $g(\cdot)$ can be obtained through simulation of DECs [17].

13.2.5 Superposition Coded Modulation (SCM)

The above IDMA scheme involves multiple signal streams from different users. We may simply allocate these signal streams to a single-user. Such scheme is referred to as superposition coded modulation (SCM) [39, 40].

We define a standard QPSK constellation as $S_{\text{QPSK}} = \{00 \to (+1, +1), 01 \to (+1, -1), 10 \to (-1, +1), 11 \to (-1, -1)\}$. Figure 13.9a is a 16-ary scheme formed by superimposing S_{QPSK} and a scaled version of S_{QPSK} with a scaling factor of 2 and a 45° phase shift [39]. Figure 13.9b is a 64-ary scheme formed by superimposing S_{QPSK} and two scaled versions of S_{QPSK} with, respectively, scaling factors of 1.18 and 1.10 plus 60° and 120° phase shifts. From the central limit theorem, the SCM signaling is more Gaussian-like when the number of streams is large. This can offer the so-called shaping gain as analyzed in [41, 42]. It has been proved that, among all possible signaling methods, an SCM constellation achieves the minimum mean squared error (MMSE) bound; that is, it minimizes the function $\psi(\cdot)$ in (13.16b) for a fixed underlying binary decoder. The details are discussed in [39].

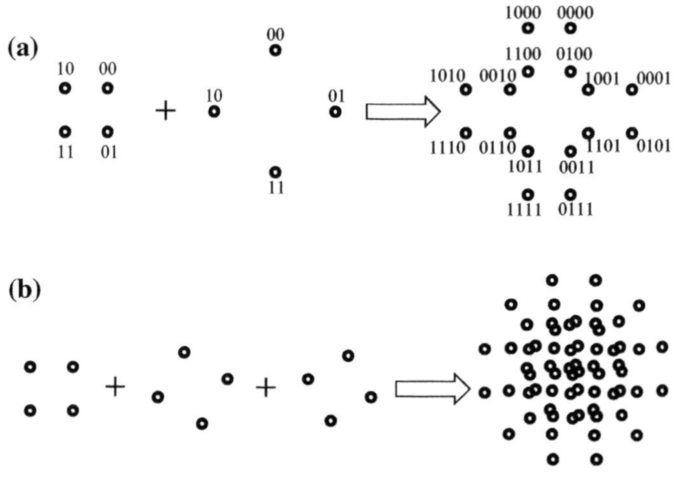

Fig. 13.9 **a** A 16-ary SCM signaling by superimposing two streams of QPSK constellations. **b** A 64-ary SCM signaling by superimposing three streams of QPSK constellations

13.3 Power Control for IDMA

At a relatively low sum-rate, such as less than 1 in a complex channel, IDMA with a GA receiver works well with equal received power. At a higher sum-rate, unequal power control is required. The situation is similar to power control for SIC in (13.2), except that iterative detection makes the problem more complicated.

13.3.1 Transmitted and Received Power Minimization

Denoting the received powers by

$$q_k = |h_k|^2 e_k. \tag{13.18}$$

We combine (13.16a) and (13.16b) into a compact form as

$$SNR_k^{(t)} = \frac{q_k}{\sum_{k'=1,k'\neq k}^{K} q_{k'} \psi\left(SNR_{k'}^{(t-1)}\right) + \sigma^2}, \forall k. \tag{13.19}$$

Let $\sum e_k$ and $\sum q_k$ be, respectively, sum transmitted power and sum received power. We can minimize them, respectively. The latter is simpler since the channel gains $\{|h_k|^2\}$ are not involved. The following Remark establishes a connection between these two problems [7, 9, 43].

Remark 1 Assume that $\{q_k^*\}$ is a minimizer for $\sum q_k$. Then, $\{e_k^* = q_k^*/|h_k|^2\}$ is a minimizer for $\sum e_k$ provided that $\{q_k^*\}$ and $\{|h_k|^2\}$ have the same order, i.e., $q_k^* \leq q_{k'}^*$ if $|h_k|^2 \leq |h_{k'}|^2$.

Based on Remark 1, we can first find the minimizer $\{q_k^*\}$ for $\sum q_k$. We then relabel $\{q_k^*\}$ such that it has the same order as $\{|h_k|^2\}$. Then, the minimizer for $\sum e_k$ can be obtained as $\{e_k^* = q_k^*/|h_k|^2\}$.

Incidentally, Remark 1 implies that a user with a higher channel gain should be assigned a higher transmitted power and vice versa. Next, we focus on minimizing $\{q_k\}$.

13.3.2 Feasible Profile

We now impose an SNR requirement Γ after T iterations. This SNR requirement can be equivalently translated into a BER requirement through (13.17). We write the received power optimization problem as follows.

$$\text{minimize} \quad \sum q_k, \tag{13.20a}$$

$$\text{subject to} \quad SNR_k^{(t)} = \frac{q_k}{\sum_{k'=1,k'\neq k}^{K} q_{k'}\psi\left(SNR_{k'}^{(t-1)}\right) + \sigma^2}, \forall k, \tag{13.20b}$$

$$SNR_k^{(T)} \geq \Gamma, \forall k. \tag{13.20c}$$

The problem in (13.20) is non-convex. We will outline two searching techniques for this problem. For convenience, we will call $\{q_k\}$ a feasible profile if it ensures the constraints in (13.20b) and (13.20c).

Incidentally, it is interesting to compare (13.2a) and (13.20b). In (13.20b), $q_{k'}\psi\left(SNR_{k'}^{(t-1)}\right)$ represents the residual interference from user k' after soft cancelation. Such terms disappear in (13.2a) for decoded users due to the error-free assumption and hard cancelation.

13.3.3 Greedy Search

We first set $T = 1$, i.e., only one iteration. Assume that approximate error-free decoding can be achieved at a sufficiently large SNR in the single-user case. We can construct an initial feasible profile $\mathcal{Q} = \{q_k\}$ according to (13.2). The sum-power for such a $\mathcal{Q}$ is typically large.

We next consider a general T. Starting from the above initial $\mathcal{Q}$, we search for a minimum value for each q_k individually to achieve (13.20c), while keeping other elements in $\mathcal{Q}$ unchanged. This involves a one-dimensional search, so its complexity is affordable. Let the search result be q_k^*. We then update $q_k \leftarrow q_k - \epsilon(q_k - q_k^*)$ in $\mathcal{Q}$, where ϵ is a damping factor (e.g., $\epsilon = 0.5$). We repeat the above process for all k iteratively. We observed reasonably good performance of this simple method for a relatively small K.

13.3.4 Approximate Linear Programming Method

Inspired by the linear program technique for LDPC code design [25], we can use the approximate technique below for a large K. The key idea is to transform the problem of finding the power for each user into finding the number of users on different given power levels, which makes the problem convex [44, 45].

Let us quantize the received power into $M + 1$ discrete values: $\{q(m), m = 0, 1, \ldots, M\}$ with $q(m - 1) < q(m)$. The received powers of all users are selected from $\{q(m)\}$. We partition K users into $M + 1$ groups according to their power levels. Let $\lambda(m)$ be the number of users assigned with power level $q(m)$ and $z(m)$ be the total power of these $\lambda(m)$ users. As such,

$$\sum_m \lambda(m) = K, \tag{13.21a}$$

$$z(m) = \lambda(m)q(m) \tag{13.21b}$$

and the sum received power

$$\sum_k q_k = \sum_m \lambda(m)q(m) = \sum_m z(m). \tag{13.21c}$$

Denote by $SNR(m)$ the SNR for the users in the mth group with power $q(m)$. Define

$$I = \sum_m z(m)\psi(SNR(m)) + \sigma^2, \tag{13.22}$$

which is the total interference power (including noise) after soft cancelation. When K is large, (13.20b) can be approximated as

$$SNR(m)^{(t)} = \frac{q(m)}{I^{(t)} - q(m)\psi\left(SNR(m)^{(t-1)}\right)} \approx \frac{q(m)}{I^{(t)}}, \tag{13.23}$$

where $I^{(t)}$ denotes the value of I at the tth iteration. Using (13.22) and (13.23), we have the update rule

$$I^{(t)} = \sum_m z(m)\psi\left(\frac{q(m)}{I^{(t-1)}}\right) + \sigma^2. \tag{13.24}$$

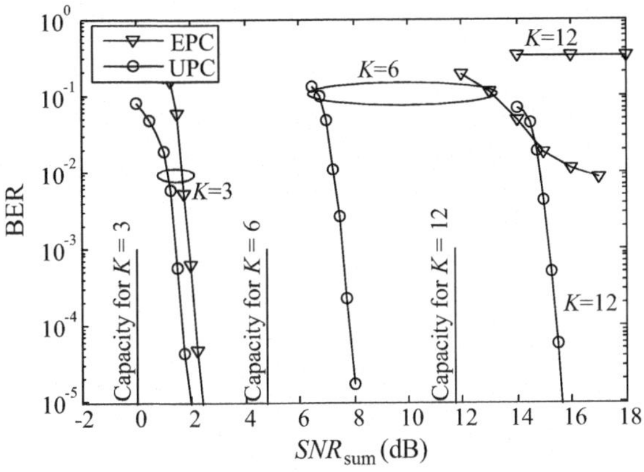

Fig. 13.10 IDMA with EPC and UPC in AWGN channels. Rate-1/3 turbo coding followed by rate-1/2 repetition coding is used for each user. Information length of each user is 1200. QPSK modulation. Sum-rate = 1, 2, and 4 for $K = 3$, 6, and 12, respectively

Equation (13.24) characterizes the evolution of the total interference at each iteration. If iterative detection converges, $I^{(t)}$ should be lower than $I^{(t-1)}$. Equivalently, we can write the convergence condition as

$$\sum_{m} z(m)\psi\left(\frac{q(m)}{I}\right) + \sigma^2 \leq (1-\delta)I, \quad I_{\min} \leq I \leq I_{\max}. \tag{13.25}$$

where $0 < \delta < 1$ is a decay factor that controls the convergence speed. $I_{\max}$ and $I_{\min}$ specify the total interference at the beginning and end of the iterative detection. In summary, we re-formulate the optimization problem as,

$$\text{minimize} \quad \sum z(m), \tag{13.26a}$$

$$\text{subject to} \quad \sum_{m} z(m)\psi\left(\frac{q(m)}{I}\right) + \sigma^2 \leq (1-\delta)I, \quad I_{\min} \leq I \leq I_{\max}, \tag{13.26b}$$

$$z(m) \geq 0, \forall m. \tag{13.26c}$$

The above optimization problem is linear with respect to $\{z(m)\}$. Hence, it can be resolved by linear programming. More details can be found in [44, 45].

Figure 13.10 is an example to illustrate the necessity of unequal power control (UPC). We can see that UPC improves the system performance for all the three cases of $K = 3$, 6, and 12 compared with equal power control (EPC). Particularly, when $K = 12$, the IDMA system does not work at all with EPC, but works well with UPC. The gap between IDMA with UPC and capacity is about 3.7 dB for $K = 12$.

Compared with the 12-user example in Fig. 13.3, we can see that IDMA recovers about 3.7 dB relative to the loss incurred by SIC. This is impressive, but there is definitely room for improvement. The gap towards capacity can potentially be further narrowed using more sophisticated techniques, such as curve-matching-based degree sequence optimization [46, 47], spatial coupling [48–57], or more sophisticated modulations [58].

A criterion called overloading, which represents the user capacity in a NOMA system, is used to assess the system performance in the recent literature. Figure 13.10 demonstrates that IDMA can offer very high overloading with centralized power control. In the next section, we will discuss IDMA techniques to achieve high throughput without centralized power control.

13.4 Random Access via IDMA

13.4.1 Limitations of Conventional Systems

A conventional uplink system with centralized control involves a connection setup procedure before data transmission. The overhead incurred by this procedure is not serious for services with long-lasting connections since it can be amortized across the connection duration. One of the main tasks envisaged for the next 5G cellular systems is to support machine-type communication (MTC) which is characterized by short and sporadic data communication. In this case, the cost of establishing centralized control can be substantial.

Random access is decentralized, in which each user makes an individual decision to transmit data packets. This avoids the overhead of connection setup, but the packets from different users may collide. In a conventional random-access scheme, such as ALOHA [59], colliding packets are discarded, which reduces throughput. For this reason, such techniques cannot satisfy the demand of high spectral efficiency in 5G cellular systems.

In a fading channel, the received powers of different users may form a feasible profile defined in Sect. 13.3.2, even without centralized control. This is captured in the multi-packet reception (MPR) model [60, 61]. Most existing MPR techniques rely on channel fading to form feasible profiles. Such a passive approach achieves limited throughput gain. In what follows, we will discuss an active approach based on the power control technique. The main idea is to optimize the probability of forming a feasible profile through decentralized power control at the transmitters.

13.4.2 Random IDMA with Decentralized Power Control

13.4.2.1 Problem Formulation

Recall from Sect. 13.3.2, a feasible received power profile $\{q_k\}$ can be formed by centralized control. Without centralized control, however, it is difficult to guarantee this. A randomized power control (RPC) technique [62–67] is discussed below to handle this difficulty.

The principles of RPC are as follows. Let $\{Q^{(l)}, l = 1, 2, \ldots, L\}$ (L is the maximum level index) be a set of pre-defined power levels and $\{P^{(l)}, l = 1, 2, \ldots, L\}$ a set of related probabilities. Upon a packet arrival, each user randomly draws a power level $Q^{(l)}$ with probability $P^{(l)}$ and uses it to transmit. Different users act individually and so their transmissions may collide. However, as long as their received powers $\{q_k\}$ form a feasible profile, their signals can still be recovered. Our aim is to optimize the probability that $\{q_k\}$ form a feasible profile.

Mathematically, $\{q_k\}$ defined above are the realizations of an underlying random variable. The random variable is characterized by a probability mass distribution $\{P^{(l)}\}$. Each q_k is independently drawn by a user from the support $\{Q^{(l)}\}$. Once the distribution is given, there is no need for centralized control.

13.4.2.2 Type-2 Collisions

The problem formulated above turns out to be difficult when K is large. So far, we have no general solution. We will discuss a sub-optimal technique below.

We say that a collision is of type-M if it involves M active users. Let $Q^{(0)} = 0$ be an element in $\{Q^{(l)}\}$. In RPC, a user will not transmit if its selected power is $Q^{(0)}$. The related $P^{(0)}$ is equivalent to the back-off probability in 802.11 Wi-fi systems. Intuitively, collisions are dominated by type-2 ones when $P^{(0)}$ is sufficiently large. Therefore, we will focus on type-2 collisions.

We consider a type-2 collision involving user i and user j. We define the union of all possible feasible profiles of the received power pair $\{q_i, q_j\}$ as a feasible region. The collision is resolvable if $\{q_i, q_j\}$ falls in this region. Figure 13.11a shows an example of the feasible region for SIC with ideal coding and decoding. The area marked by "A" in Fig. 13.11a is formed by all possible $\{q_i, q_j\}$ that meets the following conditions (see (13.2)):

$$SNR_j = \frac{q_j}{q_i + \sigma^2} \geq \Gamma, \tag{13.27a}$$

$$SNR_i = \frac{q_i}{\sigma^2} \geq \Gamma. \tag{13.27b}$$

The area marked by "B" in Fig. 13.11a is formed similarly by changing decoding order. The value of Γ here is determined by the Shannon capacity $R = \log_2(1 + \Gamma)$. Any power pair in the feasible region is resolvable by SIC.[2]

Figure 13.11b is an example of an IDMA system involving two LDPC-coded users with coding rate 0.5 per user. The receiver can achieve BER $\leq 10^{-5}$ in the feasible region, which is regarded as approximately error-free. The border of this feasible region is obtained using simulation.

[2]Figure 13.11a is for $R < 1$. If $R \geq 1$, the feasible region is divided into two disjoint sub-regions A and B symmetric to the 45° line $q_i = q_j$ [63].

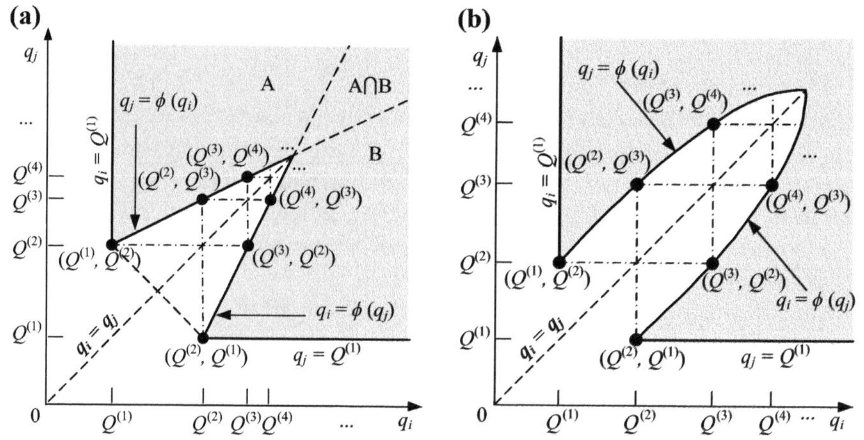

Fig. 13.11 Feasible regions for **a** a two-user ideally coded system with SIC, and **b** a two-user LDPC-coded IDMA system

Each of the two feasible regions in Fig. 13.11 is bounded by four curves $q_i = Q^{(1)}$, $q_j = \phi(q_i)$, $q_j = Q^{(1)}$ and $q_i = \phi(q_j)$. Here, the function $\phi(\cdot)$ is determined by (13.27a) taking mark of equality (for Fig. 13.11a) or by simulation (for Fig. 13.11b). $Q^{(1)}$ is the minimum power for successful single-user transmission. We construct the set $\{Q^{(l)}\}$ as follows:

$$Q^{(l)} = \begin{cases} 0, & l = 0, \\ Q^{(1)}, & l = 1, \\ \phi(Q^{(l-1)}), & l > 1. \end{cases} \tag{13.28}$$

For $\{q_k\}$ randomly selected from $\{Q^{(l)}\}$, we have the following situations:

Case 1: All $\{q_k\}$ are zeros. In this case, throughput is zero.

Case 2: Only one element in $\{q_k\}$ is nonzero. The transmission of the only active user will be successful.

Case 3: Exactly two elements in $\{q_k\}$, say q_i and q_j, are nonzero. This is a type-2 collision. It can be shown that $\{q_i, q_j\}$ falls in the feasible region (so collision is resolvable) provided that $q_i \neq q_j$.

Case 4: More than two elements in $\{q_k\}$ are nonzero. For simplicity, such events are regarded as unresolvable, which is a pessimistic assumption.

Based on the above cases, we can find an optimized probability set $\{P^{(l)}\}$ for a K user system. For convenience, we assume that the packets of all users arrive independently, following the Bernoulli process with parameter λ. The system throughput is then given by

$$T = T_1 + T_2. \tag{13.29}$$

In (13.29), T_1 is the throughput related to transmissions without collision:

$$\begin{aligned} T_1 =& \sum_{k=1}^{K} C_K^k \lambda^k (1-\lambda)^{(K-k)} C_k^1 (1-P^{(0)})(P^{(0)})^{k-1} \\ =& K\lambda(1-P^{(0)})(1-\lambda+\lambda P^{(0)})^{K-1}, \end{aligned} \tag{13.30}$$

where $C_K^k \lambda^k (1-\lambda)^{(K-k)}$ is the probability of k users among total K users having packets to transmit and $C_k^1(1-P^{(0)})(P^{(0)})^{k-1}$ the probability of only one user among these k users transmitting with nonzero power. Also in (13.29), T_2 is the throughout related to type-2 collisions. From case 3 above, a type-2 collision is unresolvable only if the two transmitting users are using the same received powers. Therefore, we have

$$\begin{aligned} T_2 =& 2\sum_{k=2}^{K} C_K^k \lambda^k (1-\lambda)^{(K-k)} C_k^2 \left((1-P^{(0)})^2 - \sum_{l>0} (P^{(l)})^2 \right) (P^{(0)})^{k-2} \\ =& K(K-1)\lambda^2 (1-\lambda+\lambda P^{(0)})^{K-2} \left((1-P^{(0)})^2 - \sum_{l>0} (P^{(l)})^2 \right), \end{aligned} \tag{13.31}$$

where $(1-P^{(0)})^2 - \sum_{l>0} (P^{(l)})^2$ is the probability that two active users transmit with different received powers.

We further consider an average transmitted power constraint $\bar{q}$ for each user. In AWGN channels with unit channel power gain, the constraint is given by

$$\sum_{l \geq 0} Q^{(l)} P^{(l)} \leq \bar{q}. \tag{13.32}$$

Under such power constraint, we can search for $\{P^{(l)}\}$ that maximize the throughput T in (13.29). It can be verified that the problem is convex if $P^{(0)}$ is fixed. The treatments for fading channels are somewhat more complicated. The details can be found in [63].

Figure 13.12 shows two examples, one for an ideally coded system and the other for an LDPC-coded IDMA system [63]. Conventional ALOHA is included as a reference. We can see that the RPC-based scheme can offer noticeably throughput gain compared with ALOHA.

It is proved in [63] that the $\{Q^{(l)}\}$ in (13.28) forms an optimal support for the decentralized power control when $K = 2$. It is sub-optimal for $K > 2$, but it can still provide excellent performance gain, as seen in Fig. 13.12.

As a short summary, we can treat collisions as NOMA cases. A conventional scheme, such as ALOHA, treats such NOMA cases as failures. The discussions in this section aim at optimizing the probability of successful detection in such NOMA cases.

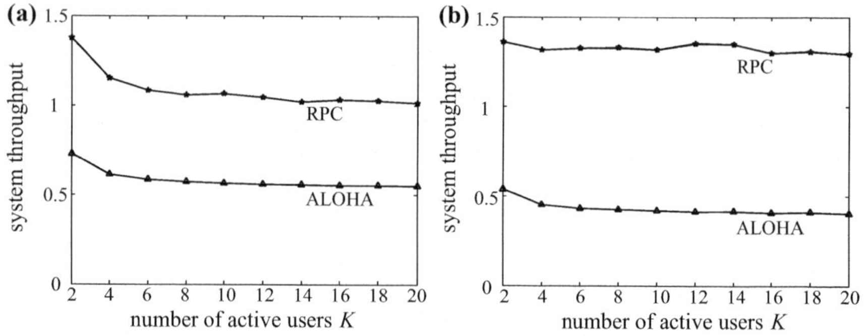

Fig. 13.12 Performance comparison of RPC and ALOHA in **a** an ideally coded system and **b** an LDPC-coded IDMA system. Rayleigh fading channel with averaged power gain 1 for all users. Same power constraint for RPC and ALOHA

13.5 IDMA in MIMO Systems

13.5.1 Multi-User Gain in MIMO Systems

MIMO is a wireless technology employing multiple transmit and receive antennas [68–74]. Multi-user gain refers to the advantage of allowing a large number of users to transmit simultaneously over the same time and same frequency in MIMO [10, 75]. This is illustrated in Fig. 13.13 by the potential sum-rate capacity gain for a single-cell system. Perfect CSI is assumed in Fig. 13.13. The curves apply to both up- and downlinks following the duality principle [3, 76, 77]. We can see that multi-user gain is very attractive. The potential gain is in the order of tens of times. Diversifying power over more users, i.e., increasing K, is a very effective way to increase sum-rate.

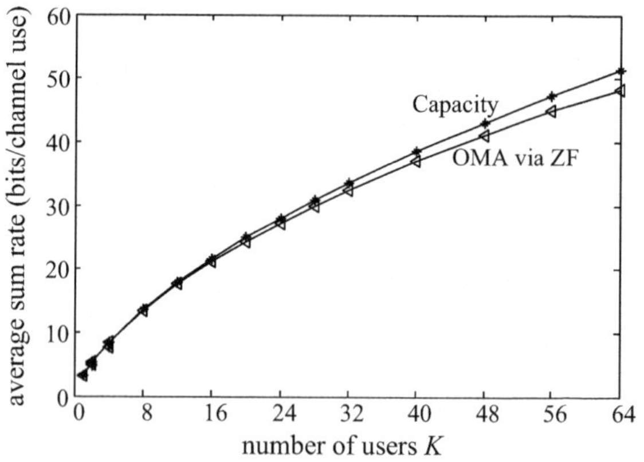

Fig. 13.13 Achievable sum-rate of ZF under perfect CSI. The number of antennas at the base station is 64. Single antenna assumed is for each user. Other system parameters are the same as those in Fig. 13.1

Figure 13.13 also includes the performance of zero-forcing (ZF) with proper power allocation. ZF is an OMA technique [75]. Different users are divided into different orthogonal subspaces in ZF, which avoids interference among users. It is seen from Fig. 13.13 that, with accurate CSI, ZF can offer very good multi-user gain. Since the gap between ZF and capacity is small, any further gain by NOMA is limited. In this case, OMA via ZF can be preferred for its low-cost SUD receiver.

However, in practice, we usually do not have reliable CSI to establish spatial orthogonality initially. ZF performance deteriorates seriously when CSI is not accurate. In the following, we will see that NOMA via IDMA offers a solution to the problem.

13.5.2 Iterative Maximum Ratio Combining (I-MRC)

We first extend the GA-based detection technique in Sect. 13.2.2 to MIMO. Consider a multi-user uplink system model with $N_{\rm BS}$ antennas at the base station. For simplicity, we assume a single antenna at each user. The IDMA principle discussed in Sect. 13.2.1 can be directly used here. We assume perfect CSI first and will return to the CSI estimation problem later.

The received signal at time j is written as

$$\boldsymbol{y}(j) = \sum_{k=1}^{K} \boldsymbol{h}_k \sqrt{e_k} x_k(j) + \boldsymbol{\eta}(j), \tag{13.33}$$

where $\boldsymbol{y}(j)$ is an $N_{\rm BS} \times 1$ signal vector received at base station antennas, h_k an $N_{\rm BS} \times 1$ complex channel coefficient vector, e_k the transmitted power of user k, $x_k(j)$ a symbol transmitted from user k, and $\boldsymbol{\eta}(j)$ an $N_{\rm BS} \times 1$ vector of complex AWGN with mean 0 and variance $\sigma^2 = N_0/2$ per dimension.

Maximum ratio combining (MRC) is a common strategy for MIMO systems. An MRC estimator is defined in a symbol-by-symbol form as

$$\hat{x}_k(j) = \boldsymbol{h}_k^H \boldsymbol{y}(j). \tag{13.34}$$

Substituting (13.33) into (13.34),

$$\hat{x}_k(j) = \lambda_k x_k(j) + \xi_k(j), \tag{13.35a}$$

where $\lambda_k \equiv \|\boldsymbol{h}_k\|^2 \sqrt{e_k} = \boldsymbol{h}_k^H \boldsymbol{h}_k \sqrt{e_k}$ is a scalar and

$$\xi_k(j) \equiv \sum_{k'=1, k' \neq k}^{K} \boldsymbol{h}_k^H \boldsymbol{h}_{k'} \sqrt{e_{k'}} x_{k'}(j) + \boldsymbol{h}_k^H \boldsymbol{\eta}(j) \tag{13.35b}$$

is an interference (from $\{x_{k'}(j), k' \neq k\}$ to $x_k(j)$) plus noise term. MRC does not involve matrix inversion and so has low cost. However, interference is a problem for MRC, especially when K is large. Iterative GA technique can alleviate this problem. Similar as that in Sect. 13.2.2, we approximate $\xi_k(j)$ in (13.35b) by a Gaussian random variable. We assume that the real and imaginary parts of $x_k(j)$ carry two bits of information in the QPSK modulation. Similar to (13.12), the real part of $x_k(j)$ can be estimated as

$$LLR_{\text{ESE}}\left(\text{Re}(x_k(j))\right) = \frac{2\lambda_k}{\text{Var}\left(\text{Re}\left(\xi_k(j)\right)\right)} \text{Re}\left(\hat{x}_k - \text{E}\left(\xi_k(j)\right)\right). \tag{13.36}$$

The mean and variance in (13.36) are updated as

$$\text{E}\left(\xi_k(j)\right) = \boldsymbol{h}_k{}^H \left(\sum_{k'=1}^{K} \boldsymbol{h}_{k'} \sqrt{e_{k'}} \text{E}(x_{k'}(j)) - \boldsymbol{h}_k \sqrt{e_k} \text{E}(x_k(j)) \right), \tag{13.37a}$$

$$\begin{aligned} &\text{Var}\left(\text{R}e\left(\xi_k(j)\right)\right) \\ &= \sum_{k'=1}^{K} \text{Var}\left(\text{Re}\left(\boldsymbol{h}_k^H \boldsymbol{h}_{k'} \sqrt{e_{k'}} x_{k'}(j)\right)\right) - Var\left(\text{Re}\left(\|\boldsymbol{h}_k\|^2 \sqrt{e_k} x_k(j)\right)\right) + \|\boldsymbol{h}_k\|^2 \sigma^2. \end{aligned} \tag{13.37b}$$

Some detailed computation techniques for (13.37) can be found in [75]. The imaginary part of $x_k(j)$ can be estimated similarly. We call the above process iterative MRC (I-MRC).

Figure 13.14 illustrates the effectiveness of I-MRC [75]. We consider three different settings:

(i) $K = 1$ and sum-rate $R_{\text{sum}} = 5$ with five signal streams (each stream with rate 1) assigned to the sole user using the SCM principle discussed in Sect. 13.2.5,
(ii) $K = 8$ and $R_{\text{sum}} = 16$ with two streams per user, and
(iii) $K = 8$ and $R_{\text{sum}} = 24$ with three streams per user.

For $K = 1$, all the signal streams see the same channel so there is no spatial diversity among them, which results in poor performance. Increasing K from 1 to 8 results in drastically enhanced rate or reduced power or both in Fig. 13.14. Figure 13.14 is a compelling evidence for multi-user gain: allowing more concurrent transmitting users is more efficient than increasing single-user rate.

13.5.3 Data-Aided Channel Estimation (DACE)

We now consider the CSI acquisition problem. Many factors may affect CSI accuracy in MIMO. In particular, the correlation among the pilots used by different users can lead to the pilot contamination problem [78].

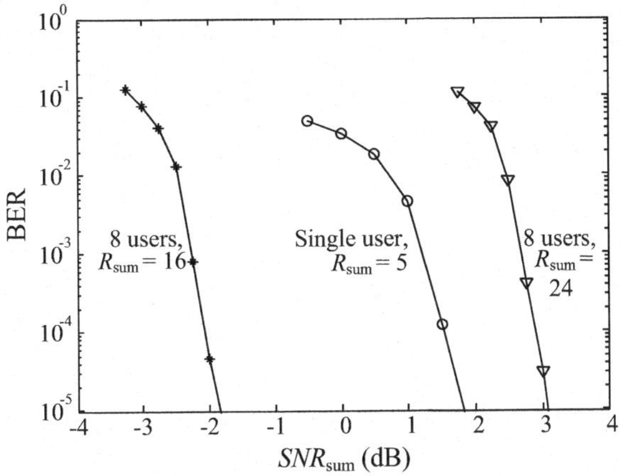

Fig. 13.14 Multi-user gain for $K = 8$ with I-MRC. Rayleigh fading. $N_{BS} = 64$. Equal transmitted power is assumed for different users. Power control is used for the streams assigned to the same user. The power allocation levels are obtained through heuristic search. Rate-1/2 turbo coding and information length = 1200 for each stream. QPSK modulation. A codeword is transmitted over ten resource blocks. Each resource block contains 180 symbols experiencing the same fading conditions

IDMA with data-aided channel estimation (DACE) [17, 79–83] technique can be used to improve CSI accuracy. The basic principle of DACE is as follows. Recall that a key difference between pilot and data is that the former is known at the receiver, while the latter is not. Therefore, if a data symbol is known, it can be used as a pilot. Furthermore, partial information of a data symbol, such as its mathematical mean, can also be used to refine the channel estimates. Such partial information is readily available in an IDMA receiver (as given in (13.8)).

DACE can be used jointly with I-MRC, which involves iterations of the following two operations [75]:

(a) using both pilots and partially decoded data information to refine CSI, and
(b) using improved channel estimates to refine data estimation by I-MRC.

The advantages of DACE are twofolds: (i) With DACE, the estimated data is gradually used for channel estimation. Pilot energy can be greatly reduced since only very coarse CSI is required initially. (ii) Data sequences are typically much longer than pilots and correlation is low among them. Therefore, DACE is robust against the pilot contamination problem. Such problem is typically caused by the correlation among the pilot sequences re-used in neighboring cells. Without DACE, longer pilot sequences will be required to reduce such correlation. Thus, DACE also reduces the time overhead related to pilots.

I-MRC and DACE can be naturally combined in an overall iterative process. After MRC and decoding operations in each iteration, partially detected data are used to refine channel estimates that are in turn used for MRC and decoding in the next iteration. This is referred to as I-MRC-DACE. Figure 13.15 compares BER

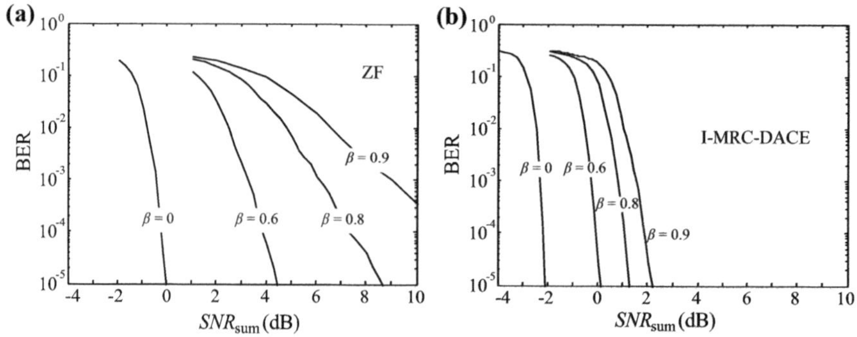

Fig. 13.15 Performance comparison of **a** ZF and **b** I-MRC-DACE with different β values. Rayleigh fading. $N_{BS} = 64$ and $K = 16$. Rate-1/3 turbo coding and information length $= 1312$ for each user. QPSK modulation. Each codeword is divided into 12 sections, and each section is transmitted over a resource block (including 16 pilot symbols). Different users in a cell are assigned different orthogonal pilots. These pilots are repeated for users in different cells. The pilot and data symbols have the same average power

performance for ZF and I-MRC-DACE, in which β is the cross-cell interference factor defined in (13.1). A larger β indicates a more serious pilot contamination problem due to more severe interference among the pilots. From Fig. 13.15, we can see that I-MRC-DACE noticeably outperforms ZF. The difference becomes very significant when β is large (e.g., $\beta \geq 0.6$).

IDMA is a natural choice for I-MRC-DACE since it is beneficial for iterative detection. Note that Fig. 13.5 can also be used to characterize IDMA in MIMO systems, if each scalar $y(j)$ in Fig. 13.5 is replaced by its vector counterpart $\boldsymbol{y}(j)$ in (13.33). The discussions on short cycles in Sect. 13.2.1 are still applicable after such replacement.

IDMA also allows a superimposed pilot scheme that can reduce the power overhead and rate loss. The related discussions can be found in [83–86].

13.6 Prospective Applications of IDMA in 5G Systems

Various approaches have been proposed recently for 5G radio link under LTE, including IDMA [87], RSMA [88], IGMA [89], PDMA [90] and SCMA [91, 92]. In the following, we will show that these schemes all share, explicitly or implicitly, the basic principle of IDMA.

We first represent these different schemes using a unified protograph framework. Assume that N resource blocks (RBs) defined in LTE are available for transmission. We label the observations from these RBs by $\{\boldsymbol{y}^{(1)}, \boldsymbol{y}^{(2)}, \ldots, \boldsymbol{y}^{(N)}\}$.

Figure 13.16 shows a scheme in which each user transmits on all available RBs as illustrated for two system settings: (a) three users over two RBs, and (b) six users

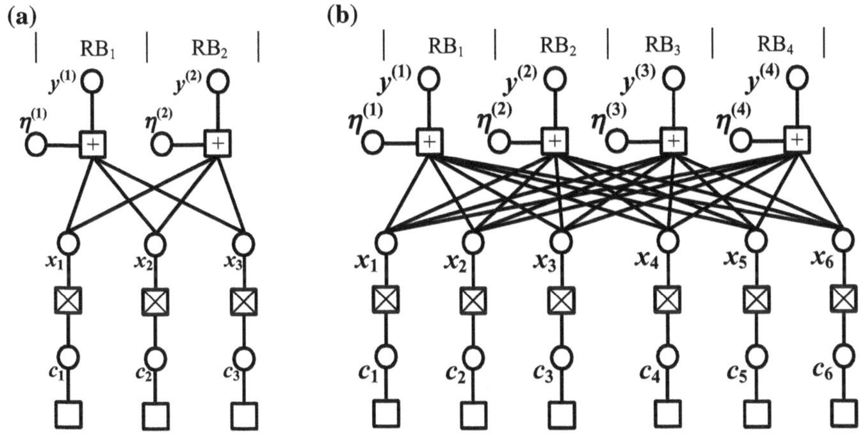

Fig. 13.16 Protograph representations of DS-CDMA and IDMA with **a** three users over two RBs, and **b** six users over four RBs

over four RBs. This can be realized by transmitting replicas of each x_k over multiple RBs. Alternatively, we may use a low-rate code to generate each x_k. Each x_k can be segmented into several blocks, with each block transmitted over an RB. The latter approach can potentially provide higher coding gain [93]. We may also use different modulations for the bits on different RBs, as for SCMA [92].

Incidentally, both DS-CDMA and IDMA can be represented using the protographs in Fig. 13.16. They are distinguished by the absence or presence of user-specific interleaving within each RB. RSMA [88] is a DS-CDMA scheme. However, user-specific interleaving is stated as an option for RSMA. If this option is used, it is equivalent to IDMA. The advantage of this option can be seen in Fig. 13.7.

Alternatively, each user can transmit over only some of the available RBs. This is referred to as sparse coding in [91]. Figure 13.17 shows an example for sparse coding. IGMA, PDMA, and SCMA all involve such treatment. Note that symbol-level interleaving as in Fig. 13.5a is not explicitly seen in Figs. 13.16 and 13.17. If such underlying interleaving is not used, size-4 cycles can be a problem in Fig. 13.16. Sparse coding in Fig. 13.17 avoids this problem. Clearly, sparse coding leads to user-specific edge connections between users and RBs. It has the same effect as symbol-level interleaving; they both reduce short cycles.

With sparse coding, each user does not fully occupy all RBs. This may cause problem for decentralized grant-free [13] or random-access applications, where each user determines its activity individually. In these cases, the number of active users, denoted by K_{active}, is a random variable. When K_{active} is small, sparse coding may lead to inefficient use of the available RBs and so low power efficiency. This implies poor scalability of user numbers. On the other hand, an IDMA system in Fig. 13.16 based on symbol-level interleaving does not have this problem, since all available RBs are fully used for any value of K_{active}.

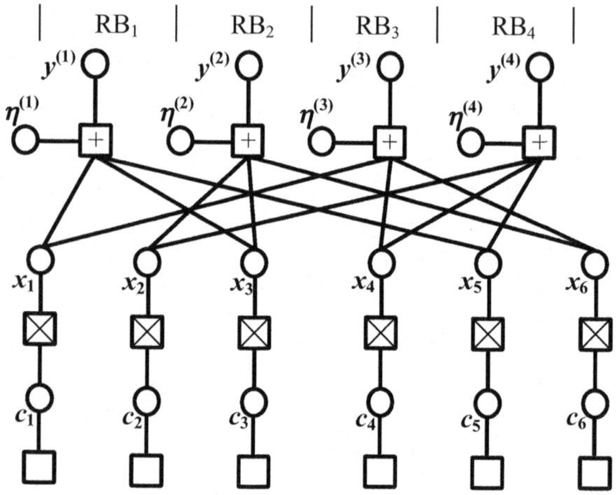

Fig. 13.17 Protograph representation of sparse coding

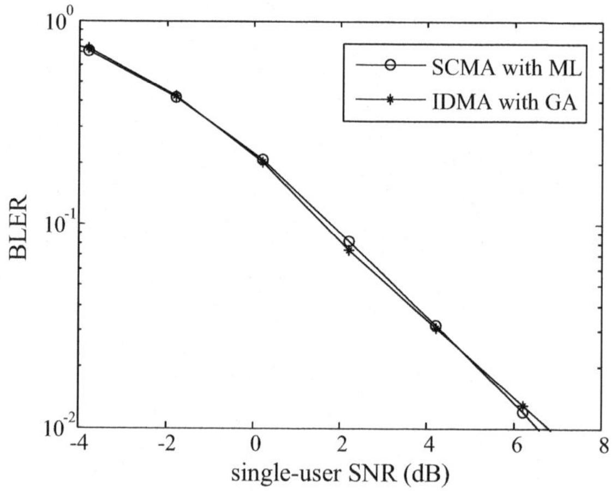

Fig. 13.18 IDMA with GA versus SCMA with ML. Ten iterations for both schemes

Also, multi-user gain in MIMO is determined by the number of users concurrently transmitting in each RB. Therefore, sparse coding may not be an efficient option in MIMO (especially in massive MIMO).

Figure 13.18 compares IDMA and SCMA in quasi-static Rayleigh fading channels. The channels remain unchanged within each transmission. Both schemes are with six users, two receiver antennas, sum-rate = 3 and equal transmitted power per user. A rate-1/2 LTE turbo code is used followed the following transmitter structures:

- IDMA is with a length-2 spreading and QPSK modulation.
- SCMA is based on Fig. 13.17 with the 16-point modulation in [92].

We can see from Fig. 13.18 that the two schemes have similar performance. However, SCMA in Fig. 13.18 is based on ML, while IDMA based on GA. The latter has much lower complexity.

13.7 Summary

We have shown that the real attractiveness of NOMA is in systems without centralized control or without accurate CSI. It is difficult or too costly to establish orthogonality in such channels, so we have to resort to NOMA. Iterative processing holds the key; interference can be gradually resolved and CSI can be gradually refined through iterative processing. IDMA is a simple implementation technique for NOMA. The features of IMDA can be seen from its sparse graphic representation. The interleaved edge connections in IDMA facilitate iterative processing at the receiver. We have demonstrated that IDMA can offer significant performance gain in random access and MIMO systems. IDMA also offers lower detection complexity compare with other alternatives.

References

1. R. Ahlswede, Multi-way communication channels, in *Second International Symposium on Information Theory* (1971), pp. 103–135
2. H. Liao, A coding theorem for multiple access communications, in *Proceedings of the International Symposium on Information Theory*, Asilomar (1972)
3. D. Tse, P. Viswanath, *Fundamentals of Wireless Communication* (Cambridge University Press, 2005)
4. P. Viswanath, D.N.C. Tse, V. Anantharam, Asymptotically optimal water-filling in vector multiple-access channels. IEEE Trans. Inf. Theory **47**(1), 241–267 (2001)
5. W. Yu, W. Rhee, S. Boyd, J.M. Cioffi, Iterative water-filling for Gaussian vector multiple-access channels. IEEE Trans. Inf. Theory **50**(1), 145–152 (2004)
6. M. Kobayashi, G. Caire, An iterative water-filling algorithm for maximum weighted sum-rate of Gaussian MIMO-BC. IEEE J. Sel. Areas Commun. **24**(8), 1640–1646 (2006)
7. R.G. Gallager, An inequality on the capacity region of multiaccess multipath channels, in *Communications and Cryptography: Two Sides of One Tapestry* (Kluwer, Boston, 1994), pp. 129–139
8. R.R. Muller, A. Lampe, J.B. Huber, Gaussian multiple-access channels with weighted energy constraint, in *IEEE Information Theory Workshop*, June 1998, pp. 106–107
9. P. Wang, J. Xiao, L. Ping, Comparison of orthogonal and non-orthogonal approaches to future wireless cellular systems. IEEE Veh. Technol. Mag. **1**(3), 4–11 (2006)
10. P. Wang, L. Ping, On maximum eigenmode beamforming and multi-user gain. IEEE Trans. Inf. Theory **57**(7), 4170–4186 (2011)
11. N. Otao, Y. Kishiyama, K. Higuchi, Performance of non-orthogonal access with SIC in cellular downlink using proportional fair-based resource allocation, in *IEEE ISWCS 2012*, August 2012, pp. 476–480
12. Z. Ding, Y. Liu, J. Choi, Q. Sun, M. Elkashlan, I. Chih-Lin, H.V. Poor, Application of non-orthogonal multiple access in LTE and 5G networks. IEEE Commun. Mag. **55**(2), 185–191 (2017)
13. L. Dai, B. Wang, Y. Yuan, S. Han, I. Chih-Lin, Z. Wang, Non-orthogonal multiple access for 5G: solutions, challenges, opportunities, and future research trends. IEEE Commun. Mag. **53**(9), 74–81 (2015)
14. G. Wunder, P. Jung, M. Kasparick, T. Wild, F. Schaich, Y. Chen, L. Mendes, 5GNOW: non-orthogonal, asynchronous waveforms for future mobile applications. IEEE Commun. Mag. **52**(2), 97–105 (2014)

15. F. Schaich, B. Sayrac, S. Elayoubi, I.P. Belikaidis, M. Caretti, A. Georgakopoulos, B. Mouhouche, FANTASTIC5G: flexible air interface for scalable service delivery within wireless communication networks of the 5th generation. Trans. Emerg. Telecommun. Technol. **27**(9), 1216–1224 (2016)
16. W. Shin, M. Vaezi, B. Lee, D.J. Love, J. Lee, H.V. Poor, Non-orthogonal multiple access in multi-cell networks: theory, performance, and practical challenges. IEEE Commun. Mag. **55**(10), 176183 (2017)
17. L. Ping, L. Liu, K. Wu, W.K. Leung, Interleave division multiple-access. IEEE Trans. Wirel. Commun. **5**(4), 938–947 (2006)
18. P.A. Hoeher, H. Schoeneich, J.C. Fricke, Multi-layer interleave-division multiple access: theory and practice. Trans. Emerg. Telecommun. Technol. **19**(5), 523–536 (2008)
19. T. Yang, J. Yuan, Z. Shi, Rate optimization for IDMA systems with iterative joint multi-user decoding. IEEE Trans. Wirel. Commun. **8**(3), 1148–1153 (2009)
20. K. Kusume, G. Bauch, W. Utschick, IDMA vs. CDMA: analysis and comparison of two multiple access schemes. IEEE Trans. Wirel. Commun. **11**(1), 78–87 (2012)
21. K. Wu, K. Anwar, T. Matsumoto, BICM-ID-based IDMA: convergence and rate region analyses. IEICE Trans. Commun. **97**(7), 1483–1492 (2014)
22. G. Song, J. Cheng, Distance enumerator analysis for interleave-division multi-user codes. IEEE Trans. Inf. Theory **62**(7), 4039–4053 (2016)
23. C. Berrou, A. Glavieux, P. Thitimajshima, Near Shannon limit error-correcting coding and decoding: turbo-codes. 1, in *IEEE ICC'93*, vol. 2, May 1993, pp. 1064–1070
24. S. Benedetto, G. Montorsi, Unveiling turbo codes: some results on parallel concatenated coding schemes. IEEE Trans. Inf. Theory **42**(2), 409–428 (1996)
25. T. Richardson, R. Urbanke, *Modern Coding Theory* (Cambridge University Press, 2008)
26. C. Douillard, M. Jzquel, C. Berrou, D. Electronique, A. Picart, P. Didier, A. Glavieux, Iterative correction of intersymbol interference: turbo-equalization. Trans. Emerg. Telecommun. Technol. **6**(5), 507–511 (1995)
27. X. Wang, H.V. Poor, Iterative (turbo) soft interference cancellation and decoding for coded CDMA. IEEE Trans. Commun. **47**(7), 1046–1061 (1999)
28. M.C. Reed, C.B. Schlegel, P.D. Alexander, J.A. Asenstorfer, Iterative multiuser detection for CDMA with FEC: near-single-user performance. IEEE Trans. Commun. **46**(12), 1693–1699 (1998)
29. F.N. Brannstrom, T.M. Aulin, L.K. Rasmussen, Iterative multi-user detection of trellis code multiple access using a posteriori probabilities, in *IEEE ICC 2001*, vol. 1, June 2001, pp. 11–15
30. M. Moher, An iterative multiuser decoder for near-capacity communications. IEEE Trans. Commun. **46**(7), 870–880 (1998)
31. L. Ping, L. Liu, W.K. Leung, A simple approach to near-optimal multiuser detection: interleave-division multiple-access, in *IEEE WCNC 2003*, vol. 1, March 2003, pp. 391–396
32. F.R. Kschischang, B.J. Frey, H.A. Loeliger, Factor graphs and the sum-product algorithm. IEEE Trans. Inf. Theory **47**(2), 498–519 (2001)
33. Y. Mao, A.H. Banihashemi, A heuristic search for good low-density parity-check codes at short block lengths, in *IEEE ICC 2001*, vol. 1, June 2001, pp. 41–44
34. R. Zhang, L. Xu, S. Chen, L. Hanzo, Repeat accumulate code division multiple access and its hybrid detection, in *IEEE ICC'08*, May 2008, pp. 4790–4794
35. Z. Chenghai, H. Jianhao, The shifting interleaver design based on PN sequence for IDMA systems, in *IEEE FGCN 2007*, vol. 2, December 2007, pp. 279–284
36. M. Noemm, T. Wo, P.A. Hoeher, Multilayer APP detection for IDM. Electron. Lett. **46**(1), 96–97 (2010)
37. D.L. Donoho, A. Maleki, A. Montanari, Message-passing algorithms for compressed sensing. Proc. Natl. Acad. Sci. **106**(45), 18914–18919 (2009)
38. J. Thorpe, Low-density parity-check (LDPC) codes constructed from protographs. IPN Prog. Rep. **42**(154), 42–154 (2003)
39. L. Ping, J. Tong, X. Yuan, Q. Guo, Superposition coded modulation and iterative linear MMSE detection. IEEE J. Sel. Areas Commun. **27**(6) (2009)

40. P.A. Hoeher, T. Wo, Superposition modulation: myths and facts. IEEE Commun. Mag. **49**(12) (2011)
41. R. Laroia, N. Farvardin, S.A. Tretter, On optimal shaping of multidimensional constellations. IEEE Trans. Inf. Theory **40**(4), 1044–1056 (1994)
42. X. Ma, L. Ping, Coded modulation using superimposed binary codes. IEEE Trans. Inf. Theory **50**(12), 3331–3343 (2004)
43. C.H.F. Fung, W. Yu, T.J. Lim, Precoding for the multiantenna downlink: multiuser SNR gap and optimal user ordering. IEEE Trans. Commun. **55**(1), 188–197 (2007)
44. L. Ping, L. Liu, Analysis and design of IDMA systems based on SNR evolution and power allocation, in *IEEE VTC2004-Fall*, vol. 2, September 2004, pp. 1068–1072
45. L. Liu, J. Tong, L. Ping, Analysis and optimization of CDMA systems with chip-level interleavers. IEEE J. Sel. Areas Commun. **24**(1), 141–150 (2006)
46. X. Yuan, L. Ping, C. Xu, A. Kavcic, Achievable rates of MIMO systems with linear precoding and iterative LMMSE detection. IEEE Trans. Inf. Theory **60**(11), 7073–7089 (2014)
47. Y. Zhang, K. Peng, J. Song, Enhanced IDMA with rate-compatible raptor-like quasi-cyclic LDPC code for 5G, in *IEEE GC Workshops 2017* (2017)
48. K. Takeuchi, T. Tanaka, T. Kawabata, Improvement of BP-based CDMA multiuser detection by spatial coupling, in *IEEE ISIT 2011*, July 2011, pp. 1489–1493
49. D. Truhachev, C. Schlegel, Spatially coupled streaming modulation, in *IEEE ICC 2013*, June 2013, pp. 3418–3422
50. S. Kudekar, T. Richardson, R.L. Urbanke, Spatially coupled ensembles universally achieve capacity under belief propagation. IEEE Trans. Inf. Theory **59**(12), 7761–7813 (2013)
51. C. Schlegel, D. Truhachev, Multiple access demodulation in the lifted signal graph with spatial coupling. IEEE Trans. Inf. Theory **59**(4), 2459–2470 (2013)
52. A. Yedla, Y.Y. Jian, P.S. Nguyen, H.D. Pfister, A simple proof of Maxwell saturation for coupled scalar recursions. IEEE Trans. Inf. Theory **60**(11), 6943–6965 (2014)
53. S. Kumar, A.J. Young, N. Macris, H.D. Pfister, Threshold saturation for spatially coupled LDPC and LDGM codes on BMS channels. IEEE Trans. Inf. Theory **60**(12), 7389–7415 (2014)
54. D.J. Costello, L. Dolecek, T. Fuja, J. Kliewer, D. Mitchell, R. Smarandache, Spatially coupled sparse codes on graphs: theory and practice. IEEE Commun. Mag. **52**(7), 168–176 (2014)
55. D.G. Mitchell, M. Lentmaier, D.J. Costello, Spatially coupled LDPC codes constructed from protographs. IEEE Trans. Inf. Theory **61**(9), 4866–4889 (2015)
56. C. Liang, J. Ma, L. Ping, Towards Gaussian capacity, universality and short block length, in *IEEE ISTC 2016*, September 2016, pp. 412–416
57. C. Liang, J. Ma, L. Ping, Compressed FEC codes with spatial-coupling. IEEE Commun. Lett. **21**(5), 987–990 (2017)
58. H.H. Chung, Y.C. Tsai, M.C. Lin, IDMA using non-Gray labelled modulation. IEEE Trans. Commun. **59**(9), 2492–2501 (2011)
59. D.P. Bertsekas, R.G. Gallager, P. Humblet, *Data Networks*, vol. 2 (Prentice-Hall, Englewood Cliffs, NJ, 1987)
60. S. Ghez, S. Verdu, S.C. Schwartz, Stability properties of slotted Aloha with multipacket reception capability. IEEE Trans. Autom. Control **33**(7), 640–649 (1988)
61. M.H. Ngo, V. Krishnamurthy, L. Tong, Optimal channel-aware ALOHA protocol for random access in WLANs with multipacket reception and decentralized channel state information. IEEE Trans. Signal Process. **56**(6), 2575–2588 (2008)
62. C. Xu, P. Wang, S. Chan, L. Ping, Decentralized power control for random access with iterative multi-user detection, in *IEEE ISTC 2012*, August 2012, pp. 11–15
63. C. Xu, L. Ping, P. Wang, S. Chan, X. Lin, Decentralized power control for random access with successive interference cancellation. IEEE J. Sel. Areas Commun. **31**(11), 2387–2396 (2013)
64. H. Lin, K. Ishibashi, W.Y. Shin, T. Fujii, A simple random access scheme with multilevel power allocation. IEEE Commun. Lett. **19**(12), 2118–2121 (2015)
65. M. Zou, S. Chan, H.L. Vu, L. Ping, Throughput improvement of 802.11 networks via randomization of transmission power levels. IEEE Trans. Veh. Technol. **65**(4), 2703–2714 (2016)

66. C. Xu, X. Wang, L. Ping, Random access with massive-antenna arrays, in *IEEE VTC2016-Spring*, May 2016, pp. 1–5
67. Y. Hu, C. Xu, L. Ping, NOMA and IDMA in random access, invited paper, in *IEEE VTC2018-Spring*, June 2018, pp. 1–5
68. J. Mietzner, R. Schober, L. Lampe, W.H. Gerstacker, P.A. Hoeher, Multiple-antenna techniques for wireless communications-a comprehensive literature survey. IEEE Commun. Surv. Tutor. **11**(2) (2009)
69. G. Caire, N. Jindal, M. Kobayashi, N. Ravindran, Multiuser MIMO achievable rates with downlink training and channel state feedback. IEEE Trans. Inf. Theory **56**(6), 2845–2866 (2010)
70. D. Gesbert, S. Hanly, H. Huang, S.S. Shitz, O. Simeone, W. Yu, Multi-cell MIMO cooperative networks: a new look at interference. IEEE J. Sel. Areas Commun. **28**(9), 1380–1408 (2010)
71. F. Rusek, D. Persson, B.K. Lau, E.G. Larsson, T.L. Marzetta, O. Edfors, F. Tufvesson, Scaling up MIMO: opportunities and challenges with very large arrays. IEEE Signal Process. Mag. **30**(1), 40–60 (2013)
72. J. Hoydis, S. Ten Brink, M. Debbah, Massive MIMO in the UL/DL of cellular networks: how many antennas do we need? IEEE J. Sel. Areas Commun. **31**(2), 160–171 (2013)
73. L. Lu, G.Y. Li, A.L. Swindlehurst, A. Ashikhmin, R. Zhang, An overview of massive MIMO: benefits and challenges. IEEE J. Sel. Top. Signal Process. **8**(5), 742–758 (2014)
74. P.A. Hoeher, N. Doose, A massive MIMO terminal concept based on small-size multi-mode antennas. Trans. Emerg. Telecommun. Technol. **28**(2) (2017)
75. C. Xu, Y. Hu, C. Liang, J. Ma, L. Ping, Massive MIMO, non-orthogonal multiple access and interleave division multiple access. IEEE Access **5**, 14728–14748 (2017)
76. S. Vishwanath, N. Jindal, A. Goldsmith, Duality, achievable rates, and sum-rate capacity of Gaussian MIMO broadcast channels. IEEE Trans. Inf. Theory **49**(10), 2658–2668 (2003)
77. W. Yu, Uplink-downlink duality via minimax duality. IEEE Trans. Inf. Theory **52**(2), 361–374 (2006)
78. E.G. Larsson, O. Edfors, F. Tufvesson, T.L. Marzetta, Massive MIMO for next generation wireless systems. IEEE Commun. Mag. **52**(2), 186–195 (2014)
79. H. Schoeneich, P.A. Hoeher, Iterative pilot-layer aided channel estimation with emphasis on interleave-division multiple access systems. EURASIP J. Appl. Signal Process. **2006**, 250–250 (2006)
80. M. Zhao, Z. Shi, M.C. Reed, Iterative turbo channel estimation for OFDM system over rapid dispersive fading channel. IEEE Trans. Wirel. Commun. **7**(8) (2008)
81. C. Novak, G. Matz, F. Hlawatsch, IDMA for the multiuser MIMO-OFDM uplink: a factor graph framework for joint data detection and channel estimation. IEEE Trans. Signal Process. **61**(16), 4051–4066 (2013)
82. J. Ma, L. Ping, Data-aided channel estimation in large antenna systems. IEEE Trans. Signal Process. **62**(12), 3111–3124 (2014)
83. P. Hoeher, F. Tufvesson, Channel estimation with superimposed pilot sequence, in *IEEE GLOBECOM'99*, vol. 4 (1999), pp. 2162–2166
84. A. Ashikhmin, T. Marzetta, Pilot contamination precoding in multi-cell large scale antenna systems, in *IEEE ISIT 2012*, July 2012, pp. 1137–1141
85. J. Ma, C. Liang, C. Xu, L. Ping, On orthogonal and superimposed pilot schemes in massive MIMO NOMA systems. IEEE J. Sel. Areas Commun. **35**(12), 2696–2707 (2017)
86. Y. Chen, Low-cost superimposed pilots based receiver for massive MIMO in multicarrier system, in *IEEE VTC2017-Spring*, June 2017
87. Nokia, Alcatel-Lucent Shanghai Bell, Performance of interleave division multiple access (IDMA) in combination with OFDM family waveforms, document R1-165021, in *3GPP TSG-RAN WG1 #85* (2016)
88. Qualcomm Incorporated, RSMA, document R1-164688, in *3GPP TSG-RAN WG1 #85* (2016)
89. Samsung, Link level performance evaluation for IGMA, document R1-166750, in *3GPP TSG RAN WG1 Meeting #86* (2016)

90. S. Chen, B. Ren, Q. Gao, S. Kang, S. Sun, K. Niu, Pattern division multiple access—a novel nonorthogonal multiple access for fifth-generation radio networks. IEEE Trans. Veh. Technol. **66**(4), 3185–3196 (2017)
91. H. Nikopour, H. Baligh, Sparse code multiple access, in *IEEE PIMRC'13*, September 2013, pp. 332–336
92. HiSilicon, Huawei, LLS results for uplink multiple access document R1-164037, *3GPP TSG RAN WG1 Meeting #85*, May 2016
93. L. Ping, L. Liu, K.Y. Wu, W.K. Leung, Approaching the capacity of multiple access channels using interleaved low-rate codes. IEEE Commun. Lett. **8**(1), 4–6 (2004)

Chapter 14
Pattern Division Multiple Access (PDMA)

Shanzhi Chen, Shaohui Sun, Shaoli Kang and Bin Ren

Generated from former research achievements on successive interference cancellation amenable multiple access (SAMA) [1–5] technology, pattern division multiple access (PDMA) [6–8] was proposed in 2014. It is a type of non-orthogonal multiple access (NOMA) technology based on the principle of the introduced reasonable diversity between multi-user to promote the capacity, which can obtain higher multiuser multiplexing and diversity gain by designing multi-user diversity PDMA Pattern matrix to implement non-orthogonal signals transmission in such domains as time, frequency, code, space and power.

This chapter provides a whole picture of PDMA, including the origination and principle, pattern design, receiver algorithms, performance evaluation, extension design, applications, challenges and trends.

14.1 Origination and Principles of PDMA

As described above, PDMA is originated from SAMA which takes successive interference cancellation (SIC) detection in the receiver. To solve the error propagation problem of SIC, transmitter and receiver joint design is a good solution [8]. Therefore, in this section, based on the explanation of error propagation problem in SIC

S. Chen (✉) · S. Sun · S. Kang · B. Ren
China Academy of Telecommunication Technologies, No. 40,
Xueyuan Road, Haidian District, Beijing, China
e-mail: chensz@datanggroup.cn

S. Sun
e-mail: sunshaohui@catt.cn

S. Kang
e-mail: kangshaoli@catt.cn

B. Ren
e-mail: renbin@catt.cn

M. Vaezi et al. (eds.), *Multiple Access Techniques for 5G Wireless Networks and Beyond*, https://doi.org/10.1007/978-3-319-92090-0_14

and the idea of transmitter receiver joint design, PDMA definition and framework are described, and also PDMA transmitting and receiving schemes are explained.

14.1.1 Error Propagation Problem in SIC

According to theoretical results of multi-user channel [9], superposition coding at a transmitter and SIC at a receiver, are able to achieve capacity boundary of multiple access channels (MAC) or degraded broadcast channels (BC) when transmitter and receiver are working together. From theoretical perspective it is rational to use SIC to achieve channel capacity, since the packet error rate tends to be zero with the increased code length as long as a user's transmission rate is below the channel capacity. However, in a real system, detection error is inevitable due to various non-ideal conditions, such as, limited code length, channel fading, and glitches, etc.

For SIC receiver, if a former user's packet is detected erroneously, it is very unlikely that the following user's packet could be detected correctly. This is the so-called error propagation problem. Since multiple users are detected one by one in serial order, the detection order of all users is usually arranged according to their signal strength. That is, the signal of the first detected user is the strongest, the signal of the second detected user is weaker, and so on. For the first detected user, it is recovered directly from the original receiving. While for the following detected users, they are recovered respectively from related cancellation receiving which should cancel those former detected users from the original receiving by user reconstruction. If a user is not correctly detected, its reconstruction is impossible to be correct. In addition, the accuracy of reconstruction also impacts on the performance of following users. For example, based on distorted channel estimation, the reconstructed signal will also be distorted. Even though the user's packet is detected correctly, it still has an adverse effect on the following users' detection.

Error propagation is a crushing blow for multi-user detection and it will deteriorate the performance of SIC-based multi-user system. In general, two approaches can be considered to alleviate the error propagation problem. The first is to enhance the reliability of those early-decoded users, the second is to adopt more advanced and complex detection algorithm than SIC. These approaches relate to joint design between transmitter and receiver, which is the origination of PDMA technology.

14.1.2 Transmitter and Receiver Joint Design

One approach to alleviate error propagation problem is to enhance the reliability of those early-decoded users, either by selecting users with good channel condition or by designing transmission parameters such that the early-decoded users have higher reliability and better channel condition.

Analytical results of multiple-input and multiple-output (MIMO) detection from [10, 11] show that the ith detected layers of SIC receiver could achieve diversity order

$$N_{div}(i) = N_R - N_T + i \tag{14.1}$$

where N_R is the receiving antenna number, N_T is number of data layers.

The diversity order increases with the detection order. PDMA design is inspired by above result [1–8]. Multi-user channel can be viewed as a virtual MIMO channel and the above result could be generalized to multi-user non-orthogonal transmission. For non-orthogonal transmission employing SIC receiver, diversity order of each user varies with the order of detection. The first detected user has the lowest diversity order, and the last detected user has the highest diversity order. In a fading channel, diversity order affects transmission reliability significantly. Increasing the diversity order typically leads to more reliable transmission. With SIC receiver, the first detected user actually determines the overall detection performance, but unfortunately its diversity order is the lowest. To optimize system performance, it is desirable to have identical pro-detection diversity order for each user.

Diversity could be obtained from transmission or reception, or from both. Assuming that transmission diversity order of the ith detected user is $D_T(i)$, the diversity order after SIC receiver can be expressed as

$$N_{div}(i) = D_T(i) - K + i \tag{14.2}$$

where K is the number of users. By joint design from transmitter and receiver, PDMA deliberately selects $D_T(i)$ so that the diversity order after SIC receiver is as close as possible.

The definition of transmission diversity means that multiple copies of a signal are transmitted from independent resources to avoid transmission error due to deep fading on one resource. The resources could be time, frequency, code, spatial or power resource.

PDMA maps transmitted data onto a group of resources according to PDMA pattern to realize disparate transmission diversity order. A PDMA pattern defines the mapping from transmitted data to a resource group. A resource group can consist of time resource, frequency resource, code resource, spatial resource, power resource or any combination of these resources. The number of mapped resources in a group determines the order of transmission diversity. Data of multiple users can be multiplexed onto the same resource group with different PDMA patterns. In this way, non-orthogonal transmission is realized. By assigning PDMA pattern with different diversity order, disparate transmission diversity orders among users could be achieved.

Another approach to alleviate error propagation problem is to adopt more advanced and complex detection algorithm such as maximum likelihood (ML) or maximum a posterior (MAP). It is anticipated that PDMA with advanced detection algorithm can alleviate error propagation effect to a substantial degree. However, ML or MAP algorithm incur tremendous detection complexity and it is hard to implement.

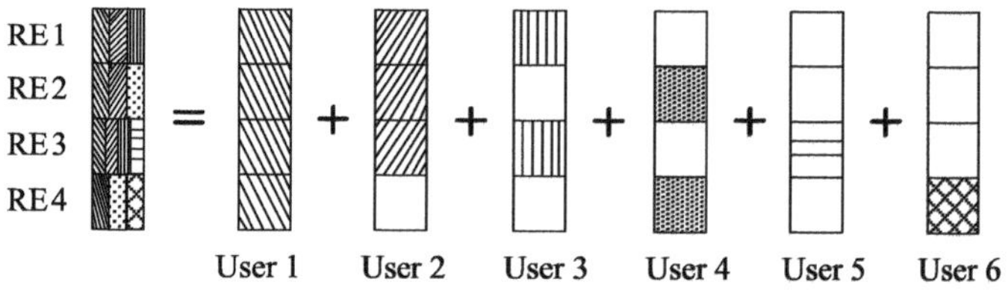

Fig. 14.1 PDMA pattern for 6 users on 4 REs

Fortunately, the detection complexity could be reduced significantly by making the PDMA pattern sparse. That is, data are only mapped to a small part of the resources in the resource group. This draws on the idea of sparse coding in low density parity check (LDPC) coding. Sparsity makes it possible to use low complexity belief propagation (BP) algorithm to approach the MAP detection. In addition, the convergence of BP algorithm could be speeded up by disparate transmission diversity of PDMA.

In summary, PDMA uses PDMA pattern to define sparse mapping from data to a group of resources. PDMA pattern could be represented by a binary vector. The dimension of the vector equals to number of resources in a group. Each element in the vector corresponds to a resource in a resource group. A "1" means that data shall be mapped to the corresponding resource. Actually, number of "1" in the PDMA pattern is defined as its transmission diversity order.

Figure 14.1 shows an example of resource mapping according to PDMA pattern. Six users are multiplexed on four resource elements (REs). A PDMA pattern is assigned to a user. User1's data are mapped to all four REs in the group, and user2's data are mapped to the first three REs, etc. The order of transmission diversity of the six users is 4, 3, 2, 2, 1, and 1 respectively.

14.1.3 PDMA Definition and Framework

PDMA is proposed as a novel NOMA scheme based on code pattern. Joint optimization of transmitting and receiving is considered with SIC amenable pattern design at the transmitter side and SIC-based detection at the receiver side. PDMA pattern is designed to offer different orders of transmission diversity, so that disparate diversity order between multiple users could be introduced to alleviate the error propagation problem of SIC receiver. PDMA pattern is also required to be sparse to facilitate advanced detection algorithm such as BP. Iterations between BP and channel decoding could further boost system performance. PDMA pattern can be also extended to include power scaling and phase shifting to harvest additional constellation shaping gain.

PDMA can design pattern for a specific user in time, frequency and space resources. Figure 14.2 shows the technical framework of the PDMA uplink application, and Fig. 14.3 shows that of the PDMA downlink application. As shown

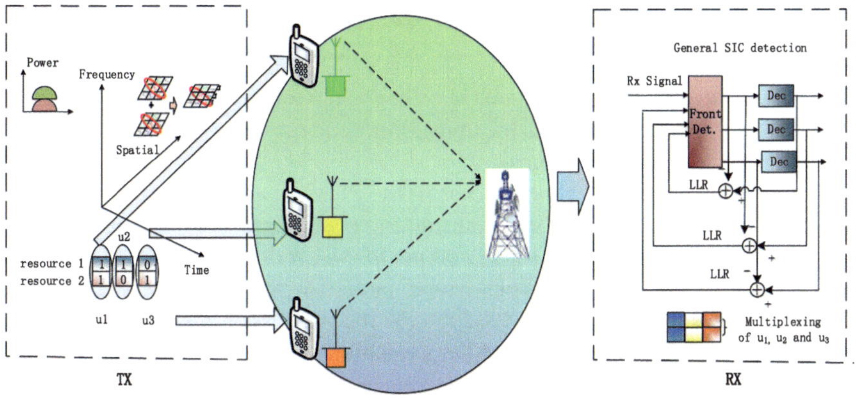

Fig. 14.2 The technical framework of the PDMA uplink application

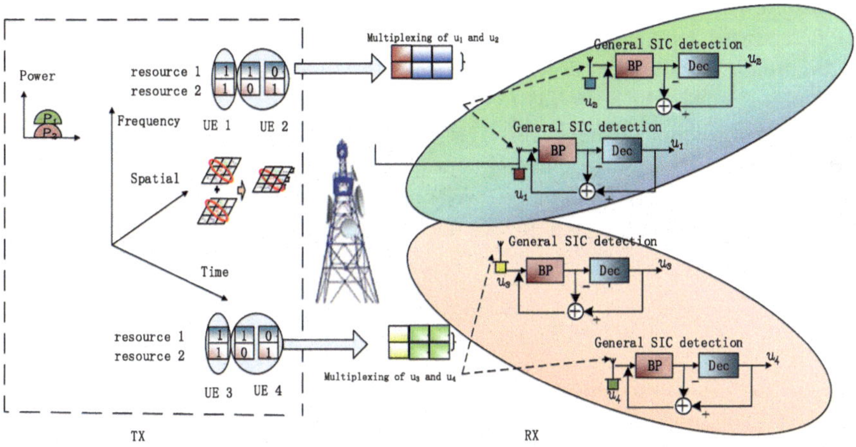

Fig. 14.3 The technical framework of the PDMA downlink application

in Figs. 14.2 and 14.3, the PDMA technical framework includes two parts: the transmitter and the receiver, which reflects that the PDMA technology considers the joint design of the transmitter and the receiver based on the optimization point of view for multi-user communication system. At transmitter side, users are distinguished by the non-orthogonal characteristic pattern based on the multiple signals domain (including time, frequency, code, power and the space domain, etc.). At the receiver side, general SIC type sub-optimal multiuser detection algorithms can be realized based on the features of the user pattern.

14.1.4 PDMA Transmitting and Receiving Scheme

With above framework of PDMA, taking orthogonal frequency division multiplexing (OFDM) waveform as a baseline, the transmitting and receiving schemes of a PDMA based system are further explained.

Figure 14.4 shows the uplink process of the PDMA based system. At the transmitting end, the system completes transmitting signal processing by multiple user data forward error correction channel coding, PDMA code modulation, PDMA subcarrier resource mapping and OFDM modulation. At the receiving end, the base station performs the opposite process, namely the system gets the transmitting data of each terminal through the OFDM demodulation and general SIC type detection like belief propagation iterative detection and decoding (BP-IDD). In the process of the PDMA modulation and coding, the symbol level mapping and spread spectrum in the frequency domain are achieved at the same time. The receiver adopts the BP-IDD algorithm which is essentially a joint iterative processing of the Turbo decoder and BP detecting.

Figure 14.5 shows the downlink process of the PDMA based system. At the transmitting end of the downlink PDMA system, the base station performs data forward error correction channel coding for multiple user like PDMA code modulation, PDMA subcarrier resource mapping and OFDM modulation. At the receiving end of the downlink PDMA system, each user performs the opposite process, including

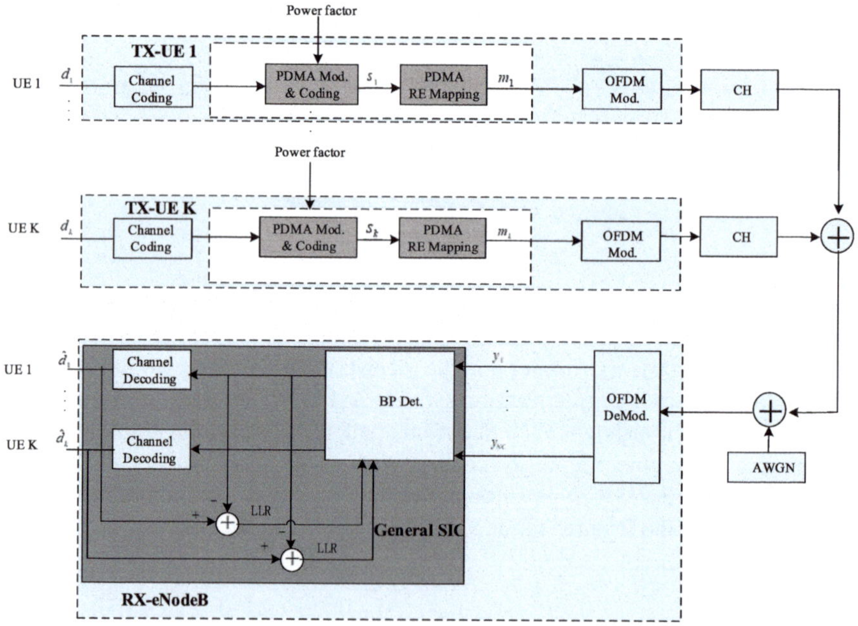

Fig. 14.4 Illustration of the transmitting and receiving scheme of the PDMA based uplink system

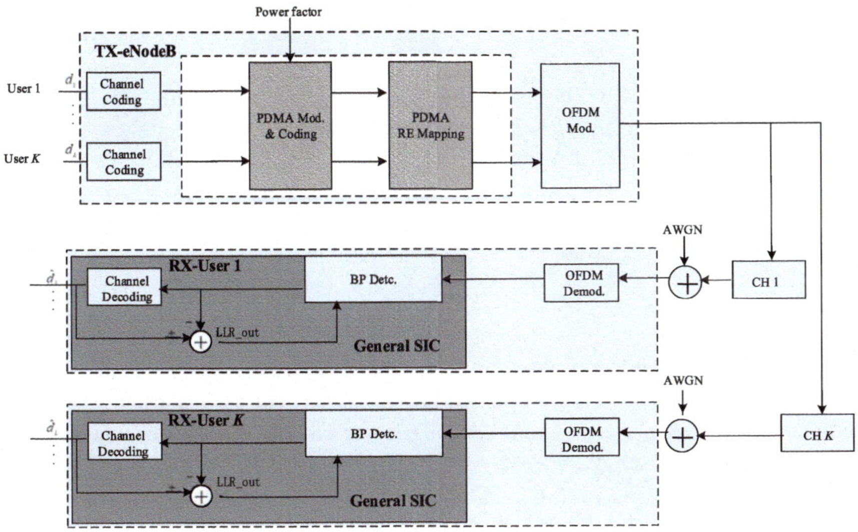

Fig. 14.5 Illustration of the transmitting and receiving schemes of the PDMA based downlink system

OFDM demodulation and general SIC type detection like BP-IDD. In this process, the downlink PDMA coding is conducted on modulation symbol level and finishes mapping on symbol level and realizes spread spectrum in frequency domain. The receiver adopts the BP-IDD algorithm which is essentially the Turbo decoding and a joint iterative processing of BP detecting.

14.2 Pattern Design of PDMA

The PDMA pattern defines the rule of mapping data to the radio resource, which can be defined as a binary vector. Each binary vector represents the PDMA pattern of one user equipment (UE). The dimension of the vector equals to the number of resources in a group. Those patterns with UEs sharing the same set of resources are arranged together to constitute the PDMA pattern matrix. Overload factor (OF) is defined as the ratio of the number of columns to the number of rows in a PDMA pattern matrix. It reflects the excessive number of UEs multiplexed on the same resources of PDMA relative to that of orthogonal multiple access (OMA) scheme. Taking resource number $N = 4$ and user number $K = 8$ as an example, the OF is then $\alpha = K/N = 200\%$, which means that PDMA supports two times the number of UEs compared with that of OMA. Properties of PDMA pattern matrix such as dimension and level of sparsity contribute to both receiver complexity and system performance.

14.2.1 Basic Pattern Matrix

Without loss of generality, both transmitter and receiver are assumed using single antenna, K UEs map onto N REs in the domains of time and frequency, in which each UE has a unique PDMA pattern. The PDMA received signal on the resource group composed by N REs at the base station (BS) is expressed as:

$$\boldsymbol{y} = \sum_{k=1}^{K} diag(\boldsymbol{h}_k)\boldsymbol{g}_k x_k + \boldsymbol{n} = \boldsymbol{H}\boldsymbol{x} + \boldsymbol{n} \tag{14.3}$$

where $\boldsymbol{y}$ denotes a vector composed by received signal on the N resources, with length N; $\boldsymbol{x} = [x_1, x_2, \ldots, x_K]^T$ represents modulated signal vector transmitted by K UEs, with length K, and x_k are the modulated signals of kth UE; $\boldsymbol{n}$ indicates the Gaussian noise vector with length N, where $\boldsymbol{n} \sim CN(0, N_0 \boldsymbol{I}_{N\times N})$; $\boldsymbol{H} = \boldsymbol{H}_{CH} \bullet \boldsymbol{G}_{PDMA}^{[N,K]}$ and $\boldsymbol{H}_{CH} = [\boldsymbol{h}_1, \boldsymbol{h}_2, \ldots, \boldsymbol{h}_K]$ are the PDMA equivalent channel response matrix and original channel response matrix of K UEs multiplexed on N REs, respectively and both have dimensions of $N \times K$, $\boldsymbol{h}_K$ is the uplink channel response of the kth UE with length N, diag($\boldsymbol{h}_k$) represents a diagonal matrix with elements from $\boldsymbol{h}_k$, the (n, k) elements of $\boldsymbol{H}_{CH}$ is the channel response from the kth UE to the BS on the nth RE, and $\bullet$ denotes element-wise product of two matrices; $\boldsymbol{G}_{PDMA}^{[N,K]}$ denotes a PDMA pattern matrix with the dimensions of $N \times K$, where $\boldsymbol{g}_k$ is the PDMA pattern used by the kth UE, corresponding to the kth column of $\boldsymbol{G}_{PDMA}^{[N,K]}$.

Given a certain overload factor, there are a number of pattern matrices available, as long as resource number N and user number K are selected properly. For example, overload factor of 150% could be achieved by a 2×3 pattern matrix, i.e., 3 users are multiplexed on 2 REs. The pattern matrix is:

$$\boldsymbol{G}_{PDMA}^{[2,3]} = \begin{bmatrix} 1 & 1 & 0 \\ 1 & 0 & 1 \end{bmatrix}$$

and another design for 150% overload is 4×6 pattern matrix:

$$\boldsymbol{G}_{PDMA}^{[4,6]} = \begin{bmatrix} 1 & 0 & 1 & 1 & 1 & 0 \\ 1 & 1 & 0 & 1 & 0 & 1 \\ 1 & 1 & 1 & 0 & 1 & 0 \\ 0 & 1 & 1 & 0 & 0 & 1 \end{bmatrix}$$

Though both pattern matrices having the same overload factor, $\boldsymbol{G}_{PDMA}^{[4,6]}$ can achieve better performance while the cost of detection complexity is higher comparing to that of $\boldsymbol{G}_{PDMA}^{[2,3]}$.

PDMA pattern matrices with different dimensions are able to achieve a given overload factor. With a higher dimension, detection complexity is also higher, and better performance is expected. Given an overload factor, the dimension of pattern matrix shall be selected to reach a tradeoff between detection complexity and system performance.

If N is the size of resource group (row number of PDMA pattern matrix), there are $2^N - 1$ possible binary vectors for a pattern matrix. Assuming K is the column number determined based on overload factor, we can thus choose K patterns out from $2^N - 1$ candidates to construct PDMA pattern matrix.

Selection of patterns also gives impacts on performance and complexity:

(1) A pattern with heavier weight (number of "1" elements in the pattern) provides higher diversity order. More reliable data transmission can be anticipated, and detection complexity is also increased. If the system can conduct complex computation, patterns with heavy weight will be preferable; otherwise, light weight patterns have to be selected, aiming at sparse PDMA pattern matrix.

(2) According to the design principle of PDMA, it is desirable to have different diversity orders in the pattern matrix to alleviate error propagation problem of SIC receiver or fasten convergence of BP receiver. Thus the selected patterns shall have as many different diversity orders as possible.

(3) For patterns with identical diversity order, smaller inner product between the patterns leads to less interference against each other. Small inner product means that the two patterns have less "1" elements in common positions. That is, the number of REs shared by the two patterns is low. Data of two users are multiplexed on only few REs. For example, if two patterns have inner product of 0, the two patterns actually map data onto a different set of REs, hence there is no interference between the two patterns. For a given diversity order, the selected patterns shall minimize the maximum inner product between any two patterns. Of course this rule is also applied to patterns with different diversity order.

The design of pattern matrix shall take overload factor, diversity order and detection complexity into account. A good pattern matrix can reach good trade-off among these aspects.

14.2.2 Pattern Optimization Method

Taking PDMA pattern matrix $\boldsymbol{G}_{PDMA}^{[2,3]}$ as an example, data of 3 users are mapped onto two REs. The transmission signal on these REs can be expressed as:

$$\begin{bmatrix} v_1 \\ v_2 \end{bmatrix} = \begin{bmatrix} 1 & 1 & 0 \\ 1 & 0 & 1 \end{bmatrix} \begin{bmatrix} x_1 \\ x_2 \\ x_3 \end{bmatrix} \tag{14.4}$$

where v_j is the transmission signal on the jth RE, and x_k is the modulation symbol of the k th user. Unlike orthogonal transmission, transmission signal on each RE is linear combination of multiple modulation symbols:

$$\begin{aligned} v_1 &= x_1 + x_2 \\ v_2 &= x_1 + x_3. \end{aligned} \tag{14.5}$$

This combination may alter the characteristics of transmission signal on each RE. For example, if all three users adopt binary phase shift eying (BPSK) modulation, the modulation symbol of user 1, user 2 and user 3 is either +1 or −1. The combined transmission signal takes value from −2, 0, +2. Assuming a noiseless channel, if the receiver receives −2, or +2 on a RE, then the receiver can infer that the transmitted symbols on the RE is [−1, −1] or [+1, +1]. But if 0 is received, it is impossible for the receiver to recover the transmitted symbols, as both [+1, −1] and [−1, +1] resulting in the same output. Furthermore, if each user adopts quadrature phase shift keying (QPSK) or 16QAM modulation, the combined constellation consists of 9 or 49 constellation points.

From the above discussions, we can see that the combined constellation has non-uniform distribution and it is no longer one-to-one map between a constellation point and an input user data, i.e., the combination leads to ambiguity.

To resolve the ambiguity, power scaling and phase shifting can be introduced in PDMA pattern matrix. Specifically, before two users symbols are mixed, a power scaling factor and a phase shifting factor shall be applied:

$$v = \sqrt{\beta} x_1 e^{j\varphi} + \sqrt{1-\beta} x_2 \tag{14.6}$$

where β is power scaling factor and φ is phase shifting factor.

As an example, by setting $\varphi = \pi/4$ and $\beta = 0.5$, i.e., only phase shifting difference is introduced between users, the combined constellation is shown in Fig. 14.6. It can be observed that by adding a phase shifting factor, the ambiguity is resolved. Moreover, the distribution of combined constellation is closer to the Gaussian distribution. It is known that, Gaussian distribution is the capacity maximizing input distribution for additive white Gaussian noise (AWGN) channel. That is, by

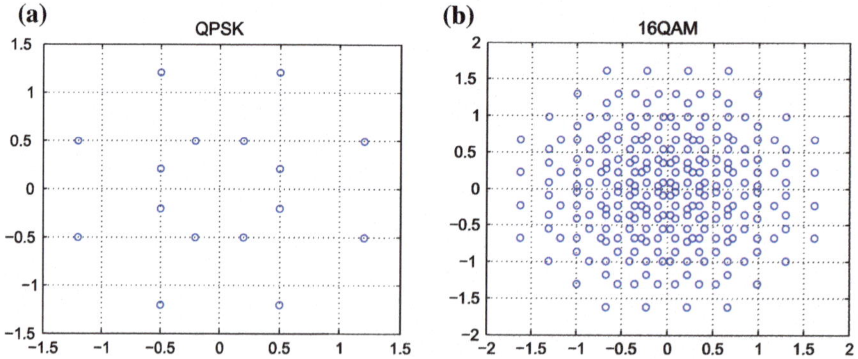

Fig. 14.6 Combined constellation from two users by phase shifting $\varphi = \pi/4$. **a** QPSK. **b** 16QAM

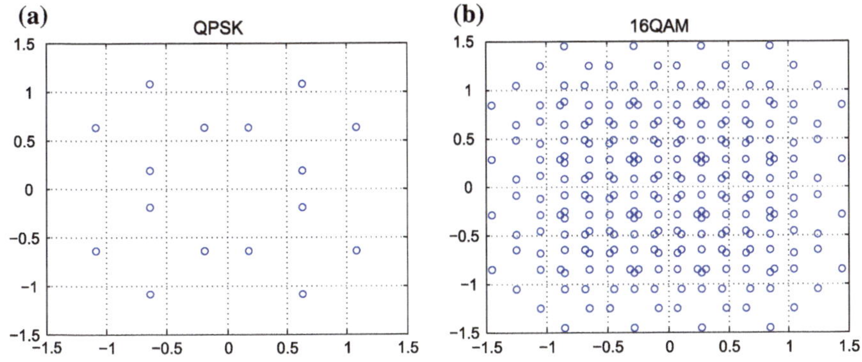

Fig. 14.7 Combined constellation from two users by power scaling and phase shifting $\beta = 0.8$ $\varphi = \pi/4$. **a** QPSK. **b** 16QAM

introducing a phase shifting factor, channel capacity gain is reaped. The gain is also called shaping gain.

When both power scaling and phase shifting are introduced, the effect is exemplified in Fig. 14.7. As expected, the shape of the constellation is changed, and it also approaches a Gaussian distribution.

For PDMA, the power scaling and phase shifting can be naturally incorporated into PDMA pattern matrix. That is, the value "1" in PDMA pattern matrix is substituted with a complex value reflecting both power scaling and phase shifting, forming an extended PDMA pattern matrix. For the PDMA pattern matrix $\boldsymbol{G}_{PDMA}^{[2,3]}$, the extended PDMA pattern matrix is given as

$$\boldsymbol{G}_{E-PDMA}^{[2,3]} = \begin{bmatrix} \alpha_{11}e^{-j\varphi_{11}} & \alpha_{21}e^{-j\varphi_{21}} & 0 \\ \alpha_{12}e^{-j\varphi_{12}} & 0 & \alpha_{32}e^{-j\varphi_{32}} \end{bmatrix} \tag{14.7}$$

where α_{kj} and φ_{kj} is the power scaling and phase shifting factor of the kth user on the jth RE.

The optimal value of power scaling and phase shifting depend on the number of users and the shape of input constellation.

14.2.3 PDMA Pattern for 5G eMBB Scenario

For 5G enhanced mobile broadband (eMBB) scenario, the key target is spectrum efficiency and experienced or peak data rate. Except for massive MIMO technology, PDMA can also be served as an efficient method. To get high performance, BP-IDD [12, 13] algorithm is preferred at the receiver. Since the complexity of BP-IDD algorithm increases exponentially with maximum row weight of PDMA pattern matrix, to reduce computation complexity at the receiver, the PDMA pattern should be designed with the sparse feature at the transmitter. Further, as the available frequency resource for eMBB is large, which provides sufficient frequency diversity, discrete subcarrier mapping can thus be adopted.

References [8, 14] proposed the criteria of maximum constellation constrained capacity (CC-Capacity) to design PDMA pattern matrix for eMBB scenario. That is, with the input information on matrix dimension and its row weight, all candidate sets of PDMA pattern matrix are calculated by CC-capacity, then the PDMA pattern matrix with maximum CC-capacity is selected.

$$\begin{aligned} &\boldsymbol{G}_{opt}^{[N,K]} = argmax\{C(N, K, \Omega)||G^{[N,K]} \subset G^{[N,M]}\}, \\ &st.\|\boldsymbol{G}^{[N,K]}(:, k)\|^2 = 1(k = 1, 2, \ldots, K), \boldsymbol{x} \in \Omega^{K\times 1} \end{aligned} \tag{14.8}$$

where $C(N, K, \Omega)$ denotes CC-Capacity for the parameter N, K and Ω, Ω denotes constellation set, $M = 2^N - 1$ and $\boldsymbol{x}$ is defined in (14.3).

As an example, assuming the matrix dimension is 4×6, the calculation of $C(N, K, \Omega)$ is as follows.

$$\begin{aligned} &C(4, 6, \Omega_{QPSK}) \\ &= I(\boldsymbol{g}_1x_1; \boldsymbol{y}) + I(\boldsymbol{g}_2x_2; \boldsymbol{y}|\boldsymbol{g}_1x_1) + I(\boldsymbol{g}_3x_3; \boldsymbol{y}|\boldsymbol{g}_1x_1, \boldsymbol{g}_2x_2) \\ &+ I(\boldsymbol{g}_4x_4; \boldsymbol{y}|\boldsymbol{g}_1x_1, \boldsymbol{g}_2x_2, \boldsymbol{g}_3x_3) + I(\boldsymbol{g}_5x_5; \boldsymbol{y}|\boldsymbol{g}_1x_1, \boldsymbol{g}_2x_2, \boldsymbol{g}_3x_3, \boldsymbol{g}_4x_4) \\ &+ I(\boldsymbol{g}_6x_6; \boldsymbol{y}|\boldsymbol{g}_1x_1, \boldsymbol{g}_2x_2, \boldsymbol{g}_3x_3, \boldsymbol{g}_4x_4, \boldsymbol{g}_5)x_5 \end{aligned} \tag{14.9}$$

where $I(.)$ represents the mutual information between the input QPSK symbol and the output of a Gaussian channel [15], and $I(\boldsymbol{g}_2x_2; \boldsymbol{y}|\boldsymbol{g}_1x_1)$ denotes the conditional mutual information between $\boldsymbol{g}_2x_2$ and $\boldsymbol{y}$ with the given value of $\boldsymbol{g}_1x_1$, where $\boldsymbol{y}$ and $\boldsymbol{g}_k$ are defined in (14.3). $I(.)$ in (14.9) can be calculated by adopting Monte Carlo integration method. Considering different row weights, the selected PDMA pattern matrix are expressed as following

$$\text{row weight 2:} \quad \boldsymbol{G}_{PDMA}^{[4,6]} = \begin{bmatrix} 1\,0\,1\,0\,0\,0 \\ 1\,0\,0\,1\,0\,0 \\ 0\,1\,0\,0\,1\,0 \\ 0\,1\,0\,0\,0\,1 \end{bmatrix}$$

$$\text{row weight 3:} \quad \boldsymbol{G}_{PDMA}^{[4,6]} = \begin{bmatrix} 1\,1\,1\,0\,0\,0 \\ 1\,0\,0\,1\,1\,0 \\ 0\,1\,0\,1\,0\,1 \\ 0\,0\,1\,0\,1\,1 \end{bmatrix}$$

As another example, in case of overload factor 200%, assuming selecting the row weight 4, the finally selected PDMA pattern matrix with dimension 3×6 is expressed as

$$\boldsymbol{G}_{PDMA}^{[3,6]} = \begin{bmatrix} 1\,1\,1\,0\,1\,0 \\ 1\,1\,0\,1\,0\,1 \\ 1\,0\,1\,1\,0\,0 \end{bmatrix}$$

And the finally selected PDMA pattern matrix with dimension 4×8 is expressed as

$$\boldsymbol{G}_{PDMA}^{[4,8]} = \begin{bmatrix} 1\,1\,1\,1\,0\,0\,0\,0 \\ 1\,1\,0\,0\,1\,1\,0\,0 \\ 1\,0\,1\,0\,1\,0\,1\,0 \\ 0\,0\,0\,1\,0\,1\,1\,1 \end{bmatrix}$$

14.2.4 PDMA Pattern for 5G mMTC Scenario

For 5G massive machine type communication (mMTC) scenario, the key target is supporting huge connections like 1 million connections per square meter. PDMA can be served as a good solution to promote the connections. To satisfy further requirements on low power consumption and low cost, usually low complexity minimum mean square error codeword level SIC (MMSE-CW-SIC) algorithm is recommended for the receiver of mMTC, so that the PDMA pattern matrix at the transmitter should be designed with the feature of low cross-correlation to improve the receiver performance. Since the available radio resource in frequency domain for typical mMTC application is too small to provide sufficient frequency diversity, localized subcarrier mapping is adopted to keep the frequency channel response unchanged within a PDMA pattern.

Assume Q is the number of possible value of a PDMA pattern $\boldsymbol{g}_i$, there are Q^N possible PDMA patterns for a PDMA pattern matrix. For a given dimension $[N, K]$, K PDMA patterns can be selected from Q^N candidates to construct a PDMA pattern matrix, where the number of combinations is $C_{Q^N}^K = (Q^N!)/((Q^N - K)!(K!))$. When N or Q becomes larger, $C_{Q^N}^K$ is too large to get the optimal PDMA pattern matrix. In [14], a suboptimal method with the criteria of minimizing sum squared correlation (SSC) combined with grant free scheme is proposed to design PDMA pattern matrix for mMTC scenario. The first step is to search a candidate set $\boldsymbol{G}_{candidate}^{[N,M]}$ in which the correlation coefficient ρ between any two PDMA patterns is lower than a predefined threshold ρ_{th}, i.e., $\rho \le \rho_{th}$. Then in the second step, the base station further searches PDMA pattern matrix $\boldsymbol{G}_{opt}^{[N,K]}$ which satisfies the minimizing SSC criterion given in (14.10) from $\boldsymbol{G}_{candidate}^{[N,M]}$, and sends the indicator of preferred $\boldsymbol{G}_{opt}^{[N,K]}$ to the UEs.

$$\boldsymbol{G}_{opt}^{[N,K]} = \arg\min\{SSC|\boldsymbol{g}_i, \boldsymbol{g}_j \in \boldsymbol{G}_{candidate}^{[N,M]} \tag{14.10}$$

$$SSC = \sum_{i=1}^{K} \sum_{j=1, i \neq j}^{K} abs(\boldsymbol{g}_i^H \boldsymbol{g}_j)^2 \tag{14.11}$$

where SSC means sum squared correlation, $abs(\boldsymbol{g}_i^H \boldsymbol{g}_j)^2, i \neq j$ represents the square of correlation coefficient between PDMA pattern pair $\boldsymbol{g}_i$ and $\boldsymbol{g}_j$, $abs(.)^2$ denotes the square of absolute value, $\boldsymbol{g}_i^H$ is the conjugate transpose of PDMA pattern $\boldsymbol{g}_i$.

For example, with resource number $N = 4$, and overload factor 150 and 200%, the preferred PDMA pattern matrices are expressed as

$$\boldsymbol{G}^{[4,6]} = \begin{bmatrix} 1 & 0 & 0 & 0.577 & -0.577 & 0 \\ 0 & 1 & 0 & 0.577 & 0.577 & 0 \\ 0 & 0 & 1 & 0.577 & 0.577 & 0 \\ 0 & 0 & 0 & 0 & 0 & 1 \end{bmatrix}$$

$$\boldsymbol{G}^{[4,8]} = \begin{bmatrix} 1 & 0 & 0 & 0.577 & -0.289 + 0.5i & -0.289 - 0.5i & -0.289 + 0.5i & 0 \\ 0 & 1 & 0 & 0.577 & 0.577 & 0.577 & 0.577 & 0 \\ 0 & 0 & 1 & 0.577 & 0.577 & 0.577 & 0.577 & 0 \\ 0 & 0 & 0 & 0 & 0 & 0 & 0 & 1 \end{bmatrix}$$

14.3 Receiver Algorithms of PDMA

Detection algorithm at receiver is the key to reap performance gain of PDMA in uplink and downlink. This section is dedicated to the details of advanced algorithms suitable for PDMA, eg., successive interference cancellation (SIC), belief propagation (BP), belief propagation with iterative detection and decoding (BP-IDD) and expectation propagation (EP). By arranging diversity order of PDMA patterns, error propagation problem of SIC receiver could be alleviated to a certain extent. Sparsity of PDMA pattern greatly reduces the complexity of BP and BP-IDD algorithm, making them suitable for PDMA system. Also, EP was designed for PDMA pattern to speed up convergence of BP and BP-IDD. However, even with BP algorithm and combined detection algorithm of BP and SIC, the receiver still has exponential complexity with respect to modulation order and PDMA pattern weight, which may become difficult for the implementation. To further reduce complexity, EP algorithm is proposed.

14.3.1 Successive Interference Cancellation

As shown in Fig. 14.8, the basic idea of SIC receiver is to reconstruct a user's signal and then subtract it from the received signal. The construction could be carried out either at symbol level or codeword level. For the symbol level SIC (SL-SIC), the construction is made from the demodulated symbols. Instead, the codeword level SIC (CW-SIC) is based on signal construction from decoded data bits. As channel

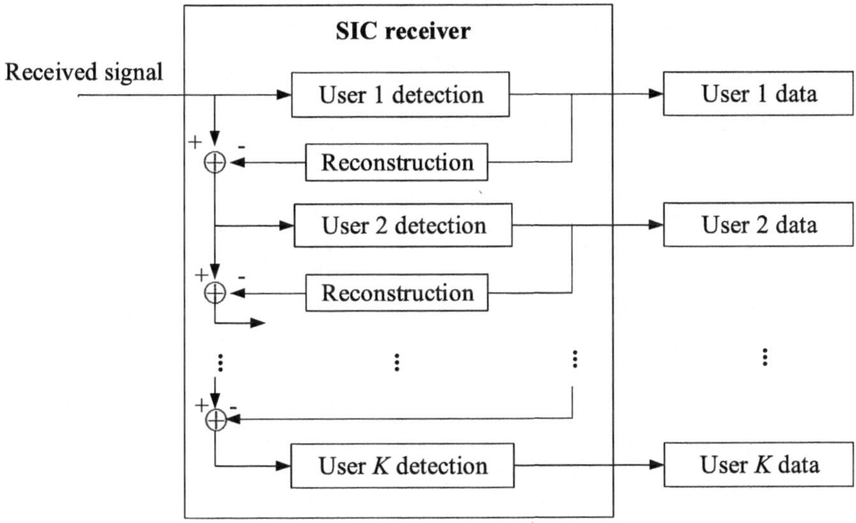

Fig. 14.8 SIC receiver

decoding is able to correct most errors, CW-SIC is expected to perform better than SL-SIC.

14.3.2 Belief Propagation (BP)

BP algorithm has been demonstrated to be able to approach MAP detection asymptotically [3, 4]. Furthermore, the sparsity of PDMA pattern reduces the complexity of BP algorithm, making it suitable for PDMA system, and disparate transmission diversity of PDMA can speed up the convergence of BP. Given the received signal vector $\boldsymbol{y}$ and the PDMA equivalent channel response matrix $\boldsymbol{H}$ in (14.3), the optimal detection of $\boldsymbol{x}$ is a joint MAP detection

$$\hat{\boldsymbol{x}} = \arg\max_{\boldsymbol{x}\in\aleph^K} p(\boldsymbol{x}\,|\,\boldsymbol{y}, \boldsymbol{H}) \tag{14.12}$$

where $\aleph^K$ represents constellation alphabet of K users.

Equation (14.12) can be approximated by a local MAP solution based on Bayesian formula

$$\hat{x}_k = \arg\max_{s\in\aleph} \sum_{\boldsymbol{x}\in\aleph^K, x_k=s} P(\boldsymbol{x}) \prod_{n\in N_v(k)} p(\boldsymbol{y}_n|\boldsymbol{x}) \tag{14.13}$$

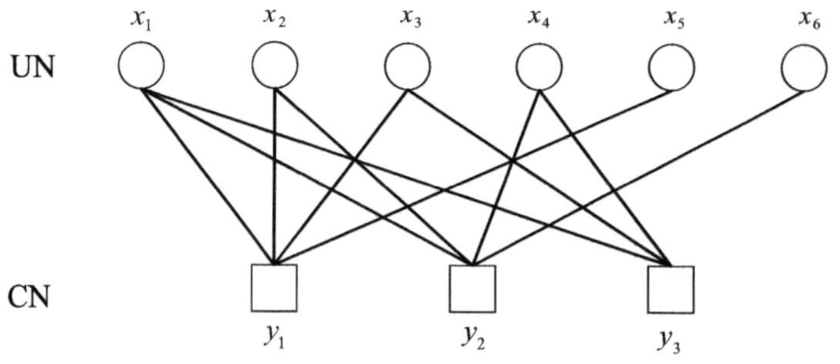

Fig. 14.9 Factor graph of PDMA with $\boldsymbol{G}_{PDMA}^{3,6}$

where $\aleph$ and $N_v(k)$ represent constellation alphabet and the RE index set corresponding to the PDMA pattern of the kth user. According to [3, 4], the problem can be solved by applying BP algorithm on the underlying factor graph.

A PDMA system with pattern matrix $\boldsymbol{G}_{PDMA}^{[N,K]}$ can be represented by a factor graph consisting of channel observation nodes (CN) and user nodes (UN). A factor graph with PDMA pattern matrix $\boldsymbol{G}_{PDMA}^{3,6}$ is shown in Fig. 14.9. The kth UN corresponds to the kth user's data symbol, and the jth CN represents the received signal on the jth RE $y_j (1 \leq j \leq N)$. If there is an edge between the kth UN and the jth CN (i.e., $\boldsymbol{G}_{PDMA}^{[N,K]}(j,k) \neq 0$), the received signal on the jth RE includes contributions from the kth user.

Here we illustrate an overview procedure of BP algorithm. As shown in Fig. 14.9, assume that x_k is the symbol at UN k, $\boldsymbol{y}_j$ is the received symbol at CN k, and let $N_C(j) = \{k|\boldsymbol{G}_{PDMA}^{[N,K]}(j,k) \neq 0, 1 \leq k \leq K\}$ be the neighboring UNs of variable nodes (VN) j, $N_v(k) = \{j|\boldsymbol{G}_{PDMA}^{[N,K]}(j,k) \neq 0, 1 \leq j \leq N\}$ be the neighboring CNs of UN k. During the lth iteration, the message passed from the kth UN to the jth CN is given by

$$I_{x_k \to y_j}^{(l)}(x_k = s) = \prod_{j' \in N_v(k) \backslash j} I_{x_k \leftarrow y_{j'}}^{(l-1)}(x_k = s) \tag{14.14}$$

where $I_{x_k \leftarrow y_{j'}}^{(l-1)}(x_k = s)$ is the message from other CNs in $N_v(k)$. The message passed from the jth CN to kth UN is given by

$$I_{x_k \leftarrow y_j}^{(l)}(x_k) = \sum_{x_{k'}, k' \in N_c(j) \backslash k} p(y_j|\boldsymbol{x}) \bullet \prod_{k' \in N_c(j) \backslash k} I_{x_{k'} \leftarrow y_j}^{(l)}(x_{k'}) \tag{14.15}$$

where $p(y_j|\boldsymbol{x})$ denotes the transition probability of receiving symbol y_j when vector $\boldsymbol{x}$ is transmitted. Detail procedure about the BP algorithm can be referred to [3, 4].

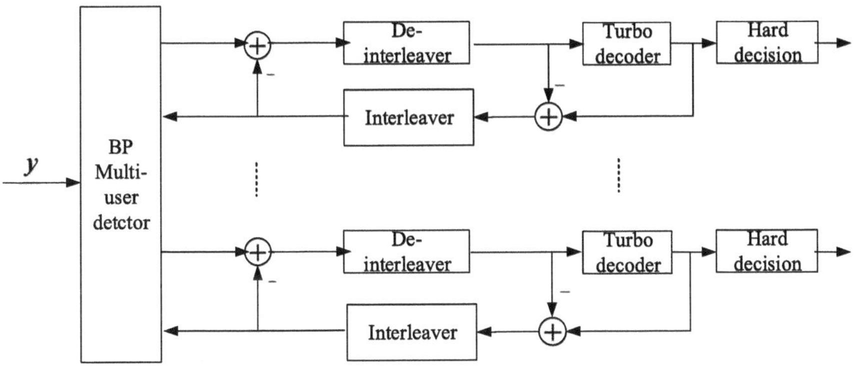

Fig. 14.10 Structure of the BP-IDD receiver for PDMA uplink system

14.3.3 BP Based Iterative Detection and Decoding (BP-IDD)

The basic principle of BP based iterative detection and decoding (BP-IDD) algorithm is that the decoded bit log likelihood ratio (bit-LLR) is fed back from Turbo decoder and converted to symbol-LLR as a priori information of the BP detector. There are two iterative processes in the BP-IDD receiver, one is inner iteration processing of BP detector and the other is outer iteration processing between BP detector and Turbo decoder. As shown in Fig. 14.10, besides a traditional BP multi-user detector, the BP-IDD receiver includes multiple parallel iterative processes, each of which is composed of modules of deinterleaver, Turbo decoder and interleaver. Here soft information is transferred between the multi-user detector and the Turbo decoder, in the form of LLR.

A factor graph of BP-IDD algorithm based on PDMA pattern matrix $\boldsymbol{G}_{PDMA}^{[3,6]}$ is shown in Fig. 14.11. Let $x_k(k = 1, \ldots, K)$ be data symbols corresponding to the UN of the kth user and associated to the VN $c_{k,i}(i = 1, \ldots, m)$, where m represents the modulation order of kth user. The connection between UN and VN shall satisfy certain condition imposed by channel encoding. According to Fig. 14.11, the iterative process between UN and CN, described in the previous section BP detection, is called inner iteration, and the iterative process between UN and VN is named outer iteration. Detail procedure about the BP-IDD algorithm can be referred to [4, 12].

14.3.4 Expectation Propagation (EP)

As shown above, the bottleneck of complexity is to calculate the (14.15). EP algorithm is a technique in Bayesian machine learning for approximating posterior beliefs with exponential family [16]. Mathematically, the projection of a particular distribution p into some distribution set Φ is defined as

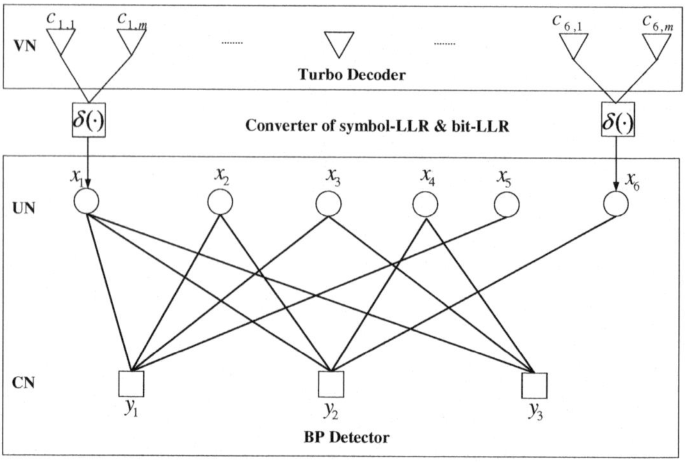

Fig. 14.11 Factor graph of PDMA with $\boldsymbol{G}_{PDMA}^{3,6}$ and Turbo decoder

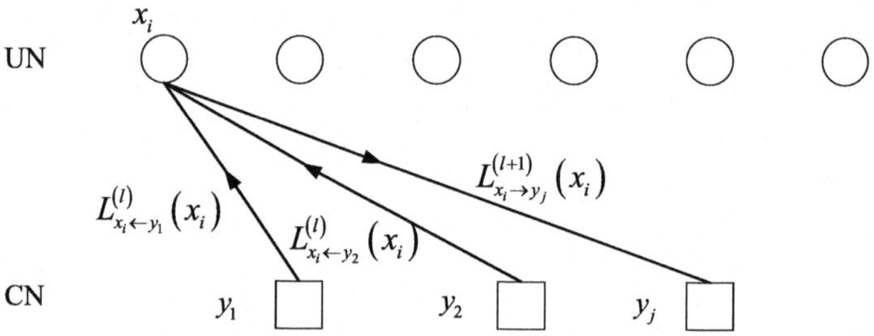

Fig. 14.12 Illustration of message process at UN with PDMA pattern matrix $\boldsymbol{G}_{PDMA}^{3,6}$

$$Proj_{\Phi}(p) = \arg\min_{q \in \Phi} D(p||q) \tag{14.16}$$

where $D(p||q)$ denotes the Kullback-Leibler divergence. In general, $p \in \Phi$ and hence the distribution projection is a nonlinear operation.

With EP algorithm [16], the message update function as shown in Figs. 14.12 and 14.13 can be written as:

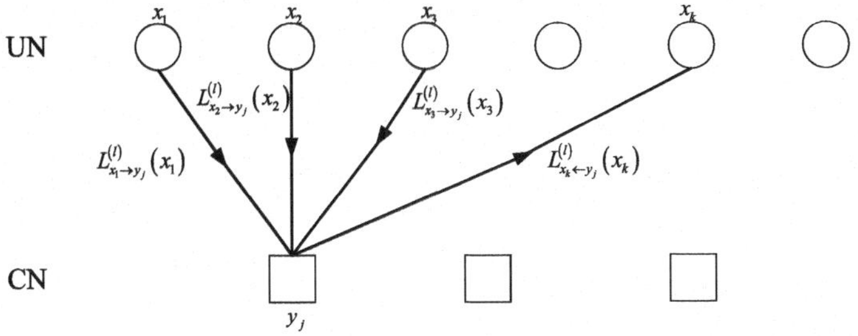

Fig. 14.13 Illustration of message process at CN with PDMA pattern matrix $\boldsymbol{G}_{PDMA}^{3,6}$

$$I_{x_k \to y_j}^{(l)}(x_k) = \frac{Proj_{\Phi}(p^{(l)}(x_k))}{I_{x_k \leftarrow y_j}^{(l-1)}(x_k)} \tag{14.17}$$

$$I_{x_k \leftarrow y_j}^{(l)}(x_k) = \frac{Proj_{\Phi}(q^{(l)}(x_k))}{I_{x_k \to y_j}^{(l)}(x_k)} \tag{14.18}$$

$$p^{(l)}(x_k) = \prod_{j' \in N_v(k) \backslash j} I_{x_k \leftarrow y_j'}^{(l-1)}(x_k) \tag{14.19}$$

$$q^{(l)}(x_k) = \sum_{x_{k'}, k' \in N_c(j) \backslash k} p(y_j | \boldsymbol{x}) \bullet \prod_{k' \in N_c(j) \backslash k} I_{x_{k'} \leftarrow y_j}^{(l)}(x_{k'}) \tag{14.20}$$

Let Φ be the set of complex Gaussian distribution, which is given by

$$\Phi = \{\rho : \rho(x_k) = \mathcal{N}_{\mathbb{C}}(x_k; \mu_k, \sigma_k)\} \tag{14.21}$$

Then, $Proj_{\Phi}(p^{(l)}(x_k))$ can be described by the mean and variance as follows:

$$\mu_k^{(l)} = \sum_{s \in S} p^{(l)}(x_k = s)s \tag{14.22}$$

$$\sigma_k^{(l)} = \sum_{s \in S} |s - \mu_k^{(l)}|^2 p^{(l)}(x_k = s) \tag{14.23}$$

Then, both $I_{x_k \to y_j}^{(l)}(x_k)$ and $I_{x_k \leftarrow y_j}^{(l)}(x_k)$ can be approximated with Gaussian distribution, which are denoted as $\mathcal{N}_{\mathbb{C}}(x_k; \overrightarrow{\mu}_{k,j}^{(l)}, \overrightarrow{\sigma}_{k,j}^{(l)})$ and $\mathcal{N}_{\mathbb{C}}(x_k; \overleftarrow{\mu}_{k,j}^{(l)}, \overleftarrow{\sigma}_{k,j}^{(l)})$, respectively.

Detail procedure about the EP algorithm can be referred to [16]. Following the same principle of BP-IDD algorithm, EP-IDD algorithm can also be achieved, where the decoded extrinsic information is fed back from Turbo decoder and converted to probability as a priori information of the EP detector.

Table 14.1 Computation complexity per modulation symbol

Algorithm	Complexity order (only dominant part is considered)
SIC	$O(KN^3)$
BP	$O(T_{in}NM^{d_f})$
BP-IDD	$O(T_{out}T_{in}NM^{d_f})$
EP	$O(T_{in}NM^{d_f})$
EP-IDD	$O(T_{out}T_{in}NM^{d_f})$

14.3.5 Comparison of Different Receivers

Among the detection algorithms described above, it is expected that BP-IDD or EP-IDD are the best of all in terms of performance, and BP or EP would be better than SIC. For a medium spectral efficiency range, EP may have similar performance with BP.

Table 14.1 summarizes computation complexity of the above five detection algorithms. Here, M denotes the size of modulation constellation, T_{in}, T_{out} and d_f represent IDD inner iteration number, outer iteration number, and maximum row weight of PDMA pattern matrix. The number of additions of IDD receiver is about T_{out} times of no IDD case [13]. It should also be noted that the computation of Turbo decoder is not accounted. It can be observed that EP has evidently decreased the computation complexity.

As the processing ability of a base station is usually more powerful compared with that of a user terminal, BP-IDD, EP-IDD, BP and EP could be used in the PDMA uplink. Detection at user terminal in the PDMA downlink can choose either EP or SIC depending on its processing ability.

14.4 PDMA Performance

Performance of PDMA is evaluated in both link level and system level, covering comparison of PDMA and OMA, effect of key factors, etc.

14.4.1 Link Level Simulation (LLS)

• ***Comparison of PDMA and OMA***

Taking orthogonal frequency division multiple access (OFDMA) in long term evolution (LTE) system as a reference, the assumptions of the link-level simulations are shown in Table 14.2, and the results are given in Figs. 14.14 and 14.15.

Table 14.2 PDMA link level simulation assumptions

Parameters	Value
Carrier	2 GHz
System bandwidth	10 MHz
Channel model	UMA-NLOS [17]
Modulation coding rate	QPSK 1/2; LTE Turbo
Antenna configuration	1Tx2Rx
Channel estimation	Perfect
HARQ	No
Uplink overload factor	150; 200; 300%
Uplink average SNR	The same of all users

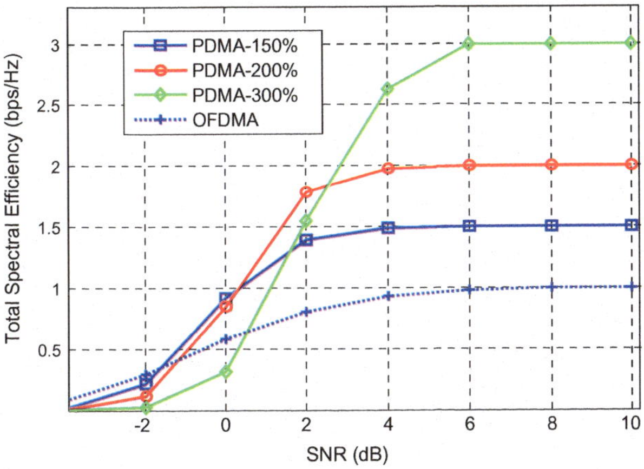

Fig. 14.14 PDMA uplink performance

Figure 14.14 shows the total spectrum efficiency (SE) of uplink PDMA under the overload factors of 150, 200, and 300% as well as SE of OFDMA. For fair comparison, a same number of source bits is assumed for PDMA and OFDMA. For the given overload factors of 150, 200 and 300%, SE gains of 50, 100 and 200% can be achieved by PDMA over OFDMA when the signal-to-noise ratio (SNR) is high enough.

In a downlink system, the SNR differences between users have tremendous influence on the performance of PDMA. The performance gain of PDMA over OFDMA gets more remarkable when the SNR difference gets larger. As shown in Fig. 14.15,

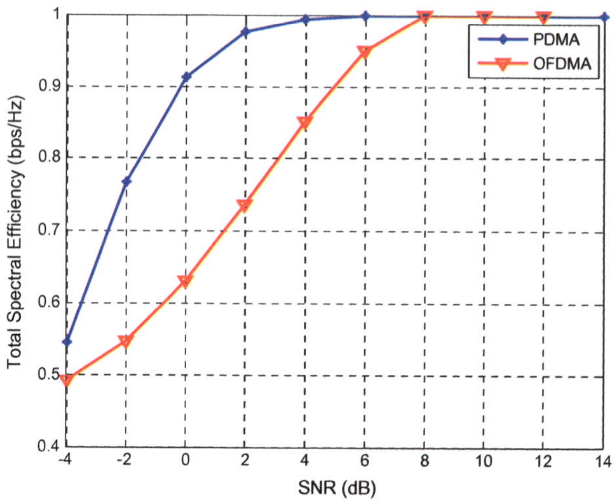

Fig. 14.15 PDMA downlink performance

when the SNR difference of two users is 12 dB, the SE gain is 14% when SNR is −4 dB, while the SE gain reaches the maximum of 50% when SNR is 0 dB, and the SE gain nearly vanishes when SNR is higher than 10 dB.

- ***Comparison of Different Receivers***

Take uplink PDMA system as an example, performance of BP and BP-IDD receivers are compared in Fig. 14.16. Also, for downlink PDMA system, performance of BP and SIC are compared in Fig. 14.17. Table 14.3 gives a comparison of the related computation complexity.

As shown in Fig. 14.16, BP-IDD receiver has evident performance gain compared with BP receiver. Especially, the gain is higher with heavier overload, e.g., the gain is 0.8 dB for overload 150% with pattern matrix [2, 3], and the gain is 1.6 dB for overload 300% with pattern matrix [4, 12]. This is because that the performance of BP algorithm converges more slowly for overload factor of 300% compared with 150% and the frequency diversity degree is higher for 300% than 150%. Whereas, from computation complexity, the addition in BP-IDD receiver is 3 times of that in BP receiver.

As shown in Fig. 14.17, BP receiver has a gain of 2 dB compared with SIC receiver in downlink where modulation coding scheme (MCS) of the near user and far user is QPSK 1/2, and power factor is 0.2 and 0.8. Whereas, the computation complexity of BP receiver is about 1.67 and 15 times higher than SIC receiver in multiplication and addition, respectively.

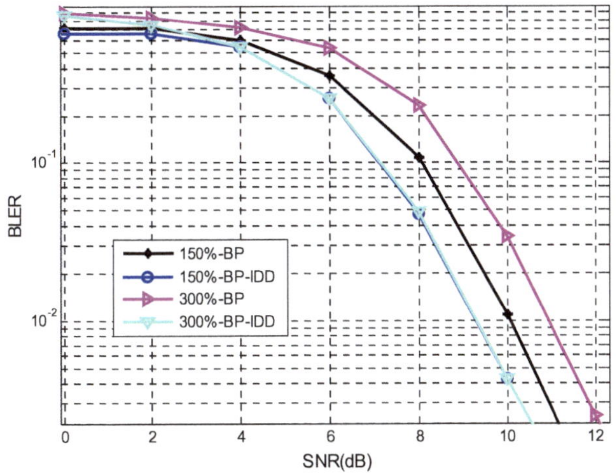

Fig. 14.16 Performance of uplink PDMA with BP or BP-IDD at receiver

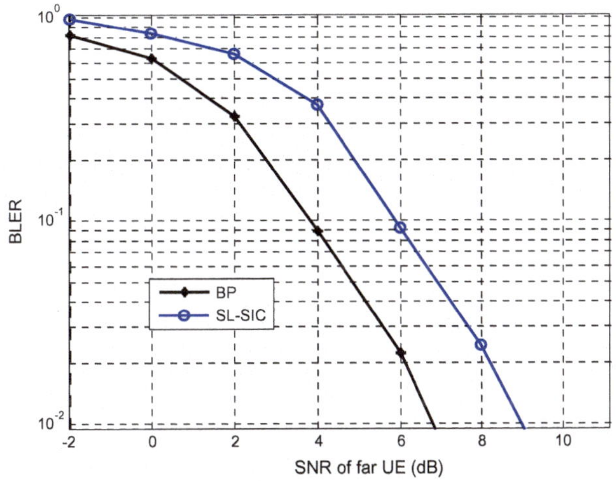

Fig. 14.17 Performance of downlink PDMA with BP or SIC at receiver

Table 14.3 Complexity of different receivers in uplink PDMA system

Link type	Receiver	Computation complexity
Uplink [2, 3]	BP-IDD	$N_{mul} \approx 64$, $N_{add} \approx 1152$
	BP	$N_{mul} \approx 64$, $N_{add} \approx 384$
Downlink [2, 3]	BP	$N_{mul} \approx 64$, $N_{add} \approx 384$
	SIC	$N_{mul} \approx 24$, $N_{add} \approx 24$

Note N_{mul} denotes the number of multiplications, N_{add} denotes the number of additions, and $\approx$ means approximately equal to

14.4.2 System-Level Simulation (SLS)

The system-level simulation assumptions are shown in Table 14.4, and the simulation results are provided in Figs. 14.18 and 14.19.

The uplink grant-free OFDMA and PDMA transmissions are evaluated under traffics with small burst packet and the latency is required to be not more than 1 ms. Gain of 500% in terms of the number of supported users under the given system packet drop rate of 1% is observed in Fig. 14.18. The gain comes from two facts. First, PDMA provides a larger resource pool than OFDMA does, so that the collision probability of PDMA is lower than that of OFDMA. Second, the BP-IDD receiver employed by PDMA is more capable of dealing with interference when collision occurs.

From the results of downlink PDMA shown in Fig. 14.19, PDMA can get about 30% gain compared with OFDMA in terms of both SE at cell edge and cell average

Table 14.4 PDMA system-level simulation assumptions

Parameters	Value
Topology	Hexagonal homogeneous network; 19 sites/57sectors
Number of users per cell	10, 20, or 30
Carrier	2 GHz
Bandwidth	Uplink: 5 MHz Downlink: 10 MHz
ISD	500 m
Channel model	ITU UMa [17]
Power control	Uplink: open-loop power control, alpha = 1, P0 = −95 dBm
The number of antenna	Uplink: 1Tx2Rx; Downlink: 2Tx2Rx
Antenna configuration	Uplink: User vertical polarization, BS cross polarization, Downlink: User cross polarization, BS cross polarization
Channel estimation	Perfect
Scheduler	Uplink: Grant-free; Downlink: PF schedule
MCS	Uplink: 160 bits @ 1PRB Downlink: Adaptive (Based on LTE downlink MCS definition)
Maximum HARQ transmission times	Uplink: 0 Downlink: 3
Traffic model	Uplink: Bursty traffic with small packet Downlink: Full buffer traffic
Receiver	Linear MMSE receiver for OFDMA, BP-IDD for PDMA

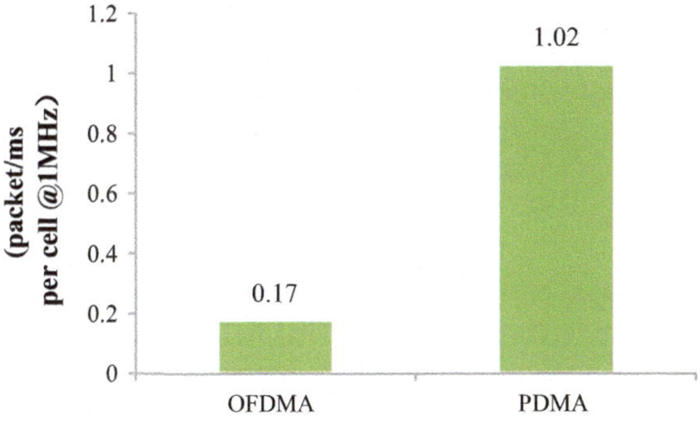

Fig. 14.18 PDMA uplink system performance (Supported packet arrival rate when packet drop rate = 1%)

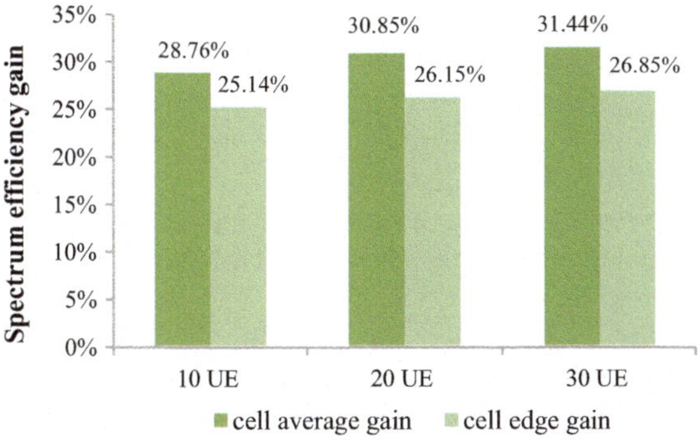

Fig. 14.19 PDMA downlink system performance

SE. The gain increases with the user number in a cell, because with more users it is easier to find suitable users for pairing in PDMA.

14.5 Extension Design of PDMA

To ensure PDMA application in 5G typical scenarios and further improve the performance of PDMA, more extension design of PDMA are proposed, such as GF-PDMA [18–20], Co-PDMA [21], LSA-PDMA [22], RIePDMA [23], PC-PDMA [24], etc.

14.5.1 PDMA Based Grant Free Transmission

Grant-free (GF) transmission is a mechanism that eliminates the dynamic scheduling request (SR) and grant signaling overhead for uplink data transmission and a usecan transmit uplink data in an "arrive-and-go" manner. With grant-free transmission, contention is usually allowed to increase the system resource utilization, i.e., the users may transmit on the same time and frequency resource as there is no coordination from the base station. In this case, NOMA based grant-free transmission will show its advantage as a solution for contention resolution with high reliability, since they are designed with high overloading capability. The design of NOMA based grant-free transmission has been proposed and discussed during Release-14 new radio (NR) Study, in which NOMA signatures are taken as part of grant-free resource besides the traditional physical resource such as time and frequency resource.

In [18–20], PDMA based grant free (GF-PDMA) transmission was proposed, and the related procedure mainly consists of following six steps as shown in Fig. 14.20:

Step 1: The base station (BS) and the users predefine the key GF parameters such as system bandwidth, time & frequency partition ration, data transmission format, basic resource unit (BRU) and the mapping rule between BRU and users.

Step 2: The users send random access request and access BS using current LTE access scheme.

Step 3: The BS builds an one to one mapping relationship according to the mapping parameters (e.g., path loss, position, UL receiving power and UL SNR), then informs the relationship to users and acquire all the candidates carried on every PDMA BRU.

Step 4: The users acquire the mapping relationship that the BS sends and transmit data and pilot on the allocated PDMA BRU at the same time when there are data need to send.

Step 5: The BS monitors all the signals of the candidates on the every PDMA BRU and determines the users whether to send data or not. Then the BS will conduct pilot channel estimation and data detection for those users sending data.

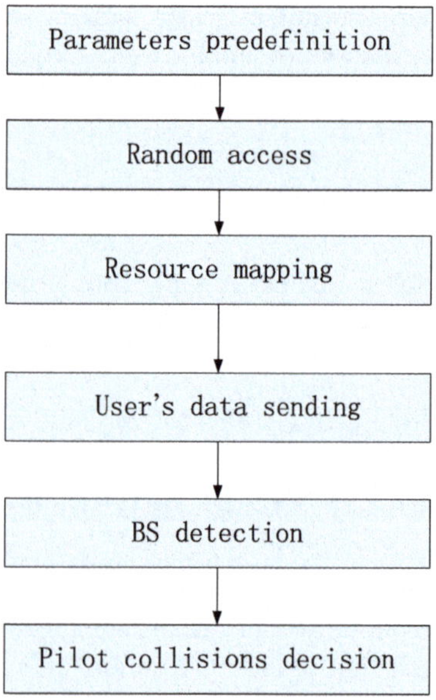

Fig. 14.20 The procedure of GF-PDMA

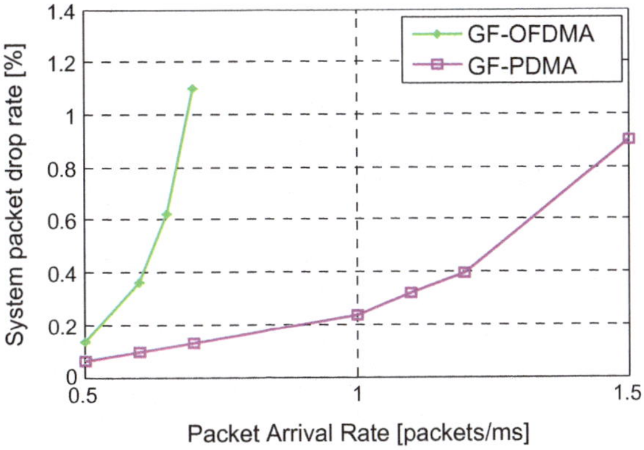

Fig. 14.21 System level simulation results of GF-PDMA compared with GF-OFDMA

Step 6: The BS determines whether there are multi-user pilot collisions (pilot collision is defined as more than two users select the same pilot on the same PDMA BRU), i.e., whether the time interval receiving the data from a user is longer than the certain fixed threshold value, if yes, the pilot collision exists otherwise non-exist.

Figure 14.21 shows comparison of GF-PDMA with OFDMA based grant-free (GF-OFDMA) transmission with the same parameter settings. It can be observed that GF-PDMA has shown the significant performance enhancement over the GF-OFDMA baseline. With the traffic load increasing, GF-OFDMA packet drop rate (PDR) degrades much faster than GF-PDMA. Specifically, at PDR 0.8% point of interest in terms of packet arrival rate, GF-PDMA demonstrates 215% gain over the GF-OFDMA baseline (or almost 2.15 times of GF-OFDMA capability).

14.5.2 Cooperative PDMA

To improve cell coverage and system performance, an uplink cooperative pattern division multiple access (Co-PDMA) scheme with half-duplex decode and forward (DF) relay is proposed in [21]. As shown in Fig. 14.22, only one half-duplex relay is considered for all transmit signals from the source. Data streams are transmitted simultaneously through both direct and relay channels. At the destination, a SIC algorithm without an ordering process is used.

Figure 14.23 compares the achievable sum data rate among three schemes, i.e., PDMA, Co-PDMA and also cooperative OMA (Co-OMA). Two conditions are assumed, one is that all users have the same target data rate (TDR), the other is that the same resource elements have the same TDR. From the figure, it can be seen that the exact analysis curves match well with the Monto Carlo simulation curves. For the three users in PDMA, each user's TDR is 0.1. For each user in Co-OMA, the TDR equals or 1.5 times as much as Co-PDMA. Clearly Co-PDMA transmission outperforms the Co-OMA and non-cooperative PDMA. When the TDR of Co-OMA

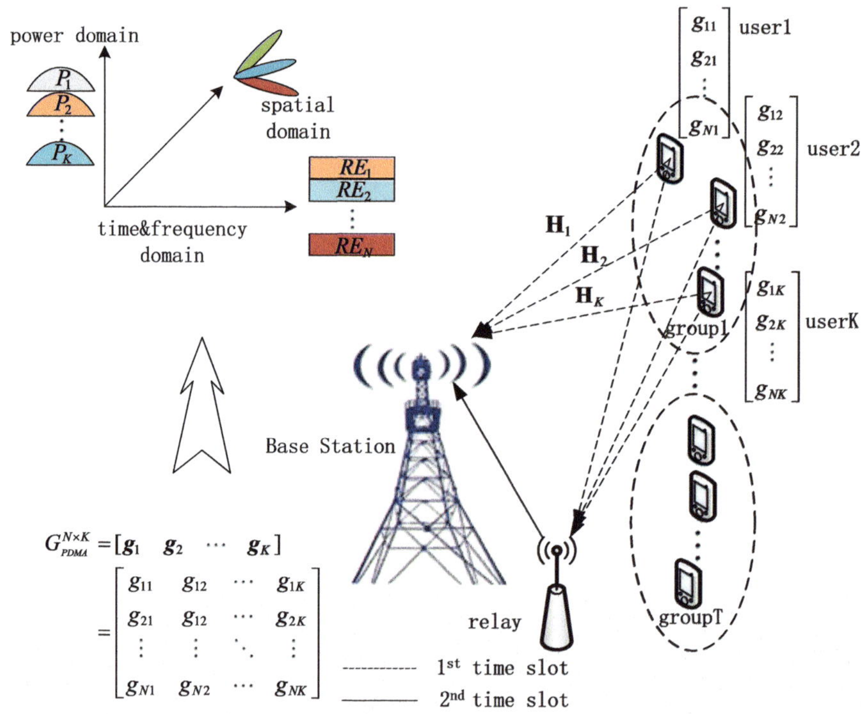

Fig. 14.22 System model of Co-PDMA

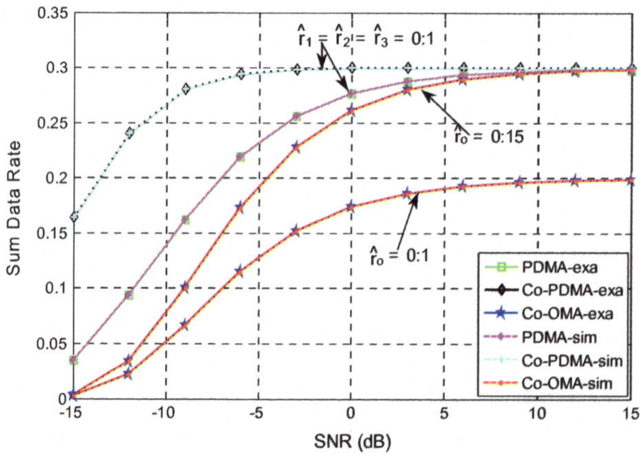

Fig. 14.23 Sum data rate comparison (-exa means exact analysis, -sim means Monte Carlo simulation)

is 1.5 times as much as Co-PDMA, Co-PDMA achieves a significant gain at lower SNR region (e.g., below 5 dB). When the TDR of Co-OMA equals Co-PDMA, Co-PDMA achieves almost 50% gain over Co-OMA at higher SNR region (e.g., above 8 dB) due to the diversity gain.

It should be noted that the performance improvement from Co-PDMA is at the expense of system complexity increasing (multiple relay stations need to be set up) and spectral efficiency degradation (duplicate transmission is conducted).

14.5.3 PDMA Combing with Massive MIMO

Pattern division multiple access with large-scale antenna array (LSA-PDMA) [22] is proposed as a novel NOMA scheme. In the proposed scheme, pattern is designed in both beam domain and power domain in a joint manner, as shown in Fig. 14.24. At the transmitter, pattern mapping utilizes power allocation to improve the system sum rate and beam allocation to enhance the access connectivity and realize the integration of LSA into multiple access spontaneously. At the receiver, hybrid detection of spatial filter (SF) and SIC is employed to separate the superposed multiple-domain signals. Furthermore, the sum rate maximization problem was formulated to obtain the optimal pattern mapping policy, and the optimization problem is proved to be convex through proper mathematical manipulations.

Figure 14.25 depicts the system sum rate versus the power gain factor where the simple pattern mapping policy is adopted for the proposed scheme. The sum transmit power is set to be 10 dB here. When $\mu > 1$, more power is allocated to the stronger user. It can be observed that the performances of the LSA-PDMA scheme

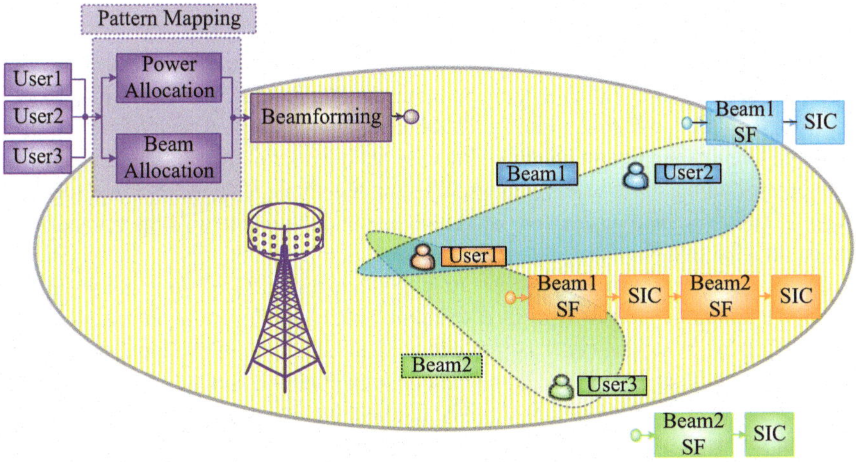

Fig. 14.24 Illustration of the proposed LSA-PDMA scheme

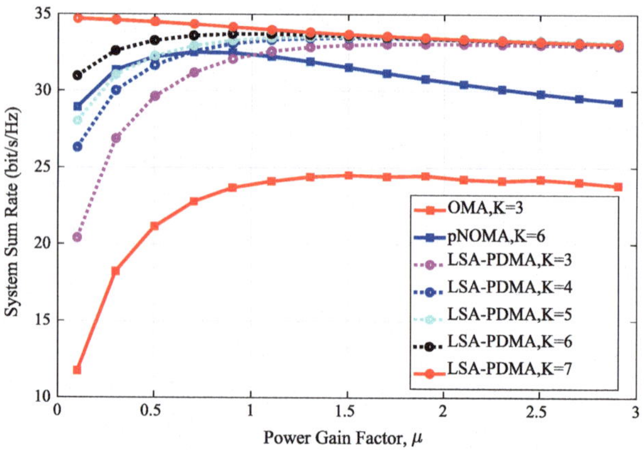

Fig. 14.25 The system sum rate versus the power gain factor, where the simple mapping policy is adopted for LSA-PDMA scheme (k is the user number)

stop increasing and tend to be constant when μ is larger than a certain threshold. This results from the design principles of beamforming. In the proposed LSA-PDMA scheme, beamforming is generated based on the channel state indication (CSI) of the weakest user within a beam for achieving the user fairness. Evidently, the proposed LSA-PDMA scheme achieves significant performance gain on system sum rate compared to both the orthogonal multiple access scheme and the power domain NOMA scheme.

14.5.4 PDMA Combing with Interleaving

In order to further improve the overload ability of PDMA, random interleaving (RI) was brought into the PDMA system to form the RIePDMA system [23]. Different from the uplink system of original PDMA, each user in a RIePDMA system is assigned to a unique interleaver after the channel encoder at the transmitter side, and the corresponding de-interleaver is used before the channel decoder at the receiver. Fig. 14.26 shows the overview of the transmitter and the receiver of the uplink of the proposed RIePDMA system.

Assuming there are K single-antenna users sharing on N resource blocks (RBs), and the base station (BS) is equipped with two antennas. The information bit stream from the kth ($1 \leq k \leq K$) user $\{d\}_k$ is encoded by the channel encoder into b_k, then the coded bits b_k are permuted by the user-specific interleaver π_k to form the chip stream. This interleaving operation can be mathematically formulated as a process by permutation matrix αI. Assuming the corresponding permutation matrix of the interleaver π_k is $a I_k$. Thus, the chip stream becomes $C_k = b_k \cdot \alpha I_k$. The chip stream C_k goes through the same processes as the original PDMA system to form the transmitted

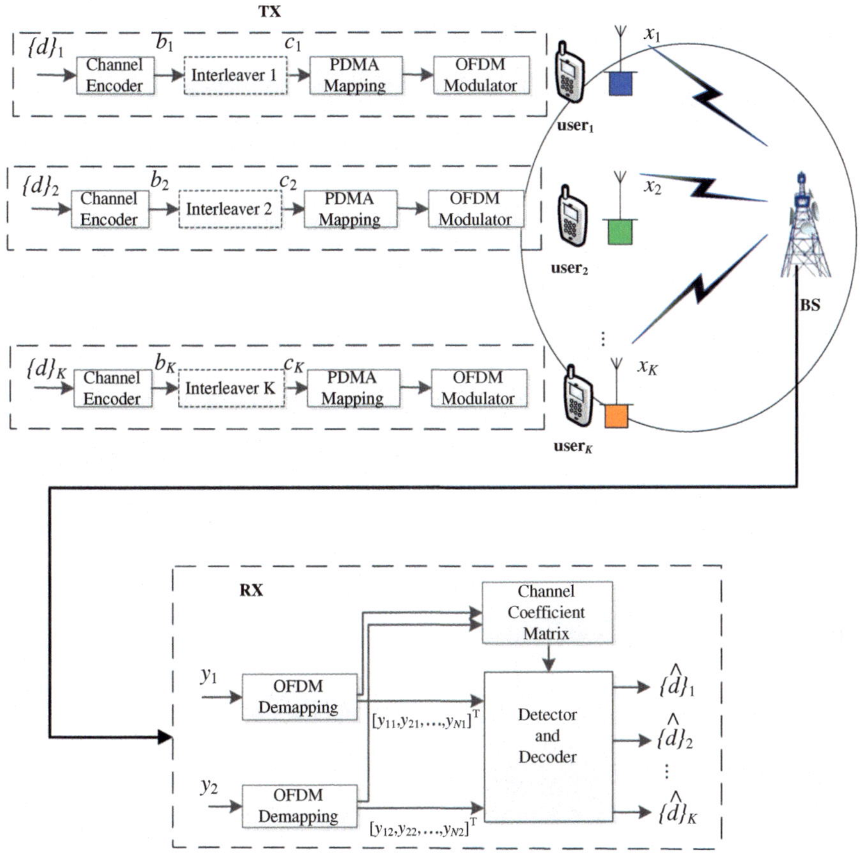

Fig. 14.26 The system model of RIePDMA

signal x_k, that is, PDMA mapping and OFDM modulator. It can be seen that the bit-level interleaving randomizes the bit sequence order, which may further bring benefit in terms of combating frequency selective fading and interference.

At the receiver side, the received signal of the mth ($m = 1, 2$) receiving antenna is $\mathbf{y}_m = [y_{1,m}, y_{2,m}, \ldots, y_{N,m}]^T$. The received signal in the nth ($1 \leq n \leq N$) RB is denoted as

$$y_{n,m} = \sum_{k=1}^{K} \mathbf{H}_{PDMA}(n,k) h_{n,m,k} x_k + n_{n,m}. \tag{14.24}$$

where $\mathbf{H}_{PDMA}(n,k)$ denotes the element at the nth line and the kth column of $\mathbf{H}_{PDMA}$, $h_{n,m,k}$ denotes the channel between the kth user and BS at the mth receiving antenna, and n_{nm} denotes the additive white Guassian noise in the nth RB at the mth receiving antenna. $\{\hat{d}\}_k$ denotes the decoded information bits of user acquired by detector and decoder.

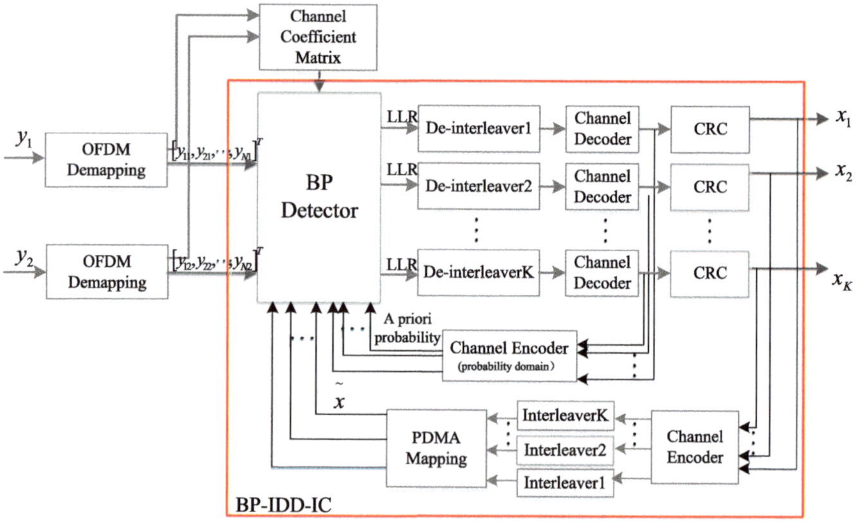

Fig. 14.27 The block diagram of BP-IDD-IC detection algorithm

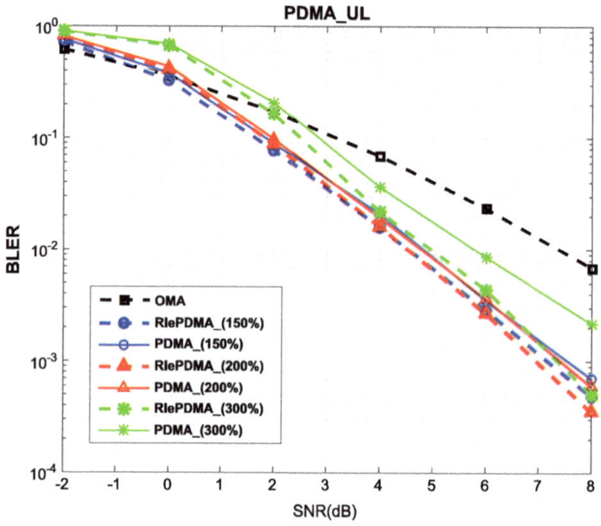

Fig. 14.28 BLER performance comparison of PDMA and RIePDMA

After signal receiving, the self-adaptive BP-IDD-IC scheme was implemented, which integrates the IC technique and the cyclic redundancy check (CRC) module into BP-IDD, to reduce complexity and improve BLER performance. As shown in Fig. 14.27, in this novel scheme, the output of BP will be sent to the turbo decoder first. The hard information from the turbo decoder goes through the CRC module.

Denote C as a set to record the correctly decoded users. When all users signals are correctly decoded, the IDD iteration and IC terminate. Otherwise, the turbo decoder will use soft information to implement turbo encoding in the probability domain and then send the coded soft information back to BP as prior information. The signals of users are also reconstituted belonging to C. The IC module cancels the reconstituted signal $\tilde{x}$ from the received signal y to get a new received signal $\tilde{y}$ in (14.25) and sends it to the BP module.

$$\tilde{y} = y - \sum_{k \in C} h_k \tilde{x}_k \tag{14.25}$$

Since some PDMA patterns are sparse, the complexity of BP is reduced and the convergence is guaranteed. Moreover, with the implementation of the CRC module, the recommended receiver scheme becomes self-adaptive. The number of iterations can be reduced, especially in high SNR region, without affecting the block error rate (BLER) performance of the receiver. With the implementation of the IC technology, the correctly decoded information can be canceled, and the inter-user interference will be reduced. Consequently, BLER performance can be improved in the self-adaptive BP-IDD-IC scheme.

As shown in Fig. 14.28, RIePDMA achieves better BLER performance than PDMA. In addition, the higher the overload is, the greater BLER performance gain RIePDMA can get. First, random interleavers can further differentiate users. Second, the interleaver disrupts the order of encoded information bits, so consecutive errors caused by the small interference and the channel fading can be avoided to some extent. From the perspective of complexity, RIePDMA only adds the process of interleaving and de-interleaving, whose complexity is negligible compared to that of the iterative decoder and detector.

14.5.5 PDMA Combing with Polar Coding

Guided by the channel-aware feature of polar coding and the generalized channel polarization idea, a polar-coded NOMA (PC-NOMA) system is designed in [24]. Compared to other coded NOMA systems, the novelty of PC-NOMA is to allow for a joint optimization of binary polar coding, signal modulation and NOMA transmission. Through a multi-stage channel transform concatenated manner, the polarization effect can be gradually enhanced. Finally, the NOMA channel will be elaborately split into a group of binary memoryless channels (BMCs), whose capacities trend to zero or one.

The proposed PC-NOMA system has been illustrated in Fig. 14.29. Specifically, three stage channel transform structure was given, which constitutes the information theoretical framework of the PC-NOMA system. The whole procedure is described as user $\rightarrow$ signal $\rightarrow$ bit partitions. In the first stage, design focuses on user partition, where the partition order and the partition structure will affect the polarization effect.

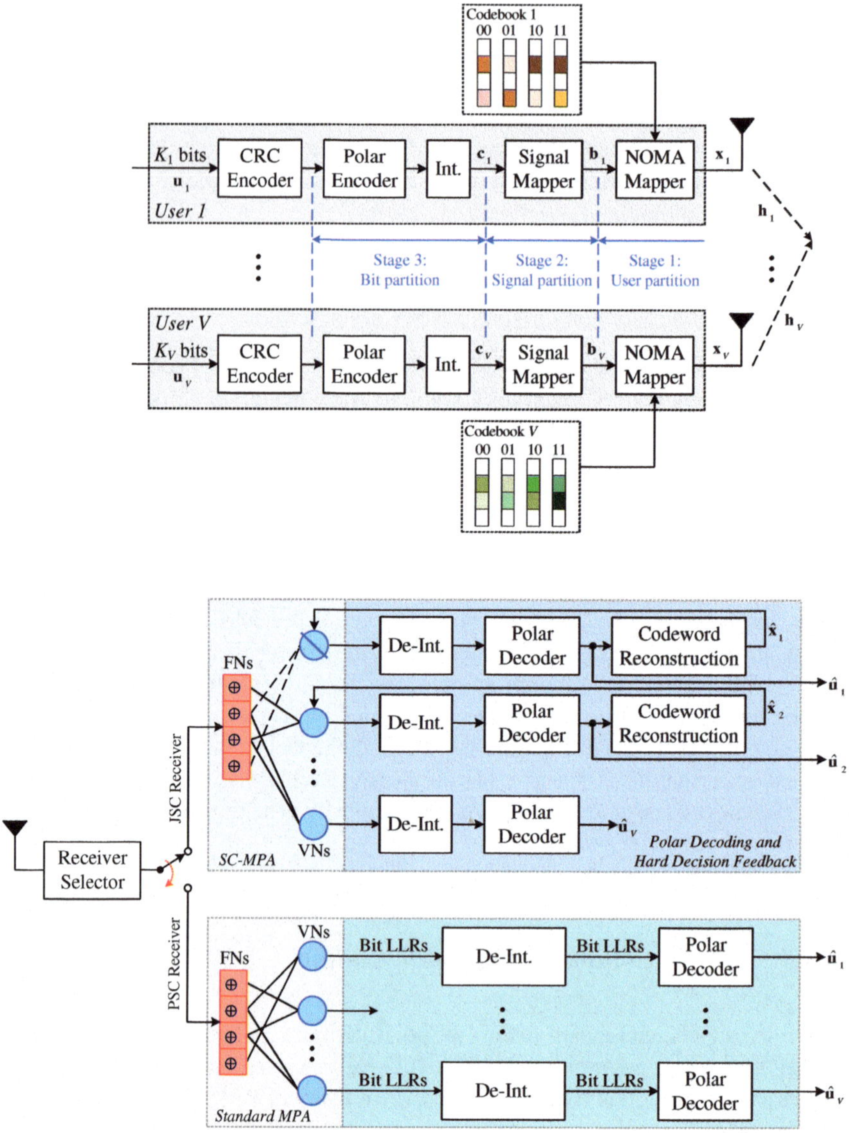

Fig. 14.29 Illustration of the uplink PC-NOMA systems

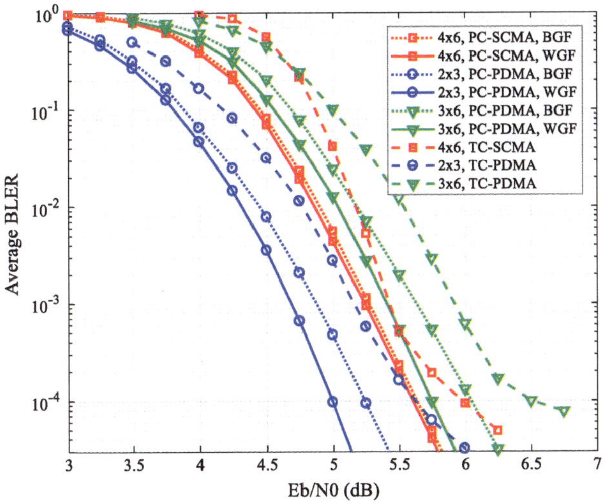

Fig. 14.30 The average BLER performance under the Rayleigh fading channel, where the SUP based PC-NOMA and the TC-NOMA are adopted

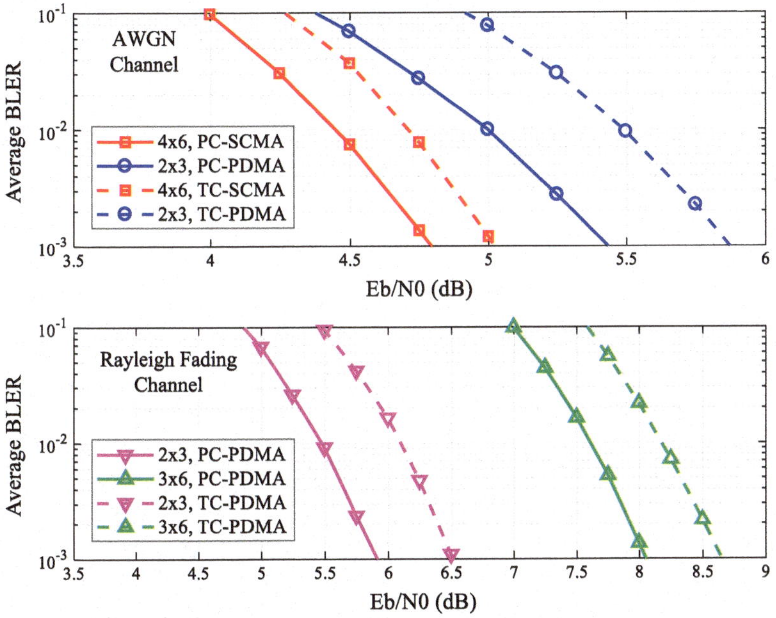

Fig. 14.31 The average BLER performance comparison under the Rayleigh fading channels, where the PUP based PC-PDMA and the TC-PDMA are adopted

Hence, two types of user partition are given, i.e., sequential user partition (SUP) and parallel user partition (PUP). In the second stage, the bit-interleaved coded modulation (BICM) scheme is used to combat the channel fading. Then by performing the binary channel polarization transform, the bit synthesized channels are split into a series of bit polarized channels in the third stage. Hence, the proposed three-stage channel transform structure facilitates the unified description of polar coding and NOMA transmission in a joint manner.

Taking PDMA and sparse code multiple access (SCMA) [25] as NOMA examples, performance of the SUP based PC-NOMA was compared with turbo code NOMA (TC-NOMA) for Rayleigh channel in Fig. 14.30. It can also be observed that the PC-NOMA systems outperform the TC-NOMA systems by 0.2–0.5 dB at various configurations. The TC-NOMA systems will also demonstrate obvious error floors in the high SNR regions, while the proposed PC-NOMA systems can effectively avoid this weakness. The proposed worst-goes-first (WGF) strategy is also efficient than best-goes-first (BGF) in fading channel. Furthermore, given overload 150%, it is worth noticing that in fading channel, the PC-PDMA system can outperform the PC-SCMA system.

Further, performance of the PUP based PC-NOMA was compared with TC-NOMA for Rayleigh channel in Fig. 14.31. With the identical overload factor, when the [2, 3] PDMA is used, this performance gain will expand to about 0.5 dB. This observation further indicates that the proposed PC-PDMA systems can better fit and exploit the irregular structure of PDMA. It will conversely guide the joint design between NOMA codebook and channel coding with the framework of PC-NOMA. Under the Rayleigh fading channel, the overload is configured as 150 and 200%, the performance gain of PC-PDMA achieves about 0.5–0.6 dB compared to the TC-PDMA systems.

14.6 PDMA Applications

14.6.1 Application Scenarios

PDMA can be applied in typical scenarios of 5G [26–35], for example, in enhanced mobile broadband (eMBB), ultra reliable low latency communication (URLLC) and massive machine type communication (mMTC), as shown in Table 14.5.

In eMBB scenario, it is desirable to improve spectrum efficiency. PDMA is expected to increase the multi-user capacity, provide better fairness against the near-far effect and improve user experience in ultra-dense networks. In addition, PDMA can be used in conjunction with massive MIMO, ultra-dense networking, and high frequency communication. For eMBB small packet, PDMA can potentially increase sum throughput and reduce latency and overhead combined with grant-free transmission.

Table 14.5 Analysis of the demand for PDMA in three 5G typical usage scenarios

Usage scenarios	Expected functionalities of 5G NOMA
eMBB	• Large network capacity
	• High user density
	• Uniform user experience
	• Easy multi user- multiple input multiple output (MU-MIMO)
	• Mixed traffic types transmission
	• Highly efficient small packets transmission
mMTC	• Massive connectivity
	• Highly efficient small packets transmission
URLLC	• Ultra low latency transmission
	• Ultra high reliability transmission
	• Highly efficient small packets transmission

In URLLC scenario, grant-free transmission can be applied to achieve ultra low latency. Although dedicate physical resources can be allocated to URLLC UEs to achieve high reliability, the spectrum efficiency is low especially for the aperiodic services. In order to achieve high reliability, low latency and improve the spectrum efficiency at the same time, PDMA can be applied to enhance the reliability in case of a collision. It is also important to point out that the application of PDMA enables efficient multiplexing of URLLC and eMBB services to further improve resource utilization.

In massive MTC scenario [31], a base station needs to provide connection to a huge number of terminals. The main challenge for massive MTC scenario is how to effectively deal with massive connection with power constraint. PDMA is a very competitive solution to address the massive connectivity issue together with the large coverage requirement.

In 4G LTE system, to transmit data, a user shall first transmit the scheduling request on periodically occurred resources which is configured by the base station. The base station then makes scheduling decisions and sends an uplink grant to the user indicating the resources on which the user can transmit data. Generally, the procedure may take 10 ms or more. For some applications, such a long latency is unacceptable. Moreover, the uplink grant is carried by downlink control signaling, and with massive number of connections the downlink control channel may become a bottleneck. In such situation, grant-free transmission is a viable option.

By means of grant-free transmission [35], a user autonomously selects a resource for transmission without SR and scheduling of base station. To avoid interfering with other traffic scheduled by the base station, the resource for grant-free transmission shall be confined within a certain set of resources. The resource set is called resource pool. For orthogonal transmission, resource pool consists of resource in time and frequency domains. A user selects a resource from the pool for transmission.

For grant-free transmission, as there is no coordination between users sharing the same resource pool, it is likely that two users select the same resource. When a collision happens, it may lead to failure in detection. The probability of collision is proportional to the number of users sharing the resource pool, and is inverse proportional to the number of resources in the pool. That is, enlarging the resource pool could reduce collision probability.

As a non-orthogonal transmission scheme, PDMA could naturally be incorporated into grant-free transmission to reduce the collision probability. That is, PDMA provides another dimension for resource sharing PDMA pattern. A traditional resource pool could be extended to include PDMA pattern. Specifically, each resource group in the pool is associated with a PDMA pattern matrix. A UE selects a time-frequency resource as well as a PDMA pattern from the pattern matrix for transmission. Even though two users may select the same time-frequency resource, as long as their PDMA patterns are different, the receiver is able to decode the two users' data successfully.

The resource pool is $\alpha - 1$ times larger than a traditional resource pool where α is the overload factor of the PDMA pattern matrix.

14.6.2 System Design Aspects

To enable PDMA in practice, a number of aspects in system design shall be considered [36, 37].

Air interface and process design

PDMA enables large number of users to transmit on the same resource, especially when PDMA is used jointly with massive MIMO. The demand of reference signals will be increased accordingly. Reference signal shall be designed carefully to meet the requirements of detecting PDMA signal and keeping the overhead incursion by the reference signal to a reasonable level.

Multi-user transmission of PDMA also leads to demanding requirements on control channel, as each user's data shall be accompanied by a control channel to provide necessary information for detection. Techniques such as multi-subframe scheduling, group control signal design and the control signaling content design could serve as a starting point to cope with the problem.

Link adaptation for downlink PDMA is based on user reporting of the channel quality information (CQI). However, the multi-user pairing nature of PDMA makes it difficult for a user to predict CQI without knowing its paring users and their PDMA patterns. Power domain optimization may further complicate the problem, as the user will not know what the transmission power is going to be before making the scheduling decision. A flexible CQI calculating and reporting mechanism is needed.

As discussed above, the grant-free transmissionv is able to reduce data latency and control overhead. Though by introduction of PDMA, a resource pool could be extended, it is still possible that two users collide on the same resource and PDMA pattern especially when the system is heavy loaded. To facilitate grant-free transmission, resource selection method and mechanism to resolve conflicts play a fundamental role.

Radio resource management

Traditionally, the radio resource management deals with the allocations of time, frequency, and spatial resources to make full use of the wireless resources. PDMA introduces another dimension - PDMA pattern. The optimization problem becomes really challenging, as more optimization variables are involved. Low complexity radio resource management algorithm, that could achieve near-optimal performance, is worthy to seek.

Cell interference management

For multi-cell networks [38], there is serious interference when different cells use the same time-frequency resources and same patterns. Therefore, it is very important to consider PDMA pattern distribution for the cells. Through rational pattern distribution between the cells, make neighboring cells using orthogonal pattern resources to coordinate neighborhood interference, can improve cell edge performance and spectrum efficiency.

14.7 Challenges and Trends

The key techniques of the PDMA include the transmitter technologies, the receiver technology, multiple antenna technology and so on. PDMA transmitter technologies include pattern matrix design, the pattern assignment scheme, power allocation scheme design, link adaptation, etc. PDMA receiver technologies include high-performance low-complexity multi-user detection algorithm, the activation detection algorithm, etc. PDMA multiple antenna technologies mainly consider the PDMA combined with multiple antenna solutions in both uplink and downlink.

The advantages of the PDMA technology are reflected as follows:

The design based on the principle of diversity

The PDMA technology improves capacity based on introducing the principle of reasonable unequal diversity. It realizes the non-orthogonal signals superposition transmission in time, frequency, space (the beam domain) and power domain by designing the sparse pattern matrix of the multi-user unequal diversity and the pattern optimized scheme of the joint coding and modulation, which can get higher multi-user multiplexing and diversity gain.

PDMA can do joint optimization of coding and modulation

PDMA can break through the tradition that the multiple access technology and coding modulation design independently, perform the scheme that the multiple access joint optimization design with modulation and channel coding, and get coding gain and constellation shaping gain at the same time.

Detection algorithm with high performance and low complexity

PDMA adopts the multi-user detection algorithm with high performance low complexity. It gets the performance approaching the maximum a posteriori (MAP) on the premise of that the complexity is reduced significantly, and gets the optimal balance point between the velocity of the convergence and the detection performance.

The main challenges of the PDMA technology include following items:

- the best way of constellation mapping based on the joint optimization design of the PDMA associated with code modulation.
- universal detection algorithms with low complexity and high performance.
- PAPR reduction in power limited case.

PDMA has been incorporated into ITU-R Report Future technology trends of terrestrial IMT systems [39] by ITU organization in 2014. Since Release 14 for 5G new radio in 2016, PDMA has been given keen concern and in hot discussion in 3GPP [26–35].

With the support of the National High Science and Technology Plan ('863' Plan, No. 2015*AA*01*A*709) by Chinese government, PDMA testbed was developed since 2015 and further updated to 5G prototypes and pre-commercial products. In 2016, PDMA passed the MIIT of China 5G 1st step verification for 5G key technologies. In 2017, PDMA passed the MIIT of China 5G 2nd step verification for 5G typical scenarios. Now, PDMA is preparing for the future 5G verification for networking.

Acknowledgements The authors would like to give special appreciates to Dr. Wang Yingmin, Dr. Zhao Zheng, Ms. Xing Yanping, Mr. Tang Wanwei and Mr. Yue Xinwei from China Academy of Telecommunication Technologies, Prof. Dai Xiaoming from University of Science & Technology Beijing, Prof. Niu Kai from Beijing University of Post and Telecommunications, Dr. Zeng Jie from Tsinghua University, and Dr. Jiang Yanxiang from Southeast University for their valuable input suggestions on PDMA extension technologies. Also, the authors give thanks to Prof. Dake Liu of Beijing Institute of Technology, and Ms. Zheng Yadan of Beijing University of Post and Telecommunications for their kind reviews and revisions.

References

1. X. Dai, Successive interference cancellation amenable space-time codes with good multiplexing-diversity tradeoffs. Wirel. Person. Commun. **55**(4), 645–654 (2010)
2. X. Dai, R. Zou, J. An, X. Li, S. Sun, Y. Wang, Reducing the complexity of quasi-maximum-likelihood detectors through companding for coded MIMO systems. IEEE Trans. Vehic. Technol. **61**(3), 1109–1123 (2012)
3. X. Dai, S. Sun, Y. Wang, Reduced-complexity performance-lossless (quasi-) maximum-likelihood detectors for S-QAM modulated MIMO systems. Electron. Lett. **49**(11), 724–725 (2013)
4. X. Dai, Z. Zhang, K. Long, S. Sun, Y. Wang, Unequal-error-correcting-capability-aware iterative receiver for (parallel) turbo-coded communications. IEEE Trans. Vehic. Technol. **63**(7), 3446–3451 (2014)
5. X. Dai, S. Chen, S. Sun, S. Kang, Y. Wang, Z. Shen, J. Xu, Successive interference cancelation amenable multiple access (SAMA) for future wireless communications, in *2014 IEEE International Conference on Communication Systems (ICCS)* (IEEE, 2014), pp. 222–226
6. S. Sun, Pattern division multiple access (PDMA). Future-Taiwan 5G Workshop (2014), http://www.future-forum.org/
7. S. Kang, X. Dai, B. Ren, Pattern division multiple access for 5G. Telecommun. Netw. Technol. **5**(5), 43–47 (2015)
8. S. Chen, B. Ren, Q. Gao, S. Kang, S. Sun, K. Niu, Pattern division multiple access a novel nonorthogonal multiple access for fifth-generation radio networks. IEEE Trans. Vehic. Technol. **66**(4), 3185–3196 (2017)
9. D. Tse, P. Viswanath, *Fundamentals of Wireless Communication*. Cambridge University Press (2005)
10. S. Loyka, F. Gagnon, Performance analysis of the V-BLAST algorithm: an analytical approach. IEEE Trans. Wirel. Commun. **3**(4), 1326–1337 (2004)
11. D. Truhachev, Universal multiple access via spatially coupling data transmission, in *2013 IEEE International Symposium on Information Theory Proceedings (ISIT)* (IEEE, 2013), pp. 1884–1888
12. J. Xu, X. Dai, W. Ma, Y. Wang, A component-level soft interference cancellation based iterative detection algorithm for coded MIMO systems, in *2014 IEEE 80th on Vehicular Technology Conference (VTC Fall)* (IEEE, 2014), pp. 1–5
13. B. Ren, X. Yue, W. Tang, Y. Wang, S. Kang, X. Dai, S. Sun, Advanced IDD receiver for PDMA uplink system, in *2016 IEEE/CIC International Conference on Communications in China (ICCC)* (IEEE, 2016), pp. 1–6
14. B. Ren, Y. Wang, X. Dai, K. Niu, W. Tang, Pattern matrix design of PDMA for 5G UL applications. China Commun. **13**(2), 159–173 (2017)
15. J. Harshan, B.S. Rajan, On two-user gaussian multiple access channels with finite input constellations. IEEE Trans. Informat. Theor. **57**(3), 1299–1327 (2011)
16. R1-166098, Discussion on the feasibility of advanced MU-detector. 3GPP RAN1#86, Huawei Technologies (2016)
17. Rep. ITU-R M.2135-1, Guidelines for Evaluation of Radio Interface Technologies for IMT-Advanced, Dec 2009
18. W. Tang, S. Kang, B. Ren, Uplink Pattern Division Multiple Access (PDMA) in 5G Systems. IET Communications, to be appeared
19. W. Tang, S. Kang, B. Ren, X. Yue, Uplink grant-free pattern division multiple access (GF-PDMA) for 5G radio access. China Commun. (to be appeared)
20. Y. Wang, S. Kang, B. Ren, PDMA grant free transmission for 5G URLLC scenario, Feb 2017
21. W. Tang, S. Kang, B. Ren, Performance analysis of cooperative pattern division multiple access (Co-PDMA) in uplink network. IEEE Acc. **5**, 3860–3868 (2017)
22. P. Li, Y. Jiang, S. Kang, F. Zheng, X. You, Pattern division multiple access with large-scale antenna array, in *IEEE VTC 2017-Spring*, Jun 2017

23. J. Zeng, D. Kong, B. Liu, X. Su, T. Lv, RIePDMA and BP-IDD-IC detection. EURASIP J. Wirel. Commun. Netw. **1**, 12 (2017)
24. J. Dai, K. Niu, Z. Si, J. Lin, Polar coded non-orthogonal multiple access, in *2016 IEEE International Symposium on Information Theory (ISIT)* (IEEE, 2016), pp. 988–992
25. L. Lei, C. Yan, G. Wenting, Y. Huilian, W. Yiqun, X. Shuangshuang, Prototype for 5G new air interface technology scma and performance evaluation. China Commun. **12**(Supplement), 38–48 (2015)
26. R1-162305, Multiple access for 5G new radio interface. 3GPP RAN1#84bis, CATT (2016)
27. R1-162306, Candidate Solution for New Multiple Access. 3GPP RAN1#84bis, CATT (2016)
28. R1-164246, Discussion on scenarios and use cases for MA. 3GPP RAN1#85, CATT (2016)
29. R1-164247, Performance on LLS of PDMA. 3GPP RAN1#85, CATT (2016)
30. R1-166466, Usage scenarios of non-orthogonal multiple access. 3GPP RAN1#86, CATT (2016)
31. R1-166467, Discussion on traffic model of mMTC. 3GPP RAN1#86, CATT (2016)
32. R1-166468, Remaining issues on evaluation assumption and methodology of MA. 3GPP RAN1#86 CATT (2016)
33. R1-166469, Update of LLS results of PDMA. 3GPP RAN1#86, CATT (2016)
34. R1-166470, Initial SLS results of PDMA. 3GPP RAN1#86, CATT (2016)
35. R1-168757, Consideration on grant-free transmission. 3GPP RAN1#86bis, CATT (2016)
36. Y. Wang, B. Ren, S. Sun, S. Kang, X. Yue, Analysis of non-orthogonal multiple access for 5G. China Commun. **13**(Supplement 2), 52–66 (2016)
37. IMT-2020(5G) Promotion Group, Whitepapers, http://www.imt-2020.org.cn/zh
38. W. Shin, M. Vaezi, B. Lee, D.J. Love, J. Lee, H.V. Poor, Non-orthogonal multiple access in multi-cell networks: theory, performance, and practical challenges. IEEE Commun. Mag. **55**(10), 176–183 (2017)
39. Report ITU-R M.2320-0, Future technology trends of terrestrial IMT systems, Nov 2014, http://www.itu.int/ITU-R/

Chapter 15
Low Density Spreading Multiple Access

Mohammed Al-Imari and Muhammad Ali Imran

15.1 Motivations for Low Density Spreading

In this section, the motivation behind the proposal of low density spreading as a multiple access technique will be presented. The aim is to provide a general understanding of the challenges/drawbacks of the conventional spreading multiple access technique that can be resolved with the low density spreading (LDS) scheme. To this end, the code division multiple access (CDMA) system is presented with focus on the multiple access interference (MAI) drawback and the approaches that have been proposed to tackle this problem. This is followed by introducing the concept of low density spreading for CDMA and highlighting its difference and advantages comparing to the conventional CDMA system.

15.1.1 Code Division Multiple Access (CDMA)

A spread-spectrum system is a system in which the transmitted signal is spread over a wide frequency band, much wider than the bandwidth required to transmit the information being sent. The power spectral density of the useful signal will be reduced to a level even maybe below the noise level. There are many applications and advantages for spreading the spectrum; anti-jamming, interference rejection, low probability of intercept and multiple access. The primary spread spectrum system for multiple access is the Direct Sequence CDMA. CDMA is an efficient multiple access

M. Al-Imari (✉)
MediaTek Wireless Ltd, Cambourne, Cambridge CB23 6DW, UK
e-mail: moh.al-imari@ieee.org

M. A. Imran
University of Glasgow, Glasgow G12 8QQ, UK
e-mail: Muhammad.Imran@glasgow.ac.uk

M. Vaezi et al. (eds.), *Multiple Access Techniques for 5G Wireless Networks and Beyond*, https://doi.org/10.1007/978-3-319-92090-0_15

technique that has been adopted in 3G mobile communication systems [1]. CDMA allows for many sources' (or users') information to be transmitted simultaneously over a single communication channel. The sources are distinguished by spreading codes. Apart from its many advantages, CDMA also has several limitations. One of the well-known disadvantages is that CDMA is an interference limited system. This means that the capacity of CDMA system is affected by the existing of MAI.

Consider an uplink synchronous CDMA system with set of users $\mathscr{K}$ and processing gain L. Each user will be assigned a unique spreading code with length L to spread its symbols. Let $a_k \in \mathscr{X}$ and $\mathbf{s}_k = [s_{k,1}, s_{k,2}, \ldots, s_{k,L}]^T$ be the modulation symbol and the spreading code of the kth user, respectively, where $\mathscr{X}$ is the constellation alphabet. The discrete-time model for the received signal on the lth chip, r_l, will be

$$r_l = \sum_{k \in \mathscr{K}} a_k s_{k,l} + z_l, \tag{15.1}$$

where z_l is Gaussian noise. In conventional CDMA systems, at the detection of a user signal, the signal of the other users is considered as interference. For the kth user, the interference component can be shown as follows

$$\mathbf{r} = a_k \mathbf{s}_k + \overbrace{\sum_{m \in \mathscr{K} \backslash k} a_m \mathbf{s}_m}^{\mathbf{mai}_k} + \mathbf{z}, \tag{15.2}$$

where $\mathbf{mai}_k$ stands for the MAI the kth user sees, $\mathbf{r} = [r_1, r_2, \ldots, r_L]^T$, and $\mathbf{z} = [z_1, z_2, \ldots, z_L]^T$. Thus, the performance of the data detection is highly affected by the number of active users in the system. Also, with conventional detection, a power control is required to mitigate the near–far effect problem. Many approaches are proposed for alleviating the MAI and improve the system performance such as

- Employing multiuser detection (MUD) techniques at the receiver by exploiting the knowledge of the user of interest as well as interferers.
- Designing suitable spreading codes so that the MAI can be decreased. It is typically done by designing the codes with good cross-correlation properties.
- Incorporating forward error correction (FEC) coding. Although the FEC is designed to remove the noise instead of MAI, when the MAI can be assumed as noise, a powerful FEC can be deployed to combat the MAI too.

15.1.1.1 Multiuser Detection

There has been great interest in improving CDMA performance through the use of MUD. Many MUD techniques are proposed to suppress the interference caused by the non-orthogonality of codes. In MUD, codes, delays, amplitudes, and phases information of all the users are jointly used to better detect each individual user. The optimal MUD has been proposed by Verdu [22] and achieves the optimal performance in terms of error probability. The optimal MUD uses the maximum likelihood criterion in detecting the users' symbols, which selects the symbols' sequence, $\hat{\mathbf{a}}$, that maximizes the likelihood function $p(\mathbf{r}|\mathbf{a})$ given the channel observation

$$\hat{\mathbf{a}} = \arg\max_{\mathbf{a}\in\mathscr{X}^K} \left(- \|\mathbf{r} - \mathbf{S}\mathbf{a}\|^2 \right), \tag{15.3}$$

where $\mathbf{S} = [\mathbf{s}_1, \mathbf{s}_2, \ldots, \mathbf{s}_K]$, and $\mathbf{a} = [a_1, a_2, \ldots, a_K]^T$. By incorporating the prior distribution of the transmitted symbols, the Maximum A Posteriori Probability (MAP) can be derived. The MAP detector maximizes the joint posterior probability, $p(\mathbf{a}|\mathbf{r})$, of the transmitted symbols, which can be implemented jointly

$$\hat{\mathbf{a}} = \arg\max_{\mathbf{a}\in\mathscr{X}^K} p(\mathbf{a}|\mathbf{r}), \tag{15.4}$$

or individually through marginalizing the joint posterior probability in (15.4) as follows

$$\begin{aligned} \hat{a}_k &= \arg\max_{b\in\mathscr{X}} \sum_{\mathbf{a}\in\mathscr{X}^K}^{a_k=b} p(\mathbf{a}|\mathbf{r}) \\ &= \arg\max_{b\in\mathscr{X}} \sum_{\mathbf{a}\in\mathscr{X}^K}^{a_k=b} \prod_{k\in\mathscr{K}} p(a_k) \prod_{l=1}^{L} p(r_l|\mathbf{a}), \quad \forall k \in \mathscr{K}, \end{aligned} \tag{15.5}$$

where $p(a_k)$ is the priori probability of the symbol of the kth user. The second equality is obtained by using Bayes' rule and the fact that the users' symbols are independent and the noise vector is independent and identically distributed (i.i.d). The optimal solution for (15.3) and (15.5) requires a search over all the possible combinations of the symbols in the vector $\mathbf{a}$. Hence, the computational complexity of the optimal MUD increases exponentially with the number of users, and a complexity order of $\mathscr{O}(|\mathscr{X}|^K)$ is required. Thus, it is clear that the optimal MUD is infeasible for practical implementation.

The high complexity of the optimal MUD has motivated the search for suboptimal MUD techniques with low computational complexity such as linear minimum mean square error, the decorrelator detector, the orthogonal multiuser detector and subtractive interference cancellation detectors. In spite of the low complexity of the linear receivers, their performance can be far from the optimal MUD performance.

15.1.1.2 Spreading Codes Design

In CDMA, the system loading is given by the ratio of the number of admissible users K and the processing gain L (number of chips per symbol), and it is denoted by $\beta = K/L$. Thus the system is called underloaded, fully-loaded, and overloaded when $\beta < 1$, $\beta = 1$, and $\beta > 1$, respectively. For underloaded and fully-loaded conditions, orthogonal codes are optimal and can be easily constructed, subsequently the multiuser channel will decouple, ideally, into single-user channels. Hence, there is no MAI and the matched filter single-user detector is optimal.

However, for an overloaded system, where K is larger than L, orthogonal codes are impossible to be constructed and non-orthogonal codes are used instead. In this case, the source of interference is due to the non-orthogonality of users' codes and the performance of the system will depend on the cross-correlation among users' codes. The codes that meet the Welch-bound equality (WBE) are known as the optimal codes that maximize the sum-rate capacity for overloading condition [24]. However, the WBE-optimized codes are constructed by using a function of a specific number of active users and spreading gain. This is the problem that fundamentally limits their practicality in the real system where the number of users dynamically changes over a period of time. Furthermore, the optimality of WBE codes can be achieved only when the optimal MUD techniques are used (e.g., MAP) [12]. Although optimal MUD techniques can effectively combat the MAI problem, its complexity grows exponentially with the number of users, and it is intractable for practical implementation.

Alternatively, another scheme for designing codes for overloaded CDMA is the hierarchy of orthogonal sequences (HOS). In HOS, a group of users ($K^{[1]} = L$) is assigned orthogonal codes and the rest of users ($K^{[2]} = K - L$) are assigned another set of orthogonal codes or pseudo-random noise (PN) sequences. Thus, the interference levels of the users are decreased significantly as compared to random spreading, since any user suffers from interference caused by the users belonging to the other group of users only. If orthogonal codes are assigned for the rest of the users, the system is referred to it as OCDMA/OCDMA, and if PN sequences are used then it is called PN/OCDMA. It has been shown that OCDMA/OCDMA has performance close to the system that is based on WBE codes.

15.1.2 *Low Density Spreading CDMA*

Motivated by the facts mentioned in the previous section and in the pursuit to develop low complexity and efficient MUD techniques, Kabashima has proposed a low complexity MUD based on belief propagation (BP) using Gaussian approximation [11]. It was shown numerically that the proposed MUD achieves near optimal performance for moderately loaded CDMA system. Improvements of the algorithm have been proposed in [18]. The same approach of MUD based on the BP was proposed to approximate the parallel interference cancellation in CDMA system [21]. Inspired by the success of low density parity check codes, the LDS was proposed as a method for guaranteeing the convergence of BP-based MUD [10, 16, 20, 26]. The LDS structure is also known as sparsely spread CDMA, sparse CDMA and low density signature CDMA.

The main idea of the LDS technique is to switch off a large number of spreading code chips, which makes the code a sparse vector. In other words, each user

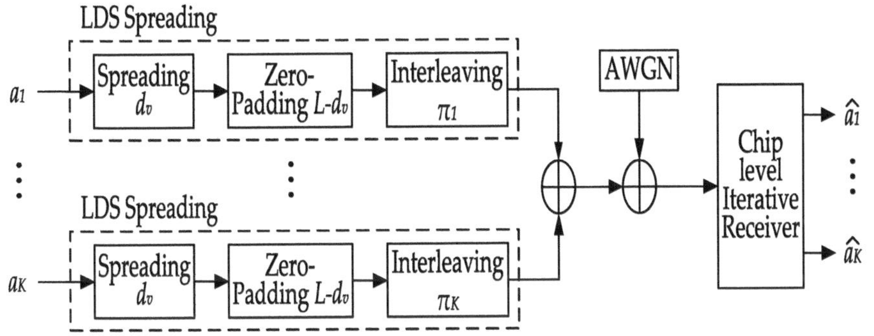

Fig. 15.1 Uplink LDS-CDMA block diagram

will spread its data over a small number of chips. Then the chips are arranged intelligently so that the number of interferers per chip is far smaller than the number of total users. It is proven in [16] that, with low density codes, the optimal MAP symbol detection (15.4) can be implemented using BP, i.e., the MUD based on BP is asymptotically optimal. The LDS-CDMA system model is depicted in Fig. 15.1. This LDS structure brings about the following advantages:

- Higher chip-level signal to interference-plus-noise ratio can be achieved which leads to a better detection process.
- At each received chip, a user will have relativity small number of interferers, so the search space should be smaller and, hence, more affordable detection techniques can be used.
- Each user will see an interference coming from different users at different chips which result in interference diversity by avoiding strong interferers to corrupt all the chips of a user.

Figure 15.2 depicts an example of the structure of the LDS-CDMA in comparison to conventional CDMA. Let d_v and d_c be the number of chips over which a single user will spread its data and the maximum number of users that are allowed to interfere within a single chip, respectively. The new spreading sequences for each user will have a maximum of d_v nonzero values and $L - d_v$ zero values. Each user will see interference coming from $d_c - 1$ users at each chip. Moreover, as shown in the figure, in LDS-CDMA the interference level is different for different chips. In conventional CDMA, each user will spread on all the L chips, so each user will see the interference coming from $K - 1$ users. LDS structure allows applying close to optimal chip-level MUD based on message passing algorithm (MPA) [13]. The complexity of the LDS detector receiver turns out to be $\mathcal{O}(|\mathcal{X}|^{d_c})$, which is significantly reduced comparing to $\mathcal{O}(|\mathcal{X}|^{K})$ for an optimal receiver for conventional CDMA system.

Figure 15.3 shows the bit error rate (BER) performance for un-coded LDS-CDMA with BPSK modulation and different system loading in comparison with single-user performance over additive white gaussian noise (AWGN) channel. As the figure

Fig. 15.2 Structure of LDS-CDMA in comparison to conventional CDMA

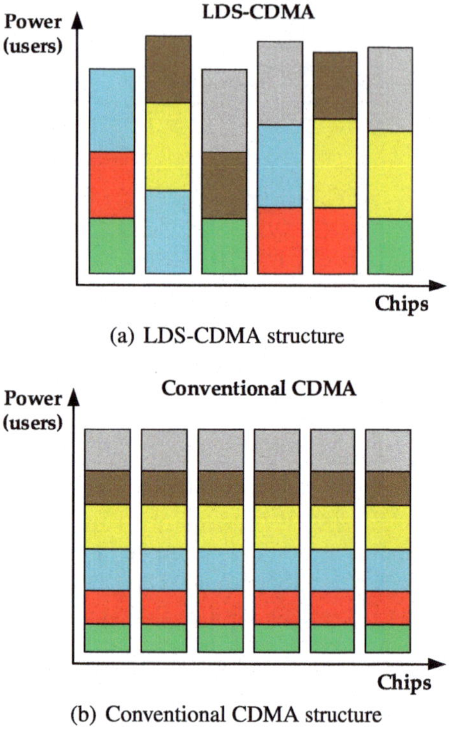

Fig. 15.3 Performance results for LDS-CDMA

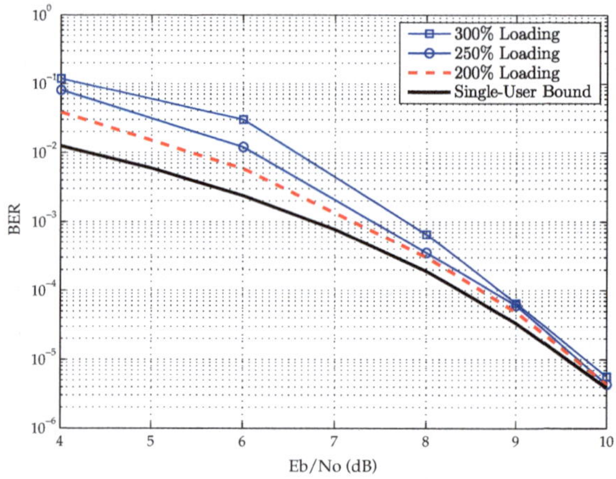

shows, LDS-CDMA can achieve an overall performance, in overloading conditions, that is close to single-user performance.

15.2 Multicarrier Low Density Spreading Multiple Access

To cope with the multipath effect of the wireless channel, a multicarrier version of LDS-CDMA has been introduced. In this section, the basic concepts of the multicarrier low density spreading multiple access (MC-LDSMA) system are introduced along with the system model and the receiver structure. Then, the properties of MC-LDSMA such as complexity and frequency diversity are discussed and compared with existing multiple access techniques.

15.2.1 MC-LDSMA System Model

Like all single-carrier communication techniques, LDS-CDMA is prone to the multipath channel conditions. The multipath propagation will result in inter-chip interference (ICI) as illustrated in Fig. 15.4, which destroys the low density structure and a dense graph will be resulted at the receiver. Therefore, the LDS structure is applied to a multicarrier system (such as orthogonal frequency division multiplexing (OFDM)) to cope with the multipath channel effect. This can be understood as the extension of LDS structure to multicarrier code division multiple access (MC-CDMA) system, in which the spreading is carried out in the frequency domain. As this technique uses multicarrier transmission and the users access the system through the LDS codes, it has been referred to as MC-LDSMA [4, 5] or LDS-OFDM [3]. The conceptual block diagram of an uplink MC-LDSMA system is depicted in Fig. 15.5. The system consists of a set of users $\mathscr{K} = \{1, \ldots, K\}$ transmitting to the same base station where the base station and each user are equipped with a single antenna. In MC-LDSMA, the user data is spread in the frequency domain before transmission. Let $\mathbf{a}_k$ be a data vector of user k consisting of M_k modulated data symbols and denoted as

$$\mathbf{a}_k = [a_{k,1}, a_{k,2}, \ldots, a_{k,M_k}]^T. \tag{15.6}$$

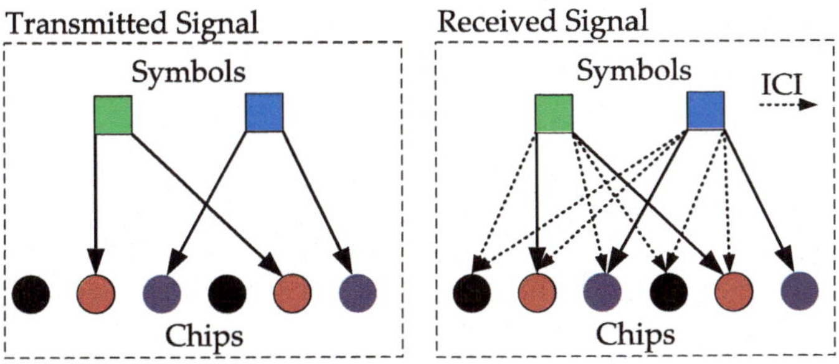

Fig. 15.4 Multipath effect on the low density structure

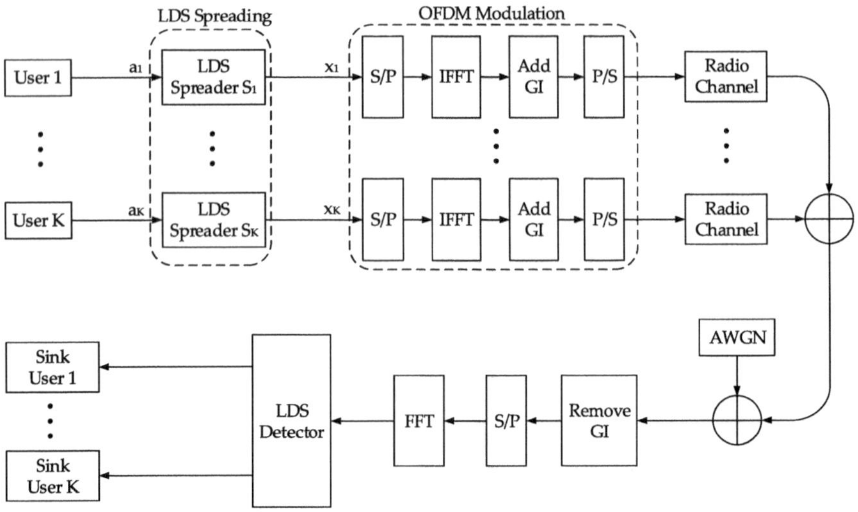

Fig. 15.5 Uplink MC-LDSMA block diagram

Without loss of generality, all users are assumed to take their symbols from the same constellation alphabet $\mathcal{X}$. Each user will be assigned a spreading matrix $\mathbf{S}_k$, which consists of M_k LDS codes

$$\mathbf{S}_k = [\mathbf{s}_{k,1}, \mathbf{s}_{k,2}, \ldots, \mathbf{s}_{k,M_k}], \tag{15.7}$$

where each LDS code, $\mathbf{s}_{k,m} \in \mathscr{C}^{N\times 1}$, is a sparse vector consisting of N chips. Among these N chips only d_v chips have nonzero values, where d_v is the effective spreading factor. Each data symbol, $a_{k,m}$, will be spread using the mth spreading sequence. Let $\mathbf{x}_k = [x_{k,1}, x_{k,2}, \ldots, x_{k,N}]^T$ denote the chips vector belonging to user k after the spreading process, which is given by

$$\mathbf{x}_k = \mathbf{S}_k \ \mathbf{a}_k. \tag{15.8}$$

Hence, the whole system code matrix has N rows and M columns each containing a unique spreading code, where M can be calculated as follows

$$M = \sum_{k=1}^{K} M_k. \tag{15.9}$$

The system loading (β), which is the ratio of the number of transmitted symbols to the total number of subcarriers, will be

$$\beta = \frac{M}{N}$$

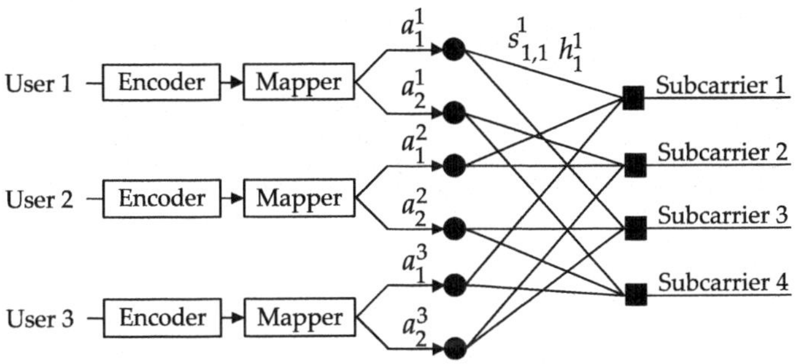

Fig. 15.6 Graphical representation of MC-LDSMA

In orthogonal transmission, the system loading cannot be more than 1, i.e., the total number of transmitted symbols is equal or less than the number of subcarriers. The main advantage of the low density spreading is that the system loading can be more than 100% ($\beta > 1$) with affordable complexity and close to the single-user performance. The case when the system loading is larger than 1 is referred to it as an overloaded system. Each user-generated chip will be transmitted over a subcarrier of the OFDM system, and the terms chip and subcarrier will be used interchangeably to refer to the same thing.

Figure 15.6 illustrates the MC-LDSMA principle by an example of a system with four subcarriers ($N = 4$), serving three users ($K = 3$) with two data symbols per user ($M_1 = M_2 = M_3 = 2$), which means 150% loading. Here the effective spreading factor is two ($d_v = 2$) and each three chips sharing one subcarrier ($d_c = 3$), where d_c denotes the number of users interfere in each subcarrier. The figure shows in more details the process of low density spreading. As it can be observed, each chip represents a subcarrier of OFDM modulation and the data symbols using the same subcarrier will interfere with each other.

The system spreading matrix can be represented by an indicator matrix $\mathbf{I}_{LDS,4\times 6}$, which represents the positions of the nonzero chips in each spreading code as follows

$$\mathbf{I}_{LDS,4\times 6} = \begin{bmatrix} 1\,0\,1\,0\,1\,0 \\ 0\,1\,1\,0\,0\,1 \\ 1\,0\,0\,1\,0\,1 \\ 0\,1\,0\,1\,1\,0 \end{bmatrix}.$$

As users are not bounded to exclusively use the subcarriers, at the receiver side, users' signals that use the same subcarrier will be superimposed. However, the number of users interfere in each subcarrier is much less than the total number of users, $d_c \ll K$.

At the receiver, after performing OFDM demodulation operation, the received signal, $\mathbf{r} \in \mathscr{C}^{N\times 1}$, is given by

$$\mathbf{r} = \sum_{k \in \mathcal{K}} \mathbf{H}_k \ \mathbf{x}_k + \mathbf{z}, \tag{15.10}$$

where $\mathbf{z} = [z_1, \ldots, z_N]^T$ is the AWGN vector and $\mathbf{H}_k$ is the frequency domain channel gain matrix of user k

$$\mathbf{H}_k = \text{diag}(\sqrt{h_{k,1}}, \sqrt{h_{k,2}}, \ldots, \sqrt{h_{k,N}}). \tag{15.11}$$

Here, $h_{k,n}$ is the channel gain of user k on subcarrier n. This signal $\mathbf{r}$ is passed to the LDS MUD to separate users' symbols which is done using chip-level iterated MUD based on MPA.

The basic form of chip-level iterated MUD can be explained as follows. Let

$$\mathcal{J}_n = \{(k, m) : s_{k,m}^n \neq 0\}, \tag{15.12}$$

identifies different users' data symbols that share the same subcarrier n. Consequently, the received signal on subcarrier n can be written as

$$r_n = \sum_{(k,m) \in \mathcal{J}_n} a_{k,m} s_{k,m}^n \sqrt{h_{k,n}} + z_n. \tag{15.13}$$

A MC-LDSMA system, with M number of symbols and N number of subcarriers, can be represented using a factor graph $\mathcal{G}(\mathcal{U}, \mathcal{V})$ where users' symbols are represented by variable nodes $u \in \mathcal{U}$ and chips are represented by function nodes $v \in \mathcal{V}$. For simplicity, it will be assumed that each user transmits one symbol only. The connection between the received chip and its related users is represented by edges. Let $e_{k,n}$ represent the edge connecting variable node u_k, $k = 1, \ldots, K$ and function node v_n, $n = 1, \ldots, N$. It is straightforward to check that the factor graph representation of the conventional MC-CDMA is fully connected, where each variable node is connected to all function nodes. However, in MC-LDSMA system, each variable node is connected to d_v function nodes only and each function node is connected to d_c variable nodes only. Figure 15.7 depicts an example of factor graph representation of LDS structure in MC-LDSMA system with $K = 8$ and $N = 6$.

Using MPA, the messages are updated and iteratively exchanged between function and variable nodes along the respective edges. Those messages are the soft-values that represent the inference or the reliability of the symbol associated to each edge. Let ζ_k be the set of chip indices over which the kth symbol is spread and ξ_n be the set of symbol indices that interfering in the nth received chip, r_n. Let $L_{v_n \leftarrow u_k}$ and $L_{v_n \rightarrow u_k}$ be the message sent along edge $e_{k,n}$ from variable node u_k and function node v_n, respectively. The message $L_{v_n \leftarrow u_k}$ gives an updated inference of a_k based on the observation taken at chips r_q, $\forall q \in \zeta_k \setminus n$. The messages of the jth iteration, sent by variable nodes, are updated using the following rule

$$L_{v_n \leftarrow u_k}^{j} = \sum_{l \in \zeta_k \setminus n} L_{v_l \rightarrow u_k}^{j-1}. \tag{15.14}$$

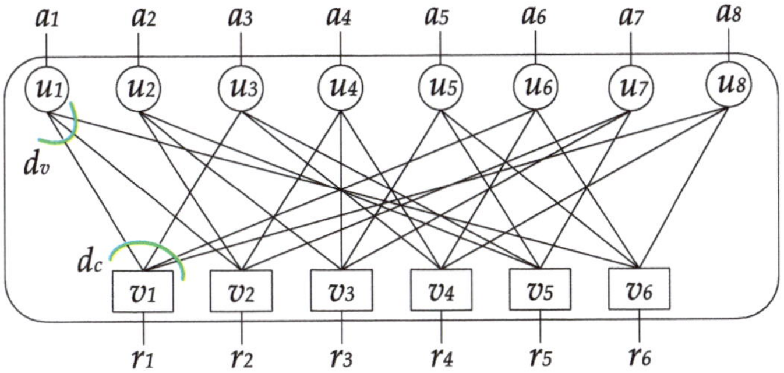

Fig. 15.7 Factor graph representation of the LDS structure in MC-LDSMA

For nth function node, the message of jth iteration is calculated as follows

$$L^{j}_{v_n \rightarrow u_k} = \mathscr{F}(a_k | r_n, L^{j}_{v_n \leftarrow u_l}, \forall l \in \xi_n \setminus k). \tag{15.15}$$

To approximate the optimal MAP detector, the function $\mathscr{F}(.)$ in (15.15) performs local marginalization and can be written as follows using the logarithmic value

$$\mathscr{F}(.) = \log \left(\sum_{\mathbf{a}_n \in \mathscr{X}^{d_c}}^{a_k} p(r_n | \mathbf{a}_{[n]}) \prod_{l \in \xi_n \setminus k} p(a_l) \right), \tag{15.16}$$

where

$$p(r_n | \mathbf{a}_{[n]}) \propto \exp\left(-\frac{1}{2\sigma^2} \| r_n - \bar{\mathbf{h}}^T_{[n]} \mathbf{a}_{[n]} \|^2\right), \tag{15.17}$$

$$p(a_k) = \exp\left(L^{j}_{v_n \leftarrow u_k}(a_k)\right). \tag{15.18}$$

Here, $\mathbf{a}_{[n]}$ and $\bar{\mathbf{h}}_{[n]}$ are the vectors that contain the symbols transmitted on the nth chip and their corresponding effective received code, respectively. Figure 15.8 illustrates the message passing process for the first function node (v_1) and the first variable node (u_1). After appropriate number of iterations or when the iterations reach the maximum limit (J), the posteriori probability of the transmitted symbol a_k will be as follows

$$L_k(a_k) = \sum_{l \in \zeta_k} L^{J}_{v_l \rightarrow u_k}. \tag{15.19}$$

If hard-decision is used, the estimated value of transmitted symbol $\hat{a}_k$ will be

$$\hat{a}_k = \arg \max_{a_k \in \mathscr{X}} L_k(a_k). \tag{15.20}$$

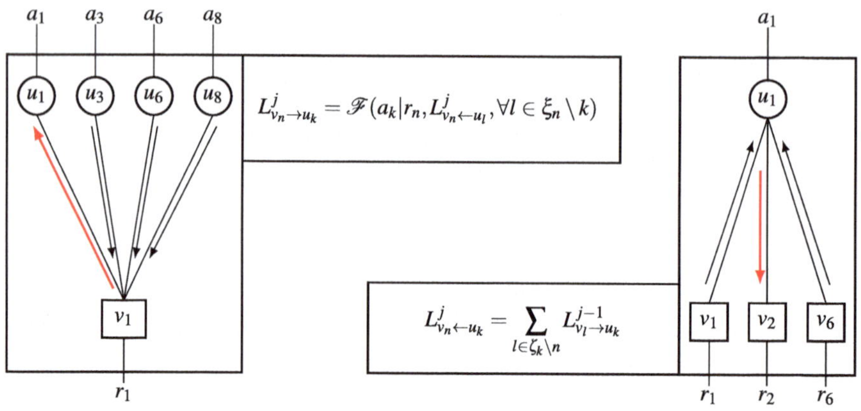

Fig. 15.8 Message passing process of the LDS detector in MC-LDSMA

Otherwise, the soft output (15.19), which is calculated at each variable node, will be sent to the channel decoder.

15.2.1.1 LDS Codes Design for MC-LDSMA

The low density spreading codes for MC-LDSMA can be implemented by first spreading the data symbol based on the effective spreading factor (d_v), then the resulted chips are mapped to specific subcarriers. The subcarrier mapping (or it can be referred to it as subcarrier allocation) can be optimized off-line, static subcarrier allocation, or it can be optimized dynamically based on the subcarriers' channel gains. In the static design of low density codes, the main aspect need to be considered is reducing the number of short-cycles in the indicator matrix. It has been shown that the MPA outputs a good approximate of the a posteriori probability if the Girth is not less than 6 [8, 25]. The Girth is the length of the smallest cycle in the factor graph associated with the indicator matrix. Thus, for better performance, short-cycle of length 4, cycle-of-four (CoF), has to be avoided in the design of the indicator matrix. There are many algorithms in the literature for designing matrices with no CoF [14, 15], which can be used for designing LDS codes.

For LDS-CDMA, these algorithms can be implemented directly. However, for MC-LDSMA, another aspect needs to be considered in the design of the indicator matrix. Due to the frequency selectivity, the subcarriers belong to each symbol has to be separated apart to achieve frequency diversity. More specifically, the subcarriers should be separated from each other by at least the channel coherence bandwidth B_c.

$$n_i - n_j \geq \left\lceil \frac{B_c}{\Delta f} \right\rceil, \quad n_i > n_j, \quad \forall n_i, n_j \in \mathscr{D} \tag{15.21}$$

where $\triangle f$ is the subcarriers' frequency separation and $\lceil x \rceil$ represent the smallest integer that is greater than x. $\mathscr{D}$ is the set of subcarriers that the symbol will be spread on, and n is the subcarrier index.

15.2.2 MC-LDSMA Properties in Comparison with Other Multiple Access Techniques

In this section, the properties of MC-LDSMA are discussed and the scheme is compared with other multiple access techniques such as orthogonal frequency division multiple access (OFDMA) and MC-CDMA. The section will focus on the link-level aspects of MC-LDSMA. The system-level aspects of MC-LDSMA and comparison with other multiple access techniques will be deferred to the chapter where radio resource allocation is considered. MC-LDSMA can represent a midway approach comparing to orthogonal multiple access techniques and conventional spread-spectrum multiple access. In the orthogonal multiple access techniques, such as OFDMA and single-carrier (SC)-FDMA, the users are orthogonal to each other, which simplifies the detection process but no frequency gain can be achieved. On the other hand, spread-spectrum multiple access methods, such as MC-CDMA, attempt to achieve the full frequency diversity gain on the cost of higher MUD complexity. In MC-LDSMA, part of the frequency diversity can be achieved by spreading on a small number of subcarriers with affordable MUD complexity. Here, the advantages of MC-LDSMA in comparison to these two conventional approaches will be highlighted.

A. Spread-Spectrum Multiple Access

In conventional MC-CDMA, each user's symbols are spread over all the subcarriers. In the synchronous MC-CDMA downlink, orthogonal spreading codes are of advantage, since they reduce the multiple access interference in comparison to non-orthogonal sequences. By contrast, in the uplink, the MC-CDMA signals received at the base station suffer from different degradations introduced by the users' independent frequency-selective channels. Consequently, users' codes are no longer orthogonal, which causes multiuser interference. Although optimal MUD techniques can effectively combat the multiuser interference, their complexity increases exponentially with the number of users, which is intractable for practical implementation. Considering that LDS structure reduces the number of interferers in each chip, it allows applying close to optimal MUD based on MPA. For MC-LDSMA, the complexity of MAP receiver will turn out to be $\mathscr{O}(|\mathscr{X}|^{d_c})$, which is significantly reduced compared to $\mathscr{O}(|\mathscr{X}|^{K})$ for optimal MUD for MC-CDMA. As MC-CDMA system the symbol is spread over all the subcarriers, full frequency diversity can be achieved if the optimal MUD is applied. However, due to the high complexity of the optimal MUD in MC-CDMA, linear MUD is usually implemented in practical systems. Therefore, MC-CDMA can't enjoy full diversity gain and MC-LDSMA outperforms the MC-CDMA performance as it will be shown in the next sections.

Furthermore, it is well known that, from frequency diversity perspective, the user does not need to transmit over all the subcarriers to achieve the maximum diversity. In fact, the frequency diversity can be achieved by transmitting on subcarriers which are separated from each other by the coherence bandwidth of the channel [7]. Thus, it is possible to reduce the MAI and the MUD complexity without sacrificing the diversity by spreading on less number of subcarriers comparing to the conventional MC-CDMA.

Comparing to sparse code multiple access (SCMA), MC-LDSMA and SCMA are based on the same concept of low density spreading. SCMA, which can be considered as an extension of MC-LDSMA, uses modulation shaping instead of the regular rectangular QAM modulation that is adopted in MC-LDSMA. Using modulation shaping may improve the performance of the LDS multiple access in AWGN. However, it is not clear if there is a gain in applying modulation shaping to LDS multiple access to frequency-selective channel, which is a more practical scenario in real systems compared to AWGN.

B. Orthogonal Multiple Access

One example of orthogonal multiple access is OFDMA. In OFDMA system, the set of subcarriers is divided into several mutually exclusive subsets and then each subset is allocated to transmission of a user signal. This approach creates frequency domain orthogonality for users' signals when the transmitter and the receiver are perfectly synchronized and hence avoids MAI. As in OFDMA user-data symbols are assigned directly to subcarriers, the frequency domain diversity will not be achievable at the modulation symbol level. Thus, it will be crucial to incorporate properly designed error correction coding and interleaving schemes to obtain this diversity at a later stage.

On the other hand, in MC-LDSMA system, each modulation symbol is spread on a number of subcarriers. Therefore, frequency diversity is achievable at the modulation symbol level in addition to the frequency diversity gained when channel coding is used. Frequency diversity can be gained by assigning distributed and spaced-enough subcarriers for spreading of a given data symbol. Consequently, even though the spreading is over a limited number of chips, the system will still be able to gain frequency diversity. However, MC-LDSMA detector has larger complexity comparing to OFDMA receiver due to the need to implement MUD. The increased computational complexity of the system in comparison to a conventional receiver used for OFDMA is practically affordable as the added complexity is in the base station side. Also, the complexity is reasonably justified considering the achieved gain in performance. Further reduction in complexity of multiuser detection for MC-LDSMA can be achieved by applying the Grouped-based technique [23]. The technique works by arranging the interfering users of a chip into two groups and approximating the interference coming from the other group as a single symmetric Gaussian distributed variable. For LDS system with 200% loading, the loss of approximately 0.3 dB compared to its brute-force counterpart can be achieved while reducing the complexity to more than half.

15.3 Challenges and Optimization Opportunities for LDS

In this section, the challenges and opportunities associated with MC-LDSMA system that can further improve the system performance will be presented.

15.3.1 Envelope Fluctuations in LDS Multiple Access

The transmitted signals in multicarrier communication systems consist of sums of trigonometric series. The major drawback of these systems is their large envelope fluctuations. It is well known that power amplifiers are peak power limited and when the input exceeds a limit the amplifier input–output characteristics are not linear any more. Therefore, when a signal with large envelope fluctuations that exceeds the dynamic range of the amplifier is passed through the amplifier it will suffer from spectrum re-growth and in-band distortion. This will cause BER degradation and out-of-band radiation. If the amplifier is designed with high de-rating to accommodate these peaks, it tends to be very inefficient. In the downlink, the base station can tolerate higher power consumption and signal processing complexity. However, in the uplink, the envelope fluctuations can be problematic where the terminals are battery powered and cannot offer high signal processing complexity. Many techniques have been proposed to reduce the envelope fluctuations in order to mitigate the associated amplifier linearity requirements, the out-of-band radiations or the required amplifier de-rating which is necessary for preventing the amplifier saturation at high input signal peaks.

As a multicarrier technique, it is expected that MC-LDSMA inherits the large envelope fluctuations drawback. As it has been explained before, in MC-LDSMA the data symbol is spread over a small number of subcarriers. This implies that the number of used subcarriers is larger than the non-spreading case (i.e., OFDMA). Hence, it is expected that MC-LDSMA has higher peak-to-average power ratio (PAPR)/cubic metric (CM) comparing to OFDMA. Thus, it is essential to investigate the envelope fluctuations of MC-LDSMA with conventional LDS codes. In the conventional LDS codes, random phases and random subcarrier allocation are used.

It is shown that MC-LDSMA with conventional LDS codes has considerably high envelope fluctuations [2]. Consequently, envelope fluctuations reduction techniques are required. At first, it may be thought that the large envelope fluctuations are caused by using more subcarriers in MC-LDSMA. However, the high envelope fluctuations for MC-LDSMA are caused by assigning random phases to the LDS codes without taking into account the correlation between the chips. In MC-LDSMA, the chips that belong to the same symbol are correlated. The phases can be tuned to reduce the envelope fluctuations. In fact, there is more flexibility in tuning the chips' phases in MC-LDSMA comparing to conventional spreading techniques such as MC-CDMA. The autocorrelation and cross-correlation properties of the LDS codes do not affect the detection process. Hence, this flexibility can be utilized and the phases can be tuned to reduce the envelope fluctuations of the MC-LDSMA signal.

However, the optimal phases that minimize the PAPR are hard to be found using classical optimization techniques [6]. Therefore, suboptimal phases to reduce the PAPR/CM of the signal have to be used. Many closely related phases have been found by different researchers that have low envelope fluctuations, such as Newman [9, 19] and Narahashi and Nojima [17] phases. These phasing schemes didn't attract significant attention within the context of PAPR reduction for MC-CDMA due to the constraints of autocorrelation and cross-correlation on the codes. Furthermore, in MC-CDMA, if the user needs to send more than one symbol, it has to be multiplexed in the code domain. Consequently, the resulted signal will have high PAPR/CM despite the single code design. On the other hand, in MC-LDSMA, there are no such constraints and the symbols can be multiplexed in the frequency domain. Hence, these phases can be employed in MC-LDSMA to reduce the envelope fluctuations. The Newman phases are given by

$$\theta_n^{Newman} = \frac{\pi(n-1)^2}{N}, \quad n = 1, 2, \ldots, N. \tag{15.22}$$

The phases of the LDS codes are adjusted according to (15.22) as follows

$$s_n = \frac{1}{\sqrt{|\zeta|}} e^{j\theta_n}. \tag{15.23}$$

These phasing schemes are designed in a way requires the subcarriers allocated to the user to be equally spaced. However, restricting all the users to have equally spaced subcarrier allocation in MC-LDSMA will result in a fully connected graph, which reduce the receiver detection efficiency, and the codes cannot be classified as low density any more. Accordingly, these phases are applied to for MC-LDSMA with resource block-based allocation, which is a structure that is already adopted in 3GPP-LTE system. This is done by dividing the total numbers of subcarriers into segments consist of N_z adjacent subcarriers. As the frequency segment can represent the frequency dimension of a resource block, it is referred to it as a resource block (RB). Therefore, instead of individual subcarriers, resource blocks are allocated to users. It is a common practice to group a number of adjacent subcarriers in both time and frequency domains in a form of a resource block. However, the time dimension of the resource block does not affect the PAPR. Figure 15.9 depicts the two types of subcarrier allocation schemes, resource block based and individual subcarriers based. It shows an example of 4 users with two symbols ($M = 2$) per user and effective spreading factor equal to two ($d_v = 2$). In resource block-based allocation, each user is allocated two resource blocks with size two ($N_z = 2$), where N_z is the number of subcarriers per resource block.

Figure 15.10 shows the results of CM for MC-LDSMA when the subcarrier allocation is resource block based and compared with OFDMA and SC-FDMA. The figure shows the CM of MC-LDSMA with Newman, Narahashi, and random phases. As the

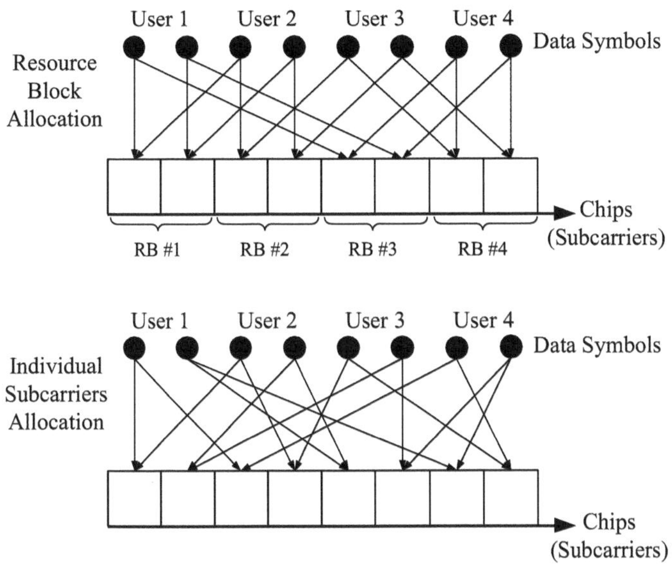

Fig. 15.9 Subcarrier allocation schemes for MC-LDSMA

CM of OFDMA and MC-LDSMA is the same for different modulation orders, only the results of 16 QAM modulation are presented in the figure. As the figure shows, Newman and Narahashi phasing schemes produce identical CM values. Both Newman and Narahashi phasing schemes significantly reduce the CM of MC-LDSMA in comparison to random phases. Random phases still suffer high CM regardless of the allocation scheme used. With phasing schemes, the CM of MC-LDSMA is reduced by 5.5 dB comparing to random phases. Comparing to SC-FDMA, MC-LDSMA has higher PAPR/CM values, especially comparing to QPSK modulation.

A major advantage of applying the phasing schemes is that it does not require modification in the MC-LDSMA system structure, and no complexity will be added to the system. In fact, generating Newman and Narahashi phases using the closed forms is less complex than the generation of random phases.

15.3.2 Radio Resource Allocation for LDS

The varying nature of the wireless channel in time domain and/or frequency domain represents a challenge for reliable communication. To cope with the varying nature of the channel, dynamic radio resource allocation is usually implemented in wireless communications. The main concept of radio resource allocation is to utilize the knowledge of channel information at the transmitter to optimize the system performance. According to the channel conditions, the transmission parameters, such as

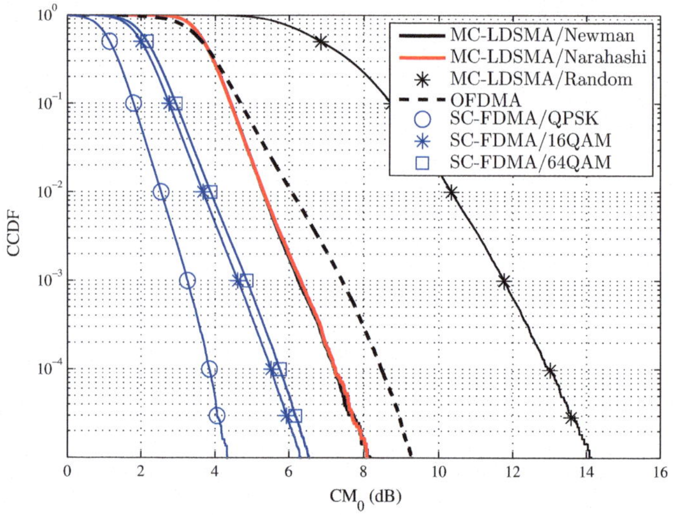

Fig. 15.10 CM comparison for MC-LDSMA with different phasing schemes: $N_z = 12$

transmitted power, allocated spectrum, modulation, and coding orders, are adjusted to improve the system performance such as data rates, fairness, delays. The low density spreading in MC-LDSMA makes the radio resource allocation more challenging compared to conventional multiple access techniques. The non-orthogonality in MC-LDSMA will couple the power allocation problem among the users, where each user's power allocation will affect other users due to the interference. Furthermore, the spreading will couple the subcarrier allocation problem among the subcarriers. Thus, the conventional radio resource allocation algorithms, optimal and suboptimal, cannot be directly applied to MC-LDSMA technique, and new analysis and algorithms need to be developed.

The single-user power allocation for LDS technique is more challenging comparing to the non-spreading case, which has a well-known water-filling solution. Assuming that N subcarriers indexed by the set $\mathcal{N} = \{1, 2, \ldots, N\}$ are allocated to a user, the user will partition this set into M subsets/groups indexed by $\mathcal{M} = \{1, 2, \ldots, M\}$. Each subset of subcarriers will be used to spread one symbol, and the number of symbols will be $M = N/d_v$. The single-user power allocation problem for LDS can be split into two parts: Firstly, what is the optimal subcarriers' partitioning that maximizes the user rate? Secondly, for a given subcarriers' partitioning, what is the optimal power allocation?

It has been shown in [5] that for a given subcarriers partitioning scheme, the optimal power allocation algorithm can be conducted in two steps: The first step is water-filling to find the power for each symbol; the second step is the maximum ratio transmission of the symbol power over the symbol's subcarriers.

In MC-LDSMA technique, the available N subcarriers should be partitioned into M subsets, where each subset will be used to transmit one symbol. In conventional

LDS codes, the subcarriers are partitioned randomly, which has shown satisfactory BER performance [4]. However, the random partitioning scheme has not been investigated from the user rate optimization perspective. The problem of partitioning the subcarriers to maximize the user rate can be formulated as follows

$$\max_{\mathscr{D}_m \in \mathscr{N}} \sum_{m \in \mathscr{M}} R_m(\mathscr{D}_m), \tag{15.24}$$

subject to

$$\mathscr{D}_m \cap \mathscr{D}_j = \emptyset, \quad \forall m \neq j,\ m, j \in \mathscr{M}. \tag{15.25}$$

$$|\mathscr{D}_m| = d_v, \quad \forall m \in \mathscr{M}. \tag{15.26}$$

This is a combinatorial optimization problem with a large search space. The number of possible LDS codes to be generated from the N subcarriers for a specific spreading factor d_v is given by

$$C_{LDS} = \frac{1}{2}\binom{2d_v}{d_v}\prod_{i=0}^{M-3}\binom{N-1-id_v}{d_v-1}. \tag{15.27}$$

In fact, the number of possible LDS codes (C_{LDS}) represents the cardinality of the feasible search space of problem (15.24)–(15.26). In order to see how large is the search space, let us consider $N = 32$ and $d_v = 2$, so the number of possible LDS codes will be $C_{LDS} = 1.92 \times 10^{17}$. It can be seen that brute-force search is unfeasible for practical systems, and partitioning schemes with low complexity need to be considered. Even though the subcarrier partitioning is a non-convex problem, there is an underlying Schur-convex structure, which can be utilized to solve the problem.

The optimal subcarrier partitioning that maximizes the user's rate with optimal power allocation is the one that gives the most majorized gain vector (MMV) and can be defined as follows

$$\mathscr{D}_m^\star = \{g_{\pi(m)}, g_{\pi(m+1)}, \ldots, g_{\pi(m+d_v-1)}\}, \quad \forall m \in \mathscr{M}, \tag{15.28}$$

where π represents a permutation of the subcarrier gains in a descending order such that $g_{\pi(1)} \geq g_{\pi(2)} \geq \cdots \geq g_{\pi(N)}$. In this scheme, the subcarriers are sorted in a descending order, then the first best d_v subcarriers are combined to create one symbol, the second best d_v for another symbol, and so on. The optimality proof of this scheme can be done by showing that this partitioning scheme will give symbols' gain vector that majorizes any other partitioning scheme [5]. Figure 15.11 shows the user spectral efficiency for different subcarriers' partitioning schemes. As it can be observed from the figure, the partitioning scheme that results in the MMV achieves the same spectral efficiency for the brute-force search and it significantly outperforms the random and least majorized vector (LMV) schemes. The random and LMV schemes only achieve

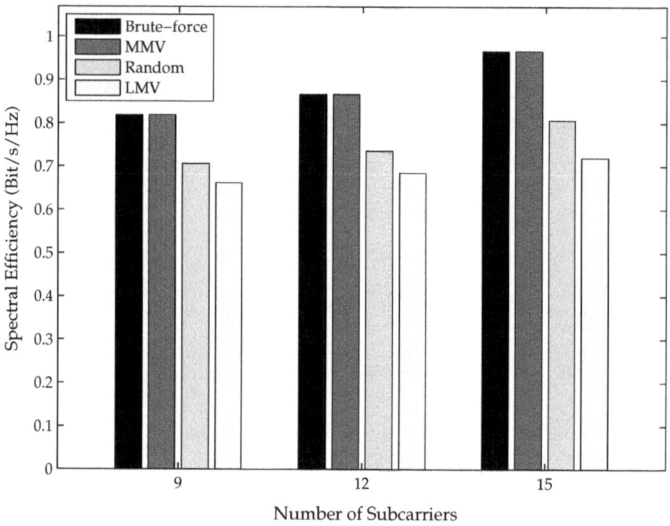

Fig. 15.11 Spectral efficiency comparison for different subcarriers partitioning schemes

85% and 72% of the optimal solution, respectively, which indicates the suboptimality of these schemes. Furthermore, the very poor performance of the LMV partitioning demonstrates the importance of vector majorization in generating the symbols gains, where it shows that a less majorized gain vector will produce very low user rate.

These results show the importance of radio resource allocation in enhancing the performance of the LDS multiple access. More analysis and radio resource optimization that consider the multiuser case can be found in [5].

15.4 Summary

In this chapter, we provided an overview of the low density spreading multiple access technique, which is considered as a promising access scheme for future wireless communication systems. The overview of the challenges in the conventional spreading techniques, due to the multiple access interference, has provided insights on the motivations for the LDS design. Low density spreading can manage the multiple access interference and offer an efficient, low complexity multiuser detection. To cope with the multipath effect of the wireless channel, low density spreading is applied to multicarrier systems, which is referred to as Multicarrier Low Density Spreading Multiple Access (MC-LDSMA) or LDS-OFDM. Compared to orthogonal access schemes (such as OFDMA), MC-LDSMA can support higher number of users in the system. Also, due the spreading nature of LDS, better frequency diversity can be achieved which results in improved link-level performance.

In addition, the chapter discussed the challenges and opportunities associated with MC-LDSMA system such as envelope fluctuations and radio resource allocation.

The impact of subcarriers' allocation schemes and the phases of the LDS codes on the PAPR/CM of MC-LDSMA signals were investigated. We have shown that with proper phasing schemes applied for the LDS codes, significant PAPR/CM reduction can be achieved. Furthermore, we presented radio resource allocation algorithms that can improve the system spectral efficiency. Numerical and simulation evaluation results have been provided to demonstrate the achieved gains by using MC-LDSMA with the optimized radio resource allocation algorithms.

References

1. F. Adachi, M. Sawahashi, H. Suda, Wideband DS-CDMA for next-generation mobile communications systems. IEEE Commun. Mag. **36**(9), 56–69 (1998)
2. M. Al-Imari, R. Hoshyar, Reducing the peak to average power ratio of LDS-OFDM signals, in *International Symposium on Wireless Communication Systems* (2010), pp. 922–926
3. M. Al-Imari, M.A. Imran, R. Tafazolli, D. Chen, Subcarrier and power allocation for LDS-OFDM system, in *IEEE Vehicular Technology Conference* (2011), pp. 1–5
4. M. Al-Imari, M.A. Imran, R. Tafazolli, Low density spreading for next generation multicarrier cellular systems, in *International Conference on Future Communication Networks (ICFCN)* (2012), pp. 52–57
5. M. Al-Imari, M.A. Imran, P. Xiao, Radio resource allocation for multicarrier low-density-spreading multiple access. IEEE Trans. Veh. Technol. **66**(3), 2382–2393 (2017)
6. S. Boyd, Multitone signals with low crest factor. IEEE Trans. Circuits Syst. **33**(10), 1018–1022 (1986)
7. X. Cai, S. Zhou, G. Giannakis, Group-orthogonal multicarrier CDMA. IEEE Trans. Commun. **52**(1), 90–99 (2004)
8. G. Colavolpe, G. Germi, On the application of factor graphs and the sum-product algorithm to ISI channels. IEEE Trans. Commun. **53**(5), 818–825 (2005)
9. D. Gimlin, C. Patisaul, On minimizing the peak-to-average power ratio for the sum of N sinusoids. IEEE Trans. Commun. **41**(4), 631–635 (1993)
10. D. Guo, C.C. Wang, Multiuser detection of sparsely spread CDMA. IEEE J. Sel. Areas Commun. **26**(3), 421–431 (2008)
11. Y. Kabashima, A CDMA multiuser detection algorithm on the basis of belief propagation. J. Phys. A: Math. General **36**, 11111–11121 (2003)
12. A. Kapur, M. Varanasi, C. Mullis, On the limitation of generalized Welch-bound equality signals. IEEE Trans. Inf. Theory **51**(6), 2220–2224 (2005)
13. F. Kschischang, B. Frey, H.A. Loeliger, Factor graphs and the sum-product algorithm. IEEE Trans. Inf. Theory **47**(2), 498–519 (2001)
14. D. MacKay, Good error-correcting codes based on very sparse matrices. IEEE Trans. Inf. Theory **45**(2), 399–431 (1999)
15. J. McGowan, R. Williamson, Loop removal from LDPC codes, in *IEEE Information Theory Workshop* (2003), pp. 230–233
16. A. Montanari, D. Tse, Analysis of belief propagation for non-linear problems: the example of CDMA (or: How to prove Tanaka's formula), in *IEEE Information Theory Workshop* (2006), pp. 160–164
17. S. Narahashi, T. Nojima, New phasing scheme of N-multiple carriers for reducing peak-to-average power ratio. Electron. Lett. **30**(17), 1382–1383 (1994)
18. J.P. Neirotti, D. Saad, Improved message passing for inference in densely connected systems. Europhys. Lett. **71**(5), 866–872 (2005)
19. D.J. Newman, An L^1 extremal problem for polynomials. Proc. Am. Math. Soc. **16**, 1287–1290 (1965)

20. J. Raymond, D. Saad, Sparsely spread CDMA–a statistical mechanics-based analysis. J. Phys. A: Math. Theor. **40**, 12315–12333 (2007)
21. T. Tanaka, M. Okada, Approximate belief propagation, density evolution, and statistical neurodynamics for CDMA multiuser detection. IEEE Trans. Inf. Theory **51**(2), 700–706 (2005)
22. S. Verdu, Minimum probability of error for asynchronous gaussian multiple-access channels. IEEE Trans. Inf. Theory **32**(1), 85–96 (1986)
23. F. Wathan, R. Hoshyar, R. Tafazolli, Dynamic grouped chip-level iterated multiuser detection based on gaussian forcing technique. IEEE Commun. Lett. **12**(3), 167–169 (2008)
24. L. Welch, Lower bounds on the maximum cross correlation of signals. IEEE Trans. Inf. Theory **20**(3), 397–399 (1974)
25. J. Yedidia, W. Freeman, Y. Weiss, Constructing free-energy approximations and generalized belief propagation algorithms. IEEE Trans. Inf. Theory **51**(7), 2282–2312 (2005)
26. M. Yoshida, T. Tanaka, Analysis of sparsely-spread CDMA via statistical mechanics, in *IEEE International Symposium on Information Theory* (2006), pp. 2378–2382

Chapter 16
Grant-Free Multiple Access Scheme

Liqing Zhang and Jianglei Ma

16.1 Motivation

The next generation (5th generation, or 5G) wireless network will target to support a wide range of applications, including enhanced mobile broadband (eMBB), ultra-reliable and low-latency communications (URLLC), and massive machine-type communications (mMTC). eMBB aims to focus on services that have requirements for high speed and capacity, such as high-definition (HD) videos and virtual reality (VR); URLLC aims to focus on latency-sensitive services such as assisted and automated driving, remote management, and e-health. mMTC aims to focus on high connection density services such as smart city and smart agriculture. Most efforts in the existing networks such as Long-Term Evolution (LTE) network are on high speed, capacity, and efficient spectrum based on information-theoretic principles with asymptotic long packets [1]. While 5G wireless network will continue to enhance the UE spectrum efficiency and high speed for eMBB services, it has also to support a variety of new types of traffic, such as small packet transmissions in machine–machine (M2M), sensors or devices, etc, and certain latency-constrained applications. These motivations and requirements will post big challenges on existing networks in terms of new and diversified demands; for example, URLLC has stringent requirements for ultra-reliable transmission (99.999%) with extremely low latency ($\leq$1 ms) [2].

This chapter is intended to address important driving technologies behind the 5G new radio (NR) system designs, which focuses on solutions to supporting 5G new services in UL transmissions with requirements such as low-latency (and high reliability), energy saving, and small packet applications. Grant-free (GF) resources in NR UL are termed as "a configured grant," which means that the preconfigured UE-

L. Zhang (✉) · J. Ma
Wireless Department, Huawei Technology Canada, Ottawa, Canada
e-mail: liqing.zhang@huawei.com

J. Ma
e-mail: jianglei.ma@huawei.com

M. Vaezi et al. (eds.), *Multiple Access Techniques for 5G Wireless Networks and Beyond*, https://doi.org/10.1007/978-3-319-92090-0_16

specific resources will be used for UE UL transmission without dynamic scheduling/grant; the term "semi-persistent scheduling (SPS)" in NR is used to indicate the DL transmissions without dynamic scheduling/assignment [3], as NR downlink (DL) SPS scheme has no significant differences compared with LTE DL SPS solution. Also, the base station (BS) in 5G NR network is referred to as "next generation NodeB" or "gNB".

16.1.1 On Grant-Free Multiple Access

To achieve the 5G network requirements, the contention-based GF multiple access scheme can be a promising solution for UL transmissions. For a GF transmission by design, a user equipment (UE) is able to send a transport block (TB) immediately upon traffic arrival without a scheduling request (SR) and a dynamic UL grant; this can be done in a way that the physical time–frequency resources, and other parameters including reference signal (or pilot) and MCS, etc., can be allocated to a UE, for example, when the UE performs an initial network entry. It is different from the grant-based (GB) UL transmission in the existing networks, where for a UL transmission, a UE has to send a scheduling request (SR) upon traffic arrival and get a dynamically scheduled UL grant from the BS before the UE starts the UL data transmission. Therefore, the GF multiple access scheme addresses low-latency and control signaling reduction, and UE energy-saving issues, thus especially applicable to small packet and latency-sensitive transmissions in 5G wireless network.

The study on the small packets and low-latency transmissions has drawn a lot of attentions to researchers lately. Early work [4] addressed the problem of short packet delivery from an information theory perspective. In [5], reliable scheme for machine-type communication with short packets was studied; and [6] discussed reliable transmission for mMTC and URLLC with short packets. In [7–9], the GF transmissions were studied to support mMTC and URLLC with short packets. The GF with non-orthogonal multiple access scheme and performance evaluation were investigated in [10–15].

As one of the promising technologies in 5G wireless network to deal with the low-latency and energy-saving issues, GF multiple access scheme will be addressed in this chapter, including key technical components, some system design details, and performance evaluations.

16.1.2 Grant-Free Key Technical Components

GF resource configurations There are two schemes for GF resource configuration. One scheme is based on higher-layer signaling only, e.g., radio resource control (RRC) message, to fully configure transmission resources and parameters, including

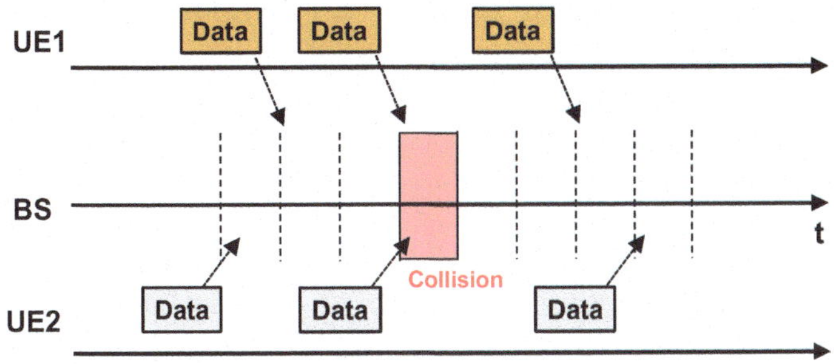

Fig. 16.1 GF contention transmissions [7]

time–frequency physical resources, pilot(s) and MCS(s), for a UE, such that the UE is able to start its data transmission in a way of "arrive-and-go."

The other scheme is operating like LTE semi-persistent scheduling (SPS) for UL where part of the UE resources and transmission parameters such as periodicity can be configured by a higher-layer signaling, while the other resources and parameters, such as frequency-domain resources, pilot, and MCS, are included in a scheduled DL control information (DCI) as an activation for the very first-time data transmission.

Hybrid automatic repeat request (HARQ) soft-combining Like the GB transmissions, the GF transmissions can also benefit from the HARQ processes with soft-combining to enhance the detection performance. In such a case, initial and one or more subsequent transmissions (e.g., repetitions or retransmissions) of a UE TB can be employed to enhance the reliability, where the signals from the initial and subsequent transmissions of the TB can be combined for enhanced signal detection at the BS.

However, as one UE packet can arrive in any time, the initial TB transmission time is uncertain. As a result, the BS has to identify the UE initial transmission (and subsequent transmissions) from the same TB of the UE in order for the signal soft-combining. For multiple HARQ processes per UE, one HARQ process identification (ID) will be used for transmissions of a UE TB where the transmissions of the UE TB are associated with redundancy versions (RVs). Thus, the initial TB transmission time and the HARQ process ID have to be determined at the receiver before the signal soft-combining.

Contention resolutions To enhance the spectrum efficiency or in some scenarios such as massive connectivity applications, two or more UEs can be configured to share the same GF resources. As a result, the contention can happen once two or more UEs transmit their packets simultaneously in the same transmission time interval (TTI), but how often the contention can happen is based on UE and system traffic loading.

As shown in Fig. 16.1, UE 1 and UE 2 can share the resources with potential contention transmissions, where there are three bursty TB transmissions for each UE while the two UEs have only one transmission contention. There are two contention scenarios in UL transmissions. The first type of contention is in the scenario where two or more UEs can content with a time–frequency resource, i.e., the collision of the time–frequency resource, but their pilots are different, each pilot being used for UE activity detection as well as channel estimation at the BS. The second type of contention is in the scenario where two or more UEs can content with not only the time–frequency resource, but their pilot(s) as well, referred to as *pilot collision*. As a result, the aggregated transmission signals from these UEs may be difficult to be detected and probably lead to a failure in decoding these received signals. Note that due to the sporadic nature of GF traffic, the resource/parameter sharing among different UEs doesn't mean that a collision will definitely happen; in fact, the collision is a probability event that depends on the UE and system traffic loading.

To handle these contention problems, different schemes can be used such as advanced reception schemes, resource allocation configuration with UE-specific hopping, which will be addressed in the following sections.

16.2 Grant-Free Transmission

To support data transmission in a wireless network, a UE has to have fundamental transmission resources such as time–frequency resources as access channel resources, and the associated parameters such as pilot(s), modulation and coding scheme. During the GF transmissions, the UE can employ HARQ procedure for the signal detection and soft-combining, while if a transmission contention happens, the BS needs to identify different UEs and resolve the contention.

16.2.1 Resource Configuration

In current LTE network, the SPS scheme has been designed to support UL and DL voice applications, where the transmission resources and parameters are configured by both higher-layer signaling (e.g., RRC) and physical layer (i.e., L1) signaling. The L1 signaling is a dynamic scheduling grant from the BS, where the DCI is included in the L1 signaling. For UL initiated traffic, a UE will need to start from a scheduling request (SR) to the BS and perform the UL voice transmissions only after the UE receives a UL grant from the BS as an activation.

In 5G network, new diverse applications and services have different requirements from LTE voice applications. For example, URLLC services have a low latency (e.g., 1 ms air-interface time) and high reliability (e.g., 99.999% success rate); mMTC services with small packets are also popular use case, with sporadic and non-periodic traffic. As a result, from at least latency and traffic characteristic perspectives for UL transmissions, the current LTE network is not capable of working efficiently to support these applications and services.

To handle these new problems in the 5G network, the newly proposed GF scheme can be a promising candidate to help resolve these issues. By the preconfigured or pregranted resources and parameters, the GF approach is able to reduce both signaling overhead and transmission latency, as well as to enhance the transmission reliability for a latency-constrained application such as URLLC services, by taking advantage of more transmission occasions (TOs) for repetitions or retransmissions, in the constrained time window. Notice that repetitions in this chapter indicate retransmissions without waiting a HARQ feedback for any previous transmission. Two types of resource configuration schemes in 5G NR are proposed for GF multiple access and transmission described in this subsection.

RRC only resource configuration In this scenario, the RRC signaling will fully configure all the required resources and parameters for a GF transmission without using dynamic DCI signaling for an activation, where the resources and parameters in the RRC signaling include

- Resource periodicity P (to indicate the time interval between any two neighboring GF resource bundles)
- Resource offset O (the resource allocation start slot relative to the system frame number #0)
- Time-domain resource parameters (a symbol offset within the transmission start slot and a TO period in number of symbols)
- Frequency resource parameters (including frequency-domain resource starting location and frequency bandwidth, as well as frequency hopping offset)
- DMRS (or pilot) parameters (including DMRS type, additional position, and sequence length)
- Modulation and coding scheme (including MCS table type and index)
- Number of repetitions K (to take a value from 1, 2, 4, or 8 by RRC configuration)
- Open-loop power control parameters (such as target received power level P_0 and fractional factor α)
- HARQ process-related parameters (e.g., the maximum number of HARQ processes per UE, RV)
- Others, including GF bandwidth part, RNTI.

A detailed list of RRC parameters for GF (and other features) can be found in 5G NR specifications [16, 17]. For RRC only UE-specific resource configuration here, a series of time-domain resources can be determined for the UE GF TOs based on the periodicity parameter. Specifically, a UE can determine its configured time-domain TOs in terms of which symbol(s) in which time slots by solving Eq. (16.1) below for $\{n_{symbol}\}$, based on the time-associated parameter set $\{n_f, O, n_{symbol.offset}, P, u\}$:

$$\lfloor 14(10(n_f)2^u + n_s - O) + n_{symbol} - n_{symbol.offset} \rfloor mod\ P = 0 \qquad (16.1)$$

where the operator $\lfloor X \rfloor$ denotes the largest integer number less than or equal to X. Based on this formulation, the UE can figure out, in a transmission frame n_f and in a slot n_s within the transmission frame n_f, if a symbol n_{symbol} is the first (starting) symbol of a configured TO or not, given configured parameters: O (# of slots relative to the system frame #0), $n_{symbol.offset}$ (a symbol offset within a slot), P (configured periodicity), and u (an index to a numerology in terms of subcarrier spacing, e.g., 0, 1, 2, 3 for 15 kHz, 30 kHz, 60 kHz, and 120 kHz, respectively). Note that a transmission frame consists of 10 sub-frames of 1 ms, and each slot consists of 14 symbols in 5G NR network for any numerology u.

For configured time-domain TOs for a GF transmission with repetitions, an initial packet transmission can start from any configured and valid (in terms of RV consideration) TO to perform data "arrive-and-go." In such a scenario, a UE doesn't need to constantly monitor dynamically scheduled DCI and wait for an activation before the first packet transmission, thus being able to reduce the transmission latency, signaling overhead, and energy usage.

RRC with L1 signaling configuration To reduce the signaling overhead and latency, the other scheme on UE-specific resource configuration for UL transmission is to employ both RRC and L1 signaling, where RRC signaling configures part of the UL transmission resources and parameters such as resource periodicity P, repetition K, RNTI; while the L1 signaling will deliver the remaining resources and parameters during the activation message by a DCI, including frequency-domain resource allocation, pilot, and MCS, before the UE starts the first packet transmission. In this case, the UL starting resource allocation for a UE is based on the DCI signaling time slot, slot offset, and symbol offset numbers, and then a series of TOs can be determined by the periodicity parameter. The detailed resource and parameter configuration in DCI can be found in the 5G NR specification [18].

This type of transmission scheme is able for the network to take advantage of dynamic activation and release on the configured GF resources. Thus, it requires a GF UE to constantly monitor the physical DL control channel (PDCCH) for a dynamic DCI signaling.

16.2.2 HARQ Procedure

HARQ process and transmission identification HARQ process for GF transmission is different from that for GB transmission in the sense that the GB transmissions have explicit initial and retransmission occasions dynamically scheduled by the BS, thus making the HARQ soft-combining straightforward.

However, the GF transmission is lack of UE scheduling request (SR) and dynamic scheduling grant (SG) by design to support sporadic and low-latency traffic with small packets, thus the BS has no idea of when a UE packet will arrive and be transmitted, and whether a transmission from the UE is new transmission or retransmission. Thus, one question of interest for a HARQ process is: How to identify the initial transmission and retransmissions from the same TB of the UE to perform HARQ signal detection and soft-combining?

To address this issue, one way is to associate (nominally) initial TOs with fixed time resource allocations, where an initial TO and the subsequent TOs (for retransmissions or repetitions) are bundled together in terms of time-domain resources in a periodicity period, and these TOs are associated with a RV pattern. Due to the sporadic and non-periodic nature for GF traffic, one packet can arrive in any time interval, thus, to reduce the transmission latency, an actual initial packet transmission may not able to be performed in a nominally configured initial TO, rather than one of other TOs among the bundled resources in a periodicity period. More limitations on an initial transmission in a TO can be applied by design; for example, by considering the RV version in the TO, the initial transmission is allowed as long as the RV element is self-decoded in order to help better signal detection and soft-combining. The initial transmission criterion associated with a RV pattern in 5G NR can be found in [17], where three RV patterns or sequences are proposed and given below.

- Sequence 1: {0, 0, 0, 0}
- Sequence 2: {0, 3, 0, 3}
- Sequence 3: {0, 2, 3, 1}

Note that Sequence 1 consists of four self-decodable elements (each denoted by 0). Sequence 3 is incrementally encoded versions, as a nonzero-value element is complementary to the zero-value element. Sequence 2 is functionally a combination of Sequences 1 and 3. The choice of a RV sequence is mainly determined by the soft-combing performance that is based on specific channel coding design. It is also noted that the configured repetition number (or the number of TOs in a bundle) can be different from the RV sequence length of 4.

HARQ process ID determination and soft-combining For multiple HARQ processes, it is challenging to determine an HARQ ID for GF transmission with repetitions. This is due to the fact that the GF traffic is sporadic and can arrive any time. Thus, HARQ process ID for a UE TB has to be determined to allow for efficient HARQ signal detection and feedback. In 5G NR network, a HARQ process ID is associated with a bundle of K repetitions (i.e., K TOs) in a periodicity period, and a RV sequence is configured and mapped over K TOs in each periodicity period [17], where for configuration of multiple HARQ processes, one HARQ process ID can be derived (by a UE and the BS) from the following equation:

$$ID_{HARQ} = \lfloor X/P \rfloor mod\, N_{HIDs} \tag{16.2}$$

where the operator $\lfloor Y \rfloor$ in Eq. (16.2) denotes the largest integer number less than or equal to Y. P is the resource periodicity in # of symbols, and N_{HIDs} is the configured number of HARQ processes. $X = (SFN * SlotPerFrame * SymbolPerSlot +$

$Slot\_index\_In\_SF * SymbolPerSlot + Symbol\_Index\_In\_Slot$), and X refers to the first-symbol index of the first TO in a repetition bundle that takes place [3].

HARQ feedback indication For GF transmission, due to its sporadic traffic nature without requesting dynamic scheduling from the BS, a HARQ-ACK indication is needed to serve as transmission acknowledgment of a TB. If a GF UE does not receive any feedback from the BS within a predefined period after a GF transmission, the GF UE may have no ideas whether the BS has miss-detected or successfully decoded the transmission data.

To acknowledge the correctly decoded data packet, UL grant or group based common PDDCH can be used, where the former scheme is UE-specific feedback mechanism; while the latter scheme is group-based feedback, which is able to reduce the signaling overhead by acknowledging a group of UEs within a single message, especially for a cell with a large number of UEs each having highly loaded traffic.

16.2.3 Contention and Resolution

For GF transmissions, more than one UE can share the transmission resources to enhance the spectrum utilization; it is reasonable due to the fact that GF traffic can be sporadic and non-periodic. As a result, the GF transmissions in such configurations may have a contention (with a probability). To resolve this potential problem, one way is to configure the GF resources with UE-specific time–frequency hopping; in this case, one UE who has a signal contention in one TO with other UEs can avoid persistent contentions in other TOs with the same group of UEs.

LTE frequency hopping LTE is cell-based hopping over time–frequency domains wherein a hopping pattern depends on time–frequency resource locations. The motivation of the hopping is mainly to take advantage of channel frequency diversity. In LTE, a carrier bandwidth will be divided into different sub-bands, and a UE can transmit a TB in a different sub-band over different time intervals (either slot based or sub-frame based). A hopping pattern for the UE can be configured either dynamically by UL grant from PDCCH or preconfigured/defined by the network.

LTE has two types of hopping configurations: Type 1 and Type 2. In Type 1, the frequency offset between the hopping time intervals is explicitly configured and determined by DCI 0. In Type 2, the frequency offset between the hopping time intervals is configured by a predefined (pseudo-random) pattern. More details can be found in LTE technical specifications in [19, 20].

The LTE frequency hopping is cell-specific, and it may incur a problem for contention-based GF transmissions in the sense that two or more UEs starting the same resource allocation can incur persistent signal collisions. Therefore, instead of cell-based hopping, UE-specific hopping in a cell can be used to support GF contention transmissions, which is described in the following.

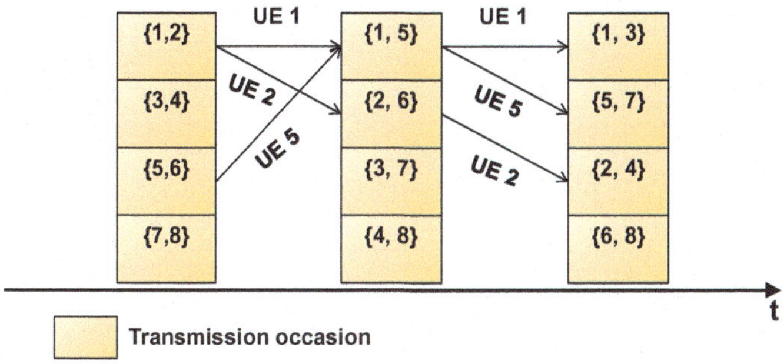

Fig. 16.2 UE-specific hopping pattern and resources

UE-specific hopping To take advantage of channel diversity and reduce potential transmission contention, UE-specific hopping scheme can be applicable to GF initial and repetition transmissions in a HARQ procedure.

For example, as shown in Fig. 16.2, multiple UEs who by configuration can share one time–frequency resource in their initial TOs may not share the resources in their subsequent redundant (repetition) transmissions, where a configured frequency hopping pattern for one UE is different from other UEs such that one collision between two UEs in one transmission time interval may not collide again in the subsequent transmission time intervals. Note that in the figure, UE1, UE2, and UE5, as an example, have explicit frequency hopping indications over the three TOs.

Other schemes can be also applied to resolve the GF collision issues. For example, a UE can choose a random back-off time from available repetitions or TOs to further randomize the retransmissions in order for the contention reduction. If severe signal contentions happen in a certain area, reconfiguring UEs to a different contention areas or increasing the GF contention resources may further help resolve the problems; or the network can switch one or more GF UE TBs to GB transmissions with dedicated resources, thus reducing the resource usage in the GF contention areas.

NOMA for contention resolution The reception procedure for a UL GF transmission consists of UE activity detection, channel estimation, and data decoding. In the contention transmissions, if the signals from multiple UEs collide in a time interval or transmission occasion, the BS typically relies on pilots for the UE differentiation. For the first type of contention, where the collisions happen in data signals but not pilots, multiple UE joint detection at receiver can be employed to enhance the detection performance. Thus, the non-orthogonal multiple access (NOMA) scheme such as sparse code multiple access (SCMA) [21] can be a promising solution to this type of the contention in UL GF transmissions. In such a case, the BS will need to first detect the UE activity (i.e., if there is any data transmission), estimate the channel, and then do the data detection, which is shown in Fig. 16.3, where a pilot is associated with a NOMA signature [32].

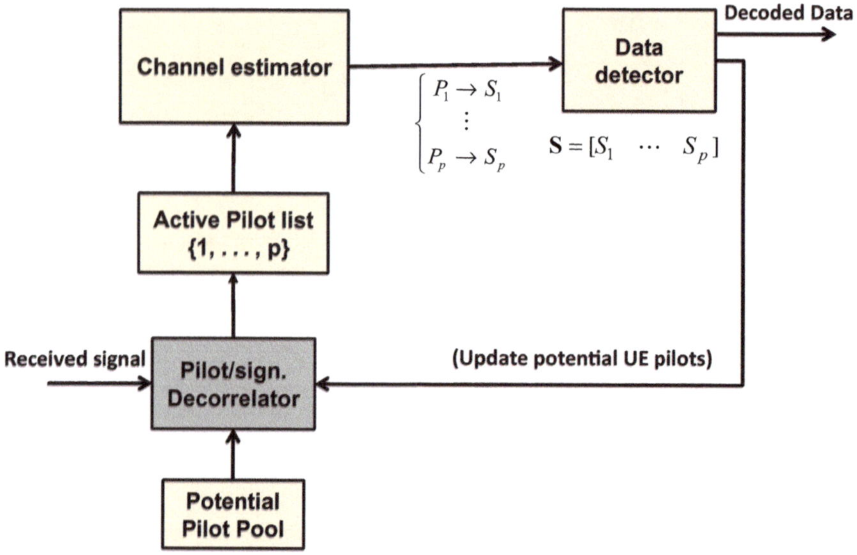

Fig. 16.3 An example of receiver procedure for UL GF scheme [32]

For multiple UE joint detection, an advanced reception scheme can be used in NOMA, where a general multi-user detection (MUD) problem can be described by Eq. (16.3).

$$\mathbf{y} = \sum_{i \in \mathscr{A}} H_i \mathbf{x}_i + \mathbf{n} \ = \ \bar{H}\bar{\mathbf{x}} + \mathbf{n} \tag{16.3}$$

where $\mathbf{y}$ is the received aggregate signal vector. $\mathscr{A}$ is the set of active UEs, $\bar{H} = [H_1, H_2, ..., H_N]$ is the channel matrix, $\bar{\mathbf{x}} = [\mathbf{x}_1^T, \mathbf{x}_2^T, ..., \mathbf{x}_N^T]^T$ is the vector of transmitted signals from N UEs, and $\mathbf{n}$ is the noise plus interference vector. The MUD receiver can support overloading scenarios for GF, e.g., mMTC services; in such a case, $\bar{H}$ is rectangular and can have more columns than rows. Many MUD receivers for NOMA have been investigated, including linear reception schemes such as zero forcing (ZF) and minimum mean squared error (MMSE) receivers, and nonlinear reception schemes such as multistage successive interference cancellation (SIC), multistage parallel interference cancellation (PIC), message passing algorithm (MPA) [21, 22] with the maximum-likelihood (ML)-like performance [23, 24] and the prior-information aided adaptive subspace pursuit (PIA-ASP) algorithm [25], where the MPA and PIA-ASP algorithms can take advantage of the sparsity of the (spreading) code and UE activity effectively. These receivers have been shown to support high overloading and provide a trade-off between detection performance and implementation complexity. The performance evaluation on NOMA with selected receiver schemes will be provided in the following section.

16.3 Performance Analysis and Evaluation

In this section, we provide numerical and evaluation performance of GF transmission on reliability with repetitions, UE activity detection with pilots, and MUD with SCMA.

16.3.1 Reliability with Repetitions

In this subsection, we provide a performance analysis on the GF reliability with repetitions and HARQ operations. We consider one initial transmission and one retransmission in single-shot for UL GF, as shown in Fig. 16.4, to support low-latency and high reliability applications such as URLLC services. At the BS, the UE activity detection based on a pilot is processed first and if successful, the channel estimation is made by the pilot, and then the signal detection is performed. The retransmission signal will be combined with initial signal for the HARQ soft-combining and detection, only when the BS has successfully detected the pilots from the two transmissions but fails in individual signal detection with each of the two transmissions.

As shown in Fig. 16.4, it can be obtained [26] that the successful detection probability, P, after the two transmissions is given below.

$$P = P_1(p_1)P_1(D_1|p_1) + P_1(p_1)(1 - P_1(D_1|p_1))P_2(p_2)P_2(D_1 + D_2|p_1, p_2, -) \\ + (1 - P_1(p_1))P_2(p_2)P_2(D_2|p_2) \quad (16.4)$$

where

- $P_1(p_1)$ and $P_1(D_1|p_1)$ are, respectively, the successful detection probabilities of the UE activity, and the successful detection of the data given the successfully detected UE activity (or pilot) in the initial transmission.
- $P_2(p_2)$ and $P_2(D_2|p_2)$ are, respectively, the successful detection probabilities of the UE activity, and the successful detection of the data given the successfully detected UE activity (or pilot) in the retransmission.
- $P_2(D_1 + D_2|p_1, p_2, -)$ is the successful detection probability of HARQ combined signal from the two transmissions of the UE, given that the UE activity successfully detected in the two receptions and first transmission data detection failed.

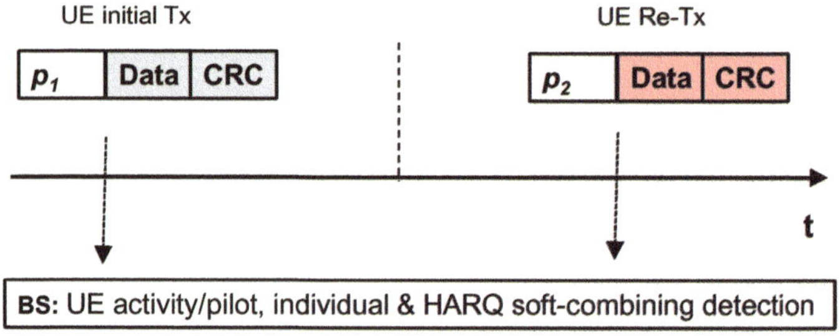

Fig. 16.4 HARQ initial transmission and retransmissions in single-shot [26]

To simplify the performance analysis, further assumptions are made below:

- The two transmissions are independent with equal error pilot detection probability on the average, defined by $p_{miss} := 1 - P_1(p_1) = 1 - P_2(p_2)$, and equal data error detection probability given the success of pilot detection, defined by $p_e := 1 - P(D_1 p_1) = 1 - P(D_2|p_2)$.
- The data error detection probability of the HARQ combined signals, given that the two pilots successfully detected in both transmissions and first transmission data detection failed, is defined by, $p_{HARQ} := 1 - P2(D1 + D2p1, p2, -)$.

As a result, using Eq. (16.4), the error detection probability after the two transmissions is given by

$$P_e \approx p_e \, p_{HARQ} + p_{miss}^2 + 2p_{miss} \, p_e \tag{16.5}$$

where p_{miss}, p_e and p_{HARQ} in Eq. (16.5) can be assumed small enough (e.g., $<10^{-2}$) to ignore its higher order terms. For the two consecutive transmissions, the GF error probably will need to satisfy $p_e \, p_{HARQ} + p_{miss}^2 + 2p_{miss} \, p_e < 10^{-5}$. It is expected that $p_{HARQ} < p_e$, however, by taking conservative considerations for simple analysis, we replace p_{HARQ} by p_e; also assuming $p_{miss} \approx p_e$, which gives

$$p_{miss} < 0.15\% \; and \; p_e < 0.15\%$$

as sufficient conditions to achieve the required URLLC reliability of 10^{-5}. From above analysis, if there is no retransmission, to support the URLLC reliability of 10^{-5}, the GF error probably has to satisfy

$$p_{miss} + p_e - p_{miss} \, p_e < 10^{-5}$$

which means that if no redundant transmissions are taken, we need the requirements of both $p_{miss} < 0.001\%$ and $p_e < 0.001\%$ to support the URLLC services that can be extremely challenging if not possible.

It is seen that HARQ retransmissions for two consecutive transmissions can significantly reduce the requirements on the error detection probabilities for both UE activity and data detection. It is expected that more retransmissions will lead to further relaxed reliability requirements; for example, within the URLLC latency constraint of 1 ms, at least 6 transmissions in single-shot can be used for 60 kHz subcarrier spacing with 7 symbols per transmission time interval. Thus, the GF scheme with HARQ retransmissions is capable of supporting the applications with low-latency and high reliability requirements such as URLLC services.

16.3.2 UE Activity Detection

In GF transmission, since the BS has no prior information of when a UE may initiate a packet transmission upon traffic arrival and possible retransmissions, it has to detect on each of GF TOs configured for the UE. In this case, UE-specific pilot can be used to perform such a UE activity detection, where the pilot actually plays as the joint functions of the UE activity detection and channel estimation.

Moreover, UE activity detection and differentiation are also required in the contention resources shared by two or more UEs. The BS has to try and decode different UE pilot candidates configured on each of contention resources as shown in Fig. 16.3; in such a sense, the user activity detection is kind of blind detection. The performance evaluation on the UE activity detection using pilots is given below.

User detection performance with pilots Pilots (as reference signals) can be used by the BS to decide which UEs transmit on the time–frequency resources for UL GF transmissions. As a pilot will be transmitted only by an active UE in a transmission resource, it is the indicator of UE activity. Due to the interference and noise, there are two types of UE activity detection error cases: some active UEs may be detected to be inactive, referred to as *missed detection*; while some inactive UEs may be detected to be active, referred to as *false alarm*. As a general detection problem, there will be a trade-off between the *missed detection* and *false alarm* '1 [27]. The *missed detection* is of more importance to us and the number of miss-detected UEs should be as small as possible, as the packets transmitted by these UEs will be dropped.

Figure 16.5 shows the user detection performance of pilot in terms of block error rate (BLER) versus received pilot signal-to-noise ratio (SNR). In the simulation and performance evaluation, four UEs with random packet arrivals share a total of 6 RBs and randomly chosen two UEs as active UEs. Detailed simulation assumptions including pilot sequences can be found in [28].

From the performance results in Fig. 16.5, we can see that pilot-based user detection works well for GF transmission as the two different types of pilot sequences (based on Comb-cyclic shift or frequency division-orthogonal cover code) all show

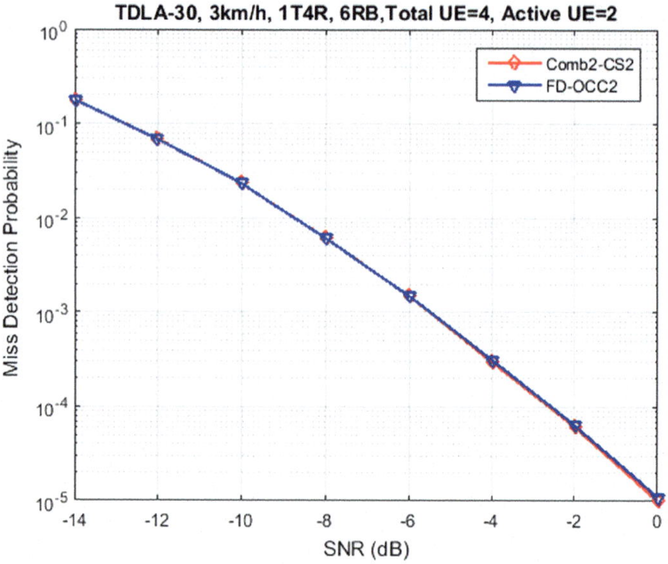

Fig. 16.5 User detection performance of pilot-based solution [28]

similar performance, where the missed pilot detection probability for one single transmission is below 10^{-3} at SNR of -5 dB.

16.3.3 Grant-Free and NOMA Performance

In this subsection, we evaluate the detection performance of GF with OFDMA and SCMA schemes [10, 11]. One key performance indicator (KPI) used is packet drop rate (PDR in %) or system UE outage over the traffic loading for the massive connection density study on mMTC services; another KPI is the percentile of UEs satisfying the latency of 1 ms and the reliability of 10^{-5} versus the system traffic loading for the URLLC study.

Connection efficiency and density The connection density/connection efficiency is the main KPI for the mMTC scenario [2]. When the connection density increases, the system traffic load increases accordingly. Assume the device number distribution density is N per km^2, and the average inter-packet arrival time per device is T_{packet} (in s), then the system traffic loading is given by $L = N/T_{packect}$ (in packets/km^2/s).

In the evaluation, the GF transmissions with NOMA and OFDMA are considered and compared for the scenarios with different antenna configurations. By varying the system traffic loading, the system outage, defined as system packet drop rate (PDR), can be captured; a user packet will be dropped if the packet transmissions fail after the retransmissions exceed some time limit in the evaluation. We'll check the performance for the supported system traffic loading at certain specific system outage points such as PDR 1% of interest.

In this study, a radio frequency resource of 4 RBs is allocated for the control/data and the packet transmissions can be in any time intervals. UE TB size is 20 bytes in the evaluation. For OFDMA, a UE will randomly choose one of the 4 RBs for GF transmissions in the UL; for SCMA, UE signal is spread over the 4 RBs for one TB transmission. The general evaluation assumptions follow the simulation methodology given in [29, 30].

Figure 16.6 provides the evaluation performance of SCMA and OFDMA with two antenna configurations: 1Tx/2Rx and 1Tx/4Rx. For 1Tx/2Rx case, the repetition of 8 for one-shot transmission is applied to some cell edge users with severe coupling loss, and no repetition (i.e., one-shot transmission includes only one transmission) to the other users; all users apply a maximum of 6 one-shot (re-)transmissions. In 1Tx/4Rx case, the repetition of 4 is applied to the cell edge users with severe coupling losses, and no repetition to the other users; all users apply a maximum of 4 one-shot (re-)transmissions. Note that the motivation for the configurations on these parameters is to support the system traffic loading while the system PDR can be around 1% or below for both OFDMA and SCMA with the GF transmissions.

It is observed that SCMA has shown the significant performance enhancement over the OFDMA. Specifically, at PDR 1% point of interest in terms of supported system traffic loading, SCMA has demonstrated 97% gain over the OFDMA system (or almost two times of OFDMA capability) in 1Tx/2Rx case, and SCMA has demonstrated 233% gain over the OFDMA system (or more than three times of OFDMA capability) in 1Tx/4Rx case.

UL URLLC support and performance The GF multiple access scheme has an advantage in the latency and control signaling reduction for small packet transmissions, and especially applicable to URLLC services with both low-latency and high reliability requirements. To support URLLC UL traffic, the GF scheme with repetition is a promising solution, while the GB scheme and LTE UL SPS schemes are also considered. We'll evaluate the performance on these options in the following.

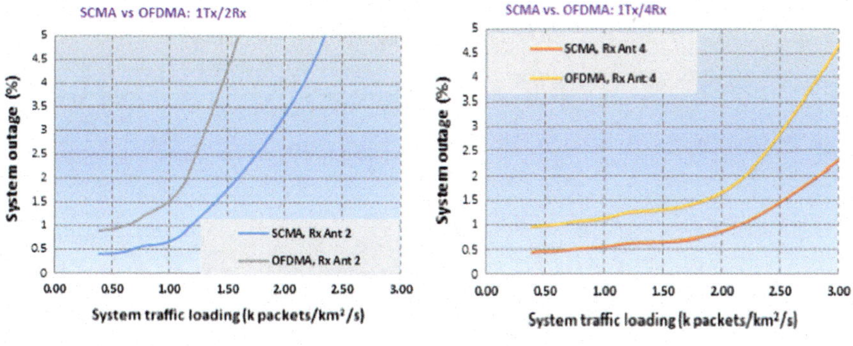

Fig. 16.6 Simulation performance on GF with SCMA and OFDMA

To evaluate the performance of the URLLC, the air-interface latency of 1 ms and high reliability of 10^{-5} are considered as the requirements to a URLLC UE [2]. As a result, the UE has a limited transmission time window for each TB upon its traffic arrival. In the evaluation, a small TB of 32 bytes is assumed and it can be transmitted in a TO with a resource unit of 7-symbol and 5 RB region, where 25 RB bandwidth is assumed with 5 partitions each being 5 RBs using subcarrier spacing of 60kHz with normal cyclic prefix (CP) [31]. Thus, there are a total of 8 TOs within 1 ms window for any UE to finish the packet transmission of one URLLC TB, from traffic arrival to the signal detection at the receiver. Moreover, a UE can take advantage of the resources with 8 TOs and frequency hopping among 5 frequency partitions to capture the channel diversity.

Due to the fact that the transmission procedures are different for GF, SPS, and GB schemes, the functional resource usage on the available TOs is different: The GF scheme can perform "arrive-and-go," where 6 TOs can be used for data transmissions, while one TO for traffic arrival, and the other one for reception and signal detection processing; multiple UEs can share the GF resources with contention as the system traffic loading increases. SPS UE each can utilize about 6 TOs (for repetitions) assuming that a total of 7 UEs will share the transmission resources, but without contention (or the allocated resources are orthogonal). To be efficient for repetition transmissions, an early acknowledgment of successful transmission is assumed, which can help increase the resource utilization as well as reduce the interference to other users. GB scheme will need to start from scheduling request (SR) and UL grant procedure and with possible BSR for more UL grant; however, the GB can take advantage of link adaption such that the spectrum efficiency can be enhanced. These evaluation assumptions are given in Table 16.1 and more details in [31].

We compare the scheme in FDD framework and show percentage of UEs satisfying the latency of 1 ms and reliability of 10^{-5} for different arrival rate per UE and different number of UEs per cell for all three schemes. The reliability of each UE is measured as average residual BLER within latency bound of all simulated packets of each UE. If the reliability for a UE is below the target reliability threshold, the UE is

Table 16.1 Usage distribution over 8 available TOs within 1 ms

Scheme	Arrival	SR/UL grant	TOs/MCS	Repetition/ decoding	Note
GF	1	0	6/Fixed MCS	1	With possible UE contention
SPS	1	0	6/Fixed MCS	1	Assumed 7 UEs sharing resources but no contention
GB	1	4	2/Adaptive MCS	1	Scheduled with no contention

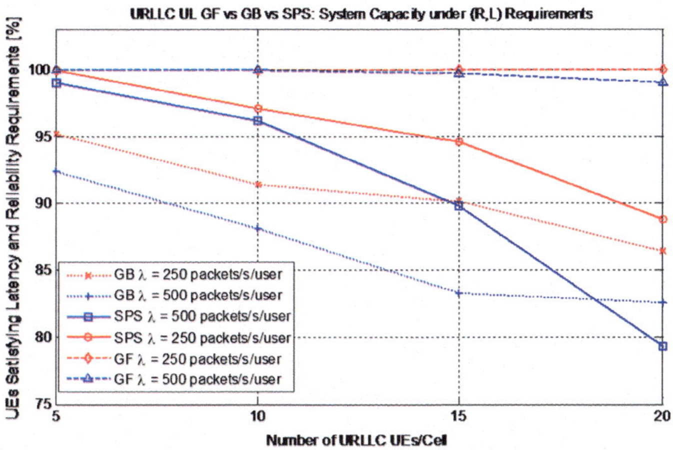

Fig. 16.7 Percentage of UEs satisfying the URLLC latency and reliability requirements for different traffic loading [31]

considered satisfied. Figure 16.7 shows the performance of contention-based GF transmission, SPS scheme, and GB transmission at different number of UEs in a cell, and each UE is with arrival rate of 250 or 500 packets/s/user. It can be seen that GF transmission provides much higher UE reliability and support higher system capacity for URLLC than both GB and SPS scheme. This is mainly because GF has comparatively more TOs to be used within the latency bound due to the "arrive-and-go" nature, which can greatly improve the reliability per packet in comparison to GB and SPS schemes. Compared to GB scheme, GF and SPS can both save latency and control overhead, due to dynamic scheduling of GB scheme. When comparing contention-based GF with SPS, contention-based GF allows multiple UEs to access the time–frequency resources and allows for more TOs for each UE (especially when the system traffic loading increases), while LTE UL SPS dedicates resources to the UEs in an orthogonal manner.

16.4 Conclusion and Summary

In this chapter, we have introduced GF multiple access scheme, which is able to send a TB in a way of "arrive-and-go" upon traffic arrival, thus reducing the transmission latency and control signaling overhead. This is done by preconfiguring the required physical time–frequency resources among other parameters including pilot and MCS for a UE. The proposed GF scheme is different from GB transmission, in which a UL transmission for a UE has to follow SR and (dynamic) scheduling grant procedure before the UE can start a data transmission.

A few key technical components for 5G wireless network have been addressed, including the GF resource configuration, UE activity detection, HARQ procedure and soft-combining, contention and resolutions. Advantages and issues have been discussed for each technical component, and some design details have been provided.

Numerical and simulation evaluation results have been provided to demonstrate the GF reliability with repetition, UE activity detection, and GF multiple access with OFDAM and NOMA. The efforts have been specifically made for the evaluations on massive connectivity applications such as mMTC services and low-latency and high reliability applications such as URLLC services.

In summary, the GF multiple access scheme is a transmission without dynamic scheduling. It is attractive to small data transmissions with control signaling reduction, energy-saving, and low-latency (with high reliability) applications such as mMTC and URLLC services in the 5G NR network.

References

1. C.E. Shannon, A mathematical theory of communication. Bell Syst. Tech. J. **27**, 379–423 (1948)
2. 3GPP TR 38.913, Study on scenarios and requirements for next generation access technologies. v14.3.0 (2017)
3. 3GPP TS 38.321, Medium Access Control (MAC) protocol specification. 3GPP TS 38.321 V15.0.0 (2017)
4. Y. Polyanskiy, H.V. Poor, S. Verdu, Channel coding rate in the finite blocklength regime. IEEE Trans. Inf. Theory **56**(5), 2307–2359 (2010)
5. P. Popovski, Ultra-reliable communication in 5G wireless network, in *2014 1st International Conference on 5G for Ubiquitous Connectivity (5GU)* (2014), pp. 146–151
6. G. Durisi, T. Koch, P. Popovski, Towards massive, ultra-reliable, and low-latency wireless communication with short packets. Proc. IEEE **104**(9), 1711–1726 (2016)
7. R1-166095, Discussion on grant-free transmission, in *3GPP TSG RAN WG1 Meeting #86*, Gothenburg, Sweden, 22–26 Aug 2016
8. A.T. Abebe, C.G. Kang, Comprehensive grant-free random access for massive & low latency communication, in *IEEE ICC 2017 Wireless Communications Symposium*, 21–25 May 2017
9. A. Azari, P. Popovski, G. Miao, C. Stefanovic, Grant-free radio access for short packet communications over 5G networks. Accepted for presentation at IEEE Globecom (2017)
10. K. Au, L. Zhang, H. Nikopour, E. Yi, A. Bayesteh, U. Vilaipornsawai, J. Ma, P. Zhu, Uplink contention based SCMA for 5G radio access, in *Globecom 2014 Workshop*, Austin (2014), pp. 900–905
11. A. Bayesteh, E. Yi, H. Nikopour, H. Baligh, Blind detection of SCMA for uplink grant-free multiple-access, in *ISWCS'2014*, Barcelona (2014)
12. Z. Zhang, X. Wang, Y. Zhang, Y. Chen, Grant-free rateless multiple access—a novel massive access scheme for Internet of Things. IEEE Commun. Lett. **20**(10), 2019–2022 (2016)
13. Z. Ding, Z. Yang, Z. Fan, H.V. Poor, On the performance of non-orthogonal multiple access in 5G systems with randomly deployed users. IEEE Signal Process. Lett. **21**, 1501–1505 (2014)
14. R. Abbas, M. Shirvanimoghaddam, Y. Li, B. Vucetic, On the performance of massive grant-free NOMA, in *IEEE 28th Annual International Symposium on Personal, Indoor, and Mobile Radio Communications (PIMRC)* (2017)
15. C. Wang, Y. Chen, Y. Wu, L. Zhang, Performance evaluation of grant-free transmission for Uplink URLLC services, in *2017 IEEE 85th Vehicular Technology Conference (VTC Spring)* (2017)
16. 3GPP TS 38.331, Radio Resource Control (RRC) protocol specification. 3GPP TS 38.331 V15.0.0 (2017)

17. 3GPP TS 38.214, Physical layer procedures for data. 3GPP TS 38.214 V15.0.0 (2017)
18. 3GPP TS 38.212, Multiplexing and channel coding. 3GPP TS 38.212 V15.0.0 (2017)
19. 3GPP TS 36.211, Physical channels and modulation. 3GPP TS 36.211 V14.4.0 (2017)
20. 3GPP TS 36.212, Multiplexing and channel coding. 3GPP TS 36.212 V14.4.0 (2017)
21. H. Nikopour, H. Baligh, Sparse code multiple access, in *2013 IEEE 24th International Symposium Personal In-door and Mobile Radio Communications (PIMRC)* (2013), pp. 332–336
22. R. Hoshyar, R. Razavi, M. AL-Imari, LDS-OFDM an efficient multiple access technique, in *VTC-Spring* (2010), pp. 1–5
23. F. Kschischang, B. Frey, H. Loeliger, Factor graphs and the sum-product algorithm. IEEE Trans. Inf. Theory **47**(2), 498–519 (2001)
24. R. Hoshyar, F.P. Wathan, R. Tafazolli, Novel low-density signature for synchronous CDMA systems Over AWGN channel. IEEE Trans. Signal Process. **56**(4), 1616–1626 (2008)
25. Y. Yang Du, B. Dong, Z. Chen, X. Wang, Z. Liu, P. Gao, S. Li, Efficient multi-user detection for uplink grant-free NOMA: prior-information aided adaptive compressive sensing perspective. IEEE J. Sel. Areas Commun. (2017)
26. R1-1611658, HARQ design for UL URLLC, in *3GPP TSG RAN WG1 Meeting #87*, Reno, Nevada, USA, 14–18 Nov 2016
27. R1-1701665, UL Grant-free transmission, in *3GPP TSG RAN WG1 Meeting #88*, Athens, Greece, 13–17 Feb 2017
28. R1-1712215, On the determination of UE ID and HARQ process for UL grant-free transmission, in *3GPP TSG RAN WG1 Meeting #90*, Prague, Czech Republic, 21–25 Aug 2017
29. IEEE 802.16m, Evaluation Methodology Document (EMD) (2008)
30. R1-166097, SLS Results for MA evaluation in mMTC, in *3GPP TSG RAN WG1 Meeting #86*, Gothenburg, Sweden, 22–26 Aug 2016
31. R1-1611223, Performance evaluation of UL URLLC schemes, in *3GPP TSG RAN WG1 Meeting #87*, Reno, Nevada, USA, 14–18 Nov 2016
32. R1-1609446, Reference signal design for UL grant-free transmission, in *3GPP TSG RAN WG1 Meeting # 86bis*, Lisbon, Portugal, 10–14 Oct 2016

Chapter 17
Random Access Versus Multiple Access

Riccardo De Gaudenzi, Oscar del Río Herrero, Stefano Cioni and Alberto Mengali

17.1 Current Random Access (RA) Schemes

Random access (RA) protocols originated in the 1970s from the need for terminal–computer and computer–computer communication. In this kind of communication, data traffic is bursty. This is the result of the high degree of randomness seen in the message generation time and size, and of the relatively low-delay constraint required by the user. Users generate traffic with a low duty cycle, but when they do, they require a fast response. As a result, there is a large peak-to-average ratio in the required data transmission rate. In this context, it is not efficient to use fixed channel allocation schemes, as they would result in a low channel utilization, that is, the percentage of the channel capacity that goes into the throughput. A more advantageous approach is to provide a single shared high-speed channel to a large number of users. However, when dealing with shared media conflicts arise, i.e., more than one user wants to access the shared resources simultaneously. Therefore, the challenge with shared media is to control the access to the common channel while providing a good level of performance and maintaining reduced implementation

R. De Gaudenzi (✉) · O. del Río Herrero
European Space and Technology Centre, European Space Agency, Keplerlaan 1,
PO Box 299 2200 AG, Noordwijk, The Netherlands
e-mail: Riccardo.De.Gaudenzi@esa.int

O. del Río Herrero
e-mail: Oscar.Del.Rio.Herrero@esa.int

S. Cioni
European Space and Technology Centre, Ajilon for European Space Agency,
Keplerlaan 1, PO Box 299 2200 AG, Noordwijk, The Netherlands
e-mail: Stefano.Cioni@esa.int

A. Mengali
Interdisciplinary Centre for Security, Reliability and Trust, University of Luxembourg,
29, rue J.F. Kennedy, 1855 Luxembourg City, Luxembourg
e-mail: a.mengali@gmail.com

M. Vaezi et al. (eds.), *Multiple Access Techniques for 5G Wireless Networks and Beyond*, https://doi.org/10.1007/978-3-319-92090-0_17

complexity. RA protocols were developed to address this multiple access scenario, and today are extensively used in terrestrial networks over wired and wireless shared media [1–3]. In this section, we critically review the main terrestrial RA techniques and their applicability in the context of 5G IoT scenarios.

17.1.1 Carrier Sensing Protocols

One of the most widely used family of distributed packet access schemes is the one comprising carrier sense multiple access (CSMA) and its variants.

17.1.1.1 Carrier Sense Multiple Access (CSMA)

In CSMA, a terminal may be at a given time either transmitting or receiving (but not both simultaneously). Transmissions by one node are generally received by all other nodes connected to the medium, such as an electrical bus or a band of the electromagnetic spectrum. When a station has a packet to transmit it senses the medium before transmitting any data, and only starts transmission if the medium is not being used by another node [4]. If it senses a transmission, the device waits for the transmission in progress to end before initiating its own transmission (persistent mode) or will differ its transmission by a random time (non-persistent mode). Once the channel is idle, the waiting device can transmit its data. However, if multiple devices access the medium simultaneously and a collision occurs, the packets will be lost and terminals have to wait for a specific random time before restarting the transmission process again.

One limitation of CSMA is that when collisions occur, the various terminals involved continue their transmissions until the end of the message despite the destructive interference, thus resulting in an increased overhead and lower throughput. Besides, to operate efficiently, CSMA assumes an environment consisting of a number of users in line of sight and within range of each other. While this can be guaranteed in wired networks, in wireless networks we can experience the hidden terminal problem, where some stations may be out of the transmission and detection range of each other. The existence of hidden terminals significantly degrades the performance of CSMA as shown in [5].

17.1.1.2 Carrier Sense Multiple Access/Collision Detection (CSMA/CD)

Carrier sense multiple access/collision detection (CSMA/CD) operates similarly to CSMA, but once the transmission has started, if the sender detects a collision it stops transmitting to reduce the overhead. CSMA/CD requires a host being able to both transmit and receive on the medium at the same time. When collisions occur, each station willing to transmit is required to back off for a random time period

[6]. The CSMA/CD mechanism performs well in wired networks and the IEEE has standardized CSMA/CD in the IEEE 802.3 standard [7]. However, collision detection cannot be used over wireless networks as wireless transceivers cannot send and receive on the same channel at the same time. The strength of its own transmissions would mask all other signals on the air.

17.1.1.3 Carrier Sense Multiple Access/Collision Avoidance (CSMA/CA)

Carrier sense multiple access/collision avoidance (CSMA/CA) is another CSMA variant that resolves contention before any data is transmitted rather than by detecting colliding packets [8]. In CSMA/CA, the sender tries to avoid a collision by waiting for an inter-frame spacing (IFS) time before contending for the channel. If the channel is found busy during the IFS interval, the station defers its transmission. The IEEE has standardized CSMA/CA with a binary exponential back-off algorithm in the IEEE 802.11 standard [9]. The back-off algorithm in CSMA/CA tries to avoid collisions, but does not remove them completely. Small values of the random back-off time cause many collisions while very large values can cause unnecessarily long delays. Secondly, CSMA/CA fails to solve the hidden terminal problem, and cannot always detect that the medium is busy, thus creating a collision in the channel.

17.1.2 Distributed Reservation Protocols

17.1.2.1 Multiple Access Collision Avoidance (MACA)

Another family of schemes is the distributed reservation one, such as the multiple access collision avoidance (MACA) [10]. In MACA, a sender transmits a request to send (RTS) message to its intended receiver before the data transmission. The data is transmitted only after reception of a clear to send (CTS) message from the receiver, which is sent after reception of a successful RTS. MACA also requires that any station that overhears an RTS or CTS packet directed elsewhere inhibits its transmitter for a specified time. The IEEE 802.11 standard [9] has also adopted an optional RTS/CTS dialogue. It is a combination of CSMA/CA and MACA.

MACA alleviates the hidden terminal problem through the use of RTS and CTS (e.g., in networks with a centralized access point or base station), but does not solve it completely (e.g., in sensor networks with direct device-to-device communications). In addition, it introduces an overhead on the channel with RTS and CTS packet transmissions. Its use is particularly useful when large message sizes are to be transmitted, but less interesting when small packets (comparable to the RTS/CTS packets) are to be transmitted.

17.1.2.2 3G Random Access Channel (RACH)

A different distributed reservation scheme is represented by the random access channel (RACH) adopted in the third-generation (3G) cellular mobile networks [11]. In case of RACH, terminals first randomly transmit short packet preambles, and then wait for a positive acquisition indicator from the base station prior to the transmission of the complete message (i.e., after successful reservation of the channel).

17.1.2.3 4G Long-Term Evolution RACH

The RACH procedure defined in Long-Term Evolution (LTE) is similar to that of 3G [12]. RACH in LTE is not used for transmission of data, but for dynamic reservation of resources on the uplink. The random access procedure consists of a four-message handshake between the user equipment (UE) and the base station (eNodeB in LTE). First, the UE transmits a short preamble (message 1) in RA Slots which are then followed by a random access response (RAR) message (message 2) by the eNodeB, in case of successful reception. The RAR message contains timing alignment instructions to synchronize uplink transmissions and uplink resource allocation that will be used by the UE to request for capacity in a connection request (message 3). The eNodeB responds with a contention resolution (message 4) which completes the access reservation process.

This mechanism suffers from a high preamble collision probability in the presence of a very large number of devices such as for machine type communications (MTC). This leads to several preamble retransmissions prior to successful completion of the random access procedure and congestion of the random access channel [13].

As shown in Sect. 17.1.3, the use of pure ALOHA protocols for channel reservation is quite inefficient in terms of throughput when the channel gets loaded with a high probability of request failure.

The distributed reservation protocols described in this section solve (or at least alleviate) the hidden terminal problem, but cannot scale to networks with a massive amount of terminals transmitting small packets, typical of the IoT scenario. The overheads of the reservation mechanism are comparable to those of the data transmissions, thus resulting in a reduced channel utilization efficiency.

17.1.3 ALOHA Protocols

17.1.3.1 ALOHA

The pure ALOHA protocol, first proposed by Abramson in [14], is one of the oldest and simplest MA protocols. In ALOHA, a terminal transmits a packet in an asynchronous fashion without checking if any other terminal is active. Within an appropriate timeout period, it receives an acknowledgment from the destination, confirming that no conflict has occurred. Otherwise, it assumes that a collision has occurred and must retransmit. To avoid repeated collisions, the retransmission time is randomized across the terminals, thus spreading the retry packets over time.

17.1.3.2 Slotted ALOHA (S-ALOHA)

A slotted version of ALOHA, referred to as slotted ALOHA (S-ALOHA), is obtained by dividing time into slots of duration equal to the duration of a fixed-length packet [15]. Users are required to synchronize the start of transmission of their packets to the slot boundaries. When two packets collide, they will overlap completely rather than partially, providing an increase on channel utilization over pure ALOHA. The S-ALOHA has the advantage of higher efficiency, but requires time slot synchronization.

17.1.3.3 Diversity Slotted ALOHA (DS-ALOHA)

The diversity slotted ALOHA (DS-ALOHA) [16] slightly improves the S-ALOHA performance at low channel loads by sending twice the same packet at random slot locations in the frame in favor of increasing the time diversity and thus reducing the packet loss rate (PLR).

It is generally assumed that whenever two packet transmissions overlap in time, they cancel each other. This assumption is pessimistic as it neglects capture or near–far effects in radio channels. Capture occurs when a terminal receives messages simultaneously from two terminals, but the signal from one of them drowns out the other, making the collision resolvable for the stronger message. The terminal with the higher received power is said to have captured the receiver. Some of these effects have been addressed in [15, 17–19]. Capture is good in the sense that it reduces the time needed in resolving collisions, but may also drive weak terminals completely out of the medium.

Figure 17.1 presents the performance results for S-ALOHA and DS-ALOHA in the presence of packets power imbalance following independent identically distributed lognormal distributions with equal mean μ and standard deviation σ, where both parameters are expressed in dB in the logarithmic domain.

The results have been obtained by detailed simulations (see the figure caption for details on the physical layer assumptions) and by using the analytical model derived in [20]. The x-axis represents the normalized average channel MAC load (G) expressed in information bits/symbol for non-spread-spectrum RA schemes and in information bits/chip for RA schemes employing spread-spectrum which will be discussed in the following section. In this way, we avoid any dependence on the modulation cardinality or coding rate used. A similar approach has been used for measuring the RA throughput shown in y-axis. The figure shows that in both schemes the throughput improves with increasing power imbalance as collisions become easier to resolve (power capture effect). However, as expected, the PLR rises quickly as the load on the channel increases.

Although for equi-powered packets the maximum channel utilization available is 18% for pure ALOHA, and 36% for S-ALOHA, low packet collision probabilities (e.g., $<10^{-3}$) typically required by satellite networks[1] are achieved at very low loads, i.e., $<10^{-3}$ bits/symbol. This results in very low channel utilization (see Fig. 17.1). It is important to remark the fact that often in the literature RA schemes are compared looking at their peak throughput rather than the value achieving the required MAC PLR for a given service or application.

17.1.3.4 Spread-Spectrum ALOHA (SS-ALOHA)

Slotted RA systems require terminals to keep time slot synchronization. The resulting network synchronization overhead reduces system efficiency, in particular for networks characterized by large number of terminals with very low transmission duty cycle (e.g., M2M applications). Moreover, a requirement of accurate network-level time synchronization increases the complexity of the terminal. Thus, slotted RA is penalizing low-cost terminal solutions. To mitigate this limitation, a pure ALOHA scheme can be employed, but its performance is worse than for S-ALOHA increasing by a factor two the packet collision probabilities [15].

Direct sequence spread spectrum (DS-SS) multiple access is the most common form of code division multiple access (CDMA), whereby each user is assigned a particular code sequence which is modulated on the carrier with the digital data modulated on top of that [23]. Users can transmit asynchronously and even when the same spreading code is used, data can be received [24, 25]. The spread spectrum ALOHA (SS-ALOHA) protocol proposed in [24] has potentially attractive features as it provides a much higher throughput capability than S-ALOHA for the same PLR target under equal power MA conditions when adopting powerful physical layer forward error correction (FEC) (e.g., coding rates $\leq 1/2$) and low order modulations (e.g., binary phase shift keying (BPSK), quadrature phase shift keying (QPSK)). Reference [26] shows that the SS-ALOHA throughput is critically dependent on the demodulator signal-to-interference-plus-noise ratio (SINR) threshold for packet decoding. Results reported in [26] indicate that, differently from S-ALOHA, SS-ALOHA shows a steep PLR increase with MAC load. Thus, SS-ALOHA can be operated with low PLR close to the peak of the throughput characteristic. As an example, using turbo codes and relatively small packets, SS-ALOHA can achieve a much higher throughput than ALOHA or S-ALOHA, in the order of $T \simeq 0.5$ bits/chip for a PLR of 10^{-3} (see Fig. 17.2 with $\sigma = 0$ dB).

---

[1] The low PLR operating point is justified by the need to avoid excessive and highly variable retransmission latencies in the satellite environment which is characterized by the very long propagation delays (in particular for satellite medium Earth orbit (MEO) and geostationary Earth orbit (GEO)), and to maximize the throughput of applications using upper-layer protocols such as the transmission control protocol (TCP) that is also highly sensitive to the packet losses [21, 22].

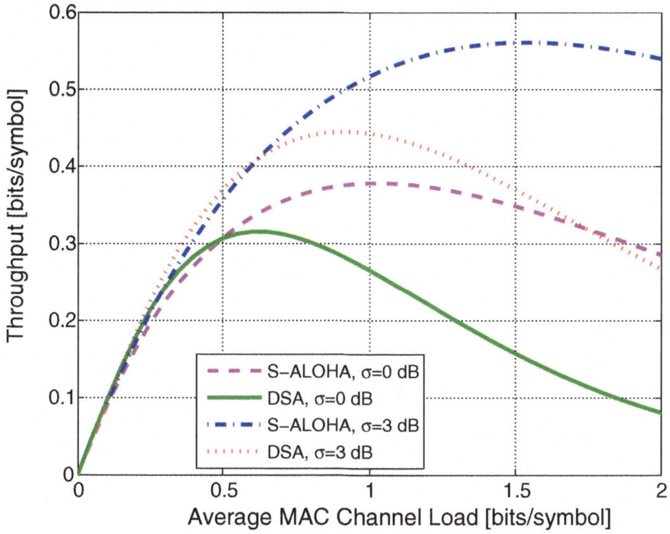

(a) Throughput

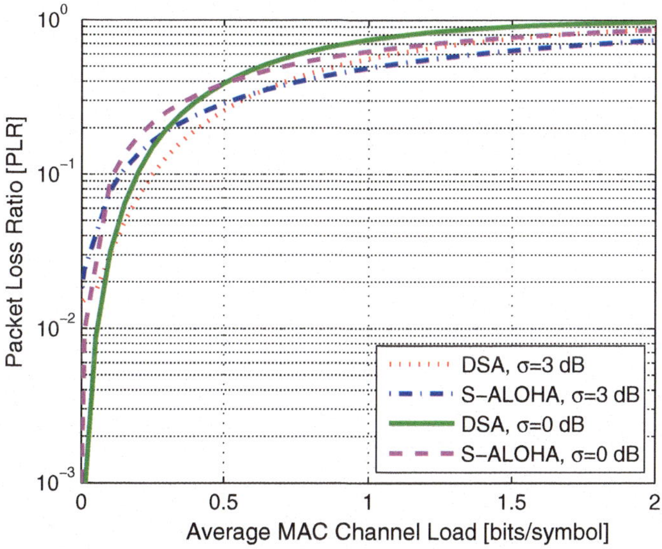

(b) PLR

Fig. 17.1 S-ALOHA and DS-ALOHA performance from [20] for QPSK modulation, 3GPP turbo code FEC $r = 1/2$, packet block size 100 bits, Energy per Symbol to Noise Power Spectral Density ratio $E_s/N_0 = 7$ dB in the presence of lognormal packets power imbalance with mean $\mu = 0$ dB, standard deviation σ and Poisson traffic (©2014 IEEE)

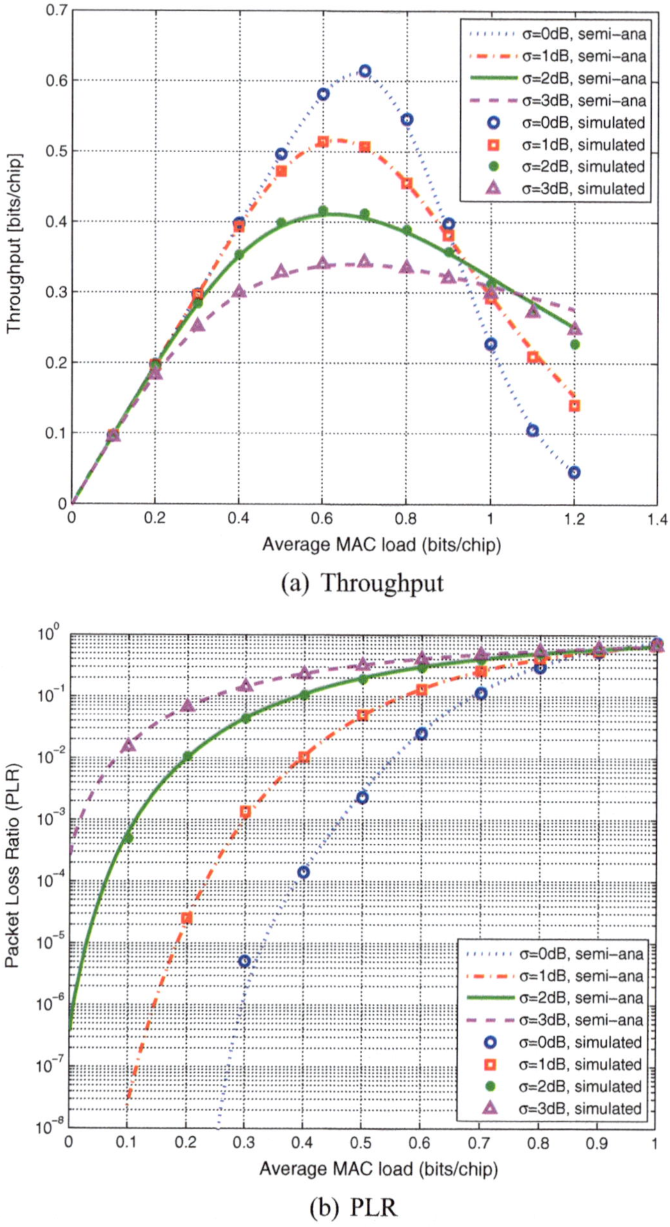

Fig. 17.2 Simulated versus analytical SS-ALOHA performance from [27] with and without power imbalance: 3GPP turbo code FEC $r = 1/3$ with block size 100 bits, BPSK modulation, spreading factor 256, $E_s/N_0 = 8.9$ dB (©2012 IEEE)

SS-ALOHA represents a very interesting RA scheme for the MTC environment, in particular when asynchronous access is required. The main reason for performance improvement of SS-ALOHA techniques with regards to pure ALOHA is that they can take advantage of a higher traffic aggregation. The average number of packet arrivals over one packet duration λ can be computed as follows:

$$\lambda = N_{\text{rep}}\, G\, G_p, \tag{17.1}$$

where N_{rep} is the number of replicas transmitted for each packet, G is the MAC load expressed in information bits/symbol in non-spread systems and bits/chip in spread systems. The processing gain is given by $G_p = SF/(r \log_2 M)$, where r is the FEC code rate, M is the modulation cardinality, and SF is the spreading factor.

It can be observed from (17.1) that large processing gain will proportionally increase the value of λ. Typical values for non-spread-spectrum systems are $\lambda \leq 5$, while for SS systems assuming $SF = 32$ and $r = 1/3$, $\lambda \approx 100$.

Figure 17.3 demonstrates that the Poisson density normalized to the mean value, approaches a Dirac delta function as λ increases. This means that the instantaneous number of interfering packets fluctuation will reduce with increasing λ. This is a favorable consequence for RA, because the system operational average MAC load can be chosen to be close to its maximum operational value (i.e., when the PLR is around 10^{-3}–10^{-4}). Furthermore, when operating with large λ values, the co-channel interference can be accurately approximated as additive white Gaussian noise (AWGN) thanks to the large number of interfering packets thus easing its analysis.

Despite the attractive features listed above, SS-ALOHA Achilles' heel resides in its high sensitivity to MA carrier power imbalance. This phenomenon is disrupting the SS-ALOHA scheme throughput because of the well known CDMA near–far problem. As shown in Fig. 17.2, for PLR of 10^{-3} or less, the SS-ALOHA throughput is diminished by several orders of magnitude when the received packets power is lognormally distributed with standard deviation of 2–3 dB, as opposed to S-ALOHA where power imbalance results in improved performance due to the power capture effect.

17.1.3.5 LoRa

There are already several technologies available to support MTC services in terrestrial low-power wide area networks (LPWAN), and most of them have been captured in [28]. Hereafter, some of them are presented, since either being pioneers in addressing such type of challenging applications or being the first answer from international standardization bodies.

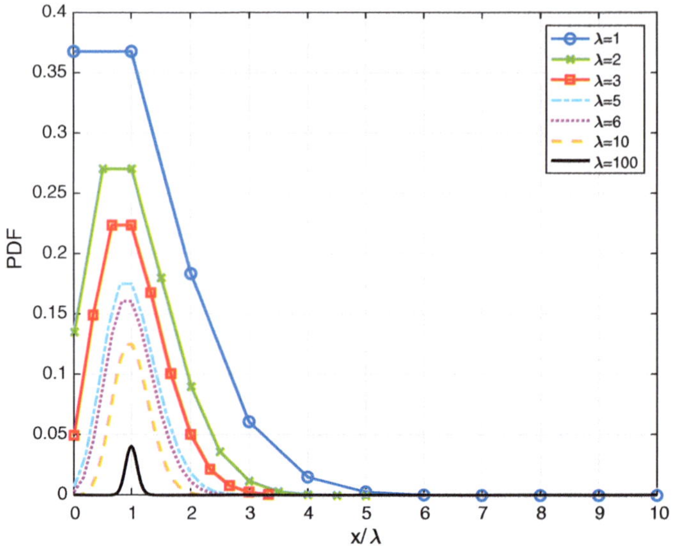

Fig. 17.3 Traffic probability distribution normalized to the mean value

One of the first solution designed for IoT applications has been LoRa [29], a proprietary waveform[2] based on spread-spectrum techniques operating in the Industrial, scientific and medical (ISM) radio bands, and commercialized by Semtech Corporation. In LoRa, the spreading of the spectrum is achieved by generating a chirp signal that continuously varies in frequency [30]. In LoRa the flow of bits to be modulated is grouped in symbols. A symbol can assume $2M$ different values. Each value corresponds to a different frequency, thus $2M$ can be considered as a sort of spreading factor. Each symbol during its duration spans the whole bandwidth available, starting from one of the $2M$ available frequencies. The information is thus not the particular frequency transmitted at each time but the initial one. Once the received signal is demodulated, a single frequency value remains which corresponds to the reconstructed information symbol. An advantage of this method is that timing and frequency offsets between transmitter and receiver are equivalent, greatly reducing the complexity of the receiver design. The frequency bandwidth of this chirp is equivalent to the spectral bandwidth of the signal. LoRa modulation is both bandwidth and frequency scalable. It can be used for both narrowband frequency hopping and wideband direct-sequence applications. Unlike existing narrowband or wideband modulation schemes, LoRa can be easily adapted for either mode of operation with only a few simple configuration register changes. The typical data rate varies from 300 bps to 30 kbps, and the trade-off is between the device throughput and the transmission range. Higher spreading factor enables long-distance commu-

[2]The LoRaWAN link layer is instead a relatively open standard with specifications available through the LoRa Alliance (https://www.lora-alliance.org).

nications at the cost of lower user data rate and overall throughput, and vice versa. The LoRa physical layer also features constant envelope combined with robustness to interference, multipath, and Doppler. The signal is also suitable to provide ranging measurement as required by localization services.

LoRa relies on a pure unslotted ALOHA access for the uplink. In a LoRaWAN network, devices are not associated with a specific base station. Instead, data transmitted by a device is typically received by multiple base stations (spatial diversity). Each base station forwards the received packet from the devices to the network server. The network server has the required intelligence for filtering the duplicate packets from different base stations.

The degrees of diversity adopted in the uplink channel are based on asynchronous and uncoordinated transmission time, on the random selection of the transmission carrier frequency (among a set of tens of allowed values), on choosing different spreading factors, and finally on the spatial diversity from the multiple base stations.

17.1.3.6 SigFox

Another proprietary solution addressing the MTC market in LPWAN is SigFox [31], an ultra narrow band (UNB) signal transmission. SigFox is using 192 kHz of the publicly available ISM band to exchange messages over the air. Each message is 100 Hz wide and transferred with a data rate of 100 or 600 bps depending on the region. The UNB technology enables very low power consumption and inexpensive terminal manufacturing. Consequently, the user data rate is rather low (i.e., approximatively 100 bps), however the communication range is in the order of 30–40 km in good propagation conditions (e.g., rural environments).

SigFox relies on unslotted diversity ALOHA with differential BPSK modulation and no FEC for the uplink. The transmission is unsynchronized between the base station and the device. The device transmits a packet on a random frequency and then sends few replicas (typically three) on different frequencies and time (time and frequency diversity). As for LoRaWAN, the devices are not associated with a specific base station, unlike cellular networks. The transmitted message is received by any base stations that are nearby (spatial diversity).

17.1.3.7 Random Phase Multiple Access (RPMA)

The RPMA INGENU proprietary LPWAN solution operates in the 2.4 GHz ISM band and exploits the more relaxed regulations on the spectrum sharing with respect to sub-GHz bands. The patented RPMA CDMA access scheme [32], is solely employed for the uplink RA. RPMA terminals and base stations are synchronized so that terminals send signal that fit inside of predefined frames of certain size. Terminals send packets with a random delay that is small enough to not exceed the frame size. In addition to choosing signal delay, each terminal calculates the optimal spreading factor to transmit at based on measurement of the downlink signal strength. The base station demodulator is demodulating in parallel packets with all possible spreading factors and delay offset. RPMA base station demodulators are constantly monitoring channel conditions and local interference levels. This information is continuously signaled back to terminal via the downlink signaling channel to optimize the RA throughput. These techniques allow RPMA to support simultaneous demodulation of up to 1200 fully overlapping signals per sector that are all in the same frequency.

INGENU RPMA provides bidirectional communications, download speeds up to 30 kb/s and upload speeds up to 15.6 kb/s. With a declared receiver sensitivity of -142 dBm (or equivalently 167 dB of maximum coupling loss), RPMA provides approximatively 10–15 dB extra link margin with respect to LoRa or SigFox. This extra margin is translated in a larger coverage, enhanced indoor penetration and increased robustness to MA interference. Furthermore, the user devices adjust their transmit power for reaching the closest base station, thus limiting the amount of co-channel interference generated.

17.2 5G NOMA-Based RA Proposals for IoT

In a massive IoT scenario, a base station needs to provide connection to a huge number of low-cost terminals. The main challenge for such a scenario is how to effectively deal with massive connection with UE tight power constraint. In a 4G LTE system (see Sect. 17.1.2.3), to transmit data, a user shall first issue scheduling request (SR) on periodically occurred resources, which is configured by the base station (BS). The base station then makes scheduling decisions and sends an uplink grant to the user indicating the resources on which the user can transmit data. Generally, the procedure may take 10 ms or more. For some IoT applications, such a long latency is unacceptable. Moreover, the uplink grant is carried by downlink control signaling, and with a massive number of connections the downlink control channel may become a bottleneck. In such a situation, grant-free transmission represents a more suitable option (see Sect. 17.1.3). By means of grant-free transmission, a user autonomously selects the resource for transmission without an SR and scheduling from the BS.

To cope with the grant-based 4G IoT approach limitations, 5G IoT is moving from orthogonal multiple access (OMA) to grant-free NOMA solutions, whereby multiple users efficiently share the same radio resource and exploit multi-user detection (MUD) to resolve collisions.

Actually, NOMA has been used in previous wireless systems, such as the 3G wideband code division multiple access (WCDMA) standard. In fact, data symbols in the uplink of 3G systems are spread by long spreading codes, and multiple users transmit their spread signals on the same frequency and time resources. Since long spreading sequences are used, only a linear detection algorithm is possible at the receiver due to the complexity of a nonlinear detection algorithm. As a result, non-orthogonal transmission of this type is demonstrated to be weak in terms of spectrum efficiency.

The rationale for moving to NOMA for 5G IoT can be summarized as:

- Removing the need for "asking-for-grant" procedure for sending packets to increase the effective throughput and reduce transmission delay. Grant-free access should be provided when there is a large number of users being served by a network as is the case for IoT. In such cases, a grant-based transmission would cause high signaling overhead and transmission delay, which would significantly reduce the spectral efficiency of the transmissions.
- Achieving higher spectral efficiencies than OMA thanks to the superimposition of several packet transmissions over the same bandwidth and collision resolution through powerful MUD techniques at the demodulator. The amount of superimposition achieved is normally called overloading.

A very good overview, categorization, and comparative performance analysis of currently proposed NOMA schemes for 5G new radio IoT is reported in [33]. In particular, the different proposals from the key wireless industry players can be grouped in three main multiple access (MA) categories: (a) Codebook-based MA; (b) Sequence-based MA; (c) Interleaver/scrambler-based MA.

Codebook-based MA maps the user data packet stream in a multi-dimensional codeword belonging to a codebook. The mapping is done in a way to achieve signal spreading and introducing zero elements to mitigate inter-user interference. Decoding is obtained through a relatively complex iterative message passing algorithm.

Sequence-based MA exploits non-orthogonal complex number sequences to separate users sharing the same spectrum and easing the MUD process. Affordable complexity linear minimum mean squared error (MMSE) plus successive interference cancellation (SIC) or parallel interference cancellation (PIC) are proposed for the packet detection.

Interleaver/scrambler-based MA utilizes interleave-division MA to separate users sharing the same bandwidth. Some repetition/scrambling is also adopted to spread the signals and achieve some interference averaging effect. Depending on the size of the interleaved bit stream, simpler MMSE-SIC or more complex soft SIC decoding techniques will be used.

Results reported in [33] show that depending on the propagation environment (outdoor/indoor) and the MA loading factor, the gain of the different proposals for 5G new radio (NR) NOMA are varying, but always performing better than the traditional OMA approach. Typically, the most complex MUD structures are outperforming simpler MMSE-SIC type of detectors, at least for high loading factors.

Figure 17.4 shows a pictorial view of the main configurable elements in the transmit-receive chain for comparing the proposed 5G NOMA techniques [34]. As far as the transmitter side is concerned, the discussed three categories have been mapped in two main operations, either at the bit or at the symbol level, in order to highlight specifically which MA signature is inserted for enabling the collision resolution process. On the receiver side, the block diagram summarizes the mostly adopted multi-user detectors iteratively combined with powerful FEC decoders. Reference [34] reports also the agreed evaluation methodology and system scenarios for the comparison among the various candidates.

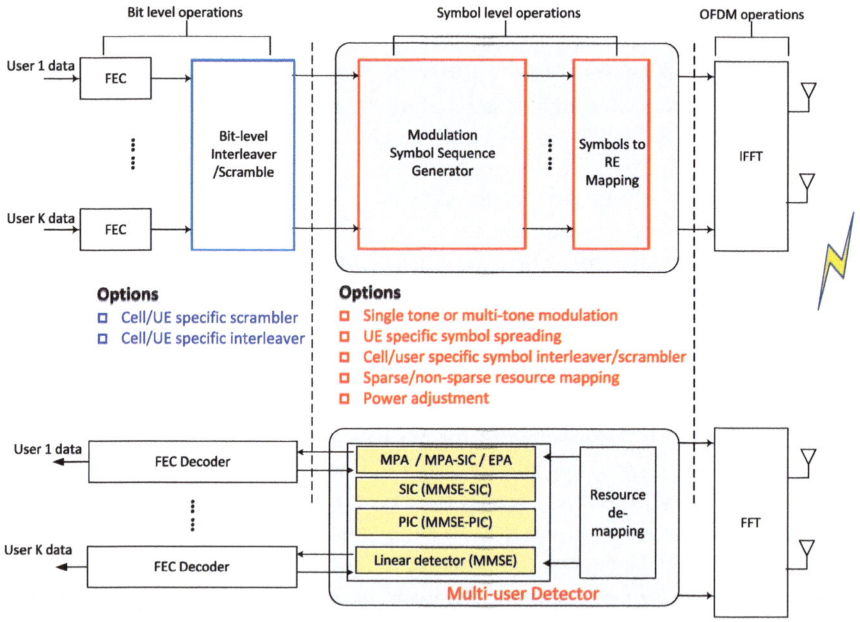

Fig. 17.4 Illustration of configurable components under the unified framework at transmitter and receiver side [34]

A more detailed discussion of each proposed 5G NOMA techniques for IoT applications is presented in the following subsections.

17.2.1 Codebook-Based MA

Two main techniques are belonging to the Codebook-based MA. The predecessor of these techniques is low-density spreading (LDS), originally proposed to reduce the complexity of MUD [35]. In the LDS system, modulated symbols are only spread over a part of the total chips and the signatures on other chips are zero. Therefore, the number of interfering users on each chip is much lower than for traditional CDMA. Similar to low-density parity-check (LDPC) code, LDS signatures can be represented by a sparse factor graph, with variable nodes (VNs) representing data symbols and function nodes (FNs) representing chips. By taking advantage of the sparsity, a message passing algorithm can be applied for multi-user detection, which has much lower complexity than the optimal maximum a posteriori (MAP) decoder but achieving almost the same performance.

17.2.1.1 Sparse Code Multiple Access (SCMA)

Sparse code multiple access (SCMA) was introduced in [36] as a generalization of the idea of LDS. At the transmitter side, for each UE, the information bits after channel coding are fed to the SCMA encoder which maps the coded bits to SCMA codewords. An SCMA codeword is a multi-dimensional constellation vector, which is spread onto the physical resource elements (REs) of a multi-carrier system (typically orthogonal frequency division multiplexing (OFDM)). SCMA codebook design is the key to ensure good performance and flexibility of the whole SCMA system. As mentioned above, the SCMA codebook is the result of the joint optimization of multi-dimensional modulation and low-density spreading. Consequently, SCMA maps the coded bits directly to complex domain multi-dimensional sparse codewords. The resulting modulation symbols on the nonzero REs are different, but dependent on among each other for each codeword. The motivation for such correlated design is to follow the multi-dimensional modulation principles of maximizing the average distance between the constellation points. Having designed the codebook, SCMA encoder simply selects the codeword corresponding to the input coded bit sequence from the SCMA codebook and then map it to the physical resource elements.

The demodulator requires to separate the overlapping OFDM packets using a message passing algorithm (MPA). To further enhance the demodulator performance, the MPA detector can be combined with the physical layer FEC decoder to operate in a turbo fashion. Performance reported in [37] shows that for an urban environment and a user moving at 3 kmph, under ideal channel estimation a similar PLR to the one of OMA can be obtained with 350% overload factor.

One potential issue of SCMA is the scalability investigated in [38], since [39] for the MPA receiver, there is lack of definite overloading performance dependence on the sparsity and the number of the sparse codes.

17.2.1.2 Pattern-Division Multiple Access (PDMA)

Pattern division multiple access (PDMA) is based on code domain superposition. To effectively separate users being multiplexed on the same resource, different types of MUD nonlinear detection algorithms have been proposed. MAP or maximum likelihood (ML) algorithms are quite complex to implement. SIC demodulation represents a good trade-off between performance and complexity aspects. However, according to [40], SIC suffers from the previous packet error propagation problem, which may degrade the NOMA overall performance.[3] Reference [40] introduced the novel PDMA NOMA scheme based on code pattern MA. The key idea of PDMA is to perform a joint optimization of the pattern design accounting for both transmitting and receiving sides. More specifically, the pattern design at the transmitter side

[3]However, Authors practical implementation experience showed that packets incorrectly detected and subtracted can be removed in later SIC processing step. This represents a kind of SIC self-healing process.

accounts for SIC-based detection at the receiver side. PDMA patterns are designed to offer different order of transmission diversity so that the disparate diversity order between multiple users could be introduced to alleviate the SIC demodulator error propagation problem. Another possibility is to use a low-complexity belief propagation (BP) algorithm for PDMA detection exploiting its sparse pattern.

The PDMA NOMA scheme could naturally be incorporated into the grant-free transmission to introduce the pattern dimension to users orthogonality and reduce collision probability. For transmission, each UE selects a time/frequency resource as well as a PDMA pattern from an assigned pattern matrix. Even though two users may select the same time/frequency resource, as long as their PDMA patterns are different, the receiver is able to decode both users' data successfully.

Simulation performance reported in [40] for a typical cellular network shows a 500% gain in terms of supported uplink users compared to the current OMA LTE for a 1% PLR using a BP demodulator.

17.2.2 Sequence-Based MA

17.2.2.1 Multi-User Shared Access (MUSA)

To alleviate the potential issues of SCMA mentioned in Sect. 17.2.1, a novel NOMA scheme named multi-user shared access (MUSA) has been proposed [39]. MUSA combines non-orthogonal transmission and grant-free access. Data of each user is spread with a family of complex tri-level real and imaginary values (i.e., $\{\pm 1,\ 0\}$) spreading sequence with short length. Signals from multiple users are superposed at the receiver, where SIC is used to cancel interference between users. The (short) spreading sequences are specifically designed to cope with heavy overloading of users and to facilitate simple SIC implementation on the receiver side.[4] Moreover, each user can choose its spreading code autonomously, therefore eliminating the need for resource coordination by BSs. With grant-free access and non-orthogonal transmission, MUSA can support a large number of connections, while minimizing signaling overhead and power consumption at the same time. The use of a simple SIC demodulator allows taking advantage of the incoming packets power imbalance due to the near–far effect. This allows eliminating the need for closed-loop power control and associated signaling overhead.

According to [41], no pilots or long preambles are required for channel estimation and MTC node (terminal) activity detection (blind detection). Instead, the successfully decoded data can be employed for channel estimation. Each user randomly chooses a spreading code from a predefined pool known by base station and UEs.

[4] In particular, it appears that the MUSA MMSE implementation through matrix inversion is limiting the spreading sequence length. Solutions like MMSE Enhanced Spread Spectrum ALOHA (ME-SSA) described in Sect. 17.3.2 are overcoming this issue with a much simpler MMSE multi-stage implementation.

Some UEs may happen to choose the same spreading code and such situation is called code collision. The base station does not know code collision occurrence before blind detection. The computational complexity of MUSA by using blind detection is not very high since a SIC receiver is applied for signal detection and decoding. The computational complexity is just linearly increasing with the number of users. This is different from MPA which is exponentially growing with the number of users in complexity and decoding time. Although in principle MUSA can operate without requiring user terminal synchronization, [39] recommends adopting simple downlink synchronization to reduce the SIC demodulator implementation complexity.

Simulation results reported in [39] for AWGN channel show that that for QPSK with rate 1/2 turbo coding MUSA with tri-level complex spreading code can achieve 225% user overloading when the spreading code length is just 4. Using longer spreading sequences provides a negligible performance advantage. Instead, if binary pseudo-noise (PN) sequences are used, 1% PLR cannot be achieved for a code length of 4, even if there is no overloading. For binary complex spreading code, the overloading ratio is reduced to 150%. It is argued that, thanks to the higher degree of freedom provided by tri-level complex spreading codes, randomly picked sequences would have lower cross-correlation.

More interesting is the comparison with LTE OMA over a single cell urban mobile channel, where MUSA can achieve an overloading factor of 400% with comparable PLR. System-level multi-cell simulations results also reported in [41] show that MUSA can achieve a very remarkable overload gain compared to LTE for both frequency reuse 1 ($\times 10$) and 3 ($\times 50$) factors. The latter is providing even a further boost in performance as there is less intra-cell interference.

17.2.2.2 Resource Spread Multiple Access (RSMA)

In resource spread multiple access (RSMA) [42], a group of different users signals are superpositioned on top of each other, and each users signal is spread to the entire frequency/time resource assigned for this group. Different users signals within the group are not necessarily orthogonal among them and they could potentially generate inter-user interference. Spreading of bits to the entire resources enables decoding at a signal level below background noise/interference. RAMA uses combination of low-rate channel codes and scrambling codes (and optionally different interleavers) with good correlation properties to separate different users signals. RAMA closely resembles to the enhanced spread-spectrum ALOHA (E-SSA) RA scheme previously devised for satellite MTC applications described in Sect. 17.3.2. Depending on the application scenarios, it can include:

- Single-carrier RAMA: optimized for battery power consumption and coverage extension for small data transactions by utilizing single-carrier waveforms, very low bit rate, low code rate and modulations with low peak-to-average power ratio (PAPR) (e.g., filtered $\pi/2$ BPSK, offset-quadrature phase shift keying (O-QPSK)

or Gaussian minimum shift keying (GMSK)). It allows grant-less transmission and potentially allow asynchronous access.
- Multi-carrier RAMA: optimized for low latency access for radio resource control (RRC) connected state users (i.e., timing with terrestrial BS already acquired) and allows for grant-less transmissions. This mode is used in good coverage areas with enough power headroom.

Although RAMA is sharing quite some similarities with the traditional Direct-Sequence CDMA, there are several differences such as:

- It supports multi-carrier mode (i.e., modulation symbols on tones) for certain use cases.
- The single-carrier mode is strictly single carrier with time division multiplexing (TDM) data and control channels, as well as using selected low PAPR modulations for coverage extension and user equipment battery power saving.
- Not relying on different Walsh codes (or spreading factors) to separate users, but utilize low-rate channel codes to fully leverage coding gains.

RAMA performance reported in [43] shows that RAMA can provide significant performance gain over orthogonal frequency division multiple access (OFDMA) especially at high spectral efficiency. It has also been shown that scrambled coded multiple access (SrCMA) has no appreciable performance gain over RAMA when every single user operates at low spectral efficiency which is the actual scenario in MTC due to the low power of each UE. Thus, sequence-based and interleave-based schemes are better solutions than codebook-based schemes for uplink IoT transmission when the complexity of the receiver is considered.

17.2.2.3 Non-Orthogonal Coded Multiple Access (NCMA)

Non-orthogonal coded multiple access (NCMA) is one of the NOMA schemes based on the spreading codes with the minimum correlation. NCMA is using non-orthogonal codewords minimizing the co-channel cross-correlation obtained by Grassmannian line packing problem [44]. In this reference, it is claimed that NCMA can provide the additional throughput or improved connectivity with a small loss of block error rate (BLER) in specific environments, by exploiting additional layers through superposed symbol while satisfying quality of service (QoS) constraints. Since the receiver of NCMA system can be based on PIC, it results in a low-complexity MUD receiver. In addition, the multi-user interference level between codeword pairs is always similar due to the specific NCMA correlation characteristics.

As far as the performance improvement is concerned, in an urban environment and slow user mobility (e.g., 3 kmph), the same target PLR $= 1\%$ can be obtained with about 180% overload factor with ideal channel estimation.

17.2.2.4 Non-Orthogonal Coded Access (NOCA)

Similarly with other spreading based non-orthogonal schemes, the basic idea of non-orthogonal coded access (NOCA) is that the data symbols are spread using non-orthogonal sequences before transmission [45]. The spreading can be applied in the frequency domain and/or time domain based on configuration.

The short 3GPP contribution [45] reports some NOCA performance results employing an MMSE-SIC demodulator in a single cell scenario with equal power. A PLR of 10^{-2} is achieved with 2 dB lower signal-to-noise ratio (SNR) compared to LTE OMA. As far as system-level simulations are concerned [34], the gain is about 516% with respect to LTE OMA at PLR $= 10^{-1}$.

17.2.2.5 Group-Orthogonal Coded Access (GOCA)

Group-orthogonal coded access (GOCA) belongs to sequence-based non-orthogonal MA. Users use two-stage structure to first generate group-orthogonal sequences, and then spread modulated symbols into shared time and frequency resources [46]. GOCA can also be seen as a variant of repetition-division multiple access (RDMA) described in the following, where the repetition is done in two stages. In GOCA, interference can only arise between users belonging to different groups. Thus, in high overload case, interference might be lower with respect to RDMA as the interference of users from the same group is orthogonal and thus not relevant. Only interference from users belonging to other groups are affecting the performances.

GOCA proponents in [46] are arguing the practical validity of the SCMA scheme. The SCMA MPA detector complexity has polynomial growth rate with the constellation size and exponential growth rate with the maximum number of users superposed at each dimension of a codeword. Therefore, the overall receiver complexity is expected to be high, if a large number of users are multiplexed on the same resource. Moreover, the multi-dimensional constellation shaping is the only difference between SCMA and LDS-CDMA.[5] The benefit from multi-dimensional constellation is probably compromised by uncontrollable independent fading channels between users in uplink transmission. The SCMA has no performance gain over other NOMA schemes in low Spectral Efficiency (SE). In IoT scenario, many devices have to operate at low SE, or low SINR, due to the transmit power limit. Thus, sequence-based and interleave-based schemes are better solutions than codebook-based schemes for uplink IoT transmission as SCMA may have marginal or even no gain over LDS-CDMA.

GOCA is considered to represent a good trade-off between transmitter and receiver complexity providing better performance than other NOMA schemes such as interleaved-division multiple access (IDMA), RAMA and RDMA. In particular,

[5]For LDS-CDMA, the nonzero positions of a user signature vector are just filled with repetition of QPSK symbols, which is a more simple way to construct signature vectors.

results reported in [46] for GOCA with perfect power control (equal average SNR), no timing offset, perfect channel estimation and user moving at 3 kmph show that it outperforms by 3.2 dB RAMA in terms of SNR for the same overloading factor. For this evaluation scenario, the gain with respect to traditional OMA schemes is about 210% at PLR $= 10^{-1}$. Initial simulation results indicate that GOCA is superior to RAMA under heavy overloading conditions thanks to the group sequence orthogonality. Instead, for reduced overloading, the performance of GOCA and RAMA are very close. In all cases, perfect UE time synchronization is required for exploiting group sequences orthogonality.

17.2.3 Interleaver/Scrambled-Based MA

A conventional random waveform CDMA system (such as IS-95) involves separate coding and spreading operations. Theoretical analysis [47, 48] shows that the optimal multiple access channel capacity is achievable when the entire bandwidth expansion is devoted to coding. This suggests combining coding and spreading using low-rate codes to maximize the coding gain. In this case, interleavers can be employed to distinguish signals from different users. This is the basic principle of Interleaver/scrambled-based MA.

17.2.3.1 Interleaved-Division Multiple Access (IDMA)

Based on literature findings that the use of different interleavers is improving CDMA performance with MUD, [49] provides an in-depth analysis of the IDMA scheme. The IDMA modulator is similar to a conventional non-spread-spectrum one. The main difference is that following a low-rate coder, coded bits are permuted by a block interleaver. The key principle of IDMA is that the interleavers should be different for different users and they are generated independently and randomly. In other words, the CDMA user unique spreading sequence here is replaced by the combination of very low-rate coding and user unique interleaving. These interleavers disperse the coded sequences so that the adjacent chips are approximately uncorrelated, which facilitates the simple chip-by-chip detection scheme. In [49] an iterative sub-optimal receiver structure has been adopted. It consists of an elementary signal estimator (ESE) and K single-user a posteriori probability (APP) decoders. The MUD complexity is considerably lower than MMSE and independent on the number of users.

IDMA results are reported in [49] for AWGN and Rayleigh fading channels considering different decoding algorithms. IDMA performance is also compared with CDMA. The IDMA performance advantages are growing with the number of active users.

17.2.3.2 Interleaved Grid Multiple Access (IGMA)

In [50] an alternative NOMA candidate derived from interleaver-based MA scheme is proposed and named interleaved grid multiple access (IGMA). Basically, the IGMA scheme could distinguish different users based on:

1. Different bit-level interleavers;
2. Different grid mapping patterns;
3. Different combinations of bit-level interleaver and grid mapping pattern.

The channel coding process can be either using simple repetition (spreading) of a moderate coding rate FEC or directly using low coding rate FEC. Compared to the need of well-designed codeword or code sequences, the sufficient source of bit-level interleavers and/or grid mapping patterns is able to not only provide enough scalability to support different connection densities, but also provide flexibility to achieve good balance between channel coding gain and benefit from sparse resource mapping. By proper selection, low correlated bit-level interleavers could be achieved. The symbol-level interleaving randomizes the symbol sequence order, which may further bring benefit in terms of combating frequency selective fading and inter-cell interference.

In the receiver side, taking advantage of the special property of interleaving, a low-complexity multi-user detector, identified as ESE [49], can be utilized with a simple de-mapping operation on the top. Note that a lower density of the grid mapping pattern could further reduce the detection complexity of ESE for IGMA implementation. In addition, MAP and MPA detectors are also applicable to IGMA, improving the detection performance a lot comparing to ESE at the cost of additional complexity.

The IGMA simulation results reported in [50] confirm that the low-complexity ESE MUD can achieve fairly good performance for low SNR. For higher SNR values a more complex chip-by-chip MAP decoder is needed to achieve the full IGMA throughput potential. More specifically, system-level simulations in an urban environment and slow user mobility (e.g., 3 kmph) show that the same target PLR = 10% can be obtained with about 860% and 680% overload factor with ideal and realistic channel estimation, respectively.

17.2.3.3 Repetition-Division Multiple Access (RDMA)

RDMA [46], which belongs to the interleave-based schemes, can separate different users signals and utilize both time and frequency diversity, just by simple cyclic-shift repetition. Consequently, at the transmitter side, RDMA is simpler than RAMA and IDMA because no scrambler or random interleaver is required. The repetition patterns reported in [46] lead to completely randomized multiple user interference and both time and frequency diversity for each repeated modulated symbol. RDMA can be seen as an unconventional form of CDMA with different spreading sequences per user.

RDMA link level simulation results are evaluated in [51]. It is shown that RDMA can provide significant gain over OFDMA with ideal channel information, and the gain can be up to 2.7 dB for the same PLR value. The RDMA performance with realistic channel estimation is still acceptable over OFDMA.

System-level RDMA simulations for a macro urban cell scenario are reported in [52]. Grant-free RDMA achieves 2.15 times the system capacity of the current benchmark OFDMA LTE Rel. 13 system at PLR equal to 10%.

17.3 Additional NOMA-Based RA Schemes for IoT

While the previous section presented the latest NOMA candidates for 5G terrestrial networks aiming at improving the deployments of massive IoT services, hereafter the most promising NOMA-based random access schemes adopted in satellite networks are reviewed. Interestingly, since long delays occur between the sender and the receiver, the satellite research community has addressed efficient NOMA protocols earlier and with more emphasis than terrestrial cellular networks.

The following schemes have been distinguished in two main categories: slotted and unslotted RA protocols. The former approach requires coarse timing synchronization at the IoT terminal side (e.g., within a defined guard time); on the contrary, the latter approach relies on a fully uncoordinated access of channel resources.

17.3.1 Slotted RA Solutions

17.3.1.1 Contention Resolution Diversity Slotted ALOHA (CRDSA)

To address the shortcomings of conventional RA techniques over satellite, the contention resolution diversity slotted ALOHA (CRDSA) technique was introduced in [53]. Similarly to DS-ALOHA (see Sect. 17.1.3), CRDSA exploits the diversity gain by transmitting multiple replicas of the same packet (number of replicas,[6] N_{rep}) in randomly selected slots within a time division multiple access (TDMA) frame of N_{slots}. To facilitate the process of solving packet collision events typical of any open-loop RA scheme, each physical layer block contains in the header the necessary information to retrieve all replicas location within the entire frame (see Fig. 17.5). At the receiver side, the whole TDMA frame is sampled and stored in a digital memory. In the following, the high-level description of the main CRDSA concepts to recover the highest number of transmitted packets in a TDMA frame are summarized.

Firstly, *clean* bursts are recovered (e.g., packet 3 in slot 5 in Fig. 17.5) and then their replicas in other slots are canceled (e.g., packet 3 in slot 4). By applying this

[6]The meaning of replicas in this context shall be understood as the number of same packet content physically repeated in the same frame.

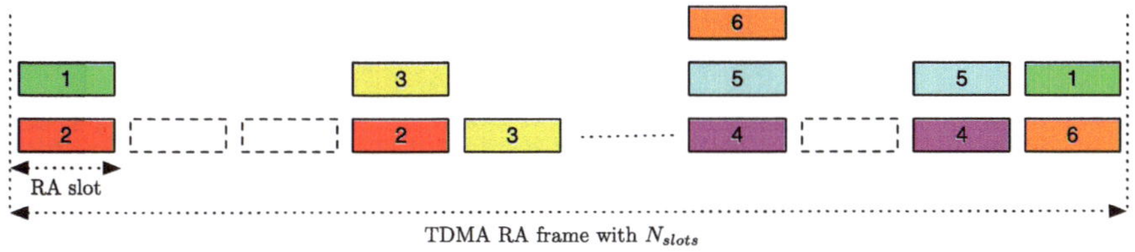

Fig. 17.5 Example of CRDSA frame structure with two replicas per packet

simple, yet efficient, replicas cancellation technique repeatedly to the TDMA frame (typically 10 times), it turns out that most of the initial collisions can be resolved at the demodulator side.

The performance boost of the CRDSA protocol is achieved thanks to the enhanced demodulator collision resolution capability on top of the packet time diversity transmission common to DS-ALOHA. Basically, CRDSA is able to localize the replica packets within the frame thanks to their location information contained in each packet. Thus, CRDSA decoded packets allow to cancel the interference generated by their replicas in other frame slots. This allows to greatly improve the maximum load at which PLR remains acceptable compared to DS-ALOHA.

In Fig. 17.6, it is shown that CRDSA largely outperforms classical S-ALOHA techniques in terms of throughput and PLR. For example, fixing the MAC packet loss probability at 1%, the CRDSA technique leads to a channel utilization of about 25% using two replicas (see Fig. 17.6b), whereas conventional S-ALOHA solutions achieve only 1% of utilization for the same packet loss probability. In other words, this represents a 25-fold throughput improvement compared to S-ALOHA. Moreover, with 3 replicas, the performance improvement is even larger, up to 58-fold throughput improvement compared to S-ALOHA.

Detailed CRDSA delay performance results have been reported in [53]. It is worthwhile noting that the steep PLR versus load CRDSA behavior leads to a remarkable end-to-end delay reduction at relatively high loads, as it avoids packet retransmissions. This benefit is particularly relevant in satellite networks where the round trip transmission time is much larger than for terrestrial networks.

A detailed discussion on how to estimate the channel parameters for packet detection and cancellation (e.g., carrier frequency and phase, symbol timing, and signal amplitude) is reported in [53, 55]. The basic idea is the introduction of a common unique word (UW) (or preamble) for all user devices [55], which also reduces the complexity of the burst search engine in the demodulator. The CRDSA performance with real detector realization was found to be very close to the one obtained with ideal channel estimation even in the presence of phase noise [53, 55].

It is important to remark that CRDSA behavior can be largely affected by a small subset of design parameters. The most relevant ones are the FEC coding rate r, the number of replicas N_{rep}, and the packets power imbalance. With the aim of optimizing its performance, in [20] an analytical framework has been developed jointly with a

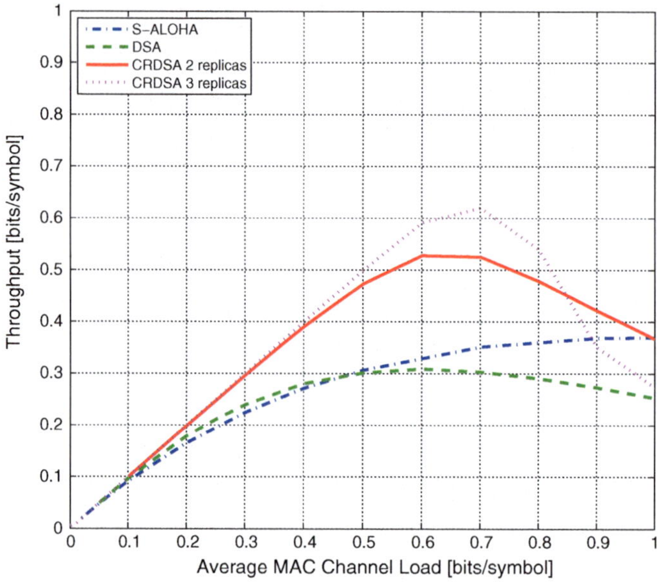

(a) Throughput

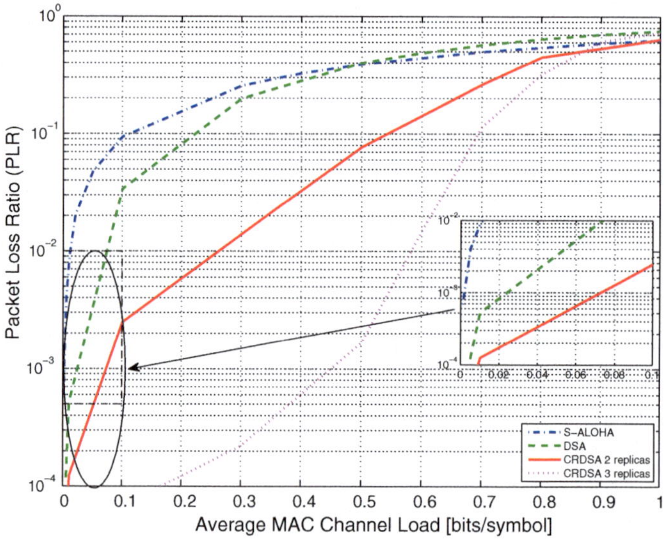

(b) PLR

Fig. 17.6 Simulated S-ALOHA, DS-ALOHA and CRDSA performance from [54] for $N_{\text{rep}} = 2$ and 3, $N_{\text{slots}} = 100$, QPSK modulation, 3GPP FEC $r = 1/2$, packet block size 100 bits, $E_s/N_0 = 7$ dB in the presence of no power imbalance and Poisson traffic (©2016 Wiley) detailed simulator for the performance assessment of slotted RA protocols. The main outcomes have been summarized below.

Concerning the coding rate, [20] highlighted the importance of the FEC protection rate to help the collision resolution process in the presence of strong co-channel interference from the colliding packet(s), thus not under pure AWGN-like conditions. Low FEC coding rates, which apparently reduce the individual packet information bit rate, end up playing an important role in enhancing the overall RA scheme throughput. In particular, [20] compared $r = 1/3$ against $r = 1/2$.

The main reason limiting the PLR performance of CRDSA with two replicas is the so-called loop phenomenon, analyzed in [20] and further refined in [56]. Essentially, a loop occurs when all replicas of a set of packets are causing an unrecoverable collision with one or more replicas. In the example of Fig. 17.5, packets 4 and 5 have formed a loop as their replicas have been transmitted in the same slots. The probability of occurrence is linked to the length of the frame, the MAC load and, more importantly, the number of replicas from each packet. Longer frames and larger number of replicas will significantly reduce the occurrence of loops at the expenses of latency and energy efficiency. As observed in Fig. 17.6b and in [57], three packet replicas gives better CRDSA performance than two, since the reduced loop occurrence probability.

As discussed for ALOHA and S-ALOHA (see Sect. 17.1.3), packets power imbalance boosts significantly the performance of CRDSA. Basically, stronger packets are decoded first and weaker packets are successively decoded following the iterative interference cancellation (IC) process. This beneficial effect is generated by several contributions: (a) effect of the time-variant atmospheric propagation; (b) open-loop power control errors (if applicable); (c) randomization of the transmitter power; (d) terminal transmit and satellite receive antenna gain variations. Figure 17.7 compares CRDSA performance with and without power imbalance, that follows a lognormal power distribution with standard deviation $\sigma = 3$ dB. It shall be noted that the peak in the throughput, $T = 1.4$ bits/symbol for a PLR$< 10^{-3}$, represents a 1400-fold improvement with respect to S-ALOHA. For completeness, the PLR floor appearing in Fig. 17.7 for $\sigma = 3$ dB is due to the presence of lognormal packet power variations which generates several packets with an overall received power well below the detector threshold, even assuming perfect interference cancellation. Further CRDSA performance improvement can be achieved optimizing the power distribution of the gateway demodulator incoming packets. In fact, [56] has demonstrated that loguniform[7] power distribution provides close to optimum performance. PLR results comparing loguniform and lognormal packet power distribution, obtained for the same mean and standard deviation in the dB domain for both distributions, are shown in Fig. 17.8. It is evident from the plots that the loguniform distribution provides a slightly higher throughput and removes the PLR floor present in the lognormal case due to the longer tail of the normal distribution.

[7]Similar to the lognormal distribution, we define as loguniform a probability distribution uniformly distributed in the log domain. In the rest of this chapter, we will discuss loguniform distribution in the dB domain, limiting ourselves to the base-10 logarithm.

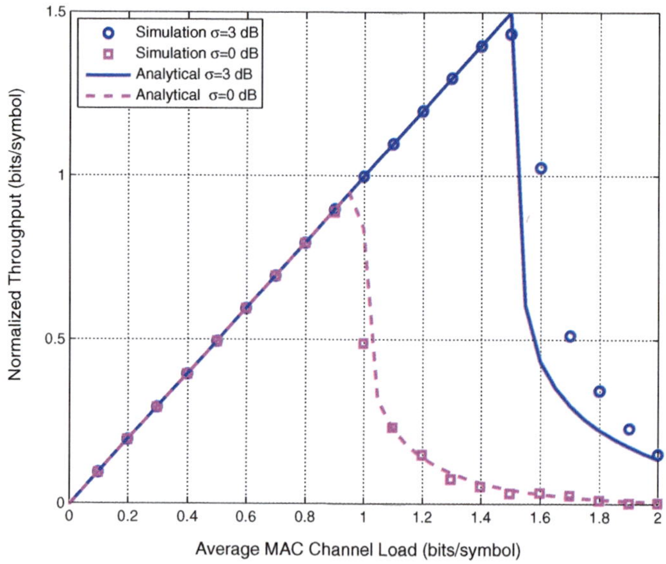

(a) Throughput

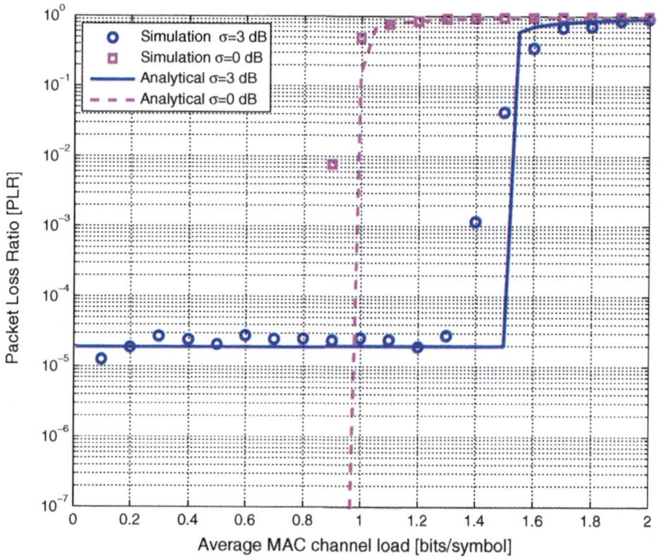

(b) PLR

Fig. 17.7 Analytical versus simulated CRDSA performance from [54] for $N_{\text{rep}} = 3$, maximum number of SIC iterations $N_{\text{max}}^{\text{iter}} = 15$, $N_{\text{slots}} = 1000$, QPSK modulation, 3GPP FEC $r = 1/3$, packet block size 100 bits, $E_s/N_0 = 10$ dB in the presence of lognormal packets power imbalance with mean $\mu = 0$ dB, standard deviation σ and Poisson traffic (©2016 Wiley)

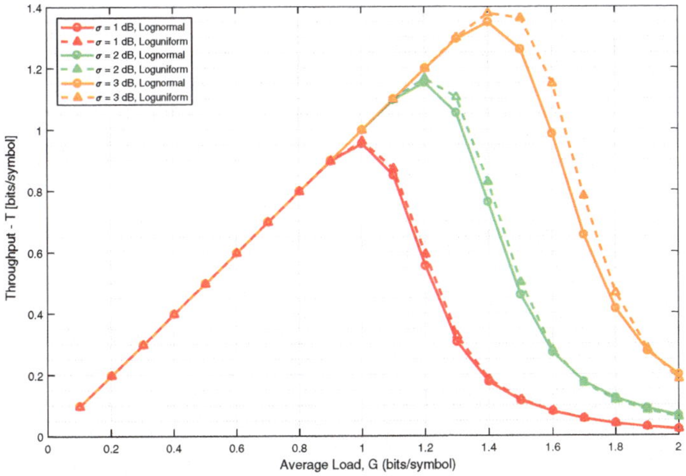

(a) Throughput

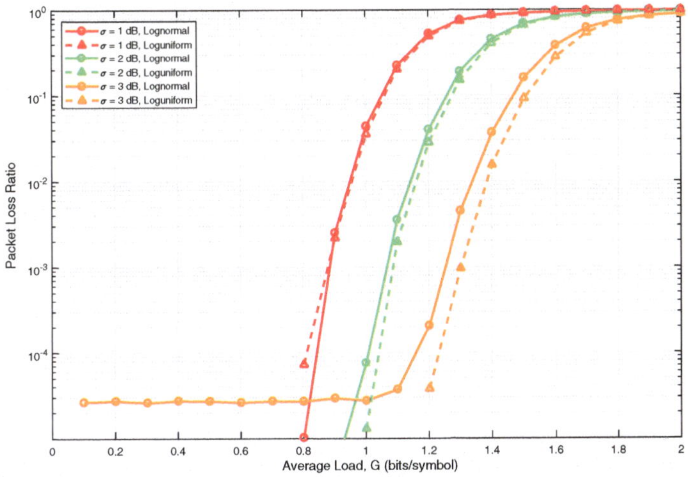

(b) PLR

Fig. 17.8 CRDSA performance for $N_{\mathrm{rep}} = 3$, $N_{\mathrm{max}}^{\mathrm{iter}} = 15$, $N_{\mathrm{slots}} = 100$, QPSK modulation, 3GPP FEC $r = 1/3$, packet block size 100 bits, $E_s/N_0 = 10$ dB in the presence of lognormal and loguniform packets power imbalance with mean $\mu = 0$ dB, standard deviation σ and Poisson traffic

17.3.1.2 Other S-ALOHA RA with Decentralized Power Control and SIC

Similarly to the optimization of CRDSA power randomization reported in [56], other publications have recently appeared exploiting NOMA slotted RA with decentralized power control and SIC [58–60]. These references, inspired by [61], suggest a combination of NOMA with SIC and RA with decentralized power control. The RA technique can be considered a CRDSA with optimized packets power distribution and $N_{\text{rep}} = 1$. Compared with other options, the scheme in [58] appears the most simple and effective one. Although it can provide higher throughput than conventional S-ALOHA, it is less performing than CRDSA with $N_{\text{rep}} = 2,\ 3$. Furthermore, no results are reported about packet error rate (PER) and delay performance.

17.3.1.3 Multi-Replica Decoding using Correlation Based Localization

A recent enhancement of the CRDSA detector, dubbed multi-replica decoding using correlation based localization (MARSALA), has been recently proposed in [62]. MARSALA enhances the CRDSA collision resolution capabilities by a further signal processing step at the end of the CRDSA detection process described before. In particular, MARSALA takes advantage of frame memory slots correlation procedures to locate replicas of packets even when all of them are affected by collisions making them not decodable. Furthermore, samples of slots containing the same packets are coherently combined, improving the symbols quality and thus enhancing the probability of detection. The transmitter side in MARSALA is exactly the same as in CRDSA, all changes are made only in the receiver architecture. The results reported in [63] show an appreciable CRDSA throughput performance increase when MARSALA is adopted. It has been presented an improvement factor of about 40% at PLR $= 10^{-3}$ for CRDSA 3 replicas with $r = 1/3$ FEC and $E_s/N_0 = 7$ dB. However, the MARSALA performance presented so far assumes the carrier phase remains constant over the packet duration, which may not be realistic in practical systems when phase noise is present. This issue is currently under investigation and results are expected to be published in the near future.

17.3.1.4 Multi-Frequency CRDSA (MF-CRDSA)

The use of time separation as a way to implement slotted RA schemes is usually associated with an undesired increase in the peak power requirements at the terminals side. This problem, common to all TDMA schemes, is caused by the necessity to transmit the packet during a small fraction ($1/N_{\text{slots}}$) of the whole frame duration. This problem is further exacerbated in RA schemes because as shown in [64], reducing the number of slots in the frame (N_{slots}) could have sizable negative effect on system performance if $N_{\text{slots}} < 60$ is selected. A more effective way to mitigate this effect was devised in satellite communications networks and dubbed multi-frequency TDMA (MF-TDMA, see Sect. 9.2.6.1 of [65]).

The multi-frequency concept can be easily applied to CRDSA leading to multi-frequency CRDSA (MF-CRDSA) [56, 66]. Adopting this approach, the packets are randomly located in a two-dimensional space composed by N_T^{MF} time slots and N_F^{MF} frequency sub-bands, for a total combined number of slots equal to $N_{\text{slots}} = N_F^{MF} \cdot N_T^{MF}$. In [56], it was found that for optimum MF-CRDSA performance the number of time slots should be minimized, leading to $N_T^{MF} = N_{\text{rep}}$ in the common case where users terminal are not allowed to transmit on more than one sub-frequency at the same time. For an equivalent number of frame slots, the reduction of the required terminal peak transmit power between MF-CRDSA and CRDSA is equal to $N_{\text{slots}}/N_T^{MF}$, which is typically $\gg$1. On the other hand, it shall be remarked that MF-CRDSA is affected by a slight increase of the loop probability. This is the consequence of imposing the transmission of replicas in different time slots, which results in a limited choice of slot combinations compared to the original CRDSA. Simulation results reported in [56] and shown in Fig. 17.9, indicate that MF-CRDSA enables a significant reduction in terminal peak power requirements at the expenses of a slight penalty in terms of PLR compared to CRDSA at small channel loads, when the performance is mainly limited by the loop probability.

17.3.1.5 Irregular Repetition Slotted ALOHA (IRSA)

The key idea behind irregular repetition slotted ALOHA (IRSA) [67] is to have a non-constant, yet random, number of packet replicas transmitted in the TDMA frame. To derive the optimized irregular packet repetition scheme probabilities, bipartite graphs techniques, typically used for the design of FEC LDPC codes, have been exploited. In IRSA, each burst is transmitted l times within the frame, and this repetition rate l changes from burst to burst according to a given mass probability distribution. It is evident that CRDSA can be seen as a special case of IRSA, where the repetition factor has a deterministic value (e.g., $l = 2$ or $l = 3$ for all users).

Simulation results reported in [67] refer to balanced packet power with constant traffic (no Poisson distribution) and simplified collision-based physical layer model. Looking at the peak value in the achievable throughput, IRSA scheme shows some advantages with respect to 2-replicas CRDSA. However, for PLR $< 10^{-3}$, the IRSA throughput is lower or comparable to CRDSA. In addition, it is noted that this randomized and variable number of replicas per frame makes the scheme implementation more complex than CRDSA. In fact, to ensure that the replicas of each user are not located in non-overlapping time slots, the required number of trial-and-error attempts could easily increase compared to CRDSA, since it may require to randomly generate a high number of non-overlapping replicas per terminal. In addition, the signaling mechanism associated with IRSA requires to be sized for the maximum number of replicas that can be generated which is larger than CRDSA.

Reference [68] extended the original IRSA graph-based model to cope with the packet capture effect. However, the model assumes that a packet is captured in a slot if the corresponding SINR is above a given fixed threshold. This approach was found to be inaccurate for predicting the PLR performance of the RA scheme. For this reason in [20], a more accurate physical layer modeling was introduced. To make a more realistic and fair comparison of CRDSA with IRSA with the real FEC and Poisson traffic, simulations have been performed in [56] and results are reported in Fig. 17.10. We can see that while the shown IRSA schemes have some advantages with a FEC code rate 1/2, the IRSA schemes show no advantages compared to the simpler and more energy-efficient CRDSA when the better performing FEC code

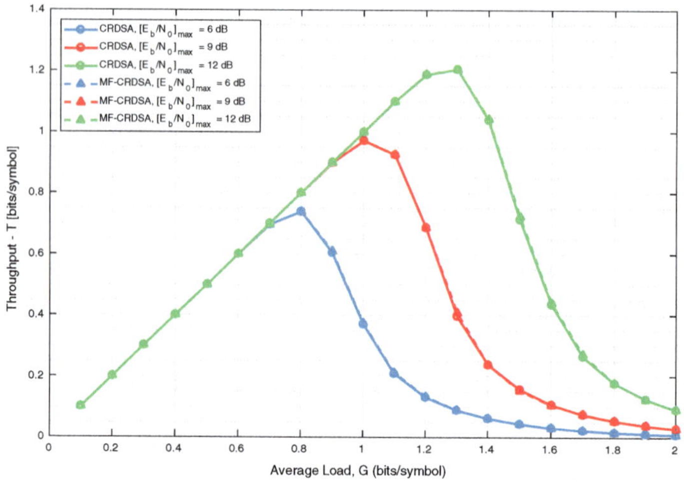

(a) Throughput

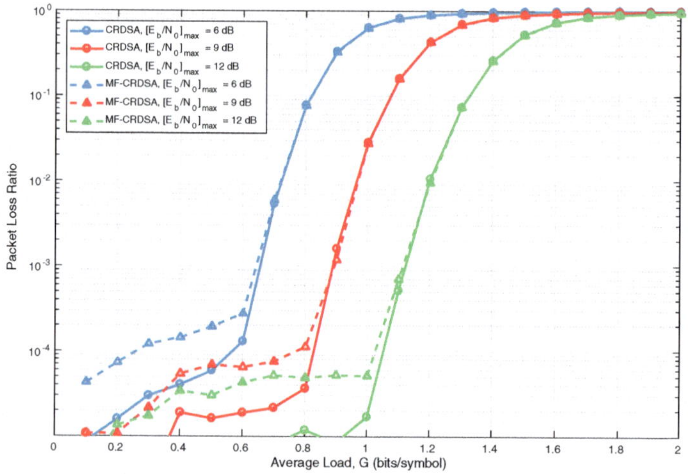

(b) PLR

Fig. 17.9 CRDSA and MF-CRDSA performance with Poisson traffic for $N_{\mathrm{rep}} = N_T^{MF} = 3$, $N_{\max}^{\mathrm{iter}} = 15$, $N_{\mathrm{slots}} = 99$, QPSK modulation, 3GPP FEC $r = 1/3$, packet block size 100 bits, in the presence of loguniform packets power imbalance with minimum power resulting in $E_b/N_0 = 2$ dB and maximum specified by $[E_b/N_0]_{\max}$

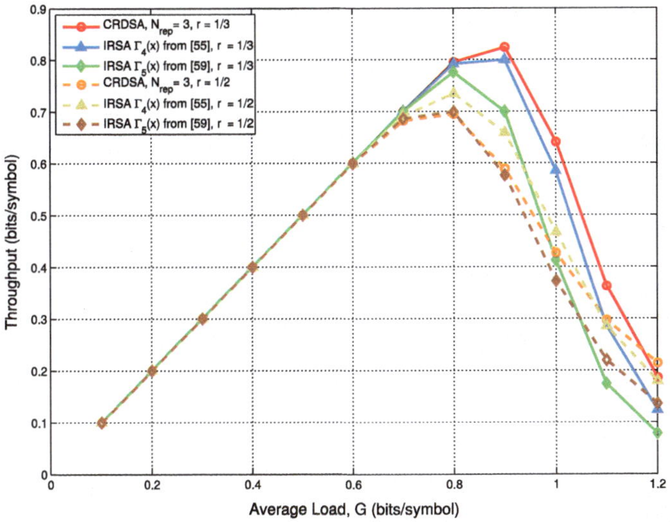

(a) Throughput

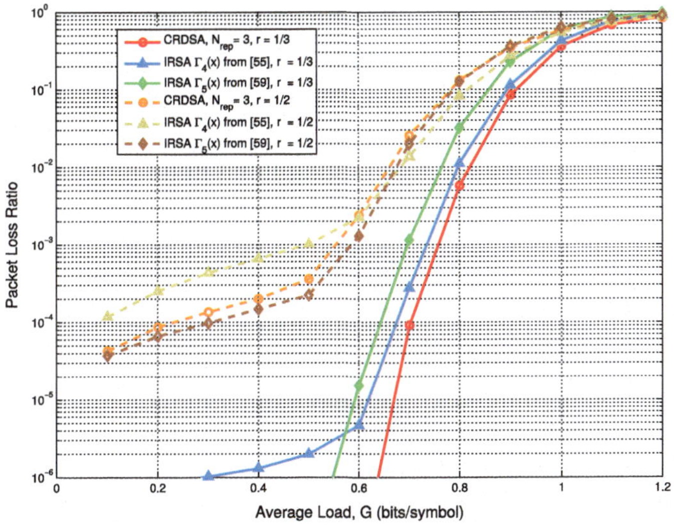

(b) PLR

Fig. 17.10 Simulated IRSA and CRDSA performance from [56] for $N_{\text{rep}} = 3$, $N_{\text{max}}^{\text{iter}} = 15$, $N_{\text{slots}} = 100$, QPSK modulation, real 3GPP FEC $r = 1/3$ and $r = 1/2$, packet block size 100 bits, $E_b/N_0 = 9$ dB in the presence of no packets power imbalance and Poisson traffic (©2017 IEEE)

rate 1/3 is used. To fully exploit the IRSA scheme potential a more accurate model of the physical layer processing at the receiver side will be required. This is subject of ongoing research activities [68].

17.3.1.6 Multi-Slots Coded ALOHA (MuSCA)

In multi-slots coded ALOHA (MuSCA) [69], differently from CRDSA, the N_{rep} slots randomly selected from the terminal in a given frame do not contain the same payload information. Instead of replicating the content, the encoded block is partitioned in sub-packets spread across two or more frame slots. This means that the MuSCA solution may exploit lower FEC coding rate for the same amount of assigned time resources. This approach has the potential to enhance the collision resolution capability of the protocol, provided that channel estimation quality is reliable at low SINR ratios. Similarly to CRDSA, each fragment contains the pointing information to retrieve the location of the other partitions.

Differently from the protocols presented so far, the introduction of this fragmentation does not allow the decoding of the single received partition. Consequently, the sub-packet location signaling field needs to be independently coded from the rest of the data payload. For this purpose, [69] suggests the introduction of a Reed–Muller block code. This represents an increased signaling overhead, particularly undesired for small size packets. Simulation results reported in [69] show that MuSCA outperforms CRDSA, however the drawback of the signaling overhead has not been accounted for. More specifically, it is shown that for PLR $= 10^{-3}$ MuSCA with 2 replicas and $r = 1/6$ FEC can achieve 100% throughput increase compared to CRDSA with 3 replicas and $r = 1/2$ FEC. For the more optimized $r = 1/3$ CRDSA configuration, the gain is close to 60%. This performance enhancement shall be scaled down by the increased overhead and by the additional complexity in the receiver for improving the channel estimation and for decoding the signaling field.

17.3.1.7 Coded Slotted ALOHA (CSA)

The coded slotted ALOHA (CSA) protocol [70] represents a generalization of both IRSA and MuSCA schemes. User packets are encoded prior to transmission, instead of being simply repeated as in IRSA. It follows that CSA is more power efficient than IRSA but less than MuSCA, since puts all the extra FEC redundancy at the physical layer level.

In CSA, prior to the transmission, the incoming user packet is divided into k information (or data) segments, equal for length in bits. The k segments are then encoded by using a packet-oriented linear block code, which generates n_h encoded segments. The level of protection for each segment (instead of the number of repetitions as in IRSA) is randomly picked up (instead of being fixed as in MuSCA) from a finite number of available codes, n_c, identical for all users. The specific code is denoted with $\mathscr{C}_h$, where $h \in 1, \ldots, n_c$. The n_h segments are further encoded through a physical layer code before being transmitted over n_h distinct channel slots.

On the receiver side, the decoding strategy is performed in two stages as follows: segments received in clean slots (i.e., segments not experiencing collisions) are firstly decoded at physical layer. Then, the relevant user-specific information, i.e., the adopted code $\mathscr{C}_h$ by the user, and the positions of the other segments in the

MAC frame are extracted. For each active user, the MAP erasure decoding of the $\mathscr{C}_h$ code is performed in order to recover as many encoded segments as possible for the user. Finally, the recovered segments may now be erased, in order to subtract their interference contribution in those time slots where collisions could have occurred. It is noted that the RA scheme based on CSA requires two layers of coding. The first is standard FEC coding in each slot. The second is erasure coding over multiple slots.

Figure 17.11 shows the performance comparison between the various CSA schemes presented in [70] and CRDSA with $N_{\text{rep}} = 3$. For the computation of CSA performance, a user has been considered decoded only if all of its original k information segments are correctly received at the end of the iterative decoding process, not considering partial decoding of user information. It is clear from the results that the proposed CSA combinations fall short of outperforming the simpler CRDSA at PLR $\leq 10^{-2}$. As for IRSA to fully exploit the CSA scheme potential a more accurate model of the physical layer processing at the receiver side will be required.

Reference [71] presents a framework for the analysis of the error floor of CSA for finite frame lengths over the packet erasure channel.

Finally, from the implementation point of view, the CSA protocol is a more complex RA scheme due to its associated signaling mechanism.

17.3.1.8 Scrambled Coded Multiple Access (SrCMA)

The SrCMA protocol (not to be confused with the previously introduced sparse code multiple access (SCMA)) has been firstly presented in [72]. The Authors got inspiration from the previously described IDMA scheme, where different users share the same bandwidth and the time slots with user-specific interleaver sequences [49]. Specifically, SrCMA simplified this user dependent aspect by adopting a single scrambling sequence with different shift factors for several users without any performance penalty.

In SrCMA, each data burst is encoded with a code rate significantly lower than the commonly used values in satellite systems. In fact, the proposed coding scheme is based on a LDPC code with $r = 1/9$. The use of a very low-rate FEC code is a distinguished feature of SrCMA. Instead of spreading the bandwidth with direct-sequence modulation, low-rate FEC provides some additional coding gain which becomes very valuable in accommodating more simultaneous users in the context of multi-user detection threshold.

The SrCMA receiver exploits the multi-user detection principles by concatenating the soft input soft output (SISO) detector and the LDPC decoder. After decoding of each user, the multi-user canceler immediately generates updated a priori information of the subsequent user, thereby incorporating the latest information from the previous decoded user. As beneficial as in the other iterative RA receivers, the presence of power variations among the users improves the SrCMA spectral efficiency. Again, this feature eliminates the need for the implementation of precise power control loop, both in the open or in the closed form.

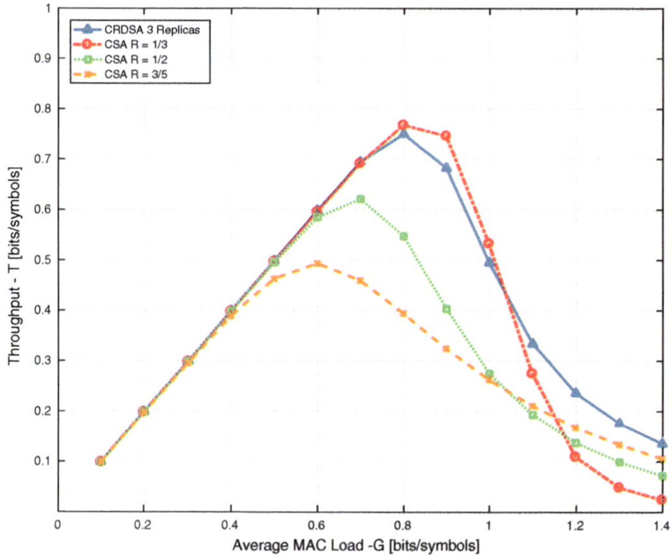

(a) Throughput

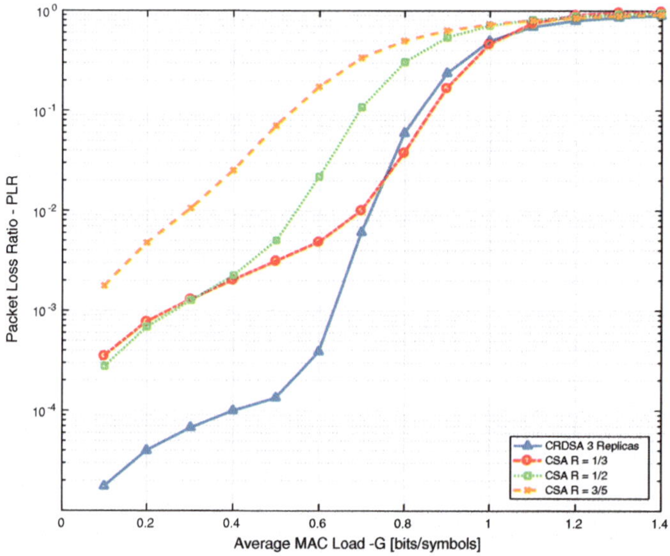

(b) PLR

Fig. 17.11 Simulated CSA and CRDSA performance for $N_{\max}^{\text{iter}} = 15$, $N_{\text{slots}} = 128$, QPSK modulation, real 3GPP FEC $r = 1/2$, packet block size 100 bits, $E_b/N_0 = 10$ dB in the presence of no packets power imbalance and Poisson traffic. CSA generating matrices are taken from [70] and CRDSA with $N_{\text{rep}} = 3$

The achievable throughput from this technique is very similar to CRDSA, and $T = 1.4$ bits/symbol might be easily reached in the presence of power imbalance with still PLR $< 10^{-3}$. The main drawback of this solution, in addition to the more complex soft SIC MUD scheme required, is the extremely different decoding performance of the selected LDPC code as a function of the aggregated traffic. In particular, as result of the FEC design algorithm in a MA channel, the PLR performance is sub-optimal in AWGN, to such an extent that the decoding threshold is several dB higher than the highly interfering channel.

17.3.2 Unslotted RA Solutions

17.3.2.1 Enhanced Spread-Spectrum ALOHA (E-SSA)

E-SSA [27] aims at solving the weakness identified for the SS-ALOHA RA scheme. In Sect. 17.1.3 it is shown how the conventional SS-ALOHA demodulator is very sensitive to packets' power imbalance. Thanks to a more advanced gateway digital signal processing, E-SSA provides remarkable enhancements in terms of robustness and absolute throughput compared to conventional SS-ALOHA.

The E-SSA demodulator main difference lies on the packet detector design exploiting iterative SIC approach customized to the asynchronous RA DS-SS scheme [73]. The central demodulator is the heart of the system, as it has to provide reliable detection of the incoming packets even under heavy MAC channel load conditions and with arbitrary packets' power distribution. Instead, the E-SSA modulator design is unchanged compared to SS-ALOHA (see Sect. 17.1.3).

The principle of the E-SSA demodulator is illustrated in Fig. 17.12. The received signal, containing the superimposition of many asynchronously generated packets, is band-pass filtered, sampled, digitally down-converted to baseband with I-Q components and stored in a digital memory. The memory size corresponds to $2WN_s^c$ real samples, where N_s^c represents the number of chips per physical layer channel symbol, and W corresponds to the iSIC memory window size expressed in symbols. The optimum window size has been found to be three times the physical layer packet length in symbols. The window memory content is shifted in time in discrete steps allowing some overlap of packets on each window step (sliding window process). The recommended window step ΔW is between $1/3$ and $1/2$ of the window length W. At each window step, the following functions are implemented:

1. Store in the detector memory the new baseband signal samples corresponding to the current window step (n);
2. Perform packets preamble detection and select the packet with highest SINR value;
3. Perform data-aided channel estimation for the selected packet over the preamble;
4. Perform FEC decoding of the selected packet;

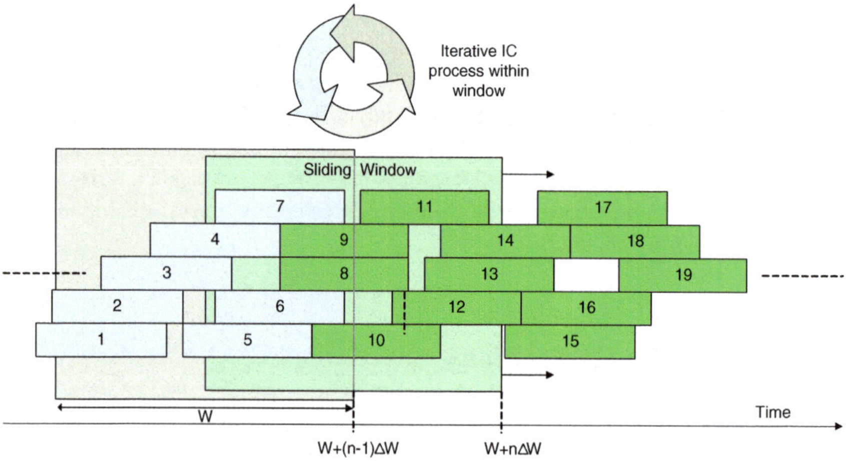

Fig. 17.12 Enhanced Spread Spectrum ALOHA algorithm description from [27] (©2012 IEEE)

5. If the decoded FEC frame is considered correct after cyclic redundancy check (CRC) then:
   a. Perform enhanced data-aided channel estimation over the whole recovered packet (carrier frequency, phase, amplitude, timing) [55];
   b. Reconstruct at baseband the detected packet for the subsequent cancellation step;
   c. Perform interference cancellation;
6. Repeat from step 2 until the maximum number of SIC iterations are performed. When the limit is reached, advance the observation window by ΔW.

The key steps for a good E-SSA detector performance are the packet preamble detection (step 2) and the IC process (step 5-c). The demodulator is first looking for the packet preambles presence in the memory. This is implemented through a conventional preamble correlator with possible parallel search in frequency to cope with the incoming packets carrier frequency uncertainty. In a good design, the preamble miss detection and false alarm probability need to be lower than the target PLR (e.g., 10^{-3}) (see Sect. IV.E in [27]). In practice, in [74] it has been found that E-SSA can operate with a single spreading sequence common to all terminals. This is largely simplifying the gateway demodulator implementation and avoid the need to randomly pick up a spreading sequence at the transmitter side.

As mentioned before, E-SSA principle closely resembles the one of RAMA 5G proposal described in Sect. 17.3.1.

Literature covers E-SSA performance by means of analysis and simulation results (see Sect. IV-D from [27]). The analytical model described in this reference accurately models the performance of spread-spectrum RA techniques, such as coded SS-ALOHA and E-SSA, with arbitrary power and traffic distributions. Figure 17.13 allows comparing E-SSA performance with SS-ALOHA ones previously reported

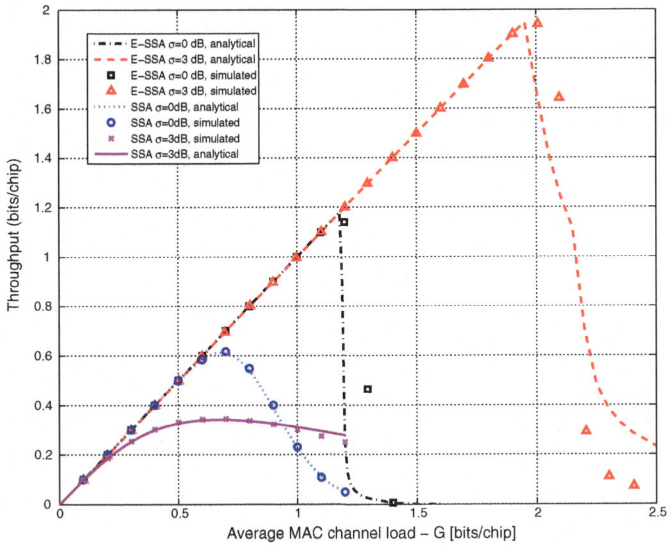

(a) E-SSA and SS-ALOHA throughput with and without power imbalance.

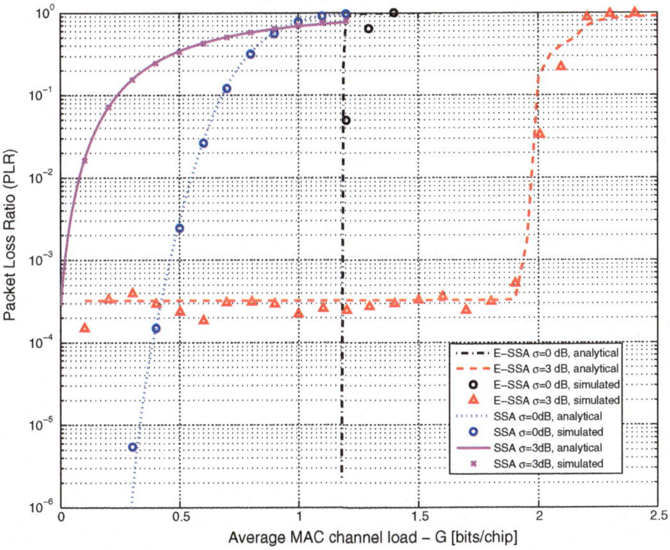

(b) E-SSA and SS-ALOHA PLR with and without power imbalance.

Fig. 17.13 Simulated versus analytical SS-ALOHA (indicated as SSA in the legend) and E-SSA throughput and PLR performance with and without power imbalance from [27], 3GPP FEC $r = 1/3$ with block size 100 bits, BPSK modulation, spreading factor 256, $E_s/N_0 = 6$ dB in Fig. 17.2. Assuming a target PLR of 10^{-3} and no power imbalance, the E-SSA throughput results to be 1.12 bits/chip, i.e., 2.4 times higher than conventional SS-ALOHA. With lognormal packet power distribution and standard deviation $\sigma = 3$ dB, the E-SSA throughput increases to 1.9 bits/chip, i.e., two order of magnitudes larger than SS-ALOHA. This amazing performance is related to the iSIC capability to combat the near–far effect heavily degrading the SS-ALOHA conventional burst demodulator performance.

An investigation about the E-SSA optimum packet power distribution is reported in [75]. It is shown that for E-SSA, a loguniform incoming hub demodulator packets power distribution closely approximate the optimum. In the same reference, semi-analytical formulation is provided for computing the optimum packets signal-to-noise ratio range as a function of the key RA system parameters. A simple open-loop power control algorithm is also proposed capable to achieve the quasi-optimum power distribution in a multi-beam satellite network. More recently, a theoretical analysis to calculate the capacity optimizing user SINR profile for an SS-ALOHA RA system adopting SIC and FEC has been reported in [76].

17.3.2.2 MMSE Enhanced Spread-Spectrum ALOHA

The ME-SSA [77], represents an extension of the E-SSA detector improving the E-SSA spectral efficiency in particular when the packets power imbalance is modest. The key ME-SSA features are: (a) adoption of QPSK modulation before spreading instead of BPSK like in E-SSA; (b) introduction of a multi-stage approximation of the MMSE filter at the demodulator preceding the iSIC. Incorporating a conventional MMSE detector in an E-SSA RA demodulator is cumbersome. In fact, E-SSA is an asynchronous access scheme particularly suited for a large population of sporadic transmissions. Thus, the active transmitters are continuously changing in time, making the co-channel interference non-stationary. This traffic feature, combined with the adoption of long (compared to the symbol duration) spreading code sequences and relatively short packets, makes infeasible the use of a conventional adaptive MMSE detector. At the same time, the implementation of MMSE through a direct matrix inversion is considered too cumbersome.

The ME-SSA proposed approach consists in the use of a multi-stage detector approximating the MMSE one [78–81]. The multi-stage approximation of the MMSE optimum detector makes the detector complexity linearly growing with the number of active transmitters and able to cope with time-variant traffic and long spreading sequences. Following Fig. 17.14, the MMSE detector is now approximated by S linear stages with each stage performing despreading (with a single user matched filter detector) and then re-spreading of the input signal. The weighting factors required for combining the S detector stages are computed off-line following [80, 81]. QPSK instead of BPSK modulation before spreading is required to maximize the number of signal dimensions, thus the MMSE performance [82]. The ME-SSA adoption of the multi-stage MMSE detector allows to operate with relatively long spreading sequences (e.g., $SF = 32–64$ chips with truly affordable complexity). This is a

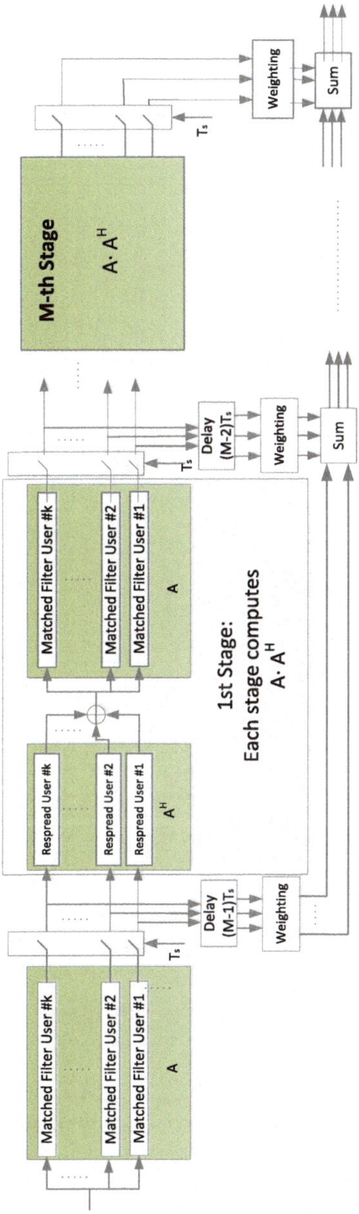

Fig. 17.14 ME-SSA MMSE multi-stage detector functional block diagram from [77]

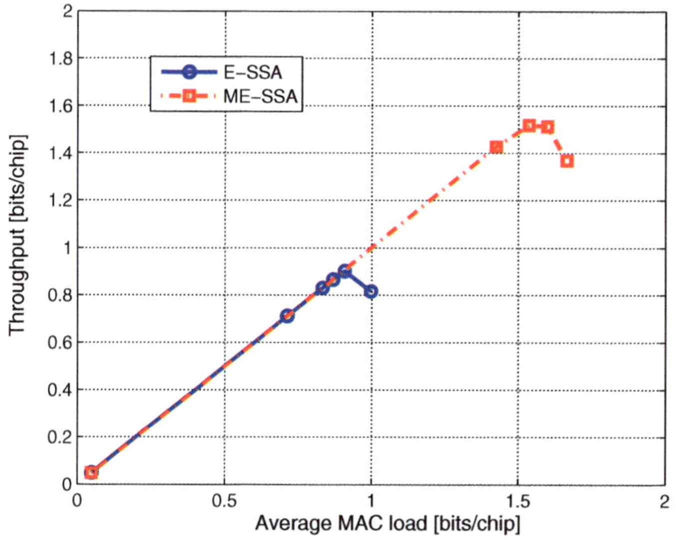

(a) Throughput

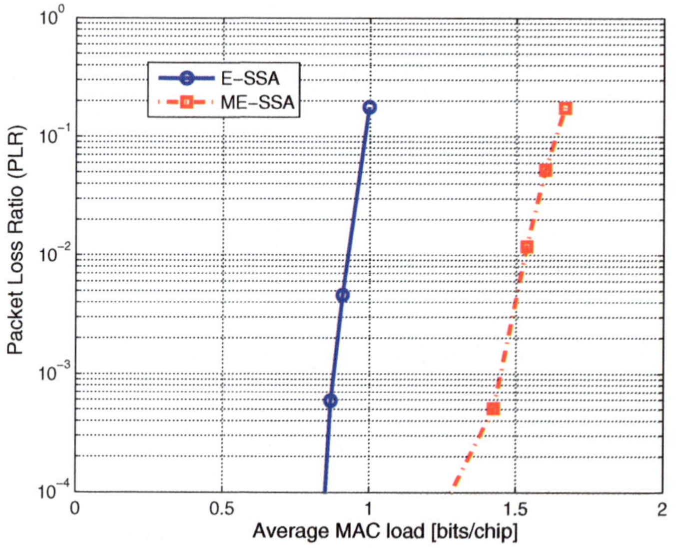

(b) PLR

Fig. 17.15 ME-SSA and E-SSA performance comparison from [77]: packet length 1200 bits, code rate 1/3, preamble length 1536 chips. ME-SSA uses QPSK modulation. E-SSA uses BPSK modulation key difference compared to the 5G MUSA scheme described in Sect. 17.2.2 which exploits short spreading sequences (e.g., $SF = 4$) to keep the MMSE detector complexity at an affordable level at the expenses of some performance reduction. Furthermore, ME-SSA as E-SSA is operating with a single spreading sequence [27]. As a consequence, it does not require a random selection of spreading sequence at the UE from a given codebook as in MUSA and allows a simplified single spreading sequence MMSE-SIC receiver design.

Simulation results reported in [77] for $SF = 16$ (see Fig. 17.15), for balanced packets power ME-SSA show gains of about 50 % over E-SSA. In case of loguniformly distributed power imbalance, the throughput gain can go up to 80%. These gains are counterbalanced by some demodulator complexity increase.

17.3.2.3 Asynchronous Contention Resolution Diversity ALOHA (ACRDA)

A possible alternative to E-SSA is represented by an asynchronous evolution of CRDSA dubbed asynchronous contention resolution diversity ALOHA (ACRDA) [83]. Differently from S-ALOHA, DS-ALOHA, CRDSA or contention resolution ALOHA (CRA) [85], ACRDA eliminates the need to maintain any kind of time slot (S-ALOHA, DS-ALOHA, CRDSA) or frame synchronization among all transmitters. Unlike SS-ALOHA and E-SSA, ACRDA does not require the use of spread-spectrum techniques. As previously discussed, the need for transmitter synchronization is a major drawback for very large networks (e.g., M2M), as the signaling

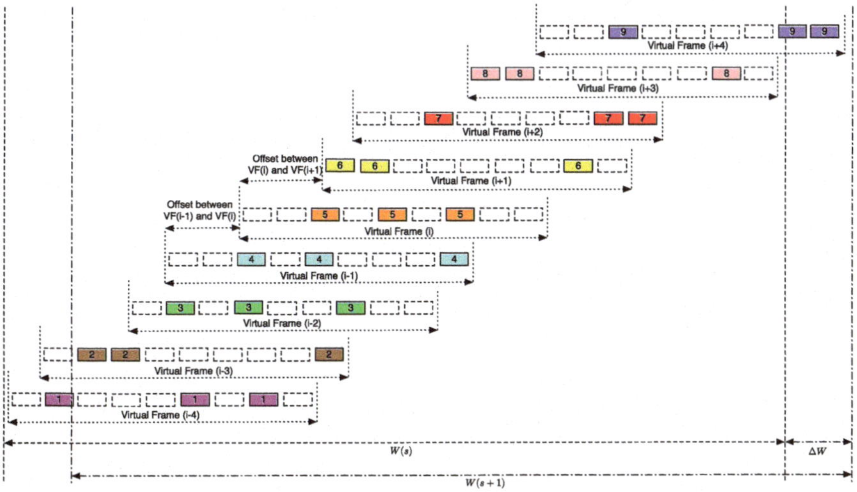

Fig. 17.16 Virtual frame compositions in ACRDA from [83]

overhead scales up with the number of transmitters independently from their traffic activity factor.

Similarly to CRDSA, ACRDA exploits packet replicas and the associated location signaling, without requiring to maintain slot synchronization among the different transmitters. Let now define the concept of the *virtual frame* (VF) as a frame composed of a number slots, only locally valid in the transmitter local time reference. While the time duration of the frame and the number slots is common to all network transmitters, the start of frame is different for each network transmitter. More precisely, each VF is composed of a number of slots N_{slots} and each slot has a duration T_{slot} with an overall frame duration $T_{\text{frame}} = N_{\text{slots}} \cdot T_{\text{slot}}$. Figure 17.16 shows an example of the ACRDA VF realization. It is apparent that the different transmitters are not time synchronized, and hence, the time offset between VF(i), VF($i-1$), VF($i+1$) result to be randomly distributed.

For mobile applications, the Doppler effect may have a non-negligible impact on the incoming packets symbol clock frequency offset. This is translated in VFs having a slightly different time duration. However, this symbol clock timing offset does not impact the localization process of the packet replicas within each VF. This is because the hub burst demodulator extracts for each VF its own local clock reference. Moreover, as the demodulator has no prior knowledge of the start of VFs, the signaling of the replicas slot location within the VF remains valid. This is because the packet replica location is signaled relative to the current packet position.

As discussed in [83], there are two alternative solutions for implementing the ACRDA modulator. The first one corresponds to transmitting all packet replicas in random locations within the VF. In the second option, the first packet replica is forced to be in the first VF slot, while the remaining replicas are randomly placed within the remaining VF slots. This second solution was found to improve the end-to-end delay performance, since the first packet replica is always transmitted at the beginning of the frame without any extra delay.

The ACRDA demodulator exploits features of CRDSA (packet replica(s) cancellation), and E-SSA (asynchronous packets processing through a sliding window). The received signal is sampled at baseband, and complex signal samples are stored in a sliding window memory of W VFs (see Fig. 17.16). A value $W = 3$ is recommended for optimal ACRDA performance (like for E-SSA). For a given window position, the demodulator performs the same iterative processing as for CRDSA to decode the clean packets and perform IC of the replica packets. The detailed ACRDA demodulator description can be found in Sect. II.B in [83].

The same reference shows that ACRDA achieves a better PLR and throughput performance than CRDSA. Figure 17.17 compares the performance of the ACRDA versus CRDSA and E-SSA, evaluated via mathematical analysis and computer simulations. For a target PLR $= 10^{-3}$ and balanced packet power, ACRDA achieves 1 bits/symbol, while CRDSA is limited to 0.75 bits/symbol and E-SSA gets up to 1.2 bits/chip. It is important to remark that the PLR floor due to loop occurrence visible at low-to-medium loads (i.e., $G < 1$ bits/symbol) is lower for ACRDA than for CRDSA. The reason for this is that the probability of loops occurrence is

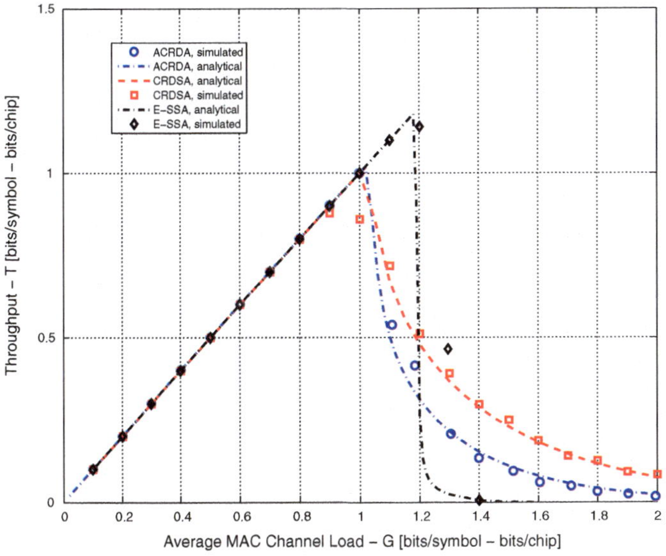

(a) Throughput

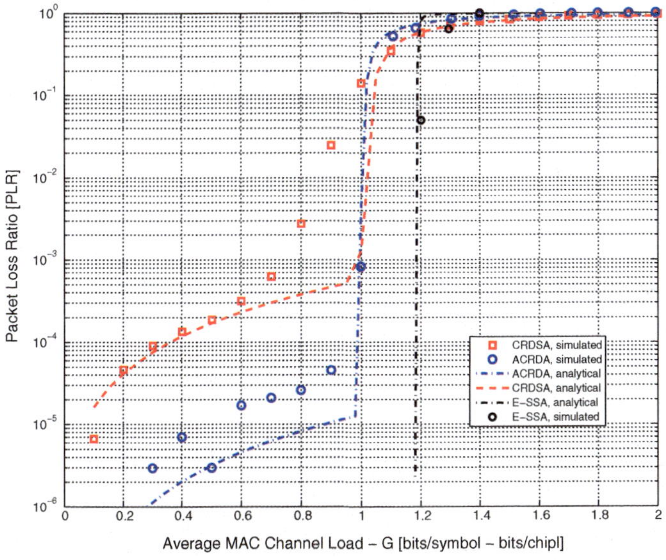

(b) PLR

Fig. 17.17 Simulation and analytical ACRDA, CRDSA and E-SSA performance. CRDSA and ACRDA with $N_{\text{rep}} = 2$, $N_{\text{slots}} = 100$ (simulations), QPSK modulation, 3GPP FEC $r = 1/3$, packet block size 100 bits, $E_s/N_0 = 10$ dB in the presence of no packets power imbalance and Poisson traffic. For ACRDA window size $W = 3$ virtual frames and a window step $\Delta W = 0.15$

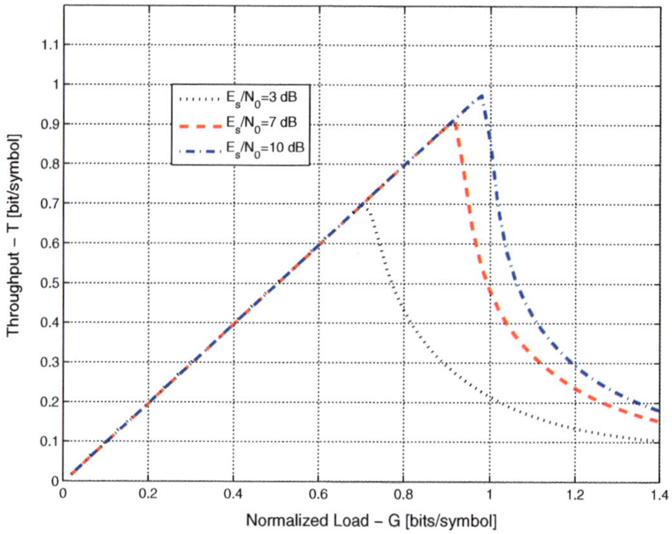

(a) Throughput

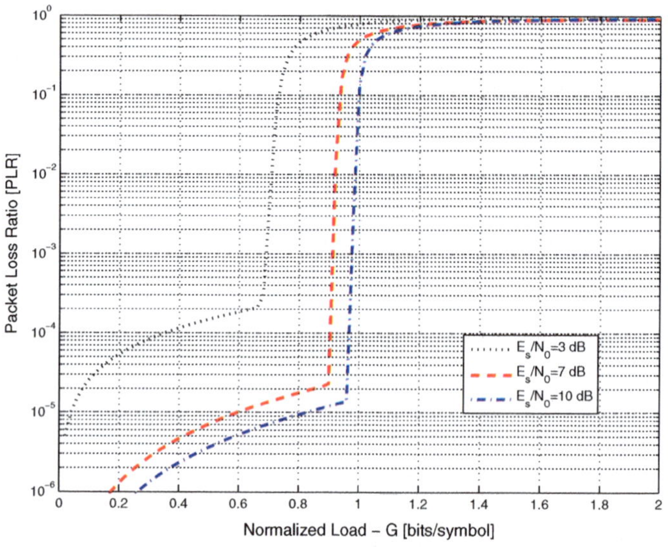

(b) PLR

Fig. 17.18 Analytical ACRDA performance for different E_s/N_0 values with $N_{\text{rep}} = 2$, $N_{\text{slots}} = 100$, QPSK modulation, 3GPP FEC $r = 1/3$, packet block size 100 bits, window size $W = 3$ virtual frames and a window step $\Delta W = 0.15$. The results are obtained in the presence of no packets power imbalance and Poisson traffic significantly lower in ACRDA than in CRDSA due to the asynchronous nature of the access.[8]

It is also observed that ACRDA with $N_{\mathrm{rep}} = 2$ is capable to achieve slightly better performance than the optimum CRDSA configuration with $N_{rep} = 3$ shown in Fig. 17.7. This is reducing the demodulator complexity which is proportional to the number of replicas used [56]. The fact that ACRDA does not need timing synchronization at network level simplifies the user terminal and the overall system complexity.

Figure 8 in [83] shows that ACRDA with FEC $r = 1/3$, $N_{\mathrm{rep}} = 2$ and power imbalance following a lognormal power distribution with standard deviation $\sigma = 3$ dB, can achieve a throughput $T = 1.5$ bits/symbol for a target PLR $< 10^{-3}$. This is similar to the observed CRDSA behavior. Finally, Fig. 17.18 shows the ACRDA analytical performance as a function of E_s/N_0 following the methodology developed in [83]. There is an evident reduction in performance changing E_s/N_0 from 7 to 3 dB.

In [83] Sect. IV analyzes and compares the delay performance between ACRDA and CRDSA. Results show that ACRDA reduces the latency by a factor of 10 compared to CRDSA for low loads (e.g., $G = 0.3$ bits/symbol) and by a factor of 2 at high loads (e.g., $G = 0.9$ bits/symbol).

17.3.2.4 Spread Asynchronous Scrambled Coded Multiple Access (SA-SrCMA)

The spread asynchronous scrambled coded multiple access (SA-SrCMA) [84] is the unslotted companion of the protocol described in Sect. 17.3.1.8. The burst structure is identical to SrCMA, and only the spreading block is added in the transmission chain. As a consequence, the iterative multi-user detection receiver exploiting a sliding window demodulation approach results to be more involved. The main difference between SA-SrCMA and E-SSA can be summarized as follows. At the transmitter side, SA-SrCMA adopts QPSK instead of E-SSA BPSK modulation along with time instead of a code multiplexed pilot transmission. SA-SrCMA exploits a very low coding rate with low spreading factor (i.e., $r = 1/9$ and $SF = 4$) instead of the low code rate and relative high spreading factor choice done in E-SSA (i.e., $r = 1/3$ and $SF = 16$ or higher values). At the receiver side, the main difference relies on the use of soft SIC techniques for SA-SrCMA instead of the hard-decision SIC approach adopted by E-SSA.

The asynchronous SA-SrCMA RA scheme outperforms the slotted SrCMA one. This is because for the slotted approach when a large number of packets arrive simultaneously, they interfere with each other for the entire burst length. This full packet overlap phenomenon significantly degrades the soft SIC convergence process. Instead, for the unslotted case, the packet overlap time span is typically shorter than

[8]The effect of loops in performance is analyzed in detail in Sect. III from [83] for ACRDA and in Appendix D from [20] for CRDSA.

the entire burst length due to the random and asynchronous arrival time among the user bursts. Therefore, even if some part of a data burst experiences a high level of interference, the remaining part may see less interference. This uneven interference level over the packet helps the soft SIC convergence.

In [84], the reported throughput performance shows a peak in the throughput at $T = 3.8$ bits/symbol for a PLR $< 10^{-3}$, with power imbalance that follows an uniform distribution of ± 3 dB around $E_s/N_0 = 3$ dB.

Acknowledgements The chapter authors would like to acknowledge the support of Gennaro Gallinaro from Space Engineering in research related to satellite random schemes and in particular to (M)E-SSA, Dr. Guray Acar ACRDA performance analysis and for Dr. Daniel Pantelis Arapouglu from European Space Agency for CRDSA optimization aspects.

References

1. F.A. Tobagi, Multiaccess protocols in packet communication systems. IEEE Trans. Commun. **28**(4), 468–488 (1980)
2. D. Bertsekas, R. Gallager, *Data Networks*, 2nd edn. (Prentice Hall, 1992). ISBN 978-0132009164
3. S. Keshav, *An Engineering Approach to Computer Networking* (Addison-Wesley Professional, 1997). ISBN 978-0201634426
4. L. Kleinrock, F.A. Tobagi, Packet switching in radio channels: Part I—Carrier sense multiple access modes and their throughput-delay characteristics. IEEE Trans. Commun. **23**(12), 1400–1416 (1975)
5. F.A. Tobagi, L. Kleinrock, Packet switching in radio channels: Part II—The hidden terminal problem in carrier sense multiple-access and the busy-tone solution. IEEE Trans. Commun. **23**(12) (1975)
6. R.M. Metcalfe, D.R. Boggs, ETHERNET: distributed packet switching for local computer networks. Commun. ACM Mag. **19**(7), 395–404 (1976)
7. IEEE Standard for Information Technology; Telecommunications and Information Exchange Between Systems Local and Metropolitan Area Networks; Specific Requirements Part 3: Carrier Sense Multiple Access with Collision Detection (CSMA/CD) Access Method and Physical Layer Specifications, in *IEEE Std 802.3-2005*, December 2005
8. A. Colvin, CSMA with collision avoidance. Comput. Commun. **6**(5), 227–235 (1983)
9. IEEE Standard for Information Technology; Telecommunications and Information Exchange Between Systems Local and Metropolitan Area Networks; Specific requirements Part 11: Wireless LAN Medium Access Control (MAC) and Physical Layer (PHY) Specifications, in *IEEE Std 802.11-2012*, March 2012
10. P. Karn, MACA—a new channel access method for packet radio, in *Proceedings of the ARRL/CRRL Amateur Radio and 9th Computer Networking Conference*, London and Ontario, Canada, 22 September 1990, pp. 134–140
11. Physical Layer Procedures (FDD); Release 1999, in *3GPP TS 25.214 v3.12.0*, March 2003
12. Evolved Universal Terrestrial Radio Access (E-UTRA) and Evolved Universal Terrestrial Radio Access Network (E-UTRAN); Overall description; Stage 2, in *3GPP TS 36.300 v14.4.0*, September 2017
13. A. Laya, L. Alonso, J. Alonso-Zarate, Is the random access channel of LTE and LTE-A suitable for M2M communications? A survey of alternatives. IEEE Commun. Surv. Tutor. **16**(1) (2014) (First Quarter 2014)

14. N. Abramson, The ALOHA system—another alternative for computer communications, in *Proceedings of 1970 Fall Joint Computer Conference AFIPS*, vol. 37 (AFIPS Press, Montvale, NJ, 1970), pp. 281–285
15. N. Abramson, The throughput of packet broadcasting channels. IEEE Trans. Commun. **25**(1), 117–128 (1977)
16. G.L. Choudhury, S.S. Rappaport, Diversity ALOHA—a random access scheme for satellite communications. IEEE Trans. Commun. **31**, 450–457 (1983)
17. L.G. Roberts, ALOHA packet systems with and without slots and capture, in *Arpanet Satellite System Note 8 (NIC Document #11290)*, 26 June 1972
18. M. Zorzi, R. Rao, Capture and retransmission control in mobile radio. IEEE J. Sel. Areas Commun. 1289–1298 (1994)
19. A. Zanella, M. Zorzi, Theoretical analysis of the capture probability in wireless systems with multiple packet reception capabilities. IEEE Trans. Commun. **60**(4), 1058–1071 (2012)
20. O. del Río Herrero, R. De Gaudenzi, Generalized analytical framework for the performance assessment of slotted random access protocols. IEEE Trans. Wirel. Commun. **13**(2), 809–821 (2014)
21. M. Mathis, J. Semke, J. Mahdavi, The macroscopic behavior of the TCP congestion avoidance algorithm. Comput. Commun. Rev. (ACM SIGCOMM) **27**(3), 67–82 (1997)
22. J. Padhye, V. Firoiu, D. Townsley, J. Kurose, Modelling TCP throughput: a simple model and its empirical validation, in *Proceedings of the SIGCOMM Symposium Communications Architectures and Protocols*, August 1998, pp. 304–314
23. G. Maral, M. Bousquet, *Satellite Communications Systems*, 5th edn. (Wiley, 2009). ISBN: 978-0-470-71458-4
24. N. Abramson, Multiple access in wireless digital networks. IEEE Proc. **82**(9), 1360–1370 (1994)
25. C. Pateros, Novel direct sequence spread spectrum multiple access technique, in *Proceedings of the MILCOM 2000, 21st Century Military Communications Conference Proceedings*, vol. 2, Los Angeles, CA, 22–25 October 2000, pp. 564–568
26. O. del Río Herrero, G. Foti, G. Gallinaro, Spread-spectrum techniques for the provision of packet access on the reverse link of next-generation broadband multimedia satellite systems. IEEE J. Sel. Areas Commun. **22**(3), 574–583 (2004)
27. O. Del Río Herrero, R. De Gaudenzi, High efficiency satellite multiple access scheme for machine-to-machine communications. IEEE Trans. Aerosp. Electron. Syst. **48**(4), 2961–2989 (2012)
28. U. Raza, P. Kulkarni, M. Sooriyabandara, Low power wide area networks: an overview. IEEE Commun. Surv. Tutor. **19**, 855–873 (2017) (2nd Quarter 2017)
29. F. Sforza, Communications system. US Patent 8'406'275, March 2013, https://www.google.com/patents/US8406275
30. LoRa Modulation Basics, in *SEMTECH Application Note AN1200.2, Revision 2*, May 2015, http://www.semtech.com/images/datasheet/an1200.22.pdf
31. C. Fourtet, T. Bailleul, Method for using a shared frequency resource, method for manufacturing terminals, terminals and telecommunication system (2013), https://www.google.com/patents/US20130142191
32. T.J. Myers, et al., Light monitoring system using a random phase multiple access system. U.S. Patent 8477830 (2013), https://www.google.com/patents/US8477830
33. H. Kim, Y.-G. Lim, C.-B. Chae, D. Hong, Multiple access for 5G new radio: categorization, evaluation, and challenges. arXiv:1703.09042
34. Study on New Radio Access Technology Physical Layer Aspects (Release 14), in *3GPP TR 38.802, v. 14.2.0*, September 2017
35. J. Choi, Low density spreading for multicarrier systems, in *Proceedings of IEEE ISSSTA*, August 2004, pp. 575–578
36. H. Nikopour, H. Baligh, Sparse code multiple access, in *Proceedings of IEEE PMRC*, September 2013

37. L. Lu, Y. Chen, W. Guo, H. Yang, Y. Wu, S. Xing, Prototype for 5G new air interface technology SCMA and performance evaluation. China Commun. **12**(Supplement), 38–48 (2015)
38. K. Au, L. Zhang, H. Nikopour, E. Yi, A. Bayesteh, U. Vilaipornsawai, J. Ma, P. Zhu, Uplink contention based SCMA for 5G radio access, in *IEEE Globecom 2014 Workshop*, December 2014, pp. 900–905
39. Z. Yuan, G. Yu, W. Li, Y. Yuan, X. Wang, X. Xu, Multi-user shared access for Internet of things, in *IEEE Vehicular Technology Conference (VTC Spring)* (2016), pp. 1–5
40. S. Chen, B. Ren, Q. Gao, S. Kang, S. Sun, K. Niu, Pattern division multiple access (PDMA)—a novel non-orthogonal multiple access for 5G radio networks. IEEE Trans. Veh. Technol. (2016)
41. 3GPP Document R1-166404: Receiver Details and Link Performance for MUSA, in *3GPP TSG RAN WG1 Meeting #86*, Gothenburg, Sweden, 22–26 August 2016
42. 3GPP Document R1-164688: Resource Spread Multiple Access (RSMA), in *3GPP TSG RAN WG1 Meeting #86*, Nanjing, China, 23–27 May 2016
43. 3GPP Document R1-1610118: Link level simulation results for RSMA vs. OMA, in *3GPP TSG RAN WG1 Meeting #86bis*, Lisbon, Portugal, 10–14 October 2016
44. 3GPP Document R1-162517: Considerations on DL/UL Multiple Access for NR, in *3GPP TSG RAN WG1 Meeting #84bis*, Busan, Korea, 11–15 April 2016
45. 3GPP Document R1-162517: Non-orthogonal Multiple Access for New Radio, in *3GPP TSG-RAN WG1 #85*, Nanjing, China, 23–27 May 2016
46. 3GPP Document R1-167535: New uplink non-orthogonal multiple access schemes for NR, in *3GPP TSG RAN WG1 Meeting #86*, Gothenburg, Sweden, 22–26 August 2016
47. A.J. Viterbi, Very low rate convolutional codes for maximum theoretical performance of spread spectrum multiple-access channels. IEEE J. Sel. Areas Commun. **8**, 641649 (1990)
48. S. Verdu′, S. Shamai, Spectral efficiency of CDMA with random spreading. IEEE Trans. Inf. Theory **45**, 622640 (1999)
49. L. Ping, L. Liu, K.Y. Wu, W.K. Leung, Interleave-division multiple-access. IEEE Trans. Wirel. Commun. **5**(4), 938–947 (2006)
50. 3GPP Document R1-163992: Non-orthogonal Multiple Access for New Radio, in *3GPP TSG-RAN WG1 #85*, Nanjing, China, 23–27 May 2016
51. 3GPP Document R1-1609333: LLS results for RDMA scheme, in *3GPP TSG RAN WG1 Meeting #86bis*, Lisbon, Portugal, 10–14 October 2016
52. 3GPP Document R1-1609334: System level evaluation result on RDMA, in *3GPP TSG RAN WG1 Meeting #86bis*, Lisbon, Portugal, 10–14 October 2016
53. E. Casini, R. De Gaudenzi, O. del Río Herrero, Contention resolution diversity slotted Aloha (CRDSA): an enhanced random access scheme for satellite access packet networks. IEEE Trans. Wirel. Commun. **6**(4), 1408–1419 (2007)
54. R. De Gaudenzi, O. del Río Herrero, G. Gallinaro, S. Cioni, P.D. Arapoglou, Random access schemes for satellite networks: from VSAT to M2M—a survey. Int. J. Satell. Commun. Netw. (2016). https://doi.org/10.1002/sat.1204
55. E. Casini, O. del Río Herrero, R. De Gaudenzi, D.M.E. Delaruelle, J.P. Choffray, Packet data transmission over a shared transmission channel. US Patent No. 8,094,672, 2 January 2012
56. A. Mengali, R. De Gaudenzi, P.D. Arapoglou, Enhancing the physical layer of contention resolution diversity slotted ALOHA. IEEE Trans. Commun. (2017). https://doi.org/10.1109/TCOMM.2017.2696952
57. R. De Gaudenzi, O. Del Río Herrero, Advances in random access protocols for satellite networks, in *Proceedings of International Workshop on Satellite and Space Communications (IWSSC)*, Siena, Italy, 10–12 September 2009
58. C. Xu, L. Ping, P. Wang, S. Chan, X. Lin, Decentralized power control for random access with successive interference cancellation. IEEE J. Sel. Areas Commun. 2387–2396 (2013)
59. H. Lin, K. Ishibashi, W.Y. Shin, T. Fujii, A simple random access scheme with multilevel power allocation. IEEE Commun. Lett. 2118–2121 (2015)
60. M. Zou, S. Chan, H.L. Vu, L. Ping, Throughput improvement of 802.11 networks via randomization of transmission power levels. IEEE Trans. Veh. Technol. 2703–2714 (2016)

61. P. Wang, J. Xiao, Li Ping, Comparison of orthogonal and non-orthogonal approaches to future wireless cellular systems. IEEE Veh. Technol. Mag. **1**(3), 4–11 (2006)
62. H.-C. Buiy, K. Zidane, J. Lacan, M.L. Boucheret, A multi-replica decoding technique for contention resolution diversity slotted Aloha, in *VTC Fall, 2015*, Boston, MA, 6–9 September 2015
63. K. Zidane, J. Lacan, M. Ginestex, C. Bes, A. Deramecourt, M. Dervin, Performance Evaluation of MARSALA with Synchronisation Errors in Satellite Communications, arXiv:1511.05359v1 [cs.IT], 17 Nov 2015
64. E. Garrido Barrabés, ACRDA Advanced Random Access Scheme for Satellite Communications. Master Thesis, Universitat Ramon Llull, La Salle Barcelona—European Space Agency (2013), https://esecretary.salle.url.edu/zona_autentica/memoriesTFC/7836_GK_2012.pdf
65. G.E. Corazza, *Digital Satellite Communications* (Springer, 2007). ISBN 9780387256344
66. G. Liva, Method for Contention Resolution in Time Hopping or Frequency Hopping. US Patent Application No. US/2011/0096795 A1, April 2011
67. G. Liva, Graph-based analysis and optimization of contention resolution diversity slotted ALOHA. IEEE Trans. Commun. **59**(2), 477–487 (2011)
68. Č. Stefanović, M. Momoda, P. Popovski, Exploiting capture effect in frameless ALOHA for massive wireless random access, in *Proceedings of IEEE WCNC 2014*, Istanbul, Turkey, April 2014, pp. 1–6
69. H.C. Bui, J. Lacan, M.-L. Boucheret, An enhanced multiple random access scheme for satellite communications, in *Proceedings of the 2012 Wireless Telecommunications Symposium (WTS)*, 18–20 April 2012
70. E. Paolini, G. Liva, M, Chiani, High throughput random access via codes on graphs: coded slotted ALOHA, in *Proceedings of the IEEE International Conference on Communications (ICC)*, Kyoto, Japan, 5–9 June 2011, pp. 1–6
71. M. Ivanov, F. Brännström, A. Graell i Amat, P. Popovski, Error floor analysis of coded slotted ALOHA over packet erasure channels. IEEE Commun. Lett. **19**(3), 419–422 (2015)
72. M. Eroz, L. Lee, Scrambled coded multiple access, in *Proceedings of the IEEE Vehicular Technology Conference*, San Francisco, CA, September 2011
73. O. del Río Herrero, R. De Gaudenzi, Methods, Apparatuses and System for Asynchronous Spread-Spectrum Communications. U.S. Patent 7,990,874 B2, 2 August 2011
74. R. De Gaudenzi, O. del Rio Herrero, G. Gallinaro, Enhanced spread Aloha physical layer design and performance. Wiley Int. J. Satell. Commun. Netw. **32**(6), 457–473 (2014)
75. F. Collard, R. De Gaudenzi, On the optimum packet power distribution for spread Aloha packet detectors with iterative successive interference cancellation. IEEE Trans. Wirel. Commun. **13**(12), 6783–6794 (2014)
76. J. Sala-Alvarez, J. Vilares, F. Rey, SINR profile for spectral efficiency optimization SIC receivers in the many-user regime, in *Proceedings of IEEE International Communication Conference (ICC)*, London, United Kingdom, 8–12 June 2015, pp. 1–6
77. G. Gallinaro, N. Alagha, R. De Gaudenzi, R. Küller, P. Salvo Rossi, ME-SSA: an advanced random access for the satellite return channel, in *Proceedings of the IEEE International Communication Conference (ICC)*, London, UK, June 2015, pp. 1–6
78. S. Moshavi, E.G. Kanterakis, D.L. Schilling, Multistage linear receivers for DS-CDMA systems. Int. J. Wirel. Inf. Netw. **3**(1), 1–17 (1996)
79. R. Müller, S. Verdù, Design and analysis of low-complexity interference mitigation on vector channel. IEEE J. Sel. Areas Commun. **19**(8), 1429–1441 (2001)
80. L. Cottatelluci, M. Debbah, R. Müller, Asymptotic design and analysis of multistage detectors for asynchronous CDMA systems, in *IEEE International Symposium on Information Theory*, 27 June–2 July 2004, Chicago, USA
81. L. Cottatellucci, R. Mùller, A systematic approach to multistage detectors in multipath fading channels. IEEE Trans. Inf. Theory **51**(9), 3146–3158 (2005)
82. D. Boudreau, G. Caire, G.E. Corazza, R. De Gaudenzi, G. Gallinaro, M. Luglio, R. Lyons, J. Romero-Garcia, A. Vernucci, H. Widmer, Wideband CDMA for the satellite component of UMTS/IMT-2000. IEEE Trans. Veh. Technol. **51**(2), 306–330 (2002)

83. R. De Gaudenzi, O. del Río Herrero, G. Acar, E. Garrido Barrabés, E. Garrido Barrabés, Asynchronous contention resolution diversity Aloha: making CRDSA truly asynchronous. IEEE Trans. Wirel. Commun. **13**(11), 6193–6206 (2014)
84. N. Becker, S. Kay, L. Lee, M. Eroz, Spread asynchronous scrambled coded multiple access (SA-SCMA)—a new efficient multiple access method, in *Proceedings of the IEEE GLOBECOM 2017*, Washington, D.C., December 2016
85. C. Kissling, Performance enhancements for asynchronous random access protocols over satellite, in *Proceedings of International Communication Conference (ICC)*, June 5–9 2011, pp. 1–6, Kyoto, Japan

Part IV
Challenges, Solutions, and Future Trends

Chapter 18
Experimental Trials on Non-Orthogonal Multiple Access

Anass Benjebbour, Keisuke Saito and Yoshihisa Kishiyama

18.1 Introduction

Non-orthogonal multiple access (NOMA) was proposed as a novel scheme where multiple users of different channel conditions are multiplexed using overlapped non-orthogonal radio resources (e.g., time/frequency/code) on the transmitter side and multi-user signal separation is conducted on the receiver side [1–11]. The transmitter and receiver designs for NOMA were considered for both closed-loop and open-loop multiple-input and multiple-output (MIMO) and for both successive interference cancellation (SIC) and non-SIC receivers [8, 9, 11]. In addition, the performance of NOMA was heavily investigated for downlink and uplink from both system-level and link-level perspectives [3–11]. In this chapter, we focus on downlink NOMA with two users combined with 2×2 open-loop single-user (SU)-MIMO. We assess its link-level performance with different types of receivers and present the experimental trials conducted in indoor and outdoor environments.

The remaining of this chapter is organized as follows. Section 18.2 describes the concept and the basic structure of the transceiver of NOMA. Section 18.3 describes the combination of NOMA with MIMO including open-loop MIMO. Section 18.4 explains about the link-level simulation parameters, experimental setup, and the transceiver structure. Section 18.5 introduces the link-level evaluation results of downlink NOMA with multiple types of receivers. Section 18.6 explains the results of experimental trials conducted in indoor and outdoor. Finally, Sect. 18.7 concludes the chapter.

A. Benjebbour (✉) · K. Saito · Y. Kishiyama
NTT DOCOMO INC., Chiyoda, Japan
e-mail: benjebbour@nttdocomo.com

M. Vaezi et al. (eds.), *Multiple Access Techniques for 5G Wireless Networks and Beyond*, https://doi.org/10.1007/978-3-319-92090-0_18

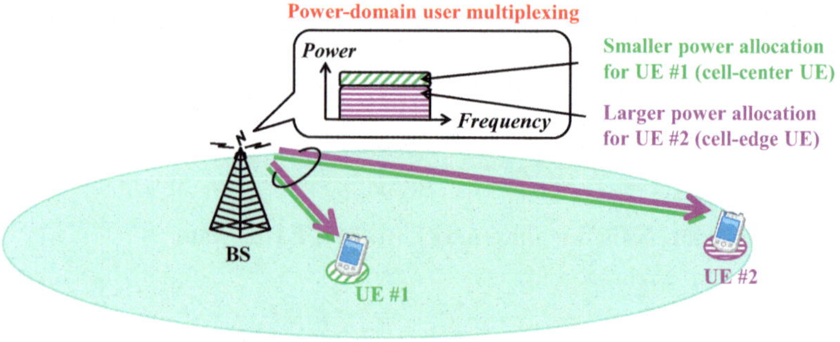

Fig. 18.1 Illustration of downlink NOMA for 2-UE multiplexing via resource overlapping in frequency domain using different transmit power allocation

18.2 Downlink NOMA

18.2.1 Concept

Figure 18.1 illustrates downlink NOMA for the case of one base station (BS) and two user equipment (UE) with resource overlapping in frequency domain and different transmit powers allocated to the two UEs to multiplex in power domain.

For simplicity, we describe here the transceiver of downlink NOMA for the case of single-input single-output (SISO) with single transmit and receive antennas. As shown in Fig. 18.1, the BS transmits a signal for UE #i (i = 1, 2), x_i, where $\mathrm{E}[|x_i|^2]$ = 1, with transmit power $P_i (P_1 < P_2, P_1 + P_2 = 1)$, and UE #1 is cell-center UE and UE #2 is cell-edge UE. The transmit signals, x_1 and x_2, are superposed using overlapped resources with different transmit powers allocated. The received signal y_i at UE #i is represented by

$$y_i = h_i \left(\sqrt{P_1}x_1 + \sqrt{P_2}x_2\right) + w_i \tag{18.1}$$

where h_i is the complex channel coefficient between UE #i and the BS. The term w_i denotes additive white Gaussian noise (AWGN) including inter-cell interference. The power spectral density of w_i is $N_{0,i}$. In downlink NOMA, multi-user multiplexing via resource overlapping using different transmit powers (power sharing among users) is conducted as shown in Fig. 18.1. The cell-center UE is allocated lower transmit power P_1, while the cell-edge UE is allocated higher transmit power P_2.

18.2.2 Receiver

For NOMA, both UEs received their own desired signal interfered with the other UE signal. In particular, the cell-center UE signal suffers from high interference from the cell-edge UE signal since $P_1 < P_2$. Therefore, in order to extract the desired signal of the cell-center UE from the received signal at the cell-center UE, advanced interference cancellation is needed for signal separation. Several types of advanced interference cancellation receivers can be applied. Examples of advanced receivers are:

1. *Codeword-level successive interference cancellation (CL-SIC)*
2. *Symbol-level successive interference cancellation (SL-SIC)*
3. *Reduced complexity maximum likelihood detection (R-ML)*
4. *R-ML with Gray-mapped composite constellation.*

CL-SIC and SL-SIC have similar structure. Figure 18.2 shows the receiver design for the CL-SIC case at the cell-center UE (UE #1). The received data symbols for the cell-edge UE (UE #2) are first demodulated by multiplying the received signal at UE #1 with the maximum ratio combining (MRC) generated weights. Then, the log-likelihood ratio (LLR) corresponding to those demodulated symbols is calculated. For the SL-SIC, those LLRs are directly used to generate a symbol replica for the cell-edge UE (UE #2). On the other hand, for the CL-SIC, a sequence of LLRs which is called codeword is input to the turbo decoder and a sequence of posteriori LLRs is generated. After interleaving the sequence of posteriori LLRs, the interleaved LLRs are used to calculate a soft symbol replica for the cell-edge UE. The soft symbol replica for the cell-edge UE is subtracted from the received data symbol. After interference cancellation, the MRC weights are applied then detection of cell-center UE (UE #1) is performed.

In the R-ML, the desired and interfering signals are jointly detected under the maximum likelihood (ML) criterion, and some complexity-reduction algorithms such as maximum likelihood detection with QR decomposition and M-algorithm (QRM-MLD) are applied. R-ML with Gray-mapped composite constellation was also considered in order to reduce signaling overhead and the receiver complexity compared to NOMA with SIC receiver [9, 11]. In such a scheme, coded bits for both

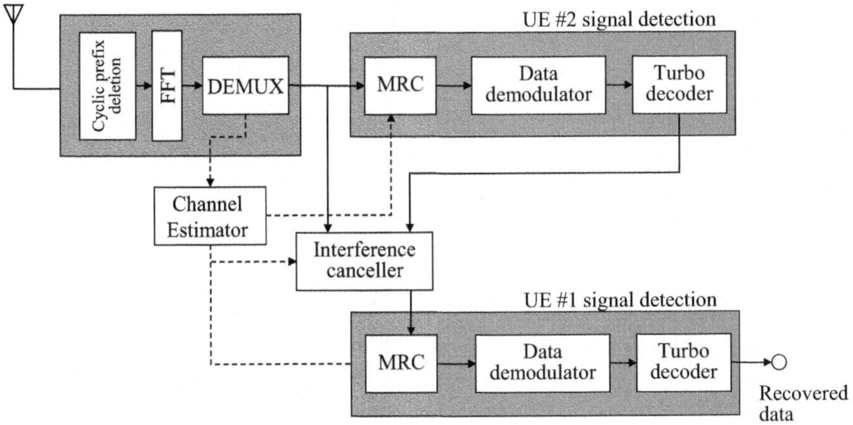

Fig. 18.2 Illustration of CL-SIC receiver for cell-center UE

the superposed UEs are jointly modulated such that the modulated signal has Gray-mapped composite constellation and reduced-maximum likelihood (R-ML) receiver is used for symbol-level interference cancellation [9].

Regarding cell-edge user (UE #2) receiver, since the transmit power of UE #2 is larger than that of UE #1, the transmit data intended to UE #2 is demodulated and decoded without interference cancellation of the signal of UE #1 and thus treating it as noise.

18.3 Combination of Downlink NOMA and MIMO

MIMO is one of the key technologies to improve spectrum efficiency in wireless communications. MIMO techniques can be categorized into single-user MIMO (SU-MIMO), where MIMO antennas are used to serve only one UE at once, and MU-MIMO, where more than one UE are served simultaneously. Since MIMO technology exploits spatial domain and NOMA overlaps the resources of multiple users in the power domain using different transmit powers, the two technologies can be combined to further boost the system performance. In the following, we explain about how NOMA can be combined with closed-loop MIMO and open-loop MIMO schemes.

18.3.1 Concept

As illustrated in Fig. 18.3, there are two major approaches to extend downlink NOMA from SISO to 2×2 MIMO.

One approach is to set multiple power levels for NOMA multiplexing and apply SU-MIMO and/or MU-MIMO technique inside each power level [6, 10]. For example, the combination of NOMA with SU-MIMO is illustrated in Fig. 18.3 on the left side for the case of 2×2 MIMO, where the number of power levels and number of multiplexed UEs are 2, respectively. UE #1 and UE #2 are NOMA multiplexed cell-center and cell-edge users, respectively and each UE receives two data streams. By combining NOMA with 2×2 SU-MIMO, with up to two user multiplexing being overlapped using two power levels, it is possible to transmit up to four data streams using only two transmit antennas.

The other approach is to convert the non-degraded 2×2 MIMO channel to two degraded 1×2 SIMO channels, where NOMA is applied over each equivalent 1×2 SIMO channel separately, as shown in Fig. 18.3 on the right side [5]. For this scheme, multiple transmit beams are created by MIMO and superposition of signals designated to multiple users is applied within each beam and signal separation is also applied within each beam. This scheme can be seen as a combination of NOMA with MU-MIMO where each user has fixed rank 1 transmission.

Moreover, NOMA can be combined with MIMO, where users use the same precoder or different precoders (not necessarily orthogonal) when multiplexed in power

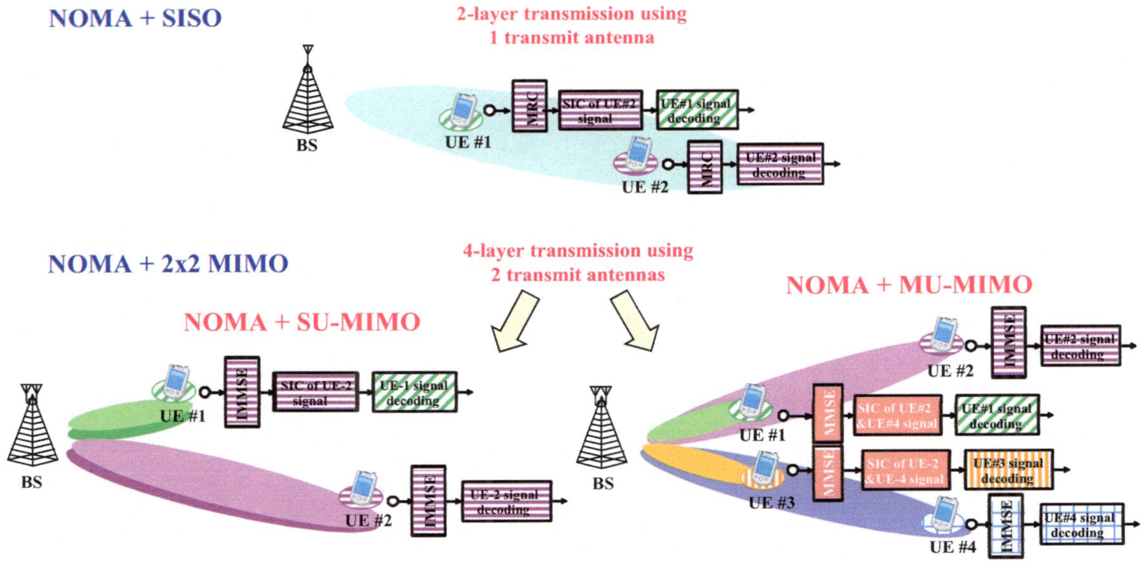

Fig. 18.3 Extension of downlink NOMA from SISO to 2 × 2 MIMO

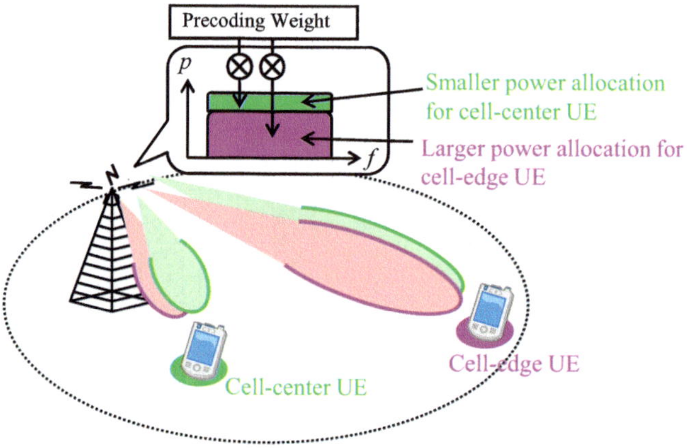

(a) Same precoder applied to downlink NOMA multiplexed UEs.

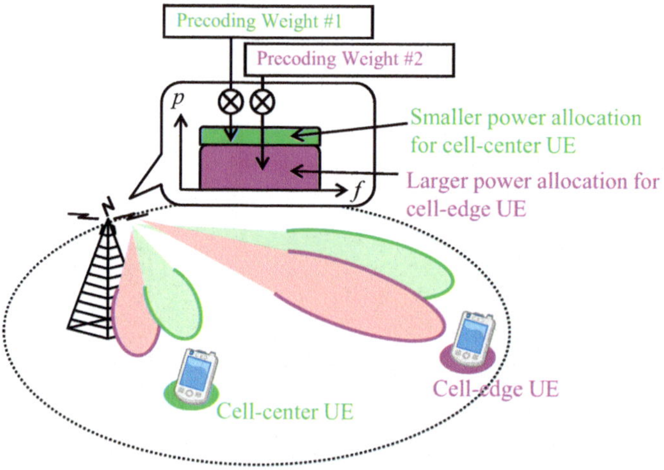

(b) Different precoders applied to downlink NOMA multiplexed UEs.

Fig. 18.4 Examples of downlink NOMA with same precoder and different precoders applied to NOMA multiplexed UEs (case of transmit rank of both UEs is two)

domain. Figure 18.4a and 18.4b illustrate the concept of downlink NOMA with precoding based SU-MIMO when applying the same precoder and different precoders for NOMA multiplexed UEs, respectively.

18.3.2 Transceiver Design for Downlink NOMA Combined with SU-MIMO

In this section, we describe the transceiver design of combining downlink NOMA with SU-MIMO. Considering LTE/LTE-Advanced MIMO transmissions, both closed-loop MIMO LTE Release 8 transmission mode 4 (TM4) [12] and open-loop MIMO LTE Release 8 transmission mode 3 (TM3) [12] are of interest.

Similar to Sect. 18.2, we assume one BS and two UEs, with one UE located at cell-center and the other UE at cell-edge. At BS, the signals of two UEs are generated using open-loop SU-MIMO, superimposed, and transmitted. At the BS transmitter, the transmit signal vector **X** is generated as follows:

$$\mathbf{X} = \sqrt{P_1}\mathbf{W}_{Tx,1}\mathbf{X}_1 + \sqrt{P_2}\mathbf{W}_{Tx,2}\mathbf{X}_2, \tag{18.2}$$

where $\mathbf{X}_i$ represents the transmit signal vector of UE #i ($i = 1, 2$), where $\mathrm{E}[|\mathbf{X}_i|^2] = 1$, and $P_i (P_1 < P_2, P_1 + P_2 = 1)$ represents the allocated transmit power to UE #i. This means that the transmit signals for UE #1 and UE #2 are multiplexed in the power domain based on the allocated transmit power P_i after applying precoding weight matrix $\mathbf{W}_{Tx,i}$ to each UE #i. At the UE receiver, **Y**, which represents the received signal vector of UE #i ($i = 1, 2$), is represented by

$$\mathbf{Y}_i = \mathbf{HX} + \mathbf{N}_i, \tag{18.3}$$

where **H** represents the complex channel matrix of UE #i, and $\mathbf{N}_i$ is the AWGN vector of UE #i, where $\mathrm{E}[\mathbf{N}_i\mathbf{N}_i^H] = \sigma^2\mathbf{I}$ (**I** is the identity matrix).

In order to combine 2-by-2 SU-MIMO transmission with NOMA, SU-MIMO is applied to each UE independently with up to 2-layer (rank 2) transmission per UE. As a result, up to 4-layer transmission is enabled by applying 2-by-2 SU-MIMO on the top of NOMA with 2 UE multiplexing. Regarding the transmit power, for the case of OMA with SU-MIMO, the transmit power of each user is split equally among transmission layers. For the case of NOMA with SU-MIMO, the transmit power of each UE is set based on the power allocation scheme such as full search power allocation (FSPA), fractional transmit power allocation (FTPA), or predefined user grouping and per-group fixed power allocation (FPA) [4]; then the transmit power of each layer within each UE is split equally among transmission layers in the same manner as OMA case. When applying NOMA with SU-MIMO, inter-user interference and inter-stream interference are caused and the amount of inter-user interference depends on the allocated transmit power among the multiplexed UEs. The larger is the difference in allocated transmit power between the UEs, the smaller is the inter-user interference.

In order to suppress the inter-stream interference effectively, UE #1 applies the CL-SIC receiver which detects and decodes the interfering data of UE #2 then cancels them. In addition, in order to mitigate the inter-user interference, we apply different receiver weights generation schemes for before successive interference cancellation

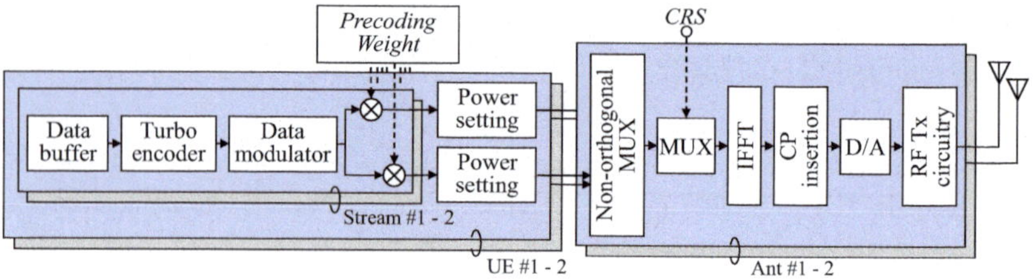

(a) BS Transmitter

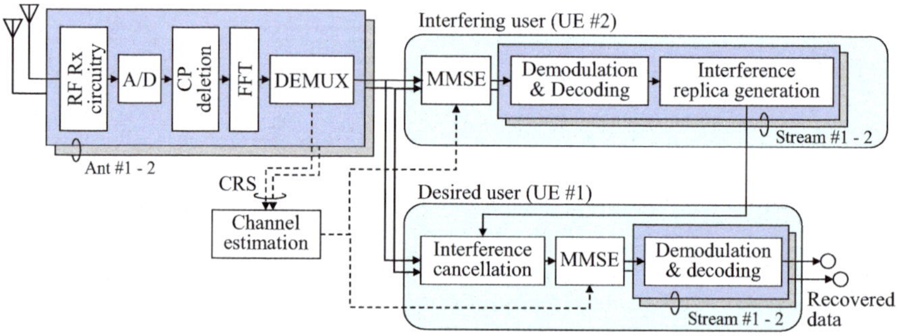

(b) UE #1 Receiver

Fig. 18.5 Structure of BS transmitter and UE #1 CL-SIC receiver for downlink NOMA combined with 2×2 MIMO

(SIC) and after SIC. Figure 18.5 shows the BS transmitter and receiver structure for cell-center UE (UE #1) using CL-SIC receiver for downlink NOMA combined with 2×2 MIMO, where channel estimation at UE side is conducted based on cell-specific reference signal (CRS).

In the following, we describe the receiver processing steps for UE #1. UE #1 receiver first calculates the receiver weight matrix $\mathbf{W}_{Rx,2}$ using the channel coefficients estimated using reference signal, and generates the symbol vector of interfering UE (UE #2) $\mathbf{S}_{rep,2}$ as follows:

$$\mathbf{S}_{rep,2} = \mathbf{W}_{Rx,2}\mathbf{Y}_1. \tag{18.4}$$

Here, we calculate the receiver weight matrix before SIC, considering the inter-user interference based on the minimum mean squared error (MMSE) criterion as follows:

$$\mathbf{W}_{Rx,2} = \left(\hat{\mathbf{H}}_2^{\mathrm{H}}\hat{\mathbf{H}}_2 + \hat{\mathbf{H}}_1^{\mathrm{H}}\hat{\mathbf{H}}_1 + \sigma^2\mathbf{I}\right)^{-1}\hat{\mathbf{H}}_2^{\mathrm{H}}, \tag{18.5}$$

where $\hat{\mathbf{H}}_i$ is defined as follows:

$$\hat{\mathbf{H}}_i = \hat{\mathbf{H}}\sqrt{P_i}\mathbf{W}_{Tx,i}, \tag{18.6}$$

where $\hat{\mathbf{H}}$ is the estimated complex channel matrix. Then, the replica signal vector of the interfering UE (UE #2), $\mathbf{X}_{rep,2}$ is generated by detecting and decoding the modulated symbol $\mathbf{S}_{rep,2}$. After generating $\mathbf{X}_{rep,2}$ and applying SIC, the received signal vector, $\mathbf{Y}_{c,1}$ is calculated as

$$\mathbf{Y}_{c,1} = \mathbf{Y}_1 - \hat{\mathbf{H}}\sqrt{P_2}\mathbf{W}_{Tx,2}\mathbf{X}_{rep,2}. \tag{18.7}$$

Finally, the receiver weights $\mathbf{W}_{Rx,1}$corresponding to desired signal UE #1 are multiplied with $\mathbf{Y}_{c,1}$, then the received signal for UE #1 is demodulated and decoded to retrieve the transmit signal sequence corresponding to UE #1. The receiver weight matrix for detecting UE #1 signal after SIC is calculated as follows:

$$\mathbf{W}_{Rx,1} = \left(\hat{\mathbf{H}}_1^{\mathrm{H}}\hat{\mathbf{H}}_1 + \sigma^2\mathbf{I}\right)^{-1}\hat{\mathbf{H}}_1^{\mathrm{H}}. \tag{18.8}$$

On the other hand, the cell-edge UE (UE #2) does not apply SIC as the interference signal from UE #1 is treated as noise; thus, only demodulation and decoding are applied to the received signal $\mathbf{Y}_2$.

18.3.3 Combination of Downlink NOMA with Open-Loop SU-MIMO

We consider applying 2-by-2 open-loop SU-MIMO transmission based on the LTE TM3. Therefore, for rank 1 transmission, space-frequency block coding (SFBC) is used to encode the same data differently and increase the SNR of the recombined data streams to obtain the transmit diversity; and for rank 2 transmission, large delay cyclic delay diversity (CDD) is used to enable spatial multiplexing using precoding vector and increase throughput [12]. Open-loop MIMO schemes when combined with NOMA are expected to provide robust performance in high mobility scenarios [3, 6, 8].

In the following, we describe how to generate transmit signals and receiver weights when downlink NOMA is combined with open-loop 2 × 2 SU-MIMO for different combinations of transmit ranks (UE #1:UE #2 = R_1:R_2).

(1) R_1:R_2 = 1:1

For this case, SFBC is applied to both UE #1 and UE #2. The transmit signal matrix [(2 transmit antennas) × (k-th and k + 1-th subcarriers)] is represented as follows:

$$\begin{pmatrix} x_i(k) & x_i(k+1) \\ -x_i(k+1)^* & x_i(k)^* \end{pmatrix}, \tag{18.9}$$

where $x_i(k)$ represents the transmit signal of k-th subcarrier of UE #i. Moreover, we extend the received signal vector as follows:

$$\tilde{\mathbf{Y}}_i = \left(y_1(k), y_2(k), y_1^*(k+1), y_2^*(k+1)\right)^T, \tag{18.10}$$

and we define the precoding weight matrix $\mathrm{W}_{Tx,i}$ in (18.2) as the identity matrix **I**. The extended complex channel matrix $\tilde{\mathbf{H}}$ is defined as follows:

$$\begin{aligned}
\tilde{\mathbf{Y}}_i &= \tilde{\mathbf{H}}\tilde{\mathbf{X}} + \tilde{\mathbf{N}}_i, \\
\tilde{\mathbf{X}} &= \sqrt{P_1}\begin{pmatrix} x_1(k) \\ -x_1^*(k+1) \end{pmatrix} + \sqrt{P_2}\begin{pmatrix} x_2(k) \\ -x_2^*(k+1) \end{pmatrix}, \\
\tilde{\mathbf{H}} &= \begin{pmatrix} h_{11}(k) & h_{12}(k) \\ h_{21}(k) & h_{22}(k) \\ h_{12} * (k+1) & -h_{11} * (k+1) \\ h_{22} * (k+1) & -h_{21} * (k+1) \end{pmatrix},
\end{aligned} \tag{18.11}$$

where $h_{ij}(k)$ represents the complex channel coefficient of k-th subcarrier of i-th receive antenna and j-th transmit antenna. Then, we calculate the receiver weight matrix by applying (18.11) to (18.5) and (18.8).

(2) $R_1{:}R_2 = 2{:}2$

The complex channel matrix $\mathbf{H}_i$ is represented as follows:

$$\mathbf{H}_i = \sqrt{P_i}\begin{pmatrix} h_{11}(k) & h_{12}(k) \\ h_{21}(k) & h_{22}(k) \end{pmatrix} \mathbf{W}'_{Tx,i} \tag{18.12}$$

where the precoding weight matrix $\mathrm{W}_{Tx,2}$is applied based on large delay CDD as specified in LTE Release 8 [12], and $\mathrm{W}_{Tx,1}$is set equal to $\mathrm{W}_{Tx,2}$ over all subcarriers. In this case, the receiver weights matrix is calculated by applying (18.12) to (18.5) and (18.8).

(3) $R_1{:}R_2 = 2{:}1$

In order to detect cell-edge UE first (UE #2, rank =1), we define the transmit signal vector and the channel matrix as in (18.11) for before SIC. On the other hand, for after SIC, we define the channel matrix as in (18.12) to detect the cell-center UE (UE #1, rank = 2).

Regarding UE #2, since the transmit power ratio of UE #2 is larger than that of UE #1, the transmit data intended to UE #2 is demodulated and decoded without applying SIC. The receiver weights are applied on the received signal before demodulation and decoding.

18.4 Link-Level Evaluation and Experiment Parameters

The radio frame structure and the parameters used in link-level simulations and experimental trials are given in Fig. 18.6 and Table 18.1, respectively.

The radio frame structure is based on LTE Release 8 specifications [12]. The control signal is multiplexed on the 1st symbol, the data signal is multiplexed on after the second symbol, and CRS [12], which is used for channel estimation, is multiplexed with the 1st, 4th, 7th, and 11th symbol in the time domain and with the every six subcarriers per resource block (RB) in the frequency domain for each transmit antenna. The system bandwidth and the number of subcarriers of the OFDM signal are 20 MHz and 1200, respectively; with a subcarrier separation of 15 kHz. The carrier frequency used in the experiment is 3.9 GHz.

The number of multiplexed UEs is 2, one cell-center UE #1 and one cell-edge UE #2. The number of transmit and receive antennas of both BS and UE is 2. At the BS transmitter, the information binary data sequence is turbo encoded with the coding rate of R and modulated using QPSK, 16QAM, or 64QAM. After modulation, the LTE TM3 codebook-based open-loop SU-MIMO transmission is applied to resultant signal sequence of each UE, and the signals of both UEs are non-orthogonally multiplexed in the power domain based using the transmit power ratio of $(P_1:P_2)$. After insertion of CRS, the signal sequence is converted into an OFDM symbol with the duration of 66.67 μs, followed by the addition of a cyclic prefix (CP) of 4.69 μs. At the UE receiver, after CP removal, the received signal is de-multiplexed into each subcarrier component using FFT. Then, MIMO signal detection is applied using the receiver weights. At the cell-center user, UE #1, advanced receiver is applied. On the other hand, for the cell-edge user, UE #2, advanced receiver is not applied. Section 18.3 describes the detection procedure for the case of SIC used as the advanced receiver for UE #1.

Finally, the sequence of likelihood values is turbo decoded using the Max-Log-MAP algorithm with six iterations to recover the transmitted binary data.

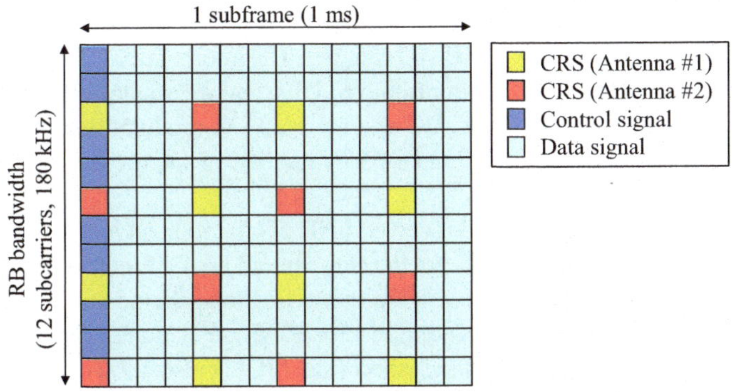

Fig. 18.6 LTE-based subframe structure used in experiment

Table 18.1 Link-Level simulation and experiment parameters

Carrier frequency	3.9 GHz
System bandwidth	20 MHz
Number of subcarriers	1200
Subcarrier spacing	15 kHz
Subframe length	1.0 ms
Symbol duration	Effective data: 66.67 s + CP: 4.69 s
Channel coding/decoding	Turbo coding (Constraint length: 4 bits)/ Max-Log-MAP decoding (six iterations)
Channel estimation	CRS-based channel estimation

LTE Release 8 radio frame structure (subcarrier spacing: 15 kHz, symbol duration: 66.67 μs with 4.69 μs cyclic prefix, channel encoding and decoding: Turbo coding with constraint length of 4 bits and Max-Log decoding with six iterations is used) is adopted, and channel estimation at UE side is conducted based on CRS.

18.5 Link-Level Performance Evaluation with Different Receivers

In this section, we present the link-level evaluation results for different types of receivers of downlink NOMA combined with SU-MIMO for two UE multiplexing case.

As described in Sect. 18.2, CL-SIC and R-ML are promising advanced receivers for downlink NOMA. In this section, we evaluate and compare the BLER performance of cell-center UE using CL-SIC and R-ML by link-level simulations.

The numbers of transmit and receive antennas are both set to two, and transmission mode 4 (TM4) [12] is assumed for closed-loop SU-MIMO transmission. The 6-path exponential power delay profile with a decaying factor of 2 dB is assumed as a fading channel for a 2×2 uncorrelated MIMO channel. The number of MIMO transmission streams is aligned among NOMA multiplexed UEs, and we assume single stream and two stream MIMO transmissions. We assume 16QAM modulation and the coding rate of 0.49 for the cell-center UE and QPSK modulation and the coding rate of 0.49 for the cell-edge UE.

Figure 18.7 shows the required received SNR of CL-SIC and R-ML for achieving the block error rate (BLER) of 10% for the cell-center UE as a function of the transmit power ratio for the cell-center UE. The joint modulation scheme with Gray-mapped composite constellation [9, 11] is applied for R-ML. For comparison, the performance for perfect interference cancellation (ideal IC) is also shown. Note that ideal SIC here refers to the ideal generation of the replica of interfering signal; however, interference cancellation may still remain imperfect due to channel estimation error.

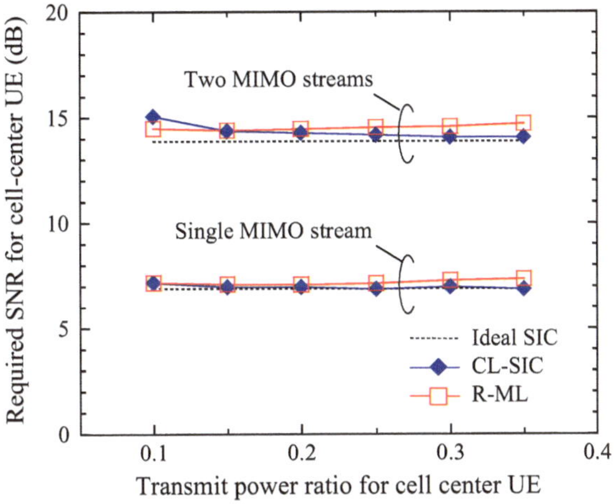

Fig. 18.7 NOMA link-level performance for CW-SIC (labeled CL-SIC) and R-ML compared with ideal SIC

From the results, we can observe that CL-SIC and R-ML have similar BLER performance and can achieve almost the same performance as Ideal IC when the transmit power ratio is set between 0.1 and 0.35.

18.6 NOMA Experimental Trials

18.6.1 Test Bed Using Fading Emulator

A test bed composed of 1 BS and 2 UEs was developed to confirm downlink NOMA performance using real hardware and in real radio environment. In order to assess downlink NOMA performance, the experiment test bed was first connected to a fading emulator. The channel model using the fading emulator was an exponentially decaying Rayleigh fading channel with 2 dB average power per decay (root mean squared (RMS) delay spread was 0.29 μs), the maximum Doppler frequency was 10 Hz, and there was no fading correlation.

In the test bed, the carrier frequency is 3.9 GHz and the total bandwidth is 20 MHz. For NOMA, the bandwidth of both UEs is 20 MHz, while for OMA the total bandwidth is split equally between the 2 UEs where each UE has 10 MHz. A 16-bit D/A and A/D converters are used in the BS and UE, respectively.

We adopt the same frame structure, turbo encoding and decoding schemes as link-level simulations. At the BS, for each UE data, turbo encoding, data modulation, and multiplication by precoding vector are applied, then the precoded signal of the two UEs is superposed according to a predefined transmit power ratio in case of NOMA, and transmitted from two antennas. For MIMO transmission, LTE TM3 is utilized for open-loop 2 $\times$ 2 SU-MIMO transmission. At the UE side, two receive antennas

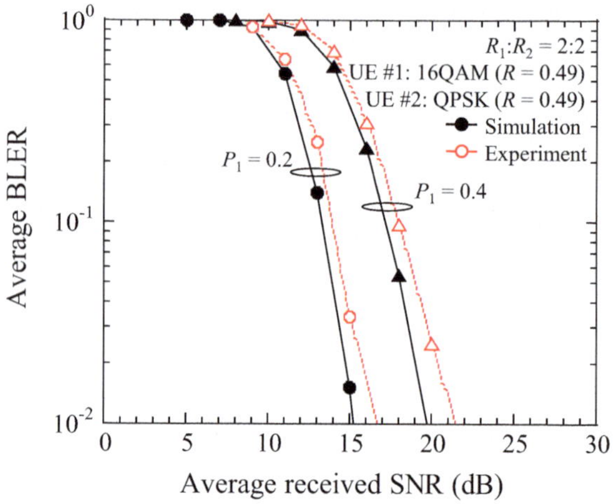

Fig. 18.8 Experiment results using fading emulator compared with link-level simulation results

are assumed. At cell-center UE (UE #1), the CL-SIC is applied to assess upper bound performance.

The experimental results using fading emulator are compared with the results obtained by computer simulations. In Fig. 18.8, $P1$:$P2$ = 0.2:0.8, and 0.4:0.6, and 2 × 2 SU-MIMO are assumed. When CL-SIC is applied, BLER performance of UE #1 versus received SNR derived based on CRS is plotted. The modulation and coding scheme (MCS) of UE #1 is 16QAM (R = 0.49), of UE #2 is QPSK (R = 0.49), and the combination of transmit rank is UE #1:UE #2 = $R1$:$R2$ = 2:2.

According to Fig. 18.8, the gap between results from experiments and simulation is within 0.8 dB in terms of required SNR to achieve BLER = 10^{-1}.

18.6.2 Configurations of Outdoor and Indoor Experimental Trials

Experimental trials are conducted as shown in Fig. 18.9 in inside (indoor) and outside (outdoor) at NTT DOCOMO R&D center, Yokosuka, Japan.

In case of the indoor experiments (Fig. 18.9a), UE #1 is located in points A (received SNR = 30 dB), B (25 dB), C (20 dB), D (15 dB), E (10 dB), and F (5 dB). UE-2 is located in points E (10 dB), F (5 dB), and G (0 dB). All points are at non-line-of-sight (NLOS) condition, and the RMS delay spread value is about 0.1–0.15 μs at points A to F, and about 0.3 μs at point G. Antenna height is 2.6 m for BS, and 1.4 m for UE. The BS transmission power is 100 mW (20 dBm). Cross-polarized antennas are applied. The antenna gain is 2 dBi for vertical and omnidirectional in horizontal.

In case of outdoor experiments (Fig. 18.9b), UE #1 is located at points A (received SNR = 30 dB), B (20 dB), and C (10 dB). UE #2 is located at points C (10 dB) and

D (0 dB). All points are at NLOS condition, and the RMS delay spread value is about 0.1 μs at points A to C, and about 0.4 μs at point D. The BS antenna height is 39.6 m for BS and 1.4 m for UE. BS antenna is 120° sectored-beam antenna with the 3 dB width of 90° in azimuth and a gain of approximately 18 dBi. BS transmission power is 10 W (40 dBm). Two antenna configurations, cross-polarized antennas and co-polarized antennas, were used at BS during the measurements. UE antenna is the same as in the case of indoor experiment. Depending on BS antenna configuration used, UE polarization was switched between cross-polarized and co-polarized antenna configuration.

We illustrate the experiment environments in Fig. 18.9 and show examples of the power delay profile of the experiment environments in Fig. 18.10. The power delay profile was measured based on CRS.

In these trials, in case of NOMA, the transmit power ratio of each UE (P_1, P_2) can be selected from (0.1, 0.9) to (0.4, 0.6) with a 0.05 step. The throughput gain

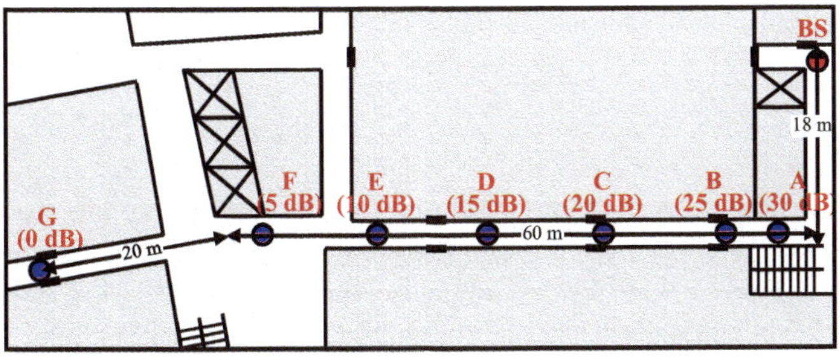

(a) Measurement course (indoor corridor).

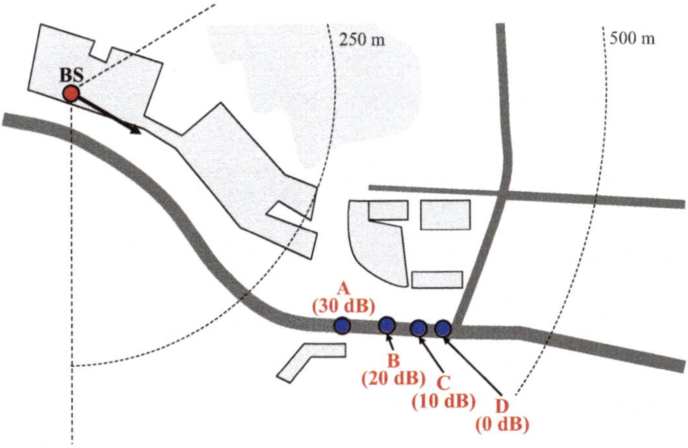

(b) Measurement course (outdoor street).

Fig. 18.9 Measurement courses of NOMA experimental trials

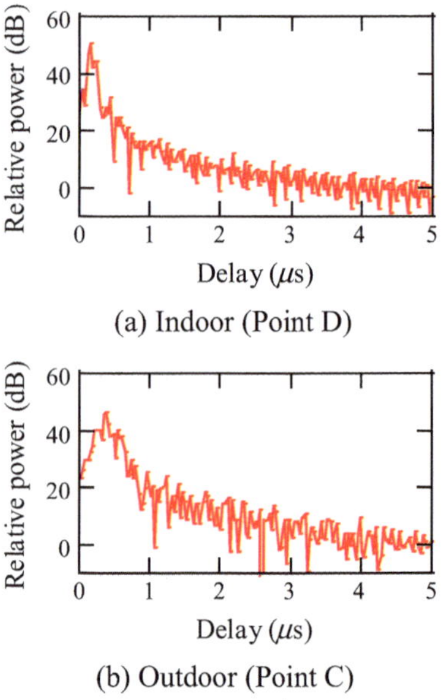

Fig. 18.10 Examples of measured power delay profile

is calculated as follows: (1) The maximum user throughput of UE #1 and UE #2 in case of OMA is measured in each measurement point by adjusting the MCS and transmission rank; (2) then, the maximum user throughput of UE #1 in case of NOMA is measured while making the user throughput of UE #2 equal to that of OMA case. By ensuring the same user-throughput of UE #2 for both OMA and NOMA tests, we were able to compare the throughput gains of cell-center UE #1 for NOMA compared to OMA in a fair manner.

18.7 Trial Results

Figures 18.11 and 18.12 show NOMA performance results in comparison with OMA for indoor and outdoor trials using 2×2 SU-MIMO, respectively.

18.7.1 Indoor Experiments

For indoor experiments, Table 18.2 shows the combination of transmit power ratio and transmit rank for each UE at each measurement point. In Table 18.2, it is possible to check that transmit power ratio and transmit rank are adjusted according to measurement point for UE #2. As the SNR of UE #2 increases, P_2 increases, this is

Fig. 18.11 NOMA performance results in indoor experiments

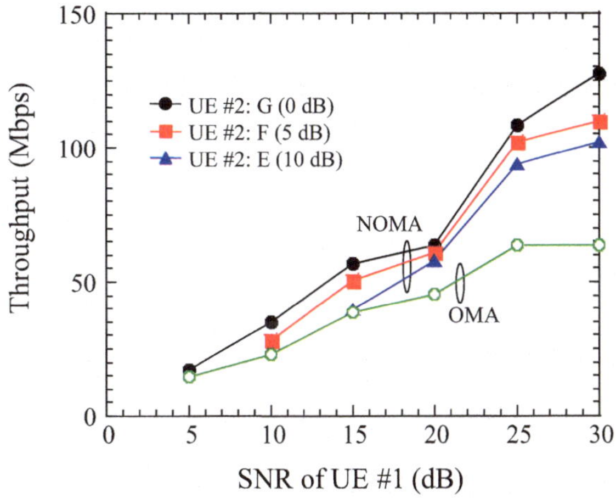

(a) UE #1 throughput performance

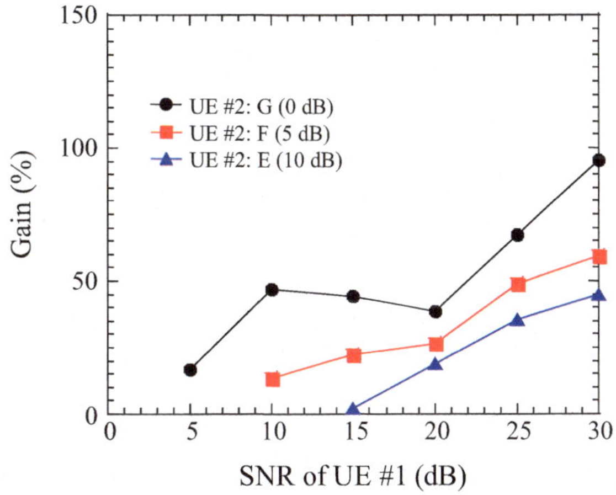

(b) NOMA total user throughput gains compared to OMA

because the user interference increases and as a result the required SNR to maintain the same throughput as that for OMA increases.

Figure 18.11a shows the throughput of UE #1 versus received SNR of UE #1, and Fig. 18.11b shows the total user throughput gain of NOMA over OMA versus received SNR of UE #1. When OMA is applied, the total bandwidth of 20 MHz is split among 2 UEs, 10 MHz per UE, and the MCS, and transmit rank that maximizes the throughput of each UE at each measurement point is applied. For the case of NOMA, 20 MHz bandwidth is applied, and the MCS and transmit rank combination are adjusted to keep UE #2 throughput about the same as that of UE #2 in case of OMA.

Fig. 18.12 NOMA performance results in outdoor experiments

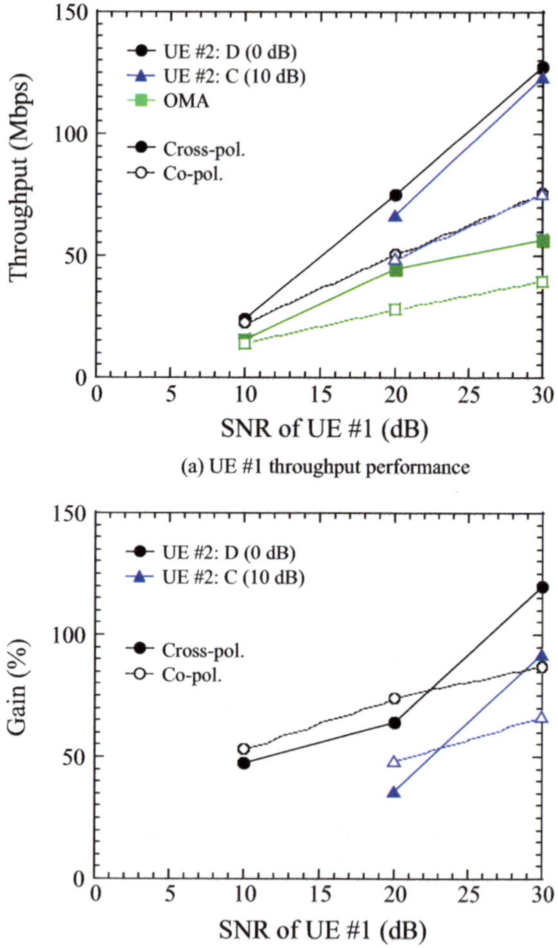

(a) UE #1 throughput performance

(b) NOMA total user throughput gains compared to OMA

Table 18.2 Combination of transmit power ratio and transmit rank (Indoor)

Point		$P1{:}P2$	$R1{:}R2$
UE #1	UE #2		
A–F	G (0 dB)	0.4:0.6	2:2
A–E	F (5 dB)	0.3:0.7	2:2
A–D	E (10 dB)	0.2:0.8	2:1

Table 18.3 Combination of transmit power ratio and transmit rank (Outdoor)

Point		$P1$:$P2$		$R1$:$R2$	
UE # 1	UE # 2	Cross-pol.	Co-pol.	Cross-pol.	Co-pol.
A–C	D (0 dB)	0.4:0.6	0.4:0.6	2:1	1:1
A, B	C (10 dB)	0.3:0.7	0.35:0.65	2:1	1:1

As shown in Fig. 18.11b, user throughput improves with NOMA over OMA for all measurement points. In particular, as the SNR of UE #1 increases, NOMA gains in terms of user throughput increases. Especially, when the SNR difference between the two UEs is larger, the performance gain increases. When UE #1 is located in point A, and UE #2 is located in point G (SNR difference of 30 dB), 95% gain in terms of total user throughput can be achieved compared to OMA.

In case of NOMA, the total transmit power is properly split among UE #1 and UE #2, while in case of OMA, both UEs transmit using maximum transmit power. As a result, for OMA UE #1 throughput achieves almost its maximum and cannot increase any further since it reaches maximum MCS. Therefore, we observed that when SNR becomes large the throughput of OMA (bandwidth sharing) is limited by the maximum MCS, while the throughput of NOMA is not limited by maximum MCS because of lowering SNR due to power sharing between the 2 UEs.

18.7.2 Outdoor Experiments

For outdoor experiments, Table 18.3 shows the combination of transmit power ratio and transmit rank for each UE at each measurement point. Figure 18.12a shows the throughput of UE #1 versus received SNR of UE #1, and Fig. 18.12b shows total user throughput gain of NOMA over OMA versus received SNR of UE #1.

In case of outdoor experiments, we adopted two antenna configurations, co-polarized and cross-polarized antennas. Other parameters, such as MCS, transmit rank, are determined in the same manner as in the indoor experiments.

In Table 18.3, irrespective of the antenna configuration used, as the SNR of UE #2 increases, P_2 increases. This result is similar as that for indoor experiments. In case of outdoor trials shown in Fig. 18.12b, the same tendency as Fig. 18.12a can be observed.

For the case of co-polarized antenna configuration, the transmit rank of the 2 UEs is 1. This is because the increase of fading correlation makes the transmission using higher transmit rank difficult. As a result, as the SNR improves the UE #1 throughput improves, however it easily reaches maximum MCS. As shown in Fig. 18.8b, when UE #1 is located in point A, and UE #2 is located in point D (SNR difference of 30 dB), 87% of throughput gain can be achieved compared to OMA. When cross-polarized antenna configuration is used, similar to indoor experiments the performance gain of NOMA increases as the SNR increases. Moreover, when UE #1 is located at point A, and UE #2 is located at point D, 120% of gain in terms of total user throughput can be achieved compared to OMA.

This is similar to the indoor experiments, when SNR becomes high, the throughput of OMA (bandwidth sharing) is limited because UE #1 throughput achieves almost its maximum and cannot increase any further since the MCS used reaches its maximum, while the throughput of NOMA is not limited by maximum MCS because of lowering SNR due to power sharing between the 2 UEs.

Based on these indoor and outdoor experiments, we show that NOMA provides performance gains compared to OMA especially when SNR difference between paired UEs is large. These gains become higher when the SNR becomes higher and higher transmit rank is used.

18.8 Conclusion

This chapter presents experimental trials of downlink NOMA where multiple users use overlapped resources but with different transmit powers. To further assess the realistic performance gains of downlink NOMA, we investigated its link-level simulations for different types of receivers, and in experimental trials conducted in both indoor and outdoor using open-loop 2×2 SU-MIMO. The results confirmed that NOMA can provide gains when the SNR difference between users is large. The results obtained from experimental trials confirmed that the system throughput gain improves significantly up to 120% when the channel gain difference between the users is large. To further improve the performance of downlink NOMA, it is important to enable smaller power ratio allocation among users. This would require more advanced interference cancellation, which may also require novel designs for the radio interface.

NOMA was studied and specified as multi-user superposition transmission (MUST) in 3GPP during LTE Release 13 and 14, respectively [13, 14]. The necessary mechanisms to enable LTE/LTE-Advanced support of MUST were specified. Up to 2 Tx common reference signal (CRS)-based transmission schemes, and up to 8 Tx DMRS (demodulation reference signal)-based transmission schemes, were considered. As an advanced receiver, only R-ML was assumed given that more advanced receivers such as CL-SIC cannot be supported with the design constraints of LTE radio interface. The RAN1 agreements and specifications of Release 14 WI are summarized in [15].

In the next step, uplink NOMA is of great interest in particular for IoT with massive number of devices [5, 7].

References

1. D. Tse, P. Viswanath, *Fundamentals of Wireless Communication* (Cambridge University Press, 2005)
2. Y. Kishiyama, A. Benjebbour, H. Ishii, T. Nakamura, Evolution concept and candidate technologies for future steps of LTE-A, in *IEEE ICCS* vol. 2012 (2012), pp. 21–23
3. A. Benjebbour, Y. Saito, Y. Kishiyama, A. Li, A. Harada, T. Nakamura, Concept and practical considerations of non-orthogonal multiple access (NOMA) for future radio access, in *IEEE ISPACS* vol. 2013 (2013), pp. 12–15
4. A. Benjebbour, A. Li, Y. Saito, Y. Kishiyama, A. Harada, T. Nakamura, System-level performance of downlink NOMA for future LTE enhancements, in *IEEE Globecom* vol. 2013 (2013), pp. 9–13
5. K. Higuchi, A. Benjebbour, Non-orthogonal multiple access (NOMA) with successive interference cancellation for future radio access. IEICE Trans. Commun. **E98-B**(3), 403–414 (2015)
6. A. Benjebbour, A. Li, Y. Kishiyama, H. Jiang, T. Nakamura, System-level performance of downlink NOMA combined with SU-MIMO for future LTE enhancements, in *IEEE Globecom*, vol. 2014 (2014), pp. 8–12
7. A. Li, A. Benjebbour, X. Chen, H. Jiang, H. Kayama, Uplink non-orthogonal multiple access (NOMA) with single-carrier frequency division multiple access (SC-FDMA) for 5G Systems. IEICE Trans. Commun. **E98-B**(8), 1426–1435 (2015)
8. K. Saito, A. Benjebbour, Y. Kishiyama, Y. Okumura, T. Nakamura, Performance and design of SIC receiver for downlink NOMA with open-loop SU-MIMO, in *IEEE ICC*, vol. 2015 (2015), pp. 8–12
9. C. Yan, A. Harada, A. Benjebbour, Y. Lan, A. Li, H. Jiang, Receiver design for downlink non-orthogonal multiple access (NOMA), in *IEEE VTC Spring*, vol. 2015 (2015), pp. 11–14
10. A. Benjebbour, A. Li, K. Saito, Y. Kishiyama, T. Nakamura, Downlink non-orthogonal multiple access (NOMA) combined with single user MIMO (SU-MIMO). IEICE Trans. Commun. **E98-B**(8), 1415–1425 (2015)
11. Y. Sano, K. Takeda, S. Nagata, T. Nakamura, X. Chen, A. Li, X. Zhang, H. Jiang, K. Fukawa, Investigation on non-orthogonal multiple access with reduced complexity maximum likelihood receiver and dynamic resource allocation. IEICE Trans. Commun. **E100-B**(8), 1301–1311 (2017)
12. 3GPP, evolved universal terrestrial radio access (E-UTRA); Physical layer procedures (Release 14). 3GPP TS 36.213, V14.1.0, January 2017
13. 3GPP, New SI proposal: study on downlink multiuser superposition transmission for LTE. 3GPP RP-150496, March 2015
14. 3GPP, Study on downlink multiuser superposition transmission (MUST) for LTE (Release 13). 3GPP TR 36.859, V13.0.0, December 2015
15. 3GPP, Summary of RAN1 agreements for Rel-14 DL MUST. 3GPP R1-1613802, November 2016

Chapter 19
Non-Orthogonal Multiple Access in LiFi Networks

Liang Yin and Harald Haas

19.1 A Brief Introduction to Visible Light Communication

Nowadays, smartphones, tablets, and smart wearables along with their data-hungry applications, such as cloud computing and augmented/virtual reality (AR/VR), are becoming more and more popular. According to Cisco's recent report, global wireless data traffic has expanded by 63% in 2016, reaching 7.2 exabytes per month [1]. Since the radio spectrum below 10 GHz is insufficient to meet such increasing demand, the wireless communications' industry has responded to this challenge by exploiting higher frequency bands. For the spectrum above 10 GHz, two main frequency bands have attracted great research interests during the last decade, including millimeter-wave communication [2] (30–300 GHz) and visible light communication (VLC) [3, 4] (430–770 THz). The rapid increase in the deployment of light-emitting diodes (LEDs) in indoor environments, such as homes, offices, and public spaces, provides a tremendous opportunity for VLC in that it can build on existing lighting infrastructures. Compared to radio frequency (RF) technology, VLC by nature exhibits a higher level of information security because light does not penetrate through walls. Moreover, the absence of electromagnetic interference to existing RF systems makes VLC particularly promising in electromagnetic sensitive areas such as aircraft cabins, hospitals, and gas stations. Extensive studies conducted during the past decade have also led to a recent standardization of VLC by the IEEE Computer Society [5] and the formation of IEEE 802.11 Light Communications Study Group [6] in 2017. A recent survey covering the latest research topics on LiFi/VLC can be found in [7].

L. Yin · H. Haas (✉)
School of Engineering, Li-Fi Research and Development Centre, Institute for Digital Communications, The University of Edinburgh, Edinburgh EH9 3JL, UK
e-mail: h.haas@ed.ac.uk

L. Yin
e-mail: l.yin@ed.ac.uk

M. Vaezi et al. (eds.), *Multiple Access Techniques for 5G Wireless Networks and Beyond*, https://doi.org/10.1007/978-3-319-92090-0_19

In VLC, data is typically modulated onto the instantaneous intensity of the emitted light, which is referred to as intensity modulation (IM) [8].[1] At the receiver side, a photodiode (PD) or an image sensor can be used to detect the varying intensities of the light, based on the principle of direct detection (DD) [8]. Therefore, the transmitted signal has to satisfy the real and nonnegative constraint, which limits the direct application of the well-researched and developed modulation schemes from the field of RF communications. Simple modulation schemes such as on–off keying (OOK), pulse amplitude modulation (PAM), pulse width modulation (PWM), and pulse position modulation (PPM) can be implemented in VLC in a straightforward manner to achieve Mbps transmission speeds. However, to achieve Gbps transmission speed, more advanced modulation schemes, for example, orthogonal frequency division multiplexing (OFDM) [12], are required. Conventional OFDM signals are both complex and bipolar, which cannot be directly implemented in VLC to drive the LED [13, 14]. Therefore, the standard RF OFDM scheme needs to be modified in order to be compatible with IM/DD. A straightforward way to obtain real-valued and unipolar OFDM signals is to impose the Hermitian symmetry constraint on the modulated subcarriers in the frequency domain and add a positive direct current (DC) bias in the time domain. The resulting modulation technique is commonly known as direct current-biased optical OFDM (DCO-OFDM) [15]. Apart from DCO-OFDM, there are many other solutions to obtain real and unipolar OFDM signals, such as asymmetrically clipped optical OFDM (ACO-OFDM) [16], unipolar OFDM (U-OFDM) [17], to name just a few.

VLC is subject to a different channel model when compared to RF communications. According to Friis equation [18], the pathloss of free-space transmission is proportional to the square of the frequency. Therefore, for the same transmission distance, VLC in general has much higher pathloss than RF communications, and this is typically compensated for by its smaller cell size [19]. Also, since the wavelength of visible light is hundreds of nanometers and the detection area of a typical PD is millions of square wavelengths, this spatial diversity essentially averages out the "multipath fading" effect in VLC. In addition, depending on the geometry and configuration of the communication link, the VLC channel is dominated by either line-of-sight (LOS) or non-line-of-sight (NLOS) components in the event that LOS is blocked by opaque objects. The LOS channel typically has a narrow power delay profile (PDP), allowing signals to be transmitted over a large bandwidth [20]. With the existence of the LOS component, the VLC channel can be generally assumed to be a flat fading channel across its 3-dB bandwidth. The NLOS channel, on the other hand, has a relatively large root mean square (RMS) delay spread, which significantly limits the usable modulation bandwidth and achievable data rate for VLC [20]. However, compared to LOS transmission, NLOS transmission is more robust to link blockages and user movement.

---

[1] Apart from IM, VLC also provides a number of unique modulation formats that are not applicable to RF, for example, color shift keying (CSK) [5], metameric modulation (MM) [9], and color intensity modulation (CIM) [10]. However, those modulation techniques are not discussed in this chapter. Interested readers can refer to [5, 9–11] and the references therein.

A general setup of the VLC link can be described as follows [3]. The information bits are first fed to the VLC transmitter, which consists of a digital signal processor (DSP), a digital-to-analog converter (DAC) and an LED. The DSP together with the DAC is responsible for modulating the information bits and transforming them into an analog signal, in the form of a continuous current, that is further used to drive the LED. After the LED, the information is carried by the intensity of the emitted light. An optical system after the transmitter is optional if extra signal gain or beam shaping is needed. For example, an optical collimator can be used to obtain parallel light beams while an optical diffuser can be used to broaden the light beams. The emitted light passes through the optical system, if there is any, and propagates through the wireless channel until it is collected by the PD. In front of the receiver, an optical system can also be implemented. For example, an optical concentrator can be used to efficiently collect and concentrate distant light beams while an optical filter can be used to remove the slow spectrum of the light so as to increase the usable modulation bandwidth and reduce the ambient noise. At the receiver, a PD can be an ordinary positive-intrinsic-negative (p-i-n) photodiode or an avalanche photodiode (APD). When the PD collects the incident light, it produces a photocurrent that is proportional to the instantaneous optical power of the received light. To enhance the received signal strength, a preamplifier in the form of a transimpedance amplifier is generally implemented, which further converts the current signal into a voltage signal. Finally, the amplified voltage pulses are passed through an analog-to-digital converter (ADC) for signal demodulation and detection.

19.2 System Model

19.2.1 Channel Model

As shown in Fig. 19.1, we consider a point-to-multipoint downlink transmission scenario, where an LED is located on the ceiling to simultaneously serve K random users that are uniformly distributed within a circular area. This model is an extension of the point-to-point VLC system to accommodate multiple users. In the following analysis, we assume that the users to be served are static or quasi-static so that their channel state information (CSI) is not outdated until the next channel estimation. The maximum cell radius is denoted by r_e, and the vertical distance from the LED to the receiving plane is denoted by L. In a polar coordinate system, the location of a random user k, $k = 1, 2, \ldots$, can be represented by (r_k, θ_k), where r_k represents its horizontal separation from the LED and θ_k represents its polar angle from the reference axis. Although a complete VLC channel consists of not only the LOS link but also the NLOS components caused by light reflections from interior surfaces, previous work has reported that in a typical indoor environment the strongest diffuse component is at least 7 dB (electrical) lower than the weakest LOS component [14]. For these reasons, only the LOS link is considered in the theoretical analysis. However, simulation

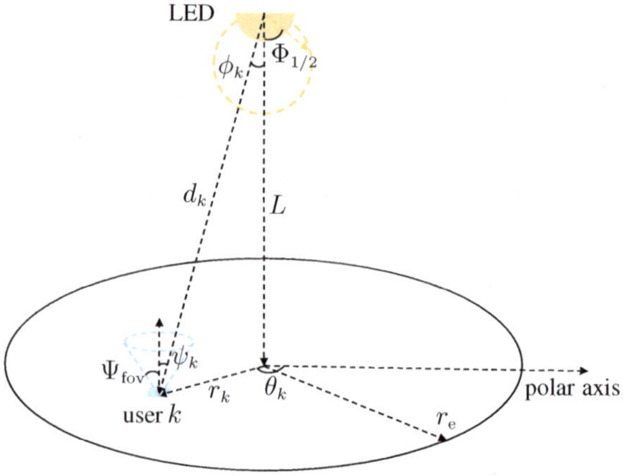

Fig. 19.1 Downlink geometry of a VLC link

results based on a complete VLC channel considering the wideband nature of VLC and the shadowing effect are presented in Sect. 19.5.2 for completeness. The LED is assumed to follow a Lambertian radiation pattern whose order is given by $m = -1/\log_2(\cos(\Phi_{1/2}))$, where $\Phi_{1/2}$ denotes the semi-angle of the LED. The PD at each user is assumed to be facing vertically upwards and its field of view (FOV) is denoted by Ψ_{FOV}. In the LOS link, the Euclidean distance between the LED and user k is denoted by d_k. The angle of irradiance and the angle of incidence are denoted by ϕ_k and ψ_k, respectively.

Without loss of generality, assume that all of the users are ordered based on their channel qualities:

$$h_1 \leq \cdots \leq h_k \leq \cdots \leq h_K, \tag{19.1}$$

where h_k denotes the direct current (DC) channel gain of the LOS link between the LED and the k-th user, given by [8]:

$$h_k = \frac{(m+1)AR_{\mathrm{PD}}}{2\pi d_k^2}\cos^m(\phi_k)g_{\mathrm{f}}(\psi_k)g_{\mathrm{c}}(\psi_k)\cos(\psi_k), \tag{19.2}$$

where A denotes the detection area of the PD; R_{PD} denotes the responsivity of the PD; $g_{\mathrm{f}}(\psi_k)$ represents the gain of the optical filter used at the receiver; and $g_{\mathrm{c}}(\psi_k)$ represents the gain of the optical concentrator, given by [8]:

$$g_{\mathrm{c}}(\psi_k) = \begin{cases} \frac{n_{\mathrm{c}}^2}{\sin^2(\Psi_{\mathrm{FOV}})}, & 0 \leq \psi_k \leq \Psi_{\mathrm{FOV}} \\ 0, & \psi_k > \Psi_{\mathrm{FOV}} \end{cases}, \tag{19.3}$$

where n_{c} is the reflective index of the optical concentrator used at the receiver front end, and it is defined as the ratio of the speed of light in vacuum and the phase velocity of light in the optical material. For visible light, the typical values for n_c are between 1 and 2.

19.2.2 Application of NOMA to LiFi

In principle, VLC also relies on electromagnetic radiation for information transmission. Therefore, existing modulation techniques that are developed for RF communication can also be applied to LiFi after necessary modifications. In this chapter, we discuss the use of DCO-OFDM in combination with NOMA for multiuser LiFi [21–24]. A simplified block diagram illustrating the transmission/reception principle is shown in Fig. 19.2. First, K parallel information streams are generated, and each information stream is sent to a modulator. According to the principle of OFDM, the information bits for each user are framed and mapped to complex symbols, $X_k(\ell)$, where $k = 1, \ldots, K$, based on the selected modulation scheme, e.g., quadrature amplitude modulation (QAM). Here, ℓ denotes the index of the subcarrier. Within each OFDM frame, the number of subcarriers that carry information is equal to $N_{\text{FFT}}/2 - 1$, where N_{FFT} is the size of the inverse fast Fourier transform (IFFT) and also the size of the fast Fourier transform (FFT). This spectral efficiency loss is because of the Hermitian symmetry constraint: ($X_k(0) = X_k(N_{\text{FFT}}/2) = 0$, $X_k(\ell) = X_k^*(N_{\text{FFT}} - \ell)$ for $\ell = 1, \ldots, N_{\text{FFT}}/2 - 1$) in order to make the time-domain signal real. In general, a cyclic prefix (CP) is included in the OFDM subcarriers to combat inter-symbol interference (ISI). However, in VLC the CP is shown to have a negligible impact on the power and bandwidth efficiencies [25]. Therefore, CP is omitted here for simplicity. The time-domain signal that carries information

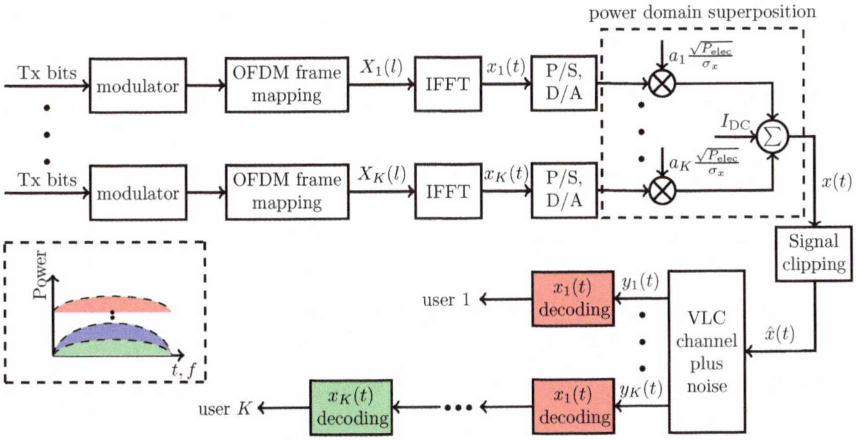

Fig. 19.2 A simplified block diagram of NOMA principle for LiFi

for user k is denoted by $x_k(t)$, and it is obtained by performing the IFFT operation on $X_k(\ell)$:

$$x_k(t) = \frac{1}{\sqrt{N_{\mathrm{FFT}}}} \sum_{\ell=0}^{N_{\mathrm{FFT}}-1} X_k(\ell) \exp\left(\frac{j2\pi \ell t}{N_{\mathrm{FFT}}}\right). \tag{19.4}$$

For a large number of subcarriers, e.g., $N_{\mathrm{FFT}} > 64$, the central limit theorem (CLT) states that $x_k(t)$ is Gaussian distributed with zero mean and a variance of σ_x^2, given by:

$$\sigma_x^2 = \mathbb{E}\left[|x_k(t)|^2\right] = \frac{N_{\mathrm{FFT}} - 2}{N_{\mathrm{FFT}}} \sigma_{\mathrm{QAM}}^2, \tag{19.5}$$

where σ_{QAM}^2 represents the variance of each QAM symbol. It can be seen from (19.5) that the value of σ_x^2 only depends on the size of the IFFT/FFT operation and the variance of QAM symbols and is independent of k.

After parallel-to-serial (P/S) conversion and digital-to-analog (D/A) conversion, the time-domain signals for all K users are superposed in the power domain: Signal $x_k(t)$ is multiplied by factor $a_k \sqrt{P_{\mathrm{elec}}}/\sigma_x$ and then summed up. With the addition of a positive DC bias I_{DC}, the overall combined signal, which consists of all the information for K users and the DC bias, can be written as:

$$x(t) = \sum_{k=1}^{K} a_k \frac{\sqrt{P_{\mathrm{elec}}}}{\sigma_x} x_k(t) + I_{\mathrm{DC}}, \tag{19.6}$$

where P_{elec}, equal to the variance of $x(t)$, represents the total available electrical power of all the message signals, and a_k represents the power allocation coefficient for user k. Because of the total power constraint, the power allocation coefficients need to satisfy:

$$\sum_{i=1}^{K} a_i^2 = 1. \tag{19.7}$$

According to the principle of NOMA, users with poorer channel conditions are allocated more signal power. This implies that $a_1 \geq \cdots \geq a_k \geq \cdots \geq a_K$.

19.2.2.1 Clipping Distortion

Note that the nonlinear characteristic of typical LEDs [26] requires the transmitted optical signal to be clipped at both the bottom and top levels, yielding the clipped signal $\hat{x}(t)$. This double-sided signal clipping is described by the following transfer function:

$$\hat{x}(t) = \begin{cases} I_{\min}, & x(t) + I_{\mathrm{DC}} \leq I_{\min} \\ x(t) + I_{\mathrm{DC}}, & I_{\min} < x(t) + I_{\mathrm{DC}} < I_{\max} \\ I_{\max}, & x + I_{\mathrm{DC}} \geq I_{\max} \end{cases}, \tag{19.8}$$

where $I_{\min}$ and $I_{\max}$ are the minimum and maximum current levels for a typical LED to operate in the linear region, respectively. According to the Bussgang theorem, the clipped signal can be modeled as [25]:

$$\hat{x}(t) = \alpha\,(x(t) + I_{\mathrm{DC}}) + z_{\mathrm{clip}}, \tag{19.9}$$

where $\alpha = Q(\rho_{\mathrm{b}}) - Q(\rho_{\mathrm{t}})$ is an attenuation factor, with $Q(\cdot)$ being the well-known Q-function; $\rho_{\mathrm{t}} = (I_{\max} - I_{\mathrm{DC}})/\sigma_x$; $\rho_{\mathrm{b}} = (I_{\min} - I_{\mathrm{DC}})/\sigma_x$; and z_{clip} is the time-domain clipping noise. The clipping of a time-domain OFDM signal modifies its mean and consequently its average optical power. Because the unclipped signal is Gaussian distributed, the clipped signal follows a truncated Gaussian distribution, whose mean is given by:

$$\mathbb{E}[\hat{x}(t)] = \sigma_x\,(\rho_{\mathrm{t}} Q(\rho_{\mathrm{t}}) - \rho_{\mathrm{b}} Q(\rho_{\mathrm{b}}) - f_{\mathrm{G}}(\rho_{\mathrm{t}}) + f_{\mathrm{G}}(\rho_{\mathrm{b}})) + I_{\min}, \tag{19.10}$$

where $f_{\mathrm{G}}(\rho) = 1/\sqrt{2\pi}\exp(-\rho^2/2)$ is the probability density function (PDF) of the standard Gaussian distribution. After the received signal being passed through the FFT circuit, z_{clip} is transformed into Gaussian noise with zero mean and variance σ^2_{clip}, given by [25]:

$$\begin{aligned}\sigma^2_{\mathrm{clip}} =& \sigma_x^2\Big(\alpha - \alpha^2 + \rho_{\mathrm{b}}^2(1 - Q(\rho_{\mathrm{b}})) + \rho_{\mathrm{t}}^2 Q(\rho_{\mathrm{t}}) + \rho_{\mathrm{b}} f_{\mathrm{G}}(\rho_{\mathrm{b}}) - \rho_{\mathrm{t}} f_{\mathrm{G}}(\rho_{\mathrm{t}}) \\ &+ (f_{\mathrm{G}}(\rho_{\mathrm{b}}) - f_{\mathrm{G}}(\rho_{\mathrm{t}}) + \rho_{\mathrm{b}}(1 - Q(\rho_{\mathrm{b}})) + \rho_{\mathrm{t}} Q(\rho_{\mathrm{t}}))^2\Big).\end{aligned} \tag{19.11}$$

The distortion of the clipping noise on the information signal can be measured by the signal-to-clipping-noise ratio (SCR), defined as:

$$\mathsf{SCR} = \frac{\alpha^2\sigma_x^2}{\sigma^2_{\mathrm{clip}}}. \tag{19.12}$$

It can be seen from (19.11) that smaller values of ρ_{b} and ρ_{t} give a lower clipping noise, that can be achieved by reducing the variance of the transmit signal. To facilitate quantifying the effect of clipping noise, we define the following variable

$$\xi = \min\left\{\frac{I_{\mathrm{DC}} - I_{\min}}{\sigma_x}, \frac{I_{\max} - I_{\mathrm{DC}}}{\sigma_x}\right\} \tag{19.13}$$

as a measure of the DC bias level relative to the standard deviation of the transmit signal. Since the clipping signal $\hat{x}(t)$ follows a truncated Gaussian distribution, setting $\xi \geq 3$ ensures that at least 99.7% of signal $x(t)$ remains unclipped, according to the three-sigma rule of thumb. This is equivalent to setting $\sigma_x \leq \bar{\sigma}_x$, where the threshold $\bar{\sigma}_x$ is given by:

$$\bar{\sigma}_x = \max\left\{\frac{I_{\mathrm{DC}} - I_{\min}}{3}, \frac{I_{\max} - I_{\mathrm{DC}}}{3}\right\}. \tag{19.14}$$

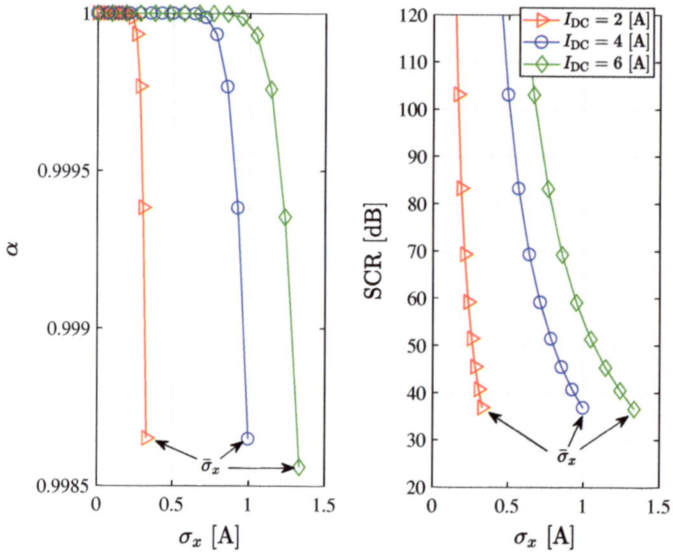

Fig. 19.3 Attenuation factor α and the SCR as a function of the standard deviation of the information signal

As an example, we plot in Fig. 19.3 the attenuation factor α and the SCR as a function of the standard deviation of the information signal under different DC bias levels. The bottom and upper clipping levels assumed in Fig. 19.3 are $I_{\min} = 1$ A and $I_{\min} = 10$ A, respectively. Note that the exact values for the bottom and upper clipping levels and the DC current are adjustable, since a high-power LED can be implemented with an array of low-power LEDs. Results confirm that for $\sigma_x \leq \bar{\sigma}_x$, the clipping noise is negligible since $\alpha \approx 1$ and $\mathsf{SCR} > 35$ dB. Therefore, the clipping noise is neglected in the following analysis.

19.2.2.2 Receiver Noise

After propagating though the VLC channel, the transmitted VLC signal is subject to random noise processes before it is detected by the PD at each receiver. Note that inside the receiver circuit, the dominant noise sources are the thermal noise and shot noise, and both noise processes can be modeled as additive white Gaussian noise (AWGN) [27] with a total variance σ^2_{noise}, given by:

$$\sigma^2_{\text{noise}} = \sigma^2_{\text{shot}} + \sigma^2_{\text{thermal}}, \tag{19.15}$$

where σ^2_{shot} and $\sigma^2_{\text{thermal}}$ are the variance of shot noise and thermal noise, respectively. Although the shot noise originates from both information signals and ambient light, the contribution of signal-induced shot noise is comparatively small and hence can

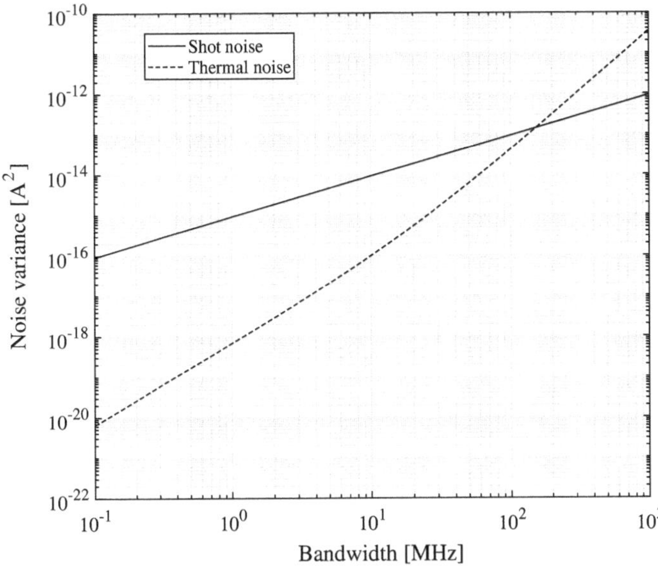

Fig. 19.4 Variance of noise as a function of the desired bandwidth. The parameters used are [27]: $I_{\text{bg}} = 5100\ \mu\text{A}$; $G = 10$; $C_{\text{PD}} = 112\ \text{pF/cm}^2$; $A = 1\ \text{cm}^2$; $I_{\text{FET}} = 1.5$; and $g_{\text{FET}} = 30\ \text{mS}$

be neglected. The variance of the ambient induced shot noise is [27]:

$$\sigma^2_{\text{shot}} = 2q_{\text{e}} I_{\text{bg}} I_2 B, \tag{19.16}$$

where q_{e} is the elementary charge; I_{bg} is the background current; $I_2 = 0.562$ is the noise bandwidth factor; and B is the desired modulation bandwidth. The thermal noise, on the other hand, is mainly caused by the preamplifier circuits at the receiver. Among preamplifier designs, the transimpedance type is generally used because of its large dynamic range and wide bandwidth. In line with previous works [27], we also assume the use of a p-intrinsic-n (p-i-n) field-effect transistor (FET) as the transimpedance amplifier. The variance of the thermal noise is given by [27]:

$$\sigma^2_{\text{thermal}} = \frac{8\pi k_{\text{B}} T_{\text{K}}}{G} C_{\text{PD}} A I_2 B^2 + \frac{16\pi^2 k_{\text{B}} T_{\text{K}} I_{\text{FET}}}{g_{\text{FET}}} C^2_{\text{PD}} A^2 I_3 B^3, \tag{19.17}$$

where k_{B} is the Boltzmann constant; T_{K} is the absolute temperature; G is the open-loop voltage gain; C_{PD} is the fixed capacitance of PD per unit area; I_{FET} is the FET channel-noise factor; g_{FET} is the transconductance of the FET; and $I_3 = 0.0868$ is another noise-bandwidth factor.

We compare the contribution of shot noise and thermal noise as a function of the bandwidth in Fig. 19.4. For the parameters considered there, it can be seen that shot noise is the dominant noise source when the bandwidth is smaller than 100 MHz. If a higher modulation bandwidth is used, thermal noise becomes the dominant noise source. As a result, for typical LEDs whose bandwidth is $10 \sim 20$ MHz, the impact of thermal noise can be neglected, and the noise power spectral density (PSD) can be assumed to be constant: $N_0 \approx N_{\text{shot}} = 2q_e I_{\text{bg}} I_2$.

19.2.2.3 Achievable Rate

After removing the DC bias, the received signal at user k can be written as:

$$y_k(t) = \frac{\sqrt{P_{\text{elec}}}}{\sigma_x} h_k \left(\underbrace{\sum_{j=1}^{k-1} a_j s_j(t)}_{\text{SIC}} + \underbrace{a_k s_k(t)}_{\text{signal}} + \underbrace{\sum_{i=k+1}^{K} a_i s_i(t)}_{\text{interference}} \right) + z_k(t), \tag{19.18}$$

where z_k denotes the AWGN whose PSD is N_0. According to the principle of NOMA, successive interference cancellation (SIC) is first carried out at all users, except user 1, to remove the message signal for other users with poorer channel conditions (the "SIC" term in (19.18)). The message signal for users whose channel gains are stronger than that of user k is treated as noise (the "interference" term in (19.18)). As a result, after optical to electrical (O/E) conversion, the achievable data rate per bandwidth, i.e., spectral efficiency, for user k is given by:

$$R_k = \begin{cases} \frac{1}{2}\log_2\left(1 + \frac{(h_k a_k)^2}{\sum_{i=k+1}^{K}(h_k a_i)^2 + \frac{1}{\rho}}\right), & k = 1, \ldots, K-1 \\ \frac{1}{2}\log_2\left(1 + \rho (h_k a_k)^2\right), & k = K \end{cases}, \tag{19.19}$$

where $\rho = P_{\text{elec}}/N_0 B$ represents the transmit SNR, and the scaling factor $1/2$ is due to the Hermitian symmetry constraint of DCO-OFDM. Note that (19.19) is conditioned on the event that user k can successfully detect the message for user j, for $\forall j \leq k$. Denote $R_{k \to j}$ as the rate for user k to detect the message intended for user j, and denote $\tilde{R}_j$ as the target data rate for successful message detection at user j. The above condition can be expressed mathematically as:

$$R_{k \to j} = \begin{cases} \frac{1}{2}\log_2\left(1 + \frac{(h_k a_j)^2}{\sum_{i=j+1}^{K}(h_k a_i)^2 + \frac{1}{\rho}}\right) \geq \tilde{R}_j, & j \leq k, j \neq K \\ \frac{1}{2}\log_2\left(1 + \rho (h_k a_j)^2\right) \geq \tilde{R}_j, & j = k = K \end{cases}. \tag{19.20}$$

If (19.20) is satisfied, we assume that perfect SIC can be performed in the decoding chain without signal error propagations.

19.3 Performance Evaluation

In this section, we study the achievable rate of users in a LiFi network based on the NOMA system model presented in Sect. 19.2. The derived results also lay the foundation for studying the impact of user pairing in the next section.

19.3.1 Distribution Functions of the Channel Gain

By substituting $d_k = \sqrt{r_k^2 + L^2}$, $\cos(\phi_k) = L/\sqrt{r_k^2 + L^2}$ and $\cos(\psi_k)=L/\sqrt{r_k^2 + L^2}$ into (19.2), the LOS channel gain can be expressed as:

$$h_k = \frac{C(m+1)L^{m+1}}{(r_k^2 + L^2)^{\frac{m+3}{2}}}, \tag{19.21}$$

where C is give by:

$$C = \frac{1}{2\pi} A R_{\mathrm{PD}} g_{\mathrm{f}}(\psi_k) g_{\mathrm{c}}(\psi_k). \tag{19.22}$$

Define a function $h = u(r) = C(m+1)L^{m+1}(r^2 + L^2)^{-\frac{m+3}{2}}$. It is evident that h is a monotonic decreasing function with respect to r. Therefore, the probability density function (PDF) of the unordered channel gain can be calculated using the "change of variable" method:

$$f_{h_k}(h) = \left| \frac{\partial}{\partial h} u^{-1}(h) \right| \cdot f_{r_k}\left(u^{-1}(h)\right), \tag{19.23}$$

where u^{-1} denotes the inverse function of u, and $f_{r_k}(r) = 2r/r_{\mathrm{e}}^2$ is the PDF of variable r_k following the uniform distribution. Consequently, the PDF of the unordered variable h_k^2 can be obtained as:

$$f_{h_k^2}(t) = \frac{1}{r_{\mathrm{e}}^2} \frac{1}{m+3} \left(C(m+1)L^{m+1}\right)^{\frac{2}{m+3}} t^{-\frac{1}{m+3}-1}, \tag{19.24}$$

for $t \in [\kappa_{\min}, \kappa_{\max}]$, where $\kappa_{\min}$ and $\kappa_{\max}$ are given as $\kappa_{\min} = \left(C(m+1)L^{m+1}\right)^2 / (r_{\mathrm{e}}^2+L^2)^{m+3}$ and $\kappa_{\max} = \left(C(m+1)L^{m+1}\right)^2 / L^{2(m+3)}$. Integrating (19.24) over $[\kappa_{\min}, \kappa_{\max}]$, the cumulative distribution function (CDF) of the unordered variable h_k^2 can therefore be obtained as:

$$F_{h_k^2}(t) = -\frac{1}{r_{\mathrm{e}}^2} \left(C(m+1)L^{m+1}\right)^{\frac{2}{m+3}} t^{-\frac{1}{m+3}} + \frac{L^2}{r_{\mathrm{e}}^2} + 1. \tag{19.25}$$

Using order statistics [28], the PDF of the ordered variable h_k^2, denoted by $f'_{h_k^2}(t)$, can be readily obtained as:

$$f'_{h_k^2}(t) = \frac{K!}{(k-1)!\,(K-k)!} F_{h_k^2}(t)^{k-1} \left[1 - F_{h_k^2}(t)\right]^{K-k} f_{h_k^2}(t)$$

$$= \frac{\Omega}{m+3} \frac{K!}{(k-1)!\,(K-k)!} \left(-\Omega t^{-\frac{1}{m+3}} + \frac{L^2}{r_{\mathrm{e}}^2} + 1\right)^{k-1} \left(\Omega t^{-\frac{1}{m+3}} - \frac{L^2}{r_{\mathrm{e}}^2}\right)^{K-k} t^{-\frac{1}{m+3}-1}, \tag{19.26}$$

in which $\Omega = \frac{1}{r_{\mathrm{e}}^2}\left(C\,(m+1)\,L^{m+1}\right)^{\frac{2}{m+3}}$. Note that (19.26) is obtained assuming that the total number of users, K, is fixed and known. If K is unknown but assumed to follow a certain point process, for example, the Poisson point process (PPP), then (19.26) becomes a conditional PDF on K. The final result should be obtained by further averaging the conditional PDF with respect to different realizations of K. For non-uniform user distribution, the results can be obtained in a similar way by plugging the specific distribution function into (19.23).

19.3.2 Case 1: Guaranteed Quality of Service

Consider the case that each user has a target data rate, which is determined by its quality-of-service (QoS) requirement. Service satisfaction at each user requires successful detection of messages not only for this user itself but also for other users with poorer channel conditions. If this constraint is met, the sum rate of the system is simply $\sum_{k=1}^{K} \tilde{R}_k$. Therefore, the sum rate is not of interest in this case. Instead, the analysis is focused on the outage probability at each user. Based on (19.20), the outage probability at the k-th user can be expressed as:

$$\begin{aligned} \mathrm{P}_k^{\mathrm{out}} &= 1 - \mathbb{P}\left[R_{k\to j} \geq \tilde{R}_j,\ \forall j \leq k\right] \\ &= 1 - \mathbb{P}\left[h_k^2 \geq \varepsilon_j,\ \forall j \leq k\right], \end{aligned} \tag{19.27}$$

where $\mathbb{P}[\cdot]$ denotes the probability of an event, and the threshold ε_j is given by:

$$\varepsilon_j = \begin{cases} \frac{\beta_j}{\rho\left(a_j^2 - \beta_j \sum_{i=j+1}^{K} a_i^2\right)}, & j \neq K \\ \frac{\beta_j}{\rho a_K^2}, & j = K \end{cases}, \tag{19.28}$$

where $\beta_j = 2^{\tilde{R}_j} - 1$ denotes the required signal-to-noise-plus-interference ratio (SINR) at the j-th user for successful message detection. Note that (19.27) is obtained based on the following condition:

$$a_j^2 > \beta_j \sum_{i=j+1}^{K} a_i^2. \tag{19.29}$$

If power allocation coefficients do not satisfy (19.29), user outage probability would always be one. Define a new threshold $\varepsilon_k^* = \min\{\max\{\varepsilon_1, \ldots, \varepsilon_k, \kappa_{\min}\}, \kappa_{\max}\}$. Using order statistics [28], the outage probability of user k can be obtained as:

$$\begin{aligned} \mathrm{P}_k^{\mathrm{out}} =& 1 - \mathbb{P}\left[h_k^2 \geq \varepsilon_k^*\right] \\ =& \sum_{i=k}^{K} \frac{K!}{i!\,(K-i)!} F_{h_k^2}(\varepsilon_k^*)^i \left[1 - F_{h_k^2}(\varepsilon_k^*)\right]^{K-i}. \end{aligned} \tag{19.30}$$

System coverage probability is defined as the probability that all of the users in the system can achieve reliable detection, which is given by:

$$\mathrm{P}^{\mathrm{cov}} = \prod_{k=1}^{K} (1 - \mathrm{P}_k^{\mathrm{out}}), \tag{19.31}$$

given that the outage event at each user is independent.

19.3.3 Case 2: Opportunistic Best-Effort Service

Consider the case where data rates for different users are opportunistically allocated based on their channel conditions, i.e., $\tilde{R}_j = R_j$. In this scenario, it can be readily verified that condition (19.20) always holds, and all of the users can be served with zero outage probability but with different data rates, depending on their SINR values. The following theorem gives a closed-form expression for the achievable sum rate of the system.

Theorem 1 *For arbitrary power allocation strategies, the ergodic sum rate of NOMA with uniformly distributed users is given by:*

$$
\begin{aligned}
R^{\mathrm{NOMA}} =& \frac{\Omega}{m+3} \sum_{k=1}^{K-1} \sum_{p=0}^{k-1} \sum_{q=0}^{K-k} \left\{ \frac{K!}{p!\,(k-1-p)!\,q!\,(K-k-q)!} \left(\frac{L^2}{r_{\mathrm{e}}^2} + 1 \right)^{k-1-p} \right. \\
& \times \left. \left(\frac{L^2}{r_{\mathrm{e}}^2} \right)^{K-k-q} (-1)^{p+K-k-q} \Omega^{p+q} \left(\varpi_1(\kappa_{\max}) - \varpi_1(\kappa_{\min}) \right) \right\} \\
& + \frac{\Omega}{m+3} \sum_{l=0}^{K-1} \left\{ \frac{K!}{l!\,(K-1-l)!} \left(\frac{L^2}{r_{\mathrm{e}}^2} + 1 \right)^{K-1-l} (-\Omega)^l \right. \\
& \times \left. \left(\varpi_2(\kappa_{\max}) - \varpi_2(\kappa_{\min}) \right) \right\},
\end{aligned} \tag{19.32}
$$

where $\varpi_1(\kappa)$ is defined as:

$$
\begin{aligned}
\varpi_1(\kappa) =& \frac{\kappa^{-\frac{p+q+1}{m+3}}}{2\left(\frac{p+q+1}{m+3}\right)^2 \ln(2)} \left\{ -\frac{p+q+1}{m+3} \ln\left(1 + \frac{(\tau_k - \tau_{k+1})\kappa}{\tau_{k+1}\kappa + \frac{1}{\rho}} \right) \right. \\
& - {}_2F_1\left(1, -\frac{p+q+1}{m+3}; -\frac{p+q+1}{m+3} + 1; -\rho\tau_{k+1}\kappa \right) \\
& \left. + {}_2F_1\left(1, -\frac{p+q+1}{m+3}; -\frac{p+q+1}{m+3} + 1; -\rho\tau_k\kappa \right) \right\},
\end{aligned} \tag{19.33}
$$

in which $\tau_k = \sum_{i=k}^{K} a_i^2$, $\tau_{k+1} = \sum_{i=k+1}^{K} a_i^2$ and ${}_2F_1$ denotes the Gauss hypergeometric function. $\varpi_2(\kappa)$ is defined as:

$$
\begin{aligned}
\varpi_2(\kappa) =& \frac{\kappa^{-\frac{l+1}{m+3}}}{2\left(\frac{l+1}{m+3}\right)^2 \ln(2)} \left\{ -1 - \frac{l+1}{m+3} \ln\left(1 + \rho a_K^2 \kappa\right) \right. \\
& \left. + {}_2F_1\left(1, -\frac{l+1}{m+3}; -\frac{l+1}{m+3} + 1; -\rho a_K^2 \kappa \right) \right\}.
\end{aligned} \tag{19.34}
$$

Proof The ergodic sum rate of NOMA can be formulated as:

$$
\begin{aligned}
R^{\mathrm{sum}} =& \sum_{k=1}^{K-1} \underbrace{\int_{\kappa_{\min}}^{\kappa_{\max}} \frac{1}{2} \log_2 \left(1 + \frac{a_k^2 t}{t \sum\limits_{i=k+1}^{K} a_i^2 + \frac{1}{\rho}} \right) f'_{h_k^2}(t)\, dt}_{Q_k} \\
& + \underbrace{\int_{\kappa_{\min}}^{\kappa_{\max}} \frac{1}{2} \log_2\left(1 + \rho a_K^2 t\right) f'_{h_K^2}(t)\, dt}_{Q_K},
\end{aligned} \tag{19.35}
$$

where Q_k denotes the ergodic data rate for the user k, $k \in \{1, \ldots, K-1\}$, and Q_K denotes the ergodic data rate for the user K. Applying binomial expansion, Q_K can be written as:

$$Q_K = \frac{\Omega K}{2(m+3)} \sum_{l=0}^{K-1} \left\{ \frac{(K-1)!}{l!(K-1-l)!} (-\Omega)^l \left(\frac{L^2}{r_{\mathrm{e}}^2} + 1 \right)^{K-1-l} \times \int_{\kappa_{\min}}^{\kappa_{\max}} \log_2\left(1 + \rho a_K^2 t\right) t^{-\frac{l+1}{m+3}-1} \mathrm{d}t \right\}. \tag{19.36}$$

Integrating by parts, the integral in (19.36) simplifies to:

$$\int \log_2\left(1 + \rho a_K^2 t\right) t^{-\frac{l+1}{m+3}-1} \mathrm{d}t = -\frac{m+3}{l+1} \log_2\left(1 + \rho a_K^2 t\right) t^{-\frac{l+1}{m+3}} + \frac{1}{\ln(2)} \frac{m+3}{l+1} \int \frac{\rho a_K^2}{1 + \rho a_K^2 t} t^{-\frac{l+1}{m+3}} \mathrm{d}t \tag{19.37}$$

After applying the geometric series, we have:

$$\begin{aligned}
&\int \log_2\left(1 + \rho a_K^2 t\right) t^{-\frac{l+1}{m+3}-1} \mathrm{d}t \\
&= -\frac{m+3}{l+1} \log_2\left(1 + \rho a_K^2 t\right) t^{-\frac{l+1}{m+3}} + \frac{1}{\ln(2)} \frac{m+3}{l+1} \int \left(1 - \sum_{n=0}^{\infty} \left(-\rho a_K^2 t\right)^n\right) t^{-\frac{l+1}{m+3}-1} \mathrm{d}t \\
&= -\frac{m+3}{l+1} \log_2\left(1 + \rho a_K^2 t\right) t^{-\frac{l+1}{m+3}} - \frac{1}{\ln(2)} \frac{m+3}{l+1} \\
&\times \int \left\{ \left(-1 + \sum_{n_1=0}^{\infty} \frac{-\frac{l+1}{m+3}(-\rho a_K^2 t)^{n_1}}{-\frac{l+1}{m+3} + n_1}\right) + \left(\sum_{n_2=0}^{\infty} \frac{n_2(-\rho a_K^2 t)^{n_2}}{-\frac{l+1}{m+3} + n_2}\right) \right\} t^{-\frac{l+1}{m+3}-1} \mathrm{d}t
\end{aligned} \tag{19.38}$$

Integrating the function in (19.38), we can obtain:

$$\begin{aligned}
&\int \log_2\left(1 + \rho a_K^2 t\right) t^{-\frac{l+1}{m+3}-1} \mathrm{d}t \\
&= -\frac{m+3}{l+1} \log_2\left(1 + \rho a_K^2 t\right) t^{-\frac{l+1}{m+3}} \\
&+ \frac{1}{\ln(2)} \left(\frac{m+3}{l+1}\right)^2 \left(-1 + \sum_{n=0}^{\infty} \frac{-\frac{l+1}{m+3}\left(-\rho a_K^2 t\right)^n}{-\frac{l+1}{m+3} + n}\right) t^{-\frac{l+1}{m+3}} \\
&= -\frac{m+3}{l+1} \log_2\left(1 + \rho a_K^2 t\right) t^{-\frac{l+1}{m+3}} \\
&+ \frac{1}{\ln(2)} \left(\frac{m+3}{l+1}\right)^2 \left(-1 + \sum_{n=0}^{\infty} \frac{(1)_n \left(-\frac{l+1}{m+3}\right)_n}{\left(-\frac{l+1}{m+3} + 1\right)_n} \frac{\left(-\rho a_K^2 t\right)^n}{n!}\right) t^{-\frac{l+1}{m+3}},
\end{aligned} \tag{19.39}$$

where $(\cdot)_n$ represents the Pochhammer symbol. After replacing the power series in (19.39) with the Gaussian hypergeometric function ${}_2F_1$, the expression of Q_K is obtained as in (19.32). In a similar way, the expression of Q_k can be obtained by applying the quotient rule of the logarithmic function. ■

Theorem 1 demonstrates that unlike OMA, the capacity of NOMA can be enhanced with an increase in the total number of users in the system. However, this performance gain is achieved at the cost of the increased computation complexity caused by the SIC process at the receivers.

With the following corollary, the ergodic sum rate gain of NOMA over OFDMA can be obtained.

Corollary 1 *The ergodic sum rate of an OFDMA-based VLC system with uniformly deployed users is given by:*

$$\begin{aligned} R^{\mathrm{OFDMA}} =& \frac{\Omega}{m+3} \sum_{k=1}^{K} \sum_{p=0}^{k-1} \sum_{q=0}^{K-k} \left\{ \frac{K!\, b_k}{p!\,(k-1-p)!\,q!\,(K-k-q)!} \left(\frac{L^2}{r_{\mathrm{e}}^2} + 1 \right)^{k-1-p} \right. \\ & \times \left. \left(\frac{L^2}{r_{\mathrm{e}}^2} \right)^{K-k-q} (-1)^{p+K-k-q}\, \Omega^{p+q} \left(\varpi_3(\kappa_{\max}) - \varpi_3(\kappa_{\min}) \right) \right\}, \end{aligned} \tag{19.40}$$

where $\varpi_3(\kappa)$ is defined as:

$$\begin{aligned} \varpi_3(\kappa) =& \frac{\kappa^{-\frac{p+q+1}{m+3}}}{2\left(\frac{p+q+1}{m+3}\right)^2 \ln(2)} \left\{ -1 - \frac{p+q+1}{m+3} \ln\left(1 + \frac{\upsilon_k}{b_k}\rho\kappa\right) \right. \\ & \left. + {}_2F_1\left(1, -\frac{p+q+1}{m+3}; -\frac{p+q+1}{m+3} + 1; -\frac{\upsilon_k}{b_k}\rho\kappa\right) \right\}. \end{aligned} \tag{19.41}$$

Proof The ergodic sum rate achieved by OFDMA is calculated as:

$$R^{\mathrm{OFDMA}} = \sum_{k=1}^{K} \int_{\kappa_{\min}}^{\kappa_{\max}} \frac{1}{2} b_k \log_2\left(1 + \frac{\upsilon_k}{b_k}\rho t\right) f'_{h_k^2}(t)\mathrm{d}t, \tag{19.42}$$

where b_k is the fraction of bandwidth occupied by the k-th user, and υ_k is the fraction of the power allocated to the k-th user. The total bandwidth constraint requires that $\sum_{k=1}^{K} b_k = 1$, and the total power constraint requires that $\sum_{k=1}^{K} \upsilon_k = 1$. The derivation can follow similar steps as the derivation of R^{NOMA} shown in the proof of Theorem 1.

With Theorem 1 and Corollary 1, the sum rate gain of NOMA over OFDMA can be obtained.

19.4 Impact of User Pairing

Selecting a subset of users to perform NOMA can effectively reduce the computational complexity of the system. This results in a hybrid MA scheme which consists a combination of NOMA and OMA techniques. As the performance of such a hybrid system is highly dependent on the user selection strategy, in this section we focus on analyzing the effect of user pairing on the system performance. In order to obtain simple but insightful results, the entity of users is divided into groups and each group consists of two users. However, it can be readily extended to the case where an arbitrary number of users is selected and paired to perform NOMA. From a qualitative point of view, the capacity region for NOMA and OFDMA in a two-user scenario is illustrated in Fig. 19.5. It can be seen that a higher performance gain can be obtained if two users with more distinctive channel conditions are paired to perform NOMA. This finding will be quantitatively validated through theoretical analysis in this section.

Assume that the user i and user j $(i \leq j)$ in the system are selected to perform NOMA so that $a_i^2 + a_j^2 = 1$. According to OFDMA, each user is allocated a fraction of subcarriers. Therefore, the achieved data rate for each user is $\bar{R}_k = \frac{1}{2} b_k \log_2 \left(1 + \frac{\upsilon_k}{b_k} \rho h_k^2\right)$, where $k = \{i, j\}$.

19.4.1 Impact of User Pairing on Individual Rates

In this subsection, the impact of user pairing on individual data rate is studied. For arbitrary power allocation strategies, the probability that both users can achieve higher individual rates in NOMA than in OFDMA is given by the following theorem.

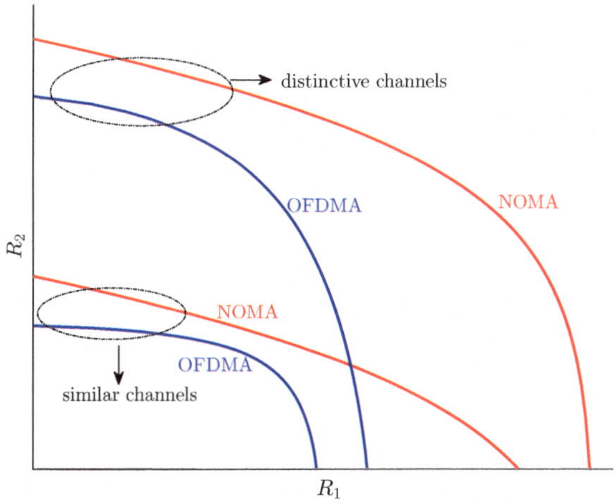

Fig. 19.5 Capacity region for NOMA and OFDMA $(\upsilon_1 = \upsilon_2)$ in a two-user scenario

Theorem 2 *Assume that user i and user j* ($i < j$) *are paired together to perform NOMA. A necessary condition for NOMA to achieve higher individual rates than OFDMA* ($b_i = b_j, \upsilon_i = \upsilon_j$) *is* $a_{\text{th1}} < a_j < a_{\text{th2}}$, *where*

$$a_{\text{th1}} = \sqrt{\frac{-1+\sqrt{1+\rho\kappa_{\max}}}{\rho\kappa_{\max}}}, \quad a_{\text{th2}} = \sqrt{\frac{-1+\sqrt{1+\rho\kappa_{\min}}}{\rho\kappa_{\min}}}.$$

If the above condition is met, the probability that NOMA can achieve higher individual rates than OFDMA is given by:

$$\begin{aligned}\mathbb{P}\left[R_i > \bar{R}_i, R_j > \bar{R}_j\right] = &\sum_{l=0}^{K-j}\sum_{p=0}^{i-1}\sum_{q=0}^{j-1-i}\left\{\frac{(-1)^{K-i-l+p-q+1}\Omega^{j+l+p}}{(j-i+l-q)(p+q+1)}\left(\frac{L^2}{r_{\text{e}}^2}+1\right)^{i-1-p}\right.\\ &\times\left(\frac{L^2}{r_{\text{e}}^2}\right)^{K-j-l}\frac{K!}{l!(K-j-l)!\,p!(i-1-p)!\,q!(j-1-i-q)!}\\ &\left.\times\left(\varsigma^{\star-\frac{p+q+1}{m+3}}-\kappa_{\min}^{-\frac{p+q+1}{m+3}}\right)\left(\kappa_{\max}^{-\frac{j-i+l-q}{m+3}}-\varsigma^{\star-\frac{j-i+l-q}{m+3}}\right)\right\},\end{aligned} \tag{19.43}$$

where $\varsigma^\star = \min\{\max\{\varsigma, \kappa_{\min}\}, \kappa_{\max}\}$, *and* $\varsigma = \frac{1-2a_j^2}{\rho a_j^4}$.

Proof The probability that NOMA achieves higher individual data rates than OFDMA can be calculated as:

$$\begin{aligned}&\mathbb{P}\left[R_i > \bar{R}_i, R_j > \bar{R}_j\right]\\ =&\mathbb{P}\left[\underbrace{1+\frac{h_i^2a_i^2}{h_i^2a_j^2+\frac{1}{\rho}} > \left(1+\frac{\upsilon_i}{b_i}\rho h_i^2\right)^{b_i}}_{\text{E}_i}, \underbrace{1+\rho h_j^2a_j^2 > \left(1+\frac{\upsilon_j}{b_j}\rho h_j^2\right)^{b_j}}_{\text{E}_j}\right].\end{aligned} \tag{19.44}$$

Note that the joint PDF of h_i^2 and h_j^2 can be obtained as [28]:

$$f_{h_i^2,h_j^2}(u,v) = \omega f_{h_k^2}(u) f_{h_k^2}(v) F_{h_k^2}(u)^{i-1}\left[1-F_{h_k^2}(v)\right]^{K-j}\left[F_{h_k^2}(v)-F_{h_k^2}(u)\right]^{j-1-i}, \tag{19.45}$$

where $\omega = \frac{K!}{(i-1)!(j-1-i)!(K-j)!}$. After some simplification, the probability of event E_i and $\text{E}_j$can be written as:

$$\mathbb{P}\left[\text{E}_i\right] = \mathbb{P}\left[h_i^2 < \varsigma\right], \tag{19.46}$$

$$\mathbb{P}\left[\text{E}_j\right] = \mathbb{P}\left[h_j^2 > \varsigma\right]. \tag{19.47}$$

Therefore, the probability that NOMA achieves higher individual rates than OFDMA can be computed as:

$$\mathbb{P}\left[R_i > \bar{R}_i, R_j > \bar{R}_j\right] = \omega \int_{\varsigma^\star}^{\kappa_{\max}} \underbrace{f_{h_k^2}(v)\left[1 - F_{h_k^2}(v)\right]^{K-j}}_{\Xi_1(v)} \left\{ \underbrace{\int_{\kappa_{\min}}^{\varsigma^\star} f_{h_k^2}(u) F_{h_k^2}(u)^{i-1} \left[F_{h_k^2}(v) - F_{h_k^2}(u)\right]^{j-1-i} du}_{\Xi_2(v)} \right\} dv, \tag{19.48}$$

where $\Xi_1(v)$ can be calculated using the binomial expansion:

$$\Xi_1(v) = \frac{1}{m+3} \sum_{l=0}^{K-j} \frac{(K-j)!}{l!\,(K-j-l)!} \Omega^{l+1} \left(-\frac{L^2}{r_e^2}\right)^{K-j-l} v^{-\frac{l+1}{m+3}-1}. \tag{19.49}$$

Again applying the binomial expansion, the inner integration in (19.48) can be calculated as:

$$\Xi_2(v) = \sum_{p=0}^{i-1} \sum_{q=0}^{j-1-i} \left\{ \frac{1}{p+q+1} \frac{(i-1)!\,(j-1-i)!}{p!\,(i-1-p)!\,q!\,(j-1-i-q)!} \Omega^{p+j-1} (-1)^{p+j-i-q} \right.$$
$$\left. \times \left(\varsigma^{\star -\frac{p+q+1}{m+3}} - \kappa_{\min}{}^{-\frac{p+q+1}{m+3}}\right) \left(\frac{L^2}{r_e^2} + 1\right)^{i-1-p} v^{-\frac{j-1-i-q}{m+3}} \right\}. \tag{19.50}$$

Combining (19.48)–(19.50), Theorem 2 is proved. ■

Theorem 2 demonstrates that given appropriate power allocation coefficients, it is nearly certain for NOMA to outperform OFDMA if two users with highly different channel qualities are paired together.

19.4.2 Impact of User Pairing on the Sum Rate

In this subsection, the impact of user pairing on the ergodic sum rate is studied. The following theorem states that the ergodic sum rate gain of NOMA over OFDMA is upper bounded in high-SNR regimes, and this upper bound remains unchanged for different power allocation strategies.

Theorem 3 *Assuming that user i and user j ($i < j$) are paired to perform NOMA ($a_i^2 + a_j^2 = 1$), as ρ increases, the sum rate gain of NOMA over OFDMA first decreases then increases until it is upper bounded in high SNR regimes, and this upper bound is given by:*

$$
\begin{aligned}
\mathbb{E}\left[R_i + R_j - \bar{R}_i - \bar{R}_j\right] \leq& \frac{1}{2}\left(b_i \log_2\left(\frac{b_i}{\upsilon_i}\right) + b_j \log_2\left(\frac{b_j}{\upsilon_j}\right)\right) \\
&+ \frac{\Omega(1-b_j)}{2\ln(2)} \sum_{p=0}^{j-1}\sum_{q=0}^{K-j} \left\{ \frac{K!(-1)^{p+K-j-q}}{p!\,(j-1-p)!\,q!\,(K-j-q)!} \Omega^{p+q} \right. \\
&\times \left.\left(\frac{L^2}{r_{\mathrm{e}}^2}+1\right)^{j-1-p} \left(\frac{L^2}{r_{\mathrm{e}}^2}\right)^{K-j-q} \left(\varpi_4(\kappa_{\max}) - \varpi_4(\kappa_{\min})\right) \right\} \\
&- \frac{\Omega b_i}{2\ln(2)} \sum_{p=0}^{i-1}\sum_{q=0}^{K-i} \left\{ \frac{K!(-1)^{p+K-i-q}\Omega^{p+q}}{p!\,(i-1-p)!\,q!\,(K-i-q)!} \right. \\
&\times \left.\left(\frac{L^2}{r_{\mathrm{e}}^2}+1\right)^{i-1-p} \left(\frac{L^2}{r_{\mathrm{e}}^2}\right)^{K-i-q} \left(\varpi_4(\kappa_{\max}) - \varpi_4(\kappa_{\min})\right) \right\}.
\end{aligned}
\tag{19.51}
$$

where

$$
\varpi_4(\kappa) = -\frac{\kappa^{-\frac{p+q+1}{m+3}}\left(m+3+(p+q+1)\ln(\kappa)\right)}{(p+q+1)^2}. \tag{19.52}
$$

Proof The sum rate gain of NOMA over OFDMA can be formulated as:

$$
\begin{aligned}
&\mathbb{E}\left[R_i + R_j - \bar{R}_i - \bar{R}_j\right] \\
&= \frac{1}{2}\int_{\kappa_{\min}}^{\kappa_{\max}} \left(\log_2\left(1 + \frac{a_i^2 t}{a_j^2 t + \frac{1}{\rho}}\right) - b_i \log_2\left(1 + \frac{\upsilon_i}{b_i}\rho t\right)\right) f'_{h_i^2}(t)\mathrm{d}t \\
&+ \frac{1}{2}\int_{\kappa_{\min}}^{\kappa_{\max}} \left(\log_2\left(1+\rho a_j^2 t\right) - b_j \log_2\left(1 + \frac{\upsilon_j}{b_j}\rho t\right)\right) f'_{h_j^2}(t)\mathrm{d}t.
\end{aligned}
\tag{19.53}
$$

It can be seen from (19.53) that $\mathbb{E}\left[R_i + R_j - \bar{R}_i - \bar{R}_j\right]$ approximates zero when ρ is extremely small. According to Leibniz integral rule, we have

$$
\begin{aligned}
\frac{\partial \mathbb{E}\left[R_i + R_j - \bar{R}_i - \bar{R}_j\right]}{\partial \rho} =& \frac{1}{2\ln(2)}\int_{\kappa_{\min}}^{\kappa_{\max}} \left(\frac{t}{1+\rho t} - \frac{a_j^2 t}{1+\rho a_j^2 t} - \frac{b_i t}{\frac{b_i}{\upsilon_i}+\rho t}\right) f'_{h_i^2}(t)\mathrm{d}t \\
&+ \frac{1}{2\ln(2)}\int_{\kappa_{\min}}^{\kappa_{\max}} \left(\frac{a_j^2 t}{1+\rho a_j^2 t} - \frac{b_j t}{\frac{b_j}{\upsilon_j}+\rho t}\right) f'_{h_j^2}(t)\mathrm{d}t.
\end{aligned}
\tag{19.54}
$$

As ρ increases, the derivative of $\mathbb{E}\left[R_i + R_j - \bar{R}_i - \bar{R}_j\right]$ first drops below zero and then increases to a positive value. Therefore, the trend of the sum rate gain of NOMA over OFDMA is proved. In high SNR regimes, it is straightforward to show:

$$\lim_{\rho\to\infty} \mathbb{E}\left[R_i\right] = \lim_{\rho\to\infty} \int_{\kappa_{\min}}^{\kappa_{\max}} \frac{1}{2}\log_2\left(1+\frac{a_i^2 t}{a_j^2 t+\frac{1}{\rho}}\right) f'_{h_i^2}(t)\mathrm{d}t$$
$$= \frac{1}{2}\log_2\left(1+\frac{a_i^2}{a_j^2}\right) = -\log_2 a_j. \tag{19.55}$$

The data rate for user j participated in NOMA can be divided into two parts:

$$\mathbb{E}\left[R_j\right] = \underbrace{\int_{\kappa_{\min}}^{\kappa_{\max}} b_i \log_2\sqrt{1+\rho a_j^2 t}\, f'_{h_j^2}(t)\mathrm{d}t}_{\mathrm{N}_1} + \underbrace{\int_{\kappa_{\min}}^{\kappa_{\max}} b_j \log_2\sqrt{1+\rho a_j^2 t}\, f'_{h_j^2}(t)\mathrm{d}t}_{\mathrm{N}_2}. \tag{19.56}$$

In high SNR regimes, the difference between N_2 and $\mathbb{E}\left[\bar{R}_j\right]$ can be calculated as:

$$\lim_{\rho\to\infty}\left(\mathrm{N}_2 - \mathbb{E}\left[\bar{R}_j\right]\right) = \lim_{\rho\to\infty}\frac{1}{2}b_j\int_{\kappa_{\min}}^{\kappa_{\max}} \log_2\frac{1+\rho a_j^2 t}{1+\frac{\upsilon_j}{b_j}\rho t} f'_{h_j^2}(t)\mathrm{d}t$$
$$= b_j\log_2 a_j + \frac{1}{2}b_j\log_2\left(\frac{b_j}{\upsilon_j}\right). \tag{19.57}$$

Applying integration by parts, N_1 in high SNR regimes can be calculated as:

$$\lim_{\rho\to\infty}\mathrm{N}_1 = b_i\log_2 a_j + \lim_{\rho\to\infty}\frac{b_i}{2}\log_2\left(1+\rho\kappa_{\min}\right) + \frac{b_i}{2\ln(2)}\int_{\kappa_{\min}}^{\kappa_{\max}}\frac{1}{t}\left(1-F'_{h_j^2}(t)\right)\mathrm{d}t, \tag{19.58}$$

where $F'_{h_j^2}(t)$ represents the CDF of the order variable h_k^2, and the integration in (19.58) can be obtained as:

$$\int_{\kappa_{\min}}^{\kappa_{\max}}\frac{1}{t}\left(1-F'_{h_j^2}(t)\right)\mathrm{d}t = -\ln(\kappa_{\min}) + \Omega\sum_{p=0}^{j-1}\sum_{q=0}^{K-j}\left\{\frac{K!(-1)^{p+K-j-q}\Omega^{p+q}}{p!(j-1-p)!q!(K-j-q)!}\right.$$
$$\left.\times\left(\frac{L^2}{r_{\mathrm{e}}^2}+1\right)^{j-1-p}\left(\frac{L^2}{r_{\mathrm{e}}^2}\right)^{K-j-q}\left(\varpi_4(\kappa_{\max})-\varpi_4(\kappa_{\min})\right)\right\}. \tag{19.59}$$

In a similar way, $\bar{R}_i$ in high SNR regimes can be calculated as:

$$\lim_{\rho\to\infty} \mathbb{E}\left[\bar{R}_i\right] = \lim_{\rho\to\infty} \frac{b_i}{2}\log_2\left(1+\frac{\upsilon_i}{b_i}\rho\kappa_{\min}\right) - \frac{b_i}{2\ln(2)}\ln(\kappa_{\min})$$
$$+ \frac{\Omega b_i}{2\ln(2)}\sum_{p=0}^{i-1}\sum_{q=0}^{K-i}\left\{\frac{K!}{p!\,(i-1-p)!\,q!\,(K-i-q)!}(-1)^{p+K-i-q}\Omega^{p+q}\right.$$
$$\left.\times\left(\frac{L^2}{r_{\mathrm{e}}^2}+1\right)^{i-1-p}\left(\frac{L^2}{r_{\mathrm{e}}^2}\right)^{K-i-q}(\varpi_4(\kappa_{\max})-\varpi_4(\kappa_{\min}))\right\}. \quad (19.60)$$

Combining (19.55)–(19.60) Theorem 3 is proved. ■

Theorem 3 gives the upper bound of the ergodic sum rate gain of NOMA over OFDMA for arbitrary i and j ($i < j$). However, it is of more interest to evaluate how the performance gain varies if i and j change. The following corollary states that the optimum sum rate gain is achieved if two users with the most distinctive channel conditions are paired together to perform NOMA.

Corollary 2 *If the i-th user and the j-th user ($i < j$) are paired to perform NOMA, the sum rate gain of NOMA over OFDMA achieves the maximum by pairing the two users with the most distinctive channel conditions, i.e., $i = 1$ and $j = K$. In high SNR regimes, this maximum gain is upper bounded by:*

$$\mathbb{E}\left[R_1 + R_K - \bar{R}_1 - \bar{R}_K\right] \le \frac{1}{2}\left(b_1\log_2\left(\frac{b_1}{\upsilon_1}\right) + b_K\log_2\left(\frac{b_K}{\upsilon_K}\right)\right)$$
$$+ \frac{\Omega}{4\ln(2)}\sum_{l=0}^{K-1}\left\{\frac{K!}{l!\,(K-1-l)!}\Omega^l\left(\varpi_5(\kappa_{\max})-\varpi_5(\kappa_{\min})\right)\right.$$
$$\left.\times\left((-1)^l\left(\frac{L^2}{r_{\mathrm{e}}^2}+1\right)^{K-1-l} - (-1)^{K-1-l}\left(\frac{L^2}{r_{\mathrm{e}}^2}\right)^{K-1-l}\right)\right\}, \quad (19.61)$$

where

$$\varpi_5(\kappa) = -\frac{\kappa^{-\frac{l+1}{m+3}}\left(m+3+(l+1)\ln(\kappa)\right)}{(l+1)^2}. \quad (19.62)$$

Proof The proof is divided into two parts. First, we prove that the maximum sum rate gain is achieved when $i = 1$ and $j = K$. Second, we prove the expression of the maximum sum rate gain in (19.61). For the first part, it is equivalent to prove the following:

$$\lim_{\rho\to\infty}\mathbb{E}\left[R_{i+1}+R_j-\bar{R}_{i+1}-\bar{R}_j\right] < \lim_{\rho\to\infty}\mathbb{E}\left[R_i+R_j-\bar{R}_i-\bar{R}_j\right], \quad (19.63)$$

for $\forall i < K$, and

$$\lim_{\rho\to\infty}\mathbb{E}\left[R_i+R_{j+1}-\bar{R}_i-\bar{R}_{j+1}\right] > \lim_{\rho\to\infty}\mathbb{E}\left[R_i+R_j-\bar{R}_i-\bar{R}_j\right], \quad (19.64)$$

for $\forall j < K$. The expression in (19.63) is equivalent to

$$\lim_{\rho\to\infty} \mathbb{E}\left[R_{i+1} - R_i - \bar{R}_{i+1} + \bar{R}_i\right] < 0. \tag{19.65}$$

From (19.55), it can be shown that $\lim_{\rho\to\infty} \mathbb{E}\left[R_{i+1}\right] = \lim_{\rho\to\infty} \mathbb{E}\left[R_i\right]$. Therefore, in order to prove (19.63), we need to prove:

$$\begin{aligned}
&\lim_{\rho\to\infty} \mathbb{E}\left[R_{i+1} - R_i - \bar{R}_{i+1} + \bar{R}_i\right] < 0 \Longrightarrow \lim_{\rho\to\infty} \mathbb{E}\left[\bar{R}_i - \bar{R}_{i+1}\right] < 0 \\
&\Longrightarrow \int_{\kappa_{\min}}^{\kappa_{\max}} \frac{1}{t}\left(1 - F'_{h_i^2}(t)\right) \mathrm{d}t < \int_{\kappa_{\min}}^{\kappa_{\max}} \frac{1}{t}\left(1 - F'_{h_{i+1}^2}(t)\right) \mathrm{d}t \\
&\Longrightarrow F'_{h_i^2}(t) > F'_{h_{i+1}^2}(t).
\end{aligned} \tag{19.66}$$

As $F'_{h_i^2}(t)$ represents the CDF of the i-th largest variable h_i^2, it is obvious that (19.66) is true. To this end (19.63) is proved, and the proof of (19.64) can be conducted in a similar way. For the second part, the expression of (19.61) can be obtained by setting $i = 1$ and $j = K$ in (19.51). ■

19.5 Simulation Results

19.5.1 Theoretical Framework

The aim of this section is to substantiate the derived analytical results through Monte Carlo simulations and to obtain insights into how the choice of LEDs affects the system performance. If not otherwise specified, the parameters used for the simulation setup are summarized in Table 19.1.

Figure 19.6 shows the system coverage probability for different power allocation coefficients in a two-user scenario. For the discussion on the two-user scenario, the user closer to the LED is referred to as the *near* user and the user further away from the LED is referred to as the *far* user, i.e., $d_{\text{near}} < d_{\text{far}}$, and their power coefficients satisfy constraint (19.7). It can be seen that the analytical results are consistent with the simulation results, and there exists an optimum set of power allocation coefficients for achieving the maximum coverage probability. For a low target data rate, the system coverage probability is nearly 100%, given that the power allocation coefficients are optimally chosen. As the target data rate increases, the achievable maximum coverage probability decreases. Also, it can be seen that a larger coverage probability can be achieved by pairing two users with more distinctive channel conditions. An interesting finding is that when the target data rate for both users increases, more signal power should be allocated to the *far* user in order to achieve optimal coverage probability.

Table 19.1 Simulation parameters

Parameter name, notation	value
Vertical separation between the LED and PDs, L	2.15 m
Cell radius, r_e	3.6 m
Total number of users, K	10
LED semi-angle, $\Phi_{1/2}$	60°
Total signal power, P_{elec}	0.25 W
PD FOV, Ψ_{fov}	60°
PD responsivity, R_p	0.4 A/W
PD detection area, A	1 cm^2
Reflective index, n	1.5
Optical filter gain, T	1
Signal bandwidth, B	20 MHz
Noise PSD, N_0	10^{-21} A^2/Hz

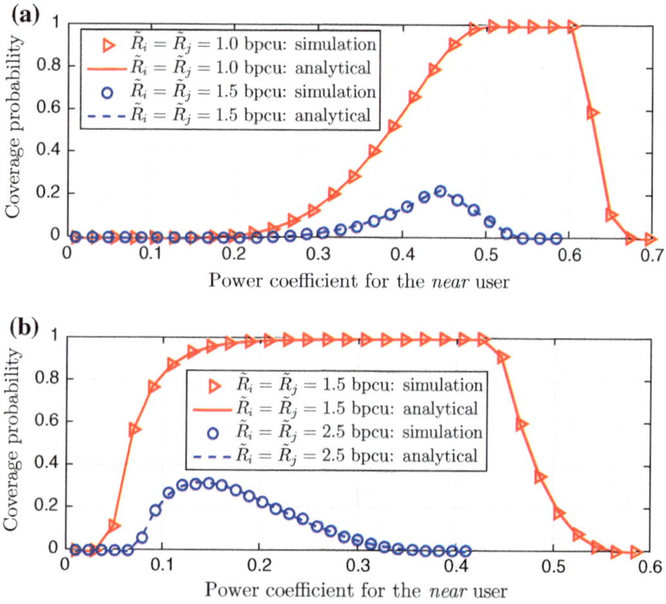

Fig. 19.6 System coverage probability for different power allocation coefficients: **a** $i = 1$ and $j = 2$ (users with similar channel conditions); **b** $i = 1$ and $j = 10$ (users with distinctive channel conditions)

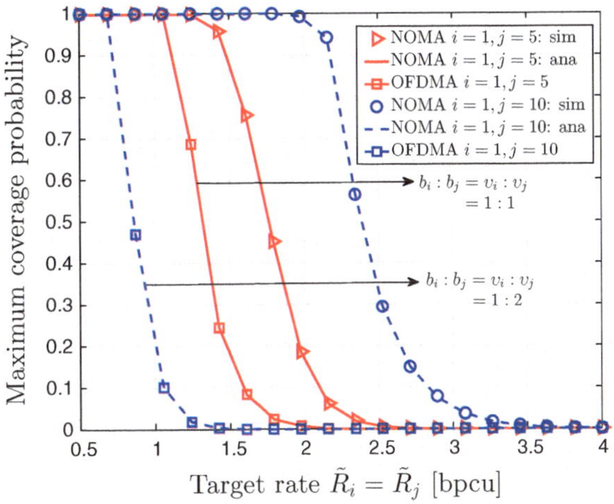

Fig. 19.7 System maximum coverage probability for different target data rates

An exhaustive search (ES) method is used to find the optimum power allocation coefficients. Specifically, a lookup table is formed, in which the system coverage probability is saved for each systematic search of the power coefficients. After this, the optimum pair of power coefficients is found by referring to the lookup table and selecting the one that gives the highest coverage probability. Figure 19.7 shows the maximum achievable coverage probability as a function of the target data rate. As expected, the maximum coverage probability would decrease as the target data rate increases. Compared with OFDMA, NOMA is shown to be able to provide a larger coverage probability, and this performance gain can be further enlarged by pairing two users with more distinctive channel conditions. For example, when $i = 1$ and $j = 10$, NOMA can provide 2.2 bpcu data rate for both users with 90% coverage probability while OFDMA can only provide 0.7 bpcu data rate for both users with the same coverage probability.

Figure 19.8 demonstrates that, in order to achieve the same QoS requirement for both users, using an LED with a larger semi-angle can provide a higher coverage probability for both NOMA and OFDMA techniques. This is because LEDs with a larger semi-angle can illuminate a wider area and produce less fluctuations in the signal strength. In Fig. 19.9, the maximum coverage probability is computed for different transmit SNR values, and the monotonically increasing characteristic of the curve is in agreement with (19.31). It can be seen that more transmit power is typically required in order to achieve a higher target rate. More specifically, when the transmit SNR is below a certain threshold, no signal coverage can be provided because the communication link is strongly limited by the interference and noise levels. As the transmit SNR increases and exceeds the threshold, the maximum coverage probability starts to increase until its value reaches unity. To investigate the impact of user pairing, the probability that NOMA can achieve higher individual rates than OFDMA is evaluated in Fig. 19.10. The developed analytical results show a good match with computer simulations. With a fixed number of users in the network,

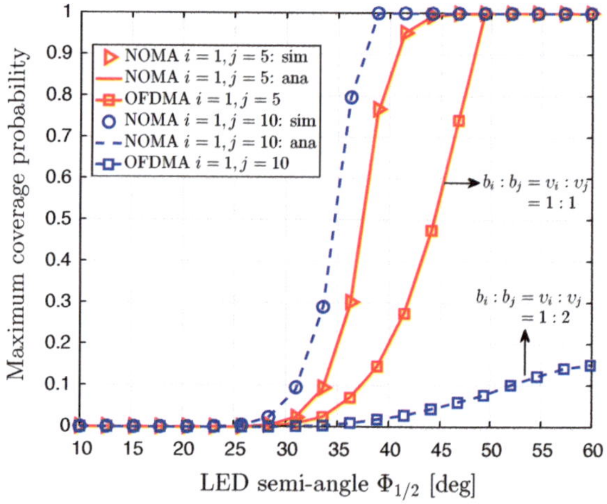

Fig. 19.8 System maximum coverage probability for different LED semi-angles ($\tilde{R}_i = \tilde{R}_j = 1$ bpcu)

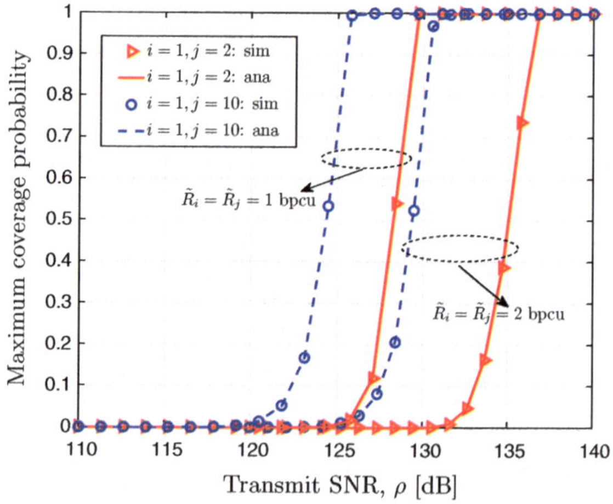

Fig. 19.9 System maximum coverage probability for different transmit SNR values

it can be seen that the performance of NOMA can be further enhanced through an efficient user pairing strategy (Theorem 2). Results shown in Figs. 19.8, 19.9, 19.10 all confirm the performance superiority of NOMA over OFDMA, especially when users have more distinct channel conditions.

In Fig. 19.11, the ergodic sum rate gain of NOMA over naive OFDMA is shown as a function of the transmit SNR. It can be seen that the derived theoretical bound shows strong consistency with the simulation results. Also, it can be seen from Fig. 19.11 that, as the transmit SNR increases, the sum rate gain of NOMA over OFDMA first decreases and reaches a minimum. As ρ continues to increase, the sum rate gain increases until it reaches the upper bound. This trend is consistent with Theorem 3.

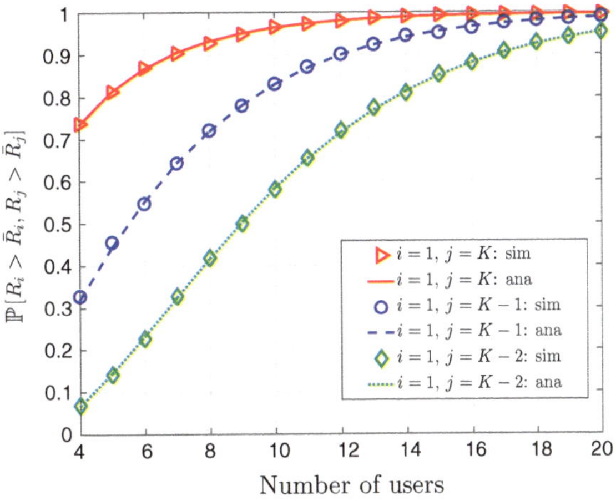

Fig. 19.10 Probability that NOMA achieves higher individual data rates than OFDMA ($a_i^2 = 9/10$, $a_j^2 = 1/10$)

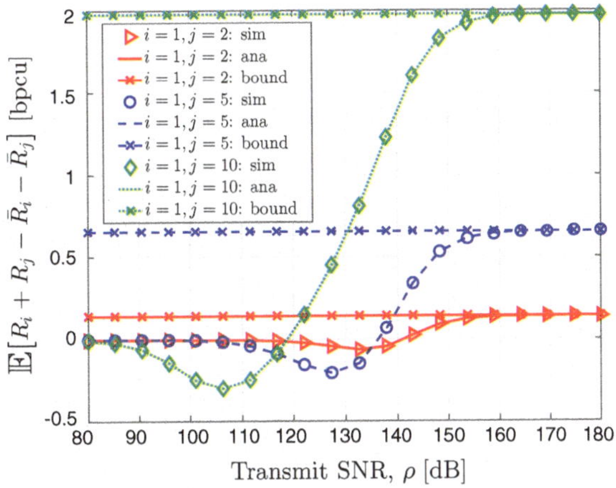

Fig. 19.11 Ergodic sum rate gain achieved by NOMA over OFDMA ($b_i : b_j = \upsilon_i : \upsilon_j = 1 : 2$)

19.5.2 Multipath Reflections and Shadowing Effect

One of the advantages of LiFi is that when the LOS is blocked by opaque objects, signal transmission is still possible via the diffuse links but with a lower data rate. Denote the LOS blockage event as X, whose probability mass function can be modeled by the Bernoulli distribution, given by:

$$\Pr[X = \chi] = \begin{cases} p, & \chi = 1 \\ 1 - p, & \chi = 0 \end{cases}, \tag{19.67}$$

where p is the probability that the LOS is blocked. When $\chi = 0$, the VLC channel consists both the LOS and diffuse components, but the dominant part is the LOS

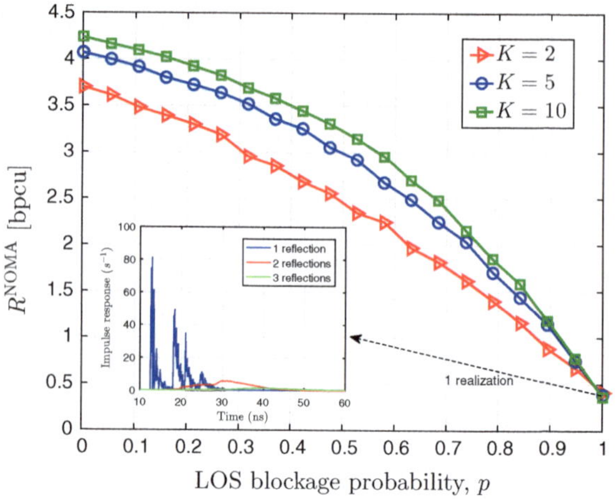

Fig. 19.12 Average sum rate of NOMA with LOS blockage

link. Theoretical analysis for this case has been presented in previous sections of this paper. When $\chi = 1$, the LOS is blocked with probability of p. In this case, the signal is transmitted via the diffuse links. Unlike narrowband infrared (IR) wireless communication, LiFi uses a wide spectrum, which ranges from 380 nm to 780 nm. This means that the wavelength-dependent properties of the PSD of the LED and the reflectance of indoor reflectors should be taken into account in the channel modeling. It has been reported that the received signal power from multipath reflections and the RMS delay spread in LiFi is generally smaller than those in IR systems [20]. In the following, the performance of NOMA is studied in a more realistic scenario where multipath reflections and LOS blockage are considered. Specifically, a room of size $5 \times 5 \times 3$ m^3 is considered. The LED is located at the center of the ceiling while receivers are randomly distributed throughout the room. Monte Carlo simulations are carried out, and the system performance is evaluated over 1000 independent trials. The channel impulse response is simulated using the recursive algorithm reported in [20], which includes both the wavelength-dependent white LED characteristics and spectral reflectance of indoor reflectors. As shown in Fig. 19.12, the performance of NOMA in a LiFi system degrades when the LOS blockage probability increases. However, due to the existence of multipath reflections in the indoor environment, signal transmission is still possible even if the LOS link is totally blocked.

19.6 Summary and Future Works

In this chapter, we have studied the application of NOMA to LiFi networks as an efficient multiuser access technique. The application takes into account unique characteristics of the VLC channel and the real and nonnegative signaling format in LiFi systems which differ from RF communications. Also, a theoretical framework for the performance evaluation of NOMA in a downlink LiFi setup has been proposed, covering both quality-of-service provisioning as well as opportunistic best-effort service provisioning. The presented studies demonstrated that NOMA is capable of providing performance gains over OMA in LiFi networks, especially for cell-edge users who generally have lower channel gains, and the performance gain can be further enlarged by selecting users with more distinct channel conditions for the power-domain multiplexing. Although high-speed VLC relies on the LOS link, it was shown that signal transmission with NOMA is still possible even if the LOS link is totally blocked.

This chapter is focused on a point-to-multipoint model. Further extensions of this work could consider the multipoint-to-multipoint model and include networked multiple-input multiple-output (MIMO) [14] to allow for parallel data transmission and to achieve higher data rates. Since the proposed framework does not assume specific power allocation strategies, it can be extended for the study of optical power allocation strategies of NOMA in LiFi networks. In addition, as the proposed framework relies on perfect CSI, a generalization of the framework to include imperfect CSI is also of interest.

References

1. Cisco visual networking index: global mobile data traffic forecast update, 2016–2021, White Paper (Cisco, Mar 2017)
2. T.S. Rappaport, S. Sun, R. Mayzus, H. Zhao, Y. Azar, K. Wang, G.N. Wong, J.K. Schulz, M. Samimi, F. Gutierrez, Millimeter wave mobile communications for 5G cellular: it will work!. IEEE Access **1**, 335–349 (2013)
3. S. Dimitrov, H. Haas, *Principles of LED Light Communications: Towards Networked Li-Fi* (Cambridge University Press, 2015)
4. S. Arnon, *Visible Light Communication* (Cambridge University Press, New York, NY, USA, 2015)
5. IEEE Std. 802.15.7-2011, *IEEE Standard for Local and Metropolitan Area Networks, Part 15.7: Short-Range Wireless Optical Communication Using Visible Light* (IEEE Std., 2011)
6. Status of IEEE 802.11 light communication SG, Meeting Update, Nov 2017, http://www.ieee802.org/11/Reports/lcsg_update.htm. Accessed 19 Dec 2017
7. P.H. Pathak, X. Feng, P. Hu, P. Mohapatra, Visible light communication, networking, and sensing: a survey, potential and challenges, IEEE Commun. Surv. Tutor. **17**(4), 2047–2077 (2015)
8. J. Kahn, J. Barry, Wireless infrared communications. Proc. IEEE **85**(2), 265–298 (1997). Feb
9. P. M. Butala, J. C. Chau, T.D.C. Little, Metameric modulation for diffuse visible light communications with constant ambient lighting, in *Proceedings of International Workshop on Optical Wireless Communications (IWOW)* (Pisa, Oct 2012), pp. 1–3
10. K.-I. Ahn, J.K. Kwon, Color intensity modulation for multicolored visible light communications. IEEE Photon. Technol. Lett. **24**(24), 2254–2257 (2012)
11. H. Haas, L. Yin, Y. Wang, C. Chen, What is LiFi? J. Lightw. Technol. **34**(6), 1533–1544 (2016)
12. S. Dimitrov, H. Haas, Information rate of OFDM-based optical wireless communication systems with nonlinear distortion. J. Lightw. Technol. **31**(6), 918–929 (2013)
13. H. Elgala, R. Mesleh, H. Haas, Non-linearity effects and predistortion in optical OFDM wireless transmission using LEDs. *Inderscience Int. J. Ultra Wideband Commun. Syst.* **1**(2), 143–150 (2009)

14. L. Zeng, D. O'Brien, H. Minh, G. Faulkner, K. Lee, D. Jung, Y.J. Oh, E.T. Won, High data rate multiple input multiple output (MIMO) optical wireless communications using white LED lighting. IEEE J. Sel. Areas Commun. **27**(9), 1654–1662 (2009)
15. J.B. Carruthers, J.M. Kahn, Multiple-subcarrier modulation for nondirected wireless infrared communication. IEEE J. Sel. Areas Commun. **14**(3), 538–546 (1996)
16. J. Armstrong, B.J.C. Schmidt, Comparison of asymmetrically clipped optical OFDM and DC-biased optical OFDM in AWGN. IEEE Commun. Lett. **12**(5), 343–345 (2008)
17. D. Tsonev, S. Videv, H. Haas, Unlocking spectral efficiency in intensity modulation and direct detection systems. IEEE J. Sel. Areas Commun. **3**(9), 1758–1770 (2015)
18. H.T. Friis, A note on a simple transmission formula. Proc. IRE **34**(5), 254–256 (1946)
19. H. Haas, High-speed wireless networking using visible light. *SPIE Newsroom* (2013)
20. K. Lee, H. Park, J.R. Barry, Indoor channel characteristics for visible light communications. IEEE Commun. Lett. **15**(2), 217–219 (2011)
21. R.C. Kizilirmak, C.R. Rowell, M. Uysal, Non-orthogonal multiple access (NOMA) for indoor visible light communications, in *Proceedings of International Workshop on Optical Wireless Communications (IWOW)* (Istanbul, Turkey, Sept 2015), pp. 98–101
22. H. Marshoud, V.M. Kapinas, G.K. Karagiannidis, S. Muhaidat, Non-orthogonal multiple access for visible light communications. IEEE Photon. J. **28**(1), 51–54 (2016)
23. L. Yin, X. Wu, H. Haas, On the performance of non-orthogonal multiple access in visible light communication, in *Proceedings of IEEE 26th Annual Symposium on Personal, Indoor and Mobile Radio Communications (PIMRC)* (Hong Kong, China, Sept 2015), pp. 1376–1381
24. L. Yin, W.O. Popoola, X. Wu, H. Haas, Performance evaluation of non-orthogonal multiple access in visible light communication. IEEE Trans. Commun. **64**(12), 5162–5175 (2016)
25. S. Dimitrov, S. Sinanovic, H. Haas, Clipping noise in OFDM-based optical wireless communication systems. IEEE Trans. Commun. **60**(4), 1072–1081 (2012)
26. H. Elgala, R. Mesleh, H. Haas, An LED model for intensity-modulated optical communication systems. IEEE Photon. Technol. Lett. **22**(11), 835–837 (2010)
27. T. Komine, M. Nakagawa, Fundamental analysis for visible-light communication system using LED lights. IEEE Trans. Consum. Electron. **50**(1), 100–107 (2004)
28. H.A. David, H.N. Nagaraja, *Order Statistics*, 3rd edn. (John Wiley, New York, 2003)

Chapter 20
NOMA-Based Integrated Terrestrial-Satellite Networks

Xiangming Zhu, Chunxiao Jiang, Linling Kuang, Ning Ge and Jianhua Lu

20.1 Background

With the rapid growth of data traffic, which is predicted to reach 1,000-fold by 2020, the wireless network will face much more challenges in the next generation wireless communication networks. Recently, non-orthogonal multiple access (NOMA) has been proposed as an efficient method for multiple access [1], which can achieve better system performance than conventional orthogonal multiple access (OMA). By means of NOMA, the challenging requirements for 5G, such as spectral efficiency and massive connectivity, can be partially fulfilled [2]. In addition, the Third-Generation Partnership Project Long-Term Evolution (3GPP-LTE) system is also doing ongoing studies upon NOMA, in the aspects of multi-user superposition transmission [3] (MUST). It can be envisioned that the technique of NOMA will play an important role in the next generation communication networks.

The NOMA schemes can be divided into two categories: power domain multiplexing and code domain multiplexing [2, 4]. In most cases, the NOMA scheme refers to the power domain multiplexing if no special explanation is provided [5–8], which is also the case that we consider in this chapter. For convenience and also without confusing, the word "NOMA" in the rest of the chapter will refer to the

X. Zhu · C. Jiang (✉) · L. Kuang · N. Ge · J. Lu
Tsinghua University, Beijing 100084, People's Republic of China
e-mail: jchx@tsinghua.edu.cn

X. Zhu
e-mail: zhuxm14@mails.tsinghua.edu.cn

L. Kuang
e-mail: kll@tsinghua.edu.cn

N. Ge
e-mail: gening@tsinghua.edu.cn

J. Lu
e-mail: lhh-dee@tsinghua.edu.cn

M. Vaezi et al. (eds.), *Multiple Access Techniques for 5G Wireless Networks and Beyond*, https://doi.org/10.1007/978-3-319-92090-0_20

power domain multiplexing. In [9], the concept of superposition coding (SC) was first proposed for broadcast channels, while the successive interference cancelation (SIC) technique has been investigated in the past systems like V-BLAST [10]. The key idea of NOMA is to serve multiple users with distinct channel conditions via SC at the transmitter and SIC at the receiver [1]. The user with poorer channel will be allocated more power, and the signals of all users will be transmitted superposed reusing the entire bandwidth. The signals of the poorer users will be decoded at all the users that experience better channels, which can then be subtracted from the received signal. As a result, the users with better channels will not suffer interference from the poorer users.

The superiority of NOMA was discussed in the initial works [1, 5], in which the author proved that NOMA could achieve superior system performance than conventional OMA. In [6], the author studied the performance of NOMA when users were randomly deployed, showing that NOMA could improve the ergodic sum rates of the 5G system. The problem of user fairness was investigated in [7], where the two cases of instantaneous channel state information (CSI) and average CSI were discussed. In [8, 11, 12], two different scenarios of cooperative NOMA were investigated. In [8], coordination between two base stations (BSs) was considered to improve the performance of the cell edge users, while the users close to the BS were utilized as relays to assist the transmission of far users with poorer channels in [11, 12]. As stated before, the NOMA scheme severs multiple users with distinct channels, and then the impact of user pairing was discussed for two NOMA-based systems in [13]. Considering the case of multiple antennas, NOMA beamforming schemes were investigated in [14–16]. In [14], a general framework of the NOMA beamforming system was proposed, while the technique of random beamforming was discussed in [15]. For the sake of energy efficiency, the problem of minimum power multicast beamforming was investigated in [16]. Then in [17], a minorization–maximization method for optimizing the sum rate of NOMA was proposed in a linearly precoded multiple-input single-output (MISO) system. The NOMA scheme was applied to multiple-input multiple-output (MIMO) system in [18–20], in which the general framework, the ergodic capacity, and the precoding designing of MIMO-NOMA systems were investigated.

While the terrestrial networks provide high bandwidth service with low cost, the satellite networks can provide the best and most comprehensive coverage for users that cannot be severed by BSs. In the next generation communication networks, the satellite has been taken into consideration to ensure ubiquitous coverage. According to [21], satellite radio access network (3GPP and non-3GPP defined) shall be supported for phase 2, and the support of satellite radio access network (3GPP defined) will be studied in conjunction with RAN activities. Up to now, quite a few works have investigated the combination of terrestrial and satellite networks in various approaches. In [22], the technique of cognitive radio (CR) was applied for the satellite to enable dynamic spectrum access [23–25]. In order to mitigate the co-channel interference (CCI) in the integrated terrestrial-satellite networks, a semi-adaptive beamforming algorithm executed at the satellite was proposed in [26]. Then, in [27, 28], the authors considered the scenario of the hybrid satellite-terrestrial relay network (HSTRN), where the symbol error and capacity performance of HSTRN were studied.

In this chapter, we consider a NOMA-based integrated terrestrial-satellite network, in which BSs and the satellite provide service for ground users cooperatively while reusing the entire bandwidth. Equipped with multi-antennas, both the terrestrial BSs and the satellite will execute beamforming for multiple access, while the NOMA scheme is applied to terrestrial networks only. The contributions of this chapter are summarized as follows:

- We propose a general downlink framework for NOMA-based integrated terrestrial-satellite networks. The satellite acts as the supplement to provide extra service for users that cannot be served by BSs. The terrestrial users are divided into groups, and in each group, the NOMA scheme is performed for multiple access. With multiple antennas, beamforming will be executed among groups and among satellite users. Based on the framework, we then formulate the optimization problem for the system capacity performance, which is decomposed into three parts: the paring scheme, the terrestrial resource allocation scheme, and the satellite resource allocation scheme.
- We propose two user pairing schemes for the satellite and the terrestrial networks separately. We select the users that to be served by the satellite based on the channel conditions. Then, a maximal minimum channel correlation criterion-based algorithm is proposed to pair users into groups for the implementation of NOMA, which is decomposed into a series of paring subproblems through an iterative algorithm. By means of the bipartite graph, we transform the paring subproblem into a maximum matching problem, which is then solved by the Hungarian method.
- We propose two schemes to solve the terrestrial resource allocation problem and the satellite resource allocation problem separately. Since the power allocation problem among groups is non-convex, the successive convex approximation (SCA) approach is utilized to transform the original problem into a series of convex subproblems, which are solved by means of the Lagrangian dual method. Then, an iterative algorithm is proposed to jointly optimize the power allocation scheme of the whole system.

The rest of the chapter is organized as follows. We first introduce the system model and formulate the optimization problem in Sect. 20.2. In Sect. 20.3, two user pairing schemes are proposed for both the satellite and the terrestrial BSs. Then, the terrestrial resource allocation problem is solved in Sect. 20.4. In Sect. 20.5, we solve the resource allocation problem of the satellite, after which an iterative algorithm is proposed for the whole system. The simulation results are shown in Sect. 20.6. Finally, Sect. 20.7 gives the conclusion.

20.2 System Model and Problem Formulation

As illustrated in Fig. 20.1, consider a downlink communication scenario of the integrated terrestrial-satellite network, in which L BSs and one satellite serve the ground users cooperatively. Each BS is equipped with N antennas for downlink transmission and can serve users within its coverage radius. The satellite can be either a low earth orbit (LEO) satellite, a medium earth orbit (MEO) satellite, or even a geosynchronous (GEO) satellite. For different types of satellite used, different techniques, such as varying channels, handover, need to be further discussed, but it is out of the range of this chapter. The satellite is equipped with M antennas and can provide downlink transmission for all users within its coverage. In this scenario, we consider that the L BSs distribute within the coverage of the satellite, and each user can be served either by the corresponding BS or by the satellite.

In the terrestrial networks, NOMA scheme is implemented for multi-user transmission, which can simultaneously serve multiple users with distinct channel conditions reusing the entire bandwidth. Considering both the system load and implementation complexity, existing studies on NOMA generally pair two users to perform NOMA [1, 16, 19]. Thus, in this chapter, we also concentrate on this typical situation, where one near user, with better channel condition, and one far user, with poorer channel condition, are paired for the implementation of NOMA. Also,

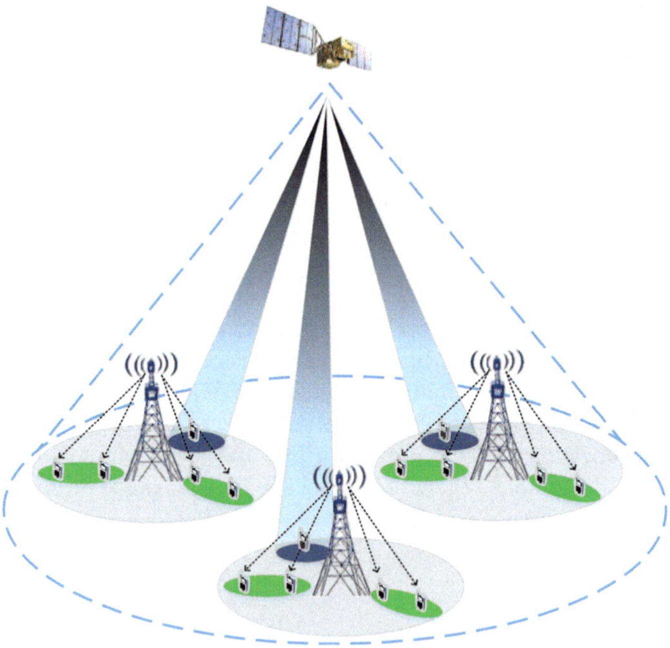

Fig. 20.1 System model of the NOMA-based integrated terrestrial-satellite network

as each BS is equipped with N antennas, beamforming is executed at the BS and thus each BS can serve multiple groups. That is, up to N group of users can be served simultaneously by one BS, when beamforming is executed to mitigate inter-group interference. Within each group, the NOMA scheme will be performed for the two users.

The user set is represented by $\{U_1, ..., U_L\}$, in which U_i represents the users within the coverage radius of BS i, and each user is equipped with one single antenna. Generally, in current communication networks, BSs-based terrestrial communication is more efficient and cheaper than the satellite communication. Thus, in the integrated terrestrial-satellite networks, we prefer to maximize the utilization of the BSs, and the satellite will act as the complement to provide service for these users that cannot be served by the BSs. By means of beamforming and NOMA, each BS can provide service for $2N$ users within its coverage radius, including N near users $\{u_{Bn,i,1}, ..., u_{Bn,i,N}\}$ and N far users $\{u_{Bf,i,1}, ..., u_{Bf,i,N}\}$, where $u_{Bn,i,j}$ is paired with $u_{Bf,i,j}$ to implement NOMA. We assume that the total number of users of BS i is $|U_i| = K_i$. If $K_i \leq 2N$, all the users will be served by the BS. When $K_i > 2N$, which is beyond the capability of BS i, in order to cover these $K_i - 2N$ users that cannot be served by BS i, the satellite is utilized to provide extra service for these users. The case of $K_i \leq 2N$ is the simplified case of $K_i > 2N$, and we consider $K_i > 2N$ for all BSs as the typical case in this chapter. Then, considering all the L BSs within the coverage of the satellite, the user set of the satellite will be $|U_S| = \sum_{i=1}^{L} K_i - 2N$. Also, as the satellite is equipped with M antennas, beamforming is executed and the $\sum_{i=1}^{L} K_i - 2N$ users will be served simultaneously using the entire bandwidth.

As stated above, for each BS, beamforming is executed among groups and NOMA is implemented within each group. The transmit signal of BS I is

$$\mathbf{x}_I = \sum_{j=1}^{N} \boldsymbol{\omega}_{I,j} \sqrt{P_{B,I,j}} (\sqrt{\alpha_{B,I,j}} s_{Bn,I,j} + \sqrt{1 - \alpha_{B,I,j}} s_{Bf,I,j}), \tag{20.1}$$

where $\boldsymbol{\omega}_{I,j}$ is the beamforming vector and $P_{B,I,j}$ is the transmit power for group j. Within each group, SC is utilized for the two users to implement NOMA, where $s_{Bn,I,j}$ and $s_{Bf,I,j}$ are the transmit signals to the near user and far user, and we have $E[|s_{Bn,I,j}|^2] = E[|s_{Bf,I,j}|^2] = 1$. Moreover, $\alpha_{B,I,j}$ is the power allocation coefficient, denoting the fraction of power allocated to the near user. The satellite executes beamforming and serves a total of $\sum_{i=1}^{L} K_i - 2N$ users:

$$\mathbf{x}_S = \sum_{i=1}^{L} \sum_{j=1}^{K_i - 2N} \boldsymbol{\nu}_{i,j} \sqrt{P_{S,i,j}} s_{S,i,j}, \tag{20.2}$$

where $\boldsymbol{\nu}_{i,j}$ is the beamforming vector, $P_{S,i,j}$ is the transmit power, and $s_{S,i,j}$ is the transmit signal to each satellite user, and we also have $E[|s_{S,i,j}|^2] = 1$. Then in group J of BS I, the received signal of the near user is

$$y_{Bn,I,J} = \mathbf{h}_{Bn,I,J}^H \sum_{j=1}^{N} \omega_{I,j}\sqrt{P_{B,I,j}}(\sqrt{\alpha_{B,I,j}}s_{Bn,I,j} + \sqrt{1-\alpha_{B,I,j}}s_{Bf,I,j}) \\ + \sum_{i=1}^{L}\sum_{j=1}^{K_i-2N} \mathbf{g}_{Bn,I,J}^H \mathbf{v}_{i,j}\sqrt{P_{S,i,j}}s_{S,i,j} + n, \tag{20.3}$$

where $\mathbf{h}_{Bn,I,J}$ is the channel vector from BS I to the near user in group J, $\mathbf{g}_{Bn,I,J}$ is the channel vector from the satellite to the near user, and n is the additive white Gaussian noise (AWGN). Also, since the main interference in the network is the inter-system interference, we mainly consider the interference from the satellite for the terrestrial users and treat the inter-cell interference as part of the AWGN. Correspondingly, the received signal of the far user is

$$y_{Bf,I,J} = \mathbf{h}_{Bf,I,J}^H \sum_{j=1}^{N} \omega_{I,j}\sqrt{P_{B,I,j}}(\sqrt{\alpha_{B,I,j}}s_{Bn,I,j} + \sqrt{1-\alpha_{B,I,j}}s_{Bf,I,j}) \\ + \sum_{i=1}^{L}\sum_{j=1}^{K_i-2N} \mathbf{g}_{Bf,I,J}^H \mathbf{v}_{i,j}\sqrt{P_{S,i,j}}s_{S,i,j} + n. \tag{20.4}$$

The SINR of the far user's signal at the near user can be calculated as

$$\varsigma_{Bf,I,J} = \frac{|\mathbf{h}_{Bn,I,J}^H \omega_{I,J}|^2 P_{B,I,J}(1-\alpha_{B,I,J})}{|\mathbf{h}_{Bn,I,J}^H \omega_{I,J}|^2 P_{B,I,J}\alpha_{B,I,J} + \sigma_{Bn,I,J} + \sigma_{Sn,I,J} + \sigma_n}, \tag{20.5}$$

where σ_n is the AWGN power, $\sigma_{Bn,I,J}$ is the interference from other groups, and $\sigma_{Sn,I,J}$ is the interference from the satellite as follows:

$$\sigma_{Bn,I,J} = \sum_{j=1, j\neq J}^{N} |\mathbf{h}_{Bn,I,J}^H \omega_{I,j}|^2 P_{B,I,j}, \\ \sigma_{Sn,I,J} = \sum_{i=1}^{L}\sum_{j=1}^{K_i-2N} |\mathbf{g}_{Bn,I,J}^H \mathbf{v}_{i,j}|^2 P_{S,i,j}. \tag{20.6}$$

We can observe that the received signal is interfered by the signal of near user, the signals from other groups, and the signals from the satellite. Correspondingly, the SINR of the far user's signal at the far user is

$$\gamma_{Bf,I,J} = \frac{|\mathbf{h}_{Bf,I,J}^H \omega_{I,J}|^2 P_{B,I,J}(1-\alpha_{B,I,J})}{|\mathbf{h}_{Bf,I,J}^H \omega_{I,J}|^2 P_{B,I,J}\alpha_{B,I,J} + \sigma_{Bf,I,J} + \sigma_{Sf,I,J} + \sigma_n}, \tag{20.7}$$

where $\sigma_{Bf,I,J}$ and $\sigma_{Sf,I,J}$ are the interference from other groups and the satellite as follows:

$$\sigma_{Bf,I,J} = \sum_{j=1,j\neq J}^{N} |\mathbf{h}_{Bf,I,J}^{H} \boldsymbol{\omega}_{I,j}|^2 P_{B,I,j}, \tag{20.8}$$

$$\sigma_{Sf,I,J} = \sum_{i=1}^{L} \sum_{j=1}^{K_i-2N} |\mathbf{g}_{Bf,I,J}^{H} \boldsymbol{\nu}_{i,j}|^2 P_{S,i,j}.$$

Within each group, the NOMA scheme is implemented by applying SIC at the user. The decoding order of SIC is in the order of the increasing channel gain, and thus, any user can decode the signals of other users that experience poorer channels than itself. As stated before, the near user experiences a better channel than the far user. The near user will first decode the signal of the far user for interference cancelation and then decode its own signal after subtracting the far user's signal from the received signal. Thus, the near user can decode its signal without the interference from the far user, and the SINR of the near user's signal at the near user is

$$\gamma_{Bn,I,J} = \frac{|\mathbf{h}_{Bn,I,J}^{H} \boldsymbol{\omega}_{I,J}|^2 P_{B,I,J} \alpha_{B,I,J}}{\sigma_{Bn,I,J} + \sigma_{Sn,I,J} + \sigma_n}. \tag{20.9}$$

Then, the capacity of the near user is

$$C_{Bn,I,J} = \log_2(1 + \gamma_{Bn,I,J}). \tag{20.10}$$

In order to perform SIC successfully, the near user needs to first decode the signal of the far user. Thus, the far user's signal has to be decodable at both the near user and the far user, and the capacity of the far user is

$$C_{Bf,I,J} = \log_2(1 + \min\{\varsigma_{Bf,I,J}, \gamma_{Bf,I,J}\}). \tag{20.11}$$

The received signal of the satellite user is

$$\begin{aligned} y_{S,I,J} = & \sum_{i=1}^{L} \sum_{j=1}^{K_i-2N} \mathbf{g}_{S,I,J}^{H} \boldsymbol{\nu}_{i,j} \sqrt{P_{S,i,j}} s_{S,i,j} \\ & + \mathbf{h}_{S,I,J}^{H} \sum_{j=1}^{N} \boldsymbol{\omega}_{I,j} \sqrt{P_{B,I,j}} \left(\sqrt{\alpha_{B,I,j}} s_{Bn,I,j} + \sqrt{1-\alpha_{B,I,j}} s_{Bf,I,j}\right) + n, \end{aligned} \tag{20.12}$$

where $\mathbf{g}_{S,I,J}$ is the channel vector from the satellite to the satellite user and $\mathbf{h}_{S,I,J}$ is the channel vector from the corresponding BS I to the satellite user. Also, we can observe that the satellite user is interfered by the signals of other satellite users and the BS users. The SINR of the satellite user is

$$\gamma_{S,I,J} = \frac{|\mathbf{g}_{S,I,J}^{H} \boldsymbol{\nu}_{I,J}|^2 P_{S,I,J}}{\sigma_{SS,I,J} + \sigma_{BS,I,J} + \sigma_n}, \tag{20.13}$$

where $\sigma_{SS,I,J}$ and $\sigma_{BS,I,J}$ are the interference received from other groups and the according BS as follows:

$$\sigma_{SS,I,J} = \sum_{i=1}^{L} \sum_{j=1,[i,j]\neq[I,J]}^{K_i-2N} |\mathbf{g}_{S,I,J}^{H} \boldsymbol{v}_{i,j}|^2 P_{S,i,j}, \tag{20.14}$$

$$\sigma_{BS,I,J} = \sum_{j=1}^{N} |\mathbf{h}_{S,I,J} \boldsymbol{\omega}_{I,j}|^2 P_{B,I,j}.$$

Then, the capacity of the satellite user is

$$C_{S,I,J} = \log_2(1 + \gamma_{S,I,J}). \tag{20.15}$$

From the above derivation, we can see that the BSs and the satellite serve multiple ground users cooperatively while beamforming and NOMA techniques are applied to reuse the entire bandwidth. Since the service capability of the BS is limited, the satellite can be utilized to provide extra service for those users that cannot be served by BSs. However, due to frequency reuse, the terrestrial BSs and the satellite will interfere with each other. Furthermore, within each network, there also exists interference among users when beamforming is executed. Thus, it is of great importance to design the total system reasonably to achieve optimal capacity performance [29, 30]. The optimization problem can be formulated as

$$\max \sum_{I=1}^{L} \sum_{J=1}^{N} (C_{Bf,I,J} + C_{Bn,I,J}) + \sum_{I=1}^{L} \sum_{J=1}^{K_I-2N} C_{S,I,J}, \tag{20.16}$$

where the optimization variables are $\boldsymbol{\omega}$, $\boldsymbol{v}$, $\mathbf{P}_B$, $\boldsymbol{\alpha}$, $\mathbf{P}_S$, and the paring scheme, and the optimization objective is the system capacity performance.

The optimization variables in (20.16) can be divided into three parts. First is the paring scheme, in which we need to decide the user set to be served by the satellite, and the pairs of near user and far user to implement NOMA for each BS. Then, the beamforming vectors $\boldsymbol{\omega}$, $\boldsymbol{v}$ for both BSs and the satellite are required to mitigate the interference between users. Finally, optimal power allocation schemes for $\mathbf{P}_B$, $\boldsymbol{\alpha}$, $\mathbf{P}_S$ are needed to implement the inter-group power allocation, the intra-group power allocation for BSs, and the inter-user power allocation for the satellite.

The optimization problem in (20.16) is a highly non-convex problem that has NP complexity, which cannot be solved directly. Also, since the satellite interferes all the BS users within its coverage, it will lead to a relatively small power allocation for the satellite user if we simply maximize the total capacity. Consider both the two above problems, we decompose the optimization problem into three parts: the paring scheme, the terrestrial resource allocation scheme, and the satellite resource allocation scheme. Then, an irrelative algorithm is proposed to optimize the capacity performance of the whole system.

20.3 User Paring Scheme

Within the coverage radius of each BS, there are $|U_i| = K_i$ users. By means of beamforming and NOMA, the BS is able to serve up to $2N$ users simultaneously while reusing the entire bandwidth. In this chapter, we assume that the total user number is beyond the service capability of BS, when $|U_i| = K_i > 2N$. In this case, the satellite is utilized to provide extra service for the remaining $K_i - 2N$ users. For each BS, we need to determine which $K_i - 2N$ users are to be served by the satellite while taking the system capacity into consideration. Also, the $2N$ BS users are divided into N groups and the NOMA scheme is applied within each group. The paring scheme for the $2N$ users will have an impact on the performance of the system since the inter-group interference is quite relative to the channel conditions of different groups when beamforming is executed, and the performance of NOMA is also influenced by the channel conditions of the two users.

20.3.1 Selection of Satellite User

First, we need to determine the $K_i - 2N$ users to be served by the satellite for each BS. Since the relation between the selection of satellite users and the system performance is indirect and complex, it is difficult to find the optimal strategy with low complexity. Thus, we turn to a suboptimal strategy taking the following paring, beamforming, and power allocation into consideration. In (20.12), we can see that the satellite user is interfered by the BS with the channel $\mathbf{h}_{S,I,J}$. It is explicit that we should choose those users that may suffer less interference from the BS, that is, the user with smaller $|\mathbf{h}_{S,I,J}|$. Also, since we consider the scenario where the BSs and the satellite serve the users cooperatively, it is better for those users with poorer channel conditions to be served by the satellite, while those users with better channel conditions are to be served by the BSs. Moreover, the channel conditions of the satellite $\mathbf{g}_{S,I,J}$ should also be taken into account for improving the capacity of the satellite users. Then, we have Algorithm 1.

Algorithm 1 Satellite User Selection Algorithm

1: **for** $I = 1$ to L **do**
2:    **for** $J = 1$ to K_I **do**
3:       Calculate $\eta_{I,J} = \frac{|g_{I,J}|}{|\mathbf{h}_{I,J}|}$
4:    **end for**
5:    Rank the K_I users of BS I by $\eta_{I,J}$
6:    Select the $K_I - 2N$ users with the largest $\eta_{I,J}$ as the satellite users $U_{S,I}$
7: **end for**

In Algorithm 1, we select the users with the largest channel condition ratio $\eta_{I,J} = \frac{|\mathbf{g}_{I,J}|}{|\mathbf{h}_{I,J}|}$ as the satellite users, where $\mathbf{g}_{I,J}$ and $\mathbf{h}_{I,J}$ represent the channels of users $J \in U_I$. In this case, they will suffer less interference from the BSs with a relatively smaller $|\mathbf{h}_{I,J}|$ and also larger received signal power with a relative larger $|\mathbf{g}_{I,J}|$. The main computation in Algorithm 1 is the ranking of $\eta_{I,J}$, whose computational complexity is $O(K_I \log_2(K_I))$ separately. Also, since there are total L BSs, the computational complexity of Algorithm 1 is $O(LK_I \log_2(K_I))$.

20.3.2 Terrestrial User Paring Scheme

After selecting the $K_I - 2N$ satellite users for each BS I, the remaining $2N$ users $U_{B,I} = \{u_{B,I,i}, ..., u_{B,I,2N}\}$ will be served by the BS. The $2N$ users will be divided into N groups, and each group consists of a near user and a far user. Within each group, the NOMA scheme is performed via superposition coding at the transmitter and SIC at the receiver. For each BS, it is important to determine how to pair the $2N$ users into N groups in order that the system can achieve better capacity performance.

In [13], the author proved that it is ideal to pair users with distinct channel conditions when performing NOMA, which is also why existing works on NOMA generally pair one near user and one far user as a group. From this point, we first divide the $2N$ users into two sets, the near user set and the far user set, according to their channel condition. For each BS I, we rank the channel conditions $|\mathbf{h}_{\mathbf{B,I,J}}|$ of the $2N$ users, and select the N users with better channel conditions as the near user set, while the N users with poorer channel conditions are selected as the far user set. For convenience, in this chapter, we assume the near user set consists of the first N users of $U_{B,I}$, while the far user set consists of the last N users of $U_{B,I}$. Each pair of users to perform NOMA will consist of one user from the near user set and one user from the far user set. Then, the paring problem is transformed into the problem how to pair the near user set with the far user set.

From (20.3) and (20.4), we can observe that the received signal of the BS users will be interfered by the signals from other groups due to frequency reuse. With only N antennas, the interference cannot be completely canceled for all the $2N$ users by means of beamforming. When performing SIC in each group, the near user needs to decode the signal of the far user before decoding its own signal. In order that the NOMA scheme can be performed correctly, it is important to protect the SINR of near users. Since zero-forcing beamforming (ZFBF) can completely cancel the co-channel interference among users, it will be a preferable choice to design the beamforming vectors using ZFBF based on the channels of nears users. Then, with respect to the far users, if the two users in one group experience the same channel or the channels of the two users are linear correlative, the two users can be treated as one user when executing beamforming, and we can simply execute ZFBF among groups, and the inter-group interference can be completely canceled. Inspired by this, it will improve the system performance to pair the users with high correlative channels into one group, which helps to mitigate the inter-group interference when executing beamforming among groups. The channel correlation coefficient is defined as follows.

$$Corr_{(I,J)} = \frac{|\mathbf{h}_{Bn,I,J}^{H}\mathbf{h}_{Bf,I,J}|}{|\mathbf{h}_{Bn,I,J}||\mathbf{h}_{Bf,I,J}|}. \tag{20.17}$$

The larger the value is, the higher the correlation of the two channels are. If $Corr_{(I,J)} = 1$, it means the two channels are linear correlative. Also, considering fairness among users, we should avoid causing too large interference to one group when optimizing the capacity of the other group, which means we should not maximize the channel correlation coefficients in some groups while ignoring the channel correlation coefficients in other groups. Then, the max–min strategy, which is typical for user fairness in many areas, is adopted, and we have the pairing scheme as

$$\max_{Pairing} \min_{J} Corr_{(I,J)} = \frac{|\mathbf{h}_{Bn,I,J}^{H}\mathbf{h}_{Bf,I,J}|}{|\mathbf{h}_{Bn,I,J}||\mathbf{h}_{Bf,I,J}|}. \tag{20.18}$$

For each BS I, the objective of the pairing scheme is to maximize the minimum channel correlation coefficient between the two users within one group. This is a non-convex problem and cannot be solved directly by typical optimization methods. The max–min problem can be equivalently transformed into

$$\begin{aligned} &\max_{Pairing} \ \rho \\ &C1: \frac{|\mathbf{h}_{Bn,I,J}^{H}\mathbf{h}_{Bf,I,J}|}{|\mathbf{h}_{Bn,I,J}||\mathbf{h}_{Bf,I,J}|} \geq \rho, J = 1, 2, ..., N. \end{aligned} \tag{20.19}$$

The transformed problem is also a non-convex problem, but it can be solved by an irrelative algorithm, in which the subproblem in each iteration is:

$$\begin{aligned} &\text{Find paring scheme subject to} \\ &\frac{|\mathbf{h}_{Bn,I,J}^{H}\mathbf{h}_{Bf,I,J}|}{|\mathbf{h}_{Bn,I,J}||\mathbf{h}_{Bf,I,J}|} \geq \rho[t], J = 1, 2, ..., N. \end{aligned} \tag{20.20}$$

If there exists a paring scheme subject to the constraints in (20.20), it means the maximum value of ρ is no less than $\rho[t]$. Otherwise, the maximum value of ρ is less than $\rho[t]$. Update $\rho[t+1]$ according to $\rho[t]$, and continue the iteration until the optimal solution is obtained.

In each iteration, the problem in (20.20) is an NP problem, which is of too high complexity if solved by exhaustive search. We solve this problem from another view. The $2N$ users of BS I can be viewed as $2N$ vertexes, in which N vertexes belong to the near user set while other N vertexes belong to the far user set. If the channel

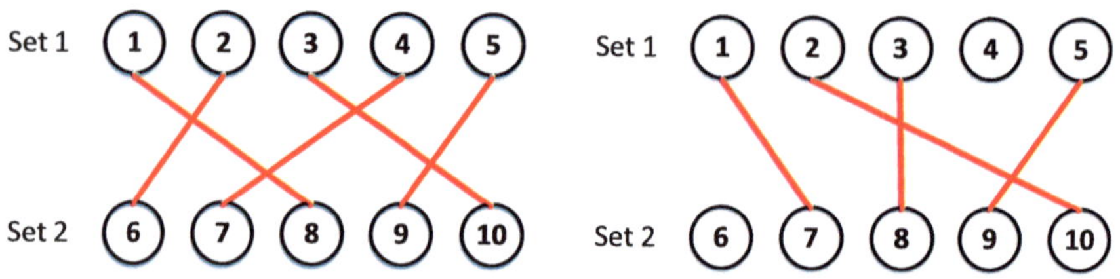

Fig. 20.2 Examples of the bipartite graph

correlation coefficient between two users $Corr_{(I,j,k)} = \frac{|\mathbf{h}_{B,I,j}^H \mathbf{h}_{B,I,k}|}{|\mathbf{h}_{B,I,j}||\mathbf{h}_{B,I,k}|} \geq \rho$, we build an edge (j, k) between the two vertexes. Then, all the $2N$ vertexes V and $|E|$ edges E form the graph $G(V, E)$. The vertexes in G can be partitioned into two sets, and each edge consists of vertexes from both sets. Thus, the graph G is a bipartite graph [31]. Then, the problem in (20.20) is to find N edges in G that exactly connect the $2N$ vertexes, which is equivalent to find the perfect matching of the bipartite graph G. An example of a perfect matching of the bipartite graph G is illustrated in Fig. 20.2a. By means of Hungarian method, one can find the maximum matching of a bipartite graph [32], which means the maximum selection of edges when no vertex is used twice, as illustrated in Fig. 20.2b. In each iteration, we can use the Hungarian method to find the maximum matching of the bipartite graph G. If the edge number of the maximum matching of is N, it means that we can find N edges in G that exactly connect the $2N$ vertexes, and the subproblem (20.20) has a solution. If the edge number of the maximum matching of G is less than N, it means that we cannot find N edges in G that exactly connect the $2N$ vertexes, and thus subproblem (20.20) has no solution.

By transforming the optimization problem into a bipartite graph problem, we can solve the subproblem in (20.20) using the Hungarian method. Then, the maximum value of ρ can be obtained via a bisection procedure through iteration. Finally, we obtain the pairing scheme that maximizes the minimum channel correlation coefficient between the two users within one group. The algorithm for user pairing is summarized as Algorithm 2. In each iteration, the computational complexity of the Hungarian method used for the subproblem is $O(N^2)$ [32], and there will be no more than $\log_2(\varepsilon)$ iterations to reach the computational accuracy ε. Thus, the total computational complexity of Algorithm 2 is $O(\log_2(\varepsilon)N^2)$.

Algorithm 2 Terrestrial User Paring Algorithm

Rank the $2N$ users of BS I by $|\mathbf{h}_{\mathbf{B,I,J}}|$
Select the first N users as the near user set $U_{Bn,I} = \{u_{B,I,1}, ..., u_{B,I,N}\}$
Select the last N users as the far user set $U_{Bf,I} = \{u_{B,I,N+1}, ..., u_{B,I,2N}\}$
Initiate $G = G(V, E)$
Calculate $Corr_{(I,j,k)} = \frac{|\mathbf{h}_{B,I,j}^H \mathbf{h}_{B,I,k}|}{|\mathbf{h}_{B,I,j}||\mathbf{h}_{B,I,k}|}$ for all $j \in U_{Bn,I}$ and $k \in U_{Bf,I}$
Initiate edge $e(j, k) = 0$ for all $j \in U_{Bn,I}$ and $k \in U_{Bf,I}$
Initiate $\rho_{LB} = 0$, $\rho_{RB} = 1$, $\varepsilon = \varepsilon_0$
repeat
  Calculate $\rho = \frac{\rho_{LB}+\rho_{RB}}{2}$
  Set edge $e(j, k) = 1$ if $Corr_{(I,j,k)} \geq \rho$ for all $j \in U_{Bn,I}$ and $k \in U_{Bf,I}$
  Find the maximum matching G_M of G using Hungarian method
  if $|G_M| = N$ **then**
    $\rho_{LB} = \rho$
  else
    $\rho_{RB} = \rho$
  end if
until $\rho_{RB} - \rho_{LB} < \varepsilon$

20.4 Terrestrial Resource Allocation

In this section, we study the resource allocation scheme for the $2N$ users when fixing the satellite parameters. The terrestrial resource allocation scheme includes terrestrial beamforming, intra-group power allocation, and inter-group power allocation.

20.4.1 Terrestrial Beamforming

Equipped with N antennas, the spatial degrees of freedom of each BS are N. However, since the NOMA technique is applied in the system, up to $2N$ users are served simultaneously by each BS. The beamforming is executed among N groups when each group consists of two users. Thus, the inter-group interference cannot be completely canceled unless the channels of the two users in one group are linear correlative.

Based on the user pairing scheme in last section, the users in one group will experience relatively high correlative channels. In this case, even if we design the beamforming vectors with only the channel of one user in each group, the other user will not suffer too large interference since its channel is correlative with the channel used for the design of beamforming. In ideal condition, the channels of the two users in one group are linear correlative, and then, the two users can be viewed as one user when executed beamforming. In addition, when performing SIC in each group, the near user needs to decode the signal of the far user before decoding its own signal. In order that the NOMA scheme can be performed correctly, we design the beamforming vectors based on the channels of near users to protect the SINR of near users. For the N near users, ZFBF can be utilized to cancel inter-group interference, and the designed beamforming vectors need to satisfy

$$\mathbf{h}_{Bn,I,j}^{H}\boldsymbol{\omega}_{I,k} = 0, j \neq k. \tag{20.21}$$

Let $\mathbf{H}$ be the channel matrix of the near user, $\mathbf{H} = [\mathbf{h}_{Bn,I,1}, \mathbf{h}_{Bn,I,2}, ..., \mathbf{h}_{Bn,I,N}]^{H}$. Then, the beamforming matrix can be obtained as

$$\mathbf{W} = [\boldsymbol{\omega}_{I,1}, ..., \boldsymbol{\omega}_{I,N}] = \mathbf{H}^{-1}\mathbf{D}, \tag{20.22}$$

where D is a diagonal matrix for normalization, and

$$\mathbf{D}^{2} = \mathtt{diag}\{\frac{1}{(\mathbf{H}^{-H}\mathbf{H}^{-1})_{1,1}}, ..., \frac{1}{(\mathbf{H}^{-H}\mathbf{H}^{-1})_{N,N}}\}, \tag{20.23}$$

where $(\mathbf{H})_{m,m}$ represents the mth element of the main diagonal of $\mathbf{H}$. In the rest of the chapter, we use $\boldsymbol{\omega}_{n,I,J}$ instead of $\boldsymbol{\omega}_{I,J}$ to represent the beamforming vector based on the channel of the near user.

20.4.2 Intra-group Power Allocation

Within each group, SIC is performed at the users for NOMA. The transmit power $P_{B,I,J}$ for group J is divided between the near user and the far user by the power allocation coefficient $\alpha_{B,I,J}$, as in (20.1). However, the selection of $\alpha_{B,I,J}$ is a dilemma problem. Generally, the optimization objective is to maximize the system capacity. It is easy to prove that the optimal value of $\alpha_{B,I,J}$ is $\alpha_{B,I,J} = 1$ if we maximize the total capacity of the two users [20]. In this case, all power is allocated to the near user that with better channel conditions. Although we can achieve higher system capacity with this choice, it ignores user fairness completely, and the NOMA scheme actually is not performed since only one user is served.

Taking user fairness into consideration, we need to find other criteria for intra-group power allocation instead of only considering the system total capacity. The reason why we apply the NOMA scheme is that NOMA can achieve better capacity performance for multiple access than traditional OMA [1], such as TDMA and FDMA. Thus, we introduce the constraint that the capacity of the far user should be no less than the capacity when conventional TDMA is applied for multiple access:

$$C_{Bf,I,J} = \log_2(1 + \gamma_{Bf,I,J}) \geq \frac{1}{2}\log_2(1 + \gamma_{conv,Bf,I,J}). \tag{20.24}$$

Since there will be no inter-group interference for the near user when ZFBF is executed based on the channel of the near user, as well as the fact that the near user experience a better channel than the far user, it is reasonable to assume that

$\varsigma_{Bf,I,J} > \gamma_{Bf,I,J}$ is naturally satisfied for all I, J. Thus, the capacity of the far user is $C_{Bf,I,J} = \log_2(1 + \gamma_{Bf,I,J})$.

When the conventional TDMA is applied, during each time slot, only one user is served within each group. Then, ZFBF will be executed among the N users served during each time slot, and thus, there will be no inter-group interference among users. The SINR of the far user in the case of TDMA is

$$\gamma_{conv,Bf,I,J} = \frac{|\mathbf{h}_{Bf,I,J}^{H}\boldsymbol{\omega}_{f,I,J}|^{2}P_{B,I,J}}{\sum_{i=1}^{L}\sum_{j=1}^{K_i-2N}|\mathbf{g}_{Bf,I,J}^{H}\mathbf{v}_{i,j}|^{2}P_{S,i,j} + \sigma_n}, \tag{20.25}$$

where $\boldsymbol{\omega}_{f,I,J}$ is the beamforming vector based on the channel of the far user, which can be obtained referring to (20.22). Also, because the time slots are divided between the two users, there exists a coefficient of $\frac{1}{2}$ in (20.24). By solving (20.24), we can obtain

$$\alpha_{B,I,J} \le \alpha_{B,I,J,1} = \frac{1}{\sqrt{1+\gamma_{conv,Bf,I,J}}} + \frac{(\sigma_{Bf,I,J} + \sigma_{Sf,I,J} + \sigma_n)(1 - \sqrt{1+\gamma_{conv,Bf,I,J}})}{|\mathbf{h}_{Bf,I,J}^{H}\omega_{n,I,J}|^{2}P_{B,I,J}\sqrt{1+\gamma_{conv,Bf,I,J}}}. \tag{20.26}$$

When $\sigma_{Bf,I,J}$ is large, the result $\alpha_{B,I,J,1}$ calculated by (20.26) may be quite small or even negative. To avoid this extreme situation and also protect the capacity of the near user, we also consider the constraint that the capacity of the near user should be no less than the capacity of conventional TDMA:

$$C_{Bn,I,J} = \log_2(1 + \gamma_{Bn,I,J}) \ge \frac{1}{2}\log_2(1 + \gamma_{conv,Bn,I,J}), \tag{20.27}$$

where

$$\gamma_{conv,Bn,I,J} = \frac{|\mathbf{h}_{Bn,I,J}^{H}\omega_{n,I,J}|^{2}P_{B,I,J}}{\sum_{i=1}^{L}\sum_{j=1}^{K_i-2N}|\mathbf{g}_{Bn,I,J}^{H}\mathbf{v}_{i,j}|^{2}P_{S,i,j} + \sigma_n}. \tag{20.28}$$

Then, we can obtain

$$\alpha_{B,I,J} \ge \alpha_{B,I,J,2} = \frac{\sigma_{Sn,I,J} + \sigma_n}{|\mathbf{h}_{Bn,I,J}^{H}\boldsymbol{\omega}_{n,I,J}|^{2}P_{B,I,J}}\left(\sqrt{1+\gamma_{conv,Bn,I,J}} - 1\right). \tag{20.29}$$

Since the channel of the near user is generally much better than the far user, the capacity of the near user will contribute much more to the total system performance than the far user. It will significantly deteriorate the system performance if we sacrifice most capacity of the near user while only little capacity gain can be achieved for the far user. Thus, we consider protecting the capacity of the near user with higher priority, and the final power allocation coefficient is selected as

$$\alpha_{B,I,J} = \min\{1, \max\{\alpha_{B,I,J,1}, \alpha_{B,I,J,2}\}\}. \tag{20.30}$$

20.4.3 Inter-group Power Allocation

After determining the beamforming vectors and the intra-group power allocation based on the above analysis, to maximize the capacity performance of the BSs, it is important to implement reasonable power allocation among groups under the constraint of total power. Since frequency reuse is considered in the system, there will be inter-group interference and also interference from the satellite. Also, in order that the NOMA scheme can be performed correctly, we implement ZFBF based on the channels of the near users to protect the SINR of the near users, and thus, only the far users will suffer from the inter-group interference. The optimization problem can be formulated as

$$\begin{aligned} \max_{\mathbf{P}_{B,I}} C_{B,I} &= \sum_{J=1}^{N} (C_{Bf,I,J} + C_{Bn,I,J}) \\ C1 &: \sum_{J=1}^{N} P_{B,I,J} \leq P_{B,I,\max}, \\ C2 &: P_{B,I,J} \geq 0, \forall J, \end{aligned} \tag{20.31}$$

where $P_{B,I,\max}$ is the maximum transmit power of the BS.

This problem is non-convex because the capacity function of the far user $\log_2(1 + \gamma_{Bf,I,J})$ is a highly non-convex function. In order to overcome this problem, we adopt the successive convex approximation (SCA) approach proposed in [33]:

1. **Step 1**: Start from a feasible point $\mathbf{P}_{B,I}[1]$ and set $t = 1$.
2. **Step 2**: In the tth iteration, use some convex function around the point $\boldsymbol{P}_{B,I}[t]$ to approximate the non-convex function, and transform the original non-convex problem into a convex subproblem.
3. **Step 3**: Solve the transformed convex subproblem in Step 2 to obtain the optimal solution $\mathbf{P}_{B,I}[t+1]$ of the subproblem.
4. **Step 4**: Increase $t = t + 1$, and continue the iteration until $\mathbf{P}_{B,I}[t]$ converges.

The convergence of the SCA approach is proved in [33], and it also proves that the SCA approach will converge to the KKT point. In each iteration, we approximate the non-convex function in the objective function by logarithmic approximation [34], which gives the lower bound of the original function:

$$\ln(1 + \gamma_{I,J}) \geq \theta_{I,J} \ln \gamma_{I,J} + \beta_{I,J}, \tag{20.32}$$

which is tight at $\gamma_{I,J} = \overline{\gamma}_{I,J}$ when approximation parameters are chosen as

$$\theta_{I,J} = \frac{\overline{\gamma}_{I,J}}{1 + \overline{\gamma}_{I,J}}, \quad \beta_{I,J} = \ln(1 + \overline{\gamma}_{I,J}) - \frac{\overline{\gamma}_{I,J}}{1 + \overline{\gamma}_{I,J}} \ln \overline{\gamma}_{I,J}. \tag{20.33}$$

Approximate the objective function by logarithmic approximation, we can obtain

$$C_{Bf,I,J} + C_{Bn,I,J} \geq \frac{1}{\ln 2}(\theta_{Bf,I,J} \ln \gamma_{Bf,I,J} + \beta_{Bf,I,J}) + \frac{1}{\ln 2}(\theta_{Bn,I,J} \ln \gamma_{BnI,J} + \beta_{Bn,I,J}), \tag{20.34}$$

in which approximation parameters are calculated referred to (20.33). Then, changing the variable $\mathbf{P}_{B,I}$ by $\widehat{\mathbf{P}}_{B,I} = \ln \mathbf{P}_{B,I}$, the optimization problem in each iteration is transformed into

$$\max_{\widehat{\mathbf{P}}_{B,I}} \sum_{J=1}^{N} \left(C^L_{Bf,I,J} + C^L_{Bn,I,J} \right) \tag{20.35}$$

$$C1 : \sum_{J=1}^{N} e^{\widehat{P}_{B,I,J}} \leq P_{B,I,\max},$$

where $C^L_{Bf,I,J} + C^L_{Bn,I,J}$ is the lower bound of $C_{Bf,I,J} + C_{Bn,I,J}$ using logarithmic approximation in (20.34) as well as the variable transformation of $\widehat{\mathbf{P}}_{B,I} = \ln \mathbf{P}_{B,I}$.

Since the log-sum-exp function is convex [35], it is easy to prove the transformed subproblem in each iteration is a standard convex optimization problem. However, by solving the subproblem, we only obtain the lower bound of the capacity. To solve the original problem (20.31), we iteratively update the approximation parameters $\boldsymbol{\theta}$ and $\boldsymbol{\beta}$ using the subproblem solution $\mathbf{P}_{B,I}[t+1]$ referring to (20.33) and use the updated parameters for the next iteration until the results converge, as stated in the SCA approach above.

In each iteration, we solve the subproblem (20.35) by means of the Lagrangian dual method. The Lagrangian function is

$$L(e^{\widehat{\mathbf{P}}_{B,I}}, \lambda) = -\sum_{J=1}^{N} \left[C^L_{Bf,I,J} + C^L_{Bn,I,J} \right] - \lambda(P_{B,I,\max} - \sum_{J=1}^{N} e^{\widehat{P}_{B,I,J}}). \tag{20.36}$$

where λ is the Lagrange multiplier for the constraint C1 in (20.35). Note that we have transformed the optimization problem into the standard form. The Lagrangian dual function is then given by:

$$D(\lambda) = \inf_{\widehat{\mathbf{P}}_{B,I}} \{L(e^{\widehat{\mathbf{P}}_{B,I}}, \lambda)\}. \tag{20.37}$$

By solving the stationary condition $\frac{\partial L}{P_{B,I,J}} = 0$, we can obtain the optimal power allocation scheme of the subproblem in the form of λ:

$$P_{B,I,J}[t+1] = e^{\widehat{P}_{B,I,J}[t+1]} = \left[\frac{\theta_{Bf,I,J} + \theta_{Bn,I,J}}{\lambda \ln 2 + \xi_{Bf,I,J}(J,t)\alpha_{B,I,J} + \sum_{j=1, j \neq J}^{N} \xi_{Bf,I,J}(j,t)} \right]^+, \tag{20.38}$$

where $(x)^+ = \max(0, x)$, and we define

$$\xi_{Bf,I,J}(j,t) = \theta_{Bf,I,j} \frac{|\boldsymbol{h}^H_{Bf,I,j} \boldsymbol{\omega}_{n,I,J}|^2}{I_{B,I,j}[t]}, \tag{20.39}$$

where

$$I_{B,I,J}[t] = |\mathbf{h}_{Bf,I,J}^{H}\omega_{n,I,J}|^2 e^{\widehat{P}_{B,I,J}[t]}\alpha_{B,I,J} + \sum\nolimits_{j=1,j\neq J}^{N} |\mathbf{h}_{Bf,I,J}^{H}\omega_{n,I,j}|^2 e^{\widehat{P}_{B,I,j}[t]} + \sigma_{Sf,I,J} + \sigma_n, \quad (20.40)$$

which is calculated based on the solution of the last iteration.

Since $D(\lambda)$ in (20.37) is not differentiable, the Lagrange multiplier λ can be obtained by using the subgradient method iteratively as follows

$$\lambda[t_\delta + 1] = \left[\lambda[t_\delta] - \delta[t_\delta + 1](P_{B,I,\max} - \sum_{J=1}^{N} P_{B,I,J})\right]^{+}, \quad (20.41)$$

where $\delta[t_\delta + 1]$ is the step size in each iteration. In addition, since $\mathbf{P}_{B,I}$ is updated in the iteration of t_δ, $I_{B,I,J}$ will also be updated for the next iteration of $t_\delta + 1$. Also, since subproblem (20.35) is a convex problem with only one inequality constraint, instead of exploiting the iterative subgradient method, there is an alternative simpler method based on the fact that $P_{B,I,J}$ is a monotonic function of λ. According to the complementary slackness in the KKT conditions, the optimal λ can be obtained as follows: First check if λ can be zero by checking whether the constraint C1 in problem (20.35) is satisfied via setting $\lambda = 0$. If satisfied, the optimal λ is zero. If not, the constraint C1 in problem (20.35) is active at the optimum and the optimal λ can be solved by the bisection method.

The inter-group power allocation scheme is summarized as Algorithm 3. In the outer loop, we update the approximation parameters $\boldsymbol{\theta}$ and $\boldsymbol{\beta}$ and transform the original optimization problem into a solvable convex problem by logarithmic approximation. In the inner loop, we solve the transformed subproblem using the Lagrangian dual method. By updating the approximation parameters iteratively, we can finally obtain the optimal solution of the original problem in (20.31). In each iteration of the inner loop, the computational complexity is O(N^2) for updating $\mathbf{P}_{\mathbf{B},\mathbf{I}}$, λ, $\boldsymbol{\alpha}_{B,I}$, and $\mathbf{I}_{\mathbf{B},\mathbf{I}}$. In each iteration of the outer loop, the computational complexity is O(N^2) for updating $\boldsymbol{\theta}_{B,I}$ and $\boldsymbol{\beta}_{B,I}$. Also, since there are total L BSs, the computational complexity of Algorithm 3 is O(LN^2).

20.5 Satellite Resource Allocation

Based on the user paring scheme in Sect. 20.3, $\sum_{i=1}^{L} K_i - 2N$ users are selected to be served by the satellite. For simplification and also without loss of typicality, we assume that the user number of the satellite equals the antenna number, that is, $\sum_{i=1}^{L} K_i - 2N = M$. By means of beamforming, the satellite serves these M users simultaneously while reusing the entire bandwidth. Similar to the case of terrestrial

Algorithm 3 Inter-group Power Allocation Algorithm

```
Initialize t = 1, θ_{Bf,I,J} = θ_{Bn,I,J} = 1, β_{Bf,I,J} = β_{Bn,I,J} = 0 and P_{B,I,J} = 0 for all J
repeat
  Initialize t_δ = 1, λ > 0
  Initialize I_{B,I} referring to (20.40)
  repeat
    Update P_{B,I} referring to (20.38)
    Update λ referring to (20.41)
    Update α_{B,I} referring to (20.30)
    Update I_{B,I} referring to (20.40)
    Set t_δ = t_δ + 1
  until λ converges
  Set P_{B,I}[t] = P_{B,I}[t + 1]
  Update θ_{B,I} and β_{B,I} referring to (20.33)
  Set t = t + 1
until P_{B,I} converges
```

resource allocation, we first study the resource allocation scheme of the satellite, including satellite beamforming vectors and power allocation, when fixing the terrestrial parameters. Then, an iterative algorithm is proposed to jointly optimize the power allocation scheme of the whole system.

20.5.1 Satellite Beamforming

For multiple access, ZFBF can be an effective method to overcome interference if multi-antennas are equipped, as in the terrestrial beamforming scheme. However, generally, there is a strong main path from the satellite to the users. Thus, the channel model of the satellite is considered as either an additive white Gaussian noise (AWGN) channel or a Rician channel in most cases, which experience small channel fluctuations. To execute ZFBF, the inverse of the channel matrix is required to obtain the beamforming vectors in (20.22). In the case of small fluctuations, this condition may not be satisfied. Due to the special characteristic of the satellite channel, a different beamforming scheme is needed.

Let G be the channel matrix of the satellite users, $\mathbf{G} = [\mathbf{g}_{1,1}, \mathbf{g}_{1,2}, \ldots, \mathbf{g}_{L,K_L-2N}]^H$. If G is invertible, it is obvious that ZFBF is the prior choice, and the beamforming vectors can be calculated by:

$$[\boldsymbol{v}_{1,1}, \ldots, \boldsymbol{v}_{L,K_L-2N}] = \mathbf{G}^{-1}\mathbf{D}, \tag{20.42}$$

where D is defined the same as (20.22).

Then, if G is not invertible, the maximum ratio transmission (MRT) beamforming will be applied instead of ZFBF [36]:

$$\boldsymbol{v}_{I,J} = \frac{\mathbf{g}_{S,I,J}}{||\mathbf{g}_{S,I,J}||}, \tag{20.43}$$

which maximizes the power of the received signal of each user.

20.5.2 Satellite Power Allocation

In this subsection, fixing the terrestrial parameters, we study the power allocation scheme for the satellite users. As stated before, the satellite will cause interference to all the BS users within its coverage. Thus, reasonable power controlling is required not only to improve the capacity performance of the satellite users, but also to avoid too large interference to the BS users.

We introduce the concept of interference temperature for the satellite to limit the interference caused to the BS users. For each BS user, the interference from the satellite should be no more than the interference temperature limit P_{th}, and then, the optimization problem can be formulated as follows:

$$\begin{aligned} &\max_{\mathbf{P_S}} \sum_{I=1}^{L} \sum_{J=1}^{K_I-2N} \log_2(1+\gamma_{S,I,J}) \\ &C1: \sum_{I=1}^{L} \sum_{J=1}^{K_I-2N} |\mathbf{g}_{Bf,l,n}^H \boldsymbol{v}_{I,J}|^2 P_{S,I,J} \le P_{th}, \forall l,n, \\ &C2: \sum_{I=1}^{L} \sum_{J=1}^{K_I-2N} |\mathbf{g}_{Bn,l,n}^H \boldsymbol{v}_{I,J}|^2 P_{S,I,J} \le P_{th}, \forall l,n, \\ &C3: \sum_{I=1}^{L} \sum_{J=1}^{K_I-2N} P_{S,I,J} \le P_{S,\max}, \\ &C4: P_{S,I,J} \ge 0, \forall I,J. \end{aligned} \tag{20.44}$$

where $P_{S,\max}$ is the maximum transmit power of the satellite. Also, the interference temperature limit P_{th} is considered for both the near users and the far users in the network.

This is also a non-convex problem due to the non-convex objective function. Similarly, we use the SCA approach introduced in Sect. 20.4.3 to solve this non-convex problem. By means of the logarithmic approximation in (20.32) and the variable transformation of $\widehat{\mathbf{P}}_S = \ln \mathbf{P}_S$, the optimization problem can be solved via iteration of subproblems. In each iteration, the subproblem is

$$\min_{\widehat{\mathbf{P}}_S} - \sum_{I=1}^{L} \sum_{J=1}^{K_I-2N} C_{S,I,J}^{L} \tag{20.45}$$

$$C1: P_{th} - \sum_{I=1}^{L} \sum_{J=1}^{K_I-2N} |\mathbf{g}_{Bf,l,n}\mathbf{v}_{I,J}|^2 e^{\widehat{P}_{S,I,J}} \geq 0, \forall l, n,$$

$$C2: P_{th} - \sum_{I=1}^{L} \sum_{J=1}^{K_I-2N} |\mathbf{g}_{Bn,l,n}\mathbf{v}_{I,J}|^2 e^{\widehat{P}_{S,I,J}} \geq 0, \forall l, n,$$

$$C3: P_{S,\max} - \sum_{I=1}^{L} \sum_{J=1}^{K_I-2N} e^{\widehat{P}_{S,I,J}} \geq 0.$$

Note that we have transform the problem into the standard form. Also, we solve this subproblem by means of the Lagrangian dual method. The Lagrangian function is

$$\begin{aligned} L(\widehat{\mathbf{P}}_S, \boldsymbol{\mu}, \lambda) = & -\sum_{I=1}^{L} \sum_{J=1}^{K_I-2N} C_{S,I,J}^{L} - \lambda(P_{S,\max} - \sum_{I=1}^{L} \sum_{J=1}^{K_I-2N} e^{\widehat{P}_{S,I,J}}) \\ & -\sum_{l=1}^{L} \sum_{n=1}^{N} \mu_{Bf,l,n}(P_{th} - \sum_{I=1}^{L} \sum_{J=1}^{K_I-2N} |\mathbf{g}_{Bf,l,n}^{H}\mathbf{v}_{I,J}|^2 e^{\widehat{P}_{S,I,J}}) \\ & -\sum_{l=1}^{L} \sum_{n=1}^{N} \mu_{Bn,l,n}(P_{th} - \sum_{I=1}^{L} \sum_{J=1}^{K_I-2N} |\mathbf{g}_{Bn,l,n}^{H}\mathbf{v}_{I,J}|^2 e^{\widehat{P}_{S,I,J}}). \end{aligned} \tag{20.46}$$

By solving the stationary condition $\frac{\partial L}{P_{S,I,J}} = 0$, we can obtain the optimal power allocation scheme of the subproblem in the form of λ and $\boldsymbol{\mu}$:

$$P_{S,I,J}[t+1] = e^{\widehat{P}_{S,I,J}[t+1]} = \frac{\theta_{S,I,J}}{(\sum_{i=1}^{L} \sum_{j=1,[i,j]\neq[I,J]}^{K_i-2N} \theta_{S,i,j} \frac{|\mathbf{g}_{S,i,j}^{H}\mathbf{v}_{I,J}|^2}{I_{S,i,j}[t]}) + \ln 2\phi_{S,I,J} + \lambda \ln 2}, \tag{20.47}$$

where $(x)^+ = \max(0, x)$, and we define

$$I_{S,i,j}[t] = \sum_{k=1}^{L} \sum_{m=1,[k,m]\neq[i,j]}^{K_k-2N} |\mathbf{g}_{S,i,j}^{H}\mathbf{v}_{k,m}|^2 e^{\widehat{P}_{S,k,m}[t]} + \sigma_{BS,i,j} + \sigma_n, \tag{20.48}$$

$$\phi_{S,I,J} = \sum_{l=1}^{L} \sum_{n=1}^{N} (\mu_{Bf,l,n}|\mathbf{g}_{Bf,l,n}^{H}\mathbf{v}_{I,J}|^2 + \mu_{Bn,l,n}|\mathbf{g}_{Bn,l,n}^{H}\mathbf{v}_{I,J}|^2),$$

which is calculated based on the solution of the last iteration. The Lagrange multipliers can be obtained by using the subgradient method iteratively as follows

$$\mu_{Bf,l,n}[t_\delta+1] = \left[\mu_{Bf,l,n}[t_\delta] - \delta_{Bf,l,n}[t_\delta+1]\left(P_{th} - \sum_{I=1}^{L}\sum_{J=1}^{K_I-2N} |\mathbf{g}_{Bf,l,n}^{H}\mathbf{v}_{I,J}|^2 e^{\widehat{P}_{S,I,J}}\right)\right]^{+},$$

$$\mu_{Bn,l,n}[t_\delta+1] = \left[\mu_{Bn,l,n}[t_\delta] - \delta_{Bn,l,n}[t_\delta+1]\left(P_{th} - \sum_{I=1}^{L}\sum_{J=1}^{K_I-2N} |\mathbf{g}_{Bn,l,n}^{H}\mathbf{v}_{I,J}|^2 e^{\widehat{P}_{S,I,J}}\right)\right]^{+},$$

$$\lambda[t_\delta+1] = \left[\lambda[t_\delta] - \delta[t_\delta+1]\left(P_{S,\max} - \sum_{I=1}^{L}\sum_{J=1}^{K_I-2N} e^{\widehat{P}_{S,I,J}}\right)\right]^{+}, \tag{20.49}$$

where $\delta[t_\delta+1]$ is the step size in each iteration.

The satellite power allocation scheme is summarized as Algorithm 4. By updating the approximation parameters iteratively, we can finally obtain the optimal solution. In each iteration of the inner loop, the computational complexity is $O(M^2+LMN)$ for updating $\mathbf{P_S}$, $\boldsymbol{\mu}$, λ, and $\mathbf{I_S}$. In each iteration of the outer loop, the computational complexity is $O(M^2)$ for updating $\boldsymbol{\theta}_S$ and $\boldsymbol{\beta}_S$. Thus, the computational complexity of Algorithm 4 is $O(M^2+LMN)$.

Algorithm 4 Satellite Power Allocation Algorithm

```
Initialize t = 1, θ_{S,I,J} = 1, β_{S,I,J} = 0 and P_{S,I,J} = 0 for all I, J
repeat
  Initialize t_δ = 1, μ > 0, and λ > 0
  Initialize I_S referring to (20.48) for all I, J
  repeat
    Update P_S referring to (20.47)
    Update μ and λ referring to (20.49)
    Update I_S referring to (20.48)
    Set t_δ = t_δ + 1
  until μ and λ converge
  Set P_S[t] = P_S[t + 1]
  Update θ_S, β_S referring to (20.33)
  Set t = t + 1
until P_S converges
```

20.5.3 *Joint Power Allocation*

In the above derivation, we obtain the power allocation schemes of the BSs and the satellite separately while fixing the parameters of another. Actually, the two network will interfere each other since the entire bandwidth is reused in the whole system. Thus, the solution of one network will affect the solution of the other. To obtain the optimal solution of the whole system, an iterative algorithm is proposed as Algorithm 5. In each iteration, the Algorithms 3 and 4 will be performed based on the power allocation scheme $\mathbf{P}_B[t]$ and $\mathbf{P}_S[t]$ obtained in the last iteration. Then, the newly obtained results will be used for the next iteration until the results converge.

Note that in each iteration, the initialization steps in Algorithms 3 and 4 will be performed based on the results of the last iteration. Since only Algorithms 3 and 4 are performed in Algorithm 5, the computational complexity of Algorithm 5 is $O(LN^2 + M^2 + LMN)$.

Then, for the global optimization of the system based on Algorithm 5, a central unit is needed for collecting the global CSI, running the algorithms, and distributing the optimized power allocation results. The framework of cloud-based integrated terrestrial-satellite networks may be one possible solution for this, which is discussed in [37–39]. Also, with small capacity loss, the suboptimal power allocation scheme may be obtained by distributed calculation, which will be discussed in the simulation.

Algorithm 5 Joint Power Allocation Algorithm

```
Initialize t = 1, P_B[t] = 0, and P_S[t] = 0
repeat
    for I = 1 to L do
        Update P_{B,I}[t + 1] referring to Algorithm 3
    end for
    Update P_S[t + 1] referring to Algorithm 4
    Set P_B[t] = P_B[t + 1]
    Set P_S[t] = P_S[t + 1]
    Set t = t + 1
until P_B and P_S converge
```

20.6 Performance Evaluation

In this section, numerical results are provided to evaluate the performance of the NOMA-based integrated terrestrial-satellite networks as well as the proposed algorithms. The carrier frequency is set as 2 GHz at the S band, and the bandwidth B is 10 MHz. Then, we can calculate the AWGN power as $\sigma_n = BN_0$, where $N_0 = -174$ dBm/Hz is the AWGN power spectral density. The satellite is assumed to be on the orbit of 1000 km, and the parameters of the satellite are defined referring to [40]. Since the satellite users are mobile users in the network, which are generally equipped with single omni antenna, the user terminal antenna gain G_R is assumed to be 0 dBi. The maximum transmit power of the BS is set as $P_{B,max} = 43$ dBm. The ground channel is assumed to be Rayleigh channel and is modeled referring to [41], while the satellite channel is modeled as Rician channel referring to [42]. The user pairing algorithms used for comparison in the simulation are as follows:

1. **Max–min Correlation**. This is the pairing scheme that we proposed in Sect. 20.3, in which we first divide the users into the near user set and the far user set according to the channel conditions and then maximize the minimum channel correlation coefficient between the two users within one group.

2. **Joint Correlation and Gain**. This algorithm was proposed in [14], which takes both the correlation and gain difference between channels into consideration. With a predefined correlation threshold ρ, each user will select the user that have the maximum channel gain difference while the channel correlation must be larger than the correlation threshold.
3. **Random Algorithm**. Neither correlation nor gain is considered, and users will be randomly paired into groups.

We also compare the three power allocation algorithms as follows:

1. **Joint Optimal Algorithm**. This is the power allocation algorithm we proposed in Sect. 20.5 as Algorithm 5. We solve the terrestrial power allocation problem and the satellite power allocation iteratively until the results converge.
2. **Distribute Suboptimal Algorithm**. Different from the centralized joint optimal algorithm, in the distributed algorithm, each BS and the satellite will solve its optimal power allocation scheme separately, which will result in a suboptimal solution.
3. **Average Algorithm**. The transmit power will be averagely allocated among groups.

Figure 20.3 shows the convergence process of Algorithm 5 when $M = 8, N = 4$, $L = 4$, and $P_{th} = -80$ dBm. We can observe that the algorithm converges fast in all cases, within less than ten iterations. In Fig. 20.3a, we can see that with smaller interference temperature limit P_{th}, the BS capacity will be larger since the BS users will receive less interference from the satellite, while the satellite capacity will decrease due to the smaller transmit power. If we increase the BS antennas from 4 to 8, as illustrated in Fig. 20.3b, the BS capacity increases by about 50%, while the satellite capacity will slightly decrease by about 15% because the satellite needs to control the interference caused to all the BS users. Then, in Fig. 20.3c, we can observe that increasing the satellite antennas from 8 to 16 will lead to larger satellite capacity, about 25% gaining, while causing no harm to the BS users if the same P_{th} is maintained.

In Fig. 20.4, we investigate the variation of the system capacity with different interference temperature limits in three cases. For comparison, we also calculate the system capacity if we do not introduce the satellite into the system, which is showed using the red dotted line in the figure. We can observe that if more antennas are equipped for the satellite, or fewer BS users are in the network, as illustrated in Fig. 20.4a, the decreasing speed of the capacity of BSs will be larger than the increasing speed of the capacity of the satellite at the first, and then the increasing speed will exceed the decreasing speed as the interference limit continues to increase. Thus, as the interference limit increases, the total system capacity will first slightly decrease, then increase, and finally a capacity gaining of 10% can be obtained. With an appropriate setting of interference limit, which should be larger than −80 dBm in the case of Fig. 20.4a, the system can serve more users simultaneously and also achieve better capacity performance compared with the no satellite case. However, if there are more BS users, either by increasing the number of BSs or by increasing the antennas of BSs, as illustrated in Fig. 20.4b, c where we increase L from

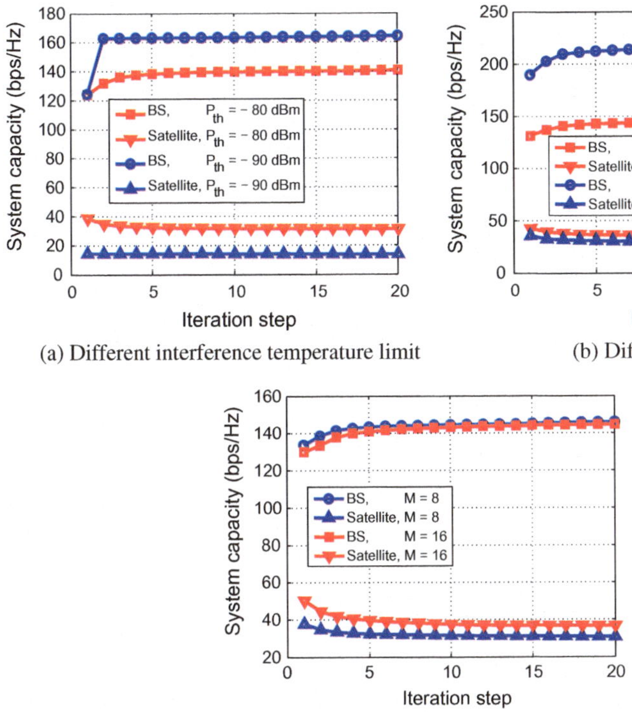

(a) Different interference temperature limit

(b) Different BS antennas

(c) Different Satellite antennas

Fig. 20.3 Convergence process over iterations

4 to 8 and N from 4 to 8 separately, the decreasing speed of the BSs will always be larger than the increasing speed of the satellite. In this case, introducing satellite into the network will lead to loss of total system capacity, about 20% in the case of Fig. 20.4b, c. However, with the help of the satellite, the network is able to provide service for more users simultaneously, especially for those users with bad BS channel conditions. As the development of wireless communication, the objective of communication is not to only provide high-speed service for high-density population. Instead, future networks aim to provide ubiquitous coverage for all ground users. Sacrificing part of the capacity of the users with good channel conditions, the system can provide better service for those users that have no access or bad access to the BSs by introducing the satellite to the system, and the total user number that can be served is also increased.

We have seen that different numbers of transmit antennas and BSs may lead to different characteristics of system performance. Thus, we then analyze the impact of different numbers of BSs, BS antennas, and satellite antennas on the total system capacity performance in Fig. 20.5. Note that we also calculate the system capacity of the case of no satellite for comparison. In Fig. 20.5a, where we set $M = 12$ and $N = 3$, we can observe that the total system capacity increases linearly as the

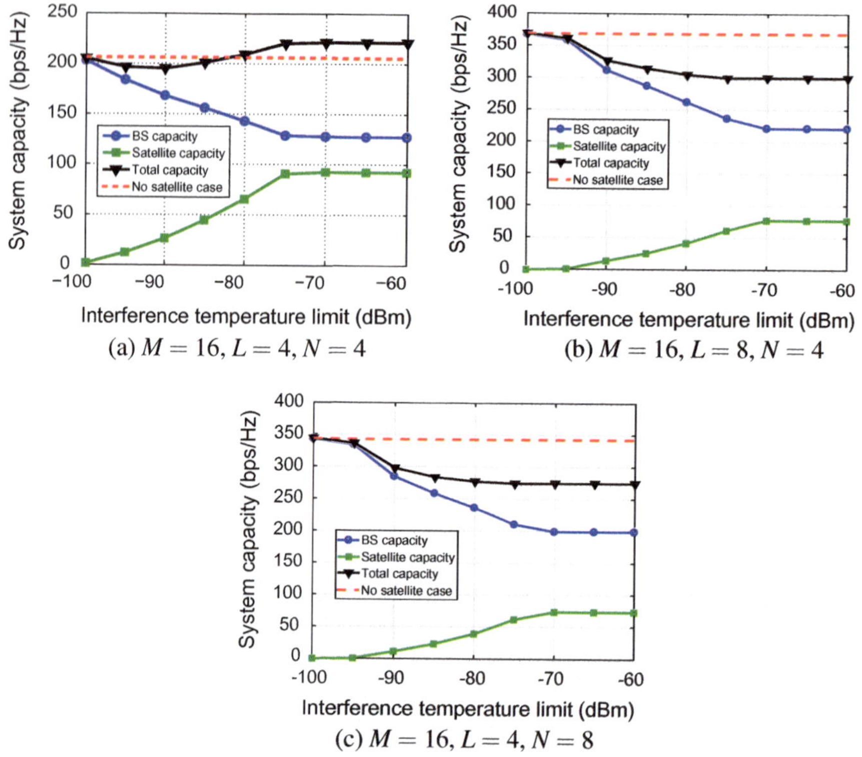

Fig. 20.4 Comparison of total system capacity, BS capacity, satellite capacity, and the case of no satellite with different interference temperature limits

number of BSs increases. For $P_{th} = -70$ dBm, the total system capacity increases from 70 to 200 bps/Hz when L increases from 1 to 6. Also, we can see that the increasing speed will be slower if the interference temperature limit is larger. When the number of BS is small, which means fewer BS users, the loss of the BS capacity due to interference from the satellite is less than the gaining of the satellite capacity. However, as illustrated in Fig. 20.4, as the number of BSs increases, which means more BS users, the BS capacity loss will exceed the gaining of the satellite capacity. For $P_{th} = -70$ dBm, there is about 87% capacity gaining compared with the case of no satellite if $L = 1$, which will turn to about 13% capacity loss when L increases to 6. In Fig. 20.5b, where $M = 9$ and $L = 3$, similar variation characteristics can be observed as the number of BS antennas increases, since either increasing L or N will lead to the increase of BS users. The major difference is that the increasing speed will be slower with more BS antennas due to less allocated power for each group and inter-group interference. Then, in Fig. 20.5c, in which $L = 2$ and $N = 4$, we can see the opposite situation. As the number of satellite antennas increases, the satellite is able to serve more users, and the gaining of the satellite capacity will be larger. For

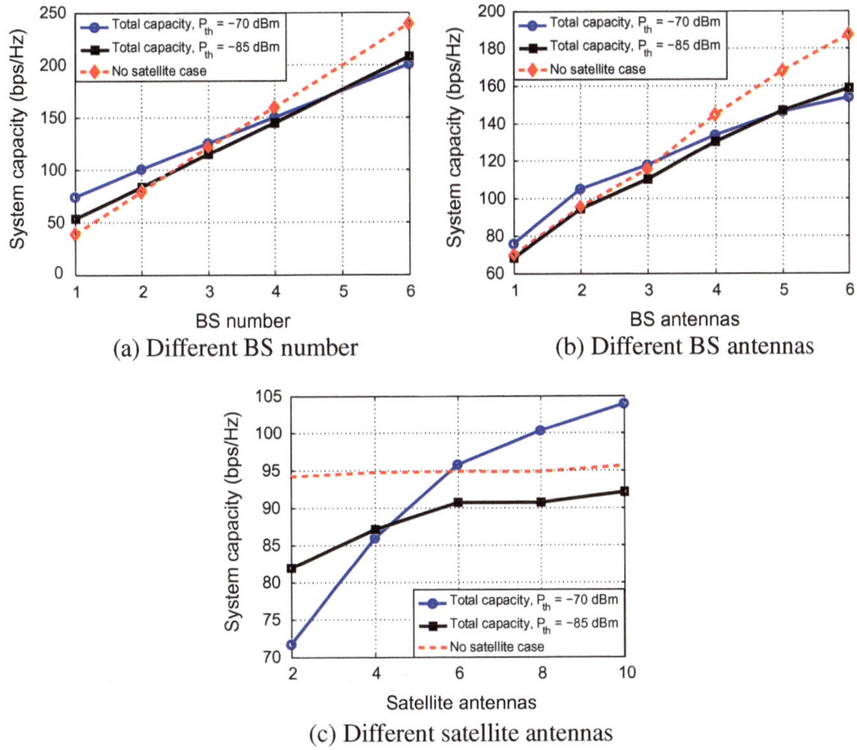

Fig. 20.5 Total system capacity of different BS numbers, BS antennas, and satellite antennas

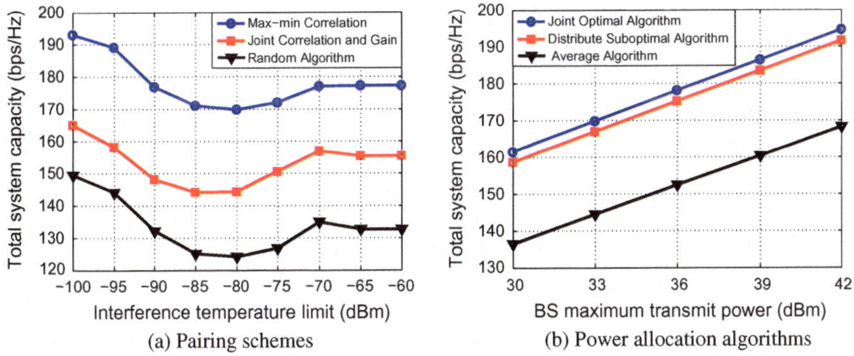

Fig. 20.6 Performance comparison of different algorithms

$P_{th} = -70$ dBm, the system capacity loss will reduce from 25% to zero and then increase to 10% gaining when M increases from 2 to 10.

Finally, we compare the performance of our algorithms with some other algorithms in Fig. 20.6, in which we set $M = 16$, $L = 4$, and $N = 4$. First, Fig. 20.6a gives the total system capacity of the three pairing schemes. It can be observed that the paring scheme significantly impacts the system performance. Compared with the random algorithm, our proposed scheme can achieve about 35% capacity gaining, while about 15% capacity gaining can be obtained compared with the "Joint Correlation and Gain" algorithm proposed in [14]. In order to achieve better system capacity, it is important to pair users that have larger channel differences into one group. At the same time, since the beamforming vectors are designed according to the channels of near users, the inter-group interference may remarkably deteriorate the performance of far users. To mitigate the inter-group interference, users that experience correlative channels should be paired into one group. Taking total system performance and user fairness into consideration, we proposed the "max–min correlation" scheme as in Algorithm 2. Figure 20.6b shows the comparison of different power allocation algorithms with different maximum transmit power of BSs, and we set $P_{th} = -90$ dBm. Our optimal power allocation scheme achieves about 15% capacity gaining compared with the average power allocation scheme, which proves the optimality of our algorithm. In addition, we can observe that there is only slight capacity gaining between the centralized optimal algorithm and the distributed suboptimal algorithm. With a small capacity loss, the suboptimal power allocation scheme can be obtained by distributed calculation.

20.7 Conclusion

In this chapter, we proposed a general downlink framework of the non-orthogonal multiple access (NOMA)-based integrated terrestrial-satellite network, in which the BSs and the satellite cooperatively provide service for ground users. The users that experience better satellite channels and worse BS channels are selected as satellite users. Then, the BS users are paired into groups when maximizing the minimum channel correlation between users, which is solved based on the bipartite graph theory. Equipped with multi-antennas, ZFBF is executed among groups for each BS, while the satellite will choose either ZFBF or MRTBF according to the channel conditions. Since the two system interfered each other, we first formulated the power allocation problem for the terrestrial networks and the satellite network separately, which are solved by means of SCA approach and the Lagrangian dual method. Then, to maximize the total system capacity, we proposed a joint iteration algorithm, in which the interference temperature limit is introduced for the satellite to control the interference caused to BS users. Numerical results showed that the proposed integrated networks can achieve good performance. With large satellite antennas, the integrated networks can serve more users and achieve higher system capacity compared with the case of no satellite. When the number of BS users increase to a large number, a trade-off between user fairness and system total capacity needs to be considered. Finally, comparison with some other proposed algorithms and existing algorithms showed the effectiveness and optimality of our proposed algorithms.

Acknowledgements This work was supported by the National Nature Science Foundation of China (Grant Nos. 91438206, 91638205, 91538203, and 61621091) and Young Elite Scientist Sponsorship Program by CAST.

References

1. Y. Saito, Y. Kishiyama, A. Benjebbour, T. Nakamura, A. Li, K. Higuchi, Non-orthogonal multiple access (NOMA) for cellular future radio access, in *Proceedings of IEEE Vehicular Technology Conference (VTC Spring)* (2013), pp. 1–5
2. L. Dai, B. Wang, Y. Yuan, S. Han, C.I, Z. Wang, Non-orthogonal multiple access for 5G: solutions, challenges, opportunities, and future research trends. IEEE Commun. Mag. **53**(9), 74–81 (2015)
3. 3rd Generation Partnership Project (3GPP), Study on downlink multiuser superposition transmission for LTE (2016)
4. W. Shin, M. Vaezi, B. Lee, D.J. Love, J. Lee, H.V. Poor, Non-orthogonal multiple access in multi-cell networks: theory, performance, and practical challenges. IEEE Commun. Mag. **55**(10), 176–183 (2017)
5. Y. Saito, A. Benjebbour, Y. Kishiyama, T. Nakamura, System level performance evaluation of downlink non-orthogonal multiple access (NOMA), in *Proceedings of IEEE 24th PIMRC* (2013), pp. 611–615
6. Z. Ding, Z. Yang, P. Fan, H.V. Poor, On the performance of non-orthogonal multiple access in 5G systems with randomly deployed users. IEEE Sig. Process. Lett. **21**(12), 1501–1505 (2014)
7. S. Timotheou, I. Krikidis, Fairness for non-orthogonal multiple access in 5G systems. IEEE Sig. Process. Lett. **22**(10), 1647–1651 (2015)
8. J. Choi, Non-orthogonal multiple access in downlink coordinated two-point systems. IEEE Commun. Lett. **18**(2), 313–316 (2014)
9. T. Cover, Broadcast channels. IEEE Trans. Inf. Theory **18**(1), 2–14 (1972)
10. P.W. Wolniansky, G.J. Foschini, G.D. Golden, R. Valenzuela, V-BLAST: an architecture for realizing very high data rates over the rich-scattering wireless channel, in *Proceedings of URSI International Symposium on Signals, Systems, and Electronics* (1998), pp. 295–300
11. Z. Ding, H.V. Poor, Cooperative non-orthogonal multiple access in 5G systems. IEEE Commun. Lett. **19**(8), 1462–1465 (2015)
12. Y. Liu, Z. Ding, M. Elkashlan, H.V. Poor, Cooperative Non-orthogonal multiple access with simultaneous wireless information and power transfer. IEEE J. Sel. Areas Commun. **34**(4), 938–953 (2016)
13. Z. Ding, P. Fan, H.V. Poor, Impact of user pairing on 5G nonorthogonal multiple-access downlink transmissions. IEEE Trans. Veh. Technol. **65**(8), 6010–6023 (2016)
14. B. Kim et al., Non-orthogonal multiple access in a downlink multiuser beamforming system, in *Proceedings of IEEE MILCOM* (2013), pp. 1278–1283
15. K. Higuchi, Y. Kishiyama, Non-orthogonal access with random beamforming and intra-beam SIC for cellular MIMO downlink, in *Proceedings of IEEE Vehicular Technology Conference (VTC Fall)* (2013), pp. 1–C5
16. J. Choi, Minimum power multicast beamforming with superposition coding for multiresolution broadcast and application to NOMA systems. IEEE Trans. Commun. **63**(3), 791–800 (2015)
17. M.F. Hanif, Z. Ding, T. Ratnarajah, G.K. Karagiannidis, A minorization-maximization method for optimizing sum rate in the downlink of non-orthogonal multiple access systems. IEEE Trans. Signal Process. **64**(1), 76–88 (2016)
18. Q. Sun, S. Han, C. I, and Z. Pan, On the ergodic capacity of MIMO NOMA systems, IEEE Wirel. Commun. Lett. **4**(4), 405–408 (2015)
19. Z. Ding, R. Schober, H.V. Poor, A general MIMO framework for NOMA downlink and uplink transmission based on signal alignment. IEEE Trans. Wirel. Commun. **15**(6), 4438–4454 (2016)

20. Z. Ding, F. Adachi, H.V. Poor, The application of MIMO to non-orthogonal multiple access. IEEE Trans. Wirel. Commun. **15**(1), 537–552 (2016)
21. 3rd Generation Partnership Project (3GPP), Study on Architecture for Next Generation System (2016)
22. E. Lagunas, S.K. Sharma, S. Maleki, S. Chatzinotas, B. Ottersten, Resource allocation for cognitive satellite communications with incumbent terrestrial networks. IEEE Trans. Cognit. Commun. Netw. **1**(3), 305–317 (2015)
23. C. Jiang, Y. Chen, K.J.R. Liu, Y. Ren, Renewal-theoretical dynamic spectrum access in cognitive radio network with unknown primary behavior. IEEE J. Sel. Areas Commun. **31**(3), 406–416 (2013)
24. C. Jiang, H. Zhang, Y. Ren, H. Chen, Energy-efficient non-cooperative cognitive radio networks: micro, meso and macro views. IEEE Commun. Mag. **52**(7), 14–20 (2014)
25. H. Yizhou, C. Gaofeng, L. Pengxu, C. Ruijun, W. Weidong, Timing advanced estimation algorithm of low complexity based on DFT spectrum analysis for satellite system. China Commun. **12**(4), 140–150 (2015)
26. A.H. Khan, M.A. Imran, B.G. Evans, Semi-adaptive beamforming for OFDM based hybrid terrestrial-satellite mobile system. IEEE Trans. Wirel. Commun. **11**(10), 3424–3433 (2012)
27. K. An, M. Lin, J. Ouyang, Y. Huang, G. Zheng, Symbol error analysis of hybrid satellite-terrestrial cooperative networks with cochannel interference. IEEE Commun. Lett. **18**(11), 1947–1950 (2014)
28. K. An, M. Lin, T. Liang, J. Wang, J. Wang, Y. Huang, A.L. Swindlehurst, Performance analysis of multi-antenna hybrid satellite-terrestrial relay networks in the presence of interference. IEEE Trans. Commun. **63**(11), 4390–4404 (2015)
29. D.T. Ngo, S. Khakurel, T. Le-Ngoc, Joint subchannel assignment and power allocation for OFDMA femtocell networks. IEEE Trans. Wirel. Commun. **13**(1), 342–355 (2014)
30. Z. Yu, K. Wang, H. Ji, X. Li, H. Zhang, Dynamic resource allocation in TDD-based heterogeneous cloud radio access networks. China Commun. **13**(6), 1–11 (2016)
31. C. Berge, Two theorems in graph theory. Proc. Nat. Acad. Sci. **43**(9), 842–844 (1957)
32. Z. Galil, Efficient algorithms for finding maximum matching in graphs. ACM Comput. Surv. **18**(1), 23–38 (1986)
33. B.R. Marks, G.P. Wright, A general inner approximation algorithm for nonconvex mathematical programs. Oper. Res. **26**(4), 681–683 (1978)
34. J. Papandriopoulos, J.S. Evans, SCALE: a low-complexity distributed protocol for spectrum balancing in multiuser DSL networks. IEEE Trans. Inf. Theory **55**(8), 3711–3724 (2009)
35. S. Boyd, L. Vandenberghe, *Convex Optimization* (Cambridge University Press, 2004)
36. T.K.Y. Lo, Maximum ratio transmission. IEEE Trans. Commun. **47**(10), 1458–1461 (1999)
37. X. Zhu, C. Jiang, W. Feng, L. Kuang, Z. Han, J. Lu, Resource allocation in spectrum-sharing cloud based integrated terrestrial-satellite network, in *Proceedings of IEEE IWCMC* (Valencia, 2017), pp. 334–339
38. J. Du, C. Jiang, Q. Guo, M. Guizani, Y. Ren, Cooperative earth observation through complex space information networks. IEEE Trans. Wireless Commun. **23**(2), 136–144 (2016)
39. J. Du, C. Jiang, Y. Qian, Z. Han, Y. Ren, Resource allocation with video traffic prediction in cloud-based space systems. IEEE Trans. Multimed. **18**(5), 820–830 (2016)
40. D. Christopoulos, S. Chatzinotas, B. Ottersten, Multicast multigroup precoding and user scheduling for frame-based satellite communications. IEEE Trans. Wirel. Commun. **14**(9), 4695–4707 (2015)
41. 3rd Generation Partnership Project (3GPP), Further advancements for E-UTRA physical layer aspects (2016)
42. E. Lutz, D. Cygan, M. Dippold, F. Dolainsky, W. Papke, The land mobile satellite communication channel-recording, statistics, and channel model. IEEE Trans. Veh. Technol. **40**(2), 375–386 (1991)

Chapter 21
Conclusions and Future Research Directions for NOMA

Zhiguo Ding, Yongxu Zhu and Yan Chen

21.1 Summary

Multiple access has always been recognized as a crucial component of modern communication systems, and a change of multiple access is the key milestone for the migration of wireless networks. To clearly illustrate the importance of multiple access techniques in mobile systems, Chaps. 1 and 2 have been provided, where the definition of multiple access was provided first and then the impact of conventional orthogonal multiple access (OMA) techniques on the past generations of cellular networks has been described. In Chaps. 3 and 4, two latest members of the OMA family have been introduced, since non-orthogonal multiple access (NOMA) is not the only type of multiple access techniques to be used in 5G and beyond. Specifically, the use of these new types of OMA can potentially yield lower system complexity compared to NOMA since users are allocated orthogonal bandwidth resource blocks, and hence, interference management among the users can be easily accomplished. It is worth pointing out that the use of new waveforms is particularly important to these new types of OMA, and hence, the details about the 5G waveform candidates have also been provided in these chapters.

The use of NOMA is to ensure that multiple users are served at the same bandwidth resource blocks and at the same time, instead of allowing a single user to solely

Z. Ding (✉)
The University of Manchester, Manchester, UK
e-mail: zhiguo.ding@manchester.ac.uk

Y. Zhu
Wolfson School of Mechanical, Electrical and Manufacturing Engineering, Loughborough University, Loughborough, UK
e-mail: Y.Zhu4@lboro.ac.uk

Y. Chen
Huawei Technologies Co., Ltd., Shanghai, China
e-mail: bigbird.chenyan@huawei.com

M. Vaezi et al. (eds.), *Multiple Access Techniques for 5G Wireless Networks and Beyond*, https://doi.org/10.1007/978-3-319-92090-0_21

occupy one block for a time as in OMA. Hence, NOMA can yield much better spectral efficiency compared to OMA, which is the reason why the majority of this book has been devoted to NOMA. In Chap. 5, the essential idea of NOMA, i.e., encouraging radio resource sharing, has been illustrated from an information theoretic perspective, which provides a foundation for the practical NOMA designs introduced in the remaining of the book. In Chap. 6, various designs of NOMA have been illustrated by focusing on the since-cell scenario with single-antenna nodes, where the use of spatial degrees of freedom to improve the performance of NOMA has also been introduced. The concept of NOMA can be extended to multi-cell scenarios with multiple-antenna nodes, but it is important to ensure interference management among the cells, which has been discussed in Chap. 7. In Chaps. 8–11, how to apply the principle of NOMA to various 5G techniques and emerging networks, such as visible light communications, full duplex and millimeter-wave communication networks, has been discussed. In Chaps. 12–17, different variants of NOMA, such as sparse code multiple access (SCMA), interleave-division multiple access, pattern division multiple access, have been introduced, since the principle of NOMA can be implemented in practical communication scenarios in different forms. The experimental trials for NOMA, the current standardization activities for NOMA in LTE and 5G as well as the potential of NOMA to satellite communications have been introduced in Chaps. 18–20 in order to show how the concept of NOMA has been included into various practical wireless networks.

21.2 Open Issues and Future Research Challenges

Although the concept of NOMA was invented only in 2013, it has quickly attracted significant attention from both academia and industry, as a paradigm shift for the design of multiple access techniques in future wireless networks [1–4]. It is worth pointing out that the NOMA principle is not new to wireless communications, and it is based on many existing ideas employing non-orthogonality. For example, the key components of NOMA systems, such as superposition coding, successive interference cancellation (SIC) and message passing algorithms, have been previously used for multi-user detection [5]. However, intentionally multiplexing different users at the same time/frequency/code/spatial direction has not been used in the previous generations of multiple access designs. Given the dynamic nature of network topologies and user traffic patterns in the next generation wireless network, there are still many open problems and research challenges for an efficient implementation of the NOMA principle as illustrated in the following subsections.

21.2.1 A New Era of Hybrid Multiple Access

Hybrid multiple access is not new to cellular networks at all, and it has been employed in early generations of mobile networks. For example, GSM relies on the combination of time division multiple access (TDMA) and frequency division multiple access (FDMA). Particularly, the principle of TDMA is used to divide one GSM frame into 8 time slots, and hence, 8 users can be supported at the same carrier frequency, which obviously is not sufficient to support a large number of users and motivates the use of FDMA in GSM. While in LTE, OFDMA is applied as another example of OMA where the time and frequency resources are gridded into orthogonal lattices and different users are scheduled on non-overlapping lattices exclusively. The future generation wireless networks are also expected to continue using such hybrid multiple access, and NOMA is envisioned to play an important role due to its superior compatibility. Particularly, NOMA can be used to improve spectral efficiency without any need for altering the fundamental resource blocks of other multiple access principles, where the integration of NOMA can be viewed as a smooth software upgrade.

While the principle of NOMA can be integrated with other multiple access techniques, there are various solutions to how such an integration can be done. In particular, there are two popular solutions. The first type of hybrid NOMA is that the principle of NOMA, spectrum sharing, is implemented on each orthogonal resource element individually and separately. If the orthogonal resource element is an OFDM subcarrier within a given time unit, this type of hybrid multiple access has also been termed single-tone NOMA. The benefit of single-tone NOMA is simple in concept as well as the transceiver design, since the implementation of NOMA on one resource element is independent to those on the others; i.e., there is no need for joint NOMA encoding or decoding across different resource elements. The most well-known form of single-tone NOMA is power-domain NOMA [6]. Particularly, power-domain NOMA invites multiple users to share the same resource element simultaneously, where the power domain is used for multiplexing; e.g., superposition coding is used at the downlink transmitter by allocating different power levels to users, and SIC is used at the receivers to remove co-channel interference. Cognitive radio-inspired NOMA (CR-NOMA) is another important form of single-tone NOMA [7]. The key difference between power-domain NOMA and CR-NOMA is that CR-NOMA recognizes the difference between users' quality-of-service (QoS) requirements, in addition to the users' channel difference. For example, one of the key features of future wireless networks is the heterogeneous traffic pattern, and such diverse QoS feature becomes more obvious after Internet of Things (IoT) is integrated with cellular networks. Compared to power-domain NOMA, CR-NOMA offers two advantages. One is that the principle of NOMA can be still implemented even if users' channel conditions are similar, since users are ordered accordingly their QoS targets. The other is that users' QoS requirements can be strictly guaranteed. It is worth noting that the concepts of power-domain NOMA and CR-NOMA are complementary to each other, where how to use the users' channel conditions as well as their QoS requirements for user ordering has been recognized as an important future research direction.

The second type of hybrid NOMA is to jointly implement the principle of NOMA across multiple orthogonal resource elements, which is thus referred as multi-tone NOMA. In such multi-tone NOMA, some types of joint coding across the multiple resource elements can be introduced to further improve NOMA performance. For instance, block-based sequence spreading, scrambling, interleaving, or multi-tone joint modulation mapping can be applied. Joint decoding across multiple resource elements and multiple users is thus needed for the implementation of multi-tone NOMA. Compared to single-tone NOMA, multi-tone NOMA offers better reception reliability and throughputs, but may suffer from increased design complexity at both transmitter and receiver sides. In another aspect, because of the joint design over multiple resource elements, the performance of multi-tone NOMA is further impacted by the way of resource elements mapping or subcarrier allocation, which can also be taken as one dimension to be optimized [8]. Currently, seeking the optimal solution for resource allocation with low complexity for multi-tone NOMA has become an important ongoing research direction.

Despite such increased design complexity, multi-tone NOMA has attracted a lot of attention from the industry, and many recently proposed industrial forms of NOMA are based on multi-tone NOMA. For example, uplink SCMA is implemented by asking each user jointly to encode its messages sent on multiple subcarriers, and using the MPA at the base station for joint decoding [9]. Regular SCMA has a strict constraint that each user can be allocated the same number of subcarriers, whereas irregular SCMA as well as pattern division multiple access uses more relaxed requirements to the number of subcarriers allocated to each user [4]. It is worth pointing out that many industrial forms of multi-tone NOMA are based on the open-loop concept; i.e., users' channel information is not used for subcarrier allocation. While this open-loop approach reduces the system complexity, the dynamic nature of users' channel conditions has not been used, which means that the performance of these industrial forms of multi-tone NOMA can be further improved by exploiting the users' channel information.

21.2.2 Combination of NOMA with Other Advanced Physical Layer Designs

Initial studies have demonstrated that the principle of NOMA is compatible not only to other types of multiple access, but also to advanced physical layer techniques to be used in the future wireless networks. Some of these examples are provided in the following:

- **mmWave-NOMA**: Both mmWave and NOMA have been recognized as key techniques to combat the spectrum crunch; i.e., there are not sufficient bandwidth for communications, although the solutions provided by the two techniques are different. A common question from the research community is why to use NOMA for mmWave networks when there is plenty of bandwidth available at mmWave bands. This question can be answered by using the following example. In 1990s, when a movie is stored in a computer by using the VCD (MPEG-1) format, the size of such a file is around 200–500 MB. When this was replaced by the DVD

format, the size of a movie file is expanded to 2–3 GB, and the size of a typical Blu-Ray file can be 20 GB. Human becomes more and more demanding to the resolution and details, which means that the amount of information to be sent in the future wireless networks will also become larger and larger. Therefore, the gains we get from mmWave bands can soon hit its ceiling, and how to efficiently use the bandwidth from the mmWave bands will become a critical issue, which motivates the use of NOMA in mmWave networks.
Furthermore, existing studies of mmWave-NOMA have revealed that mmWave transmission exhibits some features which are ideal for the application of NOMA. For example, users in mmWave networks can have strongly correlated channels, even if the antennas of these users are separated much larger than half of the signal wavelength [10]. While such correlation has been conventionally recognized as a harmful effect, the use of the quasi-degradation criterion reveals that this correlation results in an ideal situation for the application of NOMA. Another example is hardware impairments and limitations, e.g., the use of finite resolution analog beamforming, can also bring the opportunity for the integration of NOMA in mmWave networks.

- **MIMO-NOMA**: During the last two decades, MIMO has been continuously in the spotlight of the communication research and industrial activities, mainly due to its superior spectral efficiency, i.e., high data rates can be supported without using extra spectrum bandwidth but by exploring the spatial domain. At a certain stage of the development of NOMA, there was confusion about the difference between MIMO and NOMA. The reason for this confusion is that using MIMO, we can also accommodate multiple users at the same spectrum at the same time and hence yield the same non-orthogonality as NOMA. Actually, many conventional MIMO techniques aim to use the spatial domain and create spatially orthogonal channels between users, in order to avoid co-channel interference. Zero-forcing and singular value decomposition-based designs are typical examples to illustrate this orthogonal principle. On the other hand, the use of NOMA is to assume that multiple users share the same orthogonal resource unit, where one spatially orthogonal channel is just another example of such a resource unit [11]. Or in other words, conventional MIMO allows users to use the same bandwidth, but tries to create multiple orthogonal spatial directions to differentiate multiple users, whereas NOMA further supports multiple users to share the same spatially orthogonal direction.
In the context of massive MIMO, there was a concern about the feasibility for the implementation of NOMA. The rationale behind this concern is that the quasi-degradation criterion reveals that NOMA is not preferable if users' channel vectors are orthogonal to each other, but in massive MIMO, users' channel vectors are asymptotically orthogonal. However, some existing studies have demonstrated that the use of NOMA is still important to massive MIMO, where the reason is that users' channels are not completely orthogonal in a practical scenario, because of channel correlation. For example, when implementing massive MIMO at a base station, most likely this base station will be amounted at a top of a high building, without many scatters around. As a result, users from one room in this building can have highly correlated channels, instead of orthogonal channels. As discussed

in the mmWave-NOMA part, the correlation among users' channels does facilitate the implementation of NOMA [3]. Moreover, it is very costly to get accurate channel state information for massive MIMO scenarios. In the case that the channel state information is not perfect due to limited feedback quantization, channel measurement latency, or user mobility, NOMA can help to improve the system performance which will otherwise be degraded significantly.

- **Cooperative NOMA**: The importance of cooperative diversity can be easily spotted by the fact that the paper by Laneman, Tse, and Wornell has already attracted 12000+ citations, probably one of the most cited papers in the last two decades in communications [12]. The use of cooperative transmission is important to NOMA since users with poor channel conditions in current NOMA can potentially suffer some performance loss, compared to the case with OMA. By using cooperative NOMA, the reception reliability of these users can be improved. Most existing designs of cooperative NOMA can be grouped into two categories. One category is to employ NOMA users with strong channel conditions as relays to help the other users, which is also known as user cooperation. The benefit of such cooperative NOMA is that the redundant structure of NOMA can be efficiently exploited. Particularly, these so-call strong users need to decode the messages to the users with poor channel conditions in order to decode their own information, and hence, they are natural relays to help those weak users.
The use of dedicated relays is another important category of cooperative NOMA [4]. In many communication scenarios, the number of mobile devices is large, but many of them are not active in transmitting or receiving. Therefore, these idle users can be used as relays to help the active users, and the number of such relays can be quite large in practice, which is the advantage of the second category of cooperative NOMA. Given the existence of multiple relays, distributed beamforming can be designed in order to efficiently utilize the spatial degrees of freedom offered by the dedicated relays, but the system overhead caused by the coordination among the relays needs to be carefully suppressed, particularly for the scenario with a large number of relays. A low-complexity alternative to distributed beamforming is relay selection, and recent studies reveal an interesting fact that relay selection in cooperative NOMA can be fundamentally different to that in conventional cooperative networks. For example, the max-min criterion which has been shown optimal in conventional cooperative networks is no longer optimal in cooperative NOMA. This explains why relay selection becomes a quite popular area in cooperative NOMA.
- **Network NOMA**: Network MIMO has recently received a lot of attention, since the boundaries of cells are removed and base stations from different cells are encouraged to cooperate each other. There have been different forms of network MIMO, from distributed CoMP to jointly scheduled C-RAN. While the benefit of the NOMA principle in a single-cell setup has been well recognized, its benefit to the multi-cell scenarios, such as CoMP and C-RAN, has not been fully exploited and investigated. Hence recently, a lot of efforts of the NOMA research community have been devoted to thoroughly examine the impact of NOMA on multi-cell scenarios, and the resulting novel designs of NOMA can be viewed as special cases

of network NOMA, and these network NOMA designs clearly demonstrate that it is beneficial to use the NOMA principle in such network MIMO scenarios [13]. In particular, given its non-orthogonal nature, NOMA is expected to improve the CoMP transmission by relaxing the requirements of accurate time synchornization between different transmit points and joint channel state information for precise beamforming, which prohibit the boom of CoMP applications in practical systems. Without loss of general, take CoMP as an example, which is to ask multiple base stations to jointly serve a user which is at the cell boundaries. While such a design indeed helps the edge user, these base stations have to serve this user solely for a given bandwidth and time slot, which reduces the spectral efficiency since this user has poor connections to the base stations. After network NOMA is used, each base station can serve a near user while performing CoMP and helping the edge user. As a result, the overall system throughput as well as connectivity can be significantly improved. Advanced power allocation policies can be used to ensure that those near users are admitted without sacrificing the performance of the edge user. In the heterogeneous networks, the NOMA principle has also been shown to be very useful to improve the spectral efficiency as well as coverage, where not only more users can be served in each tier but also the cooperation among different tiers can be enabled. A key challenge for these network NOMA systems is how to reduce the system overhead consumed by the coordination among the cells and tiers, where sophisticated methods for low-complexity interference management are needed.

21.3 Integrating NOMA into Systems Beyond Cellular Communications

The superior compatibility of the NOMA principle can be clearly demonstrated by the fact that NOMA has found a lot of applications beyond cellular communications, as illustrated in the following:

- **Wi-Fi Networks**: While the concept of NOMA has been investigated for cellular systems, it can be straightforwardly applied to the next generation Wi-Fi systems. Conventional Wi-Fi networks still rely on orthogonal resource allocation. This leads to a difficult situation that some users cannot be admitted after all the limited orthogonal resource blocks are taken by other users. By applying the NOMA principle, more users can be simultaneously admitted, which is particularly important for the deployment of Wi-Fi in crowded areas, such as airports or sport stadiums. Different to cellular networks, user access in Wi-Fi is not realized in a centralized way, and hence, distributed designs are needed for the application of NOMA to Wi-Fi networks.
- **VLC Systems**: VLC has been recognized as an efficient method for the last mile connection in future communication networks. However, one disadvantage of VLC is that the use of narrow-band modulation and non-coherent detection limits the

number of users which can be supported. As a result, the NOMA principle is naturally compatible to VLC, and its application ensures that VLC can be used not only for small-scale smart homes, but also large-scale scenarios, such as lecture halls [14]. However, the unique channel characteristics of VLC mean that algorithms and protocols originally designed for radio frequency-based networks cannot be straightforwardly applied to VLC, and designs tailored for VLC channels are needed for the combination of VLC and NOMA.

- **TV Broadcasting**: Digital TV broadcasting is another important application of NOMA [15]. It is worth pointing out that the concept of NOMA has already been included into the next generation digital TV standard (ATSC 3.0), where it is termed Layered Division Multiplex (LDM). Particularly, a TV station will integrate several layers of video streams with different QoS requirements, and this superposition will be broadcasted to users. Each user will decode certain layers of the video streams according to its channel conditions. It is also worth pointing out that the application of NOMA to TV broadcasting can be particularly important to future wireless multicasting services, an area which has not yet been fully explored.
- **Wireless Caching**: The key idea of wireless caching is to proactively push content files to local caching infrastructure, before they are requested. As a consequence, users can fetch these files from their local caching infrastructure, without being directly served by the network controller. Existing studies have demonstrated that the NOMA concept not only helps to push the content files to local caching infrastructure timely and reliably, but also improves the spectral efficiency of content delivery from the caching infrastructure to the users [16]. However, how to integrate NOMA in wireless caching systems for coping with the dynamic changes of content popularity still remains unknown.
- **Internet of Things (IoT)**: One key feature of IoT is the diverse traffic patterns of IoT devices. Particularly, some devices have demanding bandwidth requirements, e.g., environmental monitoring cameras deliver high-resolution images or videos, whereas the others need to be served with low data rates but timely, e.g., vehicles receive incident warning messages in intelligent transportation systems. NOMA can be naturally applied to handle such a challenging situation, by integrating devices with heterogeneous QoS requirements at the same bandwidth. Furthermore, recent studies have also demonstrated that the combination of encoding with finite block length and NOMA can be a promising solution for supporting IoT as well as ultra-reliability and low latency communications (uRLLC). In particular, for contention-based grant-free transmission, NOMA is the key solution to enable reliable communications while supporting massive connectivity.

References

1. 5G radio access: Requirements, concepts and technologies, NTT DOCOMO, Inc., Tokyo, Japan, 5G Whitepaper, Jul 2014
2. 5G—A technology vision, Huawei, Inc., Shengzheng, China, 5G Whitepaper, Mar 2015

3. Z. Ding, Y. Liu, J. Choi, Q. Sun, M. Elkashlan, I. Chih-Lin, H.V. Poor, Application of non-orthogonal multiple access in LTE and 5G networks. IEEE Commun. Mag. **55**(2), 185–191 (2017)
4. Z. Ding, X. Lei, G.K. Karagiannidis, R. Schober, J. Yuan, V.K. Bhargava, A survey on non-orthogonal multiple access for 5G networks: research challenges and future trends. IEEE J. Sel. Areas Commun. **35**(10), 2181–2195 (2017)
5. S. Verdu, *Multiuser Detection* (Cambridge Press, London, 1998)
6. Y. Saito, Y. Kishiyama, A. Benjebbour, T. Nakamura, A. Li, K. Higuchi, Non-orthogonal multiple access (NOMA) for cellular future radio access, in *Proceedings of the IEEE Vehicular Technology Conference*, Dresden, Germany (2013)
7. Z. Ding, P. Fan, H.V. Poor, Impact of user pairing on 5G non-orthogonal multiple access. IEEE Trans. Veh. Tech. **65**(8), 6010–6023 (2016)
8. Y. Sun, D.W.K. Ng, Z. Ding, R. Schober, Optimal joint power and subcarrier allocation for full-duplex multicarrier non-orthogonal multiple access systems. IEEE Trans. Commun. **65**(3), 1077–1091 (2017)
9. H. Nikopour, H. Baligh, Sparse code multiple access, in *Proceedings of the IEEE International Symposium on Personal Indoor and Mobile Radio Communications*, London, UK, (2013)
10. Z. Ding, P. Fan, H.V. Poor, Random beamforming in millimeter-wave NOMA networks. IEEE Access **5**, 7667–7681 (2017)
11. Z. Chen, Z. Ding, X. Dai, G.K. Karagiannidis, On the application of quasi-degradation to MISO-NOMA downlink. IEEE Trans. Signal Process. **64**(23), 6174–6189 (2016)
12. J.N. Laneman, D.N.C. Tse, G.W. Wornell, Cooperative diversity in wireless networks: efficient protocols and outage behavior. IEEE Trans. Inform. Theory **50**(12), 3062–3080 (2004)
13. U. Vilaipornsawai, H. Nikopour, A. Bayesteh, J. Ma, SCMA for open-loop joint transmission CoMP, in *Proceedings of the IEEE VTC fall* (2015)
14. L. Yin, W.O. Popoola, X. Wu, H. Haas, Performance evaluation of non-orthogonal multiple access in visible light communication. IEEE Trans. Commun. **64**(12), 5162–5175 (2016)
15. L. Zhang, W. Li, Y. Wu, X. Wang, S.I. Park, H.M. Kim, J.Y. Lee, P. Angueira, J. Montalban, Layered-divisionmultiplexing: theory and practice. IEEE Trans. Broadcast. **62**(1), 216–232 (2016)
16. Z. Ding, P. Fan, G.K. Karagiannidis, R. Schober, H.V. Poor, NOMA assisted wireless caching: strategies and performance analysis. IEEE Trans. Commun. (submitted)

Index

M. Vaezi et al. (eds.), *Multiple Access Techniques for 5G Wireless Networks and Beyond*, https://doi.org/10.1007/978-3-319-92090-0

R

S

Printed by Printforce, the Netherlands